STATISTICS BY EXAMPLE
FIFTH EDITION

TERRY SINCICH

University of South Florida

DELLEN

an imprint of

MACMILLAN PUBLISHING COMPANY

New York

MAXWELL MACMILLAN CANADA

Toronto

MAXWELL MACMILLAN INTERNATIONAL

New York   Oxford   Singapore   Sydney

On the cover: "Aftershock" is painting measuring 56″ × 105″ by San Francisco artist Alex Baker. Its colored patterns, which incorporate fragments of organic derivation, are composed with oils, acrylics, and copper leaf on wood. Baker studied at New York's Art Students' League, the School for Visual Arts, and Parsons School of Design. Galleries in New York, New Orleans, Aspen, San Francisco, and Mexico City have shown his paintings. The artist also has done murals for clubs, restaurants, and public spaces in New York, Miami, and Aspen.

Cartoons by Tom Barnett. Technical illustrations by Kevin Tucker and Caroline Jumper.

© Copyright 1993 by Macmillan Publishing Company, a division of Macmillan, Inc. Dellen is an imprint of Macmillan Publishing Company.

Printed in the United States of America.

Macmillan Publishing Company
866 Third Avenue
New York, NY 10022

Macmillan Publishing Company is part of the
Maxwell Communication Group of Companies.

Maxwell Macmillan Canada, Inc.
1200 Eglinton Avenue East, Suite 200
Don Mills, Ontario M3C 3N1

Library of Congress Cataloging-in-Publication Data

Sincich, Terry.
    Statistics by example / Terry Sincich.—5th ed.
        p.   cm.
    Includes index.
    ISBN 0-02-410981-9
    1. Statistics.   I. Title.
QA276.12.S56   1993
519.5—dc20

92—42330
CIP

Printing: 1 2 3 4 5 6 7 8 9   Year: 3 4 5 6 7

CONTENTS

PREFACE

This introductory college statistics text is designed for students who have only a high school background in mathematics as a prerequisite. It differs from most other texts in two ways:

1. **Real data sets.** Explanations of basic statistical concepts and methodology are based on and motivated by the use of real data sets.
2. **Teaching by example.** Concepts and statistical methods are explained in examples. Many examples arise as questions posed about the data sets.

This practical orientation helps the student to relate statistics to real-life problems, and develops a pattern of thought that will persist after the student enters the working world. The American Statistical Association (ASA) sponsors an annual conference on "Making Statistics More Effective in Schools of Business," where there is a substantial consensus among the participants on the following point: Students are most effectively motivated by seeing statistics at work in real applications, problems, cases, and projects. In much of the current teaching of statistics, there is a limited opportunity for students to work with real data or to make serious use of statistical computing.

The text contains six real data sets to help motivate students. These data sets are also available on diskette (at no charge to adopters).

APPENDIX A   The first 100 observations from the set of actual starting salaries, majors, and colleges of 1,795 University of Florida graduates during the period from Fall 1989 to Spring 1991. (The complete data set is available on diskette.)

APPENDIX B   The starting salaries (extracted from Appendix A) of bachelor's degree graduates of the Colleges of Business Administration, Engineering, Liberal Arts and Sciences, Journalism, and Nursing.

APPENDIX C   The per-member per-month costs accrued by physicians in a managed-care health maintenance organization (HMO).

APPENDIX D   Length, weight, and DDT measurements for various species of fish collected from the Tennessee River, Alabama, and its creek tributaries.

APPENDIX E   Checkout times for 500 customers at a Publix supermarket.

APPENDIX F   The tar, nicotine, and carbon monoxide rankings of 372 domestic cigarette brands, as determined by the Federal Trade Commission.

Through examples and exercises, these data sets are used to develop the notion of a population and a sample, to demonstrate the need for data description, to develop the notion of a sampling distribution, and to motivate the inferential methods commonly studied in an introductory statistics course.

In addition to teaching using data sets and examples, the fifth edition retains the following features of the earlier editions:

1. **Case studies** Four case studies that detail specific and interesting current events are used in each chapter to pose questions for the student. These case studies are exctracted from journals, newspapers, and magazines. The first case study in each chapter, presented as a section in the text, demonstrates how to apply the statistical techniques learned in the chapter to solve a practical problem. The student is left to answer the questions posed by the three case studies given at the end of the chapter. (Refer to the Table of Contents for a list of the case studies.)

2. **Examples** The text, as its name implies, employs the "teaching by example" method. Each section contains several worked examples to demonstrate how to solve various types of statistical problems encountered in the real world.

3. **Key concepts highlighted** Throughout the text, key concepts are highlighted.
   a. **Definitions** are boxed.
   b. **Steps** for constructing bar graphs, performing statistical calculations, and conducting statistical tests are listed and boxed for each procedure.
   c. **Key words,** which must be added to a student's vocabulary, are listed at the end of each chapter.
   d. **Key formulas** are listed at the end of each chapter.
   e. **Warnings,** indicating situations where a student might misuse a statistical technique, are presented in boxed form. The student is directed to specific alternative methods.

4. **Real data (referenced) exercises** Since most students learn best by doing, the text contains a large number (over 1,000) of exercises. The answers for odd-numbered exercises are provided at the end of the text. Each chapter contains exercises at the end of each section and a set of Supplementary Exercises at the chapter's end. The exercises are of two types:
   a. **Learning the Mechanics** These exercises are intended to be straightforward applications of the new concepts presented in the section. They are introduced in a few words and are unhampered by a barrage of background information designed to make them "practical," but which often detracts from instructional objectives. Thus, with a minimum of labor, the student can recheck his or her ability to comprehend a concept or definition.
   b. **Applying the Concepts** The mechanical exercises are followed by realistic exercises that allow the student to see applications of statistics to the solutions of a variety of real-world problems. Almost all of these exercises contain data or information extracted from newspaper articles, magazines, and journals. Once the mechanics are mastered, these exercises develop students' skills at comprehending realistic problems that describe situations to which the techniques may be applied.

5. **Computer printouts**   To allow the instructor to emphasize interpretation of the statistical results, printouts generated from statistical computer software packages are presented throughout the text. The computer printouts for three commercial software packages—Minitab, SAS, and SPSS—are displayed.

6. **Reduced emphasis on probability formulas**   Based on the recommendations of the ASA, the emphasis on probability formulas is reduced in this introductory text. The probability chapter (Chapter 4) presents only the essential probability concepts (e.g., mutually exclusive events, conditional probability, and independent events) needed to apply the statistical inferential techniques in later chapters. Problem solving for the sake of problem solving is avoided.

7. **Reduced emphasis on analysis of variance formulas**   The text adopts a computerized rather than a "cookbook" approach to ANOVA in Chapter 13. Tedious calculation formulas are relegated to an optional section; the emphasis is on understanding designed experiments and interpretation of computer printouts.

The fifth edition contains several substantial modifications, additions, and enhancements.

1. **Computer labs**   Easy-to-follow instructions on how to use the statistical analysis commands of SAS, SPSS, and Minitab are provided at the end of most chapters. Except where noted, the commands are appropriate for both the mainframe and PC versions of the packages.

2. **Computer activities**   The text is now accompanied by ASP, a user-friendly, totally menu-driven statistical software package designed to run on IBM-compatible PCs. To encourage the use of computers in the statistical analysis of data, the student is asked to analyze a particular data set with ASP (or any other package) at the end of the relevant chapters. An ASP tutorial is provided in Appendix H.

3. **Collecting data (Chapter 1)**   The importance of collecting data with random (and representative) samples is emphasized by moving this material from its original location in Chapter 7 to Chapter 1 (Section 1.5).

4. **Descriptive methods for assessing normality (Chapter 6)**   A new section (Section 6.3) on determining whether a data set is approximately normal has been added to Chapter 6, A Continuous Probability Distribution: The Normal Distribution. In addition to the traditional graphical methods (histogram, stem-and-leaf display), we present the ratio of the interquartile range to the standard deviation as a check on normality. The emphasis on these techniques early in the text makes the student aware of the importance of checking assumptions in later chapters.

5. **The z statistic versus the t statistic in statistical inference (Chapter 8)**   We have expanded the discussion of the proper use of z and t in Chapter 8, Estimation of Population Parameters: Confidence Intervals. The importance of knowing the value of the population variance, σ^2, as well as the size of the sample (large or small), is emphasized.

6. **The relationship between confidence intervals and tests (Chapter 9)** A new section (Section 9.1) emphasizing the similarities and differences between a confidence interval and a two-tailed hypothesis test has been added to Chapter 9, General Concepts of Hypothesis Testing.

7. **Earlier presentation of p-values (Chapter 9)** The section on Reporting Test Results: p-Values (Section 9.6) is now presented much earlier in the pair of chapters on hypothesis testing. This enables the instructor to emphasize, if desired, the interpretation of results from computer printouts in Chapter 10, Hypothesis Testing: Applications.

8. **Power curves (Chapter 9)** Power curves have been added to the optional section on calculating β and the power of a test (Section 9.7).

9. **t-Test for the population coefficient of correlation (Chapter 11)** Earlier editions presented a test for zero correlation in the population based on a table of sample correlation coefficients. In this new edition, we replace the r test with the more familiar and equivalent t-test (Section 11.7) in Chapter 11, Simple Linear Regression and Correlation.

10. **Multicollinearity in multiple regression (Chapter 12)** A new section on multicollinearity (Section 12.15) has been added to Chapter 12, Multiple Regression and Model Building. The emphasis is on how to detect multicollinearity and its associated problems, and remedial measures.

11. **Analysis of variance chapter moved and expanded (Chapter 13)** The ANOVA chapter now follows the two regression chapters. This move enables us to add two new sections to Chapter 13. In new optional Section 13.9, we demonstrate the regression modeling approach to ANOVA; in new Section 13.10, we use regression residuals to check the ANOVA assumptions.

12. **Logistic regression (Chapter 14)** As an optional section (Section 14.5), we have added material on using logistic regression to model categorical probabilities in Chapter 14, Categorical Data Analysis.

13. **Nonparametric regression (Chapter 15)** In addition to Spearman's test for rank correlation, we also cover a new nonparametric test in Chapter 15: Theil's test for zero slope in regression (Section 15.8).

14. **Case studies updated** The case studies are updated, where necessary, and several new case studies are included.

15. **More exercises with "real" data** Many new "real-life" exercises have been added to each chapter. These exercises, like the case studies, are extracted from news articles, magazines, and professional journals to give students the opportunity to apply their knowledge of statistics to current practical problems.

Numerous less obvious changes in details have been made throughout the text in response to suggestions by current users of the earlier editions.

The text is also accompanied by the following supplementary material:

1. **Student's solutions manual** (by Nancy S. Boudreau) A student's exercise solutions manual presents the full solutions for half (the odd) exercises contained in the text.

2. **Instructor's solutions manual** (by Mark Dummeldinger) The instructor's exercise solutions manual presents the solutions to the other half (the even) exercises and all case studies contained in the text. For adopters, the manual is complimentary from the publisher.

3. **ASP statistical software diskette** New to this edition, the text includes a $5\frac{1}{4}''$ diskette containing the ASP program, *A Statistical Package for Business, Economics, and the Social Sciences*. ASP, from DMC Software, Inc., is a user-friendly, totally menu-driven program that contains all of the major statistical applications covered in the text, plus many more. ASP runs on any IBM-compatible PC with at least 512K of memory and two disk drives. With ASP, students with no knowledge of computer programming can create and analyze data sets easily and quickly. Appendix H contains start-up procedures and a short tutorial on the use of ASP. Full documentation is provided complimentary to adopters of the text.

4. **Minitab supplement** (by David D. Krueger and Ruth K. Meyer) The Minitab computer supplement was developed to be used with Minitab Release 6.1, a general-purpose statistical computing system. The supplement, written expecially for the student with no previous experience with computers, provides step-by-step descriptions of how to use Minitab effectively as an aid in data analysis. Each chapter begins with a list of new commands introduced in the chapter. Brief examples are then given to explain new commands, followed by examples from the text illustrating the new and previously learned commands. Where appropriate, simulation examples are included. Exercises, many of which are drawn from the text, conclude each chapter. A special feature of the supplement is a chapter describing a survey sampling project.

5. **Appendix data sets on diskette** For adopters, the complete data sets in the appendices are available on either a $3\frac{1}{2}''$ or a $5\frac{1}{4}''$ IBM PC diskette (ASCII format), complimentary from the publisher.

6. **Exercise data sets on diskette** Also for adopters, the data for all exercises containing 20 or more observations are available on either a $3\frac{1}{2}''$ or a $5\frac{1}{4}''$ IBM PC diskette (ASCII format). A list of these exercises follows this preface on pages xvii and xviii.

7. **Test bank** (by Mark Dummeldinger) This manual provides a large number of test items utilizing real data.

8. **DellenTest** This unique computer-generated random test system is complimentary to adopters. Utilizing an IBM (or compatible) PC and printer, the system will generate an almost unlimited number of quizzes, chapter tests, final examinations, and drill exercises. At the same time, the system produces an answer key and student worksheet with an answer column that exactly matches the column on the answer key.

9. **ASP Tutorial and Student Guide** (by George Blackford) Most students have little trouble learning to use ASP without documentation. Some, however, may want to purchase the *ASP Tutorial and Student Guide*. Bookstores can order the tutorial from DMC Software, Inc., 6169 Pebbleshire Drive, Grand Blanc, MI 48439.

I wish to acknowledge the many individuals who provided their invaluable assistance during the preparation of the original text and subsequent revisions. Their efforts are much appreciated. The reviewers are listed here:

Edwin F. Baumgartner (Le Moyne College)
Randal Beck (Millikin University)
Jim Bellis (Rochester Institute of Technology)
Elaine Bohanon (Bemidji State University)
John S. Bowdidge (Southwest Missouri State University)
Barbara J. Bulmahn (Indiana University at Fort Wayne)
John Cameron (Rockhurst College)
P. L. Claypool (Oklahoma State University)
Ronald L. Coccari (Cleveland State University)
Robert Cochran (University of Wyoming)
Murray Cohen (University of South Florida)
Joyce Curry-Daly (California Polytechnical)
Paul Dussere (SUNY–Oswego)
Janice Marie Dykacz (Essex Community College)
Joan Girard (Edison Community College)
Damodar Golhar (Western Michigan University)
Gavin Gregory (University of Texas at El Paso)
Derek Hart (McGill University)
Burt Holland (Temple University)
Geoffrey B. Holmewood (Hudson Valley Community College)
Rod Hurley (Hillsborough County Community College)
Thomas B. Laase (University of Southern Colorado)
Maria Ligas (Nova University)
Carolyn Likins (Millikin University)
James T. McClave (University of Florida)
William Mendenhall (University of Florida)
LaVern J. Meyer (Millikin University)
Daniel Mijalko (Western Michigan University)
Glenn W. Milligan (Ohio State University)
S. Mishra (University of South Alabama)
Amitava Mitra (Auburn University)
Maurice Monahan (South Dakota State University)
Kris K. Moore (Baylor University)
Paul Nelson (Kansas State University)
Bill Redmond (Bismarck State College)
Larry J. Ringer (Texas A&M University)
John B. Rushton (Metropolitan State College)
Dale G. Sauers (York College)
Susan Schott (University of Central Florida)
Fay Sewell (Montgomery Community College)
Brian E. Smith (McGill University)
William L. Toth (Henry Ford Community College)
Manel Wijesinha (Pennsylvania State University–York)
Charles W. Zimmerman (Robert Morris College)
Cathleen Zucco (Le Moyne College)

Caroline Jumper deserves special recognition for managing the production of this edition.

Finally, I owe very special thanks to my wife, Faith Sincich, who not only provided the necessary moral support one needs when undertaking a project like this, but also did an excellent job of typing, cutting, pasting, proofing, editing, and solving exercises. Without her, the fifth edition of this text could not have been completed.

EXERCISE DATA SETS AVAILABLE ON DISK

INTRODUCTION: STATISTICS AND DATA

CHAPTER 1

In one of the most famous experiments in psychology, Dr. Stanley Milgram (1974) studied the factors that determine the extent to which people obey authority—even if that authority is pushing them to do something they are against. In Milgram's study, subjects playing the role of "teachers" were told to give electric shocks (up to 450 volts) to "learners" who answered questions incorrectly. The object of the experiment was to see how many volts a subject would be willing to give before refusing to comply with the request.

Imagine the massive and impossible task of testing everyone in the country to see how they would respond to the shock experiment. And if we could test everyone, how would we determine what constitutes a "normal" reaction? This text describes how to use sampling and the science of statistics to solve many practical problems encountered in the real world. The chapter case in Section 1.4 discusses how Milgram measured people's obedience to authority and provides us with a glimpse of how statistics is applied to real-life data.

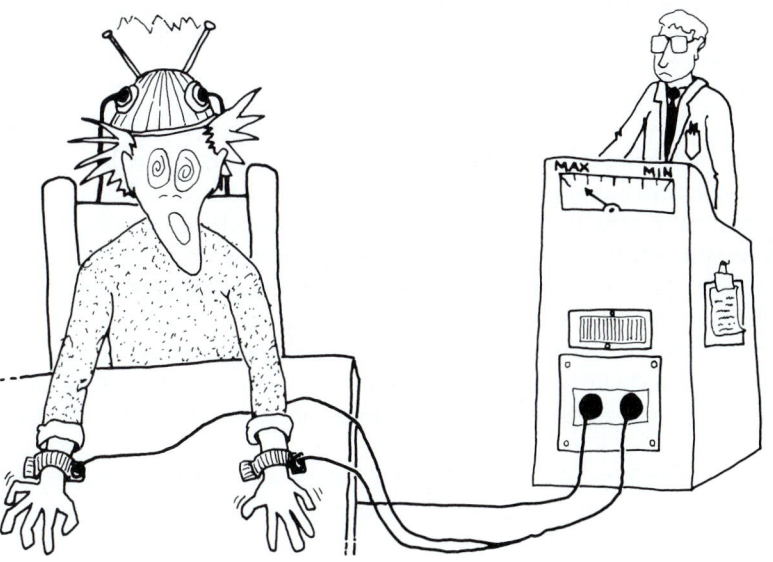

1

WHAT IS STATISTICS?

1.1 Consider the following recent items from the news media:

Tampa Tribune, Jan. 10, 1990
A computer study conducted at Columbia University demonstrated that shuffling a deck of cards seven times prevents bettors from accurately predicting the next card in blackjack. Most Atlantic City casinos will continue with their customary two shuffles of the deck, however, since the extra 5 minutes it takes to perform the additional shuffles could result in a substantial profit loss.

U.S. News & World Report, June 29, 1992
DATABASE: U.S. 18-year-olds in 1990 who had driver's licenses: 68.4%; who were registered to vote: 27.5%.

Time, Jan. 22, 1990
Nissan's ad campaign for its new Infiniti luxury cars is renowned for a novel gimmick: The autos are nowhere in sight. The Infiniti ads, which depict lushly photographed trees, boulders, lightning bolts, and ocean waves (but no cars), were found by a November 1989 Gallup poll to be the best-recalled commercial on television. Unfortunately, Infiniti dealers in the U.S. sold only 32 cars a day during the last 2 months of the year, compared to the 134 cars per day sold by its archrival, the Toyota Lexus.

San Francisco Chronicle, Apr. 27, 1992
The latest numbers on sewing add up to a lot of threads: 60% of U.S. homes have sewing machines vs. 51% with VCRs . . . about the same number of women sew for recreation as play golf or tennis . . . sewers spend an average of $646 a year on patterns, fabrics, and notions.

Sports Illustrated, June 29, 1992
A shift to the right: Because so many more Major League Baseball pitchers throw right-handed than left-handed, left-handed hitters have a tremendous advantage at the plate. That's why only once in this century have right-handed batters outhit lefties in either league over a full season. But through last Saturday, right-handed batters were dominating the American League. The American League righties were outhitting lefties by a lusty nine points (.259 to .250).

Every day we are inundated with bits of information—data—like those in the examples above, whether we are in the classroom, on the job, or at home. Many of you taking this course are studying to be (or may already be) *data producers,* but most of you will be *data users.* As such, you will need to be able to make sense out of the mass of data that others produce for you. What specialized tools will enable you to become effective data users? The answer is *statistics.*

A common misconception is that a statistician is simply a "number cruncher" or a person who calculates and summarizes numbers, like baseball batting averages or unemployment rates. Statistics involves numbers, but there is much more to it than that.

According to *The Random House College Dictionary* (1988 ed.), statistics is "the science that deals with the collection, classification, analysis, and interpretation of numerical facts or data." In short, statistics is the **science of data**—a science that will enable you to be proficient data producers and efficient data users.

> **DEFINITION 1.1**
>
> **Statistics** is the science of data. This involves collecting, classifying, summarizing, organizing, analyzing, and interpreting data.

In this chapter we explore the different types of data that you will encounter in your field of study and introduce you to some ideas on methods for collecting data. The various statistical methods for summarizing, analyzing, and interpreting that data are presented in the chapters that follow.

1.2 TYPES OF DATA

1.2 Data are obtained by measuring some characteristic or property of the objects (usually people or things) of interest to us. These objects upon which the measurements (or observations) are made are called **experimental units**, and the properties being measured are called **variables** (since, in virtually all studies of interest, the property varies from one observation to another).

> **DEFINITION 1.2**
>
> An **experimental unit** is an object (person or thing) upon which we collect data.

> **DEFINITION 1.3**
>
> A **variable** is a characteristic (property) that differs or varies from one observation to the next.

EXAMPLE 1.1   A few weeks after the end of an academic semester, the Career Resource Center (CRC) at the University of Florida (UF) mails out questionnaires pertaining to the employment status and starting salary of students who graduated that particular semester. Appendix A provides information on the annual starting salaries for 1,795 UF graduates who earned a bachelor's, master's, or doctoral degree sometime between the fall semester 1989 through the spring semester 1991. (Appendix A lists data for the first 100 graduates in the data set. The complete data set is available on disk from the publisher.) Each row of the data set given in Appendix A pertains to a single graduate and gives the following:

1. Date of graduation
2. Gender
3. Degree
4. College or School
5. Major
6. Job type
7. Starting salary

Data for 10 graduates from the data set are shown in Table 1.1. For this data set, identify the experimental units and the variables measured.

TABLE 1.1

Data for 10 University of Florida Graduates; Data Extracted from Appendix A

OBS	DATE	GENDER	DEGREE	COLLEGE	MAJOR	JOBTYPE	SALARY
1	FALL89	M	DOCTOR	LAW	LAW	CRIMINAL	$28,000
2	FALL89	M	BACHELOR	BUSINESS ADM	ACCOUNTING	AUDITING	$28,000
3	FALL89	M	BACHELOR	ENGINEERING	IND/SYSTEMS	SALES	$33,000
4	SPRG90	F	BACHELOR	HEALTH	PHYS THERAPY	THERAPY	$27,600
5	SPRG90	M	MASTERS	LIB ARTS/SCI	LATIN AM STUD	ARMY	$49,000
6	SPRG90	F	BACHELOR	HEALTH	MEDICAL TECH	THERAPY	$21,100
7	SPRG90	F	BACHELOR	LIB ARTS/SCI	POLIT SCIENCE	INS CLAIMS	$27,500
8	FALL90	M	BACHELOR	LIB ARTS/SCI	ECONOMICS	BANK/FINAN	$23,000
9	SPRG91	F	DOCTOR	DENTISTRY	DENTISTRY	DENTIST	$50,000
10	SPRG91	F	BACHELOR	NURSING	NURSING	REG NURSE	$30,000

Solution   Since the CRC collects data on each UF graduate, the UF graduates are the experimental units. The variables (properties) measured on each graduate are the seven items previously listed, i.e., graduation date, gender, degree, etc. These are called variables since their values vary from one graduate to another (i.e., they are not constant).

All data (and, consequently, the variables we measure) are either **quantitative** or **qualitative** in nature.* Quantitative data are data that can be measured on a numerical scale. In general, qualitative data take values that are nonnumerical; they can only be classified. The statistical tools that we use to analyze data depend on whether the data is quantitative or qualitative. Thus, it is important to be able to distinguish between the two types of data.

DEFINITION 1.4

Quantitative data are observations measured on a numerical scale.

---

*A finer breakdown of data types into nominal, ordinal, interval, and ratio data is possible. **Nominal** data is qualitative data with categories that cannot be meaningfully ordered. **Ordinal** data is also qualitative data, but a distinct ranking of the groups from high to low exists. **Interval** and **ratio** data are two different types of quantitative data. For most statistical applications (and all the methods presented in this introductory text), it is sufficient to classify data as either quantitative or qualitative.

EXAMPLE 1.2   Refer to the data set in Appendix A described in Example 1.1. Determine the type (quantitative or qualitative) of each of the seven variables measured.

Solution   The first six variables listed are qualitative since the data they produce are values that are nonnumerical; they can only be classified into categories or groups. For example, GENDER is either M (for male) or F (for female). Similarly, values for COLLEGE are business administration, nursing, engineering, etc. We can classify a graduate according to GENDER or COLLEGE, but we cannot represent GENDER or COLLEGE as a numerical quantity.

Caution: The values of a qualitative variable may be coded or recorded numerically. For example, gender could have been coded 1 for males and 0 for females. Nevertheless, the data are qualitative since each observation can only be classified into one of a group of categories.

Variable #7, starting salary (in dollars), is quantitative since it is measured on a numerical scale. Two additional quantitative variables not included on the data set are age (in years) and grade point average (on a 4-point scale) of the UF graduate.

EXAMPLE 1.3   The data set in Appendix D resulted from a U.S. Army Corps of Engineers study of contaminated fish inhabiting the Tennessee River (in Alabama) and its tributaries. (See Case Study 1.1.) A total of 144 fish were captured and the following variables measured for each:

1. Location of capture
2. Species
3. Length (centimeters)
4. Weight (grams)
5. DDT concentration (parts per million)

Classify each of the five variables measured as quantitative or qualitative.

Solution   The variables length, weight, and DDT are quantitative because they are all measured on a numerical scale: length in centimeters, weight in grams, and DDT in parts per million. In contrast, location and species cannot be measured quantitatively; they can only be classified (e.g., channel catfish, largemouth bass, and smallmouth buffalo for species). Consequently, data on location and species are qualitative.

EXAMPLE 1.4 Many Americans are skeptical about the information they obtain from the news media. A recent *Newsweek*/Gallup poll of 760 American adults revealed that 61% believe little or only some of the news. Surprisingly, only 13% believe that news reporters should always reveal their sources to readers or listeners. Clearly, the purpose of this survey was to examine the opinions of American adults on the reliability of the news media. The experimental units— the objects upon which observations were taken—are the individual American adults. Suppose that for each American adult we record whether he or she thinks news reporters should always reveal their sources. Describe the variable observed for each experimental unit and explain whether it is a quantitative or qualitative variable.

Solution   One way to view this situation is to note that the variable being measured, opinion on public disclosure of news sources, can assume one of two conditions. An American adult will either believe or not believe that news sources should always be revealed. From this point of view, opinion on public disclosure of news sources is a qualitative variable.

A second way to view the response would be to convert the qualitative variable into a meaningful quantitative response by assigning the number 1 to all American adults who believe that news sources should always be revealed and the number 0 to those who do not. Then the sum of all the 0's and 1's in the sample will equal the total number of American adults in the sample who believe news reporters should always reveal their sources. For example, if there were five adults in the sample with two who believe and three who do not believe news reporters should always reveal their sources, then the sample data would be 1, 1, 0, 0, 0. The sum of these observations would equal 2, the number of adults in the sample who believe news reporters should always reveal their sources.

Qualitative variables cannot always be converted into meaningful quantitative variables, but it can be done (as shown above) when the observations fall into two categories. Then for a 0–1 assignment to the two categories, the sum of the observations will equal the number of observations falling into the "1" category.

To summarize, knowing the type (quantitative or qualitative) of the data that you want to analyze is one of the keys to selecting the appropriate statistical method to use. A second key involves the concept of population and samples, as discussed in the next section.

EXERCISES
APPLYING THE CONCEPTS

1.1   In Hawaii, condemnation proceedings have been under way since 1980 to enable private citizens to own the property that their homes are built on. Before 1980, only estates were permitted to own land, and homeowners leased the land from the estate (a law that dates back to Hawaii's feudal period). The new law requires estates to sell land to homeowners at a fair market price. As part of a study to estimate the fair market value of its land (called the "leased fee" value), a large Hawaiian estate collected the data shown in the accompanying table for five properties.

PROPERTY	LEASED FEE VALUE (thousands of dollars)	LOT SIZE (thousands of sq. ft.)	NEIGHBORHOOD	LOCATION OF LOT
1	70.7	13.5	Cove	Cul-de-sac
2	52.6	9.6	Highlands	Interior
3	87.1	17.6	Cove	Corner
4	43.2	7.9	Highlands	Interior
5	144.3	13.8	Golf Course	Cul-de-sac

a. Identify the experimental units.
b. State whether each of the variables measured is quantitative or qualitative.

1.2   Many conservationists fear that Antarctic whales, particularly blue whales, will soon become extinct. State whether each of the following variables relating to Antarctic blue whales is quantitative or qualitative.

a. Length of an Antarctic blue whale
b. Cause of death of each blue whale that died last year
c. Number of blue whales sighted in the Antarctic on a particular day

1.3   A study was conducted to examine the differences in job performance of white-collar workers with type A and type B behavior (*Journal of Human Stress*, Summer 1985). Type A workers exhibit on-the-job traits such as explosiveness, accelerated speech, ambitiousness, impatience, hostility, a tendency to challenge others, and the general appearance of tension; type B behavior is generally characterized by opposite attributes and qualities. The data for several workers at a large Canadian manufacturing firm who took part in the study are given here. [*Note:* Job performance of each worker was measured on a 5-point scale (a higher score indicates better performance) based on ratings of immediate supervisors.]

WORKER	BEHAVIOR TYPE	AGE	MANAGERIAL LEVEL	NUMBER OF EMPLOYEES SUPERVISED	PERFORMANCE RATING
1	A	47	Upper	22	3
2	B	28	Middle	10	5
3	B	52	Upper	105	2
4	A	30	Lower	3	1

a. Identify the experimental units.
b. State whether each of the variables measured is quantitative or qualitative.

1.4   The *Journal of Performance of Constructed Facilities* (Feb. 1990) reported on the performance dimensions of water distribution networks in the Philadelphia area. For one part of the study, the following data were collected for a sample of water pipe sections:

1. Pipe diameter (inches)
2. Pipe material
3. Age (year of installation)
4. Location
5. Pipe length (feet)
6. Stability of surrounding soil (unstable, moderately stable, or stable)
7. Corrosiveness of surrounding soil (corrosive or noncorrosive)
8. Internal pressure (pounds per square inch)
9. Percentage of pipe covered with land cover
10. Breakage rate (number of times pipe had to be repaired due to breakage)

Identify the data as quantitative or qualitative.

1.5 White-collar criminal offenders (e.g., those convicted of tax evasion, fraud, or embezzlement) often receive preferential treatment in the criminal justice system. Michael L. Benson (1984) studied the social and economic damage incurred by convicted white-collar criminals. From the case histories of approximately seventy offenders, he collected data on several variables, listed here. Classify each of the variables as quantitative or qualitative.

   a. Age in years
   b. Length of sentence in years
   c. Type of victim (business victim, government victim, individual victim, combination of victims)
   d. Type of occupation (public/professional or private business)
   e. Type of sentence (probation, work release, or incarceration)
   f. Race
   g. Recovery time in months (i.e., time from conviction to review of the file)

1.6 List five or more variables that your family physician considers while giving you a complete physical examination. State whether each is qualitative or quantitative.

1.7 Marketers are keenly interested in the factors that motivate coupon usage by consumers. A study reported in the *Journal of Consumer Marketing* (Spring 1988) asked a sample of 290 shoppers to respond to the following questions:

   a. Do you collect and redeem coupons?
   b. Are you price-conscious while shopping?
   c. On average, how much time per week do you spend clipping and collecting coupons?

   Classify the responses to the questions as quantitative or qualitative data.

POPULATIONS AND SAMPLES

1.3 When you examine a data set in the course of your studies, you will be doing so because the data characterize some phenomenon of interest to you. In statistics, the data set that is the target of your interest is called a **population**. Notice that a statistical population does not refer to a group of people; it refers to a set of measurements on a variable. This data set, which is typically large, exists in fact or is part of an ongoing operation and hence is conceptual. Some examples of phenomena and their corresponding populations are shown in Table 1.2.

TABLE 1.2
Some Typical Populations

PHENOMENON	EXPERIMENTAL UNITS	VARIABLE MEASURED	POPULATION
(a) White blood cell count of a hemophiliac	Hemophiliacs	White blood cell count	Set of white blood cell counts of all hemophiliacs
(b) Grade point average (GPA) of a college freshman	College freshmen	GPA	Set of grade point averages of all college freshmen
(c) Profit per job in a construction company	Jobs	Profit	Set of profits for all jobs performed recently or to be performed in the near future
(d) Quality of items produced on an assembly line	Manufactured items	Quality	Set of quality measurements for all items manufactured over the recent past and future

> **DEFINITION 1.6**
> ___
> A **population** is a collection (or set) of data that describe some phenomenon of interest to you.

If you have the population in hand, i.e., if you have every measurement in the population, then statistical methodology can help you describe this set of data. In future chapters we will find graphical (Chapter 2) and numerical (Chapter 3) ways to make sense out of a large mass of data. The branch of statistics devoted to this application is called **descriptive statistics**.

> **DEFINITION 1.7**
> ___
> The branch of statistics devoted to the organization, summarization, and description of data sets is called **descriptive statistics**.

Many populations are too large to measure each observation; others cannot be measured because they are conceptual. For example, population (a) in Table 1.2 cannot be measured since it would be impossible to identify all hemophiliacs. Even if we could identify them, it would be too costly and time-consuming to measure and record their white blood cell counts. Population (d) in Table 1.2 cannot be measured because it is partly conceptual. Even though we may be able to record the quality measurements of all items manufactured over the recent past, we cannot measure quality in the future. Because of this problem, we are required to select a subset of values from a population, called a **sample**.

> **DEFINITION 1.8**
> ___
> A **sample** is a subset of data selected from a population.

___
EXAMPLE 1.5   Refer to the starting salary data of Appendix A described in Example 1.1. Recall that the Career Resource Center (CRC) surveys all University of Florida graduates at the time of their graduation. However, the CRC estimates that only about half of the UF graduates return the questionnaire, and of these, only half indicate that they have secured a job as of the date of graduation. The salaries of Appendix A, then, are the starting salaries of the fall 1989 to spring 1991 graduates who (1) returned the CRC questionnaire and (2) indicated that they had secured a job at the time of graduation. Thus, it is important to note that the data set is not a complete listing of the annual starting salaries of all recent University of Florida graduates.

a. Suppose the target of your interest is the starting salary data in Appendix A. Describe the target population.

b. Suppose the target of your interest is the first-year financial compensations of all recent University of Florida graduates. Is the data of Appendix A a population or a sample?

Solution

a. The measurements of particular interest to us are the starting salaries of UF graduates. Since Appendix A lists only the starting salaries of the UF graduates who returned the CRC survey and secured a job, the target population consists of the collection of starting salaries of UF graduates *who returned the CRC questionnaire and had secured a job at graduation time.*

b. Since we are interested in the first-year financial compensations of *all* UF graduates during this period, the target population is the set of starting salaries of all recent graduates—not only those who returned the CRC questionnaires but also those who failed to return the survey even though they had secured a job by the date of graduation. Since Appendix A is only a subset of this target population, it represents a sample.

EXAMPLE 1.6 Potential advertisers value television's well-known Nielsen ratings as a barometer of a TV show's popularity among viewers. The Nielsen rating of a certain TV program (e.g., NBC's long-running hit comedy series "Cheers") is an estimate of the proportion of viewers, expressed as a percentage, who tune their sets to the program on a given night at a given time. A typical Nielsen survey consists of 165 families selected nationwide who regularly watch television. Suppose we are interested in the Nielsen ratings for the latest episode of "Cheers" (shown at 9:00 P.M. on Thursday nights).

a. Identify the population of interest.

b. Describe the sample.

Solution

a. We want to know which TV program was watched by a family on Thursday at 9 P.M.; consequently, the population of interest is the collection of programs watched by all families in the United States during this time slot. In this example, the measurements might be recorded as 1's and 0's, where a 1 represents a family who was tuned to "Cheers" and a 0 represents a family who was not tuned to "Cheers."

b. The sample—that is, the subset of the population—consists of the set of 0's and 1's for the 165 families in the Nielsen survey.

In Chapters 2 and 3, we demonstrate several statistical methods for organizing, describing, and summarizing data sets similar to those of Examples 1.5 and 1.6. However, statistics may involve much more than data description. In succeeding chapters, we will discover statistical methods that enable us to infer the nature of the population from the information in the sample. The branch of statistics devoted to this application is called **inferential statistics**. In addition, this *statistical methodology provides measures of reliability for each inference obtained from a sample*. This is one of the major contributions of inferential statistics. Anyone can examine a sample and make a "guess" about the nature of the population. For example, we might estimate that the average grade point average of college freshmen is 2.35, or that the starting salary range of UF graduates is between $15,000 and $25,000. But statistical methodology enables us to go one step further. When the sample is selected in a specified way from the population, we can also say how accurate our estimate will be, i.e., how close the estimate of 2.35 will be to the true average GPA, or how confident we are that the typical starting salary falls between $15,000 and $25,000.

DEFINITION 1.9

The branch of statistics concerned with using sample data to make an inference about a population is called **inferential statistics**. When proper sampling techniques are used, this methodology also provides a **measure of reliability** for the inference.

EXAMPLE 1.7 Do most state lottery winners who win big payoffs quit their jobs within one year of winning? No, according to a study conducted by sociologist and professor H. Roy Kaplan (*Journal of the Institute for Socioeconomic Studies*, Sept. 1985). Kaplan mailed questionnaires to over 2,000 lottery winners who won at least $50,000 in the past 10 years. Of the 576 who responded, only 11% had quit their jobs during the first year after striking it rich. In this study, identify

a. the population
b. the sample
c. the inference made about the population

Solution

a. The researcher is interested in state lottery winners who won at least $50,000 in the past 10 years. Consequently, the experimental units in the target population are all state lottery winners who won $50,000 or more during the decade. The measurement (or variable) of interest is the winner's job status— quit job within one year of winning or still working. Note that this is a qualitative variable. In this example, the measurements could have been recorded as 1's and 0's, where a 1 would represent a lottery winner who quit his or her job within one year and a 0 would represent a winner who is still working after one year.
b. The sample consists of the collection of job status measurements (1's and 0's) for the 576 lottery winners who responded to the questionnaire.
c. Since 11% of the respondents in the sample had quit their jobs, the inference is that 11% of the state lottery winners in the population also quit their jobs

within one year of winning the lottery. This leads to the conclusion reached by the researcher, namely, that most state lottery winners *do not* quit their jobs upon winning. Note that the researcher does not provide a measure of reliability for this inference. Enough information is available, however, to calculate such a measure. We will show in Chapter 8 that, with a high degree of "confidence," the estimate of 11% is within 2.6% of the true percentage. That is, the true percentage is no lower than 8.4% and no higher than 13.6%.

The key facts to remember in this section are summarized in the accompanying boxes.

THE OBJECTIVES OF STATISTICS

1. To describe data sets (populations or samples)
2. To use sample data to make inferences about a population

THE MAJOR CONTRIBUTION OF INFERENTIAL STATISTICS

Statistical methodology allows us to provide a measure of reliability for every statistical inference based on a properly selected sample.

EXERCISES
APPLYING THE CONCEPTS

1.8 *Postpartum depression* is the term used to describe the usually short-lived period of emotional sensitivity that many women suffer following childbirth. Studies have indicated that nearly 90% of all mothers experience some symptoms of postpartum blues. However, new evidence shows that men, too, can suffer from postpartum depression. Suppose a developmental psychologist wants to estimate the proportion of fathers who suffer from postpartum blues. Fifty men who have recently fathered a child are interviewed and observed in the home, and the number experiencing some form of postpartum depression is recorded.

   a. What is the population of interest to the developmental psychologist?
   b. Describe the sample in this problem.
   c. Suppose that 31 of the 50 men are diagnosed as having postpartum blues. The psychologist then estimates that 62% of all fathers experience postpartum depression. Do you believe that this estimate is equal to the proportion for the entire population? Explain.

1.9 An assembly line that produces automobile gear shifts is considered to be operating successfully if less than 1% of the gear shifts manufactured per day are defective. If 1% or more of the gear shifts are defective, the line must be shut down and proper adjustments made. Since checking every gear shift as it comes off the line is both time-consuming and costly, quality control inspectors randomly select 100 gear shifts from a day's production and test for defects. The decision on whether to shut down the line is then made according to the proportion of defectives in the 100 gear shifts.

a. Describe the population of interest to the manufacturer of the automobile gear shifts.

b. Describe the sample.

c. If the sample proportion of defectives is larger than 1%, is it necessarily true that the actual proportion of defective gear shifts produced per day is larger than 1%? Explain.

1.10 A new precooling method has been developed for preparing Florida vegetables for market. The system employs an air–water mixture designed to yield effective cooling with much less water flow than required for conventional hydrocooling. In an effort to compare the effectiveness of the two cooling systems, researchers divided 20 batches of green tomatoes into two groups of 10 each. One group was precooled with the new method, while the other was precooled with the conventional method. The total water flow (in gallons) required to effectively cool each batch was recorded.

a. Identify the populations of interest.

b. Describe the samples.

c. How could the sample data be used to compare the cooling effectiveness of the two systems?

1.11 To evaluate the current status of the dental health of school children, the American Dental Association conducted a survey to estimate the average number of cavities per child in grade school in the United States. One thousand school children from across the country were selected and examined by a dentist; the number of cavities was recorded for each.

a. Identify the population of interest to the American Dental Association.

b. Identify the sample.

c. How could the American Dental Association use the sample information to estimate the average number of cavities per child in grade school? Will this estimate equal the average for the population? Explain.

1.12 A study was conducted to explore the relation of self-esteem and positive inequity to on-the-job productivity (*Journal of Personality*, Dec. 1985). Eighty students enrolled in an industrial psychology course at a private New England university participated. All students were asked to complete a proofreading task and were compensated for their work on an hourly basis. However, some students were overpaid for each hour they worked (positive inequity condition), while the others were given fair compensation (equity condition). The results of the study revealed that individuals of high self-esteem were more productive (i.e., completed more of the task) in the positive inequity condition than in the equity condition. However, the reverse was true for individuals of low self-esteem.

a. In this study, we can envision four experimental conditions: (1) high self-esteem/positive inequity, (2) high self-esteem/equity, (3) low self-esteem/positive inequity, and (4) low self-esteem/equity. Describe the population corresponding to the four conditions. (Recall that the variable of interest is productivity, measured as amount of the task completed.)

b. Identify the samples. Assume that 20 students were assigned to each of the four experimental conditions.

c. Do you think the samples adequately represent the populations described in part **a**?

d. Do you think the results of the study were obtained by analyzing the data in the populations, or were they derived from sample information?

1.13 *Euthanasia*, the act of painlessly putting to death a person suffering from an incurable and painful disease or condition, has long been a dilemma of medical ethics. Developments in medical technology have contributed to the dilemma. Individuals who a few years ago would have died from their affliction can now be sustained beyond the point that even they themselves would desire. Suppose you work for a major opinion pollster and you have been assigned the task of conducting a survey for the Concern for Dying (formerly the Euthanasia Society). The purpose of the survey is to estimate the proportion of American adults who support euthanasia.

a. Clearly define the population of interest to the Concern for Dying.

b. Do you think it is possible to obtain the entire population? Explain.

c. Why should the sample you select for the survey be representative of the population?

1.14 A study of merit raises at 16 U.S. corporations was conducted to discover the extent to which merit-pay policies for employees are actually tied to performance (*Personnel Journal*, Mar. 1986). One phase of the study focused on the 3,990 merit raises (measured as percentage increases in salary) awarded during a year at one of the largest of the 16 firms. The analysis revealed that over half of the merit increases were between 7% and 10%.

a. Identify the variable of interest. Is it quantitative or qualitative?

b. Do the 3,990 merit raises represent a population or a sample? Explain.

1.15 The slope of a river delta region can sometimes be accurately predicted from knowledge of the typical size of stones found there. With this in mind, a geographer studying South America would like to estimate the average size of stones found in the delta region of the Amazon River. To obtain this estimate, the geographer collects 50 stones and measures the diameter of each.

a. What is the population of interest to the geographer?

b. Describe the sample in this problem.

c. Suppose that the average diameter of the 50 stones is 7.2 inches. Do you believe that this sample average will equal the average for the population? Explain.

CHAPTER CASE:
OBEDIENCE TO
AUTHORITY—THE
SHOCKING TRUTH

1.4

A basic principle upon which our society is organized is obedience to authority. Psychologists and sociologists agree that without obedience, our society would soon be very chaotic. But what if a person is asked to obey orders that appear to be evil or malicious? Could we expect that person to disobey? Stanley Milgram (1974) conducted a series of experiments designed to isolate the psychological factors that influence a person's behavior in such a setting.

The basic set-up of the experiment was as follows. Two people were brought into a room at an appointed time and one person (the subject of the experiment) was assigned to play the role of the "teacher"; the other person (an accomplice) was assigned to play the role of the "learner." The learner's task was to learn a list of word pairs, for example, box–boat. After reviewing the entire list with the learner, the teacher went through the list one-by-one, giving the first member of the pair (e.g., box). The learner was then asked to respond with the corresponding second member of the pair (e.g., boat). If the learner answered correctly, the teacher went on to the next word pair. However, if the learner gave the wrong answer, the teacher was instructed to deliver an electric shock by depressing one of 30 levers, ranging from 15 volts to 450 volts in increments of 15 volts. In addition to the voltage, the first 28 levers were labeled in groups of four as follows: Slight shock, Moderate shock, Strong shock, Very strong shock, Intense shock, Extreme intensity shock, and Danger: Severe shock. The final two levers were simply labeled "XXX." Teachers were instructed to start with the lowest-level shock and to increase the level of shock with each error until the entire list was learned correctly. [*Note:* The shocks, of course, were imaginary. However, the learner responded in a manner that led the teacher to believe that, in fact, they were real.]

If the teacher raised questions about the experiment and whether it should be continued, the experimenter would respond with one of the following four statements:

Statement 1   Please continue or Please go on.

Statement 2   The experiment requires that you continue.

Statement 3   It is absolutely certain that you continue.

Statement 4   You have no other choice, you must go on.

Only if the teacher refused to continue after all four statements had been made was the experiment stopped. One variable measured by Milgram was how far (in the sequence of shocks) the teacher proceeded before the experiment was terminated. For this case study, we will refer to this variable as the shock level. Note that the variable is quantitative since it is measured in volts.

Milgram's shock experiment provides us with an opportunity to apply our knowledge of inferential statistics to real-life data. In this example, the *population* of interest is conceptual—the collection of shock levels administered by *all* persons, if all people were to take part in the experiment. The *sample* is the set of, say 50, shock levels administered by those persons who actually participated in the experiment. The information in the sample can be used to make several inferences about the population. For example, we can use the average of the 50 shock levels in the sample to estimate the unknown average shock level of the population, or we could use the sample proportion of shock levels above 400 volts to estimate the true corresponding population proportion.

Milgram manipulated the basic experiment in several ways to study the impact of different situational factors on obedience. There were five different settings, as described here:

Setting 1   *Predicted behavior*   Subjects (i.e., teachers) had the experimental set-up described to them and were asked to predict how far they would go (although they did not actually perform the experiment).

Setting 2   *Remote*   The learner cannot be heard or seen but pounds on the walls at 300 volts and ceases responding and pounding at 315 volts.

Setting 3   *Voice feedback*   The learner cannot be seen but vocal protest can be clearly heard.

Setting 4   *Proximity*   The learner was placed in the same room so that he could be both seen and heard.

Setting 5   *Touch proximity*   The learner received a shock only when his hand rested on a shock plate. At 150 volts, the learner demanded to be released from the experiment and refused to put his hand on the shock plate. The teacher was instructed to take the learner's hand and force it onto the plate.

The sample data for these five different settings allow us to make several interesting comparisons. For example, the differences among the average shock levels of the five samples could be used to determine whether differences exist among the five corresponding population averages. If so, one could conclude that the difference in setting influences behavior.

Do we expect the sample values we calculate (for example, the sample average or proportion) to equal the corresponding population value? The answer, of course, is no. However, the statistical procedure we use to arrive at an estimate of the population value (e.g., average shock level) also yields a measure of reliability of the estimate—that is, we will be able to assess the accuracy of the estimate. (The concept of estimation is developed more fully in Chapter 8.)

What were the results of Milgram's experiment? Surprisingly, two-thirds of the teachers continued to obey the experimenter and applied shock levels of 400 or more volts despite the pleas of the learners. From this, Milgram concluded that a high level of obedience to authority exists in our society.

COLLECTING DATA: RANDOM SAMPLING

1.5 A careful analysis of most data sets will reveal that they are samples from larger data sets that are really the object of interest. Consequently, most applications of modern statistics involve sampling and using information in the sample to make inferences about the population. For these applications, it is essential that we obtain a **representative sample**, i.e., a sample that exhibits characteristics similar to those possessed by the target population. For example, consider the problem of estimating the average price of residential properties sold last year in the United States. It would be unwise to base our estimate on data collected for a sample of properties sold in Orange County, California, since this area has one of the highest priced housing markets in the United States. Our estimate would certainly be *biased* high, and, consequently, would not be very reliable. Example 1.8 illustrates this point.

EXAMPLE 1.8 Refer to Example 1.5b. Do you think the sample of salaries in Appendix A adequately represents the first-year financial compensations of all recent University of Florida graduates?

Solution   To a certain extent the data of Appendix A do characterize first-year financial compensation, but they do not do so completely. Recall that Appendix A includes only those salaries of graduates who returned the CRC survey and had secured a job at the time of graduation. We can speculate that these particular graduates are most likely comfortable with their starting salary and thus are willing to reveal it. Many others will treat their starting salary as confidential information. Also, not all major fields of study may appear in the sample. For example, many education majors who graduate in December or in May will not have secured a job until late August or early September, when grade schools, high schools, colleges, and universities begin a new academic year. Consequently, these graduates will indicate their employment status on the CRC survey as "negotiating for a job" or "still interviewing" and their eventual starting salary will not be known. Thus, the distribution of starting salaries in the sample (Appendix A) may vary substantially from the distribution of starting salaries of all graduates in the target population.

The most common way to satisfy the requirement of representative sample is to select the sample in such a way that every different sample of size n has an equal probability (or chance) of selection from the population.* This procedure is called **random sampling** and the resulting sample is called a (simple) **random sample**.

---

*A more formal definition of "probability" will be provided in Chapter 4.

> **DEFINITION 1.10**
>
> A **random sample** of n experimental units is one selected from the population in such a way that every different sample of size n has an equal chance of selection.

How can a random sample be generated? If the population is not too large, each observation may be recorded on a piece of paper and placed in a suitable container. After the collection of papers is thoroughly mixed, the researcher can remove n pieces of paper from the container; the elements named on these n pieces of paper are the ones to be included in the sample. Lottery officials utilize such a technique in generating the winning numbers for the state of Florida's weekly 6/49 Lotto game. Forty-nine white Ping-Pong balls (the population), each identified with a number from 1 to 49 in black numerals, are placed into a clear plastic drum and mixed by blowing air into the container. The Ping-Pong balls bounce at random until a total of six balls "pop" into a tube attached to the drum. The numbers on the six balls (the random sample) are the winning Lotto numbers.

This method of random sampling is fairly easy to implement if the population is relatively small. It is not feasible, however, when the population consists of a large number of observations. Since it is also very difficult to achieve a thorough mixing (recall the *Tampa Tribune* report on shuffling the deck of cards for blackjack), the procedure provides only an approximation to random sampling. Most scientific studies, however, rely on **random number generators** to automatically generate the random sample. Random number generators can be found in a table of random numbers (such as Table 1 of Appendix G) and in computers. In fact, almost all of the commercial statistical computer software packages available today (e.g., SAS, SPSS, Minitab) have procedures for generating random samples. We illustrate the use of these random number generators in the following examples.

EXAMPLE 1.9   Consider the set of 1,795 starting salaries in Appendix A. Suppose we designate this data set as our target population. With a population this large, we will resort to sampling to make inferences about the population. Use a table of random numbers to generate a random sample of five starting salaries from the population.

Solution   A portion of the random number table in Appendix G (Table 1) is reproduced in Table 1.3 (page 18). The steps for obtaining a random sample using the table are outlined in the box below it. The five random numbers and the associated observations on starting salary are shown in Table 1.4 (page 18).

TABLE 1.3

Reproduction of a Portion of Table 1 in Appendix G

		COLUMN				
	1	2	3	4	5	6
ROW 1	10480	15011	01536	02011	81647	91646
2	22368	46573	25595	85393	30995	89198
3	24130	48360	22527	97265	76393	64809
4	42167	93093	06243	61680	07856	16376
5	37570	39975	81837	16656	06121	91782
6	77921	06907	11008	42751	27756	53498
7	99562	72905	56420	69994	98872	31016
8	96301	91977	05463	07972	18876	20922
9	89579	14342	63661	10281	17453	18103
10	85475	36857	53342	53988	53060	59533
11	28918	69578	88231	33276	70997	79936
12	63553	40961	48235	03427	49626	69445
13	09429	93969	52636	92737	88974	33488
14	10365	61129	87529	85689	48237	52267
15	07119	97336	71048	08178	77233	13916

TABLE 1.4

Random Sample of Five Starting Salaries Selected from Appendix A

RANDOM NUMBER	STARTING SALARY
1048	$11,000
1501	20,000
0153	14,500
0201	42,000
0624	27,700

USING A TABLE OF RANDOM NUMBERS TO GENERATE A RANDOM SAMPLE OF SIZE _n_ FROM A POPULATION OF _N_ ELEMENTS

STEP 1 The elements (starting salaries) in the population are numbered from 0001 to 1795 in Appendix A. This labeling implies that we will obtain random numbers of four digits from the table, selecting only those numbers with values less than or equal to 1795. Note that Table 1.3 gives 5-digit random numbers in each column. Consequently, we'll use only the first four digits of each random number.

STEP 2 Arbitrarily, let's begin in row 1, column 1 of the table. The random number entry given there is 10480. Using the first four digits, the random number generated is 1048 (shaded in Table 1.3). Thus, we'll choose the salary numbered 1048 in Appendix A as our first element of the random sample.

STEP 3 Proceeding horizontally to the right across the columns (this choice of direction is arbitrary), the next entry in the table is 1501. Therefore, the salary numbered 1501 represents the second element in the sample. Continuing in this manner, the remaining elements to be included in the sample are those numbered 0153, 0201, 8164 (skip), 9164 (skip), (proceeding to row 2) 2236 (skip), 4657 (skip), . . . , (proceeding to row 4) 4216 (skip), 9309 (skip), and 0624.

EXAMPLE 1.10 Assume, again, that our target population is the data set consisting of the 1,795 starting salaries given in Appendix A. Use the computer to select a random sample of size $n = 50$ from the population.

Solution We used the random number generator of the SAS statistical software package to obtain the sample of 50 starting salaries.* The SAS printout listing these 50 random numbers and associated salaries is shown in Figure 1.1. Consequently, the 50 salaries shown in Figure 1.1 represent the sample.

OBS	GRADUATE	SALARY	OBS	GRADUATE	SALARY
1	38	$26,500	26	966	$30,000
2	42	$32,000	27	1100	$33,100
3	62	$33,000	28	1130	$19,000
4	99	$19,000	29	1138	$20,000
5	168	$25,000	30	1141	$33,700
6	208	$18,000	31	1154	$14,000
7	318	$8,500	32	1254	$60,000
8	341	$32,000	33	1260	$29,000
9	354	$27,000	34	1277	$20,000
10	355	$24,000	35	1302	$24,600
11	374	$31,200	36	1332	$18,000
12	426	$15,000	37	1349	$30,000
13	441	$28,000	38	1380	$20,000
14	451	$27,100	39	1387	$31,500
15	480	$33,000	40	1420	$33,800
16	510	$30,000	41	1441	$26,000
17	613	$23,000	42	1450	$27,000
18	621	$17,000	43	1494	$40,000
19	673	$28,000	44	1531	$21,000
20	684	$27,900	45	1536	$24,700
21	711	$70,000	46	1563	$23,900
22	718	$22,000	47	1602	$24,900
23	774	$26,000	48	1624	$30,000
24	785	$23,000	49	1677	$26,000
25	848	$26,000	50	1685	$18,000

FIGURE 1.1

SAS-Generated Random Sample of 50 Salaries from Appendix A.

Although random sampling represents one of the simplest of the multitude of sampling techniques available for research, most of the statistical techniques presented in this introductory text assume that such a sample (or a sample that closely approximates a random sample) has been collected. We consider a few of the other, more sophisticated, sampling methods in Chapter 7.

EXERCISES
LEARNING THE MECHANICS

1.16 Appendix B contains starting salary information for University of Florida graduates in five different colleges. Use a random number generator to produce a random sample of 15 starting salaries from graduates in the College of Engineering.

1.17 Appendix E contains customer checkout times for 500 grocery shoppers at a Publix supermarket in Florida. Checkout time is measured as the total length of time (in seconds) required for service personnel to check the

*The SAS commands used to generate the sample (as well as Minitab and SPSS commands) are given in the Computer Lab section at the end of this chapter.

prices of the customer's food items, total the prices, accept payment, and return change. Use a random number generator to produce a random sample of 20 checkout times from Appendix E.

1.18   Use a random generator to produce a random sample of 10 observations on fish weights from Appendix D.

APPLYING THE CONCEPTS

1.19   Test marketing is used by companies to gauge consumer preferences for a new product. Conventional marketing tests usually involve sampling 3% of the target population over a 1-year period, a slow and expensive process to carry out. *Fortune* (Oct. 29, 1984) reports that many companies are turning to alternative methods that use a much smaller sample over a much shorter time span. One such method, called **simulated test marketing**, is described as follows: "A consumer recruited at a shopping center reads an ad for a new product and gets a free sample to take home. Later, she rates it in a telephone interview." The telephone responses are used by the test-marketing firm to predict potential sales volume. Simulated marketing tests appear to identify potential failures reasonably well, but "they don't do such a good job predicting the upside potential of products." Why might the sampling procedure yield a sample of consumer preferences that underestimates sales volume of a successful new product?

1.20   A clinical psychologist is asked to view tapes in which each of six experimental subjects is discussing his or her recent dreams. Three of the six subjects have been previously classified as "high-anxiety" individuals, and the other three as "low-anxiety." The psychologist is told only that there are three of each type and is asked to select the three high-anxiety subjects.

   a. How many different samples of three subjects may be selected by the psychologist?
   b. List them.
   c. Do you think the sample chosen by the psychologist will be random? Explain.

1.21   Many opinion surveys are conducted by mail. In such a sampling procedure, a random sample of persons is selected from among a list of people who are supposed to constitute a target population (e.g., purchasers of a product). Each is sent a questionnaire and is requested to complete and return the questionnaire to the pollster. Why might this type of survey yield a sample that would produce biased inferences?

1.22   One of the most infamous examples of improper sampling was conducted in 1936 by the *Literary Digest* to determine the winner of the Landon–Roosevelt presidential election. The poll, which predicted Landon to be the winner, was conducted by sending ballots to a random sample of persons selected from among the names listed in the telephone directories of that year. In the actual election, Landon won in Maine and Vermont but lost in the remaining 46 states. The *Literary Digest*'s erroneous forecast is believed to be the major reason for its eventual failure.

   What was the cause of the *Literary Digest*'s erroneous forecast? In other words, why might the sampling procedure described yield a sample of people whose opinion could be biased in favor of Landon?

SUMMARY

This chapter identified the types of problems for which statistical procedures are useful—namely, describing data sets and using *sample data* to make *inferences* about a sampled *population*. Basic to the application of these techniques is the identification of a population of data, *qualitative* or *quantitative*, that truly characterizes the phenomenon of interest.

Most statistical problems involve sampling and using a sample to make inferences about the sampled population. For example, the starting salaries of University of Florida bachelor's degree graduates for five colleges (given in Appendix

B) could be viewed as samples of the starting salaries of University of Florida bachelor's degree graduates for their respective colleges. Do these sample values suggest a difference in the distributions of the starting salaries among the five colleges? Statistical methods to be covered later will help us answer this question and will provide us with a measure of reliability for our decision.

The remainder of this course will examine some basic statistical procedures for describing data sets and giving them meaning. More important, we will learn how to use sample data to infer the nature of the sampled population and to do so with a known degree of *reliability*.

KEY WORDS

Data	Random sample
Experimental Unit	Reliability
Inference	Representative sample
Population	Sample
Qualitative data	Statistics
Quantitative data	Variable
Random number generator	

SUPPLEMENTARY EXERCISES

LEARNING THE MECHANICS

1.23 Use a random number table or a computer to generate a random sample of 10 observations from a population with 40,000 elements.

1.24 Use a random number table or a computer to generate a random sample of 15 observations on tar content from the cigarette data in Appendix F.

APPLYING THE CONCEPTS

1.25 State whether each of the following variables is quantitative or qualitative.

 a. Number of acres in a plot of land
 b. Mode of transportation (to and from work) for a city employee
 c. Type of residential water-heating system
 d. Time required for postoperative pain to be relieved in surgery patients

1.26 When Nissan introduced its new Infiniti luxury cars in 1989, its television ad campaign was renowned for a novel gimmick: The automobiles were nowhere in sight. The Infiniti ads, which depicted lushly photographed trees, boulders, lightning bolts, and ocean waves (but no cars) were found by a nationwide Gallup poll of 1,000 consumers to be the best-recalled commercial on television (*Time*, Jan. 22, 1990).

 a. Describe the population of interest to the pollsters.
 b. Identify the sample.
 c. What is the inference made by the Gallup poll?

1.27   Consider the customer checkout times for a sample of 500 grocery shoppers at a Publix supermarket recorded in Appendix E.

   a. Describe the population from which the sample is selected.
   b. Suppose you were to use the average of the 500 customer checkout times to estimate the average checkout time of all customers who shop at the supermarket. Would the sample average equal the average for the population? Explain.

1.28   Classify each of the following variables as quantitative or qualitative.

   a. Political affiliation of a chief executive whose firm is listed in the *Fortune 500*
   b. Geographical region with the highest unemployment rate in the United States
   c. Gas mileage attained by an automobile powered by alcohol
   d. Fee charged by an attorney to handle an uncontested divorce
   e. Highest educational degree attained by members of the faculty at a community college

1.29   A file clerk is assigned the task of selecting a random sample of 26 company accounts (from a total of 5,000) to be audited. The clerk is considering two sampling methods:

   Method A   Organize the 5,000 company accounts in alphabetical order (according to the first letter of the client's last name). Then randomly select one account card for each of the 26 letters of the alphabet.

   Method B   Assign each company account a 4-digit number from 0001 to 5000. Using a computer random number generator, choose 26 4-digit numbers (from 0001 to 5000) and match the numbers with the corresponding company account.

   Which of the two methods would you recommend to the file clerk? Which sampling method could possibly yield a biased sample?

1.30   "Possibly one of the single greatest sources of abuse and neglect of the elderly is the family, especially the . . . family in which the child has assumed a caretaking role and the parent is now in the dependent role," concludes Suzanne K. Steinmetz (*Aging*, Jan.–Feb. 1981). To investigate the problem of elder abuse, Steinmetz interviewed a sample of 60 adult children caring for a dependent elderly parent (65 years or older). During the course of the interview, Steinmetz was able to ascertain various methods used by the adult children to control their elderly parents. The results are summarized in the accompanying table.

METHOD OF CONTROL	PROPORTION OF ADULT CHILDREN
Screamed and yelled	.40
Used physical restraint	.06
Forced feeding or medication	.06
Threatened to send to nursing home	.06
Threatened with physical force	.04
Hit or slapped	.03
Used nonabusive method	.35

   State whether the variable of interest, method of controlling elderly parents, is qualitative or quantitative.

1.31   Hundreds of sea turtle hatchlings, instinctively following the bright lights of condominiums, wandered to their deaths across a coastal highway in Florida (*Tampa Tribune*, Sept. 16, 1990). This incident led researchers to begin experimenting with special low-pressure sodium lights. On one night, 60 turtle hatchlings were released on a dark beach and their direction of travel noted. The next night, the special lights were installed and the same 60 hatchlings were released. Finally, on the third night, tar paper was placed over the sodium lights. Consequently, the direction of travel was recorded for each hatchling under three experimental conditions—darkness, sodium lights, and sodium lights covered with tar paper.

a. Identify the population of interest to the researchers.
b. Identify the sample.
c. What type of data were collected, quantitative or qualitative?

CASE STUDY 1.1

CONTAMINATION OF FISH IN THE TENNESSEE RIVER

Chemical and manufacturing plants often discharge toxic waste materials into nearby rivers and streams. These toxicants have a detrimental effect on the plant and animal life inhabiting the river and the river's bank. One type of pollutant, commonly known as DDT, is especially harmful to fish and, indirectly, to people. The Food and Drug Administration sets the limit for DDT content in individual fish at 5 parts per million (ppm). Fish with DDT content exceeding this limit are considered potentially hazardous to people if consumed. A study was undertaken to examine the DDT content of fish inhabiting the Tennessee River (in Alabama) and its tributaries.

The Tennessee River flows across the northern part of the state of Alabama, through Wheeler Reservoir, a national wildlife refuge. Ecologists fear that contaminated fish migrating from the mouth of the river to the reservoir could endanger other wildlife that prey on the fish. This concern is more than academic. A manufacturing plant was once located along Indian Creek, which enters the Tennessee River 321 miles upstream from the mouth. Although the plant has been inactive for over 20 years, there is evidence that the plant discharged toxic materials into the creek, contaminating all the fish in the immediate area. Have the fish in the Tennessee River and its tributary creeks also been contaminated? And if so, how far upstream have the contaminated fish migrated? To answer these and other questions, members of the U.S. Army Corps of Engineers collected fish samples at different locations along the Tennessee River and three tributary creeks: Flint Creek, Limestone Creek, and Spring Creek. Each fish was first weighed (in grams) and measured (length in centimeters), then the fillet of the fish was extracted and the DDT concentration (in parts per million) in the fillet measured.

Appendix D contains the length, weight, and DDT measurements for a total of 144 sampled fish.* Notice that the data set also contains information on the location (i.e., where the fish were captured) and species of the fish. Three species of fish were examined: channel catfish, largemouth bass, and smallmouth buffalo. The different symbols for location are interpreted as follows. The first two characters represent the river or creek and the remaining characters represent the distance (in miles) from the mouth of the river or creek. For example, FCM5 indicates that the fish was captured in Flint Creek (FC), 5 miles upstream from the mouth of the creek (M5). Similarly, TRM380 denotes a fish sample collected from the Tennessee River (TR), 380 miles upstream from the river's mouth

---

*Source: U.S. Army Corps of Engineers, Mobile District, Alabama.

(M380). These data provide us with an opportunity to compare the DDT contents of fish at different locations and among the different species and to determine the relationship (if any) of length and weight to DDT content.

a. Suppose you view the totality of fish DDT concentrations in Appendix D as a population. What does the population characterize?
b. Suppose you view the DDT concentrations in each species of fish as a distinct population. What do the three populations characterize?
c. Randomly select 10 DDT concentrations from the listing in Appendix D. Would this constitute a sample from the population you described in part **a**? Explain.
d. Suppose you were to use the average value of the 10 DDT concentrations (part **c**) to estimate the average DDT content of the 144 fish of Appendix D. Would the sample average equal the average for the population? Explain.
e. Calculate the proportion of the 10 measurements (part **c**) with DDT levels higher than the 5 ppm maximum set by the FDA. Is this a good estimate of the true proportion in the population? Explain.

CASE STUDY 1.2

THE SAT PILL

According to *Newsweek* (Nov. 16, 1987), "Some students will do anything to inch closer to a perfect 1600 on the Scholastic Aptitude Tests (SAT). And if you think we're talking lucky shirts or a peek at a neighbor's answers, wake up to the pharmaceutical age. A researcher reports that a prescription drug may help nervous test takers improve their SATs."

The researcher is Dr. Harrison Faigel of Brandeis University. Over a period of two years Faigel experimented with giving propranolol, one of the class of heart drugs called beta blockers, to nervous high school students prior to taking their SATs. Beta blockers, which interfere with adrenaline, have been used for heart conditions and minor stress such as stage fright for over 25 years. Faigel felt that the same calming effect that beta blockers provide heart patients could also be used to reduce anxiety in test takers.

To test this theory, he selected 22 high school juniors who had not performed as well on the SAT as they should have based on IQ and other academic evaluations. Presumably, these students performed poorly because they approached the SAT with a tremendous amount of anxiety and fear. One hour before the students repeated the test in their senior year, Faigel administered each a dosage of a beta blocker. Typically, students who retake the test without special preparations will increase their scores by an average of 38 points. Faigel's students, however, improved their scores by an average of 120 points!

a. Identify the experimental units in this study.
b. Identify the measured variable.
c. Describe the population of interest to the researcher. (Give the precise statistical definition.)

d. Describe the sample.

e. Based on the sample results, what inference would you make about the use of beta blockers to increase SAT scores? (In Chapters 8–10 we show you how to assess the reliability of this type of inference.)

f. Refer to part **e**. Robert Cameron, director of research and development for the College Board (sponsors of the SAT), warns that "the findings have to be taken with a great deal of caution" and that they should not be interpreted to mean "that someone has discovered the magic pill that will unlock the SAT for thousands of teenagers who believe they do not do as well as they should because they're nervous."* Give several reasons for issuing such a warning.

CASE STUDY 1.3

THE NEW HITE REPORT—CONTROVERSY OVER THE NUMBERS

In 1968 researcher Shere Hite shocked conservative America with her now-famous "Hite Report" on the permissive sexual attitudes of American men and women. Twenty years later, Hite was surrounded by controversy again with her book, *Women and Love: A Cultural Revolution in Progress* (Knopf Press, 1988). In this new Hite report, she reveals some startling statistics describing how women feel about contemporary relationships:

- 84% of women are not emotionally satisfied with their relationship
- 95% of women report "emotional and psychological harassment" from their men
- 70% of women married 5 years or more are having extramarital affairs
- Only 13% of women married more than 2 years are "in love"

Hite conducted the survey by mailing out 100,000 questionnaires to women across the country over a 7-year period. Each questionnaire consisted of 127 open-ended questions, many with numerous subquestions and follow-ups. Hite's instructions read: "It is not necessary to answer every question! Feel free to skip around and answer those questions you choose." Approximately 4,500 completed questionnaires were returned for a response rate of 4.5%, and they form the data set from which these percentages were determined. Hite claims that these 4,500 women are a representative sample of all women in the United States, and therefore, the survey results imply that vast numbers of women are "suffering a lot of pain in their love relationships with men." Many people disagree, however, saying that only unhappy women are likely to take the time to answer Hite's 127 essay questions, and thus her sample is representative only of the discontented.

The views of several statisticians and expert survey researchers on the validity of Hite's "numbers" were presented in a recent article in *Chance* magazine (Summer 1988). A few of the more critical comments follow.†

Gainesville Sun, Oct. 22, 1987.

†*Source*: Streitfeld, D. "Shere Hite and the trouble with numbers." *Chance: New Directions for Statistics and Computing*, Vol. 1, No. 3, Summer 1988, pp. 26–31. Springer-Verlag, © 1988, the *Washington Post*. Reprinted with permission.

- Hite used a combination of haphazard sampling and volunteer respondents to collect her [data]. First, Hite sent questionnaires to a wide variety of organizations and asked them to circulate the questionnaires to their members. She mentions that they included church groups, women's voting and political groups, women's rights organizations and counseling and walk-in centers for women. These groups would not seem to be representative of women in general; there is an over-representation of feminist groups and of women in troubled circumstances. In addition, the use of groups to distribute the questionnaires meant that gatekeeprs had the power of assuring a zero response rate by not distributing the questionnaire, or conversely of greatly stimulating returns by endorsing the study in some fashion. Second, Hite also relied on volunteer respondents who wrote in for copies of the questionnaire. These volunteers seem to have been recruited from readers of her past books and those who saw interviews on television and in the press. This type of volunteer respondent is the exact opposite of the randomly selected respondent utilized in standard survey research and even more potentially unrepresentative than the group samples cited above. (**Tom Smith, National Opinion Research Center**)
- So few people responded, it's not representative of any group, except the odd group who agreed to respond. Hite has no assurance that even her claimed 4.5% response rate is correct. How do we know how many people passed their hands over these questionnaires? You don't want to fill it out, you give it to your sister, she gives it to a friend. You'll get one response, but that questionnaire may have been turned down by five people. (**Donald Rubin, Professor and Chairman, Department of Statistics, Harvard University**)
- When you get instructions to only answer those questions you wish to, you're likely to skip some. Isn't it more likely that, for example, a woman who feels strongly about affairs would be more likely to answer questions on that subject than a woman who does not feel as strongly? Thus, her finding that 70% of all women married over five years are having affairs is meaningless because she does not report how many people answered each question. I cannot tell whether this means 70% of 1,000 women or 70% of 10 women. (**Judith Tanur, Professor of Sociology and Statistical Specialist in Survey Methodology, State University of New York, Stony Brook**)
- Even in good samples, where you have a 50% or 70% response rate, you usually have some skews—say, with income, race, or region. If she can do a sample like this, she's got the Rosetta Stone, and I'll come study from her. (**Martin Frankel, Professor of Statistics & CIS, Baruch College**, commenting on Hite's claim that her sample matches that of the U.S. female population in terms of demographic balance.)
- According to Hite, whether you're 18 or 71, you're going to answer the questions the same way. Whether white, black, Hispanic, Middle Eastern, or Asian American, you're going to answer the same way. Whether you make $5,000 a year or over $75,000, you'll answer the questions the same way. I've never seen anything like this in my career—and the Kinsey Institute collects data from everybody. (**June Reinisch, Director of Kinsey Institute, Indiana University**, commenting on Hite's numbers showing that no matter what the demographic

breakdown of the women married five years or more, about 70% are having extramarital affairs.)

a. Identify the population of interest to Shere Hite. What are the experimental units?

b. Identify the variables of interest to Hite. Are they quantitative or qualitative variables?

c. Describe how Hite obtained her sample.

d. What inferences did Hite make about the population? Comment on the reliability of these inferences.

e. Discuss the difficulty in obtaining a random sample of women across the United States to take part in a survey similar to the one conducted by Shere Hite.

COMPUTER LAB 1.1
CREATING A DATA SET READY FOR ANALYSIS

In the Computer Lab sections in this text, we describe how to use any one of three statistical software packages—SAS, SPSS, and Minitab. All three were selected because of their current popularity, ease of use, and availability at most university computing centers. In addition, versions of SAS, SPSS, and Minitab are available for both large mainframe computers and for personal computers (PCs).

SAS, SPSS, and Minitab are similar in requiring the user to enter a list of commands, or instructions, in a specific form. Each utilizes the following three basic types of instructions:

1. *Data entry commands:* instructions on how the data will be entered
2. *Input data values:* the values of the variables in the data set
3. *Statistical analysis commands:* instructions on what type of analysis is to be conducted on the data

In this Computer Lab we list the steps necessary to create a data set ready for analysis. If you are using SAS, SPSS, or Minitab, this involves specifying *data entry commands* and *input data values*. (The appropriate statistical analysis commands are provided at the end of the relevant Computer Lab sections of the text.)

The data set of interest is listed in Table 1.5 (page 28). These are the starting salary data for five recent UF graduates extracted from Appendix B. (We discuss Appendix B in more detail in Chapter 2.)

Note: With few exceptions, the commands provided in the following sections are appropriate for the large mainframe and PC versions of SAS, SPSS, and Minitab. When a mainframe computer is being used, however, these statements must be preceded by the job control language (JCL) commands required at your institution.

TABLE 1.5

Data for Five UF Graduates; Data Extracted from Appendix B

GRADUATE	COLLEGE	GENDER	SALARY
1	BUS	F	$35,000
2	ENG	M	41,000
3	LAS	M	20,000
4	BUS	M	28,000
5	NUR	F	25,000

SAS

```
Command
  line
    1   DATA SAMPLE;              ⎫
    2   INPUT COLLEGE $ GENDER $ SALARY;  ⎬ Data entry instructions
    3   MONSAL = SALARY/12;       ⎪
    4   CARDS;                    ⎭
    5   BUS F 35000   ⎫
    6   ENG M 41000   ⎪   Input data values
    7   LAS M 20000   ⎬   (1 observation per line)
    8   BUS M 28000   ⎪
    9   NUR F 25000   ⎭
   10   PROC PRINT;  } Print instruction
```

COMMAND 1 SAMPLE is an arbitrarily chosen name used to identify the data set. (Data set names are restricted to a maximum length of eight characters.)

COMMAND 2 COLLEGE, GENDER, and SALARY are arbitrarily chosen names for the variables on the data set. (Variable names are also restricted to a maximum length of eight characters.) A dollar sign ($) must follow the name of any non-numeric variable on the data set.

COMMAND 3 MONSAL (monthly salary) is calculated as the yearly salary divided by 12 (months). (The standard arithmetic operation symbols, $+$, $-$, $*$, and $/$, are used for addition, subtraction, multiplication, and division, respectively.)

COMMAND 4 CARDS signals SAS that the input data values are to follow.

COMMANDS 5–9 Each data line gives the values of the variables on the data set for a single observation (UF graduate) in the order in which the variables are listed in the INPUT command. Input data values must be separated by at least one blank space; commas are not permitted in numeric values.

COMMAND 10 The PRINT procedure (PROC) will produce a listing of the entire data set. In addition to the INPUT variables, the data set will contain any variables created using the standard arithmetic operations (e.g., MONSAL) in command 3. The output from the SAS program is shown in Figure 1.2a.

FIGURE 1.2
Printouts of Data in Table 1.5

a. SAS

OBS	COLLEGE	GENDER	SALARY	MONSAL
1	BUS	F	35000	2916.67
2	ENG	M	41000	3416.67
3	LAS	M	20000	1666.67
4	BUS	M	28000	2333.33
5	NUR	F	25000	2083.33

b. SPSS

COLLEGE	GENDER	SALARY	MONSAL
BUS	F	35000.00	2916.67
ENG	M	41000.00	3416.67
LAS	M	20000.00	1666.67
BUS	M	28000.00	2333.33
NUR	F	25000.00	2083.33

Number of cases read =   5   Number of cases listed =   5

c. Minitab

ROW	COLLEGE	GENDER	SALARY	MONSAL
1	1	1	35000	2916.67
2	2	0	41000	3416.67
3	3	0	20000	1666.67
4	1	0	28000	2333.33
5	4	1	25000	2083.33

GENERAL   All SAS commands must end with a semicolon; the only exceptions to this rule are the input data values.

SPSS

```
Command
  line
    1     DATA LIST FREE/COLLEGE (A3) GENDER (A1) SALARY,  ⎫ Data entry
    2     COMPUTE MONSAL = SALARY/12,                       ⎬ instructions
    3     BEGIN DATA,
    4     BUS F 35000  ⎫
    5     ENG M 41000  ⎪
    6     LAS M 20000  ⎬  Input data values
    7     BUS M 28000  ⎪  (1 observation per line)
    8     NUR F 25000  ⎭
    9     END DATA,
   10     LIST,         ⎬ Print instruction
```

COMMAND 1   COLLEGE, GENDER, and SALARY are arbitrarily chosen names for the variables on the data set. (Variable names are restricted to a maximum length of eight characters.) An alphanumeric format of the form (An) must be specified, in parentheses, after the name of any nonnumeric variable. For example, A3 specifies that the nonnumeric variable COLLEGE will occupy three columns on the input data lines. [In the mainframe SPSS environment, an asterisk (*) separates the numeric variables from the nonnumeric variables, with the numeric variables listed first; in the PC environment, the asterisk must be omitted.]

COMMAND 2 MONSAL (monthly salary) is calculated as annual salary divided by 12. (The standard arithmetic operation symbols, $+$, $-$, $*$, $/$, are used for addition, subtraction, multiplication, and division, respectively.)

COMMAND 3 BEGIN DATA signals SPSS that the input data values are to follow.

COMMANDS 4–8 Each data line gives the values of the variables on the data set for a single observation (UF graduate) in the order in which the variables are listed in the DATA LIST command. Values in the data list must be separated by at least one blank space; commas are not permitted in numeric values.

COMMAND 9 END DATA signals SPSS that all input data values have been entered.

COMMAND 10 The LIST command will produce a listing of the data for all the variables on the data set, including the variables created using COMPUTE commands. The output from the SPSS program is shown in Figure 1.2b.

GENERAL In the PC environment, all SPSS commands must end with a command terminator (usually a period); the only exceptions to this rule are the input data values. Omit the periods when using mainframe SPSS.

MINITAB

Command line		
1	`READ C1 C2 C3`	} Data entry instruction
2	`1   1 35000`	
3	`2   0 41000`	Input data values
4	`3   0 20000`	(1 observation per line)
5	`2   0 28000`	
6	`4   1 25000`	
7	`DIVIDE C3 BY 12 PUT INTO C4`	} Data entry instruction
8	`NAME C1='COLLEGE' C2='GENDER'`	
9	`NAME C3='SALARY' C4='MONSAL'`	
10	`PRINT C1-C4`	} Print instruction
11	`STOP`	

COMMAND 1 The three variables to be read onto the Minitab "worksheet" are identified by the "columns" into which they are placed, C1, C2, and C3. (Minitab does not, in general, recognize variable names.) Thus, COLLEGE will be read in column 1, GENDER in column 2, and SALARY in column 3.

COMMANDS 2–6 Each data line gives the values of the variables read in the worksheet columns for a single observation (UF graduate). Input data values must be separated by at least one blank space; commas are not permitted. Minitab also requires that all data used in statistical analysis be numerical. Thus, values of the qualitative variables are converted to numbers in C1 and C2. For example, for COLLEGE (C1) we let 1 represent business administration, 2 represent engineering, etc., while for GENDER (C2) we let 1 represent female and 0 represent male.

COMMAND 7 Minitab uses the word commands ADD, SUBTRACT, MULTIPLY, and DIVIDE to perform the usual arithmetic operations on variables. The monthly salary (computed as annual salary divided by 12) is stored in C4.

COMMANDS 8–9 The NAME command is used to name the columns of the Minitab "worksheet" for labeling printouts. In future commands, you may refer to the columns by these names (e.g., SALARY) or by the column numbers (e.g., C3).

COMMAND 10 The PRINT command will produce a listing of the data in the Minitab worksheet for the specified variables (columns). The output of the Minitab program is shown in Figure 1.2c.

COMMAND 11 All Minitab programs terminate with the STOP command.

COMPUTER LAB 1.2
GENERATING RANDOM SAMPLES

The computer commands for generating random numbers in the interval 0 to 1 (called *uniform* random numbers) are provided in this section. These random number generators can be used (instead of Table 1 in Appendix G) to select random samples using the technique outlined in Example 1.10. For example, consider the random number .013886 generated by computer. By dropping the decimal place and retaining only the first four digits, we obtain the random number 138.

In each program, N represents the sample size (i.e., number of random numbers generated). For convenience, we selected $N = 10$.

REMINDER The commands given here are appropriate for both mainframe and PC versions of the three software packages, except where noted. When a mainframe computer is being used, however, these statements must be preceded by the JCL commands required at your computing center.

SAS

```
Command
  line
   1    DATA SELECT;              Data entry instructions:
   2    DO N = 1 TO 10;           generates N = 10 random numbers between
   3    NUMBER = RANUNI(0);       0 and 1
   4    OUTPUT;
   5    END;
   6    PROC PRINT; VAR NUMBER;   Prints the random numbers
```

COMMAND 3 RANUNI is the uniform random generator in SAS. The numerical *seed*, i.e., the value in parentheses following RANUNI, can be any integer value (including 0).

Command line		
1	`DATA LIST FREE/X.`	} Data entry instructions: generates 10
2	`COMPUTE NUMBER=UNIFORM(1).`	} random numbers between 0 and 1
3	`BEGIN DATA.`	} Input data values: one number for each
4	`1 2 3 4 5 6 7 8 9 10`	} observation in the sample, ending with $N = 10$
5	`END DATA.`	
6	`LIST.`	} Prints the random numbers

COMMAND 2 UNIFORM(1) is the uniform random number generator in SPSS.

GENERAL Remember to omit the period at the end of each command when using mainframe SPSS (e.g., SPSS^x).

Command line		
1	`RANDOM 10 NUMBERS, IN COLUMN C1;`	} Data entry instructions:
2	`UNIFORM.`	} generates $N = 10$ uniform random numbers
3	`PRINT C1`	} Prints the random numbers

COMMANDS 1–2 RANDOM with the UNIFORM subcommand is the uniform random number generator in Minitab.

COMPUTER ACTIVITIES

Refer to the data in Appendix D. As described in Case Study 1.1, Appendix D contains location, species, length, weight, and DDT measurements for a total of 144 captured fish. Enter the data into the computer using a statistical software package and obtain a listing (printout) of the data. (If you use ASP, the menu-driven software that accompanies this text, save the data set under the name FISHDATA. If you use SAS, SPSS, Minitab, or some other command-driven software package, save the program commands in a file. We will analyze this data set in future Computer Activities sections.)

Careers in Statistics. American Statistical Association and the Institute of Mathematical Statistics, 1974.

Cochran, W. G. *Sampling Techniques*, 2nd ed. New York: Wiley, 1963.

Milgram, S. *Obedience to Authority*. New York: Harper & Row, 1974.

Minitab Reference Manual, Release 8, Minitab, Inc., 1989.

Norusis, M. J. *SPSS/PC+ 4.0 Base Manual*, SPSS, Inc., 1990.

Norusis, M. J. *SPSS/PC+ Statistics 4.0*, SPSS, Inc., 1990.

Ryan, B. F., Joiner, B. L., and Ryan, T. A. *Minitab Handbook*, 2nd ed. Boston: PWS-Kent, 1990 (revised printing).

SAS Procedures Guide for Personal Computers, Version 6 ed., 1986. SAS Institute, Inc.

SAS User's Guide: Basics, Version 6 ed., 1986. SAS Institute, Inc.

SAS User's Guide: Statistics, Version 6 ed., 1986. SAS Institute, Inc.

Tanur, J. M., Mosteller, F., Kruskal, W. H., Link, R. F., Pieters, R. S., and Rising, G. R. (eds.). *Statistics: A Guide to the Unknown*, 3rd ed. San Francisco: Holden-Day, 1989.

Yamane, T. *Elementary Sampling Theory*, 3rd ed. Englewood Cliffs, N.J.: Prentice-Hall, 1967.

REFERENCES

GRAPHICAL METHODS FOR DESCRIBING DATA SETS

CHAPTER 2

Is your behavior influenced by the phases of the moon? Despite the lack of supporting scientific evidence, many people still associate aberrant behavior with a full moon. To measure the degree to which people believe in lunar effects, Rotton and Kelly (1985) administered a questionnaire to a random sample of 157 college undergraduates. How can we use statistics to make sense of the data? Graphical methods that rapidly convey information contained in a data set are the topic of this chapter. In the chapter case in Section 2.4, we apply one of these graphical methods to summarize the lunar-effects data.

CONTENTS

THE OBJECTIVE OF DATA DESCRIPTION

2.1 The objective of data description is to summarize the characteristics of a data set. Ultimately, we want to make the data set more comprehensible and meaningful. In this chapter we will show you how to construct charts and graphs that convey the nature of a data set. The procedure that we will use to accomplish this objective in a particular situation depends on the type of data, qualitative or quantitative, that you want to describe.

GRAPHICAL DESCRIPTIONS OF QUALITATIVE DATA

2.2 **Bar graphs** and **pie charts** are two of the most widely used graphical methods for describing qualitative data sets. Essentially, they show how many observations fall in each qualitative category.

The observations for brand of gasoline, a qualitative variable, could fall into one of a number of categories or **classes**. If three brands were used by a sample of race car drivers, then the number of classes would be three. If the variable brand of gasoline was observed for a group of drivers, we would find that brand 1 was used, say, n_1 times; brand 2, n_2 times; and brand 3, n_3 times; and so on.

The summary information that we seek about qualitative variables is either the number of observations falling in each class or the proportion of the total number of observations falling in each class. Bar graphs can be constructed to show either type of information. Pie charts usually show the proportions or percentages of the total number of measurements falling in the classes. Although bar graphs and pie charts can be easily constructed by hand, you will probably rely on a statistical computer software package to produce the graphs for presentations.

DEFINITION 2.1

The **frequency** for a particular class is the number of observations falling in that class.

DEFINITION 2.2

The **relative frequency** for a particular class is equal to the class frequency divided by the total number of observations.

EXAMPLE 2.1 Recall that Appendix A contains starting salaries of 1,795 University of Florida graduates who earned their degrees (bachelor's, master's, or Ph.D.) between fall 1989 and spring 1991. Appendix B lists the starting salaries for only bachelor's degree graduates of five different colleges: Business Administration, Engineering, Journalism, Liberal Arts and Sciences, and Nursing. Figure 2.1 is a computer-generated (SAS) bar graph showing the frequencies of 902 graduates who indicated they had secured a job on the questionnaire for the five colleges of Appendix B. Interpret the graph.

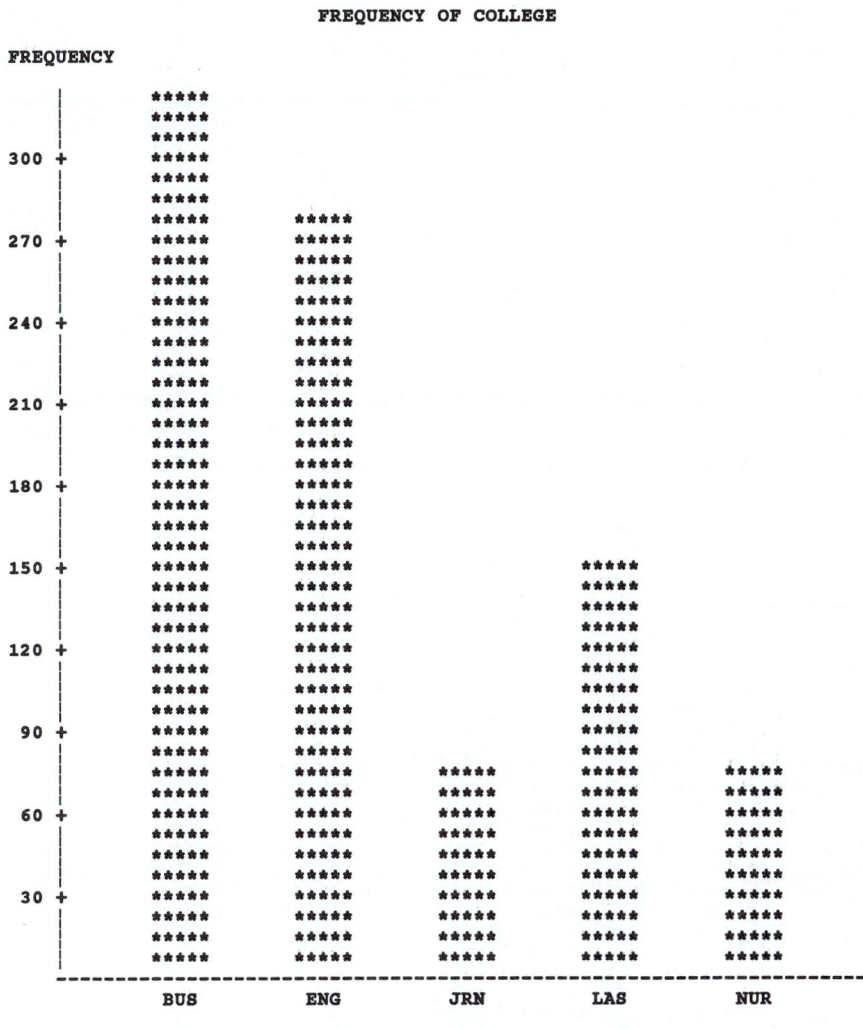

FREQUENCY OF COLLEGE

Solution The figure contains a rectangle or **bar** for each college; the height of a particular bar is proportional to the number of graduates for its college. For example, the bar graph indicates that approximately 75 employed graduates of the five colleges earned their degrees in the College of Nursing. You can rapidly compare the numbers of employed graduates for the five colleges by visually comparing the heights of the bars.

The frequencies—that is, the number of graduates who indicated they had secured a job for each of the five colleges—are shown in Table 2.1 (page 38). The class relative frequencies, obtained by dividing each class frequency by the total number of graduates, 902, are also shown.

The bar graph that permits a comparison of employed graduates can be constructed in several different ways. The heights of the bars can be measured in

TABLE 2.1
Frequencies and Relative Frequencies of Employed
Graduates for the Five Colleges of Appendix B

COLLEGE	FREQUENCY	RELATIVE FREQUENCY
Business Administration	322	.357
Engineering	281	.312
Journalism	73	.081
Liberal Arts & Sciences	149	.165
Nursing	77	.085
TOTALS	902	1.000

units of frequency (see Figure 2.1) or relative frequency (see Figure 2.2). It is also quite common to reverse the axes and display the bars in a horizontal fashion, as shown in the computer-generated (SPSS) bar graph, Figure 2.3.

FIGURE 2.2
SAS Bar Graph Showing
Relative Frequency of Employed
Graduates for Five Colleges

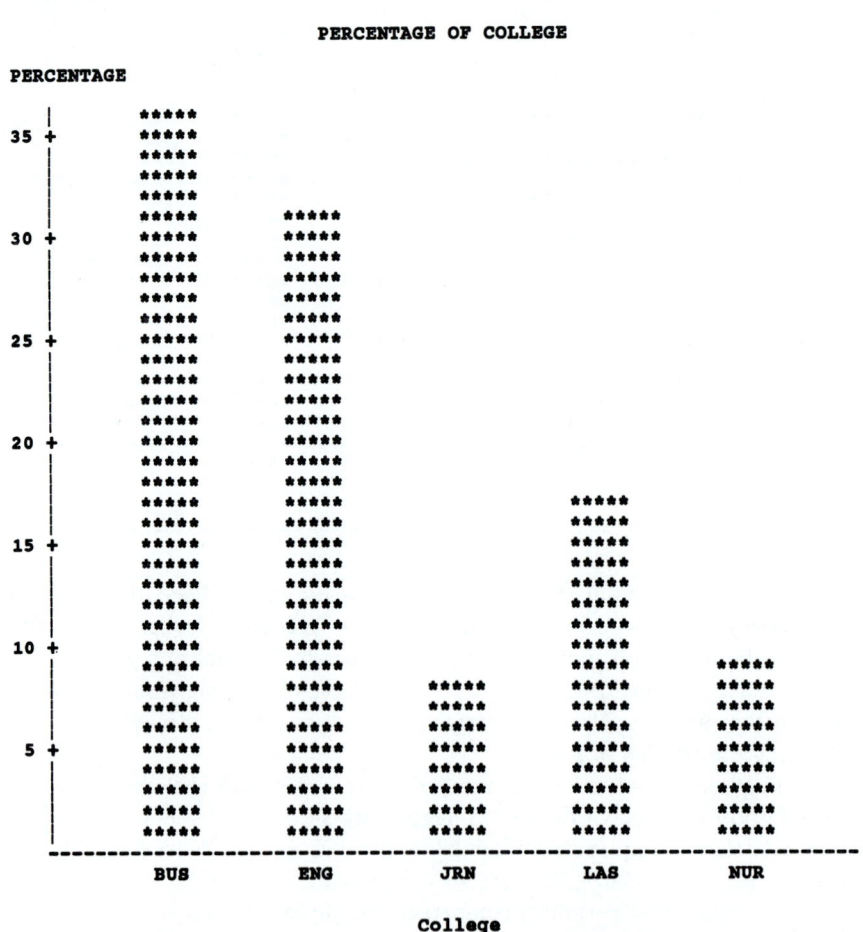

COLLEGE

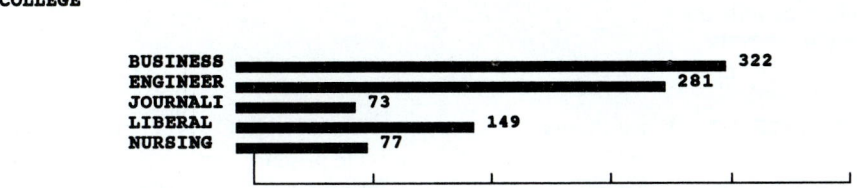

BUSINESS ████████████████████████ 322
ENGINEER ████████████████████ 281
JOURNALI ███████ 73
LIBERAL ███████████ 149
NURSING ████████ 77

    0   80   160   240   320   400

Valid cases   902   Missing cases   0

EXAMPLE 2.2   A SAS pie chart for the Appendix B salary data is shown in Figure 2.4. Confirm that the pie chart conveys the same information as the relative frequency bar chart in Figure 2.3.

Solution   The total number of employed graduates for the five colleges (the pie) is split into five pieces. The size (angle) of the slice assigned to a college is proportional to the relative frequency for that college. For example, since a complete circle spans 360°, the slice assigned to Nursing is 8.5% of 360°, or

$$.085(360) = 30.6°$$

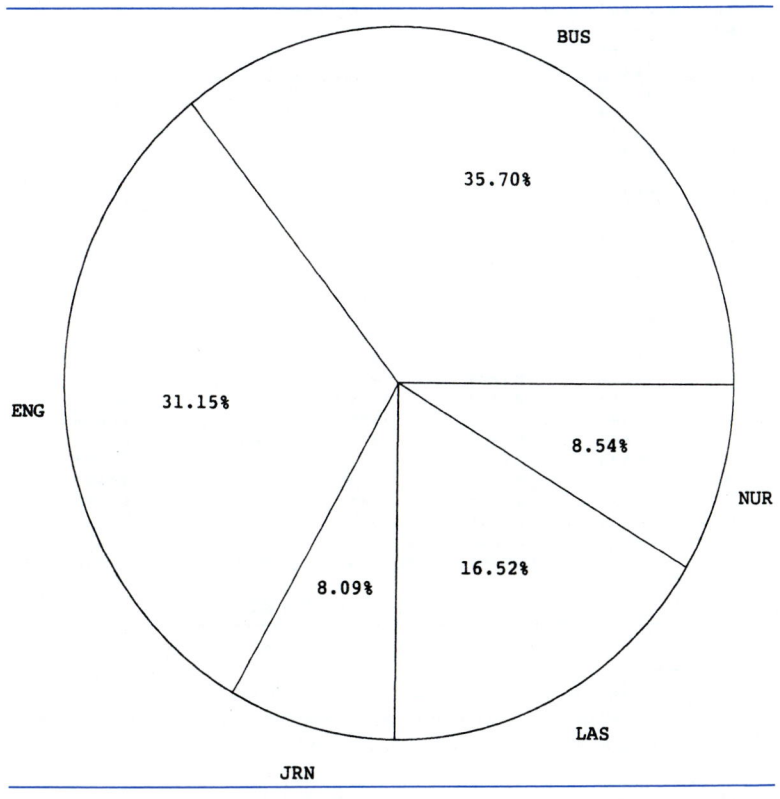

FIGURE 2.4
SAS Pie Chart Showing the
Percentages of Employed
Graduates for Five Colleges

It is common to show the percentage of measurements in each class on the pie chart as indicated.

EXAMPLE 2.3   A health maintenance organization (HMO) is an independent group of physicians whose objective is to provide primary and special health care to patients in a geographical area at low or minimal cost. Are HMO physicians truly cost-effective? To investigate this question, a study was conducted by a network of private practicing physicians in Florida, called the Tampa Bay Area Doctors (TBAD). Appendix C contains information on total costs accrued per-patient per-month by 186 TBAD physicians in 1989.* Several other variables were measured in addition to total cost, including the primary specialty of each physician. Figure 2.5 is a SAS bar graph giving the percentage breakdown according to primary specialty for the 186 physicians. Interpret the figure.

FIGURE 2.5
SAS Bar Chart Showing Percentages of Physicians in Primary Specialty Groups

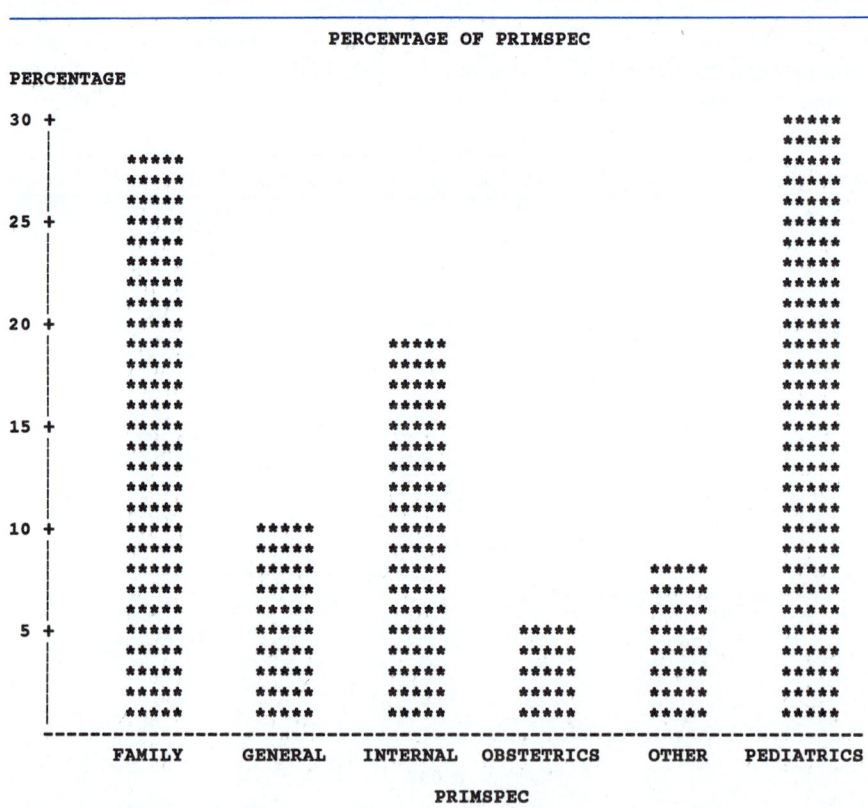

Source: Lane, W., Sincich, T., and Frazier, D. "Family physicians' HMO patients' costs compared with other primary care physician case managers." Paper presented at 43rd Annual American Academy of Family Physicians Assembly, Washington, DC, 1991.

Solution   Each of the 186 physicians in the HMO has been classified according to their primary specialty, a qualitative variable; six primary specialty groups are shown in the figure. The height of each bar is proportional to the percentage of the 186 physicians with that particular primary specialty. For example, approximately 10% of the 186 physicians (about 18 physicians) consider general practice as their primary specialty, whereas the highest percentage of these physicians (30%, or about 54) are pediatricians.

EXAMPLE 2.4   Refer to Example 2.3. Another qualitative variable measured on each physician is certification level. A certified doctor has completed training and passed the American Medical Association (AMA) Board of Physicians' examination. An uncertified physician can be one of two types: a young physician who has completed training but has not yet taken or passed the AMA board exam, or an older physician who is ineligible for certification but was granted a license to practice by the AMA. The levels of certification recorded in Appendix C are 0, 1, or 2, where 0 = uncertified, board-ineligible; 1 = board-certified; and 2 = uncertified, board-eligible. Figure 2.6 is a Minitab horizontal bar graph showing the frequencies of certification level. Interpret the graph.

```
Histogram of CERTIF    N = 186
Each * represents 5 obs.

Midpoint     Count
        0       27    ******
        1      122    *************************
        2       37    ********
```

FIGURE 2.6
Minitab Bar Graph of Physician Certification Levels

Solution   Each of the 186 HMO physicians has been classified according to certification level. The length of the horizontal bar adjacent to each category label is proportional to the number of physicians with that particular certification level. Note that each asterisk (*) in the bar represents five observations (physicians). You can see that the majority of physicians (122 of 186) are board-certified. Of the remaining (uncertified) physicians, 27 are board-ineligible and 37 are board-eligible.

EXAMPLE 2.5   *Fetal alcohol syndrome* is a group of abnormalities found in children born to chronic alcoholic mothers. Each of 60 children diagnosed as having the syndrome was examined for the abnormality of the most serious nature, with the results illustrated by the bar graph in Figure 2.7 (page 42). Discuss the information provided by the graph. Which abnormality occurs most often as the child's most serious problem?

Solution   Each of the 60 children in the study has been classified according to the most serious abnormality. Clearly, the highest bar corresponds to prenatal

FIGURE 2.7

Frequency of Fetal Alcohol
Syndrome Abnormalities

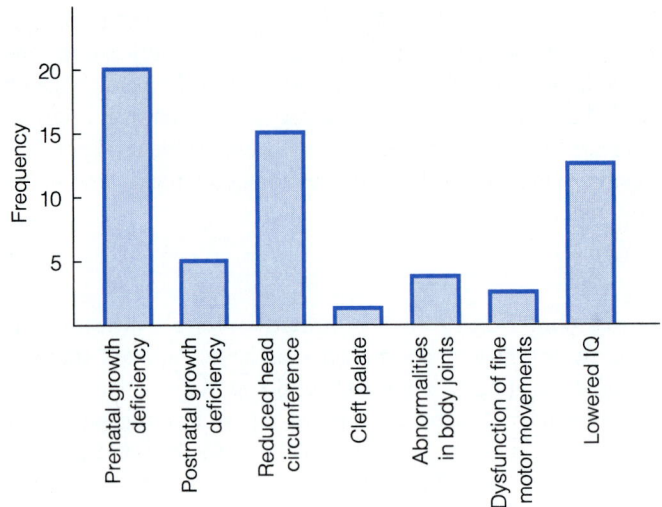

growth deficiency; thus, prenatal growth deficiency occurred most often as the children's most serious abnormality. Since the height of this bar is 20, we also conclude that 20 of the 60 children (or 33.33%) were diagnosed as having a prenatal growth deficiency as their most serious problem.

EXERCISES

LEARNING THE MECHANICS

2.1 Complete the following table:

GRADE ON STATISTICS EXAM	FREQUENCY	RELATIVE FREQUENCY
A: 90–100	16	.08
B: 80–89	36	
C: 65–79	90	
D: 50–64	30	
F: Below 50	28	
TOTAL		1.00

2.2 During August 1980, hundreds of thousands of Poland's workers walked off the job, protesting the country's poor labor conditions. One of their demands, eventually negotiated with the government through the independent Polish labor union Solidarity, was the reduction of a mandatory 6-day, 48-hour work week to a 5-day, 40-hour work week. In the United States, many corporations are considering instituting a 4-day, 40-hour work week or a 3-day, 40-hour work week. Suppose a company surveyed its employees concerning the type of work week they would prefer; a 6-day, 48-hour work week; a 5-day, 40-hour work week; a 4-day, 40-hour work week; or a 3-day, 40-hour work week. Suppose 25 employees responded as shown in the table.

EMPLOYEE	WORK WEEK	EMPLOYEE	WORK WEEK	EMPLOYEE	WORK WEEK
1	5-day, 40-hour	10	5-day, 40-hour	18	5-day, 40-hour
2	5-day, 40-hour	11	4-day, 40-hour	19	5-day, 40-hour
3	3-day, 40-hour	12	5-day, 40-hour	20	4-day, 40-hour
4	6-day, 48 hour	13	4-day, 40-hour	21	3-day, 40-hour
5	4-day, 40-hour	14	4-day, 40-hour	22	5-day, 40-hour
6	4-day, 40-hour	15	6-day, 48-hour	23	3-day, 40-hour
7	5-day, 40-hour	16	4-day, 40-hour	24	4-day, 40-hour
8	3-day, 40-hour	17	5-day, 40-hour	25	5-day, 40-hour
9	6-day, 48-hour				

a. Identify the type of variable measured.
b. Identify the classes.
c. Compute the frequency of each class.
d. Compute the relative frequency of each class.
e. Construct a relative frequency bar graph for the data on work-week preferences.

APPLYING THE CONCEPTS

2.3 Sea turtles (an endangered species) are often the subject of ecological research. However, little is known about the natural hatching success of sea turtles on nesting beaches since most studies of hatching success are conducted under artificial hatchery conditions. As part of one investigation of the natural survival rate of the green sea turtle, 350 sea turtle nests on a 4-kilometer area of the Tortuguero beach located on the Caribbean coast of Costa Rica were marked and monitored. The fate of each of the 350 marked nests was of prime interest; results are given in the table. Construct a bar graph for the data. Interpret your results.

NEST FATE	NUMBER
Undisturbed, young emerged (successful hatching)	148
Disturbed by predators, some young emerged	18
Destroyed by animal predators	122
Washed out by surf	20
Lost to human predators	23
Dead although undisturbed	19
TOTAL	350

Source: Fowler, L. E. "Hatching success and nest predation in the green sea turtle, *Chelonia mydas,* at Tortuguero, Costa Rica." *Ecology,* Oct. 1979, 60, pp. 946–955. Copyright 1979, the Ecological Society of America. Reprinted by permission.

2.4 Some scientists claim that global climate warming—caused by smoke stacks, gas-powered automobiles, power-generating stations, forest fires, etc.—will threaten the habitability of the Earth by the end of the twentieth century. *Scientific American* (July 1990) reported on computer models designed to assess the causes of global warming. The accompanying horizontal relative frequency bar chart shows a breakdown of the potential causes of global warming into five general categories: (1) energy use and production, (2) chlorofluorocarbons, (3) agriculture, (4) land-use modification, and (5) other industrial causes. Interpret the bar chart.

Human Activities That May Cause Global Warming

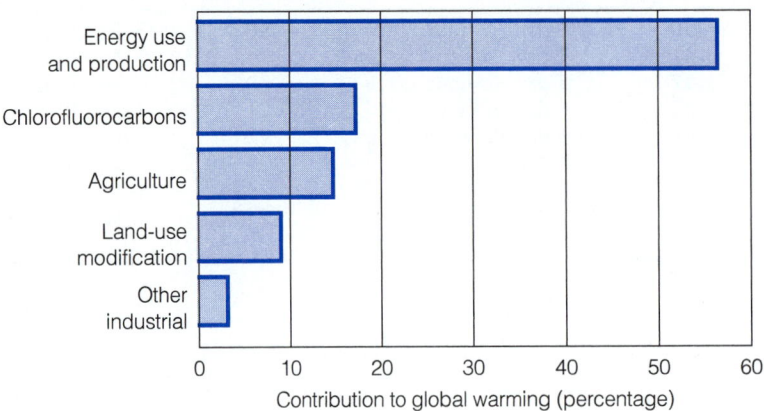

Source: White, R. M. "The great climate debate." *Scientific American*, Vol. 263, No. 1, July 1990, p. 43.

2.5 The pie chart below describes the fate of the (estimated) 242 million automobile tires that are scrapped in the U.S. each year.

    a. Interpret the pie chart.
    b. Convert the pie chart into a relative frequency bar chart.
    c. Convert the pie chart into a frequency bar chart.

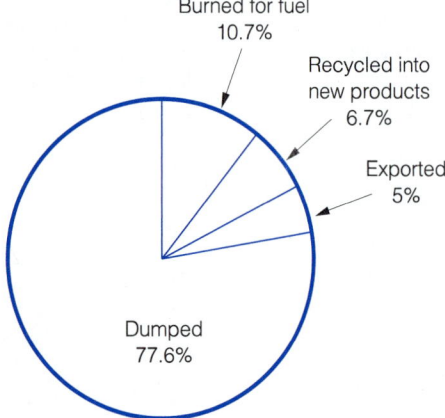

Source: U.S. Environmental Protection Agency and National Solid Waste Management Association. *Tampa Tribune*, June 29, 1992.

2.6 Casual restaurants—eateries featuring basic entrees, relaxed decors, and moderate prices—will be the major trend for restaurants in the 1990s, according to a Gallup survey for the National Restaurant Association (*Tampa Tribune*, Mar. 10, 1990). Gallup surveyed 1,000 adults about their attitudes toward casual and fine dining. One question asked how often the respondents eat out. The results are provided in the accompanying table. Use the appropriate graphical technique to describe the data.

FREQUENCY OF EATING OUT	TYPE OF RESTAURANT Casual	Fine
Once a week	29%	4%
2–3 times a month	24	9
Once a month	17	12
Once every few months	12	16
Once every 6 months	6	13
Once a year	4	15
Less than once a year	8	31
TOTALS	100%	100%

Source: 1990 National Restaurant Association Gallup Survey. *Tampa Tribune*, Mar. 10, 1990.

7 *Child Welfare* (Jan./Feb. 1992) reported on a study of 124 Canadian adoptees who recently reunited with their biological mothers and relatives. One aspect of the study dealt with how often the adoptees experienced "quest" feelings (i.e., the desire to meet or learn about their biological parents) before their reunion. The accompanying table reports the frequency of quest feelings experienced by the 124 adoptees at three different ages: before age 10, between 10 and 17, and 18 or older.

FREQUENCY OF QUEST FEELINGS	AGE CATEGORY Before 10	10–17	18 or Older
Very often, or somewhat often	19.8%	47.4%	87.5%
Occasionally	16.5	27.8	12.5
Almost never	47.2	22.7	0
Uncertain	16.5	2.1	0
TOTALS	100.0%	100.0%	100.0%

Source: Sachdev, P. "Adoption reunion and after: A study of the search process and experience of adoptees." *Child Welfare*, Vol. 71, No. 1, Jan./Feb. 1992, p. 59 (Table 2).

a. For each age category, summarize the data on frequency of "quest" feelings with a bar graph.
b. Compare and contrast the three graphs, part **a**. What inference can you make?

8 How prevalent is alcoholism among medical professionals? Does easy access to drugs tend to encourage their use? Is there a difference between the rates of addiction for nurses and physicians? To answer these and other questions, researchers conducted a survey of nurses and physicians who (1) considered themselves alcoholics, (2) were members of Alcoholics Anonymous, and (3) had been completely abstinent for at least 1 calendar year immediately prior to being interviewed.

One aspect of the study concerned the subjects' addictions to other drugs. Subjects were asked if they had used a drug outside a hospital setting and, if so, had they been addicted to it. The results shown in the table were observed. (Only the results from those subjects who responded in the affirmative are included here.)

ADDICTION	NUMBER OF NURSES	NUMBER OF PHYSICIANS
Alcohol only	65	55
Both alcohol and narcotics	1	4
Both alcohol and nonnarcotic drugs	21	24
Alcohol, narcotics, and nonnarcotic drugs	13	14
TOTALS	100	97

Source: Bissell, L., and Jones, R. W. "The alcoholic nurse." *Nursing Outlook*, Feb. 1981. Copyright by the American Journal of Nursing Company. Reprinted from *Nursing Outlook*.

a. Construct either a bar graph or a pie chart to describe addiction among the nurses interviewed.

b. Construct either a bar graph or a pie chart to describe addiction among the physicians interviewed.

c. Compare the two figures you constructed in parts **a** and **b**. Does there appear to be a difference between the rates of addiction for the two groups of subjects? Explain.

2.9 According to the American Optometric Association (AOA), about 10% of the U.S. population has normal visual acuity, while 60% suffers from farsightedness and 30% suffers from nearsightedness.

VISION PROBLEM CATEGORY	PERCENT
Nearsighted, corrected	27
Nearsighted, uncorrected	7
Farsighted, corrected	27
Farsighted, uncorrected	39
	100

Source: American Optometric Association, 1988.

a. Given this information, construct a pie chart to describe the distribution of visual acuity problems in the United States.

b. Of those that have visual acuity problems (nearsighted or farsighted), the AOA has found that many do not wear any sort of corrective lenses. Use the information provided in the table to construct a relative frequency bar chart for the four categories shown.

c. What percentage of those with vision impairments do not wear any corrective lenses?

2.10 In Florida, civil engineers are designing roads with the latest safety-oriented construction methods in response to the fact that in 1988 more people in Florida were killed by bad roads than by guns. A total of 135 traffic accidents that occurred during the year have been attributed to poorly constructed roads (*Tampa Tribune*, Nov. 14, 1989). A breakdown of the poor road conditions that caused the accidents is shown in the accompanying table. Construct and interpret a frequency bar graph for the data.

POOR ROAD CONDITION	NUMBER OF FATALITIES
Obstructions without warning	7
Road repairs/under construction	39
Loose surface material	13
Soft or low shoulders	20
Holes, ruts, etc.	8
Standing water	25
Worn road surface	6
Other	17
TOTAL	135

Source: Florida Department of Highway Safety and Motor Vehicles, 1989.

2.11 Professional purchasers are now getting tougher on their suppliers. According to *Purchasing* (Jan. 17, 1991), "suppliers are being subjected to formal and detailed monthly or quarterly performance surveys by purchasing-led teams of auditors on everything from product quality and delivery schedules to receipt of technical data sheets and timely billing paperwork." A recent survey of supplier performance evaluations revealed the results shown in the accompanying table. Summarize the data with a pie chart. Interpret the graphic.

PRIMARY REASON SUPPLIERS ARE EVALUATED	RELATIVE FREQUENCY
Quality	.42
Technical Expertise	.11
Price	.12
Service	.13
Delivery	.22
TOTAL	1.00

Source: Stundza, T. "Suppliers on the hot seat." *Purchasing,* Jan. 17, 1991, p. 92.

2.12 Reporting in the *New England Journal of Medicine* (Mar. 18, 1991), the Center for Disease Control (CDC) confirmed what many former cigarette smokers have learned from experience: People who quit smoking tend to gain weight. The CDC's research team reviewed data on 1,885 smokers and 768 former smokers who were studied over a 13-year period. Weight gain over the study period was classified as slight (3 kilograms or less), moderate (3–8 kilograms), significant (8–13 kilograms), and major (more than 13 kilograms). The smokers/quitters were also classified according to gender to compare male versus female weight gain. The percentages of men and women in the four weight-gain categories are provided in the table.

| | QUITTERS | | SMOKERS | |
WEIGHT GAIN	Men	Women	Men	Women
Slight	55	50	66	63
Moderate	22	26	24	23
Significant	14	10	8	9
Major	9	14	2	5
TOTALS	100	100	100	100

Source: *Time,* Mar. 25, 1991, p. 55.

a. Describe the data with the appropriate graphical technique. Construct one graph for each column of the table.

b. Compare the four graphs, part **a.** Do quitters tend to gain more weight than smokers? Do female quitters tend to gain more weight than male quitters?

2.13 Researchers have been creating thin films of tiny diamonds since the early 1950s, but industry has only recently utilized diamond thin films in the manufacture of such products as cutting tools, stereo loudspeaker tweeters, heat sinks, sunglasses, and components for scientific instruments. According to the Philadelphia Institute for

COUNTRY	NUMBER OF PAPERS ON DIAMOND THIN-FILM RESEARCH
United States	105
Japan	63
United Kingdom	20
Germany	17
Italy	7
Others	39
TOTAL	251

Source: Grissom, A. "U.S. and Japan sparkle in diamond thin film research." *The Scientist,* June 11, 1990, p. 17 (Table 1).

Scientific Information (ISI), the United States and Japan are preeminent in current research on diamond thin films (*The Scientist*, June 11, 1990). Each of 251 published papers was categorized according to the author's country of residence. The results are shown in the accompanying table. Use an appropriate graphical technique to summarize the data. Do you agree with ISI's assessment of the United States and Japan with regard to diamond thin-film research?

2.14   As part of a study to investigate the relationship between sexual maturation and spatial ability in college students, 274 introductory psychology students (80 males and 194 females) were asked to indicate when they reached puberty in relation to others of the same sex (*Developmental Psychology*, Mar. 1986). One of five response categories were possible: much earlier, earlier, same time, later, and much later. The proportions responding in each of the categories for the two groups of students are reported in the table.

TIME REACHED PUBERTY IN RELATION TO OTHERS	MALES	FEMALES
Much earlier	.02	.05
Earlier	.22	.20
Same time	.56	.57
Later	.20	.17
Much later	.00	.01
TOTALS	1.00	1.00

Source: Sanders, B. and Soares, M. P., "Sexual maturation and spatial ability in college students." *Developmental Psychology*, Vol. 22, No. 2, pp. 199–203, (Figure 1). Copyright 1986 by the American Psychological Association. Reprinted by permission.

a. Construct a relative frequency bar chart for the distribution of responses for the males.
b. Construct a relative frequency bar chart for the distribution of responses for the females.
c. Interpret the graphs, parts **a** and **b**.

GRAPHICAL DESCRIPTIONS OF QUANTITATIVE DATA: STEM-AND-LEAF DISPLAYS

2.3 Stem-and-leaf displays and histograms are two of the most popular graphical methods for describing quantitative data sets. Like the bar graphs and pie charts of Section 2.2, they show either the number of observations that fall into each class (class frequency) or the proportion of the total number of observations falling into each class (class relative frequency). The difference is that the classes do not represent categories of a qualitative variable; instead, they are formed by grouping the numerical values of the quantitative variable that you want to describe.

For small data sets (say, 30 or fewer observations) with measurements with only a few digits, stem-and-leaf displays can be constructed easily and quickly by hand. Histograms, on the other hand, are better suited to the description of large data sets, and they permit greater flexibility in the choice of the classes. In this section, we present stem-and-leaf displays for small data sets.

EXAMPLE 2.6   Suppose you are a realtor interested in describing and summarizing recent sales of residential properties in a particular neighborhood. The sales prices (in hundreds of dollars) of 25 recently sold (and randomly selected) properties are listed in Table 2.2. Construct a stem-and-leaf display for the sale price data of Table 2.2.

TABLE 2.2
Sale Prices for a Random Sample of 25 Residential Properties

SALE PRICE (HUNDREDS OF DOLLARS)				
660	595	1,060	500	630
899	1,295	749	820	843
710	950	720	575	760
1,090	770	682	1,016	650
425	367	1,480	945	1,120

Solution   Since all the sale prices in Table 2.2 are given to the nearest hundred dollars, we have simplified the numbers in the data set by dropping the last two zeros. Thus, the first number in the table is 660, representing a sale price of $66,000. We will designate the last two digits (60) of this number as its **leaf**; we will call the remaining digit (6) its **stem**, as illustrated here. The stem and leaf of the number 899 are 8 and 99, respectively. Similarly, the stem and leaf of the number 1,090 are 10 and 90, respectively.

STEM	LEAF
6	60

The first step in forming a stem-and-leaf display for this data set is to list all stem possibilities in a column starting with the smallest stem (3, corresponding to the number 367) and ending with the largest (14, corresponding to the number 1,480), as shown in Figure 2.8a. The next step is to place the leaf of each number in the data set in the row of the display corresponding to the number's stem. For example, for the number 660, the leaf 60 is placed in stem row 6. Similarly, for the number 899, the leaf 99 is placed in stem row 8. The usual convention is to list the leaves of each stem in increasing order. After the leaves of the 25 numbers are placed in the appropriate stem rows, the completed stem-and-leaf display will appear as shown in Figure 2.8a.

a. Stem-and-Leaf Display

b. Frequency and Relative Frequency Tabulation for the Stem-and-Leaf Display

FIGURE 2.8
Stem-and-Leaf Display of 25 Sale Prices in Table 2.2

STEM	LEAF	FREQUENCY	RELATIVE FREQUENCY
3	67	1	$\frac{1}{25}$
4	25	1	$\frac{1}{25}$
5	00 75 95	3	$\frac{3}{25}$
6	30 50 60 82	4	$\frac{4}{25}$
7	10 20 49 60 70	5	$\frac{5}{25}$
8	20 43 99	3	$\frac{3}{25}$
9	45 50	2	$\frac{2}{25}$
10	16 60 90	3	$\frac{3}{25}$
11	20	1	$\frac{1}{25}$
12	95	1	$\frac{1}{25}$
13		0	0
14	80	1	$\frac{1}{25}$
	TOTALS	25	1

You can see that the stem-and-leaf display in Figure 2.8a partitions the data set into 12 categories (called **classes**) corresponding to the 12 stems. The class corresponding to 3 would contain all numbers from 300 to 399; the class corresponding to the stem 4 would contain all numbers from 400 to 499. The number of leaves in each class gives the class frequency. Thus, a stem-and-leaf display provides the frequencies needed to construct frequency and relative frequency histograms for a data set. A **frequency histogram** is a bar graph for a quantitative data set. A **relative frequency histogram** conveys the same information, but the heights of the bars are proportional to the class relative frequencies.

EXAMPLE 2.7 Construct a frequency histogram for the data contained in the stem-and-leaf display in Figure 2.8a.

Solution   The frequencies and relative frequencies for 12 classes (corresponding to the 12 stems) are given in Figure 2.8b. Figure 2.9 shows the frequency histogram for the data.* Bars are constructed over each class, with the height of each bar proportional to the class frequency.

FIGURE 2.9

A Frequency Histogram for the Sale Price Data of Figure 2.8

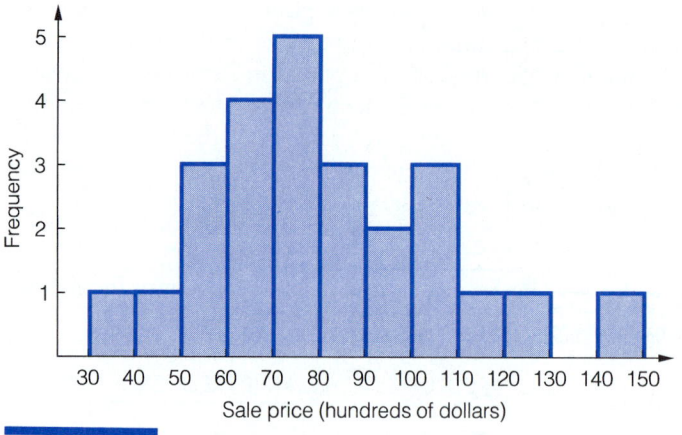

Notice that if you tip the stem-and-leaf display (Figure 2.8a) on its side, you obtain the same type of bar graph provided by the frequency histogram. Both figures show that the 25 sale prices are scattered over the interval from $30,000 to less than $150,000, with most falling between $50,000 and $110,000. Both methods provide a good visual description of the data.

*Detailed instructions for constructing a frequency histogram will be given in the next section. At this point, our goal is simply to convert the information contained in the stem-and-leaf display into graphical form.

One advantage of a stem-and-leaf display over a frequency histogram is that the original data are preserved. That is, you can look at the display and resurrect the exact values of the data. A stem-and-leaf display also arranges the data in an orderly fashion and makes it easy to determine certain numerical characteristics, which will be discussed in Chapter 3. The third advantage is that the classes and the numbers falling in them are quickly determined once we have selected the digits that we want to use for the stems and leaves. A disadvantage of the stem-and-leaf display is that there is sometimes not much flexibility in choosing the stems.

For the data in Table 2.2, two options for stems and leaves are possible. We could define the stems and leaves as shown in Figure 2.8a. Or, we could let only the last digit represent the leaf, in which case the number 660 would have the stem 66 and the leaf 0, as shown here.

STEM	LEAF
66	0

The associated stem-and-leaf display for the data of Table 2.2 would contain 113 stems, 36, 37, . . . , 148, and each of the stems would contain either a single leaf or none at all. Clearly this choice of stems and leaves would not provide as much information about the data as does the display of Figure 2.8a. Consequently, we are left with the option of using a stem-and-leaf display that produces 12 stems (and thus 12 classes) or one that produces 113 stems.*

Most statistical computer software packages have routines that will produce a stem-and-leaf display. A Minitab stem-and-leaf display for the data of Table 2.2 is shown in Figure 2.10 on page 52. The stems, 3, 4, 5, . . . , 14, are given in the second column and the leaves are given in the third column of the printout. You can see that Minitab has chosen to use 12 stems, as in Figure 2.8a. Minitab uses only a single digit—the number in the tens place—to represent a leaf.† Thus, the leaf 6 in stem row 3 of the Minitab printout (Figure 2.10) represents the value 367 in Table 2.2; the leaf 2 in stem row 4 represents the value 425; and so forth.‡

STEPS TO FOLLOW IN CONSTRUCTING A STEM-AND-LEAF DISPLAY

STEP 1 Decide how the stems and leaves will be defined.

STEP 2 List the stems in order in a column, starting with the smallest stem and ending with the largest.

STEP 3 Proceed through the data set, placing the leaf for each observation in the appropriate stem row. (You may want to place the leaves of each stem in increasing order.)

---

*By sacrificing some of the simplicity of our procedure, we could define the stems and leaves so that the number of stems falls between 12 and 113. We omit discussion of this topic.

†In Minitab, the leaf will always be the digit immediately to the right of the stem.

‡The numbers in the first column of the Minitab printout give the cumulative number of observations from the stem row to the nearest "end" of the distribution.

FIGURE 2.10
Minitab Stem-and-Leaf Display
for Sale Prices in Table 2.2

```
Stem-and-leaf of salepric   N  = 25
Leaf Unit = 10

   1        3 6
   2        4 2
   5        5 079
   9        6 3568
  (5)       7 12467
  11        8 249
   8        9 45
   6       10 169
   3       11 2
   2       12 9
   1       13
   1       14 8
```

EXERCISES

LEARNING THE MECHANICS

2.15   Consider the following sample data:

213	228	241	268	234	303
274	316	319	320	227	226
224	267	303	266	265	237
288	291	285	270	254	215

   a. Using the first two digits of each number as a stem, list the stem possibilities in order.
   b. Place the leaf for each observation in the appropriate stem row to form a stem-and-leaf display.

2.16   A sample of 20 measurements is shown here:

26	34	21	32	42	36	28	38	17	39
22	12	56	39	25	41	30	23	27	19

   a. Using the first digit as a stem, list the stem possibilities in order.
   b. Place the leaf for each observation in the appropriate stem row to form a stem-and-leaf display.

2.17   Consider the sample data shown here.

5.9	5.3	1.6	7.4	8.6	1.2	2.1
4.0	7.3	8.4	8.9	6.7	4.5	6.3
7.6	9.7	3.5	1.1	4.3	3.3	8.4
1.6	8.2	6.5	1.1	5.0	9.4	6.4

   a. Using the first digit as a stem, construct a stem-and-leaf display.
   b. Use the stem-and-leaf display from part **a** to construct a frequency distribution for the data.

APPLYING THE CONCEPTS

2.18   A team-by-team study of major league baseball players' salaries conducted by the Associated Press revealed that 152 of the 663 players earned $1 million or more in 1990. [*Note:* The major league minimum 1990 salary was

$100,000.] The 1990 salaries (recorded in $100,000s) for players on the opening-day roster of the 1989 world champion Oakland Athletics are reported in the table.

PLAYER	POSITION	SALARY	PLAYER	POSITION	SALARY
R. Henderson	OF	22.50	R. Hassey	C	7.00
J. Canseco	OF	20.00	C. Young	P	6.75
M. McGwire	1B	15.00	G. Nelson	P	6.50
C. Lansford	3B	12.75	M. Gallego	2B	5.00
M. Moore	P	11.92	S. Javier	OF	3.10
B. Welch	P	11.33	T. Burns	P	2.35
D. Stewart	P	9.50	W. Weiss	SS	2.35
D. Henderson	OF	8.50	J. Quirk	C	1.50
K. Phelps	1B	8.25	J. Corsi	2B	1.08
D. Eckersley	P	7.87	L. Blankenship	SS	1.05
R. Honeycutt	P	7.50	F. Jose	OF	1.02
S. Sanderson	P	7.50	D. Otto	3B	1.02
T. Steinbach	C	7.50	M. Norris	P	1.00

a. Construct a stem-and-leaf display for the Oakland players' 1990 salaries.
b. Locate Jose Canseco's salary of $2,000,000 on the figure constructed in part a.
c. What percentage of the Oakland A's players had salaries of $1 million or more in 1990?

2.19   Under a voluntary cooperative inspection program, all passenger cruise ships arriving at U.S. ports are subject to unannounced inspection. The purpose of these inspections is to achieve levels of sanitation that will minimize the potential for gastrointestinal disease outbreaks on these ships. Ships are rated on a 0- to 100-point scale depending on how well they meet the Center for Disease Control sanitation standards. In general, the lower the score, the lower the level of sanitation. The accompanying table (page 54) lists the sanitation inspection scores for 91 international cruise ships during 1992.

a. A Minitab stem-and-leaf display of the data is shown here. Identify the stems and leaves of the graph.
b. A score of 86 or higher at the time of inspection indicates the ship is providing an accepted standard of sanitation. Use the Minitab graph to estimate the proportion of ships that have an accepted sanitation standard.
c. Locate the inspection score of 70 (Pacific Star) on the stem-and-leaf display.

```
Stem-and-leaf of SanLevel   N  = 91
Leaf Unit = 1.0

    1      6 6
    1      6
    2      7 0
    2      7
    3      7 4
    3      7
    3      7
    3      8
    5      8 23
    7      8 44
   18      8 66666677777
   31      8 8888999999999
   42      9 00000111111
  (15)     9 222222222333333
   34      9 4444444445555555
   18      9 66666666777777
    4      9 8999
```

SHIP	SCORE	SHIP	SCORE	SHIP	SCORE
Americana	89	Hanseatic Renaissance	82	Seabourn Spirit	92
Amerikanis	97	Holiday	91	Seabourn Pride	99
Azure Seas	83	Horizon	94	Seabreeze I	96
Britanis	93	Island Princess	87	Seaward	89
Caribbean Prince	84	Jubilee	89	Sky Princess	97
Caribe I	90	Mardi Gras	92	Society Explorer	66
Carla C	90	Meridian	95	Song of America	95
Carnivale	92	Nantucket Clipper	89	Song of Flower	99
Celebration	95	New Shoreham II	95	Song of Norway	92
Club Med 1	94	Nieuw Amsterdam	97	Southward	89
Costa Classica	91	Noordam	92	Sovereign of the Seas	93
Costa Marina	91	Nordic Empress	93	Star Princess	94
Costa Riviera	91	Nordic Prince	92	Starship Atlantic	87
Crown Monarch	94	Norway	84	Starship Majestic	94
Crown Odyssey	88	Pacific Princess	88	Starship Oceanic	97
Crown Princess	88	Pacific Star	70	Starward	96
Crystal Harmony	99	Queen Elizabeth 2	98	Stella Solaris	94
Cunard Countess	96	Regent Sea	87	Sun Viking	90
Cunard Princess	89	Regent Star	74	Sunward	95
Daphne	86	Regent Sun	95	Triton	86
Dawn Princess	86	Rotterdam	92	Tropicale	93
Discovery I	93	Royal Princess	93	Universe	92
Dolphin IV	96	Royal Viking Sun	86	Victoria	96
Ecstasy	94	Sagafjord	89	Viking Princess	90
Emerald Seas	95	Scandinavian Dawn	87	Viking Serenade	96
Enchanted Isle	86	Scandinavian Song	90	Vistafjord	94
Enchanted Seas	96	Scandinavian Sun	89	Westerdam	91
Fair Princess	87	Sea Bird	86	Wind Spirit	96
Fantasy	97	Sea Goddess I	97	Yorktown Clipper	92
Festivale	94	Sea Lion	91		
Golden Odyssey	89	Sea Princess	88		

Source: Center of Environmental Health and Injury Control, Miami, Florida (reported in *Tampa Tribune*, May 17, 1992).

2.20   Many Vietnam veterans have dangerously high levels of the dioxin 2,3,7,8-TCDD in blood and fat tissue as a result of their exposure to the defoliant Agent Orange. A study published in *Chemosphere* (Vol. 20, 1990) reported on the TCDD levels of 20 Massachusetts Vietnam veterans who were possibly exposed to Agent Orange. The amounts of TCDD (measured in parts per trillion) in blood plasma and fat tissue drawn from each veteran are shown in the table. Use a graphical technique to compare the distributions of TCDD levels in plasma and fat tissue.

TCDD LEVELS IN PLASMA				TCDD LEVELS IN FAT TISSUE			
2.5	1.8	6.9	1.8	4.9	1.1	7.0	4.2
3.5	2.5	1.6	36.0	6.9	2.3	1.4	41.0
6.8	3.1	20.0	3.3	10.0	5.9	11.0	2.9
4.7	3.1	4.1	7.2	4.4	7.0	2.5	7.7
4.6	3.0	2.1	2.0	4.6	5.5	4.4	2.5

Source: Schecter, A. et al. "Partitioning of 2,3,7,8-chlorinated dibenzo-p-dioxins and dibenzofurans between adipose tissue and plasma lipid of 20 Massachusetts Vietnam veterans." *Chemosphere*, Vol. 20, Nos. 7–9, 1990, pp. 954–955 (Tables I and II).

2.21 According to *USA Today* (Aug. 8, 1985), "a baby born in the United States is nearly twice as likely to die before its next birthday as one born in Finland." The number of deaths before age 1 per 1,000 births for each of 17 countries reported in *USA Today* are given in the accompanying table.

COUNTRY	NUMBER OF DEATHS PER 1,000 BIRTHS	COUNTRY	NUMBER OF DEATHS PER 1,000 BIRTHS
Australia	10.3	Netherlands	8.4
Belgium	9.2	Norway	7.8
Canada	9.1	Spain	9.6
Denmark	8.2	Sweden	7.0
East Germany	10.7	Switzerland	7.7
Finland	6.0	United Kingdom	10.1
France	9.0	United States	10.5
Ireland	10.5	West Germany	10.1
Japan	6.2		

Source: 1985 World Population Data Sheet, Population Reference Bureau. Copyright 1985, *USA Today*. Adapted with permission.

a. Construct a stem-and-leaf display for the data.

b. Locate the infant mortality rate for the United States on the stem-and-leaf display. What percentage of the 17 countries have mortality rates greater than 10.5?

2.22 The Customer Satisfaction Index, compiled monthly by J. D. Powers & Associates, is designed to measure customer satisfaction with new automobiles in the areas of repair, reliability, and experience at the dealership. The index is based on questionnaires completed by drivers 1 year after they bought their cars. In 1992, an index of 129 was considered average, with ratings above 129 considered above average and ratings below 129 considered below average in terms of overall customer satisfaction. The accompanying table lists the 1992 Customer Satisfaction Index for the top 10 rated automobiles.

AUTO (MANUFACTURER)	FOREIGN (F) OR DOMESTIC (D)	CUSTOMER SATISFACTION INDEX
Lexus (Toyota)	F	179
Infiniti (Nissan)	F	167
Saturn (GM)	D	160
Acura (Honda)	F	148
Mercedes-Benz	F	145
Toyota	F	144
Audi (VW)	F	139
Cadillac (GM)	D	138
Honda	F	138
Jaguar (Ford)	D	137

Source: J. D. Powers & Associates, 1992.

a. Construct a stem-and-leaf display of the Customer Satisfaction Index for the top 10 ranked automobiles.

b. Circle the index values for the foreign automakers on the graph, part **a**. Are consumers more satisfied with foreign automakers?

2.23 How strong an effect do characteristics such as brand name or store name have on a buyer's perception of the quality of a product? Numerous studies have been conducted to investigate this phenomenon, but the results seem to vary depending on the method used to analyze the data, type of product, price, etc. An article in the *Journal of Marketing Research* (Aug. 1989) summarized the results of 15 recent studies that investigated the effect

EXAMPLE 2.8 Table 2.3 gives the starting salaries for 50 graduates selected from the 1,795 starting salaries of Appendix A. Construct a relative frequency histogram for the data.

TABLE 2.3
Starting Salaries for a Sample of Graduates Selected from Appendix A

$28,000	$18,600	$30,400	$12,200	$13,700
23,200	24,100	28,300	18,400	36,900
20,800	18,400	15,400	28,900	27,400
13,300	22,400	30,700	21,200	21,100
32,600	29,400	20,100	24,700	31,100
24,600	34,700	20,800	19,400	14,800
18,700	41,900	23,200	24,700	38,500
9,000	34,600	26,200	24,300	18,100
19,500	18,500	23,600	22,900	18,000
18,400	35,500	26,700	11,200	26,300

Solution

STEP 1   The first step in constructing the relative frequency histogram for this sample is to define the **class intervals (categories)** into which the data will fall. To do this, we need to know the smallest and largest starting salaries in the data set. These salaries are $9,000 and $41,900, respectively. Since we want the smallest salary to fall in the lowest class interval and the largest salary to fall in the highest class interval, the class intervals must span starting salaries ranging from $9,000 to $41,900.

STEP 2   The second step is to choose the **class interval width**; this will depend on how many intervals we want to use to span the starting salary range and whether we want to use equal or unequal interval widths. For this example, we will use equal class interval widths (the most popular choice) and 11 class intervals. (See the note in the box on page 62 on choosing the number of class intervals.)
Note that the starting salary range is equal to

$$\text{Range} = \text{Largest measurement} - \text{Smallest measurement}$$
$$= \$41,900 - \$9,000$$
$$= \$32,900$$

Since we chose to use 11 class intervals, the class intervals width should approximately equal

$$\text{Class interval width} \approx \frac{\text{Range}}{\text{Number of class intervals}}$$
$$= \frac{32,900}{11} = 2,990.9$$
$$\approx \$3,000$$

We shall start the first class slightly below the smallest observation ($9,000) and choose the starting point so that no observation can fall on a **class boundary**. Since starting salaries are recorded to the nearest hundred dollars, we can do this by choosing the lower class boundary of the first class interval to be $8,950. [*Note:* We could just as easily have chosen $8,955, $8,975, $8,990, or any one of many other points below and near $9,000.] Then the class intervals will be $8,950 to $11,950, $11,950 to $14,950, and so on. The 11 class intervals are shown in the second column of Table 2.4.

TABLE 2.4
Tabulation of Data for the Starting Salaries of Table 2.3

CLASS	CLASS INTERVAL	TALLY	CLASS FREQUENCY	CLASS RELATIVE FREQUENCY
1	8,950–11,950	II	2	.04
2	11,950–14,950	IIII	4	.08
3	14,950–17,950	I	1	.02
4	17,950–20,950	IIII IIII III	13	.26
5	20,950–23,950	IIII II	7	.14
6	23,950–26,950	IIII III	8	.16
7	26,950–29,950	IIII	5	.10
8	29,950–32,950	IIII	4	.08
9	32,950–35,950	III	3	.06
10	35,950–38,950	II	2	.04
11	38,950–41,950	I	1	.02
TOTALS			50	1.00

STEP 3   The third step in constructing a histogram is to obtain each class frequency—i.e., the number of observations falling within each class. This is done by examining each starting salary in Table 2.3 and recording by tally (as shown in the third column of Table 2.4) the class in which it falls. The tally for each class gives the class frequencies shown in column 4 of Table 2.4. Finally, we calculate the class relative frequency as

$$\text{Class relative frequency} = \frac{\text{Class frequency}}{\text{Total number of observations}}$$
$$= \frac{\text{Class frequency}}{50}$$

These values are shown in the fifth column of Table 2.4.

STEP 4   The final step is to draw the graph. Mark off the class intervals along a horizontal line, as shown in Figure 2.12 (page 60). Then construct over each class interval a bar with the height proportional to the class frequency (**frequency histogram**) or the class relative frequency (**relative frequency histogram**). Since we desire a relative frequency histogram, the bar heights will be proportional to the class relative frequencies. The resulting histogram is shown in Figure 2.13 (page 60).

FIGURE 2.12
Class Intervals for the Data of Table 2.4

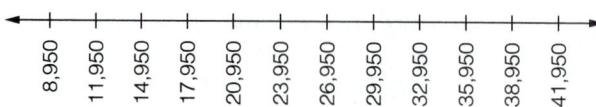

FIGURE 2.13
Relative Frequency Distribution for the Data of Table 2.3

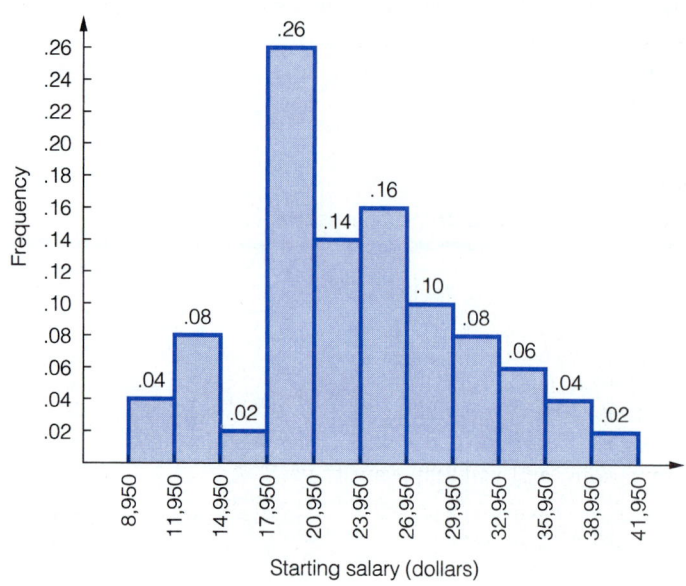

Starting salary (dollars)

EXAMPLE 2.9 Interpret the histogram for the 50 starting salaries, Figure 2.13.

Solution Remember that a histogram, like a bar graph or stem-and-leaf display, conveys a visual picture of the data. In particular, a histogram will identify a range of values where most of the data fall. You can see that most of the 50 graduates in the sample had starting salaries between $17,950 and $32,950. The five classes (bars) associated with this interval have relative frequencies of .26, .14, .16, .10, and .08, respectively. The sum of these relative frequencies is .74. Thus, 74% of the graduates in the sample had starting salaries between $17,950 and $32,950.

FIGURE 2.14
Skewed and Symmetric Distributions

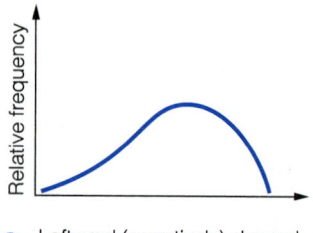

a. Leftward (negatively) skewed

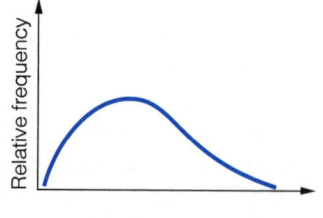

b. Rightward (positively) skewed

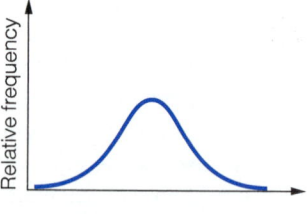

c. Nonskewed (symmetric)

Note also that none of the starting salaries was less than $8,950, but several high salaries caused the distribution to tail out to the right. We say that such a distribution is **rightward skewed** (or **positively skewed**). Similarly a **leftward** (or **negatively**) **skewed** distribution has a histogram that tails out to the left because of several unusually small values. Illustrations of rightward and leftward skewed distributions, as well as a **nonskewed** (**symmetric**) distribution, are shown in Figure 2.14.

In Examples 2.8 and 2.9, suppose we had spanned the range with three classes instead of 11. Then almost all the starting salaries would have fallen in a single class, and the resulting histogram would not have been nearly as informative as Figure 2.13. Or, suppose that we had chosen to span the interval from $9,000 to $41,900 with 50 classes. Then many classes would have contained only a few starting salaries and others would have been empty. Again, such a figure would not have been as informative as Figure 2.13. You can see that the choice of number of classes is critical when constructing a histogram.

A good rule of thumb in deciding on the number of class intervals is to use a small number when you want to describe a small amount of data—say, five or six classes for up to 25 observations. You can increase the number of classes (as we did) for 50 observations, and you may want to use 15 or 20 classes for large amounts of data. Remember that the objective is to obtain a graph that rapidly conveys a visual picture of the data. If your first choice of class interval width is not satisfactory, we recommend that you choose a different interval width and try again.

The steps used to construct a relative frequency distribution are summarized in the accompanying box.

STEPS TO FOLLOW IN CONSTRUCTING A HISTOGRAM

1. Examine the data to determine the smallest and the largest measurements.

2. Divide the interval between the smallest and the largest measurements into between 5 and 20 equal subintervals called **classes** (see next box). These classes should satisfy the following requirement:

   Each measurement falls into one and only one subinterval.

   Note that this requirement implies that no measurement falls on a boundary of a subinterval.

3. Compute the frequency or relative frequency of measurements falling within each subinterval.

4. Using a vertical axis of about three-fourths the length of the horizontal axis, plot each frequency or relative frequency as a rectangle over the corresponding subinterval.

The important concept to use in visually interpreting a histogram is as described in the next box.

EXAMPLE 2.10   Figure 2.15 is a computer-generated (SAS) relative frequency histogram that describes all 1,795 starting salaries of Appendix A.*

a. Interpret the graph.
b. Visually estimate the proportion of starting salaries between $18,000 and $38,000.

Solution

a. Note that the classes are marked off in intervals of $4,000 along the horizontal axis of the SAS histogram, Figure 2.15, with the midpoint (rather than the class boundaries) of each interval shown. You can see that the starting salaries tend to pile up near $24,000—the class from $22,000 to $26,000 has the greatest relative frequency (or percentage). Since this histogram tails out to the right, it is rightward skewed.
b. The bars that fall in the interval from $18,000 to $38,000 are shaded in Figure 2.15. You can see that this shaded portion represents approximately .80 of the total area of the bars for the complete distribution. This tells us that

*Note that the SAS-generated relative frequency histogram leaves gaps between the bars. However, the resulting figure should not be confused with a bar graph for qualitative data. When drawn by hand, the bars of a histogram are adjacent, with no gaps. (See Figure 2.13)

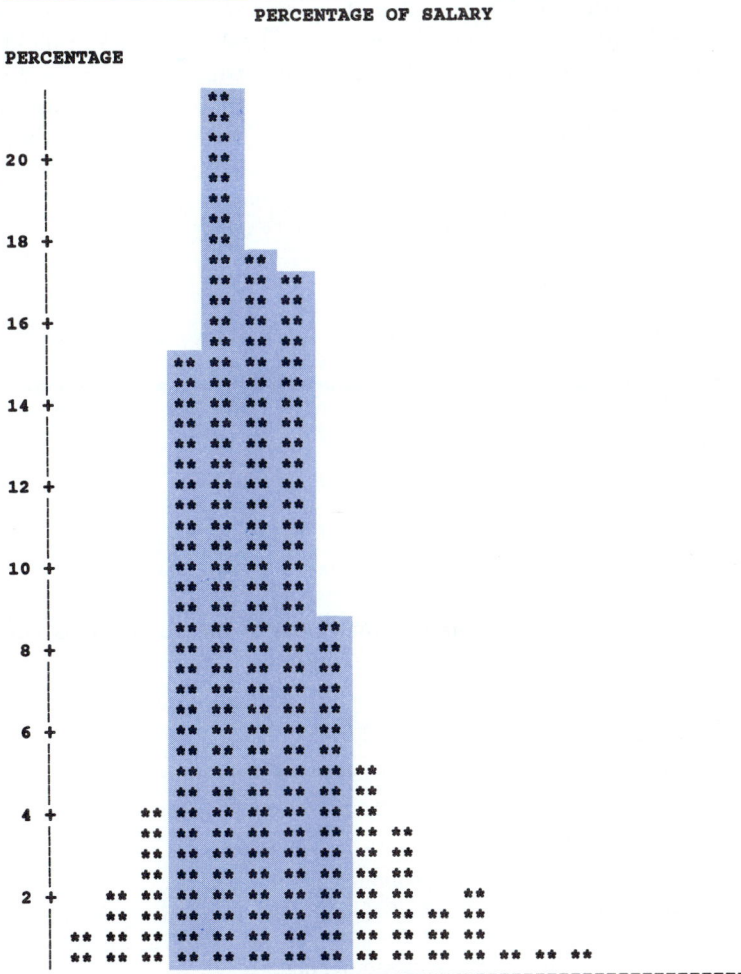

PERCENTAGE OF SALARY

PERCENTAGE

Salary (thousands) Midpoint

approximately 80% of the starting salaries were in the interval from $18,000 to $38,000. A more precise (but less rapid) answer could be obtained by recording and summing the relative frequencies for the classes in the interval from $18,000 to $38,000.

The preceding examples illustrate histograms with equal class interval widths. Histograms with unequal class widths can also be constructed. For example, you may have noticed that no bars appear atop the last four classes of the relative frequency histogram shown in Figure 2.15 (i.e., the classes 66,000–70,000;

70,000–74,000; 74,000–78,000; and 78,000–82,000). In fact, there are starting salaries in these ranges, but not enough to show on the SAS printout because of limited space. To make the histogram more visually appealing, we may choose to represent the last class interval as "Over $66,000." Then the bar over this interval would represent the sum of the relative frequencies of the four classes mentioned. By allowing the classes at the two extremes to have different widths, we obtain a more informative graphical display of the data.

EXERCISES
LEARNING THE MECHANICS

2.24   A sample of 20 measurements follows.

26 34 21 32 32 22 12 26 39 25
36 28 38 17 39 31 30 23 27 19

   a. Using a class interval width of 5, give the upper and lower boundaries for six class intervals, where the lower boundary of the first class is 10.5
   b. Determine the relative frequency for each of the six classes specified in part a.
   c. Construct a relative frequency distribution using the results of part b.

2.25   Consider the sample data shown here:

5.9 7.6 5.3 9.7 1.6 3.5 7.4
4.0 1.6 7.3 8.2 8.4 6.5 8.9
1.1 8.6 4.3 1.2 3.3 2.1 8.4
1.1 6.7 5.0 4.5 9.4 6.3 6.4

   a. Find the difference between the largest and smallest measurements.
   b. Divide the difference obtained in part a by 5 to determine the approximate class interval width for five class intervals.
   c. Specify upper and lower boundaries for each of the five class intervals.
   d. Construct a relative frequency distribution for the data.

APPLYING THE CONCEPTS

2.26   Refer to the results of the U.S. Army Corps of Engineers analysis of fish samples given in Appendix D. Consider the 50 DDT measurements corresponding to fish specimens identified on the data set by observations numbered 51–100.

   a. Construct a relative frequency histogram for the 50 DDT values. Use 10 classes to span the range.
   b. Repeat part a, but use only three classes to span the range. Compare the result with the relative frequency histogram you constructed in part a. Which is more informative? Why does an inadequate number of classes limit the information conveyed by a relative frequency histogram?
   c. Repeat part a, but use 25 classes. Comment on the information provided by this histogram and compare it with the result of part a.

2.27 Electrical engineers recognize that high neutral current in computer power systems is a potential problem. To determine the extent of the problem, a survey of the computer power system load currents at 146 U.S. sites was taken (*IEEE Transactions on Industry Applications*, July/Aug. 1990). A relative frequency histogram for the load capacities (measured as a percentage) of the 146 sites in the sample is shown here.

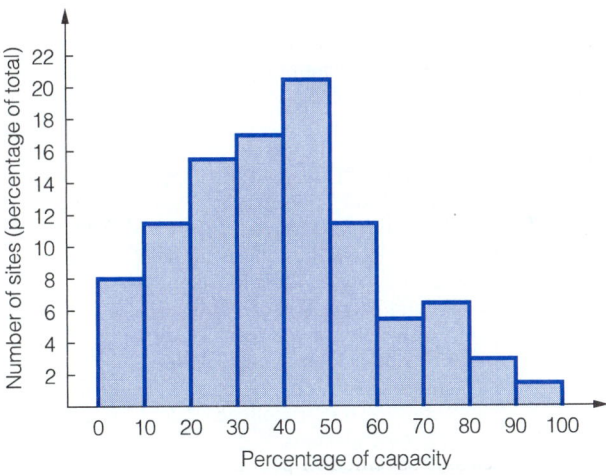

Source: Gruzs, T. M. "A survey of neutral currents in three-phase computer power systems." *IEEE Transactions on Industry Applications*, Vol. 26, No. 4, July/Aug. 1990, p. 722 (Figure 6).

  a. Approximately what proportion of the 146 computer sites had a load capacity between 20% and 30%?
  b. Approximately what proportion of the 146 computer sites had a load capacity of 50% or more?

2.28 Beginning in 1991, the nation's Department of Education began taking corrective and punitive actions against colleges and universities with high student-loan default rates. Those schools with default rates above 60% face suspension from the government's massive student-loan program, whereas schools with default rates between 40% and 60% are mandated to reduce their default rates by 5% a year or face a similar penalty (*Tampa Tribune*, June 21, 1989). A list of 66 colleges and universities in Florida and their student-loan default rates is provided in the table on page 66.
  a. Construct a relative frequency histogram for the data using 12 classes to span the range.
  b. Repeat part **a**, but use only three classes to span the range. Compare with the relative frequency distribution you constructed in part **a**. Which is more informative? Why does an inadequate number of classes limit the information conveyed by the relative frequency distribution?
  c. Repeat part **a**, but use 25 classes. Comment on the information provided by this graph as compared with that of part **a**.
  d. Refer to the histogram, part **a**. Estimate the proportion of Florida colleges and universities with a default rate of 40% or higher. Shade the bars in the histogram corresponding to this area.
  e. Note that Florida College of Business has a default rate nearly 30% higher than the next highest rate. Omit the value for Florida College of Business from the data set and reconstruct the histogram.
  f. Compare the histograms constructed in parts **a** and **e**. Which graph is more informative? Explain.

2.29 Over half of the nearly 60,000 members of the U.S. Chess Federation (USCF) have official chess ratings. A player with a rating of 1,100 or less is a "beginner"; "average" players' ratings range between 1,100 and 1,900; "experts" range between 2,000 and 2,200; "masters" range between 2,200 and 2,400; and "grand masters" have ratings

Data for Exercise 2.28

COLLEGE/UNIVERSITY	DEFAULT RATE	COLLEGE/UNIVERSITY	DEFAULT RATE
Florida College of Business	76.2	Brevard Community College	9.4
Fort Lauderdale College	48.5	College of Boca Raton	9.1
Florida Career College	48.3	Florida International University	8.7
United College	46.8	Santa Fe Community College	8.6
Florida Memorial College	46.2	Edison Community College	8.5
Bethune Cookman College	43.0	Palm Beach Junior College	8.0
Edward Waters College	38.3	Eckerd College	7.9
Florida College of Medical and		University of Tampa	7.6
Dental Careers	32.6	Lakeland College of Business	7.2
International Fine Arts College	26.5	Pensacola Junior College	6.8
Tampa College	23.9	University of Miami	6.7
Miami Technical College	23.3	Florida Institute of Technology	6.7
Tallahassee Community College	20.6	University of West Florida	6.3
Charron Williams College	20.2	Palm Beach Atlantic College	6.0
Florida Community College	19.1	University of Central Florida	5.7
Miami-Dade Community College	19.0	Seminole Community College	5.6
Broward Community College	18.4	Polk Community College	5.6
Daytona Beach Community College	16.9	Phillips Junior College	5.6
Lake Sumter Community College	16.7	Nova University	5.5
Forida Technical College	16.6	Rollins College	5.5
Florida A. & M. University	15.8	St. Leo College	5.5
Prospect Hall College	15.1	Gulf Coast Community College	5.4
Hillsborough Community College	14.4	Southern College	5.3
Pasco-Hernando Community College	13.5	Flagler College	4.7
Orlando College	13.5	Florida Atlantic University	4.4
Jones College	13.1	University of South Florida	4.2
Webber College	11.8	Manatee Junior College	4.1
Warner Southern College	11.8	Florida State University	4.0
Central Florida Community College	11.8	University of North Florida	3.9
Indian River Community College	11.8	Barry University	3.1
St. Petersburg Community College	11.3	University of Florida	3.1
Valencia Community College	10.8	Stetson University	2.9
Florida Southern College	10.3	Jacksonville University	1.5
Lake City Community College	9.8		

higher than 2,400. (Gary Kasparov, the reigning world champion from Russia has a chess rating of 2,900.) The accompanying graph, extracted from *Scientific American* (Oct. 1990), illustrates the distribution of the ratings of the 35,000 rated members of the USCF.

a. What type of graph is portrayed?

b. Visually estimate the number of USCF grand masters.

c. Is the data skewed? Explain.

2.30 According to the 1990 U.S. census, nearly half of the counties in the country exhibited a declining population trend between 1980 and 1990. In particular, the population of several major metropolitan counties also declined. For example, Wayne County (Detroit), Michigan, lost nearly 10% of its residents between 1980 and 1990. Many demographers feel this reflects "urban flight," a reversal of the 1970's trend of people migrating from rural to metropolitan counties (*USA Today*, Mar. 19, 1991). The accompanying table lists the 49 largest counties in the United States (based on 1990 population) and their percentage change in population from 1980 to 1990. A Minitab frequency histogram (with interval midpoints marked on the vertical axis) for the data follows. Interpret the graph.

Graph for Exercise 2.29

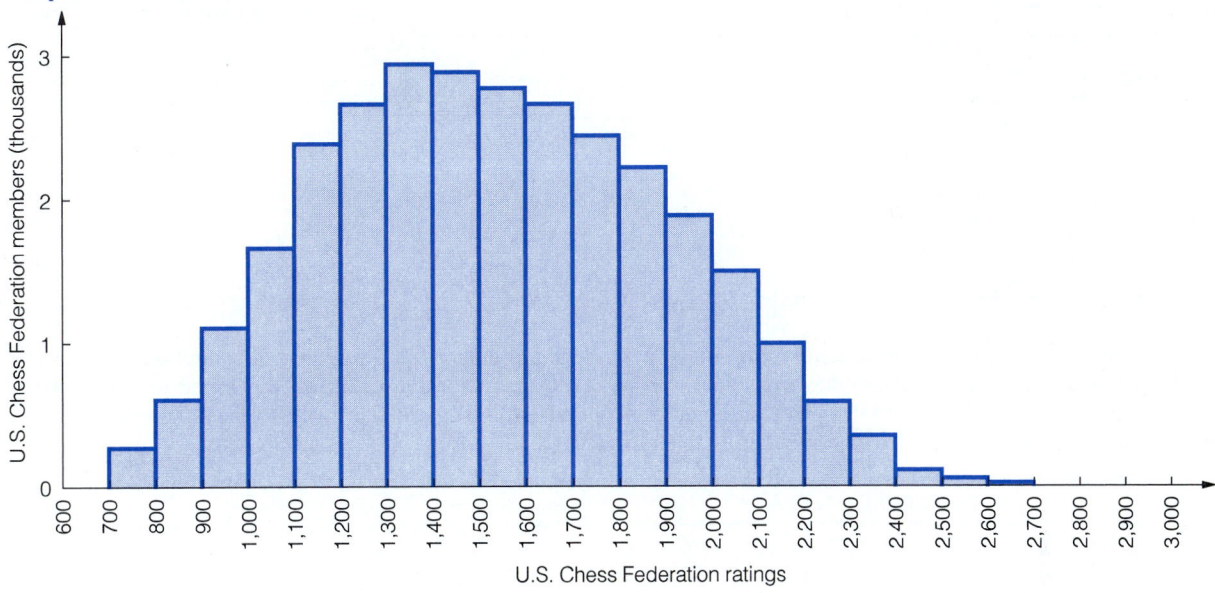

Data for Exercise 2.30

COUNTY (NEAREST BIG CITY)	PERCENT CHANGE	COUNTY (NEAREST BIG CITY)	PERCENT CHANGE
Los Angeles (Los Angeles)	18.5	Tarrant (Fort Worth TX)	35.9
Cook (Chicago)	−2.8	Oakland (Pontiac MI)	7.1
Harris (Houston)	17.0	Sacramento (Sacramento)	32.9
Orange (Los Angeles)	24.7	Hennepin (Minneapolis)	9.7
Kings (New York City)	3.1	St. Louis (St. Louis)	2.0
Maricopa (Phoenix)	40.6	Erie (Buffalo)	−4.6
Wayne (Detroit)	−9.7	Franklin (Columbus OH)	10.6
Queens (New York City)	3.2	Milwaukee (Milwaukee)	−0.6
Dade (Miami)	19.2	Westchester (White Plains NY)	1.0
Dallas (Dallas)	19.0	Hamilton (Cincinnati)	−0.8
Philadelphia (Philadelphia)	−6.1	Palm Beach (Palm Beach FL)	49.7
King (Seattle)	18.7	Hartford (Hartford CT)	5.4
Santa Clara (San Jose CA)	15.6	Pinellas (St. Petersburg FL)	16.9
New York (New York City)	4.1	Honolulu (Honolulu)	9.7
San Bernardino (Los Angeles)	58.5	Hillsborough (Tampa)	28.9
Cuyahoga (Cleveland)	−5.8	Fairfield (Bridgeport CT)	2.5
Middlesex (Boston)	2.3	Shelby (Memphis)	6.3
Allegheny (Pittsburgh)	−7.8	Bergen (Hackensack NJ)	−2.4
Suffolk (New York City)	2.9	Fairfax (VA/Washington)	37.4
Nassau (New York City)	−2.6	New Haven (New Haven CT)	5.6
Alameda (Oakland CA)	15.7	Contra Costa (San Francisco)	22.5
Broward (Ft. Lauderdale FL)	23.3	Marion (Indianapolis)	4.25
Bronx (New York City)	3.0	DuPage (Chicago)	18.6
Bexar (San Antonio)	19.9	Essex (Newark NJ)	−8.6
Riverside (Los Angeles)	76.5		

Source: U.S. Census Bureau, 1990 Census.

Minitab Printout for Exercise 2.30

```
Histogram of PctChnge    N = 49

Midpoint   Count
     -10      5   *****
       0     16   ****************
      10      7   *******
      20     13   *************
      30      2   **
      40      3   ***
      50      1   *
      60      1   *
      70      0
      80      1   *
```

2.31 *Scram* is the term used by nuclear engineers to describe rapid emergency shutdown of a nuclear reactor. The nuclear industry has made a concerted effort to significantly reduce the number of unplanned scrams and has apparently succeeded. The table gives the number of scrams at each of 56 U.S. nuclear reactor units in 1984, a year in which an unusually high number of scrams occurred.

				NUMBER OF SCRAMS					
1	0	3	1	4	2	10	6	5	2
0	3	1	5	4	2	7	12	0	3
8	2	0	9	3	3	4	7	2	4
5	3	2	7	13	4	2	3	3	7
0	9	4	3	5	2	7	8	5	2
4	3	4	0	1	7				

Source: American Nuclear Society.

a. Construct a frequency distribution for the data using 10 class intervals.
b. Construct a stem-and-leaf display. [*Note:* The leaves of all measurements will be 0.]
c. Turn the stem-and-leaf display, part b, on its side and compare its shape to the frequency distribution, part a.

2.32 Are major colleges and universities lax in hiring minorities to fill top positions in their athletic programs? A *USA Today* survey of 62 Division I schools found that only 12.5% of the jobs in the athletic department are held by minorities (blacks, Hispanics, Native Americans, and Asians). In contrast, the 1990 census shows minorities represent 19.7% of the U.S. population (*USA Today*, Mar. 19, 1991). The results of the survey are reproduced in the table on the next page. The 62 schools were selected based on the Top 25 polls for men's and women's basketball during the 1989–1990 season and the final 1990 Top 25 football poll. (Northwestern declined to respond, and Seton Hall did not supply figures.)
a. Do the data represent a sample or a population? Explain.
b. Describe the data on percentage of minority positions in the athletic department of the 62 colleges and universities with a graphical technique. Interpret the graph.

2.33 In his essay "Making Things Right," W. Edwards Deming considered the role of statistics in the quality control of industrial products.* In one example, Deming examined the quality control process for a manufacturer of steel rods. Rods produced with diameters smaller than 1 centimeter (cm) fit too loosely in their bearings and ultimately must be rejected (thrown out). To determine if the diameter setting of the machine that produces the rods is correct, 500 rods are selected from the day's production and their diameters are recorded. The distribution of the 500 diameters for one day's production is shown in the figure on the next page. Note that the symbol LSL in the figure represents the 1-centimeter lower specification limit of the steel rod diameters.

---

*From Tanur, J. M., Mosteller, F., Kruskal, W. H., Link, R. F., Pieters, R. S., and Rising, G. R. (eds.). *Statistics: A Guide to the Unknown*, 2nd ed. San Francisco: Holden-Day, 1978, pp. 279–281.

Data for Exercise 2.32

SCHOOL	TOTAL POSITIONS	POSITIONS HELD BY MINORITY	PERCENT MINORITY	SCHOOL	TOTAL POSITIONS	POSITIONS HELD BY MINORITY	PERCENT MINORITY
Georgetown	30	8	26.7	Louisiana Tech	26	3	11.5
Houston	43	11	25.6	Syracuse	52	6	11.5
Miami	55	14	25.5	Arkansas	44	5	11.4
Arizona	67	15	22.4	Northern Illinois	53	6	11.3
Long Beach State	53	11	20.8	Alabama	54	6	11.1
USC	65	13	20.0	Western Kentucky	18	2	11.1
Pittsburgh	54	10	18.5	UNLV	55	6	10.9
Oklahoma State	34	6	17.6	Connecticut	37	4	10.8
Oklahoma	63	11	17.5	South Carolina	49	5	10.2
Washington	63	11	17.5	East Tennessee State	30	3	10.0
Southern Mississippi	35	6	17.1	Texas-El Paso	30	3	10.0
Stanford	71	12	16.9	Rutgers	42	4	9.5
Iowa	72	12	16.7	North Carolina State	44	4	9.1
Georgia Tech	52	8	15.4	Texas[a]	67	6	9.0
Michigan State	60	9	15.0	Michigan	58	5	8.6
Illinois	54	8	14.8	Penn State	82	7	8.5
Kentucky	54	8	14.8	St. John's	60	5	8.3
Ohio State	64	9	14.1	Nebraska	64	5	7.8
Colorado	43	6	14.0	Mississippi State	52	4	7.7
LSU	59	8	13.6	Providence	13	1	7.7
Purdue	53	7	13.2	Mississippi	40	3	7.5
New Mexico State	38	5	13.2	Florida State	53	4	7.5
UCLA	84	11	13.1	Indiana	57	4	7.0
Clemson	48	6	12.5	Tennessee[a]	73	5	6.8
North Carolina	57	7	12.3	Duke	45	3	6.7
Kansas	49	6	12.2	Virginia	60	4	6.7
Utah	33	4	12.1	BYU	51	3	5.9
Louisville	34	4	11.8	Princeton	44	2	4.5
Georgia	60	7	11.7	Notre Dame	47	2	4.3
Florida	61	7	11.5	Stephen F. Austin	27	1	3.7

[a]Numbers combined from separate men's/women's athletic programs.

Source: *USA Today*, Mar. 19, 1991.

Graph for Exercise 2.33

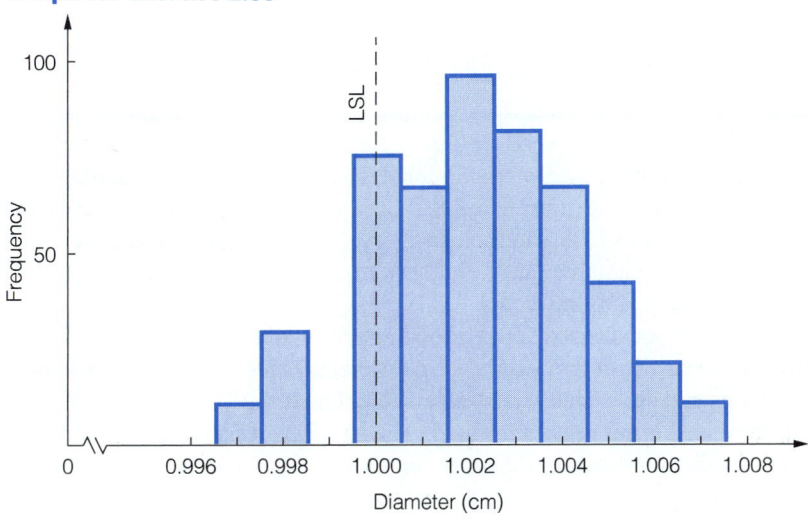

a. What type of data, quantitative or qualitative, does the figure portray?
b. What type of graphical method is being used to describe the data?
c. Use the figure to estimate the proportion of rods with diameters between 1.0025 and 1.0045 centimeters.
d. There has been speculation that some of the inspectors are unaware of the trouble that an undersize rod diameter would cause later in the manufacturing process. Consequently, these inspectors may be passing rods with diameters that were barely below the lower specification limit, and recording them in the interval centered at 1.000 centimeter. According to the figure, is there any evidence to support this claim? Explain.

2.34   Refer to the *Developmental Psychology* (Mar. 1986) study on the sexual maturation of college students, Exercise 2.14. In addition to the initial question on puberty, 75 of the male college undergraduates were asked to report the age (in years) when they began to shave regularly. The frequency distribution of age at regular shaving is shown in the accompanying figure.

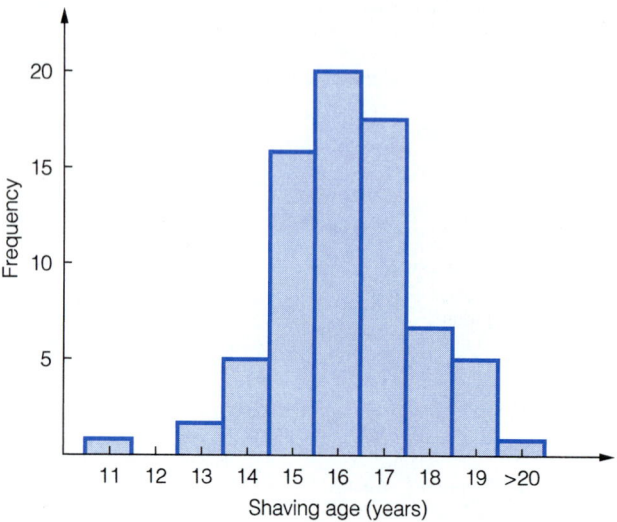

a. Approximately how many of the 75 male students began to shave regularly at age 16?
b. Use the figure to construct a relative frequency distribution for the age at which regular shaving began.
c. Approximately what proportion of the 75 male students began to shave regularly at age 16?

SUMMARY

In this chapter, we discussed methods for describing data sets of *qualitative* or *quantitative* variables. We first define the data categories and then record the number of observations falling in each category. *Bar graphs* or *pie charts* can then be constructed for qualitative data, and *stem-and-leaf displays* or *histograms* can be constructed for quantitative data.

If the data are quantitative and the number of observations is small, the categorization and the determination of class frequencies can be done automatically by constructing a stem-and-leaf display. Once the stem of a number is defined, the data *classes* are determined. The number of leaves associated with each stem in the stem-and-leaf display can easily be counted to determine the *class frequency*. The stem-and-leaf display provides a graphical picture of the data set, showing how the observations are spread from the smallest to the largest. It also enables us to reconstruct the exact values of the observations. This feature of the stem-and-leaf display will be useful in determining the values of certain numerical descriptive measures of the data to be discussed in Chapter 3.

Large sets of data are best described using *frequency* or *relative frequency histograms*. Usually, *class intervals* of equal width are defined. Bars are then constructed over each interval with the bar height proportional to the interval relative frequency. Since the bars are of equal width, the area of a particular bar, as a proportion of the total area of all bars, is equal to the *class relative frequency* of that bar. This property enables us to examine a relative frequency histogram and visually estimate the proportion of the total number of observations in the data set that fall in specific intervals. Thus, the histogram provides a good graphical description of the location and distribution of the observations in a data set.

Note the differences and similarities between the bar graph used to describe qualitative data and the histogram used to describe quantitative data. Although the classes in a bar graph are unrelated, the classes that categorize quantitative data are connected intervals on a real line—the *upper class boundary* of one interval is the *lower class boundary* of the next. The interpretations of bar graphs and histograms are similar: The height of a bar is proportional to a class frequency (or relative frequency). Thus, both convey at a glance a graphical picture of the proportions of observations falling in the data classes.

KEY WORDS

Bar graph	Leftward skewed
Class	Pie chart
Class boundaries (lower and upper)	Relative frequency histogram
Class frequency	Rightward skewed
Class interval	Skewed distribution
Class relative frequency	Stem-and-leaf display
Frequency histogram	Symmetric distribution
Histogram	

SUPPLEMENTARY EXERCISE

2.35   Refer to the fish sample data in Appendix D (see Case Study 1.1). A horizontal bar graph describing the proportion of fish specimens captured of each of three species—channel catfish, largemouth bass, and smallmouth buffalo—is shown in the figure on the next page.
   a. Interpret the graph.
   b. Use the graph to approximate how many of each species were captured. Recall from Case Study 1.1 that a total of 144 fish specimens were captured.

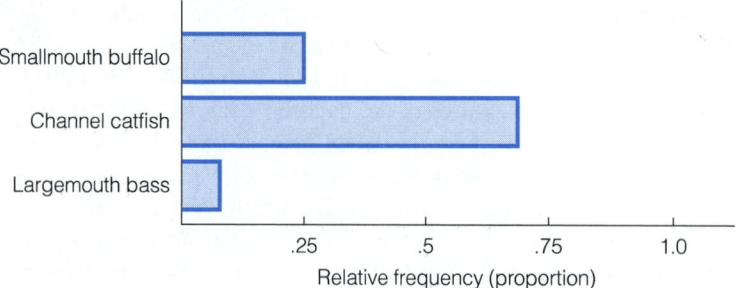

2.36 In 1986, the state of California finally declared English to be its official language. (Spanish was the official Californian language prior to 1986.) Over the past several years, however, a wave of immigrants to California brought millions of people who speak very little English. The accompanying table gives the breakdown in the percentages of current Californians who speak foreign languages and how well they speak English.

MAIN LANGUAGE SPOKEN		PERCENTAGE
English		54.2
Spanish		
Speak English very well		13.4
Speak English well		6.3
Speak little English		9.4
	TOTAL	29.2
Asian Languages (various)		
Speak English very well		4.5
Speak English well		3.0
Speak little English		2.6
	TOTAL	10.1
Other		
Speak English very well		4.4
Speak English well		1.4
Speak little English		0.7
	TOTAL	6.5

Source: U.S. Census Bureau, *San Francisco Chronicle*, May 13, 1992.

a. Use a graphical method to summarize the main languages spoken by Californians.
b. Use a graphical method to summarize how well Spanish-speaking Californians speak English.

2.37 A study was conducted to evaluate the advertisement awareness and sales effectiveness of advertising campaigns for 18 confectionary brands (*Journal of the Market Research Society*, Jan. 1986). For each brand, an ad awareness index (maximum = 100) was determined from a consumer survey, while a sales effectiveness index (maximum = 100) was estimated from market shares. The frequency distributions on the next page were used to summarize the data.

a. How many of the 18 brands had an ad awareness index of 40 or less?
b. How many of the 18 brands had a sales effectiveness index of 70 or more?
c. Use the information provided by the frequency distribution to construct a relative frequency distribution for the 18 ad awareness indexes. Interpret the graph.
d. Use the information provided by the frequency distribution to construct a relative frequency distribution for the 18 sales effectiveness indexes.

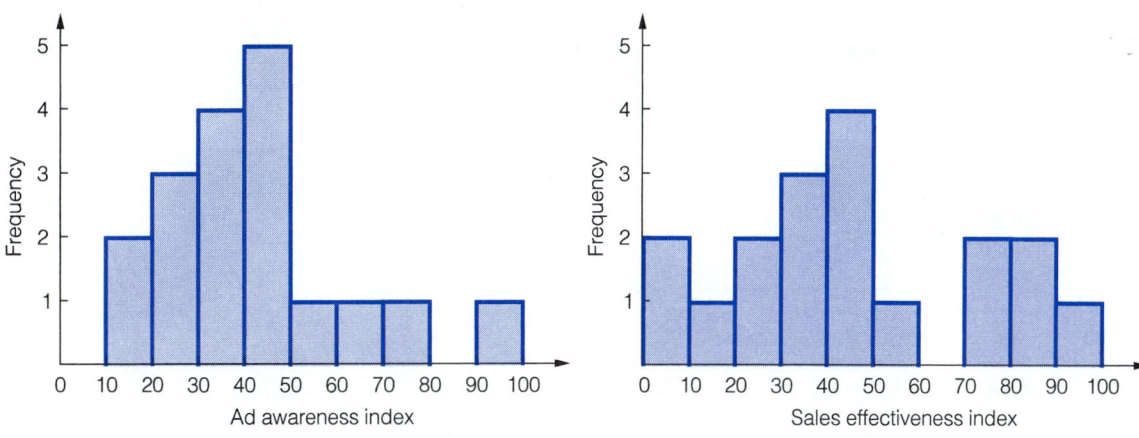

Distribution of 18 Campaigns by Ad Awareness Effectiveness (maximum = 100)

Distribution of 18 Campaigns by Sales Effectiveness (maximum = 100)

Source: Broadbent, S. and Colman, S. "Advertising effectiveness: Across brands." *Journal of the Market Research Society*, Vol. 28, No. 1, 1986, pp. 15–23.

2.38 Coca-Cola Co. and PepsiCo Inc., the makers of Coke and Pepsi, respectively, continue to be the two largest soft-drink manufacturers in the United States (*Business Week*, July 7, 1986). The pie chart below describes the U.S. soft-drink market (based on number of sales).

Soft Drinks: How the U.S. Market Splits Up

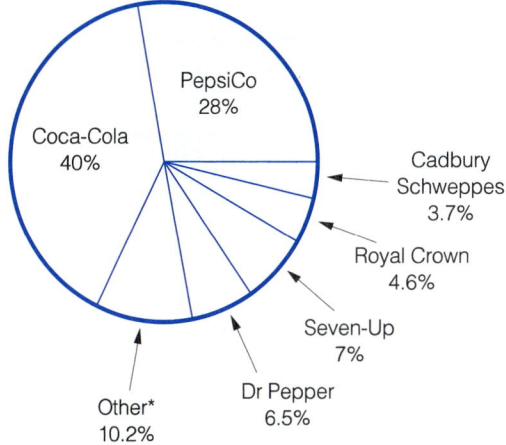

*Includes private-label and store brands

Data: *Beverage Digest*, Montgomery Securities.
Source: *Business Week*, July 7, 1986.

a. What percentage of the soft-drink market is controlled by either Coca-Cola or PepsiCo?
b. In 1986, Coca-Cola's attempted buyout of Dr Pepper Co. and PepsiCo's merger with SevenUp Co. were both blocked by the Federal Trade Commission's antitrust division. Construct a pie chart to describe the realignment of the soft-drink market if both mergers had been approved.
c. Answer the question in part **a** based on the new pie chart constructed in part **b**.

2.39 A National Cancer Institute survey of 1,580 adult women recently responded to the question, "In your opinion, what is the most serious health problem facing women?" The responses are summarized in the accompanying table.

THE MOST SERIOUS HEALTH PROBLEM FOR WOMEN	RELATIVE FREQUENCY
Breast cancer	.44
Other cancers	.31
Emotional stress	.07
High blood pressure	.06
Heart trouble	.03
Other problems	.09

Source: Nursing Outlook, Feb. 1981. Copyright by the American Journal of Nursing Company. Reprinted from Nursing Outlook.

  a. Use one of the graphical methods discussed in this chapter to describe the data.
  b. What proportion of the respondents believe that high blood pressure or heart trouble is the most serious health problem for women?
  c. Estimate the percentage of all women who believe that some type of cancer is the most serious health problem facing women.
  d. Refer to your answer to part **c**. Comment on the reliability of your inference. Do you think that the sample of 1,580 responses is representative of the set of responses for the population of all women?

2.40 The National Research Council's Committee on Underground Coal Mine Safety was established to determine "the factors that distinguish the safest from the most dangerous mines." To evaluate differences between the safest and most dangerous mines, the committee collected data on 19 of the largest underground coal companies. The "intermediate injury" rate (i.e., the number of disabling injuries resulting from falls of roof and sides, haulage, machinery, and explosive accidents per 200,000 worker hours) for each of the 19 companies is recorded in the table. Construct a stem-and-leaf display for the data.

COMPANY	INJURY RATE	COMPANY	INJURY RATE
Old Ben	2.72	American Electric Power	5.11
Bethlehem	2.89	Rochester & Pittsburgh	5.12
Island Creek	2.97	Pittston	5.39
Consolidation	2.98	Ziegler	6.19
Mapco	3.17	Freeman United	6.83
U.S. Steel	3.58	Republic	6.84
Alabama By-Product	3.88	North American	7.47
Eastern Assoc.	4.66	West Moreland	7.68
Peabody	4.81	Valley Camp	8.71
Jones & Laughlin	4.87		

Source: Spokes, E. M. "New look at underground coal mine safety." Mining Engineering, Vol. 38, No. 4, Apr. 1986, p. 267 (Table 1).

2.41 The nuclear mishap on Three Mile Island near Harrisburg, Pennsylvania, on March 28, 1979, forced many local residents to evacuate their homes—some temporarily, others permanently. To assess the impact of the accident on the area population, a questionnaire was designed and mailed to a sample of 150 households within two weeks after the accident occurred. The following two questions were asked of the sampled residents: (1) When did you learn about the accident? and (2) How did you learn about the accident? The responses to the two questions are illustrated in the accompanying frequency distributions. Based on these graphical descriptions, find each of the following:

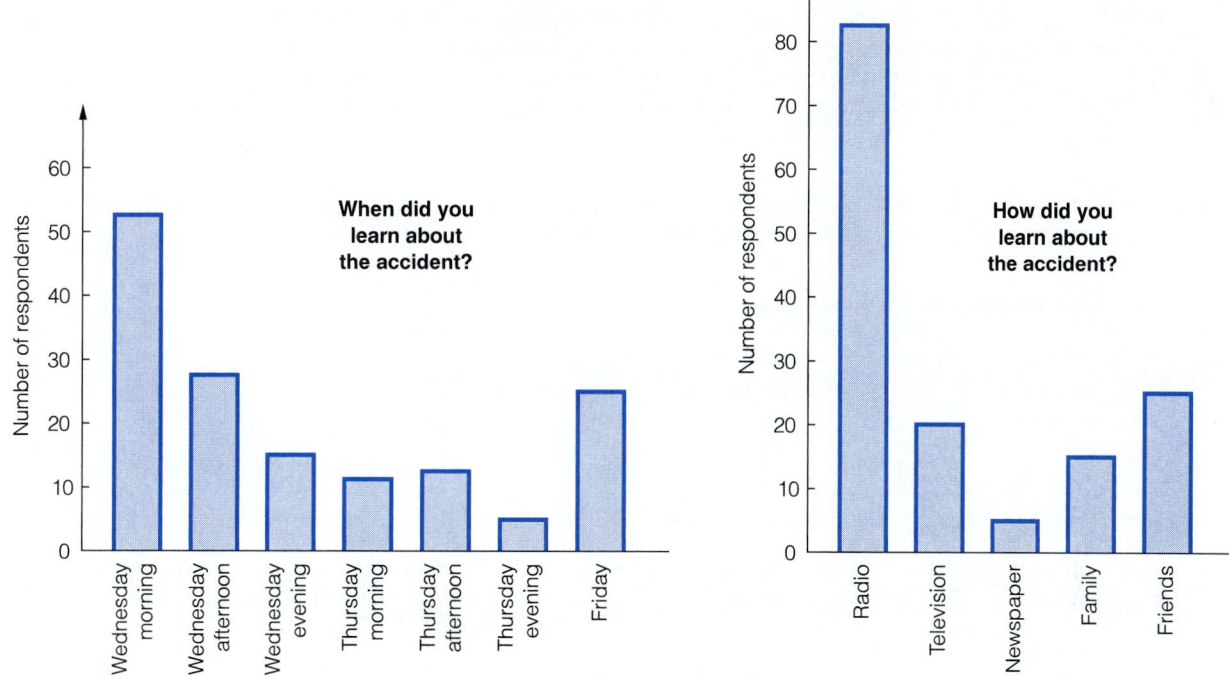

Source: Brown, S., et al. "Final Report on a Survey of Three Mile Island Area Residents." Department of Geography, Michigan State University, Aug. 1979.

    a. The number of respondents who learned about the accident on Wednesday afternoon
    b. The number of respondents who learned about the accident on Friday
    c. The number of respondents who learned about the accident from a radio report
    d. The number of respondents who learned about the accident from television

2.42 In the book *Identical Twins Reared Apart*, author Susan Farber offers a survey and reanalysis of all published cases of identical (monozygotic) twins separated and reared apart. One chapter deals extensively with the evaluation and analysis of IQ tests conducted on the twins. A portion of the data for 19 pairs of twins is reproduced in the table below.

| | IQ | | | | IQ | |
PAIR	Twin A	Twin B	PAIR	Twin A	Twin B
1	99	101	11	115	105
2	85	84	12	90	88
3	92	116	13	91	90
4	94	95	14	66	78
5	105	106	15	85	97
6	92	77	16	102	94
7	122	127	17	89	106
8	96	77	18	102	96
9	89	93	19	88	79
10	116	109			

Source: Adapted from *Identical Twins Reared Apart*, by Susan L. Farber. Copyright © 1981 by Basic Books, Inc. Reprinted by permission of Basic Books, Inc., Publishers, New York.

a. Classify the variable of interest, IQ score, as quantitative or qualitative.

b. What type of graphical method is appropriate for summarizing the data in the table?

c. Construct an appropriate graph to summarize the data for the 38 twins in the sample.

d. Use the graph to find the percentage of twins in the sample with IQ scores above 100.

e. What would be the advantages and disadvantages of using a stem-and-leaf display, rather than a relative frequency distribution, to describe the IQ data?

2.43 According to the *U.S. News and World Report Fitness Guide*, most people who start a fitness program do not stick with it at least two days a week. The table gives the number of people (in millions) who exercised with one of four activities—fitness walking, running or jogging, bicycling, and swimming—at least one day a week and at least two days a week.

ACTIVITY	DAYS PER WEEK	
	At least 1	At least 2
Fitness walking	23.2	10.3
Running/Jogging	32.9	8.1
Bicycling	28.7	4.9
Swimming	22.9	2.6

Source: *Tampa Tribune,* Aug. 14, 1988.

a. Construct a bar chart for the four categories of fitness activities for those exercising at least one day per week.

b. Construct a bar chart for the four categories of fitness activities for those exercising at least two days per week.

c. Place the bar charts, parts **a** and **b**, side by side. Do you agree with the statement made in the *U.S. News and World Report Fitness Guide*?

2.44 Shore-based marine traffic systems have been proposed to improve the safety and efficiency of marine traffic. Before the installation of the traffic system, one study was conducted to assess the current level of risk of collision to vessels operating in European waters. Data on large-vessel collisions over the 5-year period, 1978–1982, are presented in the table.

COLLISIONS BY LOCATION		COLLISIONS AT SEA BY ENCOUNTER ASPECT	
Location	Number of Ships	Aspect	Number of Ships
At sea	376	Meeting	131
Restricted waters	273	Overtaking	29
In port	478	Crossing	73
		Unknown	143
TOTAL	1,127	TOTAL	376

Source: Kemp, J. F. and Goodwin, E. M. "Risk analysis within the Cost 301 Project." *The Dock and Harbour Authority,* Vol. 66, No. 775, Dec./Jan. 1985–1986.

a. Construct a bar graph for the locations of large-vessel collisions in European waters. Interpret the bar graph.

b. Construct a bar graph for the encounter aspects of large-vessel collisions at sea in European waters. Interpret the graph.

2.45 Over the past decade, federal and state governments have proposed and implemented several new programs designed to improve automobile and highway safety standards. These include stricter driving-while-intoxicated laws and mandatory-use seat belt laws in many states and federal guidelines related to air bags in new automobiles. Recently, three Northeastern University professors conducted a survey of U.S. residents "in an attempt to gauge their opinions concerning existing and proposed automobile safety standards" (*Transportation Journal,* Summer 1986). It was hoped that the results of the survey would aid federal and state government officials in planning and designing automobile safety programs. The survey elicited usable responses from a sample of 380 residents. The responses to two of the questions in the survey are summarized in the tables here. Use a graphical method to describe the two data sets and interpret the results.

QUESTION: How often do you use your safety seat belts?

RESPONSE	PERCENTAGE
Always	40
Frequently	23
Infrequently	20
Never	17

QUESTION: Would you purchase an optional automobile safety air bag at a price of $500 or $1,000?

RESPONSE	PERCENTAGE $500	PERCENTAGE $1,000
Definitely would	16	5
Probably would	17	12
Not sure	18	17
Probably would not	21	22
Definitely would not	28	44

Source: Lieb, R. C., Wiseman, F., and Moore, T. E. "Automobile safety programs: The public viewpoint." *Transportation Journal*, Vol. 25, No. 4, Summer 1986, pp. 22–30.

2.46 Refer to the problem of elderly parent abuse introduced in Exercise 1.30. The accompanying table gives the various methods used by adults to control their elders (and the corresponding proportions), based on a sample survey of 60 adult children caring for a dependent elderly parent. Use a graphical method to summarize the survey results. Interpret the graph.

METHOD OF CONTROL	PROPORTION OF ADULT CHILDREN
Screamed and yelled	.40
Used physical restraint	.06
Forced feeding or medication	.06
Threatened to send to nursing home	.06
Threatened with physical force	.04
Hit or slapped	.03
Nonabusive method	.35

Source: Aging, Jan.–Feb. 1981.

2.47 The Global Auto Scoreboard, compiled by *Business Week*, is a comprehensive analysis of the world's 25 largest automobile manufacturers (based on world market share). The total number of units (autos and trucks) produced by each manufacturer in 1989 is provided in the table on the next page.
 a. Construct a relative frequency bar graph to describe the worldwide production of the 25 largest auto manufacturers. Interpret the graph.
 b. Nine countries are represented by the 25 largest auto manufacturers. Construct a relative frequency bar graph to describe the worldwide production of the nine countries. Interpret the graph.
 c. Construct a relative frequency bar graph to describe the worldwide production of the nine Japanese firms represented in the table. Interpret the graph.

RANK	MANUFACTURER	COUNTRY	WORLDWIDE PRODUCTION (THOUSANDS OF UNITS)
1	General Motors	United States	7,946
2	Ford	United States	6,336
3	Toyota	Japan	4,115
4	Volkswagen	Germany	2,948
5	Nissan	Japan	2,930
6	Chrysler	United States	2,382
7	Fiat	Italy	2,436
8	Peugeot	France	2,216
9	Renault	France	2,053
10	Honda	Japan	1,960
11	Mazda	Japan	1,460
12	Mitsubishi Motors	Japan	1,335
13	Hyundai	South Korea	819
14	Suzuki	Japan	875
15	Daimler	Germany	803
16	Daihatsu	Japan	600
17	Fuji Heavy	Japan	530
18	BMW	Germany	523
19	Rover	Britain	535
20	Volvo	Sweden	465
21	Isuzu	Japan	590
22	KIA Motors	South Korea	412
23	Daewoo Motors	South Korea	200
24	Lada	Soviet Union	140
25	Saab-Scania	Sweden	138

Source: Business Week, May 7, 1990, pp. 54–55.

2.48 A preliminary study was conducted to obtain information on the background levels of the toxic substance polychlorinated biphenyl (PCB) in soil samples in the United Kingdom. Such information could then be used as a benchmark against which PCB levels at waste disposal facilities in the United Kingdom could be compared. The accompanying table contains the measured PCB levels of soil samples taken at 14 rural and 15 urban locations in the United Kingdom. (PCB concentration is measured in .0001 gram per kilogram of soil.) From these preliminary results, the researchers reported "a significant difference between (the PCB levels) for rural areas . . . and for urban areas."

	RURAL					URBAN			
3.5	9.0	9.8	23.0	8.2	24.0	21.0	22.0	94.0	18.0
8.1	1.0	15.0	1.5	9.7	29.0	11.0	13.0	141.0	12.0
1.8	5.3	1.6	12.0		16.0	49.0	107.0	11.0	18.0

Source: Badsha, K. and Eduljee, G. "PCB in the U.K. environment—A preliminary survey." *Chemosphere*, Vol. 15, No. 2, Feb. 1986, p. 213 (Table 1). Reprinted with permission. Copyright 1986, Pergamon Press, Ltd.

a. Construct a stem-and-leaf display for the PCB levels of rural soil samples.
b. Construct a stem-and-leaf display for the PCB levels of urban soil samples.
c. Combine the data for rural and urban soil samples and construct a stem-and-leaf display. Identify each of the urban PCB levels on the display with a circle. Does the graph support the researchers' conclusions?

2.49 "What do you look for most in your relationship with your partner?" About 12,000 people responded to this question and others in a 1983 survey conducted by *Psychology Today*. The respondents ranged in age from 13 to

86, were well educated, and had relatively high household incomes; 41% were men and 59% were women. The responses to the question are summarized in the table.

RESPONSE	PERCENTAGE
Companionship	32
Financial security	2
Love	54
Romance	4
Sex	1
Other	7

Source: "The modern art of courtly love." *Psychology Today*, July 1983, p. 46.

a. Construct a pie chart for the survey responses.
b. What was the predominant response to the survey question?
c. What percentage of the respondents look most for love, romance, or sex in their relationships with their partners?

2.50 In 1986, McDonald's had the most retail outlets of any fast-food franchise in the United States. The accompanying bar graph describes the number of outlets for each of the ten largest U.S. fast-food franchised restaurant chains in 1986.

Ten Largest Franchised Restaurant Chains, 1986 (in number of units)

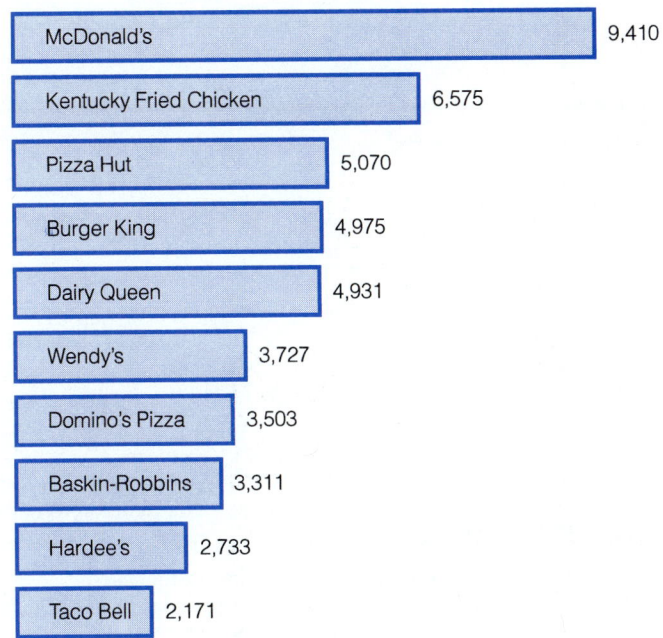

Source: 1986 Fast Food Guide.

a. Is the graph a frequency or relative frequency bar graph?
b. McDonald's, Burger King, Wendy's, and Hardee's all specialize in fast-food hamburgers. What percentage of the retail outlets of the 10 largest franchises specialize in fast-food hamburgers?

2.51   In 1985, a chemical leak from a Union Carbide plant in West Virginia led the Environmental Protection Agency (EPA) to develop tougher regulatory measures and improve emergency plans for toxic chemical substances manufactured in the United States. The accompanying table gives the 10 states with the most toxic chemical accidents from January 1983 to March 1985.

STATE	NUMBER OF ACCIDENTS	STATE	NUMBER OF ACCIDENTS
California	244	Ohio	321
Illinois	403	Pennsylvania	172
Indiana	117	Texas	451
Louisiana	289	Virginia	113
Michigan	158	West Virginia	144

Source: EPA, *USA Today*, Aug. 14, 1985. Copyright 1985, *USA Today*. Adapted with permission.

a. Construct a bar chart to describe the data.

b. Of the 10 states with the most toxic accidents, what percentage of accidents occurred in the steel industry states of Ohio, Pennsylvania, Virginia, and West Virginia?

2.52   The U.S. Chess Federation (USFC) estimates that 25 million Americans are chess enthusiasts—from beginners and casual competitors, to serious players and grand masters. Surprisingly, almost half of the nearly 14,000 junior members are age 12 and younger. Officially, there are now 59,943 dues-paying members of the USFC. The accompanying table lists the number of USFC members in each membership category. Summarize the data with a graphical method. Interpret the results.

CATEGORY	NUMBER OF USFC MEMBERS
Blind	84
Computer	2
Family	538
Full Adult	30,175
Life	9,036
Scholastic	9,002
Senior	3,463
Sustaining	1,318
Trial	1,430
Youth	4,895
TOTAL	59,943

Source: U.S. Chess Federation, Jan. 1, 1992.

2.53   *Business Week*/Harris poll conducted a survey of 1,253 adults concerning the steadily worsening U.S. trade deficit with Japan. The results were summarized in the graphic on the next page. [*Note:*   The figure conveys the same information as a bar chart, except the bars are placed horizontally side by side.]

a. What percentage of the adults surveyed believe that Japanese companies are better managed?

b. What percentage of the adults surveyed believe that U.S. products sell poorly in Japan because they are too expensive?

2.54   Many hospitals have developed their own computer system software for processing outpatient billings and other accounting services. The *1982 National Survey on Hospital Data Processing* gave the accompanying information (see next page) on the vendors who supply the computer hardware for 341 hospitals that developed their own financial system software.

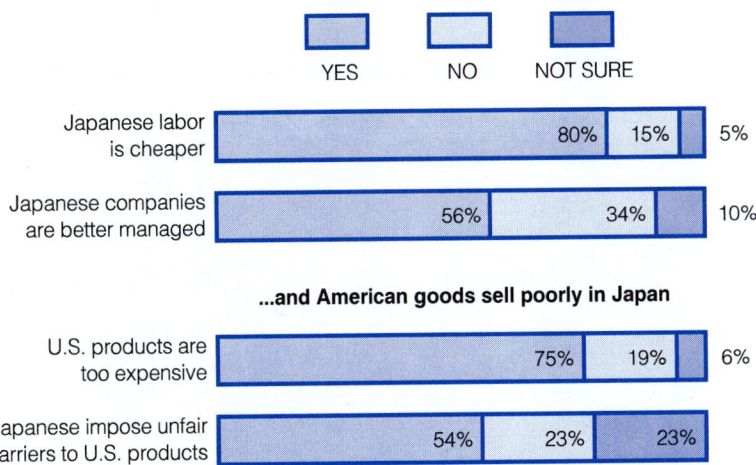

Why Americans think Japanese products sell well...

| | YES | NO | NOT SURE |

| Japanese labor is cheaper | 80% | 15% | 5% |
| Japanese companies are better managed | 56% | 34% | 10% |

...and American goods sell poorly in Japan

| U.S. products are too expensive | 75% | 19% | 6% |
| Japanese impose unfair barriers to U.S. products | 54% | 23% | 23% |

SURVEY OF 1,253 ADULTS CONDUCTED MAR. 14–17. OVERALL RESULTS SHOULD BE ACCURATE TO WITHIN THREE PERCENTAGE POINTS EITHER WAY.

Source: Jackson, S. *Business Week*/Harris Poll: Resentment of Japan Is Deepening. *Business Week*, Apr. 8, 1985, p. 53.

Data for Exercise 2.54

HARDWARE VENDOR	NUMBER OF HOSPITALS
Burroughs	45
Data General	6
Digital Equipment	8
Honeywell	8
IBM	188
NCR	50
Others	36

Source: *Modern Health Care*, May 1983, p. 188.

a. Classify the variable of interest, hardware vendor, as quantitative or qualitative.
b. What type of graphical method is appropriate for summarizing the data in the table?
c. Construct an appropriate graph to summarize the data.
d. What percentage of hospitals surveyed purchased their computer hardware equipment from Burroughs?
e. What percentage of hospitals surveyed did not purchase their computer hardware equipment from IBM?

2.55 A study of merit raises at 16 U.S. corporations was conducted to discover the extent to which merit pay policies for employees are actually tied to performance (*Personnel Journal*, Mar. 1986). One phase of the study focused on the distribution of merit raises (measured as percentage increase in salary) awarded at the companies. The frequency distribution of the 3,900 merit increases awarded at one of the largest firms in the study is reproduced in the figure on the next page.
a. Approximately how many of the firm's employees were awarded merit increases between 6% and 10%?
b. Is the distribution skewed? If so, is it skewed to the right or skewed to the left?
c. Approximately what proportion of the 3,990 merit increases were less than 5%?

Graph for Exercise 2.55

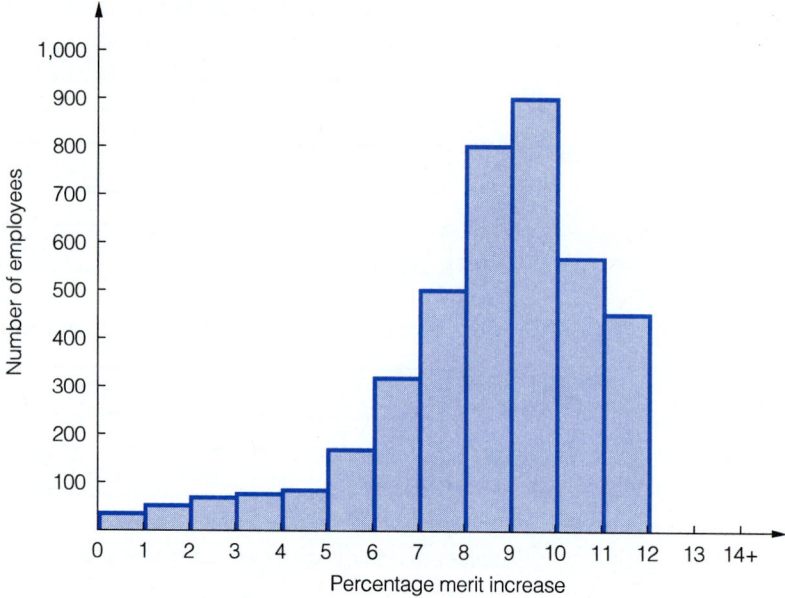

Number of employees (y-axis); Percentage merit increase (x-axis)

Source: Teel, K. S. "Compensation: Are merit raises really based on merit?" *Personnel Journal*,
Vol. 65, No. 3, Mar. 1986, p. 90. Reprinted with permission from *Personnel Journal*, Costa Mesa, Calif.
Copyright 1986. All rights reserved.

2.56   Did you know that American citizens own more guns (about 140 million) than cars? Yet, a Gallup poll (Sept. 1987) revealed that a majority of Americans favor stricter gun control laws. In one portion of the survey 500 nonowners of guns responded to the question, "In general, do you feel that the laws covering the sale of handguns should be made more strict, less strict, or kept as they are now?" The responses are summarized in the pie chart shown here. Interpret the figure.

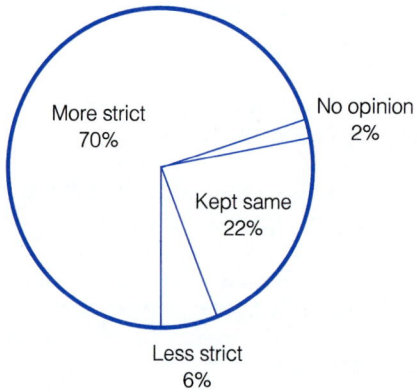

Source: *Gainesville Sun*, Nov. 15, 1987; Info Graphics,
News America Syndicate, 1987.

WHILE MOM WORKS, WHO'S MINDING THE KIDS?

According to the 1985 Current Population Survey (CPS) of the United States, "More than 29 million children under age 15 had mothers who worked; almost 19 million of these children had mothers who worked full time." With more and more mothers entering the work force, the demand for child care is growing at an unprecedented rate. Who cares for these children?

To answer this question, the Bureau of the Census conducted a special survey of working mothers with children as part of the 1985 CPS. Some of the more interesting results:

The majority of preschool-age children were cared for in their own or other homes while their mothers worked.

Full-time workers placed greater reliance on child care in the home of someone unrelated to the child and on organized child-care (day-care) facilities.

For school-age children, school is the primary source of child care.

Many children are left to care for themselves while the parents work.

A summary of these and other results are provided in the pie charts shown in Figures 2.16 (below) and 2.17 (page 84). Figure 2.16 describes the primary child-care arrangements of the approximately 8 million preschool children who have working mothers. Figures 2.17a and 2.17b describe the primary and secondary child-care arrangements, respectively, of the approximately 18 million school-age children with working mothers. ("Primary" care is the arrangement used for most of the hours the mother works; "secondary" care is the arrangement used when additional care is necessary during the mother's working hours.)

FIGURE 2.16

Primary Child-Care
Arrangements of Preschool
Children

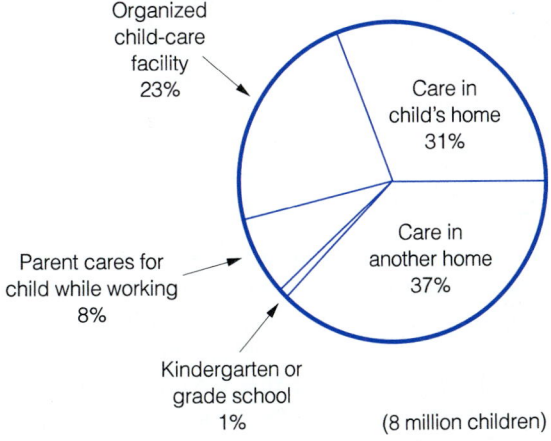

a. What percentage of the estimated 8 million preschool children with working mothers are cared for in their own or another's home?
b. How many of the estimated 8 million preschool children with working mothers are cared for at an organized child-care facility?

c. For what percentage of the estimated 18 million school-age children with working mothers is school the primary source of child care?
d. How many of the estimated 5 million school-age children who require secondary care are responsible for taking care of themselves?

FIGURE 2.17

Child-Care Arrangements of School-Age Children

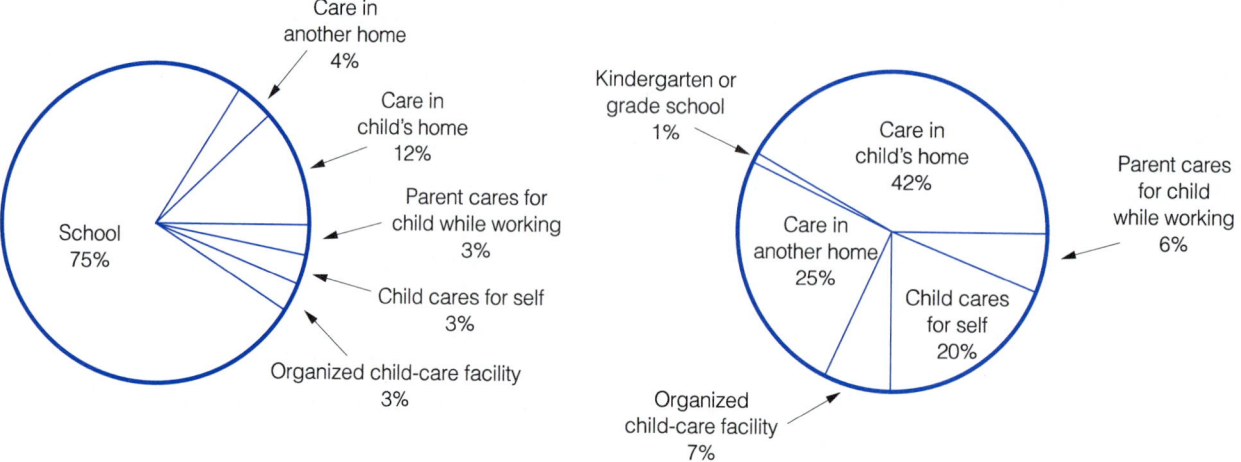

a. Primary care (18 million children)

b. Secondary care (5 million children)

Source: "Who's minding the kids?" *Data User News*, Vol. 22, No. 6, June 1987, U.S. Dept. of Commerce, Bureau of the Census.

CASE STUDY 2.2

HOW TO BEAT THE WAITING-IN-LINE BLUES

Three Boston researchers, writing in *Sloan Management Review* (Winter 1991), considered the following problem:

Historically, service businesses interested in customer satisfaction have focused on hiring and training knowledgeable, pleasant servers. Today, this approach is insufficient. Customers not only demand quality, they also demand speed. They do not tolerate waiting in line for long periods of time. Firms must respond to this change if they wish to remain competitive.

The researchers note that there are two basic ways to shorten waiting time: through operations management and through perceptions management. "The logic behind perceptions management," they go on to state, "is that when it comes to customer satisfaction, perception is reality. If customers think they are satisfied, then they are satisfied. Similarly, if customers think that their wait was short enough, then it was short enough, regardless of how long it actually was." Disneyland (California) and Disney World (Florida) are prime examples of perception management. At Disney theme parks, lines are always kept moving. In addition, "the waiting time posted by each attraction is generously overestimated, so that one comes away mysteriously grateful for having hung around 20 minutes for a 58-second twirl in the Alice in Wonderland teacups."

TABLE 2.5

Summary of Customer Interest Level

CATEGORY	PERCENTAGE
Watchers	21
Impatients	45
Neutrals	34

Source: Katz, K. L., Larson, B. M., and Larson, R. C. "Prescription for the waiting-in-line blues: Entertain, enlighten, and engage." *Sloan Management Review*, Winter 1991, p. 49 (Fig. 5).

To apply these ideas in a traditional business setting, the researchers measured the perceptions of 277 customers waiting in line at the Bank of Boston, located in downtown Boston, Massachusetts. Two video cameras filmed customers as they entered the line to see the bank teller and as they left the line. Thus, the cameras recorded the actual time the customers waited in line. As customers finished their transactions, they were interviewed by the researchers and asked about perceived waiting times. By identifying each customer on the videotape, the researchers were able to compare individual customers' perceptions of waiting time with how long they actually waited.

In addition to perceived and actual waiting time, the researchers asked customers to describe their wait in line. Customers generally fell into three categories: "watchers" who enjoy observing people and events at the bank, "impatients" who could think of nothing more boring than waiting in line, and "neutrals" who fell somewhere in the middle.

a. Table 2.5 gives the breakdown, in percentages, of the 277 customers falling into the three categories of watchers, impatients, and neutrals. Describe the data summarized in Table 2.5 graphically. Are the data quantitative or qualitative? Interpret the graph.

b. Consider the data set consisting of the differences between the perceived and actual waiting times of the 277 customers. Are the data qualitative or quantitative?

c. A graph describing the data of part b is shown in Figure 2.18. What type of graph is displayed? Interpret the graph. In particular, obtain the approximate percentage of customers who overestimated their waiting time in line at the Bank of Boston.

FIGURE 2.18

Graph Describing the Difference Between Perceived and Actual Waiting Time

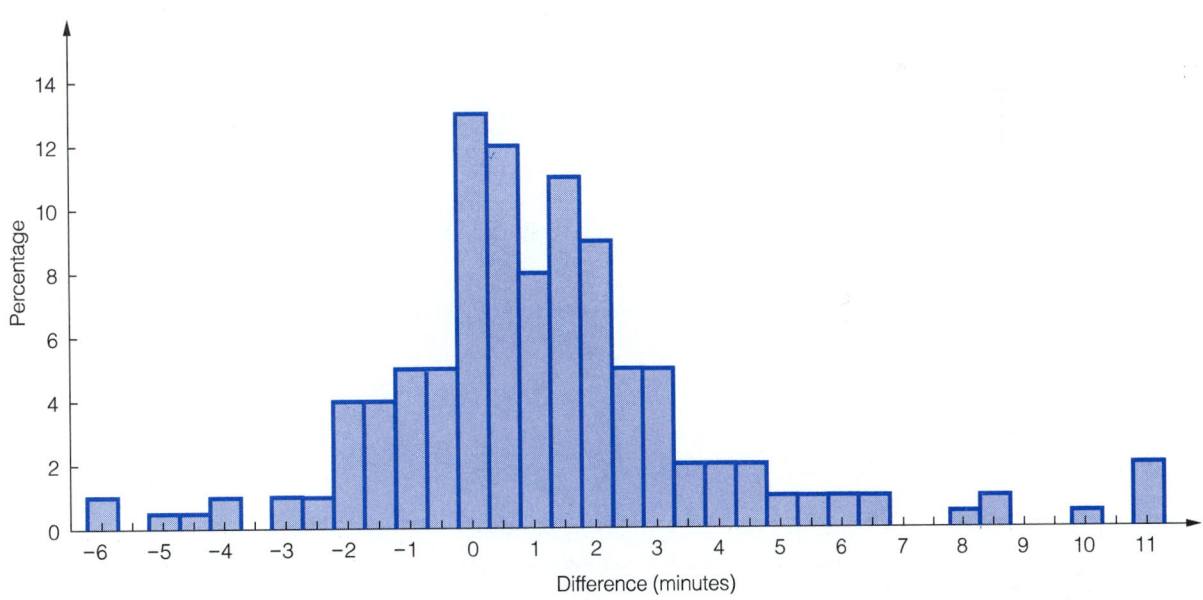

Source: Katz, K. L., Larson, B. M., and Larson, R. C. "Prescription for the waiting-in-line blues: Entertain, enlighten, and engage." *Sloan Management Review*, Winter 1991, p. 48 (Fig. 3).

WARNING: CIGARETTE SMOKE IS HAZARDOUS TO YOUR HEALTH

An ongoing campaign of heavy advertising and warnings in the media by the U.S. surgeon general has made both cigarette smokers and nonsmokers well aware of the dangers of inhaling the excess tar and nicotine from a burning cigarette. In 1980, the surgeon general added a third element to its list of hazardous substances that affect cigarette smokers—carbon monoxide. According to the surgeon general, "Breathing carbon monoxide, a product of incomplete combustion, reduces the ability of blood to carry oxygen. For smokers, this occurs at the same time that inhaled nicotine is increasing the heart's oxygen needs. Research [conducted by the surgeon general] has indicated carbon monoxide can be particularly hazardous to pregnant women smokers and heart disease patients."

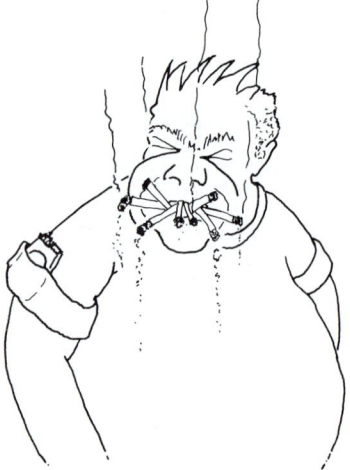

As a result of these findings, the Federal Trade Commission (FTC) annually ranks American cigarette brands in terms of the amount of carbon monoxide, tar, and nicotine in their smoke. The test results are obtained as follows: A 20-part sequential smoking machine is used to "smoke" cigarettes to a 23-millimeter butt length. Based on tests of 100 cigarettes per brand, the carbon monoxide, tar, and nicotine concentrations (rounded to the nearest milligram) in the residual "dry" particulate matter are determined.

Appendix F contains the results of the FTC's 1991 tests of over 400 cigarette brands.* In addition to carbon monoxide, tar, and nicotine content (all recorded in milligrams), the following characteristics are listed for each brand:

Light type Regular (R), Light (L), or Extra Light (E)

Filter type Filter (F) or Nonfilter (NF)

Menthol type Menthol (M) or Nonmenthol (NM)

Pack hardness Hard pack (HP) or Soft pack (SP)

Length (in millimeters) 70, 85, 100, or 120

a. Classify each of the variables in Appendix F as either quantitative or qualitative.
b. Describe each of the variables in Appendix F with a graphical technique. Interpret each graph.

[*Note:* The data in Appendix F is available on micro or floppy diskette from the publisher.]

Source: "Tar, nicotine, and carbon monoxide of the smoke of 475 varieties of domestic cigarettes." Federal Trade Commission report, 1991.

BAR CHARTS, HISTOGRAMS, AND STEM-AND-LEAF DISPLAYS

In this section we give the SAS, SPSS, and Minitab commands for producing bar charts for qualitative data, and histograms or stem-and-leaf displays for quantitative data. For purposes of illustration, we will use the computer to describe the starting salary data given in Appendix B. Recall that the data set includes the qualitative variable, college, and the quantitative variable, starting salary. In particular, the programs below will produce (1) a bar chart for college, (2) a relative frequency histogram for starting salary, and (3) a stem-and-leaf display for starting salary.

[*Note:* As in all Computer Lab sections, the SAS, SPSS, and Minitab commands given here are (unless otherwise noted) appropriate for both mainframe and personal computers. However, programs run on mainframe computers require JCL instructions. See your instructor for the appropriate JCL commands to use at your computing center.]

SAS

Command
Line

1	`DATA CRC;`	Data entry instructions
2	`INPUT COLLEGE $ SALARY;`	
3	`CARDS;`	

[Input data values (1 observation per line)]

4	`PROC CHART;`	Statistical analysis instructions:
5	`VBAR COLLEGE;`	Bar chart
6	`PROC CHART;`	
7	`VBAR SALARY/TYPE=PERCENT;`	Relative frequency histogram
8	`PROC UNIVARIATE PLOT;`	Stem-and-leaf plot
9	`VAR SALARY;`	

COMMANDS 4–5 The CHART procedure produces vertical and horizontal bar charts for qualitative data. The key word VBAR followed by the variable name COLLEGE produces a vertical bar chart for COLLEGE. (For horizontal bar charts, use the keyword HBAR.)

COMMANDS 6–7 The CHART procedure is also used to generate relative frequency and frequency histograms for quantitative data. The key word VBAR followed by the variable name SALARY produces a frequency histogram for starting salaries. Relative frequency histograms are produced by adding the option TYPE=PERCENT. SAS will automatically select suitable class intervals for the histogram.*

---

*The MIDPOINTS option allows the user to select the class intervals. Consult the SAS references given at the end of Chapter 1 for details on how to use the MIDPOINTS option in PROC CHART.

COMMANDS 8–9 The UNIVARIATE procedure with the key word PLOT will produce, among other statistics, a stem-and-leaf display for the variable (e.g., SALARY) specified in the VAR statement. [For large data sets (approximately 50 or more observations), a horizontal bar chart is produced rather than a stem-and-leaf display.]

```
Command
Line
  1    DATA LIST FREE/COLLEGE (A3) SALARY.  ⎫  Data entry instructions
  2    BEGIN DATA.                          ⎭

       [Input data values (1 observation per line)]

  3    END DATA.
  4    FREQUENCIES VARIABLES=COLLEGE/        ⎫  Statistical analysis instructions
  5                BARCHART.                 ⎪     Bar chart
  6    FREQUENCIES VARIABLES=SALARY/         ⎬
  7                HISTOGRAM=PERCENT.        ⎭  Relative frequency histogram
```

COMMANDS 4–5 The FREQUENCIES command produces horizontal bar charts for qualitative data, and horizontal relative frequency or frequency histograms for quantitative data. The subcommand BARCHART will produce a bar chart for the qualitative variable (e.g., COLLEGE) specified in the VARIABLES subcommand.

COMMANDS 6–7 The subcommand HISTOGRAM will produce a frequency histogram for the quantitative variable (e.g., SALARY) specified in the VARIABLES subcommand. Relative frequency histograms are generated by adding the =PERCENT option to the HISTOGRAM subcommand. SPSS will automatically select suitable class intervals for the histogram.*

GENERAL SPSS does not have a statistical analysis procedure specifically designed to produce stem-and-leaf displays. For certain types of data, stem-and-leaf displays may be generated through the STEMLEAF option of the MANOVA command. Consult the SPSS references for details on how to use this procedure.

REMINDER When using SPSS on a mainframe computer, place an asterisk between the numerical and alphanumeric variables in the DATA LIST command and omit the command terminators (periods).

---

*The INCREMENT option allows the user to select the width of the class intervals. Consult the SPSS references given at the end of Chapter 1 for details on how to use the INCREMENT option in FREQUENCIES.

```
Command
  Line
    1     READ C1 C2          } Data entry instruction

          [Input data values (1 observation per line)]

    2     NAME C1='COLLEGE' C2='SALARY'
    3     HISTOGRAM C1          ⎫  Statistical analysis instructions:
    4     HISTOGRAM C2          ⎬    Bar chart histogram
    5     STEM-AND-LEAF C2      ⎭    Stem-and-leaf display
```

COMMAND 3   The HISTOGRAM command will generate horizontal bar charts for qualitative data and horizontal frequency histograms for quantitative data. Since Minitab requires all data to be entered as numbers, the bar charts for the values of COLLEGE located in column 1 (C1) will look similar to a histogram with class interval midpoints identified by the values 1, 2, 3, 4, and 5 (for Business Administration, Engineering, . . . , Nursing, respectively).

COMMAND 4   Minitab will automatically select suitable class intervals for the frequency histogram of the starting salaries located in column 2 (C2).*

COMMAND 5   The STEM-AND-LEAF command produces a stem-and-leaf display for the starting salary values located in column 2 (C2).

COMPUTER ACTIVITIES

Access the data in Appendix D and generate the following graphical displays using a statistical software package. Interpret the results of each graph.

a. A bar chart for each of the two qualitative variables—location and species—in the data set.

b. A stem-and-leaf display for each of the three quantitative variables—length, weight, and DDT—in the data set.

c. A relative frequency histogram for each of the three quantitative variables—length, weight, and DDT—in the data set.

---

*The optional commands INCREMENT and START allow the user to select the class intervals. Consult the Minitab references given at the end of Chapter 1 for details on how to use these HISTOGRAM options.

REFERENCES   Chambers, J. M., Cleveland, W. S., Kleiner, B., and Tukey, P. A. *Graphical Methods for Data Analysis.* Belmont, Calif.: Wadsworth International Group, 1983.

McClave, J. T. and Dietrich, F. *A First Course in Statistics*, 4th ed. New York: Macmillan, 1991.

Koopman, L. *An Introduction to Contemporary Statistics.* Belmont, Calif.: Duxbury Press, 1981.

Rotton, J. and Kelly, I. W. "A scale for assessing belief in lunar effects: Reliability and concurrent validity." *Psychological Reports*, Vol. 57, 1985, pp. 239–245.

Tanur, J. M., Mosteller, F., Kruskal, W. H., Link, R. F., Pieters, R. S., and Rising, G. R. (eds.). *Statistics: A Guide to the Unknown*, 2nd ed. San Francisco: Holden-Day, 1978.

Tukey, J. *Exploratory Data Analysis.* Reading, Mass.: Addison-Wesley, 1977.

NUMERICAL METHODS FOR DESCRIBING QUANTITATIVE DATA

CHAPTER 3

Engineers have a term for unaided human acts of lifting, lowering, pushing, pulling, carrying, or holding and releasing a heavy object—manual materials handling activities (MMHA). Researchers are currently working to develop strength and capacity guidelines for MMHA. The obvious problem is that not every person can safely lift the same heavy object. Suppose you were to determine a guideline for the weight that can be lifted safely for a sample of 75 male college students. How could you describe the capacities of these students with a single number that would characterize the capacity norms of the sample of 75 male college students? In this chapter we will show you how numbers called **numerical descriptive measures** can be used to describe the charac-teristics of a set of measurements, and we will apply it to the chapter case in Section 3.7.

CONTENTS

WHY WE NEED NUMERICAL DESCRIPTIVE MEASURES

3.1 It is probably true that a picture is worth a thousand words, and it is certainly true when the goal is to describe a quantitative data set. But sometimes you will want to discuss the major features of a data set and it may not be convenient to produce a stem-and-leaf display or histogram for the data. When this situation occurs, we seek a few summarizing numbers, called **numerical descriptive measures**, that create in our minds a picture of the relative frequency distribution.

TYPES OF NUMERICAL DESCRIPTIVE MEASURES

3.2 Examine Figure 3.1, a reproduction of the computer-generated (SAS) histogram for the starting salaries of Appendix A. If you were allowed

FIGURE 3.1
SAS Histogram for the 1,795 Starting Salaries of Appendix A

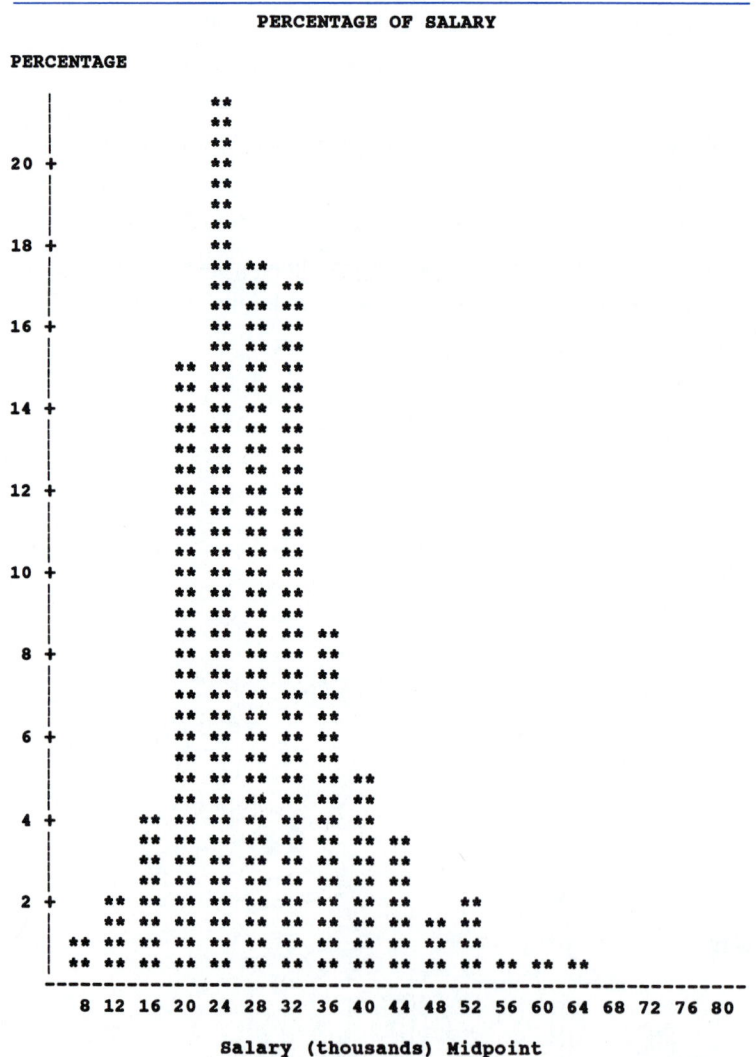

to choose two numbers that would help you construct a mental image of the distribution, which two would you choose? We think you would probably choose

1. A number that is located near the *center* of the distribution (see Figure 3.2a)
2. A number that measures the *spread* of the distribution (see Figure 3.2b)

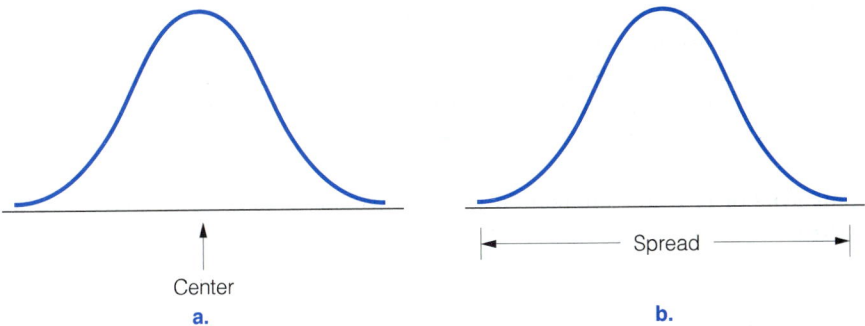

FIGURE 3.2
Numerical Descriptive Measures

For symmetric distributions of data shaped like Figure 3.2, a number that would describe the center of the distribution would be visually located near the spot where most of the data seemed to be concentrated. Consequently, numbers that fulfill this role are called **measures of central tendency**. We will define and describe several measures of central tendency for data sets in Section 3.4.

The amount of spread in a data set is a measure of the variation in the data. Consequently, numerical descriptive measures that perform this function are called **measures of variation** or **measures of dispersion**. As you will subsequently see (Section 3.5), there are several ways to measure the variation in a data set.

Measures of central tendency and data variation are not the only types of numerical measures for describing data sets. Some are constructed to measure the **skewness** of a distribution. Recall from Section 2.5 that skewness describes the tendency of the distribution to tail out to the right (or left). For example, the histogram in Figure 3.1 is *skewed to the right* (or *positively skewed*).

We will concentrate in this chapter on measures of central tendency and measures of variation. Although numerical descriptive measures of skewness are beyond the scope of this text, knowing whether a distribution of data is skewed or symmetric is important when describing the data with measures of central tendency and variation (Section 3.6). As you read this material, keep in mind our goal of using a pair of numbers to create a mental image of a distribution. Relate each numerical descriptive measure to this objective, and verify that it fulfills the role it is intended to play.

3.3 Suppose a data set was obtained by observing a quantitative variable, x. For example, x may represent the starting salary of a college graduate. By observing x (starting salary) for the 1,795 graduates of Appendix A, we obtained the data set consisting of 1,795 starting salaries. If we want to represent a particular observation in a data set—say, the 35th—we represent it by the

SUMMATION NOTATION

often generate different numbers for the same data set but all will satisfy our general objective. If we visually imagine a hump-shaped distribution, all measures of central tendency will fall near the middle of the hump.

The most common measure of the central tendency of a data set is familiar to you and is called the **arithmetic mean** of the data. The arithmetic mean, or **average**, is defined as indicated in the box.

DEFINITION 3.1

The **arithmetic mean** of a sample of n observations, $x_1, x_2, \ldots, x_n$, is denoted by the symbol $\bar{x}$ (read "x-bar"), and is computed as

$$\bar{x} = \frac{\text{Sum of the } x \text{ values}}{\text{Number of observations}} = \frac{\Sigma x}{n}$$

[*Note:* From now on, we will refer to an arithmetic mean simply as a **mean.**]

EXAMPLE 3.3 Find the mean for the data set consisting of the observations 5, 1, 6, 2, 4.

Solution   The data set contains $n = 5$ observations. Therefore,

$$\bar{x} = \frac{\Sigma x}{n} = \frac{5 + 1 + 6 + 2 + 4}{5} = \frac{18}{5} = 3.6$$

EXAMPLE 3.4 Find the mean for the 1,795 starting salaries of Appendix A. Locate it on the histogram shown in Figure 3.1. Does the mean fall near the center of the distribution?

Solution   With such a large data set, it is impractical to calculate numerical descriptive measures by hand or calculator. We will rely on one of the numerous statistical software packages available for calculating the mean. Figure 3.3 is a SAS printout giving descriptive statistics for the starting salary data. The mean of the 1,795 salaries, shaded on the printout, is

$$\bar{x} = \$28{,}475$$

This mean, or average, starting salary should be located near the center of the histogram for the 1,795 starting salaries. If you examine Figure 3.1, you will see that the mean $\bar{x}$ does indeed fall near the center of the mound-shaped portion of the distribution. If we did not have Figure 3.1 available, we could reconstruct the distribution in our minds as a mound-shaped figure centered in the vicinity of $\bar{x} = \$28{,}475$.

```
              UNIVARIATE PROCEDURE

     Variable=SALARY

                    Moments

N                   1795  Sum Wgts        1795
Mean            28474.83  Sum         51112325
Std Dev         9369.484  Variance    87787229
Skewness        1.093351  Kurtosis    2.301454
USS             1.613E12  CSS         1.575E11
CV              32.90444  Std Mean    221.1482
T:Mean=0         128.759  Prob>|T|         0.0
Sgn Rank          805955  Prob>|S|         0.0
Num ^= 0            1795
W:Normal        0.934422  Prob<W           0.0


                Quantiles(Def=5)

    100% Max      80000       99%       60000
     75% Q3       33000       95%       46000
     50% Med      27000       90%       40000
     25% Q1       22000       10%       19000
      0% Min       7100        5%       16000
                                1%       10000

    Range         72900
    Q3-Q1         11000
    Mode          20000


                   Extremes

    Lowest    Obs      Highest    Obs
      7100(   1003)     66000(    532)
      7200(   1603)     70000(    557)
      8000(   1776)     70000(    711)
      8000(    920)     70000(    884)
      8000(    849)     80000(   1433)
```

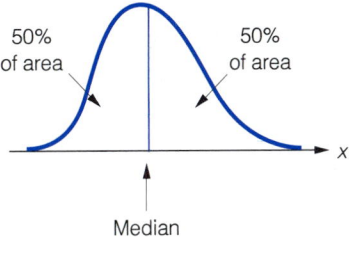

FIGURE 3.3
SAS Descriptive Measures for the 1,795 Starting Salaries of Appendix A

A second measure of central tendency for a data set is the **median**. For large data sets, the median M is a number chosen so that half the observations are less than the median and half are larger. Since the areas of the bars used to construct the relative frequency histogram are proportional to the numbers of observations falling within the classes, it follows that the median is a value of x that divides the area of the histogram into two equal portions. Half the area will lie to the left of the median (see Figure 3.4) and half will lie to the right. For example, the median for the starting salaries of Appendix A (shaded in Figure 3.3) is $27,000. You can see from Figure 3.1 that this starting salary divides the data into two sets of equal size. Half the 1,795 starting salaries are less than $27,000; half are larger. The next two examples illustrate how to find the median for small data sets.

50% of area   50% of area

Median

FIGURE 3.4
The Median Divides the Area of a Relative Frequency Distribution into Two Equal Portions

EXAMPLE 3.5   Find the median for the data set consisting of the observations 7, 4, 3, 5, 3.

Solution   We first arrange the data in increasing (or decreasing) order:

                3   3   4   5   7

than half the starting salaries and less than the remainder, then you will prefer the median. The mode is rarely the choice measure of central tendency because the measurement that occurs most often does not necessarily lie in the "center" of the distribution. There are situations, however, where the mode is preferred. For example, if the relative frequency of occurrence of values of starting salaries can be viewed as a measure of employer perception of a college graduate's first-year value (e.g., the greatest frequency of starting salaries occurred in the $22,000–$26,000 class), then the mode might be the preferred measure of central tendency. More realistically, a retailer of women's shoes would be interested in the modal shoe size of potential customers.

In making your decision, you should know *the mean is sensitive to very large or very small measurements*. Consequently, the mean will shift toward the direction of skewness and may be a misleading measure of central tendency in some situations. You can see from Figure 3.6 that the mean falls to the right of the median and that the starting salaries are skewed to the right. The high starting salaries of relatively few graduates will influence the mean more than the median. For this reason, the median is sometimes called a **resistant** measure of central tendency since it, unlike the mean, is resistant to the influence of extreme measurements. For data sets that are extremely skewed, the median would better represent the "center" of the distribution.

WARNING

For data sets that are extremely skewed, be wary of using the mean as a measure of the "center" of the distribution. In this situation, a more meaningful measure of central tendency may be the median, which is more resistant to the influence of extreme measurements.

Most of the inferential statistical methods discussed in this text are based, theoretically, on mound-shaped distributions of data with little or no skewness. For these situations, the mean and the median will be, for all practical purposes, the same. Since the mean has nicer mathematical properties than the median, it is the preferred measure of central tendency for these inferential techniques.

EXERCISES

LEARNING THE MECHANICS

3.6   Calculate the mean for samples with the following characteristics:

   a. $n = 10$ and $\Sigma x = 500$ b. $n = 20$ and $\Sigma x = 400$ c. $n = 500$ and $\Sigma x = 100$

3.7   Find the mean and median for the data set consisting of these five measurements: 3, 9, 0, 7, 4

3.8   Find the mean and median for the following sample of $n = 6$ measurements: 7, 3, 4, 1, 5, 6

3.9 Calculate the mean, median, and mode for each of the following samples:

a. 3, 4, 4, 5, 5, 5, 6, 6, 7
b. 3, 4, 4, 5, 5, 5, 6, 6, 70
c. −50, −49, 0, 0, 49, 50
d. −50, −49, 0, 9, 9, 81

3.10 The $n = 50$ starting salaries of Table 2.3 are reproduced here.

$28,000	$24,600	$18,600	$34,700	$30,400	$20,800	$12,200	$19,400	$13,700	$14,800
23,200	18,700	24,100	41,900	28,300	23,200	18,400	24,700	36,900	38,500
20,800	9,000	18,400	34,600	15,400	26,200	28,900	24,300	27,400	18,100
13,300	19,500	22,400	18,500	30,700	23,600	21,200	22,900	21,100	18,000
32,600	18,400	29,400	35,500	20,100	26,700	24,700	11,200	31,100	26,300

a. Compute the mean and median, and locate these values on the relative frequency histogram for the data set (see Figure 2.13). Notice that they fall near the center of the distribution.
b. Find the modal class (the class with the greatest relative frequency) for the relative frequency histogram shown in Figure 2.13. Compare your answer with the mean and median obtained in part **a**.
c. Suppose that a distribution of data is skewed to the right. Would you expect the mean of this data set to be larger or smaller than the median? Does your answer agree with the results of part **a**?

APPLYING THE CONCEPTS

3.11 Refer to the Center for Disease Control study of sanitation levels for 72 international cruise ships, Exercise 2.19. A Minitab printout of the descriptive statistics for the data is shown here. (Recall that sanitation scores range from 0 to 100.) Interpret the numerical descriptive measures of central tendency displayed in the printout.

	N	MEAN	MEDIAN	TRMEAN	STDEV	SEMEAN
sanlevel	91	91.044	92.000	91.580	5.566	0.583

	MIN	MAX	Q1	Q3
sanlevel	66.000	99.000	89.000	95.000

3.12 Refer to the *Journal of Marketing Research* article on the effect of brand name and store name on product quality, Exercise 2.23. The stem-and-leaf displays showing the distribution of effect size (an index ranging from 0 to 1) for the 15 brand-name studies are reproduced here.

Brand Name (15 studies)   Store Name (17 studies)

STEM	LEAF
.6	0
.5	7
.4	
.3	4
.2	5 5
.1	0 1 1 2 4
.0	3 3 5 5 7

STEM	LEAF
.6	
.5	
.4	3 4
.3	
.2	
.1	2
.0	0 0 0 1 1 2 2 3 3 4 6 7 8 8

Source: Rao, A. R. and Monroe, K. B. "The effect of price, brand name, and store name on buyers' perceptions of product quality: an integrative review." *Journal of Marketing Research*, Vol. 26, Aug. 1989, p. 354 (Table 2).

a. Calculate the mean, median, and mode of effect size for the 15 brand-name studies. Which measure of central tendency best describes the data set?

b. Calculate the mean, median, and mode of effect size for the 17 store-name studies. Which measure of central tendency best describes the data set?

c. Combine the data for the two studies, and compute the mean, median, and mode of effect size. Which measure of central tendency best describes the combined data set?

3.13 According to one study, "The majority of people who die from fire and smoke in compartmented fire-resistive buildings—the type used for hotels, motels, apartments, and other health care facilities—die in the attempt to evacuate" (*Risk Management*, Feb. 1986). The accompanying data represent the numbers of victims who attempted to evacuate for a sample of 14 recent fires at compartmented fire-resistive buildings reported in the study.

FIRE	NUMBER OF VICTIMS
Las Vegas Hilton (Las Vegas)	5
Inn on the Park (Toronto)	5
Westchase Hilton (Houston)	8
Holiday Inn (Cambridge, Ohio)	10
Conrad Hilton (Chicago)	4
Providence College (Providence)	8
Baptist Towers (Atlanta)	7
Howard Johnson (New Orleans)	5
Cornell University (Ithaca, New York)	9
Westport Central Apartments (Kansas City, Missouri)	4
Orrington Hotel (Evanston, Illinois)	0
Hartford Hospital (Hartford, Connecticut)	16
Milford Plaza (New York)	0
MGM Grand (Las Vegas)	36

Source: Macdonald, J. N. "Is evacuation a fatal flaw in fire fighting philosophy?" *Risk Management*, Vol. 33, No. 2, Feb. 1986, p. 37.

a. Construct a stem-and-leaf display for the data.

b. Compute the mean, median, and mode for the data set. Which measure of central tendency appears to best describe the center of the distribution of data?

3.14 Appendix B contains the starting salaries for University of Florida bachelor's degree graduates in five specific colleges. The mean and median starting salaries (to the nearest dollar) for the five data sets are shown in the accompanying table. Use these measures of central tendency to construct a mental picture of the relative locations of the relative frequency histograms for the five data sets.

COLLEGE	MEAN STARTING SALARY	MEDIAN STARTING SALARY
Business Administration	$24,814	$24,000
Engineering	30,877	32,000
Journalism	19,995	20,000
Liberal Arts/Sciences	21,321	21,000
Nursing	28,147	27,000

3.15 Recently, organizational behaviorists and social psychologists have begun to study the process by which decision makers escalate their commitment to an ineffective course of action. This phenomenon has been labeled many things, including the "sunk cost" effect, the "knee-deep-in-the-big-muddy" effect, and the "too-much-invested-to-quit" effect, but is most commonly known as "entrapment." Fifty-two introductory psychology students took part in a laboratory experiment designed to explore whether individuals' tendencies to view prior outcomes as revealing of their self-identity would heighten entrapment (*Administrative Science Quarterly*, Mar. 1986). The experiment consisted of 30 trials in which points were "awarded" based on the accuracy of students' judgments of geometric patterns of various shapes. The total points awarded on each trial are listed in the table.

5	5	4	7	24	6
10	12	11	15	11	10
3	23	4	20	5	4
7	5	6	6	15	5
15	10	13	9	4	6

Source: Brockner, J., et al. "Escalation of commitment to an ineffective course of action: The effect of feedback having negative implications for self-identity." *Administrative Science Quarterly*, Vol. 31, No. 1, Mar. 1986, p. 115. Reprinted by permission of *Administrative Science Quarterly*. Copyright 1986.

a. Construct a stem-and-leaf display for the data.
b. Compute the mean, median, and mode for the data set and locate them on the graph, part a. Do these measures of central tendency appear to locate the center of the distribution of data?

3.16   In many medical experiments, the "success" or "failure" of a treatment or drug is determined by a follow-up study. Biographical and medical histories are collected on the subjects, treatment is administered for a specified time period, and then a follow-up study is conducted to compare pre- and post-treatment statistics. In a recent follow-up study of alcoholics treated at Maudsley Hospital in London, the data in the accompanying table were collected at the time of admission for each of seven subjects suffering from alcohol addiction.

CASE NUMBER	AGE	YEARS OF EXCESSIVE DRINKING
1	41	10
2	42	8
3	47	13
4	26	3
5	28	2
6	40	5
7	35	6

Source: Edwards, G. "A later follow-up of a classic case series: D. L. Davies's 1962 report and its significance for the present." *Journal of Studies on Alcohol*, Vol. 46, No. 3, 1985, pp. 181–190. Copyright by Alcohol Research Documentation, Inc., Rutgers Center of Alcohol Studies, New Brunswick, N.J. 08903. Reprinted with permission.

a. Find the mean and median of the age of the seven patients.
b. Find the mean and median of the number of years of excessive drinking reported by the seven patients.
c. Which measure of central tendency, the mean or the median, better describes the age distribution of the seven subjects?
d. Which measure of central tendency, the mean or the median, better describes the distribution of number of years of excessive drinking for the seven subjects?

3.17   What are the top corporate executives being paid? To answer this question, *Business Week* magazine conducts a survey of corporate executives each year. *Business Week*'s 1990 survey of executives at 356 companies revealed only a 3.4% increase in salaries and bonuses from 1988 to 1989, the smallest increase since 1970. The top 20 corporate executives and their 1989 total cash compensations (salary plus bonus plus long-term compensation) are shown in the table. Assume that these represent a sample of the highest-paid corporate executives in the United States.

If a sample data set contains n observations, the **sample variance** is equal to the "average" of the squared deviations of all n observations. That is, the sample variance, denoted by the symbol s^2, is*

$$s^2 = \frac{\Sigma(x - \bar{x})^2}{n - 1}$$

The larger the value of s^2, the more spread out (i.e., the more variable) the sample data.

DEFINITION 3.5

The **variance**, s^2, of a set of n sample measurements is equal to the sum of squares of deviations of the measurements about their mean, divided by ($n - 1$):

$$s^2 = \frac{\Sigma(x - \bar{x})^2}{n - 1}$$

Note: A shortcut formula for calculating the numerator of s^2 is:

$$\Sigma(x - \bar{x})^2 = \Sigma x^2 - \frac{(\Sigma x)^2}{n} = \Sigma x^2 - n(\bar{x})^2$$

The greater the value of s^2, the more variation in the sample data.

EXAMPLE 3.8 Find the variance for the sample measurements: 3, 7, 2, 1, 8.

Solution The five observations are listed in the first column of Table 3.1. You can see that $\Sigma x = 21$ and, therefore,

$$\bar{x} = \frac{\Sigma x}{n} = \frac{21}{5} = 4.2$$

TABLE 3.1

Data and Computation Table

	OBSERVATION		
	x	$x - \bar{x}$	$(x - \bar{x})^2$
	3	−1.2	1.44
	7	2.8	7.84
	2	−2.2	4.84
	1	−3.2	10.24
	8	3.8	14.44
TOTALS	21	0	38.8

---

*We use ($n - 1$) rather than n in the denominator of s^2 in order to obtain a mathematically good estimator of the true population variance. When n is used in the denominator, the value of s^2 tends to underestimate the population variance. Dividing by ($n - 1$) adjusts for the underestimation problem.

This value of $\bar{x}$, 4.2, is subtracted from each observation to determine how much each observation deviates from the mean. These deviations are shown in the second column of Table 3.1. A *negative deviation* means that the observation fell *below* the mean; a *positive deviation* indicates that the observation fell *above* the mean. *Notice that the sum of the deviations equals 0. This will be true for all data sets.*

The squares of the deviations are shown in the third column of Table 3.1. The total at the bottom of the column gives the sum of squares of deviations,

$$\Sigma(x - \bar{x})^2 = 38.8$$

Then the **sample variance** is

$$s^2 = \frac{\Sigma(x - \bar{x})^2}{n - 1} = \frac{38.8}{4} = 9.7$$

The procedure illustrated in Example 3.8 for calculating a variance is tedious and often leads to rounding errors in finding the sum of squares of deviations, $\Sigma(x - \bar{x})^2$. A shortcut formula for calculating the sum of squares of deviations and the variance (shown in the box) is illustrated in the following example.

EXAMPLE 3.9 Use the shortcut procedure to calculate the sum of squares of deviations, $\Sigma(x - \bar{x})^2$, and the sample variance, s^2, for the data set of Example 3.8.

Solution The shortcut procedure provides an easy way to compute $\Sigma(x - \bar{x})^2$, the numerator of the sample variance calculation. Instead of calculating the deviation of each measurement from the mean, we calculate the squares of the observations, as shown in Table 3.2.

Then it can be shown (proof omitted) that

$$\Sigma(x - \bar{x})^2 = \Sigma x^2 - \frac{(\Sigma x)^2}{n}$$

Substituting the sum of squares Σx^2 and the sum Σx of the observations into this formula, we obtain

$$\Sigma(x - \bar{x})^2 = \Sigma x^2 - \frac{(\Sigma x)^2}{n} = 127 - \frac{(21)^2}{5} = 127 - 88.2 = 38.8$$

This is exactly the same total that you obtained for the sum of squares of deviations in Table 3.1. Finally, we have

$$s^2 = \frac{\Sigma(x - \bar{x})^2}{n - 1}$$

$$= \frac{38.8}{4} = 9.7$$

as we obtained in Example 3.8.

TABLE 3.2
Table for Calculating a Standard Deviation: The Shortcut Procedure

	OBSERVATION	
	x	x^2
	3	9
	7	49
	2	4
	1	1
	8	64
TOTALS	21	127

How can we interpret the value of the sample variance calculated in Example 3.8? We know that data sets with large variances are more variable (i.e., more spread out) than data sets with smaller variances. But what information can we obtain from the number, $s^2 = 9.7$? One interpretation is that the average squared deviation of the sample measurements from their mean is 9.7. However, a more practical interpretation can be obtained by calculating the square root of this number.

A third measure of data variation, the **standard deviation**, is obtained by taking the square root of the variance. This results in a number with units of measurement equal to the units of the original data. That is, if the units of measurement for the sample observations are feet, dollars, or hours, the standard deviation of the sample is measured in feet, dollars, or hours (instead of feet^2, dollars^2, or hours^2). Like the variance, the standard deviation measures the amount of spread in a quantitative data set.

> **DEFINITION 3.6**
>
> The **standard deviation** of a set of n sample measurements is equal to the square root of the variance:
>
> $$s = \sqrt{s^2} = \sqrt{\frac{\Sigma(x - \bar{x})^2}{n - 1}}$$

The standard deviation of the five sample measurements in Examples 3.8 and 3.9 is

$$s = \sqrt{s^2} = \sqrt{9.7} = 3.1$$

Now that you know how to calculate a standard deviation, we will demonstrate in the next section how it can be used to measure the spread or variation of a distribution of data.

INTERPRETING THE STANDARD DEVIATION

3.6 In this section, we give two rules for interpreting the standard deviation. Both rules use the mean and standard deviation of a data set to determine an interval of values within which most of the measurements fall. For samples, the intervals take the form

$$\bar{x} \pm (k)s$$

where k is any positive constant (usually 1, 2, or 3). The particular rule you apply will depend on the shape of the relative frequency histogram for the data set, as the following examples illustrate.

EXAMPLE 3.10 Recall that each year, the Federal Trade Commission (FTC) ranks domestic cigarette brands according to the amount of carbon monoxide (CO), in milligrams, emitted in smoke (Case Study 2.3). The 1990 CO rankings of a sample of 372 brands are given in Appendix F. Suppose we want to describe the distribution of CO measurements for this sample. To do so, we require the mean and standard deviation of the measurements.

a. Calculate $\bar{x}$ and s for the data set.
b. Form an interval by measuring 1 standard deviation on each side of the mean, i.e., $\bar{x} \pm s$. Also, form the intervals $\bar{x} \pm 2s$ and $\bar{x} \pm 3s$.
c. Find the proportions of the total number (372) of CO measurements falling within these intervals.

Solution

a. Rather than compute $\bar{x}$ and s by hand, we utilize the computer. A Minitab printout giving numerical descriptive measures for the data set is shown in Figure 3.8. The mean and standard deviation, shaded on the printout, are

$$\bar{x} = 11.66 \quad \text{and} \quad s = 4.09$$

	N	MEAN	MEDIAN	TRMEAN	STDEV	SEMEAN
CO	372	11.660	12.000	11.772	4.089	0.212

	MIN	MAX	Q1	Q3
CO	0.500	22.000	9.250	14.000

FIGURE 3.8
Minitab Printout: Descriptive Statistics for 372 CO Measurements in Appendix F

b. The intervals $\bar{x} \pm s$, $\bar{x} \pm 2s$, $\bar{x} \pm 3s$, are formed as follows:

$$\bar{x} \pm s = 11.66 \pm 4.09$$
$$= (11.66 - 4.09, \ 11.66 + 4.09)$$
$$= (7.57, \ 15.75)$$

$$\bar{x} \pm 2s = 11.66 \pm 2(4.09)$$
$$= 11.66 \pm 8.18$$
$$= (11.66 - 8.18, \ 11.66 + 8.18)$$
$$= (3.48, \ 19.84)$$

$$\bar{x} \pm 3s = 11.66 \pm 3(4.09)$$
$$= 11.66 \pm 12.27$$
$$= (11.66 - 12.27, \ 11.66 + 12.27)$$
$$= (-.61, \ 23.93)$$

c. It is too tedious to check each of the CO measurements in Appendix F by hand to determine whether it falls within the three intervals, so we did it by computer. The proportions of the total number of CO measurements falling within the three intervals are shown in Table 3.3 on the following page. The three intervals—$\bar{x} \pm s$, $\bar{x} \pm 2s$, and $\bar{x} \pm 3s$—are also shown on a SAS relative frequency histogram for the CO data displayed in Figure 3.9. If you visually estimate the proportions of the total area in the histogram that lie over the three intervals, you will obtain proportions approximately equal to those given in Table 3.3.

TABLE 3.3
Proportions of the Total Number of CO Measurements
in Intervals $\bar{x} \pm s$, $\bar{x} \pm 2s$, and $\bar{x} \pm 3s$

INTERVAL		NUMBER OF OBSERVATIONS IN INTERVAL	PROPORTION IN INTERVAL
$\bar{x} \pm s$ or	(7.57, 15.75)	244	.656
$\bar{x} \pm 2s$ or	(3.48, 19.84)	353	.949
$\bar{x} \pm 3s$ or	(−.61, 23.93)	372	1.000

FIGURE 3.9
SAS Histogram for CO
Measurements of Appendix F

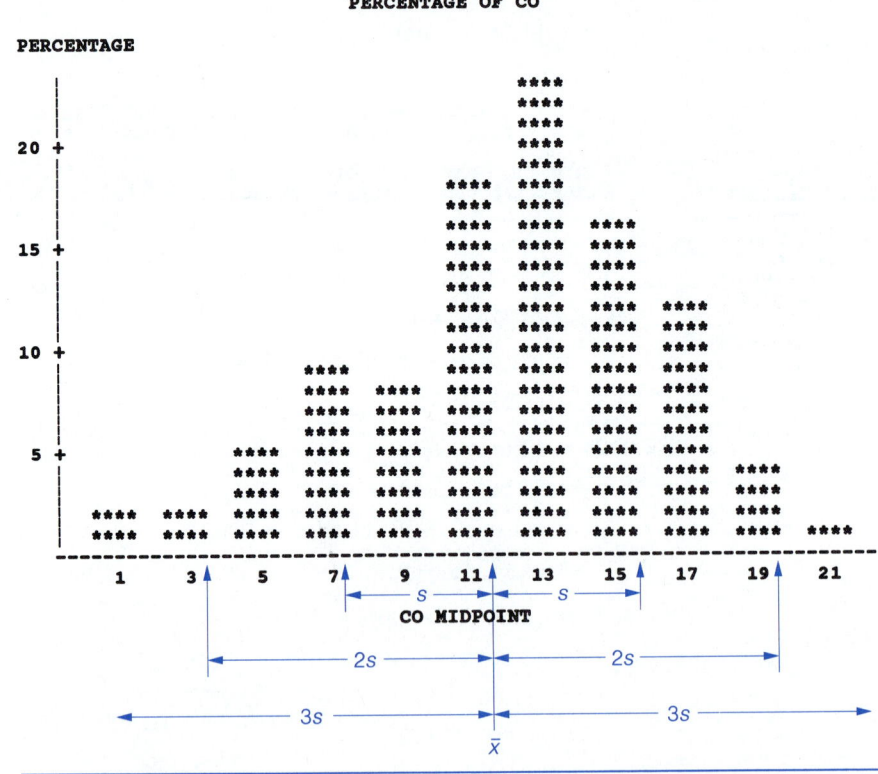

Will the proportions of the total number of observations falling within the intervals $\bar{x} \pm s$, $\bar{x} \pm 2s$, and $\bar{x} \pm 3s$ remain fairly stable for most distributions of data? To examine this possibility, consider the next example.

EXAMPLE 3.11 Calculate the mean and standard deviation of each of the following data sets from the appendixes.

a. The 1,795 starting salaries for UF graduates in Appendix A
b. The 186 per-member, per-month costs of physicians in Appendix C
c. The 144 weights of fish specimens in Appendix D
d. The 144 DDT levels of fish specimens in Appendix D
e. The 372 tar contents of cigarettes in Appendix F

Solution Because of the large amounts of data involved, we computed the means and standard deviations on a computer. They are shown in Table 3.4. The SPSS relative frequency histograms for the five data sets are shown in Figures 3.10–3.14.

TABLE 3.4
Means and Standard Deviations for Five Data Sets

DATA SET	APPENDIX	MEAN	STANDARD DEVIATION
a. Starting salaries	A	$28,475	$9,369
b. Physician costs	C	$102.96	$366.31
c. Fish weights	D	1,049.7 grams	376.6 grams
d. Fish DDT measurements	D	24.36 ppm	98.38 ppm
e. Tar contents of cigarettes	F	11.60 mg	4.97 mg

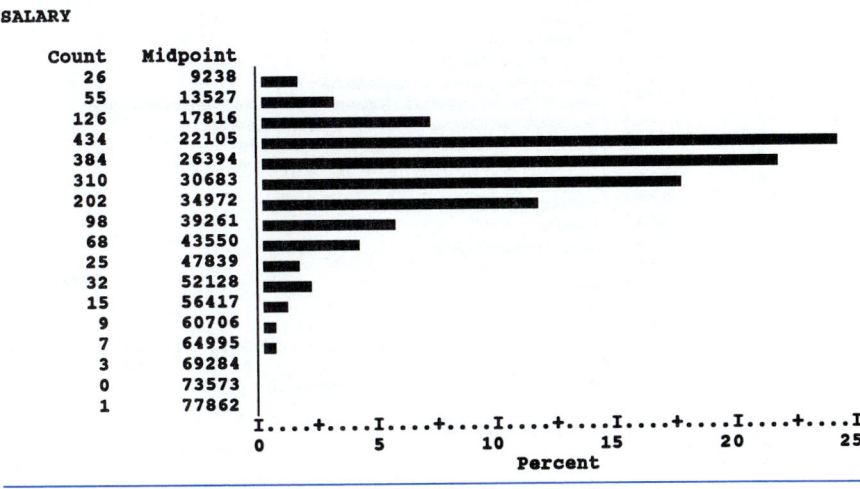

FIGURE 3.10
SPSS Relative Frequency Histogram for the 1,795 Starting Salaries of UF Graduates in Appendix A

FIGURE 3.11

SPSS Relative Frequency Histogram for the 186 Per-Member, Per-Month Costs of Physicians in Appendix C

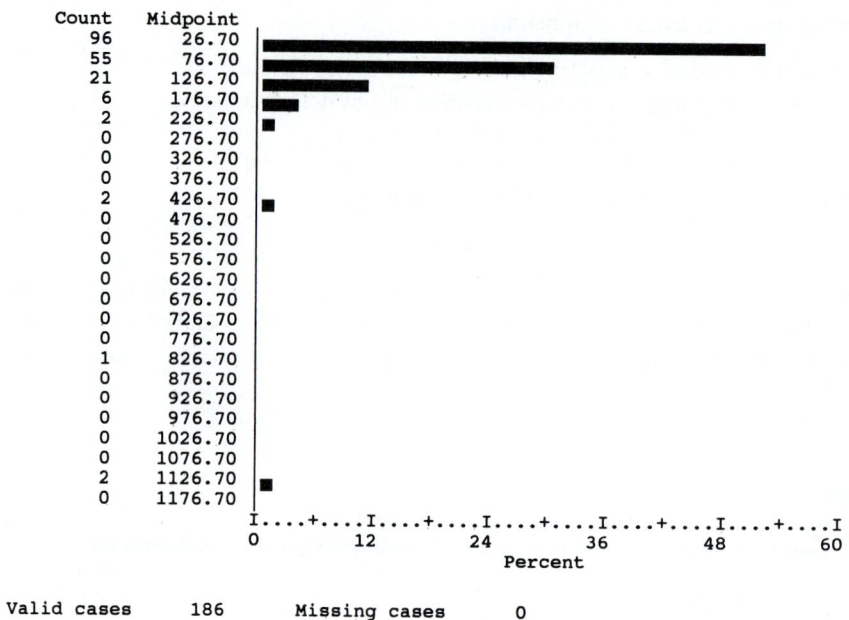

COST

Count	Midpoint
96	26.70
55	76.70
21	126.70
6	176.70
2	226.70
0	276.70
0	326.70
0	376.70
2	426.70
0	476.70
0	526.70
0	576.70
0	626.70
0	676.70
0	726.70
0	776.70
1	826.70
0	876.70
0	926.70
0	976.70
0	1026.70
0	1076.70
2	1126.70
0	1176.70

```
I....+....I....+....I....+....I....+....I....+....I
0        12        24        36        48        60
                    Percent
```

Valid cases   186   Missing cases   0

FIGURE 3.12

SPSS Relative Frequency Histogram for the 144 Weights of Fish Specimens in Appendix D

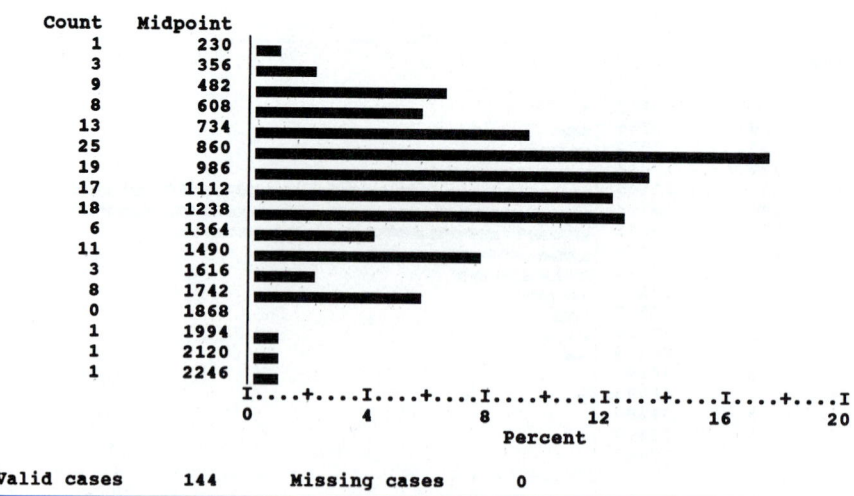

WEIGHT

Count	Midpoint
1	230
3	356
9	482
8	608
13	734
25	860
19	986
17	1112
18	1238
6	1364
11	1490
3	1616
8	1742
0	1868
1	1994
1	2120
1	2246

```
I....+....I....+....I....+....I....+....I....+....I
0         4         8        12        16        20
                    Percent
```

Valid cases   144   Missing cases   0

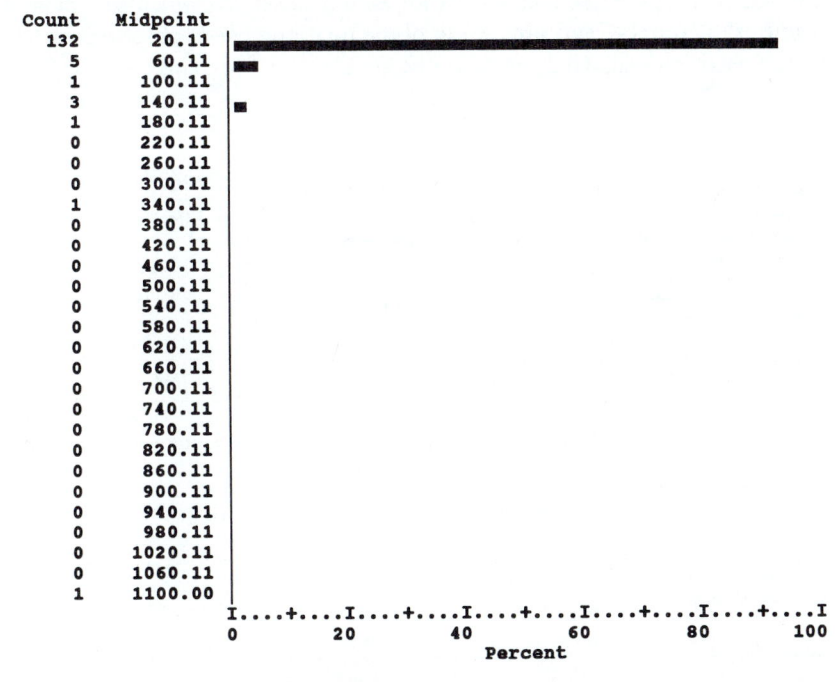

FIGURE 3.13

SPSS Relative Frequency Histogram for the 144 DDT Measurements of Fish Specimens in Appendix D

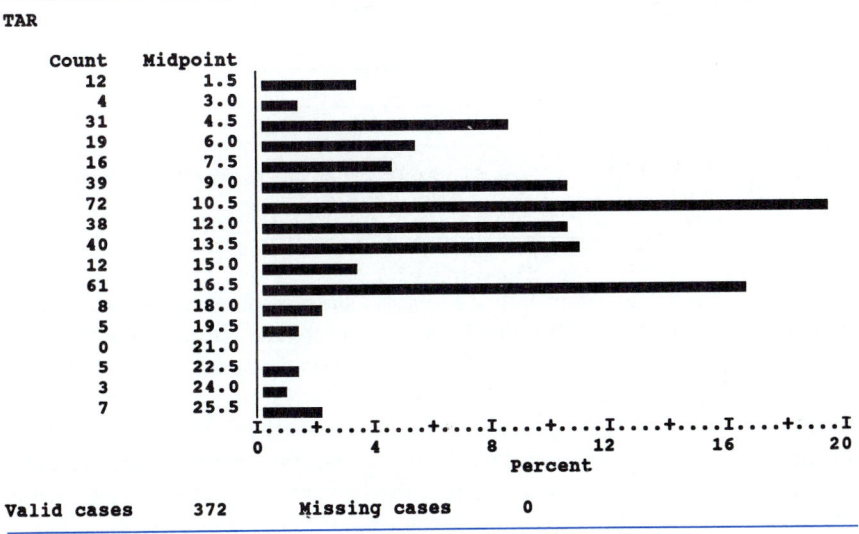

FIGURE 3.14

SPSS Relative Frequency Histogram for the 372 Tar Contents of Cigarettes in Appendix F

The means and standard deviations of Table 3.4 were used to calculate the intervals $\bar{x} \pm s$, $\bar{x} \pm 2s$, and $\bar{x} \pm 3s$ for each data set. We obtained a computer count of the number and proportion of the total number of observations falling within each interval. These proportions are presented in Tables 3.5a–e.

TABLE 3.5

Proportions of the Total Number of Observations
Falling within $\bar{x} \pm s$, $\bar{x} \pm 2s$, $\bar{x} \pm 3s$

a. Starting Salaries

	Interval	Proportion in Interval
$\bar{x} \pm s$ or	($19,105, $37,844)	.758
$\bar{x} \pm 2s$ or	($9,736, $47,214)	.947
$\bar{x} \pm 3s$ or	($366, $56,583)	.986

b. Physician Costs

	Interval	Proportion in Interval
$\bar{x} \pm s$ or	($-263.35, $469.27)	.978
$\bar{x} \pm 2s$ or	($-629.66, $835.58)	.984
$\bar{x} \pm 3s$ or	($-995.97, $1,201.89)	.995

c. Fish Weights

	Interval (grams)	Proportion in Interval
$\bar{x} \pm s$ or	(673.1, 1,426.3)	.681
$\bar{x} \pm 2s$ or	(296.5, 1,802.9)	.972
$\bar{x} \pm 3s$ or	(−80.1, 2,179.5)	.986

d. Fish DDT Measurements

	Interval (ppm)	Proportion in Interval
$\bar{x} \pm s$ or	(−74.02, 122.74)	.958
$\bar{x} \pm 2s$ or	(−172.40, 221.12)	.986
$\bar{x} \pm 3s$ or	(−270.78, 319.50)	.986

e. Tar Contents of Cigarettes

	Interval (mg)	Proportion in Interval
$\bar{x} \pm s$ or	(6.63, 16.57)	.659
$\bar{x} \pm 2s$ or	(1.66, 21.54)	.932
$\bar{x} \pm 3s$ or	(−3.31, 26.51)	1.000

Tables 3.3 and 3.5 demonstrate a property that is common to many data sets. The percentage of observations that lie within one standard deviation of the mean $\bar{x}$, i.e., in the interval $\bar{x} \pm s$, is fairly large and variable, usually from 60% to 80% of the total number, but the percentage can reach 90% or more for highly skewed distributions of data. The percentage within two standard deviations of $\bar{x}$, i.e., in the interval $\bar{x} \pm 2s$, is close to 95% but, again, the percentage will be larger for highly skewed sets of data. Finally, the percentage of observations within three standard deviations of $\bar{x}$, i.e., in the interval $\bar{x} \pm 3s$, is almost 100%, meaning that almost all of the observations in a data set will fall within this interval. This property, which seems to hold for most data sets that contain at least 20 observations and are *mound-shaped*, is called the **Empirical Rule**. The Empirical Rule provides a very good rule of thumb for forming a mental image of a distribution of data when you know the mean and standard deviation of the data set. Calculate the intervals $\bar{x} \pm s$, $\bar{x} \pm 2s$, and $\bar{x} \pm 3s$ and then picture the observations grouped as described in the box and shown in Figure 3.15.

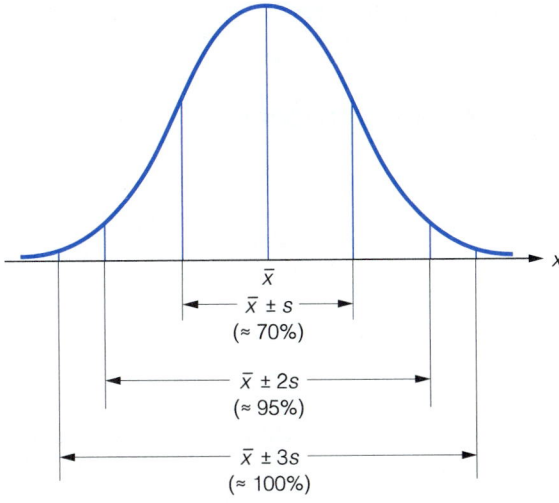

FIGURE 3.15
Illustration of the Empirical Rule

THE EMPIRICAL RULE

If a distribution of sample data is mound-shaped with mean $\bar{x}$ and standard deviation s, then the proportions of the total number of observations falling within the intervals $\bar{x} \pm s$, $\bar{x} \pm 2s$, and $\bar{x} \pm 3s$ are as follows:

$\bar{x} \pm s$: Usually between 60% and 80%. The percentage will be approximately 70% for distributions that are nearly symmetric, but larger (near 90%) for highly skewed distributions.

$\bar{x} \pm 2s$: Close to 95% for symmetric distributions. The percentage will be larger (near 100%) for highly skewed distributions.

$\bar{x} \pm 3s$: Near 100%

Note that the relative frequency histograms of physician costs (Figure 3.11) and DDT levels (Figure 3.13) are highly skewed. Consequently, actual percentages of observations falling within the intervals $\bar{x} \pm s$, $\bar{x} \pm 2s$, $\bar{x} \pm 3s$ for these data sets will tend to be on the high side of the range of values given by the Empirical Rule. On the other hand, the mound-shaped distributions of starting salaries (Figure 3.10), fish weights (Figure 3.12), and tar contents (Figure 3.14) are nearly symmetric; consequently, the percentages falling within the three intervals are very close to the values given by the Empirical Rule.

Can the Empirical Rule be applied to data sets with non–mound-shaped histograms or histograms of unknown shape? The answer, unfortunately, is no. However, in these situations we can apply a more conservative rule, called **Tchebysheff's theorem**.

The theorem states (proof omitted) that at least $1 - 1/k^2$ of the total number of observations in a sample data set will fall within the interval $\bar{x} \pm ks$, where k is a constant. We present the theorem in the following box for two useful values of k, $k = 2$, and $k = 3$.*

TCHEBYSHEFF'S THEOREM

For any set of sample measurements with mean $\bar{x}$ and standard deviation s, the proportions of the total number of observations in the sample falling within the intervals $\bar{x} \pm 2s$ and $\bar{x} \pm 3s$ are as follows:

$\bar{x} \pm 2s$:   At least 75%

$\bar{x} \pm 3s$:   At least 89%

Note that Tchebysheff's theorem applies to any set of sample measurements, regardless of the shape of the relative frequency histogram. The rule is conservative in the sense that the specified percentage for any interval is a lower bound on the actual percentage of measurements falling in that interval. For example, Tchebysheff's theorem states that at least 75% of the 372 CO measurements in Appendix F will fall in the interval $\bar{x} \pm 2s$. We know (from Table 3.3) that the actual percentage (94.9%) is closer to the Empirical Rule's value of 95%. Consequently, whenever you know that a relative frequency histogram for a data set is mound-shaped, the Empirical Rule will give more precise estimates of the true percentages falling within the intervals $\bar{x} \pm s$, $\bar{x} \pm 2s$, and $\bar{x} \pm 3s$.

The last example in this section demonstrates the use of these rules in statistical inference.

EXAMPLE 3.12   Travelers who have no intention of showing up often fail to cancel their hotel reservations in a timely manner. These travelers are known, in the parlance of the hospitality trade, as "no-shows." To protect against no-shows and late cancellations, hotels invariably overbook rooms. A recent study reported in the *Journal of Travel Research* examined the problems of overbooking rooms in the hotel industry. The data in Table 3.6, extracted from the study, represent daily numbers of late cancellations and no-shows for a random sample of 30 days at a large (500-room) hotel. Based on this sample, how many rooms, at minimum, should the large hotel overbook each day?

---

*For $k = 1$, $1 - 1/k^2 = 1 - 1/(1)^2 = 0$. Thus, Tchebysheff's theorem states that at least 0% of the observations fall within $\bar{x} \pm s$. Consequently, no useful information is provided about the interval.

TABLE 3.6

Hotel No-Shows for a Sample of 30 Days

18	16	16	16	14	18	16	18	14	19
15	19	9	20	10	10	12	14	18	12
14	14	17	12	18	13	15	13	15	19

Source: Toh, R. S. "An inventory depletion overbooking model for the hotel industry." *Journal of Travel Research*, Vol. 23, No. 4, Spring 1985, p. 27. The *Journal of Travel Research* is published by the Travel and Tourism Research Association (TTRA) and the Business Research Division, University of Colorado at Boulder.

Solution   To answer this question, we need to know a range of values where most of the daily numbers of no-shows fall. This requires that we compute $\bar{x}$ and s, and examine the shape of the relative frequency distribution for the data.

Figure 3.16 is a Minitab printout that shows a stem-and-leaf display and descriptive statistics of the sample data. Notice from the stem-and-leaf display that the distribution of daily no-shows is mound-shaped, and only slightly skewed on the low (top) side of Figure 3.16. Thus, the Empirical Rule should give a good estimate of the percentage of days that fall within one, two, and three standard deviations of the mean.

```
Stem-and-leaf of noshows     N = 30
Leaf Unit = 0.10

    1     9 0
    3    10 00
    3    11
    6    12 000
    8    13 00
   13    14 00000
   (3)   15 000
   14    16 0000
   10    17 0
    9    18 00000
    4    19 000
    1    20 0
```

	N	MEAN	MEDIAN	TRMEAN	STDEV	SEMEAN
noshows	30	15.133	15.000	15.231	2.945	0.538

	MIN	MAX	Q1	Q3
noshows	9.000	20.000	13.000	18.000

FIGURE 3.16

Minitab Printout: Describing the No-Show Data, Example 3.12

The mean and standard deviation of the sample data, shaded on the Minitab printout, are $\bar{x} = 15.133$ and $s = 2.945$. From the Empirical Rule, we know that about 95% of the daily number of no-shows fall within two standard deviations of the mean, i.e., within the interval

$$\bar{x} \pm 2s = 15.133 \pm 2(2.945)$$
$$= 15.133 \pm 5.890$$

or between 9.243 no-shows and 21.023 no-shows. (If we count the number of measurements in this data set, we find that actually 29 out of 30, or 96.7%, fall in this interval.)

From this result, the large hotel can infer that there will be at least 9.243 (or, rounding up, 10) no-shows per day. Consequently, the hotel can overbook at least 10 rooms per day and still be highly confident that all reservations can be honored.

EXERCISES
LEARNING THE MECHANICS

3.18 Calculate the variance and standard deviation of samples for which

    a. $n = 10$ $\Sigma x^2 = 331$ $\Sigma x = 50$
    b. $n = 25$ $\Sigma x^2 = 163{,}456$ $\Sigma x = 2{,}000$
    c. $n = 5$ $\Sigma x^2 = 26.46$ $\Sigma x = 11.5$

3.19 Calculate the range, variance, and standard deviation of each of the following examples:

    a. 0, 2, 4, 6, 8, 10   **b.** 0, 4, 5, 5, 6, 10   **c.** 4, 4, 4, 4, 4, 4

3.20 Find the range, variance, and standard deviation for the following data set:

    3, 9, 0, 7, 4

    Use the shortcut procedure to calculate s^2.

3.21 Find the variance and standard deviation for the following $n = 25$ measurements:

    2, 1, 7, 6, 5, 3, 8, 5, 2, 4, 5, 6, 3, 4, 4, 6, 9, 4, 3, 4, 5, 5, 7, 3, 5

3.22 Refer to the data given in Exercise 3.21. Construct the intervals $\bar{x} \pm s$, $\bar{x} \pm 2s$, and $\bar{x} \pm 3s$. Count the number of observations falling within each interval and find the corresponding proportions. Compare your results to the Empirical Rule and Tchebysheff's theorem.

3.23 Suppose a set of data has a mound-shaped, symmetric distribution. Make a statement about the percentage of measurements contained in each of the following intervals:

    a. $\bar{x} \pm s$ **b.** $\bar{x} \pm 2s$ **c.** $\bar{x} \pm 3s$

3.24 Suppose a data set has a non–mound-shaped (skewed) distribution. Make a statement about the percentage of measurements contained in each of the following intervals:

    a. $\bar{x} \pm s$ **b.** $\bar{x} \pm 2s$ **c.** $\bar{x} \pm 3s$

APPLYING THE CONCEPTS

3.25 Refer to the 500 supermarket checkout times in Appendix E. The SAS printout at the top of the next page provides numerical descriptive measures for customer checkout time (in seconds).
    a. Use the information in the printout to describe the relative frequency histogram for customer checkout time.
    b. Is the data set better described by the Empirical Rule or Tchebysheff's theorem?

3.26 Refer to the data on student-loan default rates for 66 Florida colleges, Exercise 2.28. An SPSS printout giving descriptive statistics for the data set is displayed on the next page.

UNIVARIATE PROCEDURE

Variable=CHKTIME

Moments

N	500	Sum Wgts	500
Mean	50.118	Sum	25059
Std Dev	49.06435	Variance	2407.311
Skewness	2.541635	Kurtosis	9.731435
USS	2457155	CSS	1201248
CV	97.89767	Std Mean	2.194225
T:Mean=0	22.84087	Prob>|T|	0.0001
Sgn Rank	62625	Prob>|S|	0.0001
Num ^= 0	500		

Quantiles(Def=5)

100% Max	353	99%	241
75% Q3	65	95%	135
50% Med	35	90%	110
25% Q1	18	10%	9.5
0% Min	2	5%	5.5
		1%	3
Range	351		
Q3-Q1	47		
Mode	30		

Extremes

Lowest	Obs	Highest	Obs
2 (	409)	245 (	81)
2 (	404)	292 (	262)
2 (	378)	340 (	214)
3 (	472)	350 (	215)
3 (	436)	353 (	280)

```
                     Histogram                       #        Boxplot
   350+*                                              3           *
      .
      .
      .*                                              1           *
      .
      .*                                              1           *
      .*                                              3           *
      .*                                              2           *
      .*                                              2           0
      .*                                              2           0
      .***                                            9           0
      .*******                                       20           |
      .*********                                      26           |
      .**********                                     29           |
      .*****************                              48        +-----+
      .******************************                 89        |  +  |
      .**********************************************133        *-----*
   10+**********************************************  132       +-----+
      ----+----+----+----+----+----+----+----+----+
      * may represent up to 3 counts
```

Number of Valid Observations (Listwise) =   66.00

Variable   DEFRATE

Mean	14.682	S.E. Mean	1.741
Std Dev	14.141	Variance	199.974
Kurtosis	5.427	S.E. Kurt	.582
Skewness	2.204	S.E. Skew	.295
Range	74.700	Minimum	1.50
Maximum	76.20	Sum	969.000

Valid Observations -   66   Missing Observations -   0

a. Locate the mean default rate on the printout.

b. Locate the variance and standard deviation of the default rates on the printout.

c. What proportion of measurements would you expect to find within two standard deviations of the mean?

d. Determine the proportion of measurements (default rates) that actually fall within the interval of part **c**. Compare this result with your answer to part **c**.

e. Suppose the college with the highest default rate (Florida College of Business, 76.2%) was omitted from the analysis. Would you expect the mean to increase or decrease? Would you expect the standard deviation to increase or decrease?

f. Calculate the mean and standard deviation for the data set with Florida College of Business excluded. Compare these results with your answer to part **e**.

g. Answer parts **c** and **d** using the recalculated mean and standard deviation. This problem illustrates the dramatic effect a single observation can have on the analysis.

3.27 Refer to the *Journal of Performance of Constructed Facilities* (Feb. 1990) study of water distribution networks, Exercise 1.4. The internal pressure readings (measured in pounds per square inch, psi) for a sample of pipe sections had a mean of 7.99 psi and a standard deviation of 2.02 psi.

a. Use this information to construct an interval that captures about 95% of the pressure readings sampled.

b. Would you expect to observe an internal pressure reading of 20 psi? Explain.

3.28 Refer to the data on population change of the 50 largest U.S. counties, Exercise 2.30. For convenience, the data are reproduced in the accompanying table.

COUNTY (NEAREST BIG CITY)	PERCENT CHANGE	COUNTY (NEAREST BIG CITY)	PERCENT CHANGE
Los Angeles (Los Angeles)	18.5	Tarrant (Fort Worth TX)	35.9
Cook (Chicago)	−2.8	Oakland (Pontiac MI)	7.1
Harris (Houston)	17.0	Sacramento (Sacramento)	32.9
Orange (Los Angeles)	24.7	Hennepin (Minneapolis)	9.7
Kings (New York City)	3.1	St. Louis (St. Louis)	2.0
Maricopa (Phoenix)	40.6	Erie (Buffalo)	−4.6
Wayne (Detroit)	−9.7	Franklin (Columbus OH)	10.6
Queens (New York City)	3.2	Milwaukee (Milwaukee)	−0.6
Dade (Miami)	19.2	Westchester (White Plains NY)	1.0
Dallas (Dallas)	19.0	Hamilton (Cincinnati)	−0.8
Philadelphia (Philadelphia)	−6.1	Palm Beach (Palm Beach FL)	49.7
King (Seattle)	18.7	Hartford (Hartford CT)	5.4
Santa Clara (San Jose CA)	15.6	Pinellas (St. Petersburg FL)	16.9
New York (New York City)	4.1	Honolulu (Honolulu)	9.7
San Bernardino (Los Angeles)	58.5	Hillsborough (Tampa)	28.9
Cuyahoga (Cleveland)	−5.8	Fairfield (Bridgeport CT)	2.5
Middlesex (Boston)	2.3	Shelby (Memphis)	6.3
Allegheny (Pittsburgh)	−7.8	Bergen (Hackensack NJ)	−2.4
Suffolk (New York City)	2.9	Fairfax (VA/Washington)	37.4
Nassau (New York City)	−2.6	New Haven (New Haven CT)	5.6
Alameda (Oakland CA)	15.7	Contra Costa (San Francisco)	22.5
Broward (Ft. Lauderdale FL)	23.3	Marion (Indianapolis)	4.25
Bronx (New York City)	3.0	DuPage (Chicago)	18.6
Bexar (San Antonio)	19.9	Essex (Newark NJ)	−8.6
Riverside (Los Angeles)	76.5		

Source: U.S. Census Bureau, 1990 Census.

a. Compute $\bar{x}$ and s.

b. Calculate the intervals $\bar{x} \pm s$, $\bar{x} \pm 2s$, and $\bar{x} \pm 3s$.

c. Calculate the percentage of observations falling in the interval $\bar{x} \pm 2s$.

d. Does the Empirical Rule adequately describe the distribution of percentage change in population for the U.S. counties?

3.29 Nevada continues to be the leading gold producer in the United States. According to the U.S. Bureau of Mines, it ranks among the top four regional producers worldwide (trailing South Africa, Russia, and Australia). The data in the table represent the production (in thousands of ounces) for the top 30 gold mines in the state.

1,467.8	228.0	111.3	76.0	55.1	40.0
318.0	222.6	89.1	72.5	54.1	32.4
296.9	214.6	82.0	66.0	50.0	30.9
256.0	207.3	81.5	60.4	50.0	30.3
254.5	120.7	78.8	60.0	44.5	30.0

Source: Engineering & Mining Journal, June 1990, p. 38.

a. Summarize the data with a graphical technique.

b. Calculate the mean, median, and standard deviation of the data.

c. What proportion of Nevada mines have production values that lie within two standard deviations of the mean?

d. Note the extremely large production value, 1.467.8, for the first mine listed in the table. Recalculate the mean, median, and standard deviation with the production measurement for this mine deleted.

e. Explain how the three numerical descriptive measures (mean, median, and standard deviation) are affected by the deletion of the measurement 1,467.8.

3.30 The Trail Making Test (TMT) is frequently used in neuropsychological assessment to provide a quick estimate of brain damage in humans. Subjects taking the TMT are asked to perform a certain task as quickly as possible. The test is sensitive to the effects of age—an older person normally takes longer to complete the task. To investigate the neuropsychological deficits in alcoholics, 50 problem drinkers (25 drinkers under the age of 40 and 25 drinkers 40 years or older) were given the TMT and their performance scores observed (the higher the score, the more extensive the brain damage). The results are reported in the accompanying table.

	ALCOHOLICS UNDER AGE 40	ALCOHOLICS 40 OR OLDER
Mean performance score	39.6	49.7
Standard deviation	19.7	19.1

a. Use the information in the table, in conjunction with either the Empirical Rule or Tchebysheff's theorem, to sketch your mental images of the relative frequency histograms of TMT performance scores for the two groups of alcoholics.

b. Estimate the fraction of alcoholics under age 40 who score between 19.9 and 59.3 on the TMT.

c. Approximately what percentage of alcoholics aged 40 or older score between 11.5 and 87.9 on the TMT?

3.31 The data on unplanned rapid emergency shutdowns of nuclear reactors (scrams), Exercise 2.31, are reproduced in the table.

NUMBER OF SCRAMS									
1	0	3	1	4	2	10	6	5	2
0	3	1	5	4	2	7	12	0	3
8	2	0	9	3	3	4	7	2	4
5	3	2	7	13	4	2	3	3	7
0	9	4	3	5	2	7	8	5	2
4	3	4	0	1	7				

Source: American Nuclear Society.

a. Compute $\bar{x}$, s^2, and s for this data set.

b. What percentage of the measurements would you expect to find in the interval $\bar{x} \pm 2s$?

c. Count the number of measurements that actually fall within the interval of part **b** and express the interval count as a percentage of the total number of measurements. Compare this result with the answer to part **b**.

d. Suppose the nuclear reactor unit that had the most scrams (13) was omitted from the analysis. Would you expect $\bar{x}$ to increase or decrease? Would you expect s to increase or decrease?

CHAPTER CASE: HOW MUCH WEIGHT CAN YOU LIFT—AND STILL AVOID INJURY?

3.7 Most people either have lifted or know people who have lifted (or at least attempted to lift) a heavy load (e.g., a refrigerator, a wheelbarrow filled with bricks, a dining room table) by themselves and, as a result, were forced to remain in bed the following day with severe back pain or a slipped disk. An unusually large number of injuries, both at work and at home, arise from handling (or mishandling) heavy materials, and most of these injuries result from slipping and falling, dropping the load, or straining to lift the load. Engineers have a term for unaided human acts of lifting, lowering, pushing, pulling, carrying, or holding and releasing an object—*manual materials handling activities (MMHA)*.

M. M. Ayoub et al. (1980) have attempted to develop strength and capacity guidelines for MMHA.* These include norms for lifting and lowering, pushing, pulling, and carrying. The authors point out that a clear distinction between strength and capacity must be made. "Strength implies what a person can do in a single attempt, whereas capacity implies what a person can do for an extended period of time. Lifting strength, for example, determines the amount that can be lifted at infrequent intervals."

Ayoub et al. arrive at these norms by combining the results of other researchers. Much of the successful research in determining norms for MMHA relies on the psychophysical methodology. In a study of MMHA using psychophysics, subjects are required to adjust one of the task variables (e.g., weight of the load) according to their own perception of muscular effort or force. For example, "Switzer (1962) instructed 75 male college students to find reasonable weights that could be lifted without excessive strain or discomfort for a single lift. Subjects varied the weight lifted by adding or subtracting 2.26-, 4.53-, and 9.06-[kilogram] bags of lead shot." In another study, Kroemer (1974) had 73 male subjects maintain (what they perceived to be) a maximum push force steadily over a 5-second period while standing in various postures on a slippery floor. (An interesting result of this experiment was that body support, not body weight or body size, was most useful in predicting force output. "A physically strong subject," conclude Ayoub et al., "may be able to exert only weak push or pull forces due to lack of body support.") A third example of the psychophysical approach [Snook (1974)] involved both male and female workers who determined their own work load and the walking speed with which they carried the load.

---

*Ayoub, M. M., Mital, A., Bakken, G. M., Asfour, S. S., and Bethea, N. J. "Development of strength capacity norms for manual materials handling activities: The state of the art." *Human Factors*, June 1980, Vol. 22, pp. 271–283. Copyright 1980 by the Human Factors Society, Inc., and reproduced by permission.

In Table 3.7 we present a portion of the recommendations of Ayoub et al. for the lifting capacities of males and females based on the psychophysical technique. The table gives the means and standard deviations of the maximum weight (in kilograms) of a box 30 centimeters wide that can be safely lifted from the floor to knuckle height at two different lift rates: 1 lift per minute and 4 lifts per minute. Thus, for any given sex–lift rate combination, Ayoub et al. envision a distribution of maximum weight of lift that will vary from one individual to another.

TABLE 3.7
Mean and Standard Deviation of the Maximum Weight of Lift (in Kilograms)

SEX	LIFTS PER MINUTE	MEAN	STANDARD DEVIATION
Male	1	30.25	8.56
	4	23.83	6.70
Female	1	19.79	3.11
	4	15.82	3.23

Using the means and standard deviations given in Table 3.7, we can reconstruct and compare the distributions of maximum weights of lift. While we do not know the exact shapes of the distributions, we know from the preceding sections that the center of a given distribution will be located (approximately) over its mean and that most of the distribution will lie within two standard deviations of its mean. Using this information, and *assuming the distributions are mound-shaped and symmetric*, we can reconstruct approximations to the four distributions, one for each of the four sex–lift rate combinations, as shown in Figure 3.17.

The four distributions shown in Figures 3.17a–d provide some interesting information about the comparison of maximum weight of lift for men and women. The distribution for males (Figures 3.17a and 3.17c) is shifted to the right of the distribution for females (Figures 3.17b and 3.17d) for both weight lifts, but the shift is less for 4 lifts per minute. The maximum weight of lift is much more variable for males than females, as indicated by the spreads of the distributions.

Examining the distributions for 1 lift per minute (Figures 3.17a–b), we can see that the maximum weight of lift for almost all women is less than 30 kilograms. In comparison, the maximum weight of lift for men was as high as 50. However, you will note that the smallest maximum weight of lift for women is approximately equal to the smallest maximum weight of lift for men.

The distributions for 4 lifts per minute (Figure 3.17c–d) show less difference between males and females. Once again, most females were comfortable with maximum weights of lift of less than 30 kilograms, compared to an upper limit of approximately 42 kilograms for males. Perhaps surprisingly, it appears that a higher proportion of men than women score less than 10 kilograms.

The preceding discussion shows how you can use the means and standard deviations of sets of measurements to create mental images of their distributions. These distributions provide descriptions of the data sets and enable us to compare one data set with another.

FIGURE 3.17

Approximate Distributions of the
Maximum Weight (in Kilograms)
of Lift for Males and Females at
Lift Rates of 1 and 4 Lifts/Minute

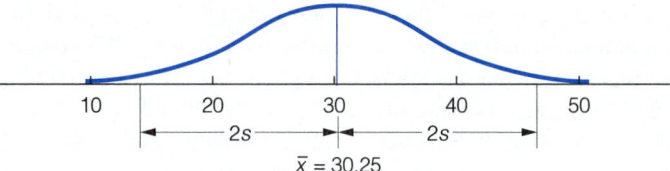

a. Male; 1 lift/minute

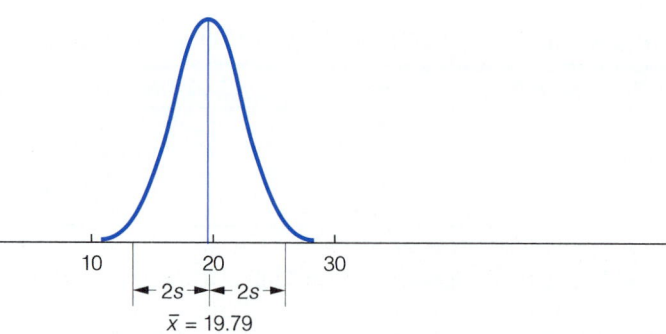

b. Female; 1 lift/minute

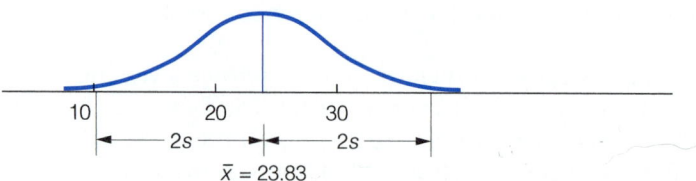

c. Male; 4 lifts/minute

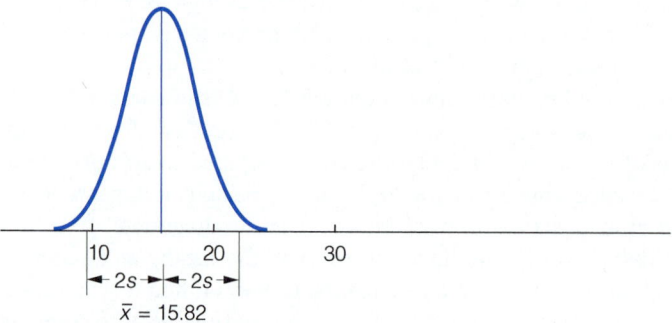

d. Female; 4 lifts/minute

3.8 In Sections 3.4 and 3.5, we gave formulas for computing the mean and standard deviation of a data set. However, these formulas apply only to **raw data sets**, i.e., those in which the value of each of the individual observations in the data set is known. If the data have already been grouped into classes of equal class width and arranged in a frequency table, you must use an alternative method to compute the mean and standard deviation.

CALCULATING A MEAN AND STANDARD DEVIATION FROM GROUPED DATA (OPTIONAL)

EXAMPLE 3.13 Refer to Example 2.10. Calculate the mean and standard deviation for the starting salary data of Table 2.3, using the grouping shown in the frequency table (Table 2.4).

Solution Since the data of Table 2.3 are raw (that is, the starting salaries for each of the 50 college graduates selected from Appendix A are given), we could compute the sample mean starting salary and sample standard deviation of starting salaries directly, using the formulas of Sections 3.4 and 3.5. For the purposes of illustration, however, we will assume that we have access only to the grouped data of Table 2.4. The formulas for calculating $\bar{x}$, s^2, and s from **grouped data** are given in the box.

FORMULAS FOR CALCULATING A MEAN AND STANDARD DEVIATION FROM GROUPED DATA

x_i = Midpoint of the ith class

f_i = Frequency of the ith class

k = Number of classes

n = Total number of observations in the data set

$$\bar{x} = \frac{\sum x_i f_i}{n}$$

$$s^2 = \frac{\sum (x_i - \bar{x})^2 f_i}{n - 1}$$

$$= \frac{\sum x_i^2 f_i - \frac{(\sum x_i f_i)^2}{n}}{n - 1} \quad \text{Shortcut formula}$$

$$s = \sqrt{s^2}$$

The 11 class intervals, midpoints, and frequencies of Table 2.4 are reproduced in Table 3.8. Substituting the class midpoints and frequencies into the formulas, we obtain

$$\bar{x} = \frac{\sum x_i f_i}{n} = \frac{(10,450)(2) + (13,450)(4) + (16,450)(1) + \cdots + (40,450)(1)}{50}$$

$$= \frac{1,191,500}{50} = 23,830$$

TABLE 3.8
Class Intervals, Midpoints, and Frequencies for the Starting Salaries of Table 2.4

CLASS	CLASS INTERVAL	CLASS MIDPOINT x_i	CLASS FREQUENCY f_i
1	8,950–11,950	10,450	2
2	11,950–14,950	13,450	4
3	14,950–17,950	16,450	1
4	17,950–20,950	19,450	13
5	20,950–23,950	22,450	7
6	23,950–26,950	25,450	8
7	26,950–29,950	28,450	5
8	29,950–32,950	31,450	4
9	32,950–35,950	34,450	3
10	35,950–38,950	37,450	2
11	38,950–41,950	40,450	1
			$n = 50$

Since we found $\sum x_i f_i = 1,191,500$ when calculating $\bar{x}$, we have (using the shortcut formula for s^2)

$$s^2 = \frac{\sum x_i^2 f_i - \frac{(\sum x_i f_i)^2}{n}}{n - 1}$$

$$= \frac{[(10,450)^2(2) + (13,450)^2(4) + \cdots + (40,450)^2(1)] - \frac{(1,191,500)^2}{50}}{49}$$

$$= \frac{30,845,225,000 - 28,393,445,000}{49} = 50,036,327$$

and

$$s = \sqrt{50,036,327} = 7,073.6$$

Thus, using the grouped data method, we have $\bar{x} = \$23,830$ and $s = \$7,073.6$.

The values of $\bar{x}$, s^2, and s based on the formulas for grouped data will usually not agree exactly with those obtained using the raw or ungrouped data. Applying the formulas of Sections 3.4 and 3.5 to the raw data of Table 2.3, we obtain $\bar{x} = 23,708$ and $s = 7,338$. These values are different from those computed in Example 3.13 because, in the grouped data method, we have substituted the value of the class midpoint for each value of x, starting salary, in a class interval. Only when every value of x in each class is equal to its respective class midpoint (which is rarely the case) will the formulas for grouped and for ungrouped data give identical values of $\bar{x}$, s^2, and s. Therefore, *the formulas for grouped data are approximations to these numerical descriptive measures.*

LEARNING THE MECHANICS

3.32 Consider the data summarized in the accompanying table.

CLASS INTERVAL	CLASS FREQUENCY f_i
5–10	8
10–15	12
15–20	10
20–25	6
25–30	4
	$n = 40$

a. Find the midpoint x_i for each of the five classes.
b. Compute $\Sigma\, x_i f_i$ and $\Sigma\, x_i^2 f_i$.
c. Use the results of part **b** to calculate $\bar{x}$, s^2, and s.

3.33 Calculate $\bar{x}$, s^2 and s for the data summarized in the frequency table.

CLASS INTERVAL	MIDPOINT x_i	FREQUENCY f_i
.715–2.115	1.415	8
2.115–3.515	2.815	7
3.515–4.915	4.215	5
4.915–6.315	5.615	4
6.315–7.715	7.015	3
7.715–9.115	8.415	3
		$n = 30$

APPLYING THE CONCEPTS

3.34 *Stores* magazine, with the help of Ernst & Young, recently conducted a survey of distribution, transportation, and warehousing trends in the retail industry (*Stores*, Feb. 1991). Over 50 companies, ranging from major department store chains to specialty apparel chains, participated in the survey. One of the survey questions concerned the turnaround time (in days) for moving merchandise received by the warehouse to the stores. The accompanying table gives the distribution of turnaround time for the 53 stores surveyed.

TURNAROUND TIME (days)	NUMBER OF STORES
0–2	11
3–4	16
5–10	17
11–40	9
	$n = 53$

Source: "Ernst & Young 1990 survey of distribution, transportation, and warehousing." *Stores*, Feb. 1991, p. A12.

a. Use the information in the table to compute the mean and standard deviation of the turnaround times for the 53 stores.

b. The actual mean turnaround time reported in the article is 8.66 days. Explain why this number differs from your answer to part **b**.

3.35 Refer to the hotel no-show data of Example 3.12. The data is reproduced here for convenience.

18	16	16	16	14	18	16	18	14	19
15	19	9	20	10	10	12	14	18	12
14	14	17	12	18	13	15	13	15	19

Source: Toh, R. S. "An inventory depletion overbooking model for the hotel industry." *Journal of Travel Research*, Vol. 23, No. 4, Spring 1985, p. 27. The *Journal of Travel Research* is published by the Travel and Tourism Research Association (TTRA) and the Business Research Division, University of Colorado at Boulder.

a. Using the seven class intervals 7.5–9.5, 9.5–11.5, . . . , 19.5–21.5, construct a frequency table similar to Table 3.8.

b. Compute $\bar{x}$ and s using the grouped data formulas. Compare these values of $\bar{x}$ and s obtained in Example 3.12 using the raw data.

3.36 Do urban residents have the ability to accurately comprehend and evaluate the services delivered by local governments? This question was the topic of research conducted by Stephen L. Percy (*Urban Affairs Quarterly*, Sept. 1986). In particular, Percy focused on residents' perceptions of police response time for a sample of 625 calls to the Fort Worth (Texas) Police Department. The accompanying frequency table describes the absolute difference (in minutes) between citizen-reported and police-record measures of response time.

DIFFERENCE BETWEEN MEASURES OF POLICE RESPONSE TIME (Minutes)	NUMBER OF CASES
0	38
1–5	262
6–10	139
11–20	102
21–30	46
30	38

Source: Percy, S. L. "In defense of citizen evaluations as performance measures." *Urban Affairs Quarterly*, Vol. 22, No. 1, Sept. 1986, p. 74.

a. Calculate the approximate mean and standard deviation of the absolute differences between citizen-reported and police-record measures of police response time.

b. Form the interval $\bar{x} \pm 2s$. Using the frequency table, determine the percentage of cases in the sample with time differences in this interval.

c. Compare the results of part **b** with Tchebysheff's theorem.

3.37 How well are market research questionnaires designed? A recent study examined almost 570 items obtained from questionnaires developed by 12 research companies. One of the major objectives of the study was to analyze the frequency of word use. In the 570 questions analyzed, a total of 1,298 different words were used. The table gives the breakdown on the frequency of word use. Calculate $\bar{x}$, s^2, and s for frequency of word use in the sample questionnaires, using the grouping given in the table. (Assume that the midpoint for the "100 or more" interval is 120.)

FREQUENCY OF WORD USE (NUMBER OF TIMES USED)	NUMBER OF DIFFERENT WORDS
1–4	964
5–9	152
10–19	80
20–49	52
50–99	26
100 or more	24
	$n = 1,298$

Source: O'Brien, J. "How do market researchers ask questions?" *Journal of the Market Research Society,* Apr. 1984, Vol. 26, No. 2, p. 102.

3.38 The accompanying data were collected by a national polling service in April 1983 as part of a telephone survey on deterrence of alcohol-impaired driving. Respondents, all licensed drivers age 16 or older, were asked to give their annual family income and their preferred alcoholic beverage.

ANNUAL FAMILY INCOME	PREFERRED BEVERAGE (NUMBER OF RESPONDENTS)		
	Beer	Wine	Spirits
Below $10,000	26	12	15
$10,000–$20,000	67	36	36
$20,000–$30,000	79	47	42
$30,000–$40,000	43	35	31
$40,000+	31	50	32
	$n = 246$	180	156

Source: Berger, D. E. and Snortum, J. R. "Alcoholic beverage preferences of drinking-driving violators," *Journal of Studies on Alcohol,* Vol. 46, No. 3, 1985, pp. 232–239. Copyright by Alcohol Research Documentation, Inc., Rutgers Center of Alcohol Studies, New Brunswick, N.J. 08903. Reprinted by permission.

a. Compute $\bar{x}$, s^2, and s for the annual family income of those respondents preferring beer. (Assume that the "$40,000+$" category has a midpoint of $50,000.)
b. Repeat part **a** for those respondents preferring wine.
c. Repeat part **a** for those respondents preferring spirits.

3.9 In some situations, you may want to describe the relative position of a particular measurement in a data set. For example, suppose a college graduate in the data set of Appendix A has a starting salary of $46,000. You might want to know whether this is a relatively low or high starting salary, etc. What percentage of the starting salaries were less than $46,000; what percentage were larger? Descriptive measures that locate the relative position of a measurement—in relation to the other measurements—are called **measures of relative standing**. One measure that expresses this position in terms of a percentage is called a **percentile** for the data set.

MEASURES OF RELATIVE STANDING

Let $x_1, x_2, \ldots, x_n$ be a set of n measurements arranged in increasing (or decreasing) order. The **pth percentile** is a number x such that $p\%$ of the measurements fall below the pth percentile and $(100 - p)\%$ fall above it.

A starting salary of \$46,000 falls at the 95th percentile of the starting salary data given in Appendix A. This tells you that 95% of the starting salaries were less than \$46,000 and $(100 - 95)\% = 5\%$ were greater.

The median, by definition, is the 50th percentile. The 25th percentile, the median, and the 75th percentiles are often used to describe a data set because they divide the data set into four groups, with each group containing one-fourth (25%) of the observations. They would also divide the relative frequency histogram for a data set into four parts, each containing the same area (.25), as shown in Figure 3.18. Consequently, the 25th percentile, the median, and the 75th percentile are called the **lower quartile**, the **mid-quartile**, and the **upper quartile**, respectively, for a data set.

FIGURE 3.18
Locations of the Lower and Upper Quartiles

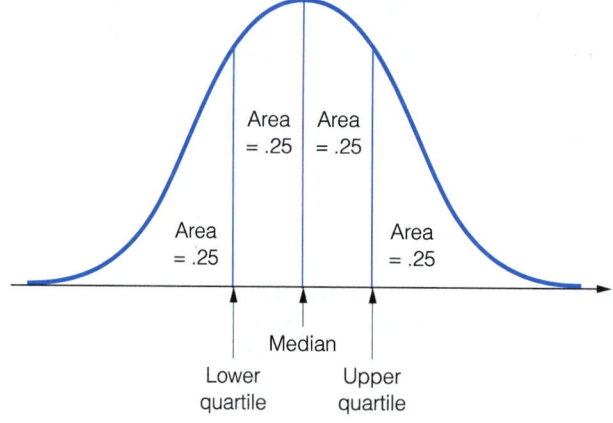

The **lower quartile**, Q_L, for a data set is the 25th percentile.

The **mid-quartile (median)**, M, for a data set is the 50th percentile.

DEFINITION 3.10

The **upper quartile**, Q_U, for a data set is the 75th percentile.

For large data sets, percentiles can be found by locating the corresponding areas under the relative frequency distribution; however, they are usually found via the computer. Figure 3.19 is a reproduction of the SAS printout describing the starting salary data set in Appendix A. The values of Q_L, M, Q_U, and the 90th percentile are shaded on the printout. From these values, we know that 25% of the 1,795 starting salaries fall below the lower quartile, \$22,000; 50% fall below the median, \$27,000; 75% fall below the upper quartile, \$33,000; and 90% fall below the 90th percentile, \$40,000.

```
                    UNIVARIATE PROCEDURE

Variable=SALARY

                          Moments

N                1795   Sum Wgts        1795
Mean         28474.83   Sum         51112325
Std Dev      9369.484   Variance    87787229
Skewness     1.093351   Kurtosis    2.301454
USS          1.613E12   CSS         1.575E11
CV           32.90444   Std Mean    221.1482
T:Mean=0      128.759   Prob>|T|      0.0
Sgn Rank       805955   Prob>|S|      0.0
Num ^= 0         1795
W:Normal     0.934422   Prob<W        0.0


                     Quantiles(Def=5)

  100% Max      80000         99%      60000
   75% Q3       33000         95%      46000
   50% Med      27000         90%      40000
   25% Q1       22000         10%      19000
    0% Min       7100          5%      16000
                               1%      10000

  Range         72900
  Q3-Q1         11000
  Mode          20000


                          Extremes

    Lowest     Obs        Highest      Obs
     7100(     1003)       66000(      532)
     7200(     1603)       70000(      557)
     8000(     1776)       70000(      711)
     8000(      920)       70000(      884)
     8000(      849)       80000(     1433)
```

FIGURE 3.19
SAS Descriptive Statistics for Starting Salary Data in Appendix A

When the sample data set is small, it may be impossible to find a measurement in the data set that exceeds, say, *exactly* 25% of the remaining measurements. Consequently, percentiles for small data sets are not well defined. The box describes a procedure for finding quartiles and other percentiles with small data sets.

	FINDING QUARTILES (AND PERCENTILES) WITH SMALL DATA SETS
STEP 1	Rank the n measurements in the data set in increasing order of magnitude.
STEP 2	Calculate the quantity $\frac{1}{4}(n+1)$ and round to the nearest integer. The measurement with this rank represents the lower quartile or 25th percentile. [*Note:* If $\frac{1}{4}(n+1)$ falls halfway between two integers, round up.]
STEP 3	Calculate the quantity $\frac{3}{4}(n+1)$ and round to the nearest integer. The measurement with this rank represents the upper quartile or 75th percentile. [*Note:* If $\frac{3}{4}(n+1)$ falls halfway between two integers, round down.]
GENERAL	To find the pth percentile, calculate the quantity $p(n+1)/100$ and round to the nearest integer. The measurement with this rank is the pth percentile.

One advantage of a stem-and-leaf display is that the display makes it easy to locate the median and the upper and lower quartiles for a data set. We will illustrate with an example.

EXAMPLE 3.14   Find the lower quartile, the median, the upper quartile, and the 90th percentile for the 25 sale prices in Table 2.2.

Solution   The stem-and-leaf display for the data of Table 2.2 is reproduced for convenience in Figure 3.20.

Since there are 25 observations in the data set, $\frac{1}{4}(n+1) = \frac{1}{4}(26) = 6.5$. Since this value is halfway between 6 and 7, we round *up* to 7. Thus, the lower quartile Q_L will be the seventh observation when the data are arranged in order from smallest to largest. We can locate Q_L by proceeding down the stem-and-leaf display until we reach the stem (6) that contains the seventh leaf. Of the leaves in this stem, the value 50 represents the seventh leaf; therefore, $Q_L = 650$ (corresponding to \$65,500).

The median of the data set is the 13th observation when the data are arranged in order. Counting leaves from the top of the display, you can see that the 13th leaf is the next-to-largest leaf in stem 7—namely, 60. Therefore, the median $M = 760$ (or \$76,000).

To find the upper quartile Q_U, we calculate $\frac{3}{4}(n+1) = \frac{3}{4}(26) = 19.5$. Since this value is halfway between 19 and 20, we round *down* to 19. Thus, the upper quartile is the 19th leaf from the top (or the seventh leaf from the bottom) of the display: the leaf 50 in stem 9. Therefore, $Q_U = 950$ (or \$95,000).

Finally, we find the 90th percentile by computing $90(n+1)/100 = 90(26)/100 = 23.4$ (or 23, after rounding). Thus, the 90th percentile is the 23rd leaf from the top (or the third leaf from the bottom) of the stem-and-leaf display. This value is 1,120, corresponding to a sale price of \$112,000.

STEM	LEAF
3	67
4	25
5	00 75 95
6	30 50 60 82
7	10 20 49 60 70
8	20 43 99
9	45 50
10	16 60 90
11	20
12	95
13	
14	80

FIGURE 3.20

Stem-and-Leaf Display for the Data of Table 2.2

When a data set is large, the locations of the quartiles relative to the median help us detect possible skewness in the distribution for a data set. For example, if Q_L is farther away from the median than Q_U, then the distribution is likely to be skewed to the left, as shown in Figure 3.21a. If Q_U is farther away from the median than Q_L, then the distribution is likely to be skewed to the right, as in Figure 3.21b. Lack of skewness is suggested when Q_L and Q_U are approximately equidistant from the median, as depicted in Figure 3.21c.

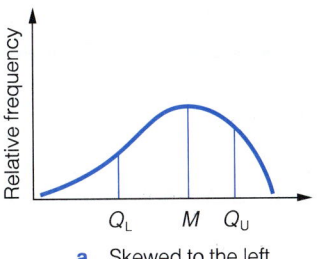

a. Skewed to the left

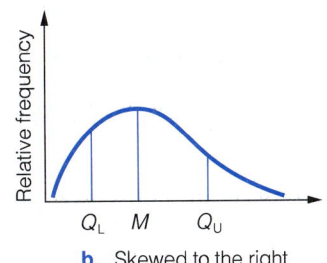

b. Skewed to the right

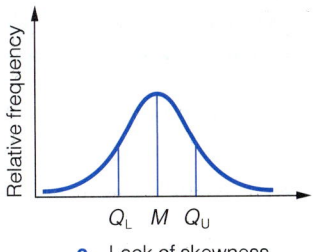

c. Lack of skewness

Another measure of relative standing is the **z score** for a measurement. For example, suppose you were told that $42,000 lies 1.44 standard deviations above the mean of the 1,795 starting salaries of Appendix A. Knowing that most of the starting salaries will be less than two standard deviations from the mean and almost all will be within three, you would have a good idea of the relative standing of the $42,000 starting salary. The distance that a measurement x lies above or below the mean $\bar{x}$ of a data set, measured in units of the standard deviation s, is called the **z score** for the measurement. A negative z score indicates that the observation lies to the left of the mean; a positive z score indicates that the observation lies to the right of the mean.

DEFINITION 3.11

The sample **z score** for the measurement x is

$$z = \frac{x - \bar{x}}{s}$$

EXAMPLE 3.15 We have noted that the mean and standard deviation for the 1,795 starting salaries of Appendix A are $\bar{x} = \$28,475$ and $s = \$9,369$, respectively. Use these values to find the z score for a starting salary of $42,000.

Solution Substituting the values of x, $\bar{x}$, and s into the formula for z, we obtain

$$z = \frac{x - \bar{x}}{s} = \frac{42,000 - 28,475}{9,369} = 1.44$$

a. Calculate $\bar{x}$ and s for the TCDD levels in plasma.

b. Calculate $\bar{x}$ and s for the TCDD levels in fat tissue.

c. Calculate the z score for a TCDD level in plasma of 20.

d. Calculate the z score for a TCDD level in fat tissue of 20.

e. Based on the results of parts **c** and **d**, is a TCDD level of 20 or more likely to occur in plasma or fat tissue? Explain.

3.50   Refer to the data on percentage change in population for U.S. counties given in Exercise 3.28. Compute and interpret the values of Q_L, M, and Q_U.

METHODS FOR DETECTING OUTLIERS

3.10 Sometimes inconsistent observations are included in a data set. For example, when we discuss starting salaries for college graduates with bachelor's degrees, we generally think of traditional college graduates—those near 22 years of age with four years of college education. But suppose one of the graduates is a 34-year-old Ph.D. chemist who has returned to the university to obtain a bachelor's degree in metallurgy. Clearly, the starting salary for this graduate could be much larger than the other starting salaries because of the graduate's additional education and experience, and we probably would not want to include it in the data set. An unusual observation that lies outside the range of the data values that we want to describe is called an **outlier**.

Outliers are often attributable to one of several causes. First, the measurement associated with the outlier may be invalid. For example, the experimental procedure used to generate the measurement may have malfunctioned, the experimenter may have misrecorded the measurement, or the data might have been coded incorrectly in the computer. Second, the outlier may be the result of a misclassified measurement. That is, the measurement belongs to a population different from that from which the rest of the sample was drawn, as in the case of the chemist's salary described in the preceding paragraph. Finally, the measurement associated with the outlier may be recorded correctly and from the same population as the rest of the sample, but represents a rare (chance) event. Such outliers occur most often when the relative frequency histogram of the sample data is extremely skewed, because such a distribution has a tendency to include extremely large or small observations relative to the others in the data set.

DEFINITION 3.12

An observation (or measurement) that is unusually large or small relative to the other values in a data set is called an **outlier**. Outliers typically are attributable to one of the following causes:

1. The measurement is observed, recorded, or entered into the computer incorrectly.
2. The measurement comes from a different population.
3. The measurement is correct, but represents a rare (chance) event.

The most obvious method for determining whether an observation is an outlier is to calculate its z score (Section 3.9), as the following example illustrates.

EXAMPLE 3.16 The fish specimen identified by observation 115 in Appendix D has a DDT measurement of 1,100 ppm. Is this observation an outlier?

Solution Recall from Table 3.4, Example 3.11, that $\bar{x} = 24.36$ ppm and $s = 98.38$ ppm for the 144 DDT measurements in Appendix D. Therefore, the z score for the DDT measurement of fish 115 is

$$z = \frac{x - \bar{x}}{s} = \frac{1{,}100 - 24.36}{93.38} = 10.93$$

Both the Empirical Rule and Tchebysheff's theorem (Section 3.6) tell us that almost all the observations in a data set will have z scores less than 3 in absolute value. Since a z score as large as 10.93 is highly improbable, the DDT measurement of 1,100 ppm is called an outlier. Some research by the Army Corps of Engineers revealed that the DDT value for this fish specimen was correctly recorded, but that the fish was one of the few found at the exact location where the manufacturing plant was discharging its toxic waste materials into the river.

Another procedure for detecting outliers is to construct a **box plot** of the data. With this method, we construct intervals similar to the $\bar{x} \pm 2s$ and $\bar{x} \pm 3s$ intervals of the Empirical Rule; however, the intervals are based on a quantity called the **interquartile range** instead of the standard deviation s.

DEFINITION 3.13

The **interquartile range**, **IQR**, is the distance between the upper and lower quartiles:

$$\mathrm{IQR} = Q_U - Q_L$$

You can see from Figure 3.22 that the interquartile range is a measure of data variation. The larger the interquartile range, the more variable the data tend to be.

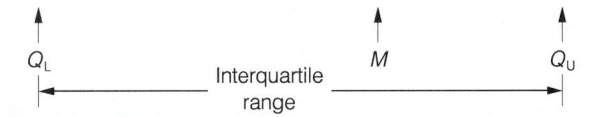

FIGURE 3.22
The Interquartile Range

The box plot procedure is especially easy to use for small data sets because the quartiles and interquartile range can be quickly determined. We will illustrate the procedure in Example 3.17.

EXAMPLE 3.17 Refer to the sale price data of Table 2.2. Construct a box plot for the data and check for outliers.

Solution

STEP 1 Find M, Q_L, Q_U, and IQR. From Example 3.14, $M = 760$, $Q_L = 650$, and $Q_U = 950$; therefore, the interquartile range for the data set is

$$IQR = Q_U - Q_L = 950 - 650 = 300$$

STEP 2 Construct a box with Q_L and Q_U located at the lower corners (see Figure 3.23). The base width will then be equal to the interquartile range.

STEP 3 Locate the **inner fences**, which lie a distance of $1.5(IQR) = 1.5(300) = 450$ below Q_L and above Q_U. These values, $Q_L - 1.5(IQR) = 650 - 450 = 200$ and $Q_U + 1.5(IQR) = 950 + 450 = 1,400$, are located on the box plot shown in Figure 3.23.

STEP 4 Locate the **outer fences**, which lie a distance of $1.5(IQR) =$ below the lower inner fence and above the upper inner fence. Thus, the outer fences for this data set are located at -250 and $1,850$, as indicated in the figure.

STEP 5 Observations that fall between the inner and outer fences (usually indicated by asterisks) are deemed to be **suspect outliers**. Observations falling outside the outer fences (usually indicated by small circles) are judged **highly suspect outliers**. Checking the data set in Table 2.2, you can see that only the observation 1,480 (representing a sale price of $148,000) falls outside the inner fences. Since it lies between the outer fences, it would be judged a suspect outlier, and is located on the box plot with an asterisk (*).

FIGURE 3.23

A Box Plot for the Data of Table 2.2

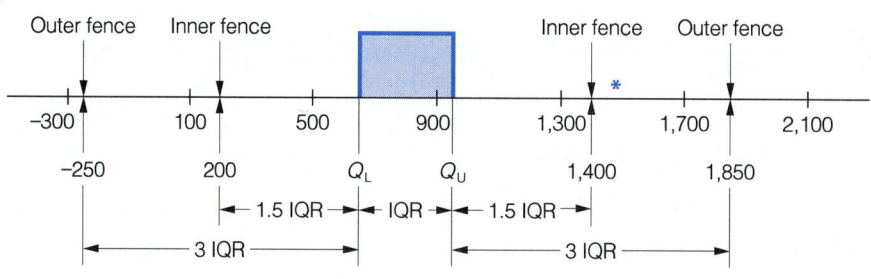

For large data sets, box plots can be constructed using an available statistical computer software package. A computer-generated (Minitab) box plot for the data in Table 2.2 is shown in Figure 3.24. Note that Q_L and Q_U are indicated on the box plot with the symbol I (called a **hinge**). The plus (+) symbol locates the median on the box plot and the asterisk (*) identifies the suspect outlier at 1,480. To further highlight extreme values, Minitab adds dashed lines—called **whiskers**—to the box plot of Figure 3.24. The left whisker extends to the smallest sale price in the region between Q_L and the lower inner fence, and the right whisker extends to the largest price in the region between Q_U and the upper inner fence.

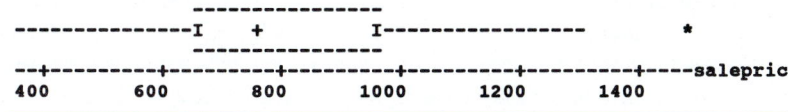

FIGURE 3.24
Minitab Box Plot for the Data of Table 2.2

The z score and box plot methods both establish rule-of-thumb limits outside of which a measurement is deemed to be an outlier (see the accompanying box).

Usually, the two methods produce similar results. However, the presence of one or more outliers in a data set can inflate the computed value of s. Consequently, it will be less likely that an errant observation would have a z score larger than 3 in absolute value. In contrast, the values of the quartiles used to calculate the intervals for a box plot are not affected by the presence of outliers.

EXAMPLE 3.18   Each county in the state of Florida negotiates an annual contract for bread to supply the county's public schools. Sealed bids are submitted by vendors, and the lowest bid (price per pound of bread) is selected as the bid winner. This process works extremely well in competitive markets, but it has the potential to increase the cost of purchasing if the markets are noncompetitive or if collusive practices are present. The latter occurred in the early 1980s in the Florida bread market. In several markets the suppliers of white bread were found guilty of price-fixing, i.e., setting the price of bread several cents above the fair, or competitive, price.

For this example, we have obtained the winning (or low) bid prices for a random sample of 303 white-bread contracts awarded in eight geographic markets over a 6-year period in Florida. (For confidentiality, the specific years and markets are not identified.) Descriptive statistics and a histogram for the data are shown in the SAS printouts, Figures 3.25a and 3.25b.

a. Use the z score method to identify any outliers in the bid-price data.

b. Use the box plot method to identify any outliers in the bid-price data.

Solution

a. The mean and standard deviation of the sample bid prices are shaded on the SAS printout, Figure 3.25a. To check for outliers we would use these values, $\bar{x} = .243$ and $s = .052$, to calculate z scores for all $n = 303$ bid prices in the data set. For the purposes of this example, we will focus on only the largest bid prices in the sample. Notice that the five highest prices are given at the bottom of the printout, Figure 3.25a, in the section titled "Extremes." The z scores for these five prices (corresponding to contracts #17, #303, #233, #224, and #295) are computed as follows:

$$x_{17} = .364 \qquad z = \frac{(.364 - .243)}{.052} = 2.33$$

$$x_{303} = .375 \qquad z = \frac{(.375 - .243)}{.052} = 2.54$$

$$x_{233} = .405 \qquad z = \frac{(.405 - .243)}{.052} = 3.12$$

$$x_{224} = .410 \qquad z = \frac{(.410 - .243)}{.052} = 3.21$$

$$x_{295} = .440 \qquad z = \frac{(.440 - .243)}{.052} = 3.79$$

All five prices have z scores that exceed 2; the prices for contracts #233, #224, and #295 have z scores that exceed 3.

---

a. Descriptive Statistics for Bid-Price Data

FIGURE 3.25

SAS Printouts

UNIVARIATE PROCEDURE

Variable=PRICE

Moments

N	303	Sum Wgts	303
Mean	0.242777	Sum	73.56131
Std Dev	0.051701	Variance	0.002673
Skewness	0.674997	Kurtosis	0.243131
USS	18.66622	CSS	0.807249
CV	21.29577	Std Mean	0.00297
T:Mean=0	81.73874	Prob>|T|	0.0001
Sgn Rank	23028	Prob>|S|	0.0001
Num ^= 0	303		
W:Normal	0.948543	Prob<W	0.0001

Quantiles(Def=5)

100% Max	0.44	99%	0.375
75% Q3	0.2793	95%	0.328889
50% Med	0.23	90%	0.312
25% Q1	0.2052	10%	0.186667
0% Min	0.145	5%	0.173333
		1%	0.153333

Range	0.295
Q3-Q1	0.0741
Mode	0.2

Extremes

Lowest	Obs	Highest	Obs
0.145(	145)	0.364(	17)
0.150267(	1)	0.375(	303)
0.153333(	288)	0.405333(	233)
0.153333(	282)	0.41(	224)
0.153333(	193)	0.44(	295)

b. Histogram and Box Plot for Bid-Price Data

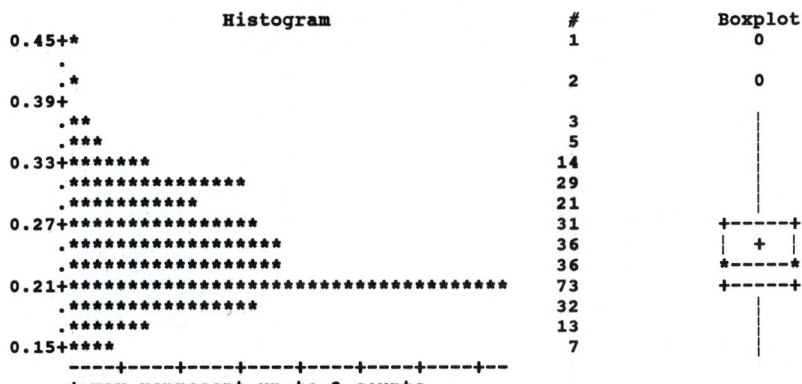

UNIVARIATE PROCEDURE

Variable=PRICE

```
         Histogram                              #    Boxplot
0.45+*                                          1       0
    .
    .*                                          2       0
0.39+
    .**                                         3       |
    .***                                        5       |
0.33+*******                                    14      |
    .***************                            29      |
    .***********                                21      |
0.27+*****************                          31    +-----+
    .********************                       36    |  +  |
    .********************                       36    *-----*
0.21+*****************************************  73    +-----+
    .*****************                          32      |
    .*******                                    13      |
0.15+****                                       7       |
    ----+----+----+----+----+----+----+--
    * may represent up to 2 counts
```

Which of these cutoffs, 2 or 3, should be used with this data set? Since the (horizontal) histogram shown in Figure 3.25b is mound-shaped and only slightly skewed to the high (top) side, the Empirical Rule best describes the distribution of bid prices. Thus, we expect nearly all the bid prices to fall within three standard deviations of the sample mean. Those falling beyond this range are considered outliers and should be investigated for cause. In this example, the bid prices corresponding to contracts #233, #224, and #295 are outliers. Further investigation of these bread contracts may reveal that they were not from the population of competitively bid prices, but were "fixed" during collusion.

b. A SAS box plot for the bid price data set is shown in Figure 3.25b. Like the Minitab box plot, the plus (+) symbol locates the median bid price. However, SAS uses zeros (0) to locate suspect outliers. You can see from the box plot in Figure 3.25b that the two highest bid prices are identified as suspect outliers.

EXERCISES
LEARNING THE MECHANICS

3.51 Construct a box plot for the 24 sample measurements reproduced in Exercise 3.41.

3.52 Construct a box plot for the 28 sample measurements reproduced in Exercise 3.43.

APPLYING THE CONCEPTS

3.53 Recall that Appendix C contains the per-member, per-month costs of a sample of 186 physicians in a managed-care HMO. The mean and standard deviation of the costs are (from Table 3.4) $\bar{x} = \$102.96$ and $s = \$366.31$. One of the physicians in the data set had a per-member, per-month cost of $4,725.10.

  a. Calculate and interpret the z score for this physician's cost. Is the observation an outlier?
  b. A careful examination of this physician's data in Appendix C (observation #166) reveals a total member–months value of 1. In other words, the physician treated only a single patient during a single month of the year at a cost of $4,725.10. Based on this information, how would you classify this outlier?

3.54 Refer to the hotel no-show data of Example 3.12 and Exercise 3.46.

  a. Use the stem-and-leaf display and the results of Exercise 3.46 to form a box plot for the data.
  b. How would you classify the observations that lie between the inner and outer fences of the box plots?
  c. How would you classify the observations that lie outside the outer fences of the box plot?
  d. Use z scores to detect suspect outliers in the data. Do your results agree with the box plot? Explain.

3.55 The accompanying table lists the lymphocyte (LYMPHO) and white blood cell (WBC) count results from hematology tests administered to a sample of 50 black (West Indian or African) workers.
  a. Construct a box plot for the white blood cell count data. Do you detect any outliers or suspect outliers?
  b. Repeat part a for lymphocyte cell counts.

CASE NUMBER	WBC	LYMPHO	CASE NUMBER	WBC	LYMPHO
1	4,100	14	26	4,300	9
2	5,000	15	27	5,200	16
3	4,500	19	28	3,900	18
4	4,600	23	29	6,000	17
5	5,100	17	30	4,700	23
6	4,900	20	31	7,900	43
7	4,300	21	32	3,400	17
8	4,400	16	33	6,000	23
9	4,100	27	34	7,700	31
10	8,400	34	35	3,700	11
11	5,600	26	36	5,200	25
12	5,100	28	37	6,000	30
13	4,700	24	38	8,100	32
14	5,600	26	39	4,900	17
15	4,000	23	40	6,000	22
16	3,400	9	41	4,600	20
17	5,400	18	42	5,500	20
18	6,900	28	43	6,200	20
19	4,600	17	44	4,900	26
20	4,200	14	45	7,200	40
21	5,200	8	46	5,800	22
22	4,700	25	47	8,400	61
23	8,600	37	48	3,100	12
24	5,500	20	49	4,000	20
25	4,200	15	50	6,900	35

Source: Royston, J. P. "Some techniques for assessing multivariate normality based on the Shapiro–Wilk W." *Applied Statistics*, Vol. 32, No. 2, 1983, pp. 121–133.

3.56 Construct a box plot for the data on nuclear scrams discussed in Exercise 3.31. Do you detect any suspect outliers? Highly suspect outliers?

3.57 A computer-generated (Minitab) box plot for the executive salary data of Exercise 3.17 is shown here.

```
         ----------
   -I +  I---------                *                    o
         ----------
+---------+---------+---------+---------+---------+------totalpay
0     10000     20000     30000     40000     50000
```

a. Do you detect any outliers in the sample of 25 corporate executive salaries?
b. Calculate the z scores for any outliers identified in part a. Do the results agree with the box plot approach?

3.58 The data in the accompanying table represent sales, in thousands of dollars per week, for a random sample of 24 fast-food outlets located in four cities.

CITY	WEEKLY SALES (Thousands of Dollars)
A	6.3, 6.6, 7.6, 3.0, 9.5, 5.9, 6.1, 5.0, 3.6
B	2.8, 6.7, 5.2
C	82.0, 5.0, 3.9, 5.4, 4.1, 3.1, 5.4
D	8.4, 9.5, 8.7, 10.6, 3.3

A computer-generated (Minitab) box plot for the data is reproduced here.

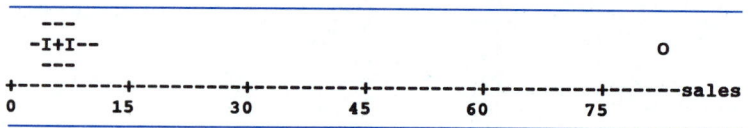

a. Examine the box plot. Do you detect any outliers? If so, identify the city and the weekly sales measurement associated with the outlier.
b. Calculate $\bar{x}$ and s for the sample data. Use this information to compute the z score for the outlier(s) identified in part **a**. Is the result consistent with the box plot constructed in part **a**? Explain.
c. A careful check of the sales records revealed that the weekly sales value for the first fast-food outlet in city C was actually 8.2, but was incorrectly recorded as 82.0. When this recording error is corrected, the Minitab box plot appears as follows. Repeat parts **a** and **b** for the corrected sales data set.

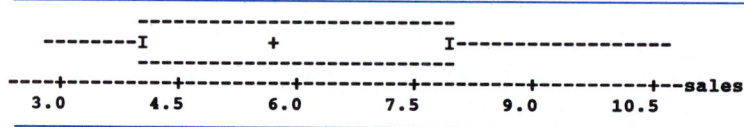

NUMERICAL DESCRIPTIVE MEASURES FOR POPULATIONS

3.11 When you analyze a sample, you are doing so for a reason. Presumably, you want to use the information in the sample to infer the nature of some larger set of data—the population. For example, we might want to know something about the complete set of fall 1989 through spring 1991 graduates of the University of Florida. If we consider the starting salary for each graduate to be the financial compensation that the graduate *would have received* during the year following graduation (if the graduate had secured a job), then the conceptual set of starting salaries for *all* recent University of Florida graduates would be the population of interest to us. However, many of these graduates *did not* secure jobs immediately after graduation, and consequently we can never obtain all the starting salaries that constitute the population. We do know that the entire population of starting salaries has a relative frequency histogram (the exact form of which is unknown to us). We will want to infer the nature of this distribution based on the sample of 1,795 starting salaries contained in Appendix A. It is natural that we would want to use the descriptive measures of this sample to infer the nature of the population.

Numerical descriptive measures that characterize the distribution for a population are called **parameters**. Since we will often use numerical descriptive measures of a sample to estimate the corresponding unknown descriptive measures of the population, we need to make a distinction between the numerical descriptive measure symbols for the population and for the sample.

In our previous discussion, we used the symbols $\bar{x}$ and s to denote the mean and standard deviation, respectively, of a sample of n observations. Similarly, we will use the symbol μ (mu) to denote the mean of a population and the symbol σ (sigma) to denote the standard deviation of a population. As you will subsequently see, we will use the sample mean $\bar{x}$ to estimate the population mean μ, and the sample standard deviation s to estimate the population standard deviation σ. In doing so, we will be using the sample to help us infer the nature of the population relative frequency distribution.

SAMPLE AND POPULATION NUMERICAL DESCRIPTIVE MEASURES

Sample mean: $\bar{x}$ Population mean: μ

Sample standard deviation: s Population standard deviation: σ

Sample z score: $z = \dfrac{x - \bar{x}}{s}$ Population z score: $z = \dfrac{x - \mu}{\sigma}$

SUMMARY

Numerical descriptive measures enable us to construct a mental image of the relative frequency histogram for a data set. The two most important types of numerical descriptive measures are those that measure *central tendency* and *data variation*.

Three numerical descriptive measures are used to locate the "center" of a relative frequency histogram: the *mean*, the *median*, and the *mode*. Each conveys a special piece of information. In a sense, the mean is the balancing point for the data. The median, which is insensitive to *outliers*, divides the data; half of the observations will be less than the median, and half will be larger. The mode is the value of x that occurs with greatest frequency. It is the value of x that locates the point where the relative frequency histogram achieves its maximum relative frequency.

The *range* and the *standard deviation* measure the spread of a distribution. In particular, we can obtain a very good notion of the way data are distributed by constructing the intervals $\bar{x} \pm s$, $\bar{x} \pm 2s$, and $\bar{x} \pm 3s$ and referring to the *Empirical Rule* or *Tchebysheff's theorem*. The percentages of the total number of observations falling within these intervals will be approximately as shown in the accompanying table (page 146).

INTERVAL	EMPIRICAL RULE (MOUND-SHAPED)	TCHEBYSHEFF'S THEOREM
$\bar{x} \pm s$	60% to 80%	At least 0%
$\bar{x} \pm 2s$	95%	At least 75%
$\bar{x} \pm 3s$	Almost 100%	At least 89%

Percentiles, quartiles, and *z scores* measure the relative position of a measurement within a data set. The *lower* and *upper quartiles* and the distance between them can also help us visualize a data set. *Box plots* constructed from intervals based on the *interquartile range* and *z* scores provide an easy way to detect possible outliers in the data.

KEY WORDS

Box plots
Empirical Rule
Grouped data*
Inner fences
Interquartile range
Mean
Measures of central tendency
Measures of data variation or spread
Measures of relative standing
Median
Mode
Numerical descriptive measures
Outer fences
Outliers

Parameters
Percentile
Quartiles
Range
Raw data*
Skewness
Standard deviation
Tchebysheff's theorem
Variance
z score

KEY SYMBOLS

Sample mean: $\bar{x}$
Sample variance: s^2
Sample standard deviation: s
Population mean: μ
Population variance: σ^2
Population standard deviation: σ

---

*From the optional section

KEY FORMULAS

Sample mean: $\bar{x} = \dfrac{\sum x}{n}$

Sample variance: $s^2 = \dfrac{\sum(x - \bar{x})^2}{n-1} = \dfrac{\sum x^2 - \dfrac{(\sum x)^2}{n}}{n-1}$

Sample standard deviation: $s = \sqrt{s^2}$

Interquartile range: $\text{IQR} = Q_U - Q_L$

Sample z score: $z = \dfrac{x - \bar{x}}{s}$

Population z score: $z = \dfrac{x - \mu}{\sigma}$

Sample mean (grouped data):* $\bar{x} = \dfrac{\sum x_i f_i}{n}$

Sample variance (grouped data):* $s^2 = \dfrac{\sum x_i^2 f_i - \dfrac{(\sum x_i f_i)^2}{n}}{n-1}$

*From the optional section

SUPPLEMENTARY EXERCISES

[*Note:* Starred (*) exercises refer to the optional section in this chapter.]

3.59   Research has indicated that the inhabitants of some older cities in the United States may ingest small, but potentially harmful, amounts of lead that is introduced into their drinking water by the use of lead-lined pipes installed in some of the early city water systems. The data reported in the table are the mean lead, copper, and iron contents (milligrams per liter) for samples of water collected on each of 23 days from the Boston water system. The data were collected in 1977 after a sodium hydroxide water treatment system was installed. Each mean is based on approximately 40 measurements taken at different sites in the Boston water system where lead pipe was still being used.

LEAD				COPPER				IRON			
.035	.038	.015	.021	.12	.16	.04	.08	.20	.17	.14	.15
.060	.049	.015	.036	.18	.14	.04	.05	.33	.17	.12	.13
.055	.073	.022	.016	.10	.07	.05	.05	.22	.23	.12	.14
.035	.047	.043	.010	.07	.07	.07	.04	.17	.18	.16	.11
.031	.031	.030	.020	.08	.08	.10	.04	.15	.25	.17	.11
.039	.016	.019		.09	.07	.04		.19	.14	.13	

Source: Karalekas, P. C., Jr., Ryan, C. R., and Taylor, F. B. "Control of lead, copper, and iron pipe corrosion in Boston." *Journal of American Water Works Association*, Vol. 75, No. 2 (Feb. 1983), pp. 92–95. Reprinted by permission. Copyright © 1983, American Water Works Association.

a. Construct a relative frequency histogram for the 23 daily mean lead concentration measurements.

b. Calculate the mean and standard deviation for the sample of part **a**.

c. What proportion of the mean lead concentrations lie in the interval $\bar{x} \pm 2s$?

3.60 Repeat Exercise 3.59 using the copper concentration means.

3.61 Repeat Exercise 3.59 using the iron concentration means.

3.62 What is the relationship between ants and plants? Do ants protect certain plants from insect seed predators? To investigate the phenomenon of ants protecting plants, a naturalist sampled 50 flower heads of a certain sunflower plant that attracts ants. (Each flower head was on a different sunflower plant.) The flower heads were divided into two groups of 25 each. Ants were prevented from reaching the flower heads of one group (this was accomplished by painting ant repellent around the flower stalks of 25 plants). After a specified period of time, the number of insect seed predators on each flower head was counted. The results are summarized in the table. Sketch mental images of the two relative frequency distributions. (The Empirical Rule will help you to do this.) Does it appear that ants protect sunflower plants from insect seed predators? [*Note:* We will present a statistically valid method for answering this question in Chapters 9–10.]

	PLANTS WITH ANTS	PLANTS WITHOUT ANTS
Mean number of predators per flower head	2.9	7.6
Standard deviation	2.4	4.4

3.63 Refer to the fire deaths data given in Exercise 3.13. The MGM Grand Fire in Las Vegas was treated separately in the *Risk Management* analysis because of the size of the high-rise hotel and other unique factors. Do the data support treating the MGM Grand fire deaths differently than the other measurements in the sample? Explain.

3.64 Industrial engineers periodically conduct "work measurement" analyses to determine the time used to produce a single unit of output. At a large processing plant, the total number of man-hours required per day to perform a certain task was recorded for 50 days. This information will be used in a work measurement analysis. The total number of man-hours required for each of the 50 days is listed here, accompanied by a SAS printout (page 149) summarizing and describing the data.

128	119	95	97	124	113	109	124	132	97
146	128	103	135	114	124	131	133	131	88
100	112	111	150	117	128	142	98	108	120
138	133	136	120	112	109	100	111	131	113
118	116	98	112	138	122	97	116	92	122

a. Locate the mean, median, and mode of the data set on the printout and interpret their values.

b. Locate the range, variance, and standard deviation of the data set on the printout, and interpret their values.

c. Construct the intervals $\bar{x} \pm s$, $\bar{x} \pm 2s$, and $\bar{x} \pm 3s$. Count the number of observations that fall within each interval and find the corresponding proportions. Compare the results to the Empirical Rule. Do you detect any outliers?

3.65 The variation of the rates of return on a bond is often used to measure the level of risk associated with buying the bond—the greater the variation, the higher the level of risk. The accompanying table presents a portion of the results of a simulation study to compare the performance of zero-coupon bonds (those for which the interest coupons have been removed and therefore pay no interest) to bonds with coupon payments attached. The rates of return on zero-coupon bonds are determined solely by the changes in bond prices, while the return on coupon bonds depends also on the market interest rate. The simulation study yielded means and standard deviations of the rates of return for both zero-coupon and coupon bonds at a fixed interest rate. The results for a market interest rate of 14% are shown in the table on the next page.

UNIVARIATE PROCEDURE

Variable=MANHOURS

Moments

N	50	Sum Wgts	50
Mean	117.82	Sum	5891
Std Dev	15.01114	Variance	225.3343
Skewness	0.00906	Kurtosis	-0.69123
USS	705119	CSS	11041.38
CV	12.74074	Std Mean	2.122896
T:Mean=0	55.49967	Prob>$|T|$	0.0001
Sgn Rank	637.5	Prob>$|S|$	0.0001
Num ^= 0	50		

Quantiles(Def=5)

100% Max	150	99%	150
75% Q3	131	95%	142
50% Med	117.5	90%	137
25% Q1	109	10%	97
0% Min	88	5%	95
		1%	88
Range	62		
Q3-Q1	22		
Mode	97		

Extremes

Lowest	Obs	Highest	Obs
88(	44)	138(	7)
92(	40)	138(	49)
95(	21)	142(	16)
97(	42)	146(	3)
97(	31)	150(	35)

```
Stem Leaf                        #     Boxplot
 15 0                            1        |
 14 26                           2        |
 13 1112335688                  10     +-----+
 12 0022444888                  10     |     |
 11 1122233466789               13     *--+--*
 10 003899                       6     +-----+
  9 2577788                      7        |
  8 8                            1        |
    ----+----+----+----+
Multiply Stem.Leaf by 10**+1
```

Table for Exercise 3.65

	ZERO-COUPON BONDS	COUPON BONDS
Mean	12.48%	12.48%
Standard deviation	20.80%	14.56%

Source: Potter, T. "Your finances." Reprinted with permission from the August 1984 issue of the *ABA Journal, The Lawyer's Magazine*, published by the American Bar Association.

a. Use the Empirical Rule to sketch the relative frequency histograms for the rates of return of both zero-coupon bonds and coupon-attached bonds.
b. Compute the z score for a zero-coupon bond with a rate of return of -20%. Interpret this value.
c. Compute the z score for a coupon-attached bond with a rate of return of -20%. Interpret this value.

3.66 Neuropsychologists often measure the impact of brain damage on a patient's verbal skills by means of a verbal fluency test. One such test requires subjects to produce as many words as they can that begin with a particular letter. The production figures for one such study are summarized in the table.

	MEAN PRODUCTIVITY	STANDARD DEVIATION
Normal patients	19.30	5.73
Brain-damaged patients	9.20	2.50

a. A particular brain-damaged patient produced eight words. Find the z score for this observation and interpret the value.

b. Would you expect a normal patient to produce a total of eight words? Explain.

3.67 A simulation study was conducted to investigate the rounding accuracies of several new algorithms for functions in the FORTRAN computer program library (*IBM Journal of Research and Development*, Mar. 1986). For each new FORTRAN function, the rounding error for each of 10,000 trials was calculated.

a. The 99th percentile of the rounding errors for one particular FORTRAN function was found to be .53. Interpret this value.

b. For the FORTRAN function of part **a**, the mean rounding error is $\bar{x} = .22$ and the standard deviation (estimated) is $s = .07$. Use this information to calculate the z score for the 99th percentile of .53. Interpret the z score.

*3.68 The time patients spend waiting in a physician's office before they receive health care services, i.e., *patient waiting time*, plays an important role in the efficient operation of the physician's practice. The number of minutes (after their appointment times) that each of a sample of 55 patients had to wait in a dentist's office before being served is shown in the table.

WAITING TIME (Minutes)	NUMBER OF PATIENTS
0– 3.5	12
3.5– 7.0	11
7.0–10.5	7
10.5–14.0	15
14.0–17.5	6
17.5–21.0	3
21.0–24.5	1
	$n = 55$

a. Calculate approximate values for the mean waiting time and standard deviation of waiting times from this sample.

b. Form the interval $\bar{x} \pm 2s$. Using the frequency table, determine the number of patients in the sample with waiting times in this interval.

c. Compare the results of part **b** with the Empirical Rule.

3.69 Laws to protect infants and children as occupants of motor vehicles, i.e., *child restraint laws*, have been in effect in all 50 states since 1985. However, there is little uniformity among the laws, and in each state some children are excluded from coverage because of age, vehicle type, seating position, unregistered vehicles, etc. As part of a study to show the limiting effects of exemptions to the coverage of child restraint laws, the number of children (0–5 years) killed in motor vehicles in each state over a 5-year period was recorded, and the percentage of deaths not covered because of exemptions was calculated. The state percentages are given in the table.

a. Calculate the mean and standard deviation for the data.

b. Calculate the proportion of states with percentages falling within one standard deviation of the mean.

STATE	PERCENTAGE	STATE	PERCENTAGE	STATE	PERCENTAGE	STATE	PERCENTAGE
AL	41	IN	27	NE	45	SC	37
AK	14	IA	10	NV	11	SD	37
AZ	49	KS	70	NH	9	TN	41
AR	44	KY	38	NJ	5	TX	30
CA	27	LA	37	NM	40	UT	50
CO	47	ME	55	NY	27	VT	50
CT	29	MD	42	NC	76	VA	47
DE	29	MA	8	ND	53	WA	18
FL	15	MI	32	OH	26	WV	26
GA	50	MN	36	OK	17	WI	46
HI	8	MS	68	OR	20	WY	60
ID	46	MO	50	PA	36		
IL	17	MT	57	RI	0		

Source: Teret, S. P., Jones, A. S., Williams, A. F., and Wells, J. K. "Child restraint laws: An analysis of gaps in coverage." *American Journal of Public Health*, Vol. 76, No. 1, Jan. 1986, p. 33 (Table 3).

c. Calculate the proportion of states with percentages falling within two standard deviations of the mean.

d. Calculate the proportion of states with percentages falling within three standard deviations of the mean.

e. Which rule, Tchebysheff's theorem or the Empirical Rule, best describes the distribution of state percentages of child motor vehicle deaths?

3.70 It is well known that worker absenteeism is costly and leads to decreased production efficiency at most firms. A study was recently conducted to investigate employee absenteeism at a medium-size assembly and packaging plant in Great Britain. Workers in each of three departments—packaging, assembly, and maintenance—were monitored over a 2-year period. The number of days of unanticipated absences because of sickness was recorded each week. The accompanying table gives the mean and standard deviation of number of days absent per 100 employees for each department.

	PACKAGING DEPARTMENT	ASSEMBLY DEPARTMENT	MAINTENANCE DEPARTMENT
Mean number of days per 100 employees	39.24	17.38	14.56
Standard deviation	9.88	5.16	6.88

Source: Moch, M. K. and Fitzgibbons, D. E. "The relationship between absenteeism and production efficiency: An empirical assessment." *Journal of Occupational Psychology*, Vol. 58, 1985, pp. 39–47.

a. Use the information in the table to sketch mental images of the three relative frequency histograms. Construct them on the same graph so that you can see how they appear relative to each other.

b. Estimate the proportion of weeks in which between 19.48 and 59.00 days of unanticipated absences per 100 workers occur in the packaging department.

c. In a typical week, how many days of unanticipated absences per 100 workers would you expect to occur in the assembly department?

d. Repeat part c for the maintenance department.

3.71 *Muck* is a rich, highly organic type of soil that serves as a growth medium for most vegetation in the Florida Everglades. Because of its high concentration of organic material, muck can be destroyed by drought, fire, and windstorms. During a recent drought in south Florida, the Everglades lost a considerable amount of muck. To assess the loss, members of the Florida Fish and Game Commission marked 40 plots with stakes at various locations in the Everglades, and measured the depth of the muck (in inches) at each stake. The data are given on the next page.

27	30	40	21	41	23	24	30	26	40
35	32	35	7	37	19	30	28	39	33
33	23	35	22	15	32	26	38	26	30
45	15	29	57	27	27	18	16	31	36

a. Compute the mean, median, and mode of the data set.
b. Find the range, variance, and standard deviation of the data set. (Use the shortcut procedure to calculate s^2.)
c. Construct the intervals $\bar{x} \pm s$, $\bar{x} \pm 2s$, and $\bar{x} \pm 3s$. Count the number of observations falling within each interval and find the corresponding proportions. Compare the results to the Empirical Rule.
d. Find the 10th percentile for the 40 muck depths.
e. Construct a box plot for the data. Use the plot to detect any suspect outliers in the data.

3.72   Behavior geneticists are scientists who study the ways in which genetic factors influence behavior. One characteristic that has been studied is the "emotional" behavior of different strains of rats. Emotional behavior is sometimes defined as the tendency to "freeze"—i.e., not move—when presented with a new situation. For rats from different strains that are raised under identical circumstances, differences in emotional behavior may suggest a genetic base. Summary statistics on the number of meters traversed by 500 rats of a particular strain when put in a box with a bright light and white noise are listed here.

$$\bar{x} = 7.65 \qquad s = .88 \qquad M = 7.3$$

a. Find the 50th percentile of the distribution of the number of meters traversed by this sample of rats.
b. Compute the interval $\bar{x} \pm 2s$. Use the Empirical Rule to approximate the percentage of rats that had distance measurements within this interval.
c. Would you expect to observe a rat of this particular strain traverse a distance of 9.50 meters?

3.73   Each year Video Storyboard Tests conducts a national survey of approximately 30,000 television viewers. Each person surveyed is asked to name the "most outstanding" television commercial they have seen. The products are then ranked according to both popularity of commercials and advertising expenditure. In 1986, Video Storyboard Tests also conducted a separate survey of approximately 4,000 television viewers who were regular users of a product. They were asked to recall a commercial they had seen for that product category in the past week. The television advertising budget for a brand was then divided by the number of times the brand was mentioned in the survey to develop a measure that Video Storyboard calls "cost per 1,000 retained impressions," or "cost efficiency." The table on the next page shows the advertising expenditures (in millions of dollars) and cost efficiencies for the top 25 product brands in 1986. The brands are ranked according to the popularity of their television commercials. Assume that the 25 brands represent a sample of products popularized by their television commercials.

a. Calculate the mean and standard deviation of the television advertising expenditures.
b. Calculate the mean and standard deviation of the cost efficiencies. [*Note:* The cost efficiencies for 12 companies were not available (N.A.)]
c. Calculate a measure of relative standing for Miller Lite's $70.9 million television advertising budget. Interpret this value.
d. Calculate a measure of relative standing for Miller Lite's cost efficiency of $42.89 per 1,000 retained impressions. Interpret this value.
e. According to the *Wall Street Journal*, ". . . Adolph Coor's debut in the top 25 indicates that public-service type ads may be a good marketing strategy. In the Coor's spot that viewers praised most, a bartender takes the car keys from his buddy who has had one too many beers." Repeat parts c and d for Coor's television advertising expenditures and cost efficiencies, respectively.

3.74   Research shows that children born to couples in their mid or late 30's have a greater risk of becoming senile in old age. A study of 80 elderly patients with atrophy of the prefrontal brain areas—the most prevalent cause of senility in the United States—indicated that the average age of the patients' mothers at the time of their births was 36.5 years, 10 years older than the average for first-time parents. Assume that the standard deviation of the 80 ages was 7.2 years.

RANK	BRAND (AGENCY)	TV SPENDING (IN MILLIONS)	COST EFFICIENCY
1	Coca-Cola (McCann-Erickson/SSC&B)	$ 48.3	$ 12.07
2	McDonald's (Leo Burnett)	321.2	64.67
3	California raisins (Foote, Cone & Belding)	5.8	N.A.
4	Miller Lite (Backer & Spielvogel)	70.9	42.89
5	Pepsi-Cola (BBDO)	42.1	10.29
6	Bartles & Jaymes (Hal Riney & Partners)	30.3	N.A.
7	Bud Light (DDB Needham)	52.3	42.99
8	Burger King (J. Walter Thompson)	165.2	43.40
9	Jell-O (Young & Rubicam)	44.5	N.A.
10	Isuzu (Della Femina, Travisano)	29.1	N.A.
11	Kibbles'n Bits (J. Walter Thompson)	10.0	23.13
12	Levi's (Foote, Cone & Belding)	18.1	7.47
13	Snuggle (SSC&B)	22.0	N.A.
14	Coors (Foote, Cone & Belding)	72.1	73.00
15	Honda scooters (Wieden & Kennedy)	12.7	N.A.
16	Wendy's (DFS Dorland)	79.6	40.26
17	Seagram's wine cooler (Ogilvy & Mather)	23.6	N.A.
18	Taco Bell (Tracy-Locke)	50.2	102.76
19	Huggies (Ogilvy & Mather)	20.2	N.A.
20	Sprite (Lowe Marschalk/Carden & Cherry)	24.0	34.40
21	French's mustard (J. Walter Thompson)	2.1	N.A.
22	Acutrim (Ally Gargano/MCA)	8.3	N.A.
23	Pizza Hut (Chiat/Day)	71.3	53.52
24	Kodak (J. Walter Thompson)	43.4	N.A.
25	Hefty (Wells, Rich, Greene)	10.5	N.A.

Source: Alsop, R. "New Coke is a smash success with consumers in this poll." *Wall Street Journal*, Feb. 26, 1987, p. 25.

a. Use the Empirical Rule to describe the relative frequency distribution of the ages of the 80 elderly patients' mothers at the time of the patients' births.

b. Approximately what percentage of the patients in the study had mothers who were between 22.1 and 50.9 years old at the time of the patients' births?

c. The median age of the 80 patients' mothers at the time of their births was 35.5 years. What percentage of the mothers were younger than 35.5 years at the time of the patients' births?

d. One of the 80 senile patients is selected arbitrarily. Would you expect the age of this patient's mother at the time of the patient's birth to be 20 years or less? Explain.

*3.75 Paper and lumber companies pay for timber by the weight per truckload of 16-foot logs. Consequently, an investor in forest land needs to estimate the total weight of logs that can be produced by a property. This is done by *cruising the property* and counting the total number[†] of trees capable of producing 16-foot logs. A sample of trees (usually 10% of the total number) is selected from this group, and the diameter at chest height and the number of logs per tree (a visual guess) are recorded for each. A forester can then use the diameter and logs-per-tree measurements to calculate the *approximate* weight for each tree. The data in the table were obtained from "cruising" an actual 40-acre tract of short-leaf pine timber located in western Arkansas.[‡] The chest-height diameters and logs per tree were measured for each of 117 trees on the tract. The trees were then grouped according to diameter (10, 11, . . . , 15, or 16 inches), and an estimate of the weight per tree in each diameter group (based on the

---

[†]This number is usually close to, but not actually equal to, the exact number of trees. For the purpose of this discussion, we will assume that it is an exact one.

[‡]Timber data and information courtesy of Delton F. Price, Fort Smith, Arkansas.

average of the logs-per-tree measurements) was calculated. Assuming that the weights in the table represent the midpoints of various weight class intervals (e.g., 580 pounds could represent the midpoint of the class interval 579.5–580.5, 750 pounds could represent the midpoint of the class interval 749.5–750.5, etc.), compute the approximate sample mean weight $\bar{x}$ and the sample standard deviation s for the 117 trees. [*Note:* There are numerous class intervals (e.g., 580.5–581.5, 581.5–582.5, ... , 748.5–749.5) that have a class frequency of 0.]

DIAMETER AT CHEST HEIGHT (Inches)	ESTIMATED (AVERAGE) WEIGHT (Pounds per Tree)	NUMBER OF TREES
10	580	38
11	750	34
12	1,100	21
13	1,800	15
14	2,000	5
15	2,660	3
16	3,000	1
		$n = 117$

3.76 Many people living in metropolitan areas receive impressions of what is happening in their area primarily through their major newspapers. A study was conducted to determine whether the *Uniform Crime Report*, compiled by the U.S. Federal Bureau of Investigation, and the daily newspaper give consistent information about the trend and distribution of crime in a metropolitan area. An attention score, based on the amount of space devoted to a story, was calculated for each paper's coverage of murders, assaults, robberies, etc. Suppose the murder attention scores of metropolitan newspapers across the country had a mean of 62 and a standard deviation of 7.

a. Use the Empirical Rule to determine the approximate percentage of metropolitan newspapers that had murder attention scores between 41 and 83.

b. Suppose your metropolitan newspaper had a murder attention score of 76. Describe its rank among all metropolitan newspapers.

CASE STUDY 3.1

ARE CEOs REALLY WORTH THEIR PAY?

How much are the top corporate executives being paid, and are they worth it? To answer these questions, *Business Week* magazine compiles its Executive Compensation Scoreboard each year based on a survey of chief executive officers (CEOs) at the highest ranking companies listed in the *Business Week 1000*. Among the 356 companies surveyed in 1990, over half of the executives "joined the million-dollar club by earning total compensation of more than $1 million" in 1989.*

The top spot in the survey was claimed by Craig McCaw, the 40-year-old chairman and founder of McCaw Cellular Communications. According to *Business Week*, McCaw earned "$53.9 million in salary, bonuses, and stock options—even though his company has never earned a profit!" Interestingly, only $298,000 of McCaw's 1989 compensation was in cash. The remainder is from the exercise of stock options acquired by McCaw before he took the company public.

Source: "Executive pay: Who made the most . . . and are they worth it?" *Business Week*, May 7, 1990.

To determine which executives are worth their pay, *Business Week* calculates several useful benchmarks as part of the Executive Compensation Scoreboard. One of these, the focus of this case study, is the ratio of total shareholder return (measured by the dollar value of a $100 investment in the company made 3 years earlier) to total executive pay (in thousands of dollars). For example, a $100 investment in Walt Disney Corporation in 1987 was worth $162 at the end of 1989. When this shareholder return ($162) is divided by Disney CEO Michael Eisner's total 1989 compensation in thousands ($56,413), the result is a shareholder return-to-pay ratio of only .003—one of the lowest among all other executives in the survey. The lower the return-to-pay ratio, the less likely the CEO is worth his pay to prospective investors in the firm.

The CEOs in the 1990 scoreboard are categorized by nine industry groups: industrial high-tech, industrial low-tech, consumer products, financial, transportation, telecommunications, services, utilities, and resources. The data for the 36 CEOs in the industrial high-tech group are listed in Table 3.9. A SAS printout describing the return-to-pay ratios of CEOs in the industrial high-tech (IHT) industry group is shown in Figure 3.26 on the following page. Parts a–f refer to this printout.

TABLE 3.9

Executive Compensation Scoreboard: Industrial High-Tech Industrial Group

OBS	CEO	COMPANY	RATIO	OBS	CEO	COMPANY	RATIO
1	Shrontz	Boeing	0.0626	19	Sculley	Apple Computer	0.0129
2	Pace	General Dynamics	0.0208	20	Weston	Automatic Data Processing	0.0374
3	Tellep	Lockheed	0.0492	21	Canion	Compaq Computer	0.0771
4	Augustin	Martin Marietta	0.0303	22	Rollwage	Cray Research	0.0109
5	McDonnel	McDonnell Douglas	0.0304	23	Haverty	Deluxe	0.0604
6	Daniell	United Technologies	0.0317	24	Olsen	Digital Equipment	0.0066
7	Raab	AMP	0.0507	25	Young	Hewlett-Packard	0.0162
8	Cizik	Cooper Industries	0.0250	26	Akers	IBM	0.0172
9	Knight	Emerson Electric	0.0269	27	Shirley	Microsoft	0.0881
10	Hartley	Harris	0.0417	28	Exley	NCR	0.0179
11	Grove	Intel	0.0316	29	Bills	Novell	0.3266
12	Hoch	Litton Industries	0.0159	30	Ellison	Oracle Systems	0.2365
13	Weisz	Motorola	0.0405	31	Harvey	Pitney Bowes	0.0291
14	Kirschne	National Service Industries	0.0769	32	McNealy	Sun Microsystems	0.0846
15	Junkins	Texas Instruments	0.0301	33	Treybig	Tandem Computers	0.0737
16	Marous	Westinghouse Electric	0.0214	34	Roach	Tandy	0.0359
17	Renier	Honeywell	0.0425	35	Blumenth	Unisys	0.0053
18	Keyes	Johnson Controls	0.0452	36	Kearns	Xerox	0.0207

a. Find and interpret the measures of central tendency give on the printout.
b. Find and interpret the measures of variation given on the printout.
c. Find and interpret the measures of relative standing given on the printout.
d. Is the distribution of return-to-pay ratios for industrial high-tech CEOs best described by the Empirical Rule or Tchebysheff's theorem? Explain.
e. Do you detect any outliers in the data set? If so, identify them.

INDUSTRY = IHT

UNIVARIATE PROCEDURE

Variable=RATIO

Moments

N	36	Sum Wgts	36
Mean	0.050848	Sum	1.830513
Std Dev	0.061868	Variance	0.003828
Skewness	3.433158	Kurtosis	12.91788
USS	0.227046	CSS	0.133968
CV	121.6737	Std Mean	0.010311
T:Mean=0	4.931223	Prob>$|T|$	0.0001
Sgn Rank	333	Prob>$|S|$	0.0001
Num ^= 0	36		

Quantiles(Def=5)

100% Max	0.326558		99%	0.326558
75% Q3	0.055599		95%	0.236545
50% Med	0.031646		90%	0.084606
25% Q1	0.020777		10%	0.012893
0% Min	0.005286		5%	0.006588
			1%	0.005286
Range	0.321272			
Q3-Q1	0.034822			
Mode	0.005286			

Extremes

Lowest	Obs	Highest	Obs
0.005286(	35)	0.077095(	21)
0.006588(	24)	0.084606(	32)
0.010857(	22)	0.088113(	27)
0.012893(	19)	0.236545(	30)
0.015875(	12)	0.326558(	29)

```
Stem Leaf                              #        Boxplot
   3 3                                 1           *
   2
   2 4                                 1           *
   1
   1
   0 5556678889                       10        +--+--+
   0 111122222222333333344444         24        *-----*
     ----+----+----+----+----
     Multiply Stem.Leaf by 10**-1
```

f. Identify the CEOs that fall below the 10th percentile of return-to-pay ratios for this industry group.

g. Delete the outliers identified in part e and recalculate the mean, median, mode, range, variance, and standard deviation. Which measure of central tendency is most affected by the deletion of outliers? Which measure of variation?

CONSUMER COMPLAINTS: DUE TO CHANCE OR SPECIFIC CAUSES?

Statistics plays a key role in monitoring the quality of a manufactured product and in controlling the quality of products shipped to customers. The term *statistical quality control* is now an integral part of any successful firm's day-to-day operation.

In this case study, we present a simple, yet powerful quality control procedure. J. Namias (1964) was one of the first to suggest using the technique to control the level of consumer complaints.* In Namias' application, the rate of consumer complaints about a product is used to determine when—and when not—to conduct a search for specific causes of consumer complaints.

The consumer complaint rate may change or vary merely as a result of chance or fate, or it may be due to some specific cause, such as a decline in the quality of the product. According to Namias,

> As long as the results exhibit chance variability, the causes are common, and there is no need to attempt to improve the product by making specific changes. Indeed, this may only create more variability, not less, and may inject trouble where none existed, with waste of time and money. . . . On the other hand, time and money are again wasted through failure to recognize specific conditions when they arise. It is therefore economical to look for a specific cause when there is more variability than is expected on the basis of chance alone.

Namias determined that the complaint rate of a product (e.g., the number of customer complaints per 10,000 units sold) has a relative frequency histogram that is approximately mound-shaped. This leads to a *decision rule* with which to determine when the observed variation in the rate is due to chance and when it is due to specific causes.

DECISION RULE

If the observed complaint rate is within two standard deviations of the mean rate of complaint, it is attributed to chance. If the observed rate is farther than two standard deviations above the mean rate, it is attributed to a specific problem in the production or distribution of the product.

The reasoning is that if there are no problems with the production and distribution of the product, then 95% of the complaint rates should be within two standard deviations of the mean rate. If the production and distribution process were operating normally, it would be very unlikely for a rate higher than two standard deviations above the mean to occur. Instead, it is more likely that the

*Namias, J. "A method to detect specific causes of consumer complaints." *Journal of Marketing Research*, Aug. 1964, pp. 63–68. By permission of the American Marketing Association.

high complaint rate is caused by abnormal operation of the production or distribution process—that is, something specific is wrong with the process.

Namias collected data from the records of a beverage company for a 2-week period to demonstrate the effectiveness of the rule. Consumer complaints primarily concerned chipped bottles that looked dangerous. For one of the firm's brands, the complaint rate was determined to have a mean of 26.01 per 10,000 bottles sold and a standard deviation of 11.28. The complaint rate observed during the two weeks under study was 93.12 complaints per 10,000 bottles sold.

a. Which rule, the Empirical Rule or Tchebysheff's theorem, is Namias' decision rule based on? Why?

b. Compute the z score for the observed rate of 93.12.

c. Give a general interpretation of the z value computed in part **b**.

d. Use the Namias decision rule to determine whether the observed rate is due to chance or whether it is due to some specific cause. (In actuality, a search for a possible problem in the bottling process led to a discovery of rough handling of the bottled beverage in the warehouse by newly hired workers. As a result, a training program for new workers was instituted.)

CASE STUDY 3.3

BID-COLLUSION IN THE HIGHWAY CONTRACTING INDUSTRY

Many products and services are purchased by governments, cities, states, and businesses on the basis of competitive bids, and frequently contracts are awarded to the lowest bidders. As we learned in Example 3.18, such a process is prone to price-fixing or bid-collusion by the bidders. Recall that price-fixing involves setting the bid price above the fair, or competitive, price to increase profit margin.

Numerous methods exist for detecting the possibility of collusive practices among bidders. According to Rothrock and McClave (1979), these procedures involve detecting significant departures from normal market conditions, such as (1) systematic rotation of the winning bid, (2) stable market shares over time, (3) geographic market divisions, (4) lack of relationship between delivery costs and bid levels, (5) high degree of uniformity and stability in bid levels over time, and (6) presence of a baseline point pricing scheme.*

In this case study, we examine a data set collected during the late 1970s and early 1980s involving road construction contracts in the state of Florida. During this time period, the Office of the Florida Attorney General suspected numerous contractors of practicing bid-collusion. The investigation led to an admission of guilt by several of the contractors. Although these contractors were heavily fined, they avoided harsher punishment by identifying which road construction contracts were competitively bid and which involved fixed bids. By comparing the bid prices (and other important bid variables) of the fixed contracts to the competitive contracts, the Florida Attorney General was able to establish invaluable

---

*Rothrock, T. P. and McClave, J. T. "An analysis of bidding competition in the Florida school bidding competition using a statistical model." Paper presented at the TIMS/ORSA Joint National Meeting, Chicago, 1979.

TABLE 3.10

Low-Bid-Estimate Ratios for Competitive and Fixed Contracts

COMPETITIVE CONTRACTS					
0.84706	0.81294	0.84947	0.82045	0.74058	0.95659
0.79604	0.94029	0.83473	1.01686	0.86738	0.83417
0.99504	0.96421	0.56458	1.20942	0.98283	0.85788
0.75519	0.55926	0.82667	0.79416	0.80052	0.53256
0.80993	0.61726	0.87136	0.70124	0.84679	0.58660
0.85818	1.02779	1.02066	0.82872	0.56154	0.80604
0.83678	0.88429	0.67427	0.85418	0.80769	0.82356
0.90790	0.84337	0.97863	0.85910	0.95524	0.99704
0.93263	0.98418	1.01218	0.91837	0.75132	0.56958
0.76891	0.98350	0.90747	0.86848	0.87175	1.06275
1.03847	0.81453	0.97092	0.84639	0.91569	0.80036
0.74178	1.15942	0.92574	1.05129	0.93190	0.94623
1.02971	1.00238	0.99094	1.08052	0.82088	0.86044
0.91729	0.99481	0.90621	1.04081	0.96586	0.93358
0.75446	0.92454	0.90958	0.81181	1.04067	0.85952
1.02410	0.91999	0.66151	0.77619	0.91446	0.99429
0.75085	0.94714	0.79414	1.23987	0.91706	1.17211
0.94463	0.95872	0.86617	0.70829	0.63516	0.88898
0.94505	0.86283	0.89214	1.03969	0.81636	0.76920
1.28562	0.89310	0.78920	0.86720	0.72732	1.00800
0.76011	0.74089	1.12640	0.84523	0.90192	0.76520
1.06395	0.96832	1.05070	0.86668	0.81801	0.95358
0.79852	1.17036	1.05659	0.96512	1.04657	0.78969
0.91536	0.87965	0.99660	0.93633	0.79439	0.98118
0.98819	0.99483	1.10669	1.03936	1.08561	1.11340
0.92861	1.07797	1.03524	1.12084	1.00052	1.02635
1.35009	0.96836	1.10043	0.91859	1.02384	0.86765
0.91293	1.30215	0.92623	0.88246	0.96999	0.97556
0.87291	0.91385	0.87678	0.91367	0.89729	1.05454
0.87506	1.02810	0.79883	0.98221	0.99722	0.85552
0.80171	0.83699	0.79261	0.77651	0.66907	0.82356
1.01985	1.04099	0.93329	0.99935	0.98071	0.94421
0.86350	0.80633				

FIXED CONTRACTS					
1.05701	1.18705	0.99505	1.56376	1.19798	1.34676
1.11814	1.07163	0.84640	0.93359	1.14827	1.09177
0.86693	1.52402	1.10792	1.23126	1.23735	0.88570
1.21388	0.99578	1.20061	1.16075	0.83974	1.03118
1.34423	1.17919	1.25092	1.02069	1.05933	0.96487
1.01136	1.16030	0.89721	0.94805	1.33058	1.33976
1.19888	1.17608	1.17053	1.10921	1.29868	1.26369
1.10506	1.12001	0.97225	1.03805	1.34698	1.18351
1.33447	1.04488	1.00582	0.87470	1.21535	0.98325
1.05964	1.41907	0.98426	1.07672	1.03658	1.10761
1.01960	1.13592	1.29874	1.10312	1.15079	1.09186
1.17556	1.28360	1.07151	1.02573	0.99354	1.22666
1.08694	0.99070	1.08237	0.81352	1.07066	1.09208
0.96967	1.05283	1.21457	0.91678	1.12155	1.04438
1.03024					

benchmarks for detecting bid-rigging in the future. (In fact, the benchmarks led to a virtual elimination of bid-rigging in road construction in the state.)

Table 3.10 (on the previous page) shows two of the many variables measured for each contract in the data set. STATUS is a qualitative variable representing the bid status (fixed or competitive) of the contract. LBERATIO is a quantitative variable representing the ratio of the winning (low) bid to the Department of

```
----------------------------- STATUS = COMPET -----------------------------
                           UNIVARIATE PROCEDURE

        Variable=LBERATIO        LOW BID/ESTIMATE(DOT) RATIO

                              Moments

                N             194   Sum Wgts          194
                Mean     0.907248   Sum          176.0062
                Std Dev  0.136927   Variance     0.018749
                Skewness 0.034461   Kurtosis     0.918657
                USS      163.2998   CSS           3.61855
                CV       15.09254   Std Mean     0.009831
                T:Mean=0 92.28655   Prob>|T|       0.0001
                Sgn Rank   9457.5   Prob>|S|       0.0001
                Num ^= 0      194
                W:Normal 0.978221   Prob<W         0.2297

                         Quantiles(Def=5)

             100% Max   1.350088        99%   1.302152
              75% Q3     0.99483        95%   1.120844
              50% Med   0.913298        90%   1.054544
              25% Q1    0.820453        10%   0.754458
               0% Min   0.532556         5%   0.669066
                                         1%   0.559258
             Range      0.817532
             Q3-Q1      0.174377
             Mode       0.532556

                              Extremes

             Lowest      Obs      Highest      Obs
           0.532556(      24)  1.209417(       16)
           0.559258(      20)  1.239872(      100)
           0.561543(      35)  1.285623(      115)
           0.564582(      15)  1.302152(      164)
           0.569583(      54)  1.350088(      157)


Stem Leaf                                         #          Boxplot
  13 5                                            1             0
  13 0                                            1             0
  12 9                                            1             0
  12 14                                           2
  11 677                                          3
  11 01123                                        5
  10 5555666889                                  10
  10 0000000112222233334444444444               25
   9 5555666677777788888888899999               27          +-----+
   9 00111111112222222223333333444              29          *--+--*
   8 555555666666666777777777788888999          31          |     |
   8 000000111111112222223333444                26          +-----+
   7 5556677788999999                            16
   7 013444                                       6
   6 677                                          3
   6 24                                           2
   5 66679                                        5
   5 3                                            1             0
     ----+----+----+----+----+----+-
Multiply Stem.Leaf by 10**-1
```

Transportation engineer's estimate of the "fair" bid price. Theoretically, the ratio will be near 1 for competitive bids. The larger the ratio, the more evidence that an unusually high bid (and possibly price-fixing) has occurred.

The data set of Table 3.10 contains 194 competitively bid and 85 fixed contracts—a total of 279 contracts. Descriptive statistics for both sets of contracts are provided in the SAS printouts in Figure 3.27 (competitive) and Figure 3.28

FIGURE 3.28
Descriptive Statistics for Fixed Contracts

```
--------------------------- STATUS = FIXED ---------------------------
                        UNIVARIATE PROCEDURE

        Variable=LBERATIO       LOW BID/ESTIMATE(DOT) RATIO

                            Moments

        N                      85    Sum Wgts             85
        Mean              1.11232    Sum            94.54722
        Std Dev          0.148423    Variance       0.022029
        Skewness         0.513358    Kurtosis       0.511053
        USS              107.0173    CSS            1.850478
        CV               13.34358    Std Mean       0.016099
        T:Mean=0         69.09346    Prob>|T|         0.0001
        Sgn Rank           1827.5    Prob>|S|         0.0001
        Num ^= 0               85
        W:Normal         0.973421    Prob<W           0.2884

                       Quantiles(Def=5)

        100% Max    1.563757         99%    1.563757
         75% Q3     1.198876         95%    1.346765
         50% Med    1.092076         90%    1.330577
         25% Q1     1.019604         10%    0.933592
          0% Min     0.81352          5%    0.874703
                                      1%     0.81352

        Range       0.750236
        Q3-Q1       0.179272
        Mode         0.81352

                           Extremes

          Lowest      Obs      Highest      Obs
        0.81352(       76)   1.346765(       6)
        0.839736(      23)   1.346982(      47)
        0.846403(       9)   1.419068(      56)
        0.866931(      13)   1.524021(      14)
        0.874703(      52)   1.563757(       4)


        Stem Leaf                        #       Boxplot
          15 6                           1          0
          15 2                           1          0
          14
          14 2                           1
          13 55                          2          |
          13 003344                      6          |
          12 568                         3          |
          12 000112334                   9       +-----+
          11 5566788889                 10       |     |
          11 011112224                   9       |  +  |
          10 5666777889999              13       *-----*
          10 0011223334444              13       +-----+
           9 56778899                    8          |
           9 023                         3          |
           8 5779                        4          |
           8 14                          2          |
             ----+----+----+----+
          Multiply Stem.Leaf by 10**-1
```

(fixed). In addition, side-by-side box plots for the two sets of contracts are shown in Figure 3.29.

Use the methods in Chapters 2 and 3 to analyze the road construction contract data. In particular, comment on the belief that fixed bids result in higher low bid-to-estimate ratios than competitively bid contracts.

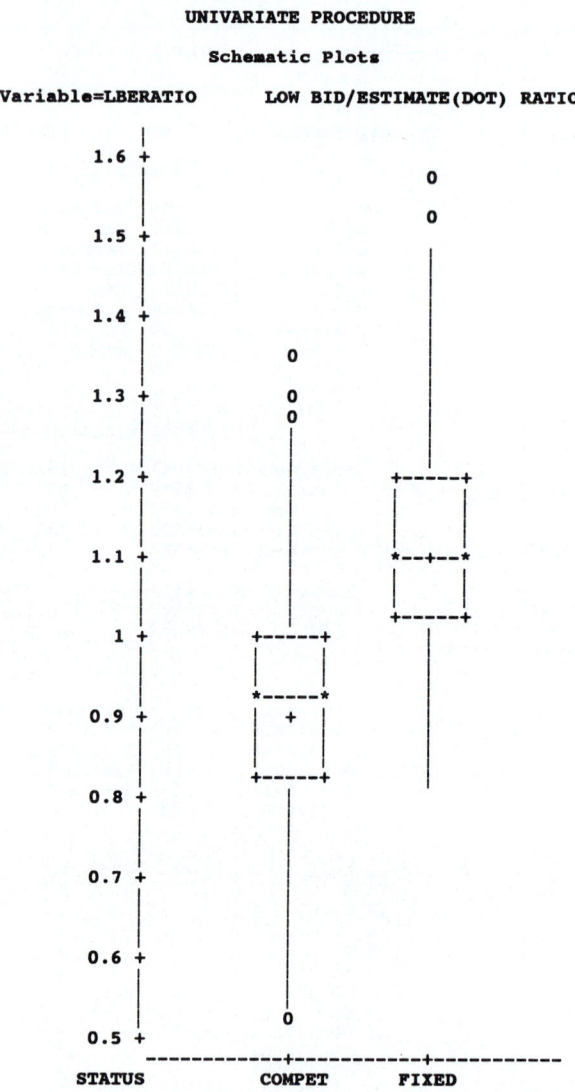

FIGURE 3.29

Box Plots for Competitive and Fixed Contracts

DESCRIPTIVE STATISTICS

The SAS, SPSS, and Minitab commands for producing numerical descriptive measures are provided in this section. The data set of interest for the sample programs is the set of 50 starting salaries given in Table 2.3.

REMINDER The SAS, SPSS, and Minitab commands given here are appropriate for both the mainframe and PC versions of the three software packages, except where noted. When a mainframe computer is being used, however, these statements must be preceded by the JCL commands required at your computing center.

SAS

Command
line

```
1    DATA SAMPLE;        ⎫
2    INPUT SALARY @@;    ⎬  Data entry instructions
3    CARDS;              ⎭

     [Input data values]

4    PROC UNIVARIATE PLOT;  ⎫ Descriptive statistics and box plot
5    VAR SALARY;            ⎭
```

COMMAND 2 The symbols @@ placed at the end of the INPUT statement allow multiple observations to be read on a single input data line as long as the values of the variable SALARY are separated by blanks.

COMMANDS 4–5 The UNIVARIATE procedure in SAS generates descriptive statistics (mean, median, mode, variance, standard deviation, lower quartile, upper quartile, etc.) for the variable specified in the VAR statement (command 5). The PLOT option (command 4) will generate a box plot for the data.

SPSS

Command
line

```
1    DATA LIST FREE/SALARY.  ⎫ Data entry instructions
2    BEGIN DATA.             ⎭

     [Input data values]

3    END DATA.
4    FREQUENCIES VARIABLES=SALARY/                        ⎫ Descriptive
5                STATISTICS=ALL/PERCENTILES=25 75.         ⎭ statistics
```

COMMAND 1 The FREE subcommand permits multiple observations of SALARY to be read on each input data line.

COMMAND 4 The FREQUENCIES command in SPSS produces descriptive statistics for the variable (SALARY) specified.

COMMAND 5 The STATISTICS=ALL subcommand computes the values of the mean, median, mode, range, variance, and standard deviation. The PERCENTILES subcommand calculates the percentiles (25th and 75th) specified after the equals sign.

GENERAL Recall that the periods at the end of the SPSS commands are required when using a personal computer. However, they should be omitted when using mainframe SPSS.

MINITAB

```
Command
  line
    1    SET C1           } Data entry instruction

         [Input data values]

    2    DESCRIBE C1      } Descriptive statistics
    3    BOXPLOTS C1      } Box plot
    4    STOP
```

COMMAND 1 The SET command allows multiple observations of starting salary to be read on the input data lines. These values are stored in C1.

COMMAND 2 The DESCRIBE command prints descriptive statistics (mean, median, standard deviation, lower quartile, and upper quartile) for the values in the specified columns.

COMMAND 3 The BOXPLOTS command generates one box plot for the data in each specified column.

COMPUTER ACTIVITIES

Access the data in Appendix D and perform the following using a statistical software package:

a. Generate numerical descriptive measures (mean, median, standard deviation, etc.) for each of the three quantitative variables LENGTH, WEIGHT, and DDT.
b. Estimate the number of the 144 fish specimens that fall within two standard deviations of the mean for each variable. Which rule, the Empirical Rule or Tchebysheff's theorem, did you apply? Why? Compare your estimates with the actual number obtained by counting.
c. Generate a box plot for each of the three variables LENGTH, WEIGHT, and DDT. Do you detect any outliers?
d. Delete the outliers you detected in part c from the data set. Repeat parts a–c for the new data set. What effect does the deletion of the outliers have on the results?

Freedman, D., Pisani, R., and Purves, R. *Statistics*. New York: W. W. Norton and Co., 1978.

Mendenhall, W. *Introduction to Probability and Statistics*, 8th ed. Boston: Duxbury, 1990.

Tukey, J. *Exploratory Data Analysis*. Reading, Mass.: Addison-Wesley, 1977.

PROBABILITY: BASIC CONCEPTS

CHAPTER 4

How would you like to win a state lottery (Case Study 4.1), choose the grand prize in "Let's Make a Deal" (Case Study 4.3), or obtain a winning edge in blackjack (Section 4.4)? What do each of these ventures have to do with statistics? The answer is **uncertainty**. The return in real dollars on most investments cannot be predicted with certainty. Neither can we be certain that an inference about a population, based on the partial information contained in a sample, will be correct. In this chapter we shall learn how probability can be used to measure uncertainty, and we shall take a brief glimpse at its role in assessing the reliability of statistical inferences.

CONTENTS

THE ROLE OF PROBABILITY IN STATISTICS

4.1 If you play blackjack, a popular gambling game, you know that whether you win in any one game is an outcome that is very uncertain. Similarly, investing in bonds, stock, or a new business is a venture whose success is subject to uncertainty. (In fact, some would argue that investing is a form of educated gambling—one in which knowledge, experience, and good judgment can improve the odds of winning.)

Much like playing blackjack and investing, making inferences based on sample data is also subject to uncertainty. A sample rarely tells a perfectly accurate story about the population from which it was selected. There is always a margin of error (as the pollsters tell us) when sample data are used to estimate the proportion of people in favor of a particular political candidate, some consumer product, or some political or social issue. There is always uncertainty about how far the sample estimate will depart from the true population proportion of affirmative answers that you are attempting to estimate. Consequently, a measure of the amount of uncertainty associated with an estimate (which we called *the reliability of an inference* in Chapter 1) plays a major role in statistical inference.

How do we measure the uncertainty associated with events? Anyone who has observed a daily newscast can answer that question. The answer is **probability**. For example, it may be reported that the probability of rain on a given day is 20%. Such a statement acknowledges that it is uncertain whether it will rain on the given day and indicates that the forecaster measures the likelihood of its occurrence as 20%.

Probability also plays an important role in decision making. To illustrate, suppose you have an opportunity to invest in an oil exploration company. Past records show that for 10 out of 10 previous oil drillings (a sample of the company's experiences), all 10 resulted in dry wells. What do you conclude? Do you think the chances are better than 50–50 that the company will hit a producing well? Should you invest in this company? We think your answer to these questions will be an emphatic "no." If the company's exploratory prowess is sufficient to hit a producing well 50% of the time, a record of 10 dry wells out of 10 drilled is an event that is just too improbable. Do you agree?

In this chapter we will examine the meaning of probability and develop some properties of probability that will be useful in our study of statistics.

EXPERIMENTS, EVENTS, AND THE PROBABILITY OF AN EVENT

4.2 In the language employed in a study of probability, the word *experiment* has a very broad meaning. In this language, an **experiment** is a process of making an observation or taking a measurement on the experimental unit. For example, suppose you are dealt a single card from a standard 52-card bridge deck. Observing the outcome (i.e., the number and suit of the card) could be viewed as an experiment. Counting the number of U.S. citizens who live in a particular state or county is an experiment. Similarly, recording a voter's opinion on an important political issue is an experiment. Observing the fraction of insects killed by a new insecticide is an experiment. Note that most experiments result in outcomes (or measurements) that cannot be predicted with certainty in advance.

EXAMPLE 4.1 Consider the following experiment. You are dealt one card from a standard 52-card bridge deck. List some possible outcomes of this experiment that cannot be predicted with certainty in advance.

Solution Some possible outcomes of this experiment that cannot be predicted with certainty in advance are as follows (see Figure 4.1):

a. You draw an ace of hearts.
b. You draw an eight of diamonds.
c. You draw a spade.
d. You do not draw a spade.

  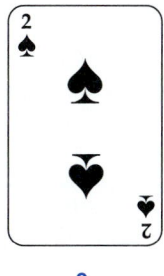 

a.   b.   c.   d.

FIGURE 4.1

Possible Outcomes of Card-Drawing Experiment

EXAMPLE 4.2 Consider the following experiment. Five hundred urban residents with children are selected from a large number of urban residents with children to determine the proportion who favor a busing plan which would, in theory, racially balance public schools in the city. The response of each urban resident is recorded. List some possible outcomes of this experiment that cannot be predicted with certainty in advance.

Solution Since we are observing the responses of 500 urban residents with children (where urban residents are the experimental units), this experiment can result in a very large number of outcomes. Three of the many possible outcomes are listed here.

a. Exactly 387 of the 500 urban residents favor the busing plan.
b. Exactly 388 favor the busing plan.
c. A particular resident, the Jones family, favors the busing plan.

Clearly, we could define many other outcomes of this experiment that cannot be predicted in advance.

In the language of probability theory, outcomes of experiments are called **events**.

<blockquote>

DEFINITION 4.2

Outcomes of experiments are called **events**. [*Note:* To simplify our discussion, we will use italic capital letters, A, B, C, ..., to denote specific events.]

</blockquote>

The outcome for each of the experiments described in Examples 4.1 and 4.2 is shrouded in uncertainty; that is, prior to conducting the experiment, we could not be certain whether a particular event would occur. This uncertainty is measured by the **probability** of the event.

EXAMPLE 4.3 Suppose we perform the following experiment: Toss a coin and observe whether the upside of the coin is a head or a tail. Define the event H by

H: Observe a head.

What do we mean when we say that the probability of H, denoted by $P(H)$, is equal to $\frac{1}{2}$?

Solution Stating that the probability of observing a head is $P(H) = \frac{1}{2}$ does *not* mean that exactly half of a number of tosses will result in heads. (For example, we do not expect to observe exactly 1 head in 2 tosses of a coin or exactly 5 heads in 10 tosses of a coin.) Rather, it means that, in a very long series of tosses, we believe that approximately half would result in a head. Therefore, the number $\frac{1}{2}$ measures the likelihood of observing a head on a single toss of the coin.

The "relative frequency" concept of probability discussed in Example 4.3 is illustrated in Figure 4.2. The graph shows the proportion of heads observed after $n = 25$, 50, 75, 100, 125, ..., 1,450, 1,475, and 1,500 computer-simulated repetitions of a coin-tossing experiment. The number of tosses is marked along the horizontal axis of the graph, and the corresponding proportions of heads are plotted on the vertical axis above the values of n. We have connected the points by line segments to emphasize that the proportion of heads moves closer and closer to .5 as n gets larger (as you move to the right on the graph).

Although most people think of the probability of an event as the proportion of times the event occurs in a very long series of trials, some experiments can never be repeated. For example, if you invest $50,000 in starting a new business, the probability that your business will survive 5 years has some unknown value that you will never be able to evaluate by repetitive experiments. The probability of this event occurring is a number that has some value, but it is unknown to us. The best that we could do, in estimating its value, would be to attempt to determine the proportion of similar businesses that survived 5 years and take this as an approximation to the desired probability. In spite of the fact that we may not be able to conduct repetitive experiments, the relative frequency definition for probability appeals to our intuition.

DEFINITION 4.3

The **probability of an event** A, denoted by $P(A)$, is a number between 0 and 1 that measures the likelihood that A will occur when the experiment is performed. $P(A)$ can be approximated by the proportion of times that A is observed when the experiment is repeated a very large number of times.

In the next section we give two rules for calculating the exact probability of an event.

FIGURE 4.2

The Proportion of Heads in n Tosses of a Coin

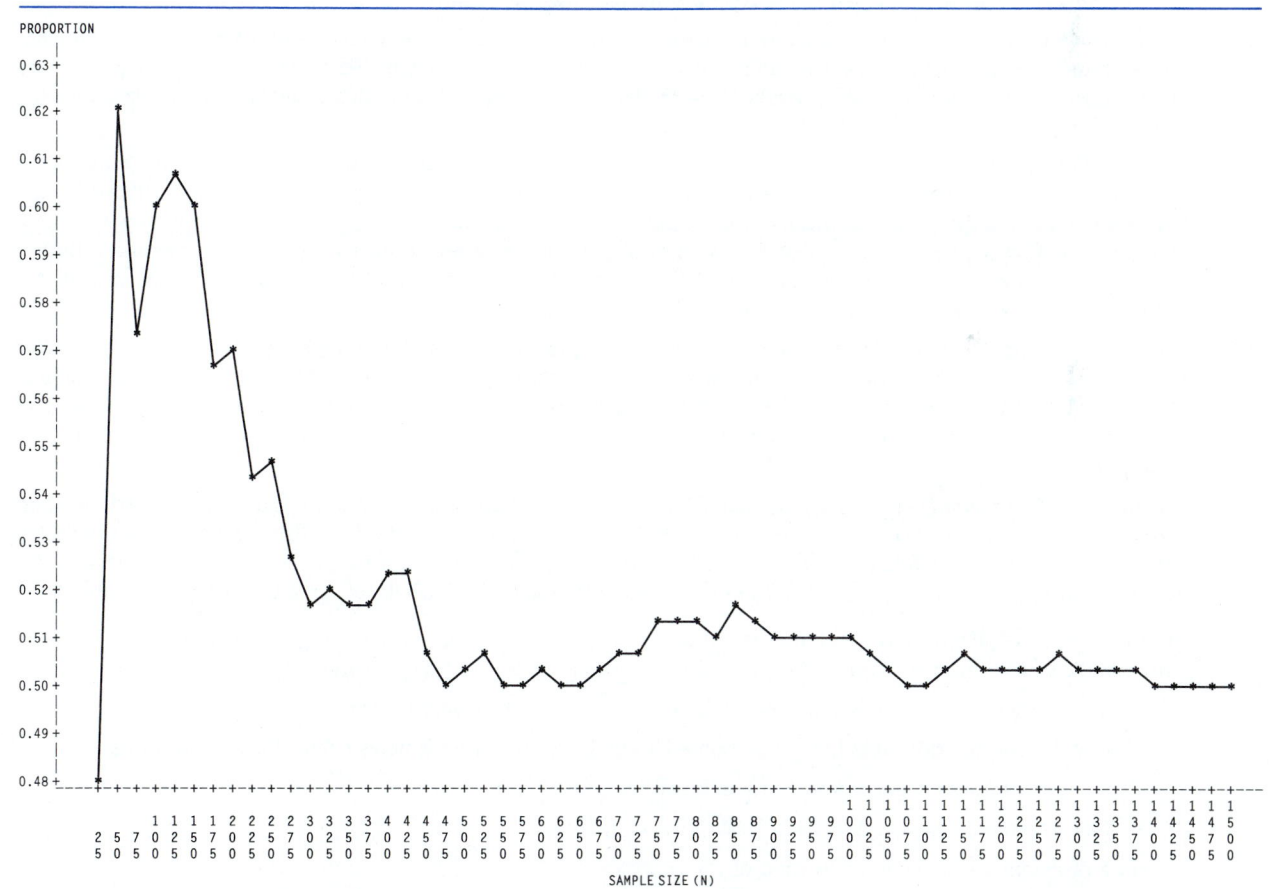

EXERCISES
APPLYING THE CONCEPTS

4.1 Consider the following experiment: Toss a single die and observe the number showing on the upper face.

   a. If this experiment were to be repeated over and over again in a very long series of trials, what proportion of the experimental outcomes do you think would result in a 5?
   b. What does it mean to say, "The probability that the outcome is a 5 is $\frac{1}{6}$?"
   c. Perform the experiment a large number of times and calculate the proportion of outcomes that result in a 5. Note that as the number of repetitions becomes larger and larger, this proportion moves closer and closer to $\frac{1}{6}$.

4.2 During the 1989 U.S. Open, four professional golfers (Doug Weaver, Jerry Pate, Nick Price, and Mark Wiebe) made holes in one (aces) on the sixth hole at Oak Hill Country Club—all on the same day! How unlikely is such a feat? According to *Golf Digest* (Mar. 1990), the probability of a Professional Golf Association (PGA) tour pro making an ace on a given hole is approximately 1/3,000. The estimate is based on the ratio of the number of aces made on the PGA tour to the total number of rounds played.

   a. Interpret the probability of 1/3,000.
   b. *Golf Digest* also estimates that the probability of any four players getting aces on the same hole on the same day during the next U.S. Open is 1/150,000. Interpret this probability.

4.3 A Gallup Youth Survey of teenagers across the United States found that, of those teens who attend church regularly, 55% admit to having cheated on examinations in school (*Tampa Tribune*, Aug. 6, 1988). Approximate the probability that a teenager who attends church regularly has cheated on an exam. Why is this value an approximation to the probability?

4.4 A power plant that discharges its waste into a nearby gulf performed the following experiment each day for a period of 1 year (365 days): A water sample was selected from an area near the plant's discharge, analyzed for the presence of PCB (a dangerous chemical), and the amount of PCB (in parts per million) in the sample was recorded. Suppose the power plant observed that the amount of PCB in the water samples exceeded government pollution standards on 2 days. Find the approximate probability of A, where A is the event that, on any given day, the amount of PCB in the water sample will exceed the standard.

4.5 U.S. Department of Labor estimates showed that the unemployment in the labor force reached 6.0% in November 1990. That is, of all those who were considered eligible workers in November 1990, 6.0% were unemployed. Suppose we selected one eligible worker from the labor force in November 1990 and determined that person's employment status. Based on the Department of Labor estimate, what is the probability that this person was unemployed?

4.6 A survey of 1,035 family doctors found that 70% have increased tests in an effort to detect cancer early in their patients (*Ca—A Cancer Journal for Clinicians*, July/Aug. 1985). Suppose that one of the 1,035 doctors is selected at random. Consider the following statement: "The probability that this doctor has increased tests to detect cancer early in patients is only $\frac{1}{1,035}$ = .00097." Do you agree with the statement? If not, give the correct probability.

4.7 In his book *100% American* (Poseidon Press, 1988), Daniel Weiss presents "fascinating facts about who we are and what some of us—or even most of us—are thinking, doing, believing today." Some excerpts:*

   8% of American 17-year-olds think Abraham Lincoln wrote *Uncle Tom's Cabin*.

   43% of Americans think it is likely that some UFOs are really space vehicles from other civilizations.

   71% of American men think women should call men for a date.

   95% of American peanut eaters eat at least nine at a sitting.

   Form a probability statement about each of these events.

100% American. Copyright © 1988 by Daniel Weiss. Reprinted by permission of Poseidon Press, a division of Simon & Schuster, Inc.

4.3 One special property of events can be seen in the examples of the preceding section. Two events are said to be **mutually exclusive** if, when one occurs, the other cannot occur. The events listed under parts **c** and **d** of Example 4.1 are mutually exclusive. You cannot conduct an experiment and "draw a spade" (the event listed under part **c**) and at the same time "not draw a spade" (the event listed under part **d**). If one of these two events occurs when an experiment is conducted, the other event cannot have occurred. Therefore, we say that they are mutually exclusive events.

PROBABILITY RULES
FOR MUTUALLY
EXCLUSIVE EVENTS

DEFINITION 4.4

Two events are said to be **mutually exclusive** if, when one of the two events occurs in an experiment, the other cannot occur.

EXAMPLE 4.4   Refer to Example 4.2 and define the following events:

A: Exactly 387 of the 500 urban residents favor the busing plan.

B: Exactly 388 favor the busing plan.

C: A particular resident, the Jones family, favors the busing plan.

State whether the following pairs of events are mutually exclusive:

a. A and B   **b.** A and C   **c.** B and C

Solution

a. Events A and B are mutually exclusive. If you have observed exactly 387 residents who favor the busing plan, then you could not, at the same time, have observed exactly 388.
b. Events A and C are *not* mutually exclusive. The Jones family may be one of the residents among the 387 residents in event A who favor busing. Therefore, it is possible for both events A and C to occur simultaneously.
c. Events B and C are not mutually exclusive for the same reason given in part **b**.

EXAMPLE 4.5   Suppose an experiment consists of selecting two electric light switches from an assembly line for inspection. Define the following events:

A: The first switch is defective.

B: The second switch is defective.

Are A and B mutually exclusive events?

Solution   Events A and B are not mutually exclusive because both occur when the inspection is made. That is, both the first and the second switch could be defective.

If two events A and B are mutually exclusive, then the probability that *either* A *or* B occurs is equal to the sum of their probabilities. We will illustrate with an example.

EXAMPLE 4.6 Consider the following experiment: You toss two coins and observe the upper faces of the two coins. What is the probability that you toss exactly one head?

Solution The experiment can result in one of four mutually exclusive events. One possibility is that you will observe a head on coin 1, call it H_1, and a head on coin 2, H_2. We could denote this event as H_1H_2. Similarly, we could observe a head on coin 1 and a tail on coin 2; call this H_1T_2. The other two possible outcomes are a tail on coin 1 and a head on coin 2, T_1H_2; or tails on both coins, T_1T_2. These four events represent the most basic outcomes of the experiment and are called **simple events**. The simple events are shown diagrammatically in Figure 4.3.*

FIGURE 4.3
Four Mutually Exclusive Outcomes (Simple Events) and Associated Probabilities when Tossing a Pair of Coins

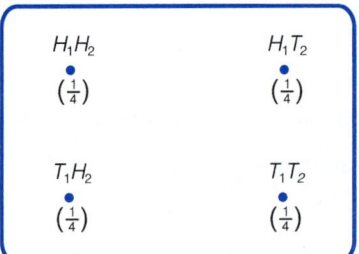

Since the chance of tossing either a head or a tail on each coin is the same $\left(\frac{1}{2}\right)$, we would expect each of the four simple events of Figure 4.3 to occur with approximately equal relative frequency $\left(\frac{1}{4}\right)$ if the coin-tossing experiment were repeated a large number of times. Since you will observe exactly one head only if T_1H_2 occurs or if H_1T_2 occurs, and these simple events are mutually exclusive, either one or the other of these events will occur $\left(\frac{1}{4} + \frac{1}{4}\right) = \frac{1}{2}$ of the time. Therefore, the probability of observing exactly one head in the toss of two coins is equal to the probability of observing *either* T_1H_2 *or* H_1T_2, which is $\frac{1}{2}$. You can verify this result experimentally, using the procedure employed in Section 4.2.

> **DEFINITION 4.5**
>
> **Simple events** are mutually exclusive events that represent the most basic outcomes of an experiment.

---

*The graphical representation that shows all possible simple events in an experiment is called a **Venn diagram**.

Although Example 4.6 utilized the concept of simple events, the additive probability rule shown in the next box applies to any two mutually exclusive events. You can now see why the concept of mutually exclusive events is important. We will illustrate with several more examples.

PROB

The P
of Eq
result
and t*
A is

P(*/

PROBABILITY RULE #1

The Additive Rule for Mutually Exclusive Events   If two events A and B are mutually exclusive, then *the probability that either A or B occurs* is equal to the sum of their respective probabilities:

$$P(A \text{ or } B) = P(A) + P(B)$$

EXAMP
chance,
prizes b
at least

Solution
the two
mutuall

$(D_1,$

STEP 1

STEP 2

STEP 3

Ex*
summ
Bef
exclu*
ment

A:
$\overline{A}$:

EXAMPLE 4.7   A local weather reporter forecasts the following three mutually exclusive weather conditions, with their associated probabilities:

Overcast with some rain: 40%

No rain, but partially or completely overcast: 30%

No rain and a clear day: 30%

What is the probability that the day will be overcast?

Solution   The forecaster suggests that the experiment of observing the day's weather can result in one (and only one) of three mutually exclusive events:

A: Overcast with some rain

B: No rain, but partially or completely overcast

C: No rain and a clear day

where

$$P(A) = .4 \qquad P(B) = .3 \qquad P(C) = .3$$

The event of interest, that the day will be overcast, will occur if either event A or event B occurs. Therefore, using Probability Rule #1, we have:

$$P(A \text{ or } B) = P(A) + P(B)$$
$$= .4 + .3 = .7$$

The implication is that if we were to forecast the weather for a large number of days with similar weather conditions, 70% of the days would be overcast.

EXAMPLE 4.8   Consider the following experiment: Two dice are tossed and the number of dots on the upper faces of the dice are observed. Find the probability that the sum of the two numbers is equal to 7 (a winning number in the casino game of craps).

FIGURE 4

The Sum
Mutually
the Tossin

4.14   Environmental engineers classify U.S. consumers into five mutually exclusive groups based on consumers' feelings about environmentalism:

1. *Basic browns* claim they don't have the knowledge to understand environmental problems.
2. *True-blue greens* use biodegradable products.
3. *Greenback greens* support requiring new cars to run on alternative fuel.
4. *Sprouts* recycle newspapers regularly.
5. *Grousers* believe industries, not individuals, should solve environmental problems.

The proportion of consumers in each group is shown in the table. Suppose a U.S. consumer is selected at random and his or her feelings about environmentalism determined.

Basic browns	.28
True-blue greens	.11
Greenback greens	.11
Sprouts	.26
Grousers	.24

Source: The Orange County (Calif.) *Register*, Aug. 7, 1990.

a. List the simple events for the experiment.
b. Assign reasonable probabilities to the simple events.
c. Find the probability that the consumer is either a basic brown or a grouser.
d. Find the probability that the consumer supports environmentalism in some fashion (i.e., is a true-blue green, greenback green, or sprout).

4.15   Local area merchants often use "scratch off" tickets with promises of grand prize giveaways to entice customers to visit their storerooms. One such game, called Jackpot, was recently used in Tampa, Florida. Residents were mailed game tickets with 10 "play squares" and the instructions: "Scratch off your choice of ONLY ONE Jackpot Play Square. If you reveal any 3-OF-A-KIND combination, you WIN. Call NOW to make an appointment to claim your prize." All 10 play squares, of course, have winning 3-of-a-kind combinations. On one such ticket, three cherries (worth up to $1,000 in cash or prizes) appeared in play squares numbered 1, 2, 3, 4, 5, 6, 7, and 9; three lemons (worth a 10-piece microwave cookware set) appeared in play square number 8; and three sevens (worth a whirlpool spa) appeared in play square 10.

a. If you play Jackpot, what is the probability that you win a whirlpool spa?
b. If you play Jackpot, what is the probability that you win $1,000 in cash or prizes?

4.16   The effect of race in the selection of political ideology (conservatism versus liberalism) was recently investigated. Demographic data used in the analysis were collected by the National Opinion Research Center and included the race and state of residence at age 16 of each respondent (*Phylon, A Review of Race and Color*, June 1985). Consider the following events:

A: Respondent is black.

B: Respondent is white.

C: Respondent resides in a southern state.

D: Respondent resides in a nonsouthern state.

Which of the events are mutually exclusive?

4.17 An improved method for measuring the electrical resistivity of concrete has been developed that eliminates difficulties caused by polarization effects and capacitive resistance (*Magazine of Concrete Research*, Dec. 1985). The method was tested on concrete specimens with different water–cement mixes. Three different water-weight ratios (40%, 45%, and 50%) and three different amounts of cement (300, 350, and 400 kilograms per cubic meter) were examined.

    a. List all possible water–cement mixes for this experiment.

    b. Suppose we determine the water–cement mix that yields the highest electrical resistivity. Before the experiment is performed, should equal probabilities be assigned to the simple events? Why or why not?

4.18 The Commission of the European Communities initiated a research program to determine the influence of traffic noise on sleep, subjective assessment, and psychomotor performance. For one portion of the study, a team of German acoustical engineers monitored the sleep of 10 couples (one male and one female per couple) during 12 consecutive nights. All 10 couples slept under usual conditions on seven of the nights. (This represents the *control phase* of the study.) For the other five nights (the *experimental phase*), the 10 couples were divided into two groups of equal size. One group slept with the windows open and the other slept with earplugs. The experimental setup is described in the table.

NIGHTS	CONTROL PHASE 1 2 3 4 5	EXPERIMENTAL PHASE 6 7 8 9 10	CONTROL PHASE 11 12
Earplugs group (couples 1, 4, 6, 9, 10)	Without earplugs Windows open	With earplugs Windows open	Without earplugs Windows open
Windows group (couples 2, 3, 5, 7, 8)	Windows closed No earplugs	Windows open No earplugs	Windows closed No earplugs

Source: Griefahn, B. and Gros, E. "Noise and sleep at home, a field study on primary and aftereffects." *Journal of Sound and Vibration*, Vol. 105, No. 3, Mar. 1986, p. 376 (Figure 2).

Suppose we randomly select one couple from the experiment on one randomly selected night and note whether the couple is wearing earplugs and whether the windows are open or closed.

    a. List the simple events for the experiment.

    b. Assign probabilities to the simple events.

    c. What is the probability that the couple is wearing earplugs?

    d. What is the probability that the windows are closed?

4.19 The American Association for Marriage and Family Therapy (AAMFT) is a group of professional therapists and family practitioners who treat many of the nation's couples and families. The AAMFT released the findings of a 5-year study that has tracked the postdivorce history of 98 pairs of former spouses with children. Each divorced couple was classified into one of four groups, nicknamed "perfect pals," "cooperative colleagues," "angry associates," and "fiery foes." The proportions classified into each group are given here.

"Perfect pals": .12
    Joint-custody parents who get along well, do not remarry.

"Cooperative colleagues": .38
    Occasional conflict, likely to be remarried.

"Angry associates": .25
    Cooperate on issues related to children only, conflicting otherwise.

"Fiery foes": .25
    Communicate only through children, hostile toward each other.

Source: Gainesville Sun, Nov. 12, 1985.

Suppose one of the 98 couples is selected at random.

a. What is the probability that the former spouses are "cooperative colleagues"?
b. What is the probability that the former spouses are "angry associates"?
c. What is the probability that the former spouses are not "perfect pals"?

4.20 A Northwestern University Department of Pediatrics study of bicycle-related injuries over a 7-year period found that more than 2,500 babies and toddlers were hurt while riding in seats mounted on an adult's bike. The accompanying table gives a breakdown of the causes of the injuries.

CAUSE OF BIKE-RELATED INJURIES TO CHILDREN	PERCENT
Fell out of seat	39
Accident with car	10
Stationary bike fell over	24
Seat fell off bike	6
Extremity caught in spoke	21
TOTAL	100

Source: Tampa Tribune, July 10, 1988.

Suppose a child is injured in a bicycle accident.

a. What is the probability that the injury results from an extremity getting caught in the bicycle spoke?
b. What is the probability that the injury results from either the child falling out of the seat or the seat falling off the bike?
c. What is the probability that the injury did not occur as a result of an accident with a car?

4.21 "Employee Burnout: America's Newest Epidemic" is the name of a 1991 survey of American workers commissioned by Northwestern National Life Insurance Company. The main topic of the study was job-related stress. The origins of job stress, according to the responses of 600 workers, are summarized in the pie chart.

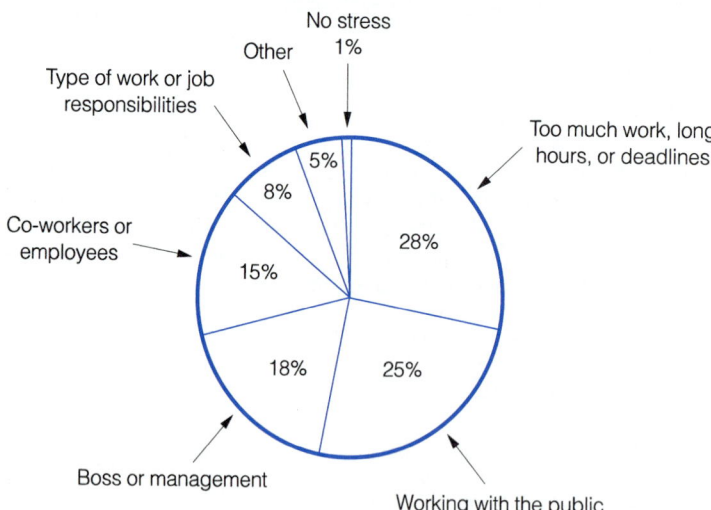

Source: Tampa Tribune, May 8, 1991.

Suppose one of the 600 workers in the survey is selected at random.

a. What is the probability that the worker attributes on-the-job stress to his or her boss or management?
b. What is the probability that the worker attributes on-the-job stress to too much work, long hours, or deadlines?

c. What is the probability that the worker has on-the-job stress?

d. What is the probability that the worker attributes on-the-job stress to people (e.g., co-workers, employees, boss, management, or the public)?

4.22 The YES/MVS (Yorktown Expert System/MVS Manager) is an experimental expert system designed to exert active control over a computer system and provide advice to computer operators. YES/MVS is designed with a knowledge base consisting of 548 rules that are triggered in response to messages or queries from the computer operator. The table gives the number of rules allocated to different subdomains of the operator's actions. Periodically, the rules in the YES/MVS knowledge base are tested and, if necessary, adjusted. Suppose a rule is selected at random for testing and its type (operator action/query) noted.

OPERATOR'S ACTION/QUERY	NUMBER OF RULES
Batch scheduling	139
JES queue space	104
C-to-C links	68
Hardware errors	87
SMF management	25
Quiesce and IPL	52
Performance	41
Background monitor	32
TOTAL	548

Source: Ennis, R. L., et al. "A continuous realtime expert system for computer operations." *IBM Journal of Research and Development*, Vol. 30, No. 1, Jan. 1986, p. 19. Copyright 1986 by International Business Machines Corporation; reprinted with permission.

a. List the simple events for this experiment.

b. Assign probabilities to the simple events based on the information contained in the table.

c. What is the probability that the rule is a C-to-C link or hardware error rule?

d. What is the probability that the rule is not a performance rule?

4.23 In 1987 Congress enacted the Surface Transportation and Uniform Act, which allowed states to increase the speed limit to 65 miles per hour (mph) on interstate highways located outside of an urban area of 50,000 or more persons. In a study of traffic fatalities on interstate highways, the Fatal Accident Reporting System discovered that 96% of the interstate highway miles eligible to be posted at 65 mph are rural interstates, and 97% of these eligible miles were actually posted at 65 mph (*American Journal of Public Health*, Oct. 1989).

a. For a particular 1-mile stretch of interstate highway eligible to be posted at 65 mph, estimate the probability that the 1-mile stretch is posted at 65 mph.

b. For a particular 1-mile stretch of interstate highway eligible to be posted at 65 mph, estimate the probability that the 1-mile stretch is not rural.

c. Are the events in parts a and b mutually exclusive? Explain.

4.24 Entomologists are often interested in studying the effect of chemical attractants (*pheromones*) on insects. One common technique is to release several insects equidistant from the pheromone being studied and from a control substance. If the pheromone has an effect, more insects will travel toward it than toward the control. Otherwise, the insects are equally likely to travel in either direction. Suppose five insects are released. If we are interested in which insects travel toward the pheromone, how many outcomes (simple events) are possible? [*Hint:* If the five insects are denoted as *A, B, C, D,* and *E,* one possible outcome is that insects *A* and *B* travel toward the pheromone. Another possible outcome is that insects *A, B, C,* and *D* travel toward the pheromone.]

4.25 Refer to Exercise 4.24. Suppose the pheromone under study has no effect and, therefore, it is equally likely that an insect will move toward the pheromone or toward the control—i.e., the possible outcomes are equiprobable. Find the probability that both insect *A* and insect *E* travel toward the pheromone.

4.26 After Evelyn Marie Adams won the New Jersey weekly lottery twice within 4 months in 1986, the event was widely reported as an amazing feat that beat the odds of 1 in 17 trillion. Although the probability of the event, 1/17,000,000,000,000, is technically correct, it does not take into account a fundamental law of statistics called "the law of very large numbers." In the words of Harvard statisticians Perci Diaconis and Frederick Mosteller (*Journal of the American Statistical Association*, Dec. 1989), the law states that "with a large enough sample, any outrageous thing is apt to happen." Diaconis and Mosteller go on to explain that "one in 17 trillion is the odds that a given person (e.g., Ms. Adams) who buys a single ticket for exactly two New Jersey lotteries will win both times." The true question, they say, is "What is the chance that some person, out of all the millions and millions who buy lottery tickets in the United States, hits a lottery twice in a lifetime?" Based on a 7-year study of state lottery winners, Purdue statisticians Stephen Samuels and George McCabe estimated the "odds are better than even that there will be a double lottery winner somewhere in the United States" (*Wall Street Journal*, Feb. 27, 1990).* Let *A* be the event that you buy a New Jersey lottery ticket for exactly two different weeks and win both times. Let *B* be the event that any person wins a state lottery twice.

a. Are the events *A* and *B* mutually exclusive?
b. What is the probability of *A*?
c. What is the probability of *B*?
d. Explain why the probabilities, parts **b** and **c**, are so drastically different.

**CHAPTER CASE:
THE PROBABILITY
OF BLACKJACK**

FIGURE 4.5

Some Typical Two-Card Draws
and Their Values

4.4 Blackjack (or twenty-one, as it is sometimes called) is a game played by two or more players using an ordinary 52-card bridge deck. Each card is assigned a value. A card numbered from 2 to 10 is assigned the value shown on the card. For example, a seven of spades has a value of 7; a three of hearts has a value of 3. Face cards (kings, queens, and jacks) are each valued at 10, and an ace can be assigned a value of either 1 or 11, at the discretion of the player holding the card.

The essence of the game is that you are dealt two cards and you may request more. The objective is to acquire cards whose total value is as near as possible to, but not exceeding, 21. If the sum of the initial two cards is equal to 21, you have drawn a blackjack, a winning hand in most circumstances. Each player bets and plays against one of the players who is designated as the *dealer*. You win the bet if the total value of your hand exceeds the total value of the dealer's hand. If the total value of the cards in your hand exceeds 21, you automatically lose; if you hold five cards and their total value is 21 or less, you automatically win;

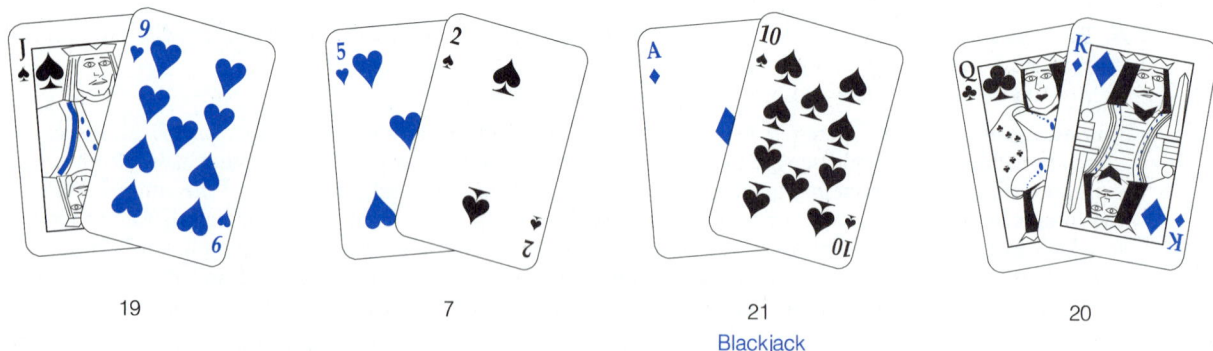

19   7   21   20
                                      Blackjack

---

*Reprinted by permission of the *Wall Street Journal.* © 1990 Dow Jones & Company, Inc. All rights reserved worldwide.

if you tie the dealer, you lose. (At some casinos, a tie is a *standoff* and the bet is canceled.) Some typical initial draws are shown in Figure 4.5.

The game is played as follows. The dealer distributes two cards, face down, to each player and two cards to himself, one face up and one face down. You see the total value of your hand and, seeing *one* of the dealer's two cards, you have some information on the total value of the cards in the dealer's hand. The next move belongs to you. You can decide to face the dealer with the total value of the two cards contained in your initial draw, or you can elect to draw one or more new cards, hoping to push the total value of your hand nearer to, but not exceeding, 21. If additional draws lead to a total exceeding 21, you lose. If not, presumably at some draw, you decide to face the dealer with the total value you then hold in your hand.

After you and the other players finish playing, the dealer must decide whether to face you with the total value contained in his initial pair of cards (if the initial pair is a blackjack, the dealer automatically wins) or whether to draw additional cards in an attempt to increase the total value of his hand.* In doing so, he risks the possibility of drawing a card whose value will push the total value of his hand over 21. If this were to occur, the dealer would lose to all players still in the game. When the dealer concludes his draw, he displays his cards face up. If the total value of your cards exceeds the total value of the dealer's cards, you win. If not, you lose. Naturally, you hope to draw an ace and a 10 or a face card (blackjack) on the initial draw. This draw is an automatic win for you unless the dealer matches it.

We conducted an experiment similar to that described in Example 4.3 to determine the approximate probability of drawing a blackjack when two cards are dealt from a well-mixed standard deck of bridge cards. Figure 4.6 (page 186) shows the proportion of times a blackjack is observed when the deal is repeated $n = 100, 200, 300, 400, \ldots, 9{,}800, 9{,}900$, and $10{,}000$ times. The proportions are plotted in Figure 4.6 for each value of n. Notice that the proportion of times a blackjack is observed moves closer and closer to .048 as n gets larger (as you move to the right along the horizontal axis).

Since every possible pair of cards is as likely to be drawn as any other and the outcomes are mutually exclusive, we can use Probability Rule #2 to find the probability of drawing a blackjack. The calculation is described in the following steps.

STEP 1   First, we find M, the number of distinctly different combinations of two cards that can be selected from a deck of 52. Since there are too many outcomes to list, we will use the formula contained in the box on page 187 (proof omitted), which gives the number of ways for selecting n objects from a total of N.

---

*Different variations of the game are played in casinos throughout the world. For example, blackjack is an automatic win for a player in Las Vegas. Also, according to Las Vegas rules, the dealer has no decision on a *hit*. If the total value of the dealer's cards is 16 or less, he or she must take a *hit* (draw another card). If the dealer's hand has a total value between 17 and 21 (inclusive), he or she must *stick* (play the cards in hand).

APPLYING THE CONCEPTS

4.31 A National Merit Scholarship award committee recently claimed that each of five grant applications received equal consideration for two grants and that, in fact, the recipients were randomly selected from among the five. Three of the applicants were from a majority group and two were from a minority group. Suppose that both grants were awarded to members of a majority group.

   a. What is the probability of this event occurring if, in fact, the committee's claim is true?

   b. Is the probability computed in part **a** inconsistent with the committee's claim that the selection was at random?

4.32 Market researchers are keenly interested in consumer preferences for the various benefits offered by a product. Researchers at the University of Pennsylvania conducted a study of consumer preferences for shampoo benefits. Part of the study involved a survey of 186 undergraduate business students. Each respondent was shown a list of 13 benefits and asked to select up to four benefits that he or she most strongly desires in a shampoo brand. The group of benefits selected by a respondent is termed a *benefit bundle*. The list of 13 benefits is shown in the accompanying table.

SHAMPOO BENEFITS

Body	Thickness	Natural ingredients
Bounciness	Softness	Repairs split ends
Control	Manageability	Conditions hair
Luster	Gentle action	
Protection against dandruff	Contains protein	

Source: Journal of the Market Research Society, Jan. 1984.

   a. How many bundles of four shampoo benefits can be specified?

   b. Assuming that the benefit bundles of part **a** are equally likely, what is the probability that a respondent will select the "Body, Bounciness, Luster, Protein" bundle as most desirable?

   c. Assuming that the benefit bundles of part **a** are equally likely, what is the probability that a respondent will select "Softness" as one of the four benefits? [*Hint:* The number of bundles that include "Softness" is $\binom{12}{3}$.]

4.33 To evaluate the air traffic control systems of facilities relying on computer-based equipment, the Federal Aviation Administration (FAA) formed a 16-member task force. If the FAA wants to assign four task force members to a specific facility, how many different assignments are possible?

4.34 According to the *Wall Street Journal* (Sept. 23, 1985) "the rise in recent years of high-powered betting syndicates has become one of the hot controversies in sports betting." A syndicate is a group of high-stakes bettors who pool their resources and cover every combination of a pari-mutuel bet. That is, the syndicate collectively bets on every possible jackpot outcome to assure a winning bet. For example, in jai alai, the pick-six jackpot requires bettors to name the winners in six straight games (where each game involves competition among eight players). One Miami syndicate recently bet every possible combination in a pick six jackpot—a total of 262,144 different combinations (or bets) at a cost of $2 per bet—to win a total of $752,778. Consider a less complicated jai alai bet—the quinella. To win a quinella, a bettor must pick the two players (of the eight) who finish in the top two positions ("win" and "place") in a single game without regard to their order of finish. Suppose your betting syndicate wants to cover all possible outcomes of a quinella bet. How many different bets must be made, i.e., how many combinations of two players from the eight are possible? List them.

CONDITIONAL PROBABILITY AND INDEPENDENCE

4.5 The event probabilities we have discussed thus far give the relative frequencies of occurrence of the events when an experiment is repeated a very large number of times. They are called **unconditional probabilities** because no special conditions are assumed other than those that define the experiment.

Sometimes we may want to revise the probability of an event when we have additional knowledge that might affect its outcome. To give a simple example, we found that the probability of observing a 7 when two dice are tossed is $\frac{1}{6}$ (see Example 4.8). But suppose you are given the information that the sum of the two numbers showing on the dice is even. Would you still believe that the probability of observing a 7 on that particular toss is $\frac{1}{6}$? Intuitively, you will realize that the probability of observing a 7 is now 0. Since you know that an even number occurred, the outcome 7 cannot have occurred (because 7 is an odd number). The probability of observing a 7, *given that you know some other event has already occurred*, is called the **conditional probability** of the event.

DEFINITION 4.7

The probability of an event A, given that an event B has occurred, is called the **conditional probability of A given B** and is denoted by the symbol

$P(A \mid B)$

[*Note:* The vertical bar between A and B is read "given."]

EXAMPLE 4.10   A box contains three fuses, one good and two defective. Two fuses are drawn in sequence, first one and then the other.

a. What is the probability that the second fuse drawn is defective?
b. What is the probability that the second fuse drawn is defective if you know, for certain, that the first fuse drawn is defective?

Solution

a. We will denote the good fuse by G and the two defective fuses as D_1 and D_2. If the fuses are drawn at random from the box, the six possible orders of selection, i.e., the six simple events, are

(G, D_1) (G, D_2) (D_1, G) (D_1, D_2) (D_2, G) (D_2, D_1)

STEP 1   Since these six mutually exclusive events are equally likely and comprise all possible outcomes of the draw, we have $M = 6$.

STEP 2   Next, we must find the number of selections in which a defective fuse is selected in the second draw. You can see from the listed draws that $m = 4$.

STEP 3   Using Probability Rule #2, we conclude that the unconditional probability of obtaining a defective fuse on the second draw is

$$P(\text{defective fuse on the second draw}) = \frac{m}{M} = \frac{4}{6} = \frac{2}{3}$$

b. The probability of observing a defective fuse on the second draw, given that you have observed a defective fuse on the first draw, is the conditional probability $P(A \mid B)$, where

      A: Observe a defective fuse on the second draw.

      B: Observe a defective fuse on the first draw.

If the first fuse drawn from the box is defective, then the box now contains only two fuses, one defective and one nondefective. This means that there is a 50% chance of drawing a defective fuse on the second draw, given that a defective fuse has already been drawn. That is,

$$P(A \mid B) = \frac{1}{2}$$

The probability obtained in part **a**, (the unconditional probability of event A) was equal to $\frac{2}{3}$. Clearly, the probability has changed when we know that event B has occurred.

EXAMPLE 4.11   A balanced coin is tossed 10 times, resulting in 10 tails. If the coin is tossed one more time, what is the probability of observing a head?

Solution   We are asked to find the conditional probability of event A, given that event B has occurred, where

      A: The 11th toss results in a head.

      B: The first 10 tosses resulted in 10 heads.

Intuitively, it may seem reasonable to expect the probability of observing a head on the 11th toss (given that the 10 previous tosses resulted in heads) to be greater than $\frac{1}{2}$, but such is not the case. If the coin is truly balanced and is tossed in an unbiased manner, then the probability of observing a head on the 11th toss is still $\frac{1}{2}$. (This has been verified both theoretically and experimentally.) Therefore, this is a case where the conditional probability of an event A is equal to the unconditional probability of A.

Example 4.11 illustrates an important relationship that exists between some pairs of events. If the probability of one event does not depend on whether a second event has occurred, then the events are said to be **independent**.

The notion of independence is particularly important when we want to find the probability that *both* of two events will occur. When the events are independent, the probability that both events will occur is equal to the product of their unconditional probabilities.

DEFINITION 4.8

Two events A and B are said to be **independent** if

$$P(A \mid B) = P(A)$$

or if

$$P(B \mid A) = P(B)$$

[*Note*: If one of these equalities is true, then the other will also be true.]

PROBABILITY RULE #3

The Probability That Both of Two Independent Events A and B Occur   If two events A and B are independent, then *the probability that both A and B occur* is equal to the product of their respective unconditional probabilities:

$$P(A \text{ and } B) = P(A)P(B)$$

Probability Rule #3 can be extended to apply to any number of independent events. For example, if A, B, and C are independent events, then

$$P(\text{all of the events, } A, B, \text{ and } C, \text{ occur}) = P(A \text{ and } B)P(C)$$
$$= P(A)P(B)P(C)$$

EXAMPLE 4.12   Find the probability of observing two heads in two tosses of a balanced coin.

Solution   Define the following events:

A: Observe a head on the first toss.

B: Observe a head on the second toss.

Since we know that events A and B are independent and that $P(A) = P(B) = \frac{1}{2}$, the probability that we observe two heads, i.e., both events A and B, is

$$P(\text{observe two heads}) = P(A)P(B)$$
$$= \left(\frac{1}{2}\right)\left(\frac{1}{2}\right) = \frac{1}{4}$$

You can see that this answer agrees with our reasoning in Example 4.6.

We now consider a problem in statistical inference.

EXAMPLE 4.13 Experience has shown that a manufacturing operation produces, on the average, only one defective unit in 10. These are removed from the production line, repaired, and returned to the warehouse. Suppose that during a given period of time you observe five defective units emerging in sequence from the production line.

a. If experience has shown that defective units usually emerge randomly from the production line, what is the probability of observing a sequence of five consecutive defective units?
b. If the event in part **a** really occurred, what would you conclude about the process?

Solution

a. If the defectives really occur randomly, then whether any one unit is defective should be independent of whether the others are defective. Second, the unconditional probability that any one unit is defective is known to be $\frac{1}{10}$. We will define the following events:

D_1: The first unit is defective.
D_2: The second unit is defective.
$\vdots$ $\qquad\qquad$ $\vdots$
D_5: The fifth unit is defective.

Then

$$P(D_1) = P(D_2) = P(D_3) = P(D_4) = P(D_5) = \frac{1}{10}$$

and the probability that all five are defective is

$$P(\text{all five are defective}) = P(D_1)P(D_2) \cdots P(D_5)$$
$$= \left(\frac{1}{10}\right)\left(\frac{1}{10}\right)\left(\frac{1}{10}\right)\left(\frac{1}{10}\right)\left(\frac{1}{10}\right)$$
$$= \frac{1}{100,000}$$

b. We do not need a knowledge of probability to know that something must be wrong with the production line. Intuition would tell us that observing five defectives in sequence is highly improbable (given the past), and we would immediately infer that past experience no longer describes the condition of the process. In fact, we would infer that something is disturbing the stability of the process.

Example 4.13 illustrates how you can use your knowledge of probability and the probability of a sample event to make an inference about some population. The technique, called the **rare event approach**, is summarized in the box.

EXERCISES
LEARNING THE MECHANICS

4.35 Consider an experiment that consists of two trials and has the nine possible outcomes (simple events) listed here:

AA	BA	CA
AB	BB	CB
AC	BC	CC

where *AC* indicates that *A* occurs on the first trial and *C* occurs on the second trial. Suppose the following events are defined:

 D: Observe an *A* on the first trial.

 E: Observe a *B* on the second trial.

a. List the simple events associated with event *D* and find $P(D)$.
b. List the simple events associated with event *E* and find $P(E)$.
c. Find $P(E \mid D)$.
d. Are *D* and *E* independent?
e. Find P(both *D* and *E* occur).

4.36 Assume that $P(A) = .6$ and $P(B) = .3$. If *A* and *B* are independent, find $P(A \text{ and } B)$.

4.37 Assume that $P(A) = .6$, $P(B) = .4$, $P(C) = .5$, $P(A \mid B) = .15$, $P(A \mid C) = .5$, and $P(B \mid C) = .3$.

a. Are events *A* and *B* independent?
b. Are events *A* and *C* independent?
c. Are events *B* and *C* independent?

4.38 Nightmares about college exams appear to be common among college graduates. In a recent survey of 30- to 45-year-old graduates from Transylvania University, 50 of 188 respondents admitted they had recurring dreams about college exams (*Tampa Tribune*, Dec. 12, 1988). Of these 50, 47 felt distress, anguish, fear, or terror in their dreams. (For example, some dreamers "couldn't find the building or they walked in and all the students were different." Other dreamers either overslept or didn't realize they were enrolled in the class.)

    a. Calculate the approximate probability that a 30- to 45-year-old graduate of Transylvania University has recurring dreams about college exams. Why is this probability approximate?

    b. Refer to part **a**. Given that the graduate has recurring dreams, what is the approximate probability that the dreams are unpleasant (i.e., that the graduate feels distress, anguish, fear, or terror in the dreams)?

    c. Are the events {graduate has recurring dreams} and {dreams are unpleasant} independent?

4.39 Refer to the U.S. Army Corps of Engineers study on the DDT contamination of fish in the Tennessee River in Alabama (see Case Study 1.1). Part of the investigation focused on how far upstream the contaminated fish have migrated. (A fish is considered to be contaminated if its measured DDT concentration is greater than 5.0 parts per million.) Recall that Appendix D gives the DDT concentration, species, and capture location (in miles from the river's mouth) for each in a sample of 144 fish specimens. The accompanying table gives the number of contaminated fish found for each species–location combination. Suppose a contaminated fish is captured from the river.

		CAPTURE LOCATION		
		275–300	305–325	330–350
	Smallmouth Buffalo	9	7	0
SPECIES	Largemouth Bass	0	0	1
	Channel Catfish	31	23	6

    a. Given that the fish is a channel catfish, what is the probability that it is captured 330–350 miles upstream?

    b. Given that the fish is captured 275–300 miles upstream, what is the probability that it is a smallmouth buffalo?

4.40 According to a recent report from the Newspaper Advertising Bureau, 40% of all primary car maintainers are women (*American Demographics*, June 1985). Consequently, advertisements for car care products, traditionally geared toward men, are now being aimed at women also. Consider the population consisting of all primary car maintainers.

    a. What is the probability that a primary car maintainer, selected from the population, is a woman? A man?

    b. What is the probability that both primary car maintainers in a sample of two selected from the population are women?

4.41 The transport of neutral particles in an evacuated duct is an important aspect of nuclear fusion reactor design. In one experiment, particles entering through the duct ends streamed unimpeded until they collided with the inner duct wall. Upon colliding, they are either scattered (reflected) or absorbed by the wall (*Nuclear Science and Engineering*, May 1986). The reflection probability (i.e., the probability a particle is reflected off the wall) for one type of duct was found to be .16.

    a. If two particles are released into the duct, find the probability that both will be reflected.

    b. If five particles are released into the duct, find the probability that all five will be absorbed.

    c. What assumption about the simple events in parts **a** and **b** is required to calculate the probabilities?

4.42 According to the National Highway Traffic Safety Administration, "as many as nine out of ten heavily used cars, such as those from the leasing companies, may have had their mileage rolled back when resold" (*Orlando Sentinel*, Apr. 13, 1984). Officials estimate that an altered odometer adds $750 to the price of a used car. Suppose you are considering buying a used car from an auto leasing company that has three cars available. Assume also that nine out of every ten used cars sold by the company have falsified odometer readings and that the three available cars represent a random sample of used cars sold by the company.

a. What is the probability that all three cars have falsified odometer readings?
b. What is the probability that none of the three cars has a falsified odometer reading?
c. Suppose a salesman claims that none of the three cars has a falsified odometer reading. What would you infer about the claim?

4.43 Periodically, magazine publishers promote their products by mailing sweepstakes packets to consumers. These packets offer the chance to win a grand prize of $1 million or more, with no obligation to purchase any of the advertised products. Despite the low odds of winning, marketing experts have found that the sweepstakes contests dramatically increase consumer interest and orders. Recently, the U.S. government investigated the legitimacy of popular sweepstakes conducted by Publishers Clearing House, American Family Publishers, and Reader's Digest. On a nationwide basis, the odds of winning the grand prize are 1 in 181,795,000 for the current Publishers Clearing House sweepstakes, 1 in 200,000,000 for the American Family Publishers sweepstakes, and 1 in 84,000,000 for the Reader's Digest sweepstakes (*Gainesville Sun*, Jan. 24, 1985).

a. Calculate the probability of winning the grand prize in the Publishers Clearing House sweepstakes.
b. Repeat part a for the American Family Publishers sweepstakes.
c. Repeat part a for the Reader's Digest sweepstakes.

4.44 Refer to Exercise 4.43. Suppose you enter the sweepstakes contests of all three companies.

a. What is the probability that you win the grand prize in all three contests?
b. What is the probability that you do not win any of the three grand prizes?
c. Use the probability computed in part b to calculate the probability of winning at least one of the three grand prizes. [*Hint:* The complement of "at least one" is "none."]

4.45 An article in *IEEE Computer Applications in Power* (Apr. 1990) describes "an unmanned watching system to detect intruders in real time without spurious detections, both indoors and outdoors, using video cameras and microprocessors." The system was tested outdoors under various weather conditions in Tokyo, Japan. The numbers of intruders detected and missed under each condition are provided in the table.

	WEATHER CONDITION				
	Clear	Cloudy	Rainy	Snowy	Windy
Intruders detected	21	228	226	7	185
Intruders missed	0	6	6	3	10
TOTALS	21	234	232	10	195

Source: Kaneda, K., et al. "An unmanned watching system using video cameras." *IEEE Computer Applications in Power*, Apr. 1990, p. 24.

a. Under cloudy conditions, what is the probability that the unmanned system detects an intruder?
b. Given that the unmanned system missed detecting an intruder, what is the probability that the weather condition was snowy?

4.46 Roadside breath-testing surveys conducted throughout the country indicate that 6% of the people driving at night have a blood alcohol concentration (BAC) of at least .10%, the legal definition of driving under the influence of alcohol (DUI) in most states. Thus, the probability of a police patrol officer detecting a person DUI from a random stop at night is only .06. However, research has shown that certain visual clues can aid the patrol officer in discriminating between DUI and driving while sober (DWS). These visual clues serve to increase the DUI detection rate above the chance probability of .06. As a result, Harris (1980) developed the Drunk Driver Detection Guide for patrol officers. The guide, a portion of which is shown in the table on page 196, gives the percentage of nighttime drivers stopped for a traffic violation (based on various characteristics that serve as visual clues for the patrol officer) with a BAC of at least .10%.

VISUAL CLUE	PERCENTAGE OF NIGHTTIME DRIVERS STOPPED FOR TRAFFIC VIOLATION (BASED ON VISUAL CLUE) WHO HAVE BAC OF AT LEAST .10%
Stopping (without cause) in traffic lane	70%
Following too closely	60%
Slow response to traffic signals	50%
Headlights off (at night)	50%
Weaving	45%
Accelerating or decelerating rapidly	45%
Braking erratically	35%
Driving into opposing or crossing traffic	30%

Source: Harris, D. H. "Visual detection of driving while intoxicated." *Human Factors*, Dec. 1980, Vol. 22, pp. 725–732. Copyright 1980 by the Human Factors Society, Inc., and reproduced by permission.

Suppose a patrol officer has stopped a nighttime driver for a traffic violation.

a. Given that the driver has been charged with following too closely behind another vehicle, what is the probability that he or she is DUI?

b. Given that the driver has been charged with stopping (without cause) in a traffic lane, what is the probability that he or she is DUI?

c. Consider the following events:

   A: A nighttime driver has a BAC of at least .10%.

   B: A nighttime driver is driving with headlights off.

   C: A nighttime driver is weaving.

   Special adjustments must be made to the Drunk Driver Detection Guide when two or more visual clues are detected. If two clues are detected, add 5% to the larger of the two corresponding percentages to obtain the conditional probability of DUI; if three or more clues are detected, add 10% to the largest of the corresponding percentages. Use this information to find the conditional probability of event *A*, given that both event *B* and event *C* have occurred.

4.47 How would you like to be an oil baron? Since 1960, parcels of land in Wyoming and New Mexico that may contain oil have been placed in a lottery with the winner receiving leasing rights for 10 years. Any U.S. citizen older than 21 can play by paying a $10 filing fee to the Bureau of Land Management. If you win—and if an oil company is interested in drilling on the parcel you obtain—you may strike it rich. The *Orlando Sentinel* (Oct. 12, 1980) reported on some suspected cases of "hanky-panky" in the conduct of the lottery, and, for several months in 1980, the lottery was suspended. One reported case involved a player who won three times in one month. The three parcels had 1,836 entries, 1,365 entries, and 495 entries, respectively. In this particular case, an Interior Department audit stated that "federal workers did a poor job of shaking the drum before the drawing."

a. Find the probability that a player would win on three parcels involving 1,836 entries, 1,365 entries, and 495 entries.

b. Is this probability consistent with the Interior Department's explanation of this particular event?

c. Based on your knowledge of probability and rare events, would you make the same inference as that made by the auditor? Explain.

THE ADDITIVE AND MULTIPLICATIVE LAWS OF PROBABILITY (OPTIONAL)

4.6 In this optional section, we define some standard probability notation and give two laws for finding probabilities. Although these laws are not required for a study of the remaining material in the text, they are needed to complete an introductory coverage of probability.

The event of Section 4.5 that occurs when both of two events A and B occur is called the **intersection of A and B**, and is denoted by $A \cap B$. The event of Section 4.3 that occurs when either A or B occur is called the **union of A and B**, and is denoted $A \cup B$.

> **DEFINITION 4.9**
>
> The **intersection** of A and B, denoted by $A \cap B$, is the event that both A and B occur.

> **DEFINITION 4.10**
>
> The **union** of A and B, denoted by $A \cup B$, is the event that either A or B occurs.

Probability Rule #3 (Section 4.5) gave a formula for finding the probability that both events A and B occur (i.e., $A \cap B$) for the special case where A and B are independent events. We now give a formula, called the **Multiplicative Law of Probability**, that applies in general—that is, regardless of whether A and B are independent events.

> **THE MULTIPLICATIVE LAW OF PROBABILITY**
>
> The probability that **both** of two events A and B occur is
>
> $$P(A \text{ and } B) = P(A \cap B) = P(A)P(B \mid A)$$
> $$= P(B)P(A \mid B)$$

EXAMPLE 4.14   Refer to Example 4.10, where we selected two fuses from a box that contained three, two of which were defective. Use the Multiplicative Law of Probability to find the probability that you first draw defective fuse D_1 and then draw D_2.

Solution   Define the following events:

   A: The second draw results in D_2.

   B: The first draw results in D_1.

The probability of event B is $P(B) = \frac{1}{3}$. Also, from Example 4.10, the conditional probability of A given B is $P(A \mid B) = \frac{1}{2}$. Then the probability that both events A *and* B occur is

$$P(A \text{ and } B) = P(B)P(A \mid B) = \left(\frac{1}{3}\right)\left(\frac{1}{2}\right) = \frac{1}{6}$$

You can verify this result by rereading Example 4.10.

An additive probability rule, Probability Rule #1, was given for the event that either A or B occurs (i.e., $A \cup B$), but it applies only to the case where A and B are mutually exclusive events. A rule that applies in general is given by the **Additive Law of Probability**.

THE ADDITIVE LAW OF PROBABILITY

The probability that **either** an event A or an event B **or both** occur is

$$P(A \text{ or } B) = (P(A \cup B) = P(A) + P(B) - P(A \cap B)$$

EXAMPLE 4.15   Suppose an experiment consists of tossing a pair of coins and observing the upper faces. Define the following events:

  $A:$ Observe at least one head.

  $B:$ Observe at least one tail.

Use the Additive Law of Probability to find the probability of observing either A or B or both.

Solution   We know the answer to this question before we start because the probability of observing at least one head or at least one tail is 1—i.e., the event is a certainty. To obtain this answer using the Additive Law of Probability, we could use the method of Example 4.6 to find

$$P(A) = P(\text{at least one head}) = \frac{3}{4}$$

$$P(B) = P(\text{at least one tail}) = \frac{3}{4}$$

The event that both A and B occur—observing at least one head and at least one tail—is the event that you observe exactly one head and exactly one tail. We found this probability in Example 4.6 to be $\frac{1}{2}$. Therefore,

$$P(\text{either } A \text{ or } B \text{ or both occur}) = P(A) + P(B) - P(\text{both } A \text{ and } B \text{ occur})$$

$$= \frac{3}{4} + \frac{3}{4} - \frac{1}{2} = 1$$

This answer confirms what we already knew, that the probability of the event is equal to 1.

In Examples 4.14 and 4.15, two key words helped us identify which probability law to employ. In Example 4.14, the key word was *and*, as in "find the probability that both *A and B* occur." The word *and* implies intersection; therefore, we use the Multiplicative Law of Probability. Alternatively, the key word was *or* in Example 4.15, as in "find the probability that either *A or B or both* occur." The word *or* implies union; therefore, we use the Additive Law of Probability.

EXAMPLE 4.16 Psychologists believe that there is a relationship between aggressiveness and order of birth. To test this belief, a psychologist randomly chose 1,000 elementary school children and administered to each a test designed to measure the student's aggressiveness. Each student was then classified according to aggressiveness (aggressive or unaggressive) and order of birth (firstborn, secondborn, or other). The percentages of students falling in the six categories are shown in Table 4.1.

TABLE 4.1
Results of Aggressiveness Test

		ORDER OF BIRTH			
		Firstborn	Secondborn	Other	TOTALS
AGGRESSIVENESS	Aggressive	6	10	8	24
	Unaggressive	12	33	31	76
TOTALS		18	43	39	100

Suppose we use the percentages contained in the cells of the table to give the approximate probabilities that a single elementary school student would fall in the respective categories. We will define the following events:

A: The student is aggressive.

B: The student was firstborn.

a. Find the probability that both *A* and *B* occur.
b. Find the conditional probability that *A* will occur given that *B* has occurred.
c. Find the probability that *A* will not occur.
d. Find the probability that either *A* or *B* or both occur.

Solution

a. We can see from the table that 6% of the students were classified as both aggressive (*A*) and firstborn (*B*). Therefore,

$$P(A \text{ and } B) = P(A \cap B) = .06$$

b. Again, examining Table 4.1, we find that 24% of all the students were classified as aggressive and 18% were the firstborn in their family. Therefore,

$$P(A) = .24 \quad \text{and} \quad P(B) = .18$$

To find $P(A \mid B)$, we substitute the answer to part **a** and the value of $P(B)$ into the formula for the Multiplicative Law of Probability. Thus,

$$P(A \text{ and } B) = P(A \cap B) = P(B)P(A \mid B)$$

or

$$.06 = (.18)P(A \mid B)$$

Solving for $P(A \mid B)$ yields

$$P(A \mid B) = \frac{.06}{.18} = .333$$

c. The event that A does not occur is the complement of A, denoted by the symbol $\overline{A}$. Since A is the event that a student is aggressive, $\overline{A}$ is the event that a student is unaggressive. Recall that $P(A)$ and $P(\overline{A})$ bear a special relationship to each other:

$$P(A) + P(\overline{A}) = 1$$

From part **b**, we have $P(A) = .24$. Therefore,

$$P(\overline{A}) = 1 - .24 = .76$$

which can be verified by examining Table 4.1.

d. The probability that either A or B or both occur is given by the Additive Law of Probability. From parts **a** and **b** we know that

$$P(A) = .24$$
$$P(B) = .18$$
$$P(A \cap B) = .06$$

Then,

$$P(A \text{ or } B) = P(A) + P(B) - P(A \cap B)$$
$$= .24 + .18 - .06 = .36$$

EXERCISES

LEARNING THE MECHANICS

4.48   For two events A and B, $P(A) = .3$, $P(B) = .5$, and $P(A \cap B) = .2$. Find $P(A \cup B)$.

4.49   For two events A and B, $P(A) = .5$, $P(B) = .6$, and $P(A \cap B) = .4$. Find $P(A \cup B)$.

4.50   Consider the experiment of tossing a pair of dice. Define events A and B as follows:

   A: Observe a sum of 7.

   B: Observe a 4 on at least one die.

   a. List the possible outcomes in event A and find $P(A)$.
   b. List the possible outcomes in event B and find $P(B)$.
   c. Find $P(A \cap B)$ using the Multiplicative Law of Probability. Then list the possible outcomes associated with this event, and find $P(A \cap B)$ by summing the probabilities of these outcomes.
   d. Find $P(A \cup B)$ using the Additive Law of Probability. Then list the possible outcomes associated with this event, and find $P(A \cup B)$ by summing the probabilities of these outcomes.

4.51   Refer to the study of automobile safety discussed in Exercise 2.45. The researchers also investigated the frequency of safety seat belt usage among automobile owners. Each in a sample of 387 drivers was classified according to frequency of use (always, frequently, infrequently, and never) and state of residence (states with mandatory safety seat belt laws, states with pending mandatory laws, and states without mandatory laws). The results are shown in the table.

STATE OF RESIDENCE	SEAT BELT USAGE				TOTALS
	Always	Frequently	Infrequently	Never	
Mandatory Seat Belt Law	67	24	18	19	128
Pending Mandatory Seat Belt Law	27	20	23	8	78
No Mandatory Seat Belt Law	63	42	38	38	181
TOTALS	157	86	79	65	387

Source: Lieb, R. C., Wiseman, F., and Moore, T. E. "Automobile safety programs: The public viewpoint." *Transportation Journal*, Vol. 25, No. 4, Summer 1986, p. 25. Permission to reprint this material for educational purposes has been granted by the American Society of Transportation and Logistics, publisher of the *Transportation Journal*.

Suppose we select one of the 387 drivers in the study.

a. Find the probability that the driver resides in a state with no mandatory seat belt law.
b. Find the probability that the driver uses safety seat belts infrequently.
c. Find the probability that the driver resides in a state with a pending mandatory seat belt law and never uses seat belts.
d. Find the probability that the driver either resides in a state with a mandatory seat belt law or always uses seat belts.
e. Given that the driver never uses seat belts, what is the probability that the driver resides in a state with no mandatory seat belt law?
f. Given that the driver resides in a state with a pending mandatory seat belt law, what is the probability that the driver frequently uses seat belts?

4.52   The National Acid Precipitation Assessment Program (NAPAP) has recently concluded a 10-year study of acid rain. In its report, NAPAP estimates the probability of an Adirondack lake being acidic at .14. Given that the Adirondack lake is acidic, the probability that the lake comes by its acidity naturally is .25 (*Science News*, Sept. 15, 1990). Use this information to find the probability that an Adirondack lake is naturally acidic.

4.53   Refer to the Jackpot scratch-off game described in Exercise 4.15. Recall that eight of the 10 play squares reveal three cherries, "worth up to $1,000 in cash or prizes." Given that you scratch off three cherries, you will receive one of the following prizes (with associated probabilities):

$1,000 cash (.0000125)   Grandfather clock (.0000625)
$500 cash (.0000125)   Designer watch (.0000625)
VCR (.0000125)   Assorted prizes (.9998375)

a. Given that you scratch off three cherries, find the probability that you win cash.
b. What is the unconditional probability that you win cash in Jackpot?

4.54   A study conducted by the National Opinion Research Center found that Hispanics who are proficient in English do better in school than those who communicate primarily in Spanish (*Demographics*, June 1983). Of those students who speak only Spanish at home, 41% earn a high school grade average of B or better. In contrast, 53% of all students earn a grade of B or better. It is also known that 5% of all students speak only Spanish at home, while 89% speak only English at home. Consider the following events:

A: A student earns a grade of B or better.

B: A student speaks only Spanish at home.

C: A student speaks only English at home.

   a. Find $P(A)$.
   b. Find $P(B)$.
   c. Find $P(C)$.
   d. Find $P(A \mid B)$.
   e. Find $P(A \text{ and } B)$.
   f. Which (if any) of the pairs of events, A and B, A and C, or B and C, are mutually exclusive?

4.55 As part of a study of phonological dyslexia in a highly literate subject, researchers asked 22 undergraduate students to spell a list of familiar, but commonly misspelled words (*Quarterly Journal of Experimental Psychology*, Aug. 1985). The purpose of the study was to compare the misspellings of the control group (i.e., the 22 undergraduates) with the misspellings of the dyslexic subject. Several words on the list, along with the percentage of students in the control group who misspelled the words, are given in the accompanying table.

WORD	PERCENT MISSPELLING
sacrilegious	95
Gandhi	86
professor	36
khaki	23

Suppose one of the 22 students in the control group is selected at random.

   a. What is the probability that the student misspelled the word *sacrilegious*?
   b. What is the probability that the student misspelled the word *khaki*?
   c. What is the probability that the student spelled the word *professor* correctly?
   d. Suppose that 14% of the students in the control group misspelled both the words *Gandhi* and *khaki*. What is the probability that the selected student misspelled either *Gandhi* or *khaki*, or both?

4.56 In the summer of 1941, two incredible and well-publicized events occurred during the major league baseball season: (1) Ted Williams of the Boston Red Sox became the last player to record a batting average over .400, and (2) Joe DiMaggio of the New York Yankees had at least one hit in 56 consecutive games—the longest hitting streak in major league baseball history. Baseball fans and experts alike continually argue about which accomplishment is more likely to be duplicated first. *Chance* (Fall 1991) assessed the probabilities of the events occurring within the next 50 years as $P(\text{bat over } .400) = .035$ and $P(\text{56-game hitting streak}) = .016$ under specific conditions (i.e., a league batting average of .270).

   a. Given a current major league baseball star (e.g., Tony Gwynn of the San Diego Padres) has just completed a 56-game hitting streak, what is the probability that he will end the season with a batting average of over .400? Assume the two events, {bat over .400} and {56-game hitting streak}, are dependent.
   b. Explain why it is not possible to calculate the conditional probability of part **a**.
   c. Calculate the conditional probability of part **a**, assuming the two events are independent.
   d. Based on your knowledge of baseball, batting averages, and hitting streaks, assess the likelihood of the two events being independent.

4.57 According to the 1985 National Crime Survey, "Household burglary ranks among the more serious felony crimes, not only because it involves the illegal entry of one's home, but also because a substantial proportion of violent crimes that occur in the home take place during a burglary incident." The survey found that 3.8% of all household burglaries over the 1973–1982 period resulted in a violent crime. A breakdown of these violent burglaries by type of crime and type of entry is shown in the table. (Entries in the table are percentages.)

| | TYPE OF ENTRY | | | |
TYPE OF CRIME	Forcible Entry	Unlawful Entry	Attempted Forcible Entry	TOTALS
Rape	3.5	6.3	0.3	10.1
Robbery	9.6	17.2	1.7	28.5
Aggravated Assault	4.9	12.7	5.0	22.6
Simple Assault	8.1	25.0	5.7	38.8
TOTALS	26.1	61.2	12.7	100.0

Source: *Household burglary*, Bureau of Justice Statistics Bulletin, U.S. Department of Justice, Jan. 1985, NCJ-96021.

a. What is the probability that a household burglary committed between 1973 and 1982 resulted in a violent crime?

Assuming a household burglary committed between 1973 and 1982 did, in fact, result in a violent crime, answer parts b–e.

b. What is the probability that an assault (simple or aggravated) was committed?
c. What is the probability that a robbery with forcible entry occurred?
d. Given an unlawful entry, what is the probability a rape was committed?
e. What is the probability that either a forcible entry or an aggravated assault occurred?

4.58 The merging process from an acceleration lane to the through lane of a freeway constitutes an important aspect of traffic operation at interchanges. From a study of parallel and tapered interchange ramps in Israel, the accompanying table provides information on traffic lags (where a lag is defined as an interval of time between arrivals of major streams of vehicles) accepted and rejected by drivers in the merging lane.

TYPE OF INTERCHANGE LANE	TRAFFIC CONDITION ON FREEWAY	NUMBER OF MERGING DRIVERS ACCEPTING THE FIRST AVAILABLE LAG	NUMBER OF MERGING DRIVERS REJECTING THE FIRST AVAILABLE LAG
Tapered	Heavy traffic	16	115
	Little traffic	67	121
Parallel	Heavy traffic	40	139
	Little traffic	144	331

Source: Polus, A. and Livneh, M. "Vehicle flow characteristics on acceleration lanes." *Journal of Transportation Engineering*, Vol. III, No. 6, Nov. 1985, pp. 600–601 (Table 4).

a. What is the probability that a driver in a tapered merging lane with heavy traffic will accept the first available lag?
b. What is the probability that a driver in a parallel merging lane will reject the first available lag in traffic?
c. Given that a driver accepts the first available lag in little traffic, what is the probability that the driver is in a parallel merging lane?

4.59 A survey of 300 children in the San Francisco Bay area showed that 90% lived in families owning pets. Of these, 30% owned dogs only, 18% owned dogs and other pets, 22% owned cats only, 22% owned cats and other pets, and 8% owned only pets other than dogs and cats (*Psychological Reports*, 1985, Vol. 57). For a randomly selected child, find the following probabilities:

a. The child owns a pet. b. The child owns a dog only.
c. The child owns a cat only. d. The child owns a dog.

4.60 In a study of 15,000 overweight women (*Science*, Feb. 5, 1982), researchers at the Medical College of Wisconsin found that approximately 25% were upper-body obese (excess weight distributed in the waist, chest, neck, and arms). National statistics have shown that 40% of all women in the United States are obese. Suppose a woman is chosen at random in the United States.

a. Find the probability she is obese.
b. Given that she is obese, find the probability that she is upper-body obese.
c. Find the unconditional probability that the woman is upper-body obese.

SUMMARY

In this chapter we introduced the notion of *experiments* whose outcomes could not be predicted with certainty in advance. The uncertainty associated with these outcomes (events) was measured by their probabilities—the relative frequencies of their occurrence in a very large number of repetitions of the experiment.

We presented four rules for finding the probabilities of events. The first two rules enable us to find the probability that either one or the other of two events will occur when the events are *mutually exclusive* (Probability Rule #1), and when all possible outcomes of the experiment are both mutually exclusive and

---

PROBABILITY RULES

1. *Additive Rule:* If two events A and B are mutually exclusive, then

$$P(A \text{ or } B) = P(A \cup B) = P(A) + P(B)$$

In general,*

$$P(A \text{ or } B) = P(A \cup B) = P(A) + P(B) - P(A \text{ and } B)$$

2. *Modified Additive Rule for Equally Likely Mutually Exclusive Events:* If an experiment results in one and only one of M equally likely mutually exclusive (simple) events, of which m of these result in an event A, then

$$P(A) = \frac{m}{M}$$

3. *Multiplicative Rule:* If two events A and B are independent, then

$$P(A \text{ and } B) = P(A \cap B) = P(A)P(B)$$

In general,*

$$P(A \text{ and } B) = P(A \cap B) = P(A)P(B \mid A) = P(B)P(A \mid B)$$

4. *Rule of Complements:* $P(A) = 1 - P(\overline{A})$

5. *Combinatorial Rule:* The number of different ways to select n objects from a total of N is

$$\binom{N}{n} = \frac{N!}{n!(N - n)!} \qquad \text{where} \quad N! = N(N - 1)(N - 2) \cdots (2)(1).$$

---

*From the optional section in this chapter.

equiprobable (Probability Rule #2). Probability Rule #3 provides a formula for finding the probability that both of two events will occur when the two events are *independent*. A fourth rule gives the probability of the *complement* of an event. In the optional section of this chapter we gave probability rules that apply in general—the Multiplicative Law of Probability and the Additive Law of Probability. These probability rules, as well as the Combinatorial Rule, are summarized in the box. Finally, we used the *rare event* approach to illustrate how probability plays a role in statistical inference. We drew a sample from a population and then, based on the probability of observing the sample under various assumptions about the population, we made a decision concerning the nature of the sampled population.

We will not be using the probability rules to solve probability problems in the succeeding chapters because the sample probabilities that we need are too difficult to obtain (that is, their calculation is beyond the scope of this text). Nevertheless, the basic concepts of probability covered in this chapter will be of considerable benefit in understanding how probability plays a role in the inferential methods that follow.

KEY WORDS

Complementary events	Mutually exclusive events
Conditional probability	Probability
Event	Rare event approach
Experiment	Simple events
Independent events	Uncertainty
Intersection*	Union*

*From the optional section in this chapter.

SUPPLEMENTARY EXERCISES

[*Note:* Starred (*) exercises refer to the optional section in this chapter.]

4.61 A child psychologist is interested in the ability of 5-year-old children to distinguish between imaginative and nonfiction stories. Two 5-year-old children are selected from a kindergarten class and are given a test to determine if they can distinguish imagination from reality. Consider the following events:

A: The first child can distinguish imagination from reality.

B: Neither child can distinguish imagination from reality.

C: The second child can distinguish imagination from reality, but the first child cannot.

Explain whether the pairs of events, A and B, A and C, B and C, are mutually exclusive.

4.62 Californians living along the San Andreas fault constantly fear the "Big One," i.e., a devastating major earthquake. According to the U.S. Geological Survey, "the question is not whether a [major] earthquake is coming. The question is when." The accompanying table gives survey estimates of the probability of a major earthquake striking a particular area along the fault line before the year 2018.

LOCATION	PROBABILITY OF EARTHQUAKE
North Coast	Less than .10
Northern East Bay	.20
San Francisco Peninsula	.20
Southern East Bay	.20
South Santa Cruz Mountains	.30
Parkfield	.90
Cholame	.30
Carrizo	.10
Mojave	.30
San Bernardino Mountain	.20
San Bernardino Valley	.20
San Jacinto Valley	.10
Anza	.30
Coachella Valley	.40
Borrego Mountain	Less than .10
Imperial	.50

Source: U.S. Geological Survey; *Time,* Oct. 30, 1989.

a. Are the events listed in the table mutually exclusive? Why?
b. Do you believe the events listed in the table are independent? Why?
c. Explain why the probabilities listed in the table do not sum to 1.
d. Northern East Bay, San Francisco Peninsula, Southern East Bay, and South Santa Cruz Mountains are fault line segments in the San Francisco area. What is the probability that a major earthquake will strike the San Francisco area before the year 2018? What did you assume about the four locations to answer this question?

4.63 In business, how often do corporate managers commit a "courageous" act, i.e., voice an unpopular position, speaking out despite the possibility that the action might hurt their careers? A *Psychology Today* (Sept. 1986) study of the actions of American managers from 1969 to 1983 revealed that managers were six times more likely to act courageously during the years 1981–1983 than during the recessionary period of 1972–1974. The table gives the probability of a courageous act for each of the periods studied. [*Note:* These probabilities were obtained by dividing the number of courageous acts reported in a given period by the total number reported over the years 1969–1983.]

PERIOD	PROBABILITY OF A COURAGEOUS ACT
1969–1971	.21
1972–1974	.07
1975–1977	.13
1978–1980	.15
1981–1983	.44

Source: Hornstein, H. A. "When corporate courage counts." *Psychology Today,* Sept. 1986, p. 58. Copyright © 1986 (PT Partners, L.P.). Reprinted with permission.

a. Verify that the probabilities in the table sum to 1.
b. What is the probability that a courageous act committed by an American manager in the study period occurred prior to 1975?
c. What is the probability that a courageous act committed by an American manager in the study period did not occur during 1981–1983?

4.64 To find the probability that a randomly selected United States citizen was born in a given state—Virginia, for example—an introductory statistics student divides the number of favorable outcomes, one, by the total number of states, 50. Thus, the student reports that the probability that John Q. Citizen was born in Virginia is $\frac{1}{50}$. Explain why this probability is incorrect. If you had access to the complete 1990 census, how could the correct probability be obtained?

4.65 A recent study on shoplifting for the National Mass Retailing Institute (*Orlando Sentinel*, Dec. 11, 1983) revealed that retailers blame shoplifting for only 30% of their shrinkage (disappearance of inventory). Fifty percent of the shrinkage is attributed to employee theft and 20% to poor paperwork control.

a. Find the probability that a missing sales item was stolen.
b. Find the probability that a missing sales item was not stolen.

4.66 The illegal "numbers game," one of organized crime's largest sources of revenue, operates in the following manner: A player selects a three-digit number, such as 987 or 243, and then places a bet with an agent, who might be a shopkeeper, an office worker, a newsstand operator, etc. These bets are picked up by a "runner" (who receives a commission of between 10% and 25%) who passes the money on to the numbers "bank" that finances the operation and pays off the winners. In New York, the winning number is usually based on "the handle," the total dollar amount bet on each of the third, fifth, and seventh races at one of the local racetracks. The winning number is obtained by taking the last digit of the handle for each of the three races. Bets can be as small as 50¢ or $1, and the payoff to a winner is 500 to 1. For example, if you bet $1 and win, your return is $499 ($500, less the dollar placed with the agent).

a. What is the probability that the winning first digit will be a 1?
b. What is the probability that the winning second digit will be a 3? The winning third digit will be a 9?
c. What is the probability that a person selecting one three-digit number will win?
d. Considering your answer to part c, do you think the payoff rate is reasonable for the player?

4.67 There are very few (if any) tests for pregnancy that are 100% accurate. Sometimes a test may indicate that a woman is pregnant even though she really is not. This is known as a *false positive* test. Similarly, a *false negative* result occurs when the test indicates that the woman is not pregnant even though she really is. Suppose a woman submits to a certain pregnancy test. Define the events *A*, *B*, and *C* as follows:

A: The woman is really pregnant.

B: The test gives a false positive result.

C: The test gives a false negative result.

Which pair(s) of events are mutually exclusive?

4.68 Managers of oil exploration portfolios make decisions on which prospects to pursue based, in part, on the level of risk associated with each venture. Kinchen (1986) examined the problem of risk analysis in oil exploration using the outcomes and associated probabilities for a single prospect shown in the table.

a. What is the probability that a single oil well prospect will result in no more than 100,000 barrels of oil?
b. What is the probability that a single oil well prospect will strike oil?

OUTCOME (BARRELS)	PROBABILITY
0 (dry hole)	.60
50,000	.10
100,000	.15
500,000	.10
1,000,000	.05

Source: Kinchen, A. L. "Projected outcomes of exploration programs based on current program status and the impact of prospects under consideration." *Journal of Petroleum Technology*, Vol. 38, No. 4, Apr. 1986, pp. 461–467. Copyright 1986 Society of Petroleum Engineers.

c. Kinchen also considered two identical oil well prospects. List the possible outcomes if the two wells are drilled. Assume that the outcomes listed in the table are the only possible outcomes for any one well. [*Hint:* One possible outcome is two dry holes.]

d. Use the information in the table to calculate the probabilities of the outcomes listed in part **c**. (Assume that the individual outcomes of the two wells are independent of each other.)

e. Refer to part **d**. Find the probability that at least one of the two oil prospects strikes oil.

4.69 The Department of Housing collects data on loan size, loan value, default status, and various other characteristics of all FHA home mortgages. Housing planners and credit institutions use this information to assess the risk of default in home mortgages. The accompanying table gives the probability of default for several different types of home loans at 95%–100% of value.

LOAN	RACE OF HOME OWNER	HOME LOCATION	PROBABILITY OF DEFAULT
Above $20,750	White	City	.0344
Below $20,750	White	City	.0389
Above $20,750	Black	City	.1250
Above $20,750	Black	Suburb	.1695

Source: FHA Cross Reference File, 1978. Data also reported in Evans, R. D., Mans, B. A., and Weinstein, R. I. "Expected loss and mortgage default risk." *Quarterly Journal of Business and Economics*, Winter 1985, Vol. 24, No. 1, p. 77.

a. What is the probability that a white home owner, located in the city, will default on a loan of greater than $20,750?

b. What is the probability that a black home owner, located in the city, will default on a loan of greater than $20,750?

c. Suppose that 70% of the blacks who own homes are city dwellers and 30% reside in the suburbs. What is the probability that a black home owner will default on a loan of greater than $20,750?

4.70 The New York State Bureau of Fisheries has reported that acid rain and snowfall—originating from oxides of nitrogen and sulfur deposited in the atmosphere from industrial burning of coal and from automobile exhaust—have killed off all fish and many plants in 50% of the high elevation lakes in the Adirondack Mountains. Of all 2,800 lakes in the entire American Adirondack area (including the high-elevation lakes), at least 10% have been found to contain no fish.

a. If one of the high-elevation lakes in the Adirondack Mountains is randomly selected and tested, what is the probability that it contains fish?

b. If one of the 2,800 lakes in the entire American Adirondack area is randomly selected and tested, what is the probability that it will be found to contain no fish?

*4.71 Despite penicillin and other antibiotics, bacterial pneumonia still kills thousands of Americans every year. An antipneumonia vaccine, called Pneumovax, is designed especially for elderly or debilitated patients, who are usually the most vulnerable to bacterial pneumonia. Suppose the probability that an elderly or debilitated person will be exposed to these bacteria is .45. If exposed, the probability that an elderly or debilitated person inoculated with the vaccine acquires pneumonia is only .10. What is the probability that an elderly or debilitated person inoculated with the vaccine does acquire bacterial pneumonia?

4.72 A daily newspaper operates with two high-speed printing presses (presses #1 and #2). The manufacturer of these high-speed presses claims that, when operating properly, the machines shut down for repairs on only 1% of the operating days. Suppose that the presses operate independently—that is, the chance of one press breaking down is in no way influenced by the current operating condition of the other. One operating day is randomly selected and the performance of the presses is observed.

a. What is the probability that press #1 will be shut down for repairs?
b. What is the probability that press #2 will not need to be shut down for repairs?
c. What is the probability that both presses will be shut down for repairs?
d. Suppose that both presses actually do need to be shut down for repairs during the operating day. Based on this observation, what would you infer about the press manufacturer's claim?

4.73 One of the most popular card games among Americans is the game of poker. Each player in the game is dealt five cards from a standard 52-card bridge deck. The player with the best (as defined by the rules of the game) five-card hand is declared the winner. Use the Combinatorial Rule to determine the number of different five-card poker hands that can be dealt from a 52-card deck.

4.74 Recent survey research reveals a tremendous variability in consumer purchasing experience and behavior in less developed countries (LDCs). For example, one large multinational corporation failed in its attempt to sell baby food in Africa using a package designed for its home country—Africans interpreted the labels to mean the jars contained ground-up babies! How do multinational corporations market and adapt consumer products from developed countries in LDCs? To answer this question, Hill and Still (1984) surveyed subsidiaries of consumer goods manufacturers with operations in LDCs. The survey results were used to assess the likelihood that a specific factor (for example, consumer preference or social custom) caused the change in a product characteristic (for example, brand name). The probabilities for three product characteristics are shown in the table.

| | PRODUCT CHARACTERISTIC | | |
FACTOR	Brand Name	Labeling	Package Aesthetics
Consumer preferences	.363	.314	.530
Competition	.018	.205	.269
Literacy and education	.200	.051	.000
Legal considerations	.309	.340	.000
Sociocultural customs and taboos	.072	.051	.037
Other	.038	.039	.164

Source: Adapted from Hill, J. S. and Still, R. R. "Adapting products to LDC tastes." *Harvard Business Review*, Mar.–Apr. 1984, Vol. 62, No. 2, p. 96.

a. Show that the factor probabilities sum to 1 for each product characteristic.
b. Given that the brand name of a product is changed, what is the probability that the change is caused by a marketing factor (i.e., either consumer preference or competition)?
c. Repeat part b for the product characteristic labeling.
d. Repeat part b for the product characteristic package aesthetics.
e. For each of the three product characteristics, find the probability that the factor causing the change is not an LDC's customs and taboos.

4.75 The most common data-collection method in consumer and market research is the telephone survey. However, a major problem with consumer telephone surveys is nonresponse. How likely are consumers to be at home to take the call and, if at home, how likely are they to take part in the survey? To answer these and other questions, R. A. Kerin and R. A. Peterson directed a study of over 250,000 random-digit dialings of both listed and unlisted telephone numbers across the United States. The study enabled the researchers to assess the probabilities of various outcomes on the first dialing attempt, as shown in the table on the next page.
a. What is the probability that a single call will result in no answer, a busy signal, or an out-of-service number?
b. What is the probability that an eligible person will be at home to take the call?
c. Given that an eligible person is at home to take the call, what is the probability that he or she will refuse to participate in the interview? (This probability is known as the *refusal rate*.)

RESULT OF DIALING ATTEMPT	PROBABILITY
No answer	.347
Busy signal	.020
Out-of-service	.203
No eligible person at home	.291
Business number	.041
Eligible person at home—refusal	.014
Eligible person at home—completed interview	.084

Source: Kerin, R. A. and Peterson, R. A. *Journal of Advertising Research*, Apr./May 1983.

4.76 A panel studying emergency evacuation plans for Florida's Gulf Coast in the event of a hurricane has determined that it would take between 14 and 17 hours to evacuate people living in low-lying land, with the probabilities shown in the table.

EVENT	TIME TO EVACUATE	PROBABILITY
A	14 hours	.27
B	15 hours	.42
C	16 hours	.18
D	17 hours	.13

a. What is the probability that the residents of low-lying coastal areas will take at least 16 hours to evacuate their homes?

b. What is the probability that the residents of low-lying coastal areas will take less than 17 hours to evacuate their homes?

c. Weather forecasters say they cannot accurately predict a hurricane landfall more than 14 hours in advance. If the civil engineering department of the Gulf Coast waits until the 14-hour warning before beginning evacuation, what is the probability that all residents of low-lying areas are evacuated safely (i.e., before the hurricane hits the Gulf Coast)?

4.77 The game of craps is played with two dice. A player throws both dice, winning unconditionally if he produces either of the outcomes 7 or 11 (the sum of the numbers showing on the two dice), which are designated as *naturals*. If the player casts the outcome 2, 3, or 12—referred to as *craps*—he loses unconditionally.

a. Find the probability of a player throwing a natural.

b. Find the probability of a player throwing craps.

c. Suppose that a "hot" player has thrown five naturals in a row. What is the probability that the player throws a natural on his next toss?

d. Suppose that a "cold" player has thrown five craps in a row. What is the probability that the player throws craps on his next toss?

4.78 Refer to Exercise 4.77. In the two-dice game of craps, a player wins if he throws a *natural* (a 7 or 11) and loses if he throws *craps* (a 2, 3, or 12). However, if the sum of the two dice is 4, 5, 6, 8, 9, or 10 (each of these is known as a *point*), the player continues throwing the dice until the same outcome (point) is repeated (in which case the player wins) or the outcome 7 occurs (in which case the player loses). For example, if a player's first toss results in a 6, the player continues to toss the dice until a 6 or 7 occurs. If a 6 occurs first, the player wins. If a 7 occurs first, the player loses.

a. What is the probability that a player throws a point on the first toss? [*Hint:* Find P(4 or 5 or 6 or 8 or 9 or 10).]

b. If a player throws a point of 6 on the first toss, what is the probability that the player wins the game on the next toss?

c. If a player throws a point of 6 on the first toss, what is the probability that the player loses the game on the next toss?

4.79 Refer to Exercises 4.77 and 4.78. From the information provided by these exercises, it can be seen that there are basically three events that result in a win for the craps player:

> A: The player throws a 7 on the first toss.
>
> B: The player throws an 11 on the first toss.
>
> C: The player throws a point on the first toss, and throws the same point on a subsequent toss before throwing a 7.

Since the events A, B, and C form pairs of mutually exclusive events, the probability that the player wins the game—that is, the probability of *making a pass*—is simply $P(A) + P(B) + P(C)$. It can be shown (proof omitted) that the probability of a player making a pass is .493.

a. Interpret this win probability.

b. In most casinos, betting that a player makes a pass pays off at *even odds*—i.e., for every $1 bet, you win $1 if the player makes a pass. Considering the .493 probability of winning, do you think that the even payoff odds are fair, that is, if you repeatedly bet on a player to make a pass, would you expect to win as much money as you lose? [*Hint:* When the payoff odds for a winning bet are even, the game is deemed fair if the probability of winning the bet is .50.]

*4.80 Most national advertisers use mixed-media advertising campaigns that include both television and magazines. The key to a successful campaign is the correct apportionment of the advertising budget to be spent in each medium. The *Journal of Marketing Research* (Feb. 1984) reported on a model for evaluating mixed-media advertising schedules, based on the probabilities of the following events:

> A: A consumer views a typical television show in the schedule.
>
> B: A consumer reads a typical magazine issue in the schedule.
>
> C: A consumer views a typical television show and reads a typical magazine issue in the schedule.
>
> D: A consumer is *not* exposed to either a typical television show or a typical magazine issue in the schedule.

Assume that an individual is selected from the population of potential buyers of a product. The advertisers of the product have developed a mixed-media campaign, with the following probabilities:

$$P(A) = .20 \qquad P(B) = .08 \qquad P(C) = .03$$

a. Find $P(A \text{ or } B)$.

b. Describe, in words, the probability of part **a**.

c. Describe the event $\bar{D}$.

d. Use your answers to parts **a**–**c** and the probability relationship for complementary events to find $P(D)$.

*4.81 The U.S. Command, Control, Communication and Intelligence (C^3I) System includes sensors (e.g., satellites and radars), communication links, and computer systems that allow gathering and processing of information that a missile attack on the continental United States may be on the way. The C^3I System contains two basic components: a warning system, designed to detect missile attacks on the United States, and a response system, designed to launch a counterattack. The response system cannot launch a missile unless a signal is detected by the warning system. However, if a warning signal is received, it is possible the response system will fail to launch a counterattack.

Paté-Cornell and Neu conducted a probabilistic analysis of the reliability of the C^3I System (*Risk Analysis*, Vol. 5, No. 2, 1985). In particular, the researchers were interested in the probabilities of an accidental strike by the C^3I System, i.e., the probability that the system launches a nuclear missile toward the USSR based on a false alert.

Suppose the conditional probability that the response system will launch a missile toward the (former) USSR, given a false alert from the warning system, is .90. Suppose also that the probability of false alert (i.e., the probability that the warning system sends a signal to the response system, given no Soviet attack is made) is .02. If the probability of a Soviet missile attack on the continental United States is .01, find the probability of an accidental strike by the C^3I System. [Hint: For three events, A, B, and C, the formula

$$P(A \cap B \cap C = P(C \mid A \cap B) \cdot P(B \mid A) \cdot P(A)$$

gives the probability of their intersection.]

CASE STUDY 4.1
LOTTERY BUSTER!

"Welcome to the Wonderful World of Lottery Bu$ters." So begins the premier issue of *Lottery Buster*, a monthly publication for players of the state lottery games. *Lottery Buster* provides interesting facts and figures on the 27 state lotteries currently operating in the United States and, more importantly, tips on how to increase a player's odds of winning the lottery.

New Hampshire, in 1963, was the first state in modern times to authorize a state lottery as an alternative to increasing taxes. (Prior to this time, beginning in 1895, lotteries were banned in America because of corruption.) Since then, lotteries have become immensely popular for two reasons. First, they lure you with the opportunity to win millions of dollars with a $1 investment, and second, when you lose, at least you know your money is going to a good cause.

The popularity of the state lottery has brought with it an avalanche of "experts" and "mathematical wizards" (such as the editors of *Lottery Buster*) who provide advice on how to win the lottery—for a fee, of course! These experts—the legitimate ones, anyway—base their "systems" of winning on their knowledge of probability and statistics.

For example, most experts would agree that the "golden rule" or "first rule" in winning lotteries is *game selection*. State lotteries offer three types of games: Instant (scratch-off) tickets, Daily Numbers (Pick-3 and Pick-4), and the weekly Pick-6 Lotto game.

The Instant game involves scratching off the thin opaque covering on a ticket with the edge of a coin to determine whether you have won or lost. The cost of a ticket is 50¢, and the amount won ranges from $1 to $100,000 in most states, and to as much as $1 million in others. *Lottery Buster* advises against playing the Instant game because it is "a pure chance play, and you can win only by dumb luck. No skill can be applied to this game."

The Daily Numbers game permits you to choose either a three-digit (Pick-3) or four-digit (Pick-4) number at a cost of $1 per ticket. Each night, the winning number is drawn. If your number matches the winning number, you win a large sum of money, usually $100,000. You do have some control over the Daily Numbers game (since you pick the numbers that you play) and, consequently, there are strategies available to increase your chances of winning. However, the Daily Numbers game, like the Instant game, is not available for out-of-state play.

For this reason, and the fact that payoffs are relatively small, lottery experts prefer the weekly Pick-6 Lotto game.

To play Pick-6 Lotto, you select six numbers of your choice from a field of numbers ranging from 1 to N, where N depends on which state's game you are playing. For example, Florida's Lotto game involves picking six numbers ranging from 1 to 49 (denoted 6/49) as shown on the Florida Lotto ticket, Figure 4.7; Delaware's Lotto is a 6/30 game; and Pennsylvania's is a 6/40 game. The cost of a ticket is $1 and the payoff, if your six numbers match the winning numbers drawn at the end of each week, is $1 million or more, depending on the number of tickets purchased. (To date, Florida has had the largest weekly payoff of $48 million.) In addition to the grand prize, you can win second-, third-, and fourth-prize payoffs by matching five, four, and three of the six numbers drawn, respectively. And you don't have to be a resident of the state to play the state's Lotto game. Anyone can play by calling a toll-free "hotline" number.

FIGURE 4.7

Reproduction of Florida's 6/49 Lotto Ticket

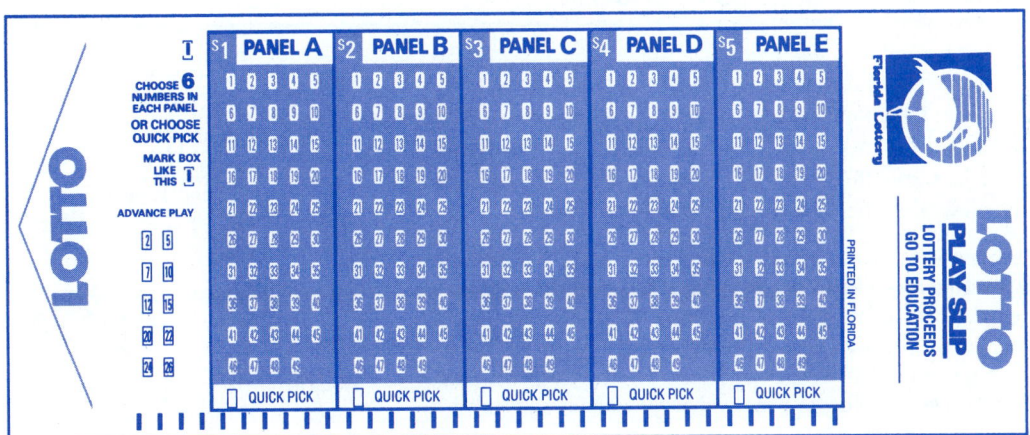

a. Consider Florida's 6/49 Lotto game. Calculate the number of possible ways in which you can choose the six numbers from the 49 available. If you purchase a single $1 ticket, what is the probability that you win the grand prize (i.e., match all six numbers)?

b. Repeat part **a** for Delaware's 6/30 game.

c. Repeat part **a** for Pennsylvania's 6/40 game.

d. Since you can play any state's Lotto game, which of the three, Florida, Delaware, or Pennsylvania, would you choose to play? Why?

e. One strategy used to increase your odds of winning a Lotto is to employ a *wheeling system*. In a complete wheeling system, you select more than six numbers, say, seven, and play every combination of six of those seven numbers. Suppose you choose to "wheel" the following seven numbers in a 6/40 game: 2, 7, 18, 23, 30, 32, 39. How many tickets would you need to purchase to have every possible combination of the seven numbers? List the six numbers on each of these tickets.

f. Refer to part **e**. What is the probability of winning the 6/40 Lotto when you wheel seven numbers? Does the strategy, in fact, increase your odds of winning?

g. Another strategy is to play neighboring pairs. Neighboring pairs are two consecutive numbers that come up together on the winning ticket. In one state lottery, for example, 79% of the winning tickets had at least one neighboring pair. Thus, some "experts" feel that you have a better chance of winning if you include at least one neighboring pair in your number selection. Calculate the probability of winning the 6/40 Lotto with the six numbers: 2, 15, 19, 20, 27, 37. [*Note:* 19, 20 is a neighboring pair.] Compare this probability to the one in part **c**. Comment on the neighboring pairs strategy.

CASE STUDY 4.2

DR. CRYPTON, BRIDGE, AND THE BIRTHDAY PROBLEM

Science Digest magazine features an interesting and informative columnist, known only by the eerie name of Dr. Crypton. Each month Dr. Crypton presents mind-twisters, riddles, puzzles, and enigmas that very often can be solved using the laws of probability. In the January 1982 issue, Dr. Crypton strayed from his usual column and offered his view of coincidences and probability theory:

> Much too much is often made of the uncanniness of rare events. For every Dame Rebecca West whose servants find a hedgehog in the garden while she happens to be writing a story about such a discovery, there are scores of novelists whose stories do not play themselves out in reality as they are being written. For every Freud or Jung who hears a poltergeist while discussing extrasensory perception, there are undoubtedly numerous people who discuss parapsychology without currently experiencing paranormal phenomena.
>
> In each of these cases, if you could count the number of "bizarre" episodes and the number of "normal" episodes, I am convinced you would find that they obey the laws of probability. If the odds against an event are a billion to one, the event can still happen. Indeed, you would *expect* to observe the event—at an average rate of one event for every billion nonevents.
>
> In other cases, the odds can be precisely calculated, and these calculations should undermine any feelings of uncanniness caused by rare events. From time to time you've probably come across a short newspaper item describing someone who was dealt a bridge hand of 13 cards of one suit. You should not react by groveling in awe at the power of the card gods.
>
> The odds of being dealt all the cards of a suit are 158,753,389,899 to 1. In *A Mathematician's Miscellany*, J. E. Littlewood points out that roughly 2 million Englishmen play an average of 30 hands of bridge a week; that's 3,120,000,000 bridge hands a year. If you add in the bridge hands that the rest of the world is playing, it is not surprising that you've heard of someone being dealt an entire suit.
>
> Another source of eerie coincidences is shared birthdays. Surely you have been in a situation in which a small group of people compared birthdays and found, to their surprise, that at least two of them were born on the same day of the same

month. Suppose there are 10 people in the group. Intuition may suggest that the odds of two sharing a birthday are quite poor. Probability theory, however, shows the odds are better than one in nine.

If you like to win bets, you should keep in mind that for a group of 23 people the odds are in favor of at least two of them sharing a birthday. (For 22 people, the odds are slightly against this.) Warren Weaver, the celebrated mathematician who wrote *Lady Luck: The Theory of Probability*, once explained these odds at a dinner party of Army and Navy officers. The group included 22 people, and one proposed that they test Weaver's explanation. "We got all around the table without a duplicate birthday," Weaver recalled. "At which point a waitress remarked, "Excuse me. But I am the twenty-third person in the room, and my birthday is May 17, just like the general's over there."

This all goes to show that you need look no further than elementary probability theory in order to understand the mundaneness of coincidences.

—*Dr. Crypton*

Consider Dr. Crypton's reference to the perfect bridge hand of 13 cards of one suit. The probability of being dealt this hand can be computed using the rules presented in this chapter.

a. Use the Combinatorial Rule to find M, the number of different bridge hands containing 13 cards that can be selected from the total of 52 cards in a standard deck.

b. Find m, the number of different bridge hands resulting in 13 cards of the same suit. (Recall that there are four suits in a standard bridge deck.)

c. Use your answers to parts **a** and **b** to compute the probability of being dealt the perfect bridge hand. [*Note:* This probability is equivalent to Dr. Crypton's odds of 158,753,389,899 to 1.]

Now refer to the birthday problem. Dr. Crypton states that "for a group of 23 people the odds are in favor of at least two of them sharing a birthday." We can compute the probability of this event using the concepts introduced in this chapter.

d. Find the probability of the complementary event—namely, that no two of a group of 23 people share the same birthday. [*Hint:* Since there are 365 days in a year, any of which may be the birthday of one of the 23 people, the probability that none of the people in the group share a birthday is

$$\underbrace{\left(\frac{365}{365}\right)}_{\substack{\text{Person} \\ \text{\#1}}} \cdot \underbrace{\left(\frac{364}{365}\right)}_{\substack{\text{Person} \\ \text{\#2}}} \cdot \underbrace{\left(\frac{363}{365}\right)}_{\substack{\text{Person} \\ \text{\#3}}} \cdots \underbrace{\left(\frac{343}{365}\right)}_{\substack{\text{Person} \\ \text{\#23}}}$$

Compute this probability.]

e. Using the Rule of Complements, find the probability that at least two of 23 people will share the same birthday.

f. Use the steps outlined in parts **d** and **e** to find the probability that at least two of 22 people will share the same birthday.

TO SWITCH OR NOT TO SWITCH

Marilyn vos Savant, who is listed in *Guinness Book of World Records Hall of Fame* for "Highest IQ," writes a monthly column in the Sunday newspaper supplement *Parade Magazine*. Her column, "Ask Marilyn," is devoted to games of skill, puzzles, and mind-bending riddles. In a recent issue, vos Savant posed the following question:*

> Suppose you're on a game show, and you're given a choice of three doors. Behind one door is a car; behind the others, goats. You pick a door—say, #1—and the host, who knows what's behind the doors, opens another door—say, #3—which has a goat. He then says to you, "Do you want to pick door #2?" Is it to your advantage to switch your choice?

Marilyn's answer: "Yes, you should switch. The first door has a $\frac{1}{3}$ chance of winning [the car], but the second has a $\frac{2}{3}$ chance [of winning the car]." Needless to say, vos Savant's surprising answer led to thousands of critical letters disagreeing with her. Many of the letters were from Ph.D. mathematicians; some of the more interesting and critical letters (that were printed in her next column) are condensed here:

- "May I suggest you obtain and refer to a standard textbook on probability before you try to answer a question of this type again?" (University of Florida)
- "Your logic is in error, and I am sure you will receive many letters on this topic from high school and college students. Perhaps you should keep a few addresses for help with future columns." (Georgia State University)
- "You are utterly incorrect about the game-show question, and I hope this controversy will call some public attention to the serious national crisis in mathematical education. If you can admit your error you will have contributed constructively toward the solution of a deplorable situation. How many irate mathematicians are needed to get you to change your mind?" (Georgetown University)
- "I am in shock that after being corrected by at least three mathematicians, you still do not see your mistake." (Dickinson State University)
- "You are the goat!" (Western State University)
- "You're wrong, but look on the positive side. If all the Ph.D's were wrong, the country would be in serious trouble." (U.S. Army Research Institute)

The logic employed by those who disagree with vos Savant is as follows: Once the host shows you door #3 (a goat), only two doors remain. The probability of the car being behind door #1 (your door) is $\frac{1}{2}$; similarly, the probability is $\frac{1}{2}$ for door #2. Therefore, in the long run (i.e., over a long series of trials), it doesn't matter whether you switch to door #2 or keep door #1. Approximately 50% of the time you will win a car, and 50% of the time you will "win" a goat.

---

*Parade Magazine, Feb. 17, 1991.

Who is correct, the Ph.D. mathematicians or Marilyn? By answering the following series of questions, you will arrive at the correct solution.

a. Before taping of the game show, the host randomly decides behind which of the three doors to put the car; the goats will go behind the remaining two doors. List the simple events for this experiment. [*Hint:* One simple event is $C_1G_2G_3$, i.e., car behind door #1 and goats behind door #2 and door #3.]

b. Randomly choose one of the three doors, as the contestant does in the game show. Now, for each simple event in part **a**, circle the selected door and put an X through one of the remaining two doors that hides a goat. (This is the door that the host shows the contestant—always a goat.)

c. Refer to the altered simple events in part **b**. Assume your strategy is to keep the door originally selected. Count the number of simple events for which this is a "winning" strategy (i.e., you win the car). Assuming equally likely simple events, what is the probability that you win the car?

d. Repeat part **c**, but assume your strategy is to always switch doors.

e. Based on the probabilities of parts **c** and **d**, "Is it to your advantage to switch your choice?"

REFERENCES

Epstein, R. A. *The Theory of Gambling and Statistical Logic*, revised ed. New York: Academic Press, 1977.

Feller, W. *An Introduction to Probability Theory and Its Applications*, 3rd ed. Vol. I. New York: Wiley, 1968.

Mendenhall, W., Wackerly, D., and Scheaffer, R. *Mathematical Statistics with Applications*, 3rd ed. Boston: PWS-Kent, 1989.

Mosteller, F., Rourke, R., and Thomas, G. *Probability with Statistical Applications*, 2nd ed. Reading, Mass.: Addison-Wesley, 1970.

Parzen, E. *Modern Probability Theory and Its Applications*. New York: Wiley, 1960.

A DISCRETE PROBABILITY DISTRIBUTION: THE BINOMIAL DISTRIBUTION

CHAPTER 5

I n 1979, the U.S. military attempted a daring helicopter mission to rescue the American hostages held by Iran. Was it doomed from the start, or was there a reasonable chance for the mission's success? The Department of Defense claimed that each of the eight helicopters used in the attempt flew 20 practice missions without malfunction. Consider the number x of the eight helicopters that malfunction during a mission. What value of x should we expect to observe, and what values of x would be considered unusual or rare events, in view of the preraid training record? In this chapter we will learn how to answer questions of this type, and will specifically address the questions concerning the Iran rescue mission in Section 5.5.

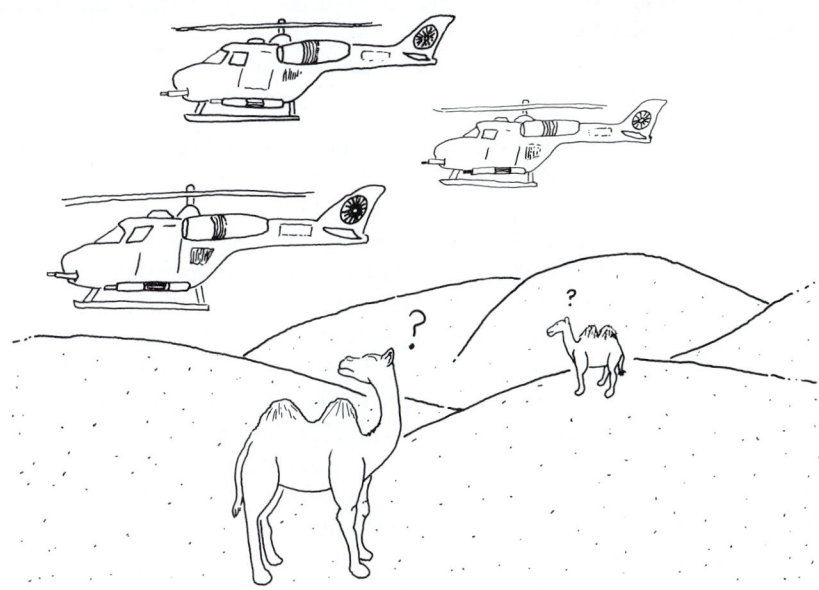

INTRODUCTION

5.1

In Chapter 4, we introduced the concept of probability and gave several useful rules for finding probabilities of events. If the experiment that generated the events has certain characteristics, the event probabilities follow a specific pattern, and can be expressed in an easy-to-use mathematical formula called a **probability distribution**. One such experiment, called a **binomial experiment**, has broad applications in the real world and is the focus of this chapter.

In Sections 5.2 and 5.3, we introduce some general concepts of probability distributions, then devote the remainder of the chapter to the binomial distribution.

RANDOM VARIABLES

5.2

In a practical setting, an experiment (as defined in Chapter 4) involves selecting a sample of data consisting of one or more observations on some variable. For example, we might survey 1,000 physicians concerning their preferences for aspirin and record x, the number who prefer a particular brand. Or, we might randomly select a single supermarket customer from Appendix E and record his or her total checkout time, x. Since we can never know with certainty in advance the exact value that we will observe when we record x for a single performance of the experiment, we call x a **random variable**.

DEFINITION 5.1

A **random variable** is a variable that assumes numerical values associated with events of an experiment.

The random variables described above are examples of two different types of random variables—**discrete** and **continuous**. The number x of physicians in a sample of 1,000 who prefer a particular brand of aspirin is said to be a discrete random variable because it can assume only a countable number of values—namely, 0, 1, 2, 3, . . . , 999, 1,000. In contrast, the checkout time of a supermarket customer is a continuous random variable because it could theoretically assume any one of an infinite number of values—namely, any value from 0 seconds upward. Of course, in practice we record checkout time to the nearest second but, *in theory*, the checkout time of a supermarket customer could assume any value in an interval of real numbers—say 137.21471 seconds.

A good way to distinguish between discrete and continuous random variables is to imagine the values that they may assume as points on a line. Discrete random variables may assume any one of a countable number (say, 10, 21, or 100) of values corresponding to points on a line. In contrast, a continuous random variable can theoretically assume *any* value corresponding to the points in one or more intervals on a line. For example, the checkout time of a supermarket customer could be represented by any of the infinitely large number of points on some portion of the positive half of a line.

EXAMPLE 5.1   Suppose you randomly select a student attending your college or university. Classify each of the following random variables as discrete or continuous:

a. Number of credit hours taken by the student this semester
b. Current grade point average of the student

Solution

a. The number of credit hours taken by the student this semester is a discrete random variable because it can assume only a countable number of values (for example, 15, 16, 17, and so on). It is not continuous since the number of credit hours cannot assume values such as 15.2062, 16.1134, and 17.0398 hours.
b. The grade point average for the student is a continuous random variable since it could theoretically assume any value (for example, 2.87355) corresponding to the points on the line interval from 0 to 4.

The focus of this chapter is discrete random variables. Continuous random variables are the topic of Chapter 6.

5.3 PROBABILITY MODELS FOR DISCRETE RANDOM VARIABLES

We learned in Chapter 4 that we make inferences based on the probability of observing a particular sample outcome. Since we never know the *exact* probability of some events, we must construct probability models for the values assumed by random variables. For example, if we toss a die, we assume that the values 1, 2, 3, 4, 5, and 6 represent equiprobable events, i.e., $p(1) = p(2) = p(3) = \cdots = p(6) = \frac{1}{6}$. In doing so, we have constructed a probabilistic model for the relative frequency distribution for the number x of dots that we would observe if we tossed the die thousands and thousands of times and recorded x for each toss. It is unlikely that a perfectly balanced die exists, but most dice would produce relative frequencies very close to $\frac{1}{6}$, and a bar graph of the results of a large number of tosses would appear as shown in Figure 5.1.

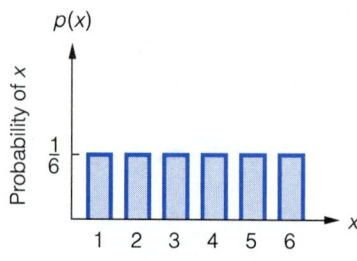

FIGURE 5.1
The Probability Distribution for x, the Number of Dots Observed on a Balanced Die

Figure 5.1, which gives the relative frequency for each value of x in a very large number of tosses of a die, is called the **probability distribution for the discrete random variable** x.

DEFINITION 5.4

The **probability distribution for the discrete random variable** x is a table, graph, or formula that gives the probability of observing each value of x. If we denote the probability of x by the symbol $p(x)$, the probability distribution has the following properties:

1. $0 \leq p(x) \leq 1$ for all values of x
2. $\sum_{\text{all } x} p(x) = 1$

EXAMPLE 5.2 Consider the following sampling situation: Draw a random sample of $n = 5$ physicians from a very large number—say, 10,000—and record the number x of physicians who favor aspirin brand A. Suppose that 2,000 of the physicians actually prefer brand A. Replace the five physicians in the population and randomly draw a new sample of $n = 5$ physicians. Record the value of x again. Repeat this process over and over again 100,000 times.

a. Construct a relative frequency histogram for the 100,000 values of x.
b. Assuming that the relative frequencies of Table 5.1 are good approximations to the probabilities of x, show that the properties of a probability distribution are satisfied.

TABLE 5.1
Relative Frequencies for 100,000 Observations on x, the Number of Physicians in a Sample of $n = 5$ Who Prefer Aspirin Brand A

x	FREQUENCY	RELATIVE FREQUENCY
0	32,807	.32807
1	40,949	.40949
2	20,473	.20473
3	5,122	.05122
4	645	.00645
5	4	.00004

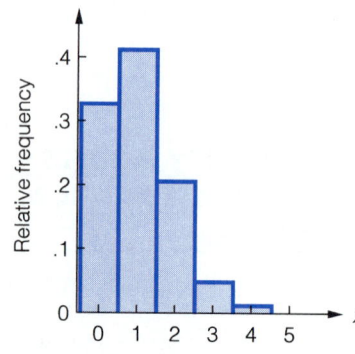

FIGURE 5.2
Relative Frequency Histogram for Example 5.2

Solution

a. The relative frequency histogram for the values of x is shown in Figure 5.2. This figure provides a very good approximation to the probability distribution for x, the number of physicians in a sample of $n = 5$ who prefer brand A (assuming that 20% of the physicians in the population prefer brand A).

b. We use the relative frequencies of Table 5.1 as approximations to the probabilities of $x = 0$, $x = 1$, ..., $x = 5$. Note that each probability (relative frequency) is between 0 and 1. Summing these probabilities, we obtain

$$.32807 + .40949 + .20473 + .05122 + .00645 + .00004 = 1$$

Thus, the properties of a discrete probability distribution are satisfied.

Often, the sample collected in an experiment (such as the physician survey of Example 5.2) is quite large. For example, the Gallup and Harris survey results reported in the news media are usually based on sample sizes from $n = 1,000$ to $n = 2,000$ people. Since we would not want to calculate $p(x)$ for values of n this large, we need an easy way to describe the probability distribution for x. To do this, we need to know the mean and standard deviation for the distribution. Then we can describe it using either the Empirical Rule or Tchebysheff's theorem from Chapter 3.

The mean μ and standard deviation σ for a discrete random variable are found using the following definitions:

DEFINITION 5.5

The **mean** μ (or **expected value**) of a discrete random variable x is equal to the sum of the products of each value of x and the corresponding value of $p(x)$:

$$\mu = \Sigma\, xp(x)$$

DEFINITION 5.6

The **variance** σ^2 of a discrete random variable x is equal to the sum of the products of $(x - \mu)^2$ and the corresponding value of $p(x)$:

$$\sigma^2 = \Sigma(x - \mu)^2 p(x) = \Sigma\, x^2 p(x) - \mu^2$$

DEFINITION 5.7

The **standard deviation** σ of a random variable x is equal to the positive square root of the variance.

EXAMPLE 5.3 Consider the sample survey of $n = 5$ physicians, described in Example 5.2. The graph of the probability distribution for x, the number of physicians in the sample who favor aspirin brand A, is reproduced in Figure 5.3.

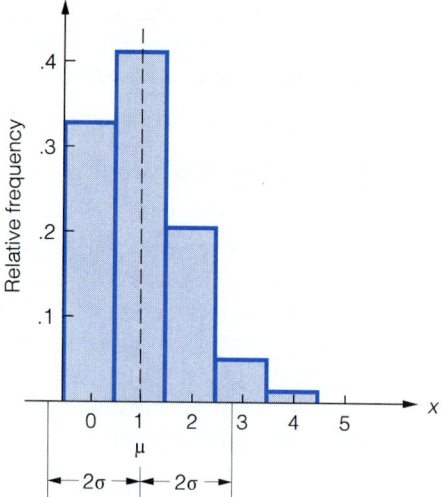

FIGURE 5.3

Probability Distribution for x, the Number of Physicians in a Sample of $n = 5$ Who Prefer Aspirin Brand A

a. Find the mean μ for this distribution. That is, find the expected value of x.
b. Interpret the value of μ.
c. Find the standard deviation of x.

Solution

a. By Definition 5.5, the mean μ is given by

$$\mu = \Sigma \, xp(x)$$

where $p(x)$ is given in Table 5.1. Since x can take values $x = 0, 1, 2, \ldots, 5$, we have

$$
\begin{aligned}
\mu &= 0p(0) + 1p(1) + 2p(2) + \cdots + 5p(5) \\
&= 0(.32807) + 1(.40949) + 2(.20473) + 3(.05122) \\
&\quad + 4(.00645) + 5(.00004) \\
&= 1.0
\end{aligned}
$$

b. To gain insight on the value of μ, note that μ can also be obtained by adding the 100,000 values of x shown in Table 5.1 and dividing the sum by $n = 100{,}000$:

$$\mu = \frac{\Sigma \, x}{n} = \frac{32{,}807(0) + 40{,}949(1) + 20{,}473(2) + 5{,}122(3) + 645(4) + 4(5)}{100{,}000}$$

$$= 1.0$$

Consequently, the value $\mu = 1.0$ implies that *over a long series of surveys* similar to the one described in Example 5.2, the average number of physicians in the sample of 5 who favor aspirin brand A will equal 1. The key to the interpretation of μ is to think in terms of repeating the experiment over a long series of trials (i.e., in the long run). Then μ represents the average x value of this large number of trials.

c. By Definition 5.6, we obtain

$$\sigma^2 = \Sigma\, x^2 p(x) - \mu^2 = (0)^2 p(0) + (1)^2 p(1) + \cdots + (5)^2 p(5) - (1)^2$$
$$= (0)(.32807) + (1)(.40949) + (4)(.20473) + (9)(.05122)$$
$$\quad + (16)(.00645) + (25)(.00004) - (1)$$
$$= .79$$

Then by Definition 5.7, the standard deviation σ is given by

$$\sigma = \sqrt{\sigma^2} = \sqrt{.79} = .89$$

EXAMPLE 5.4 Refer to Example 5.3. Locate the interval $\mu \pm 2\sigma$ on the graph of the probability distribution for x. Confirm that most of the (theoretical) population falls within this interval.

Solution Recall that $\mu = 1.0$ and $\sigma = .89$ for this distribution. Then

$$\mu - 2\sigma = 1.0 - 2(.89) = -.78$$
$$\mu + 2\sigma = 1.0 + 2(.89) = 2.78$$

The interval from $-.78$ to 2.78, shown in Figure 5.3, includes the values of $x = 0$, $x = 1$, and $x = 2$. Thus, the probability (relative frequency) that a population value falls within this interval is

$$p(0) + p(1) + p(2) = .32807 + .40949 + .20473 = .94229$$

This certainly agrees with the Empirical Rule, which states that approximately 95% of the data will lie within 2σ of the mean μ.

In the remainder of this chapter, we present a special discrete probability distribution that provides a useful model for many types of data encountered in the real world.

EXERCISES
LEARNING THE MECHANICS

5.1 Consider the probability distribution shown in the table.

x	−5	0	2	5
p(x)	.2	.3	.4	.1

a. List the values that x may assume.
b. What value of x is most probable?
c. Find the probability that x is greater than 0.
d. What is the probability that $x = -5$?
e. Verify that the sum of the probabilities equals 1.

5.2 Find μ and σ for the probability distribution given in Exercise 5.1.

5.3 A discrete random variable can assume five possible values, as listed in the accompanying probability distribution.

x	1	2	3	4	5
p(x)	.20	.25	—	.30	.10

a. Find the missing value for $p(3)$.
b. Find the probability that $x = 2$ or $x = 4$.
c. Find the probability that x is less than or equal to 4.

5.4 Find μ and σ for the probability distribution given in Exercise 5.3.

5.5 The probability distribution for a discrete random variable x is given by the formula,

$$p(x) = (.8)(.2)^{x-1}, \ x = 1, 2, 3, \ldots$$

a. Calculate $p(x)$ for $x = 1$, $x = 2$, $x = 3$, $x = 4$, and $x = 5$.
b. Sum the probabilities of part **a**. Is it likely to observe a value of x greater than 5?
c. Find $P(x = 1 \text{ or } x = 2)$.

APPLYING THE CONCEPTS

5.6 Refer to Exercise 4.21. The pie chart describing the origins of job stress for American workers is reproduced below. Let x be the number of workers that must be sampled until the first worker with no stress is found.

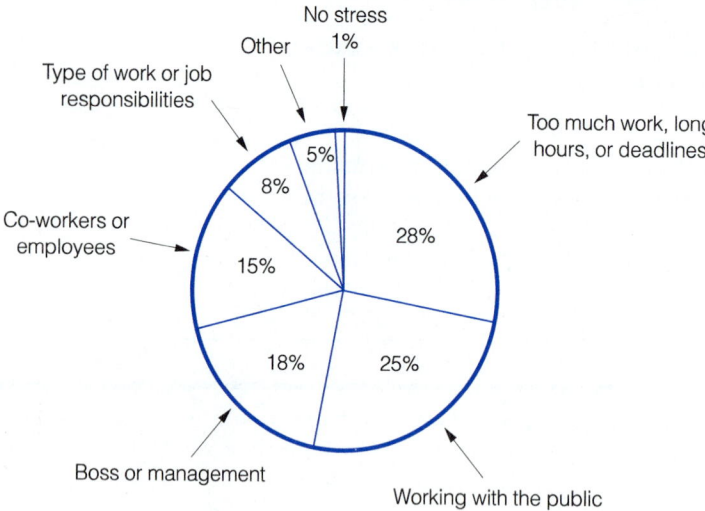

Source: Tampa Tribune, May 8, 1991.

a. List the possible values of x.
b. Refer to part **a**. If you sum the probabilities for these x values, what result will you obtain?
c. Find $p(1)$.
d. Find $p(2)$. [*Hint:* Use Probability Rule #3 for two independent events (Chapter 4).]
e. It can be shown that the probability distribution of x is given by the formula $p(x) = \pi(1 - \pi)^{x-1}$, where π is the probability that a single worker has no on-the-job stress. Use this formula to find $p(12)$.

5.7   Refer to the *Developmental Psychology* (Mar. 1986) study on the sexual maturation of college students, Exercises 2.14 and 2.34. A probability distribution for the age *x* (in years) when males begin to shave regularly is shown in the figure.

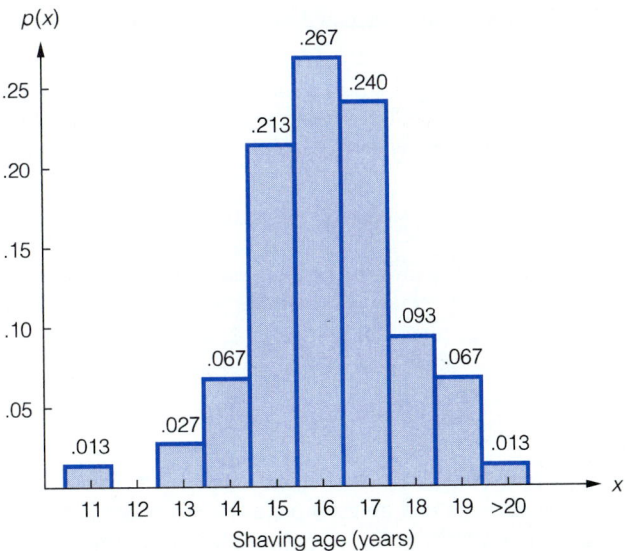

a. Is this a valid probability distribution? Explain.
b. Display the probability distribution in tabular form.
c. What is the probability that a randomly selected male college student began shaving at age 16?
d. What is the probability that a randomly selected male college student began shaving before age 15?

5.8   Refer to Exercise 5.7.

a. Calculate μ and σ.
b. Interpret the value of μ.
c. Locate the interval $(\mu - 2\sigma, \mu + 2\sigma)$ on the graph constructed in Exercise 5.7.
d. Find the probability that *x* falls within the interval of part **c**. Compare your results to the Empirical Rule.

5.9   According to a *Wall Street Journal* survey of businesses that use personal computers (PCs) in the office, 70% use IBM PCs. If two surveyed businesses are selected, find the probability distribution of *x*, the number of businesses that use IBM PCs in the office. [*Hint:* The possible outcomes of the experiment are (*I, I*) (both companies use IBM PCs), (*I, N*) (the first company uses IBM PCs and the second does not), (*N, I*), and (*N, N*). For each outcome, determine the value of *x* and then calculate its probability using Probability Rule #3 (Chapter 4).]

5.10   A panel of meteorological and civil engineers studying emergency evacuation plans for Florida's Gulf Coast in the event of a hurricane has estimated that it would take between 13 and 18 hours to evacuate people living in low-lying land, with the probabilities shown in the table.

TIME TO EVACUATE (nearest hour)	PROBABILITY
13	.04
14	.25
15	.40
16	.18
17	.10
18	.03

a. Calculate the mean and standard deviation of the probability distribution of the evacuation times.

b. Within what range would you expect the time to evacuate to fall?

c. Weather forecasters say they cannot accurately predict a hurricane landfall more than 14 hours in advance. If the Gulf Coast Civil Engineering Department waits until the 14-hour warning before beginning evacuation, what is the probability that all residents of low-lying areas are evacuated safely (i.e., before the hurricane hits the Gulf Coast)?

5.11   In Exercise 4.47, we introduced the oil-lease lottery conducted by the Bureau of Land Management. Given various probabilities of winning, it is interesting to compute the expected gain associated with a single $10 entry. For example, it is estimated that the probability of winning on a parcel worth $25,000 to $150,000 is approximately $\frac{1}{5,000}$.

a. Use the values stated above to calculate the expected gain associated with a single $10 entry. Interpret this result.

b. Suppose you enter the lottery for three parcels of land. If the probability that you will win $25,000 on a single entry is also $\frac{1}{5,000}$, what is the expected gain for the $30 investment in the three parcels? [*Hint:* The gain can assume one of four values, depending on whether the number x of wins is 0, 1, 2, or 3.]

THE BINOMIAL PROBABILITY DISTRIBUTION

5.4 Consumer preference and opinion polls (i.e., sample surveys) are conducted so frequently in political, psychological, sociological, medical, and business situations that it is useful for us to know the probability distribution of the number x in a random sample of n experimental units (people) who prefer some specific proposition. This probability distribution, known as a **binomial probability distribution**, is applicable when the sample size n is small relative to the number N of experimental units in the population.

Strictly speaking, the binomial probability distribution applies only to sampling that satisfies the conditions of a **binomial experiment**, as listed in the box.

CONDITIONS REQUIRED FOR A BINOMIAL EXPERIMENT

1. A sample of n experimental units is selected from the population **without replacement** (i.e., once an experimental unit is selected, it cannot be selected again).

2. Each experimental unit possesses one of two mutually exclusive characteristics. We conventionally call the characteristic of interest a *success* and the other a *failure*.

3. The probability that a single experimental unit possesses the success characteristic is equal to π. This probability is the same for all experimental units.

4. The outcome for any one experimental unit is independent of the outcome for any other experimental unit (i.e., the draws are independent).

5. The random variable x counts the number of successes in n trials.

In real life, there are probably few experiments that satisfy exactly the conditions for a binomial experiment. However, there are many that satisfy approximately—at least for all practical purposes—these conditions. Consider, for example, a sample survey. When the number N of elements in the population is large and the sample size n is small relative to N, the sampling satisfies, approximately, the conditions of a binomial experiment. The next two examples illustrate the point.

EXAMPLE 5.5 Suppose that a sample of $n = 2$ elements is randomly selected from a population containing $N = 10$ elements, three of which are designated as successes and seven as failures. Explain why this sampling procedure violates the conditions of a binomial experiment.

Solution The probability of selecting a success on the first draw is equal to $\frac{3}{10}$—that is, the number of successes in the population (3) divided by the total number of elements in the population (10). In contrast, the probability of a success on the second draw is either $\frac{2}{9}$ or $\frac{3}{9}$, depending on whether a success was or was not selected on the first draw. In other words, selecting a success on the second draw is dependent on the outcome of the first draw and this is a violation of condition 4 required for a binomial experiment.

EXAMPLE 5.6 Suppose that a sample of size $n = 2$ is randomly selected from a population containing $N = 1,000$ elements, 300 of which are successes and 700 of which are failures. Explain why this sampling procedure satisfies, approximately, the conditions required for a binomial experiment.

Solution The probability of success on the first draw is the same as in Example 5.5—namely, $300/1,000 = \frac{3}{10}$. The probability of a success on the second draw is either $\frac{299}{999} = .2993$ or $\frac{300}{999} = .3003$, depending on whether the first draw resulted in a success or a failure. But since these conditional probabilities are approximately equal to the unconditional probability $\left(\frac{3}{10}\right)$ of drawing a success on the second draw, we can say that, *for all practical purposes*, the sampling satisfies the conditions of a binomial experiment. Thus, when N is large and n is small relative to N (say, n/N less than .05), we can use the binomial probability distribution to calculate the probability of observing x successes in a survey sample.

The binomial probability distributions for a sample of $n = 10$ and $\pi = .1$, $\pi = .3$, $\pi = .5$, $\pi = .7$, and $\pi = .9$ are shown in Figure 5.4 (page 230). Note that the probability distribution is skewed to the right for small values of π, skewed to the left for large values of π, and symmetric for $\pi = .5$. [*Note:* As you work the exercises in this chapter, you will encounter several nonsurvey applications of the binomial experiment. Check each to convince yourself that the five characteristics of a binomial experiment are satisfied.]

The formula used for calculating probabilities of the binomial probability distribution is shown in the box on page 230.

THE BINOMIAL PROBABILITY DISTRIBUTION

$$p(x) = \binom{n}{x} \pi^x (1 - \pi)^{n-x} \qquad x = 0, 1, 2, \ldots, n$$

where n = Sample size (number of trials)

             x = Number of successes in n trials

             π = Probability of success on a single trial

$$\binom{n}{x} = \frac{n!}{x!(n - x)!}$$

Assumption: The sample size n is small relative to the number N of elements in the population (say, n/N smaller than 1/20).

[*Note:* $x! = x(x - 1)(x - 2)(x - 3) \cdots (3)(2)(1)$, and $0! = 1$.]

FIGURE 5.4
Binomial Probability
Distributions for $n = 10$,
$\pi = .1, .3, .5, .7, .9$

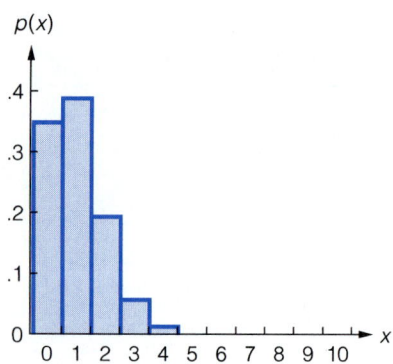

a. $\pi = .1$

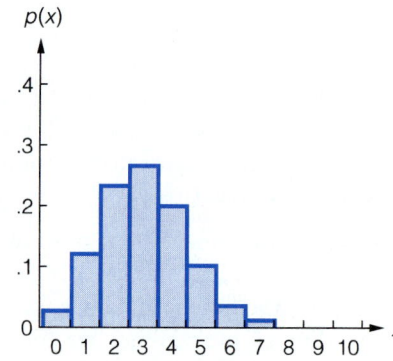

b. $\pi = .3$

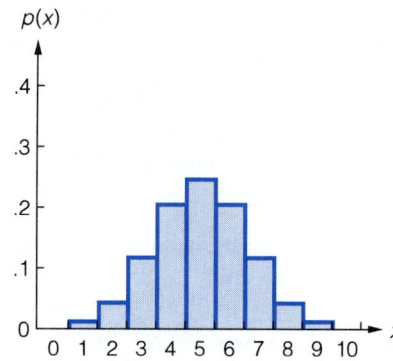

c. $\pi = .5$

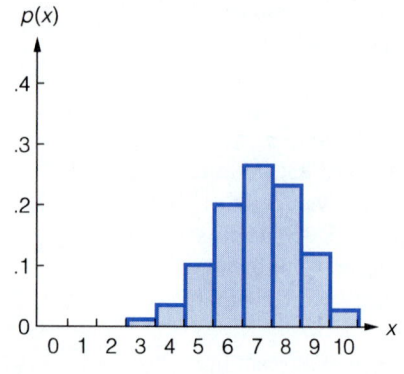

d. $\pi = .7$

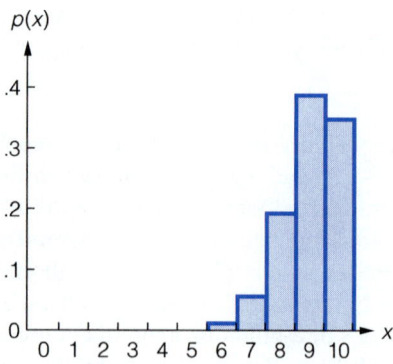

e. $\pi = .9$

EXAMPLE 5.7 Verify that the physician survey described in Example 5.2 is a binomial experiment. Then use the binomial probability distribution to calculate the probabilities of $x = 0, x = 1, \ldots, x = 5$.

Solution The sample survey satisfies the five requirements for a binomial experiment given in the box:

1. A sample of $n = 5$ physicians (experimental units) is selected from a population.
2. Each physician surveyed possesses one of two mutually exclusive characteristics: favors aspirin brand A (a *success*) or does not favor aspirin brand A (a *failure*).
3. The proportion of physicians in the population who prefer aspirin brand A is .2; thus, the probability of a success is $\pi = .2$. Since the sample size $n = 5$ is small relative to the population size $N = 10,000$, this probability remains the same for all trials.
4. The response for any one physician is independent of the response for any other physician.
5. We are counting $x =$ the number of physicians in the sample who favor aspirin brand A.

To calculate the binomial probabilities, we will substitute the values of $n = 5$ and $\pi = .2$ and each value of x into the formula for $p(x)$:

$$p(x) = \binom{n}{x} \pi^x (1 - \pi)^{n-x} = \frac{n!}{x!(n-x)!} \pi^x (1 - \pi)^{n-x}$$

Thus, remembering that $0! = 1$, we have

$$P(x = 0) = p(0) = \binom{5}{0}(.2)^0(.8)^5$$

$$= \frac{5!}{0!5!}(.2)^0(.8)^5 = (1)(1)(.32768)$$

$$= .32768$$

Similarly,

$$P(x = 1) = p(1) = \binom{5}{1}(.2)^1(.8)^4 = \frac{5!}{1!4!}(.2)^1(.8)^4 = .40960$$

$$P(x = 2) = p(2) = \binom{5}{2}(.2)^2(.8)^3 = \frac{5!}{2!3!}(.2)^2(.8)^3 = .20480$$

$$P(x = 3) = p(3) = \binom{5}{3}(.2)^3(.8)^2 = \frac{5!}{3!2!}(.2)^3(.8)^2 = .05120$$

$$P(x = 4) = p(4) = \binom{5}{4}(.2)^4(.8)^1 = \frac{5!}{4!1!}(.2)^4(.8)^1 = .00640$$

$$P(x = 5) = p(5) = \binom{5}{5}(.2)^5(.8)^0 = \frac{5!}{5!0!}(.2)^5(.8)^0 = .00032$$

Note that these probabilities are approximately equal to the relative frequencies reported in Table 5.1.

EXAMPLE 5.8 Refer to Example 5.7. Find the probability that three or more physicians in the sample prefer brand A.

Solution The values that a random variable x can assume are always mutually exclusive events—i.e., you could not observe $x = 2$ and, at the same time, observe $x = 3$. Therefore, the event "x is 3 or more" (the event that $x = 3$ or $x = 4$ or $x = 5$) can be found using Probability Rule #1 (Chapter 4). Thus,

$$P(x = 3 \text{ or } x = 4 \text{ or } x = 5) = P(x = 3) + P(x = 4) + P(x = 5)$$
$$= p(3) + p(4) + p(5)$$

Substituting the probabilities found in Example 5.7, we obtain

$$P(x = 3 \text{ or } x = 4 \text{ or } x = 5) = .05120 + .00640 + .00032 = .05792$$

EXERCISES
LEARNING THE MECHANICS

5.12   A coin is tossed ten times and the number of heads is recorded. To a reasonable degree of approximation, is this a binomial experiment? Check to determine whether each of the five conditions required for a binomial experiment is satisfied.

5.13   Compute each of the following:

a. $4!$ b. $\dfrac{4!}{1!3!}$ c. $\binom{5}{3}$ d. $(.4)^3$ e. $\binom{5}{3}(.4)^3(.6)^2$

5.14   Four coins are selected without replacement from a group of five pennies and five dimes. Let x equal the number of pennies in the sample of four coins. To a reasonable degree of approximation, is this a binomial experiment? Determine whether each of the five conditions required for a binomial experiment is satisfied.

5.15   Consider a binomial experiment with $n = 4$ trials and probability of success $\pi = .5$.

a. Use the formula for the binomial probability distribution to find the probabilities for $x = 0, 1, 2, 3,$ and 4. Construct a graph, similar to Figure 5.4, of the probability distribution.
b. Find the probability that x is less than 2.
c. Find the probability that x is less than or equal to 2.
d. Locate the probabilities computed in parts **b** and **c** on the graph that you constructed in part **a**.
e. Verify that (except for rounding) the sum of the probabilities for $x = 0, 1, 2, 3,$ and 4 equals 1.

5.16   Let x be a binomial random variable with parameters $n = 4$ and $\pi = .2$.

a. Find the probability that x is less than 2.
b. Find the probability that x is equal to 2 or more.
c. How are the events in parts **a** and **b** related?
d. What relationship must the probabilities of the events in parts **a** and **b** satisfy?

APPLYING THE CONCEPTS

5.17   According to the American Hotel & Motel Association, women are expected to account for half of all business travelers by the year 2000. To attract these women business travelers, hotels are providing more amenities that women particularly like, such as shampoo, conditioner, and body lotion. A recent survey of American hotels found that 86% now offer shampoo in their guest rooms (*Wall Street Journal*, Oct. 14, 1988). Consider a random sample of five hotels and let x be the number that provide shampoo as a guest room amenity.

a. To a reasonable degree of approximation, is this a binomial experiment?
b. What is a "success" in the context of this experiment?
c. What is the value of p?
d. Find the probability that $x = 4$.
e. Find the probability that $x \geq 4$.

5.18 A study of 5-year trends in the logistics information systems of industries found that the greatest computerization advances were in transportation (*Industrial Engineering*, July 1990). Currently, 90% of all industries contain shipping open-order files in their computerized database. In a random sample of 10 industries, let x equal the number that include shipping open-order files in their computerized database.

a. Verify that the probability distribution of x can be modeled using the binomial distribution.
b. Find $P(x) = 7$.
c. Find $P(x) > 5$.

5.19 In a recent study, *Consumer Reports* (Feb. 1992) found widespread contamination and mislabeling of seafood in supermarkets in New York City and Chicago. One alarming statistic: 40% of the swordfish pieces available for sale had a level of mercury above the Food and Drug Administration (FDA) minimum amount. For a random sample of three swordfish pieces, find the probability that:

a. All three swordfish pieces have mercury levels above the FDA minimum.
b. Exactly one swordfish piece has a mercury level above the FDA minimum.
c. At most one swordfish piece has a mercury level above the FDA minimum.

5.20 Would most wives marry the same man again, if given the chance? According to a poll of 608 married women conducted by *Ladies Home Journal* (June 1988), 80% would, in fact, marry their current husbands. Assume the women in the sample were randomly selected from among all married women in the United States. Does the number x in the sample who would marry their husbands again possess (approximately) a binomial probability distribution? Explain.

5.21 The *Wall Street Journal* (Nov. 2, 1983) reported on a survey conducted to evaluate the public's opinion concerning the honest and ethical behavior of business executives. The survey included a representative sample of 1,558 adults selected from among all adults in the United States. Each person in the sample was asked to respond "yes" or "no" to each of a series of questions. If this sample were randomly selected from among all adults in the United States, would the number x responding "yes" to a particular question possess (approximately) a binomial probability distribution? Explain.

5.22 Refer to the neutral particle transport problem described in Exercise 4.41. Recall that particles released into an evacuated duct collide with the inner duct wall and are either scattered (reflected) with probability .16 or absorbed with probability .84 (*Nuclear Science and Engineering*, May 1986).

a. If four particles are released into the duct, what is the probability that all four will be absorbed by the inner duct wall? Exactly three of the four?
b. If 20 particles are released into the duct, what is the probability that at least 10 will be reflected by the inner duct wall? Exactly 10?

5.23 The dairy industry is capitalizing on new medical research in the field of osteoporosis (an age-related disorder characterized by decreased bone mass and increased susceptibility to fractures) to promote its products. According to the National Institute of Health, by the age of 90, 32% of women and 17% of men will suffer a hip fracture because of osteoporosis (*American Demographics*, Oct. 1985).

a. Find the probability that in a random sample of five women aged 90, exactly three have suffered a broken hip due to osteoporosis.
b. Refer to part **a**. Find the probability that at least two of the five women have suffered a broken hip due to osteoporosis.

c. In a random sample of five men aged 90, find the probability that at least two have suffered a fractured hip due to osteoporosis.

5.24 A particular system in a space vehicle must work properly in order for the space ship to gain reentry into the earth's atmosphere. One component of the system operates successfully only 85% of the time. To increase the reliability of the system, four of the components will be installed in such a way that the system will operate successfully if at least one component is working successfully.

a. What is the probability that the system will fail? Assume the components operate independently.
b. If the system does in fact fail, what would you infer about the claimed 85% success rate of a single component?

5.25 One of the most complete and comprehensive studies of drug use in junior high school was conducted in Pinellas County (Florida). The survey of 1,532 eighth grade students found that 25.8% believe they will use marijuana and 10.8% believe they will use cocaine by the time they enter high school (*Alligator*, Sept. 27, 1984). A representative of the community group that conducted the study claims that these "results are applicable nation-wide." Consider a random sample of 10 eighth-graders selected from your school district, and assume that the results do apply nationwide.

a. What is the probability that exactly five of the eighth-graders believe they will use marijuana before entering high school?
b. What is the probability that exactly four of the eight-graders believe they will use cocaine before entering high school?
c. What is the probability that at least two of the eighth-graders believe they will use marijuana before entering high school?
d. What is the probability that at most two of the eighth-graders believe they will use cocaine before entering high school?

5.26 A study of vehicle flow characteristics on acceleration (i.e., merging ramps) at a major freeway in Israel found that one out of every six vehicles use less than one-third of the acceleration lane before merging into traffic (*Journal of Transportation Engineering*, Nov. 1985). Suppose we monitor the location of the merge for the next five vehicles that enter the acceleration lane.

a. What is the probability that none of the vehicles will use less than one-third of the acceleration lane?
b. What is the probability that exactly two of the vehicles will use less than one-third of the acceleration lane?

CHAPTER CASE: THE ILL-FATED IRANIAN HOSTAGE RESCUE MISSION

5.5

In 1979, President Jimmy Carter authorized American commandos to undertake a daring attempt to rescue the U.S. hostages that were being held in Iran at the time. Before embarking on the mission, the military estimated that at least six helicopters were needed to provide a reasonable probability of success. Although eight helicopters were used on the mission, three failed. Since only five remained, the mission was aborted.

In April of the following year, the U.S. House Armed Services Investigations Subcommittee announced that it would investigate the aborted rescue attempt. According to the subcommittee chairman, Representative Samuel S. Stratton, the purpose of the inquiry would be to find out why three out of eight helicopters in the mission failed.

Specifically, Stratton said, "The failure rate of three out of eight doesn't match the record. President Carter said this was practiced 20 times and the whole thing ran perfectly. We need to find out what the problem was and how it can be prevented in the future."

To summarize, the Department of Defense claimed that each of the eight helicopters flew 20 missions without malfunction, for a total of 160 missions. On the day of the rescue mission, however, three of the eight helicopters that embarked on the mission failed. It is clear that Representative Stratton believed that the data collected on the day of the rescue attempt disagreed with the government's claim. In fact, it seemed extremely doubtful that helicopters capable of flying 160 successful missions out of a total of 160 could produce three malfunctions in a sample of $n = 8$ flights.

We can show just how doubtful three malfunctions in $n = 8$ missions would be, using our knowledge of the binomial probability distribution. The statement that the helicopters flew 160 successful missions in 160 attempts suggests that the probability that a single helicopter would malfunction during a mission is low. The more reliable the helicopter (and the smaller the value of this probability), the less likely would be the observance of as many as three malfunctions in eight missions.

To show that the observance of three malfunctions in eight helicopter missions is inconsistent with the helicopter reliability demonstrated during the training missions, we will assume that the helicopter reliability is relatively low and that the probability π that a helicopter will malfunction during a single mission is as large as .05. Then the probability of observing as many as three (i.e., three or more) malfunctions in $n = 8$ missions is

$$P(x = 3 \text{ or more}) = p(3) + p(4) + \cdots + p(8)$$

where $p(x)$ is the binomial probability distribution with $n = 8$ and $\pi = .05$. Using the Rule of Complements (Chapter 4),

$$P(x = 3 \text{ or more}) = 1 - P(x = 2 \text{ or less})$$
$$= 1 - [p(0) + p(1) + p(2)]$$
$$= 1 - p(0) - p(1) - p(2)$$

To solve this problem, we calculate

$$p(x) = \binom{n}{x} \pi^x (1 - \pi)^{n-x}$$

for

$$p(0) = \binom{8}{0}(.05)^0(.95)^8 = .663420$$

$$p(1) = \binom{8}{1}(.05)^1(.95)^7 = (8)(.05)(.95)^7 = .279335$$

$$p(2) = \binom{8}{2}(.05)^2(.95)^6 = \frac{8!}{2!6!}(.05)^2(.95)^6 = .051456$$

Then

$$P(x = 3 \text{ or more}) = 1 - p(0) - p(1) - p(2)$$
$$= 1 - .663420 - .279335 - .051456$$
$$= .005789$$

Thus, the probability of observing three or more malfunctions in $n = 8$ missions, given that the probability of malfunction for a single helicopter is as large as .05, is only $.005789 \approx .006$, or less than 6 chances in 1,000. This small probability indicates a marked inconsistency between the helicopter malfunction rate during training missions and that observed during the actual Iranian rescue mission. It is very likely that the value of π on the day of the rescue mission was much larger than .05, implying that the mission may have been doomed from the start.

5.6 TABLES OF THE BINOMIAL PROBABILITY DISTRIBUTION

As you saw in the preceding sections, the calculation of binomial probabilities can be very tedious. In practice, we would refer to one of the many tables that give the computed values of $p(x)$ for a wide range of values of n and π. To aid you in using these tables, we have included binomial probability tables for $n = 5, 6, 7, 8, 9, 10, 15, 20$, and 25 in Table 2 of Appendix G. We illustrate the use of this table in the next two examples.

EXAMPLE 5.9 Find the probability that $x = 1$ for a binomial random variable with $n = 5$ and $\pi = .2$.

Solution The tabulated binomial probability distribution for $n = 5$ (from Table 2 of Appendix G) is reproduced in Table 5.2. The table gives $P(x = k)$ for some value k and some value of π. Note that the values of π are given in the top row of the table, and the values that x can assume ($k = 0, 1, 2, 3, 4, 5$) are shown in the first column. To find the binomial probability for some value of π—in this example, $\pi = .2$—move across the top row of the table to $\pi = .2$. The values of $p(x)$ appear in the column beneath this value of π. We desire $p(1) = P(x = 1)$. Thus, $k = 1$ in this example. Move down the column beneath $\pi = .2$ until you reach the row corresponding to $k = 1$. The value of $p(1)$ is given as .4096 (shaded in Table 5.2). Note that this value is identical to the value that we computed in Example 5.7 using the formula for $p(x)$.

TABLE 5.2

A Reproduction of a Portion of Table 2 in Appendix G: Values of the Binomial Probability $P(x = k)$, $n = 5$

	π												
k	.01	.05	.1	.2	.3	.4	.5	.6	.7	.8	.9	.95	.99
0	.9510	.7738	.5905	.3277	.1681	.0778	.0313	.0102	.0024	.0003	.0000	.0000	.0000
1	.0480	.2036	.3280	.4096	.3601	.2592	.1563	.0768	.0283	.0064	.0005	.0000	.0000
2	.0010	.0214	.0729	.2048	.3087	.3456	.3125	.2304	.1323	.0512	.0081	.0011	.0000
3	.0000	.0011	.0081	.0512	.1323	.2304	.3125	.3456	.3087	.2048	.0729	.0214	.0010
4	.0000	.0000	.0004	.0064	.0283	.0768	.1563	.2592	.3601	.4096	.3280	.2036	.0480
5	.0000	.0000	.0000	.0003	.0024	.0102	.0313	.0778	.1681	.3277	.5905	.7738	.9510

In some situations, we will want to compare an observed value of x obtained from a binomial experiment with some theory or claim associated with the sampled population. In particular, we will want to see if the observed value of x represents a **rare event**, assuming that the claim is true.

EXAMPLE 5.10 A manufacturer of O-ring seals used to prevent hot gases from leaking through the joints of rocket boosters claims that 95% of all seals that it produces will function properly. Suppose you randomly select 10 of these O-ring seals, test them, and find that only six prevent gas from leaking. Is this sample outcome highly improbable (that is, does it represent a *rare event*), if in fact the manufacturer's claim is true?

Solution If π is in fact equal to .95 (or some larger value), then observing a small number, x, of O-ring seals that function properly would represent a rare event. Since we observed $x = 6$, we want to know the probability of observing a value of $x = 6$ or some other value of x even more contradictory to the manufacturer's claim, i.e., we want to find the probability that $x = 0$ or $x = 1$ or $x = 2$ or . . . or $x = 6$. Using the additive rule for values of $p(x)$, we obtain

$$P(x = 0 \text{ or } x = 1 \text{ or } x = 2 \text{ or } \ldots \text{ or } x = 6)$$
$$= p(0) + p(1) + p(2) + p(3) + p(4) + p(5) + p(6)$$

when $n = 10$ and $\pi = .95$. Referring to Table 2 of Appendix G for these values of $p(x)$ and substituting, we find

$$P(x = 0 \text{ or } x = 1 \text{ or } x = 2 \text{ or } \ldots \text{ or } x = 6)$$
$$= .0000 + .0000 + .0000 + .0000 + .0000 + .0001 + .0010$$
$$= .0011$$

This small probability tells us that observing as few as six good O-ring seals out of 10 is indeed a rare event, if in fact the manufacturer's claim is true. Such a sample result suggests either that the manufacturer's claim is false or that the 10 O-ring seals tested do not represent a random sample from the manufacturer's total production. Perhaps they came from a particular production line that was temporarily malfunctioning.

Solving practical problems of the type illustrated in Example 5.10 requires us to sum values of $p(x)$. Partial sums of the values of $p(x)$, called **cumulative probabilities**, are given for $n = 5, 6, 7, 8, 9, 10, 15, 20,$ and 25 in Table 3 of Appendix G. A reproduction of the cumulative binomial probability table for $n = 10$ is shown in Table 5.3. We show how to use this table in the next two examples.

TABLE 5.3
Reproduction of a Portion of Table 3 in Appendix G: Cumulative Binomial Probabilities $P(x \le k)$, $n = 10$

k	.01	.05	.1	.2	.3	.4	.5	.6	.7	.8	.9	.95	.99
							π						
0	.9044	.5987	.3487	.1074	.0282	.0060	.0010	.0001	.0000	.0000	.0000	.0000	.0000
1	.9957	.9139	.7361	.3758	.1493	.0464	.0107	.0017	.0001	.0000	.0000	.0000	.0000
2	.9999	.9885	.9298	.6778	.3828	.1673	.0547	.0123	.0016	.0001	.0000	.0000	.0000
3	1.0000	.9990	.9872	.8791	.6496	.3823	.1719	.0548	.0106	.0009	.0000	.0000	.0000
4	1.0000	.9999	.9984	.9672	.8497	.6331	.3770	.1662	.0473	.0064	.0001	.0000	.0000
5	1.0000	1.0000	.9999	.9936	.9527	.8338	.6230	.3669	.1503	.0328	.0016	.0001	.0000
6	1.0000	1.0000	1.0000	.9991	.9894	.9452	.8281	.6177	.3504	.1209	.0128	.0010	.0000
7	1.0000	1.0000	1.0000	.9999	.9984	.9877	.9453	.8327	.6172	.3222	.0702	.0115	.0001
8	1.0000	1.0000	1.0000	1.0000	.9999	.9983	.9893	.9536	.8507	.6242	.2639	.0861	.0043
9	1.0000	1.0000	1.0000	1.0000	1.0000	.9999	.9990	.9940	.9718	.8926	.6513	.4013	.0956

EXAMPLE 5.11   Refer to Example 5.10. Use Table 5.3 to find the probability of six or fewer properly functioning O-ring seals (i.e., $P(x \le 6)$), when $n = 10$ and $\pi = .95$.

Solution   In general, the cumulative probabilities in Table 3 of Appendix G are obtained as follows. For specified values of n and π, the table gives the cumulative sum

$$\sum_{x=0}^{k} p(x) = p(0) + p(1) + \cdots + p(k) = P(x \le k)$$

for some value of k (see Figure 5.5). To find this sum, locate the table entry corresponding to the row k under the appropriate column for π. The cumulative sum of probabilities for this example is given in the column corresponding to $\pi = .95$ and the row corresponding to $k = 6$ in the $n = 10$ table reproduced in Table 5.3. Therefore,

$$P(x \le 6) = p(0) + p(1) + \cdots + p(6) = .001$$

This value, shaded in Table 5.3, agrees (except for rounding) with the value found in Example 5.10.

FIGURE 5.5
Illustration of Cumulative Probabilities given in Table 3, Appendix G

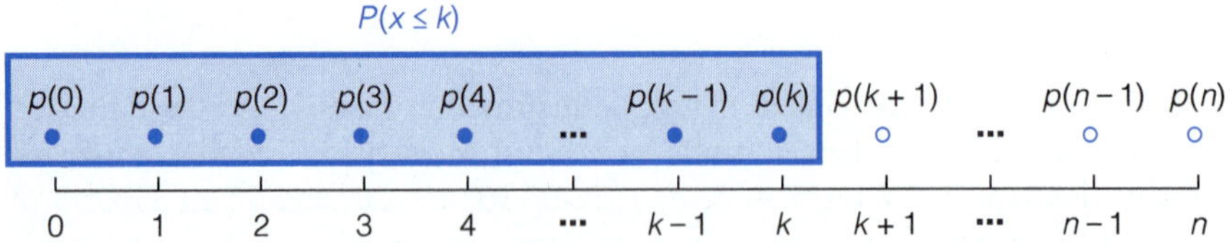

EXAMPLE 5.12 If you toss a balanced coin 10 times, what is the probability that you will observe eight or more heads?

Solution First, suppose that we define a success to be a head. Then, the probability of a success is .5 and

$$P(\text{eight or more heads}) = P(x = 8 \text{ or } x = 9 \text{ or } x = 10)$$
$$= p(8) + p(9) + p(10)$$

Remember that Table 3 in Appendix G gives cumulative sums, $P(x \le k)$ for some k, and that the sum of the values of $p(x)$ over all values of x is equal to 1. Therefore,

$$P(x = 8 \text{ or } x = 9 \text{ or } x = 10) = p(8) + p(9) + p(10)$$
$$= 1 - [p(0) + p(1) + \cdots + p(7)]$$

We now turn to the cumulative binomial probability table for $n = 10$ (reproduced in Table 5.3) and find

$$P(x \le 7) = \sum_{x=0}^{7} p(x) = p(0) + p(1) + \cdots + p(7)$$

The tabulated value in the $\pi = .5$ column and the row corresponding to $k = 7$ is .9453. Then, the probability of tossing eight or more heads in 10 tosses of a balanced coin is

$$P(x = 8 \text{ or } x = 9 \text{ or } x = 10) = 1 - \sum_{x=0}^{7} p(x) = 1 - .9453 = .0547$$

EXAMPLE 5.13 Refer to Example 5.12. Find the probability of tossing eight or more heads by defining a success as observing a tail.

Solution If a success is observing a tail, then the probability of a success is still $\pi = .5$, and x is the number of tails in 10 tosses of the coin. The next step is to define the event "observe eight or more heads" in terms of x, the number of tails. This event will occur if the number of tails is 0, 1, or 2. Therefore,

$$P(\text{eight or more heads}) = P(x = 0 \text{ or } x = 1 \text{ or } x = 2) = \sum_{x=0}^{2} p(x) = P(x \le 2)$$

We can read this cumulative sum directly from Table 5.3. Looking in the column corresponding to $\pi = .5$ and the row corresponding to $k = 2$, we read

$$P(\text{eight or more heads}) = P(x = 0 \text{ or } x = 1 \text{ or } x = 2)$$
$$= P(x \le 2) = .0547$$

This is exactly the same answer as was obtained in Example 5.12.

EXERCISES

LEARNING THE MECHANICS

5.27 Refer to the binomial probability distribution for $n = 10$ given in Table 2 of Appendix G.

a. Find $p(0)$ when $\pi = .1$.
b. Find $p(1)$ when $\pi = .1$.
c. Find $p(2)$ when $\pi = .1$.

5.28 Refer to the binomial probability distribution for $n = 6$ in Table 2 of Appendix G.

a. Find $p(2)$ when $\pi = .30$.
b. Find $p(3)$ when $\pi = .50$.
c. Find $p(5)$ when $\pi = .10$.

5.29 Refer to the cumulative binomial probability table for $n = 5$, given in Table 3 of Appendix G.

a. Find $\sum_{x=0}^{2} p(x) = p(0) + p(1) + p(2)$ when $\pi = .3$.

b. Find $\sum_{x=0}^{4} p(x)$ when $\pi = .3$.

c. Find $\sum_{x=4}^{5} p(x)$ when $\pi = .3$. $\left[Hint: \sum_{x=4}^{5} p(x) = 1 - \sum_{x=0}^{3} p(x). \right]$

d. Find $p(2)$ when $\pi = .3$. $\left[Hint: \quad p(2) = \sum_{x=0}^{2} p(x) - \sum_{x=0}^{1} p(x). \right]$

5.30 Refer to the cumulative binomial probability table for $n = 10$ and $\pi = .4$, given in Table 3 of Appendix G.

a. Find the probability that x is less than or equal to 8.
b. Find the probability that x is less than 8.
c. Find the probability that x is larger than 8.

APPLYING THE CONCEPTS

5.31 *Organic Gardening* magazine conducted a poll to determine whether consumers would prefer organically grown fruits and vegetables over those grown with fertilizers and pesticides (*New York Times*, Mar. 21, 1989). If the costs of the two food types were the same, 85% said they would prefer the organic food. Surprisingly, 50% said they would prefer the organic food even if they had to pay more for it. Consider the preferences of a random sample of $n = 25$ consumers.

a. Assuming the percentage in the poll are reflective of the population, find the probability that at least 20 of the 25 consumers would prefer the organically grown food, if the costs were the same.
b. Assuming the percentages in the poll are reflective of the population, find the probability that at least 20 of the 25 consumers would prefer the organically grown food, even if the costs were higher than for food grown with fertilizers and pesticides.

5.32 Refer to the *IEEE Computer Applications in Power* study of an outdoor unmanned watching system designed to detect trespassers, Exercise 4.45. In snowy weather conditions, the system detected seven out of 10 intruders; thus, the researchers estimated the system's probability of intruder detection in snowy conditions at .70.

a. Assuming the probability of intruder detection in snowy conditions is only .50, find the probability that the unmanned system detects at least seven of the 10 intruders.

b. Based on the result, part **a**, comment on the reliability of the researcher's estimate of the system's detection probability in snowy conditions.

5.33 A "Louis Harris and Associates poll on public attitudes toward water pollution cleanup found overwhelming support for tougher clean water standards, even if it means more expensive goods and services or fewer jobs" (*Engineering News-Record*, Dec. 23, 1982). Forty-six percent of those surveyed believe that government regulations regarding pollution control are "not protective enough," while 60% felt that "Congress should make the Clean Water Act stricter." In response to a question about whether industry should be required to install best available technology (BAT) for pollution control, over 50% of those surveyed said they would rather endure factory shutdowns and lost jobs than waivers from BAT standards. Suppose 10 people are randomly selected and asked to give their opinion regarding BAT pollution control. Assuming $\pi = .5$, find the probability that:

  a. None would prefer factory shutdowns and lost jobs to waiving BAT standards.
  b. At least five would prefer factory shutdowns and lost jobs to waiving BAT standards.
  c. At least one would prefer factory shutdowns and lost jobs to waiving BAT standards.

5.34 In a nationwide poll of 2,052 adults, conducted by the American Association of Retired Persons, approximately 60% stated the opinion that the current version of the federal income tax is unfair (*USA Today*, Aug. 8, 1985). Suppose we randomly sample 20 of the 2,052 adults surveyed and record x, the number who think the federal income tax is unfair.

  a. To a reasonable degree of approximation, is x a binomial random variable? If yes, specify the values of n and π.
  b. What is the probability that x is less than or equal to 15?
  c. What is the probability that x is greater than 10?

5.35 Do you have the basic skills necessary to succeed in college? Most likely, your college professor does not think so. According to a survey conducted by the Carnegie Foundation for the Advancement of Teaching, over 70% of college professors consider their students "seriously unprepared in basic skills" that should have been learned in high school (*Tampa Tribune*, Nov. 6, 1989). In a sample of 25 college faculty at your institution, let x represent the number who agree that students lack basic skills. Assume that $\pi = .70$ at your institution.

  a. What is the probability that x is less than 20?
  b. What is the probability that x is less than nine?
  c. What is the probability that x is more than nine?
  d. If, in fact x is less than nine, make an inference about the value of π at your institution.

5.36 When radar was first introduced during World War II, it was very difficult for an operator manning the screen to distinguish a static interference blip from an actual enemy aircraft blip. Although the operator did not want to sound an alarm needlessly, failure to alert the defenses could have serious consequences. Records indicate that 60% of all observed blips represented enemy aircraft. Suppose that during a particular siege there were five blips spotted on the screen at different points in time and the radar operator alerted the defenses on each occasion. Assume that the events are independent and compute the probability of each of the following events:

  a. Radar operator made the correct decision on all five occasions.
  b. Radar operator made the correct decision on at least three occasions.
  c. Radar operator was incorrect all five times (and therefore sounded five false alarms).

5.37 According to a Harvard School of Public Health Survey, nearly 30% of children 6–11 years old are considered obese (*American Journal of Diseases of Children*, May 1987), a sharp increase from the 18% considered obese in 1963. Researchers speculate that the increased prevalence of obesity in children is probably due to a combination of "too little exercise and too much or more fatty foods." Suppose you randomly sample 25 children ages 6 to 11 and examine each for obesity. If the proportion of children in this age group who are considered obese is .3, what is the probability that:

  a. More than 10 children in the sample will be obese?
  b. Five or fewer children will be obese?

5.38 Are today's college graduates willing to work long, hard hours on the job? In a survey of over 2,000 college students conducted by the College Placement Council, approximately 87% indicated they would, in fact, work long hours on the job (*Personnel Journal*, July 1984). Suppose the true percentage of college graduates who expect to work long, hard hours is .90. Let x be the number of college graduates in a sample of 10 who say they would be willing to work long, hard hours on the job.

    a. Find the probability that x is less than 5.
    b. Find the probability that $x = 9$.
    c. Find the probability that x is 9 or more.

5.39 Bird watchers and observers of asteroids, mineral deposits, caterpillars, or other objects of natural history or geology are keenly interested in the following question: "What is the optimum number of observers to use in a search party, to be reasonably sure of seeing 95% of the objects (e.g., birds or caterpillars) that are available to be seen?" Frank Preston considered the problem of "bird spotting" from a probabilistic point of view (*Ecology*, June 1979). Assuming that π, the probability that a single observer sees a particular bird, is .5, and that this probability is the same for all observers in the party, Preston used a binomial probability model to estimate the proportion of birds that would be missed for various numbers of observers. The estimate is simply the probability that all observers in the party miss seeing a particular bird. The results are shown in the table.

NUMBER OF OBSERVERS	PROPORTION OF BIRDS MISSED
1	.50
2	.25
3	.13
4	.06
5	.03
6	.02
7	.01
8	.00

    a. Use the binomial tables in Appendix G to verify the probabilities in the table.
    b. Preston calculated similar proportions for different values of the probability that a single observer sees a bird. Show that if π is as low as .2, two observers will miss an estimated 64% of the birds available to be seen.

THE MEAN AND STANDARD DEVIATION FOR A BINOMIAL PROBABILITY DISTRIBUTION

5.7 In Section 5.3, we gave formulas for finding the mean μ, variance σ^2, and standard deviation σ of any discrete probability distribution. Substituting $p(x)$ for the binomial distribution into the formulas in Definitions 5.5, 5.6, and 5.7, you can show (proof omitted) that μ, σ^2, and σ for a binomial probability distribution are as listed in the accompanying box.

MEAN, VARIANCE, AND STANDARD DEVIATION FOR A BINOMIAL PROBABILITY DISTRIBUTION

$$\mu = n\pi \qquad \sigma^2 = n\pi(1 - \pi) \qquad \sigma = \sqrt{n\pi(1 - \pi)}$$

where n = Sample size

          π = Probability of success on a single trial

            = Proportion of experimental units in a large population that are successes

---

For large samples, the mean and standard deviation of the binomial probability distribution (in conjunction with the Empirical Rule or Tchebysheff's theorem) can be used to describe and make inferences about the sampled population.

EXAMPLE 5.14   Find μ and σ for a binomial probability distribution with $n = 10$ and $\pi = .1$, and find the theoretical proportion of the population that lies within the interval $\mu \pm 2\sigma$. Does this result agree with the Empirical Rule, which states that approximately 95% of the measurements in the distribution should lie within this interval?

Solution   Using the formulas for μ and σ, we obtain

$$\mu = n\pi = (10)(.1) = 1$$
$$\sigma = \sqrt{n\pi(1 - \pi)} = \sqrt{(10)(.1)(.9)} = .949$$

Then

$$\mu - 2\sigma = 1 - 2(.949) = -.898$$
$$\mu + 2\sigma = 1 + 2(.949) = 2.898$$

The values of x in the interval from $-.898$ to 2.898 are $x = 0, 1,$ and 2. To find the sum of the probabilities in the interval $\mu \pm 2\sigma$ (i.e., from $-.898$ to 2.898), we need to find $p(0) + p(1) + p(2)$. The easy way to find this partial sum is to refer to the cumulative binomial probability table for $n = 10$ and $\pi = .1$. This gives

$$\sum_{x=0}^{2} p(x) = p(0) + p(1) + p(2) = .9298$$

You can see that the sum of the probabilities for values of x in the interval $\mu \pm 2\sigma$ agrees very closely with the Empirical Rule. It can also be seen graphically in Figure 5.6.

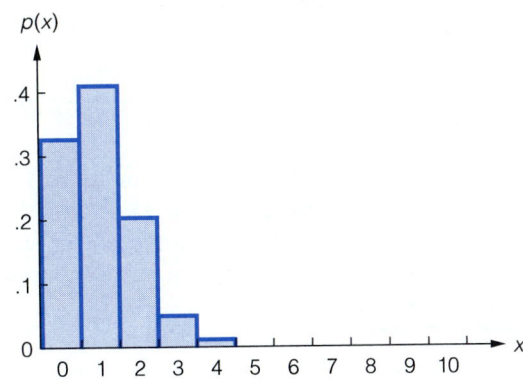

FIGURE 5.6
The Binomial Probability Distribution with $n = 10$ and $\pi = .1$

EXAMPLE 5.15 People who work at high-stress jobs frequently develop stress-related problems (for example, high blood pressure, ulcers, and irritability). In a recent study it was found that 40% of the large number of business executives surveyed have symptoms of stress-induced problems. Consider a group of 1,500 randomly selected business executives and assume that the probability of an executive with stress-induced problems is $\pi = .40$. Let x be the number of business executives in the sample of 1,500 who develop stress-related problems.

a. What is the mean and standard deviation of x?
b. Based on the Empirical Rule, within what limits would you expect x to fall?
c. Suppose you observe $x = 800$ executives with symptoms of stress-induced problems. What can you infer about the value of π?

Solution

a. Since $n = 1,500$ executives in the sample is small relative to the large number of business executives in the population, the number x of executives with stress-induced problems is a binomial random variable with $\pi = .4$. We use the formulas for μ and σ given in the box to obtain the mean and standard deviation of this binomial distribution:

$$\mu = n\pi = (1,500)(.4) = 600$$
$$\sigma = \sqrt{n\pi(1 - \pi)} = \sqrt{(1,500)(.4)(.6)} = \sqrt{360} = 18.97$$

b. According to the Empirical Rule, most (about 95%) of the x values will fall within two standard deviations of the mean. Thus, we would expect the number of sampled executives with stress-induced problems to fall in the interval from

to
$$\mu - 2\sigma = 600 - 2(18.97) = 562.06$$
$$\mu + 2\sigma = 600 + 2(18.97) = 637.94$$

c. Using the rare event approach of Chapter 4, we want to determine whether observing $x = 800$ executives with stress-induced problems is unusual, assuming $\pi = .40$. You can see that this value of x is highly improbable when $\pi = .40$, since it lies a long way outside the interval $\mu \pm 2\sigma$. The z score for this value of x is

$$z = \frac{x - \mu}{n} = \frac{800 - 600}{18.97} = 10.54$$

Clearly, if $\pi = .40$, the probability that the number x of executives with stress-induced problems in the sample is 800 or larger is almost 0. Therefore, we are inclined to believe (based on the sample value of $x = 800$) that the proportion of executives with stress-induced problems is much higher than $\pi = .40$.

5.40   Calculate μ and σ for a binomial probability distribution with

   a. $n = 15$ and $\pi = .1$ b. $n = 15$ and $\pi = .5$

5.41   Since a probability distribution is a theoretical model for a population relative frequency distribution, what proportion of the distribution would you expect to lie within the interval $\mu \pm 2\sigma$?

5.42   Construct a graph of the binomial probability distribution for $n = 15$ and $\pi = .5$. Locate the interval $\mu \pm 2\sigma$ on the graph. Find the probability that x lies within the interval $\mu \pm 2\sigma$. Do these probabilities agree with your answer to Exercise 5.41?

5.43   Find the mean and standard deviation for each of the following binomial probability distributions:

   a. $n = 100$ and $\pi = .99$ b. $n = 100$ and $\pi = .8$ c. $n = 100$ and $\pi = .5$
   d. $n = 100$ and $\pi = .2$ e. $n = 100$ and $\pi = .01$

5.44   Use the values of π and σ calculated in Exercise 5.43 to construct rough sketches of the five binomial probability distributions.

5.45   Find the mean and standard deviation for each of the following binomial probability distributions:

   a. $n = 900$ and $\pi = .99$ b. $n = 900$ and $\pi = .8$ c. $n = 900$ and $\pi = .5$
   d. $n = 900$ and $\pi = .2$ e. $n = 900$ and $\pi = .01$

APPLYING THE CONCEPTS

5.46   Electrical engineers recognize that high neutral current in computer power systems is a potential problem. A recent survey of computer power system load currents at U.S. sites found that 10% of the sites had high neutral to full-load current ratios (*IEEE Transactions on Industry Applications*, July/Aug. 1990). In a sample of 20 computer power systems selected from the large number of sites in the country, let x be the number with high neutral to full-load current ratios.

   a. Find and interpret the mean of x.
   b. Find and interpret the standard deviation of x.

5.47   As a college student, are you frequently depressed or overwhelmed by the pressure to succeed? According to the American Council on Education (ACE), the level of stress among college freshmen is rising rapidly. In a 1988 survey of 380,007 full-time college freshmen conducted for the ACE by UCLA's Higher Education Research Institute, more than 10% reported frequently "feeling depressed," compared with 8.2% in 1985 (*Tampa Tribune*, Jan. 9, 1989). Assume that 10% of all college freshmen frequently feel depressed.

   a. On average, how many of the 308,007 college freshmen surveyed would you expect to report frequent feelings of depression if the true percentage in the population is 8.2% (the 1985 figure)?
   b. Find the standard deviation of the number of the 308,007 freshmen surveyed who frequently feel depressed if the true percentage is 8.2%.
   c. Suppose that 30,800 of the college freshmen surveyed reported frequent feelings of depression. (The number 30,800 is approximately the number observed in the actual study.) Calculate the z score for 30,800.

5.48   *Psychological Reports* (Vol. 57, 1985) examined sex differences in unique and common first names of infants. A name was judged to be unique if it was given only once a year at the participating study hospital. Results showed that 45% of the females and 27% of the males received unique first names. Assuming these percentages can be generalized to the entire U.S. population, consider a nationwide random sample of 500 newborn males and 450 newborn females.

a. What is the mean number of males in the sample to receive a unique first name? Within what limits would you expect x, the number of males receiving a unique first name, to fall?

b. Repeat part **a** for females.

5.49 During the 1950s, a number of atomic weapons tests were conducted in the desert in Nevada. Since that time, estimates of radiation exposure to off-site populations, especially in Utah, have been the subject of considerable scientific effort. The Surveillance, Epidemiology, and End Results (SEER) Registry collected data on incidence of thyroid cancer among Utah residents over the period 1973–1977. SEER found that the incidence rate of thyroid cancer among 50-year-old males is 3.89 per 100,000 population (*Health Physics*, Jan. 1986). This implies that the probability of a 50-year-old Utah male developing thyroid cancer is .0000389. In a random sample of 1,000 50-year-old Utah males, let x equal the number developing thyroid cancer.

a. Calculate the mean and variance of x.

b. Would you expect to observe at least one 50-year-old male with thyroid cancer among the 1,000? [*Hint:* Use Tchebysheff's theorem and the results of part **a**.]

5.50 An Associated Merchandising Corporation survey of shoppers at malls revealed that 82% of the shoppers feel there is more credibility in buying a similar or identical piece of merchandise at regular discount stores than on sale at department stores (*The Discount Merchandiser*, Aug. 1983). Suppose the survey consisted of a sample of 1,000 shoppers and assume that, in fact, $\pi = .82$.

a. Compute μ, the mean number of shoppers in a sample of 1,000 who feel there is more credibility in buying merchandise at regular discount stores.

b. Compute σ.

c. Construct the interval $\mu \pm 2\sigma$.

d. Would you expect to see as few as 600 shoppers in 1,000 who feel there is more credibility in buying at regular discount stores? Explain.

5.51 Travelers who fail to cancel their hotel reservations when they have no intention of showing up are commonly referred to as *no-shows*. In anticipation of no-shows and late cancellations, most hotels overbook (i.e., accept reservations for more rooms than the number of rooms in their inventory). (See Example 3.12.) The *Journal of Travel Research* (Spring 1985) reported that six major hotels in the Seattle area had a no-show rate of 10%. Consider a random sample of 200 travelers with reservations at one of these six Seattle hotels. Let x equal the number of no-shows.

a. Find the mean of x.

b. Find the standard deviation of x.

c. Within what limits would you expect x to fall?

d. Use your answer to part **c** to assess the likelihood of observing 30 or more no-shows in the sample of 200.

5.52 According to the *Training and Development Journal* (Aug. 1984), statistics show that between 10% and 20% of the labor force are considered "troubled employees" who suffer from mental health problems such as alcohol, drug abuse, and psychological stress disorders. Suppose that 20% of a firm's 10,000 employees are "troubled employees." Let x be the number of "troubled employees" in a sample of 25 randomly selected employees.

a. Find the mean of x.

b. Find the standard deviation of x.

c. Within what limits would you expect x to fall?

d. If none of the 25 employees admits to being a "troubled employee," what would you infer? [*Hint:* Refer to your answer to part **c**.]

5.53 In a national study, it was discovered that four out of every 10 full-time employees claim they participate in their companies' purchasing decisions on a regular basis (*Journal of Advertising Research*, Aug./Sept. 1984). Let x be the number of full-time employees in a sample of $n = 500$ who regularly participate in their companies' purchasing decisions.

a. Is this a binomial experiment? Explain.
b. What is π, the probability of success?
c. What is the expected value of x?
d. Within what interval would you expect x to fall?
e. Would the value $x = 185$ contradict the statement that "four out of every 10 employees regularly participate in their firms' buying decisions"? Explain.

SUMMARY

This chapter introduces the notion of *random variables* and then focuses on *discrete random variables*, those that can assume a countable number of values. A complete list (or graph or formula) that gives the probabilities associated with each value of a discrete random variable x is called its *probability distribution*.

A useful discrete probability distribution encountered in many real-world problems is the *binomial probability distribution*. The binomial distribution provides a good model for the probability distribution of the number x of favorable (or unfavorable) responses when the number N of events in the population is large relative to the sample size n—a situation that generally occurs with public opinion polls and sample surveys. Other random variables that have the same type of probability distribution are those that satisfy the five conditions required for a binomial experiment. The formula for a binomial probability distribution and the *binomial probability tables* enable us to calculate probabilities about x. Particularly, in a practical situation, we can determine whether an observed sample value of x is an improbable or rare event.

When the sample size n is large, it is difficult to calculate the values of $p(x)$ but it is easy to describe the binomial probability distribution by finding the mean μ and standard deviation σ of the distribution. Then we can use the Empirical Rule to describe $p(x)$ and to identify values of x that are improbable.

KEY WORDS

Binomial experiment

Binomial probability distribution

Binomial random variable

Continuous random variable

Cumulative binomial probabilities

Discrete random variable

Random variable

Rare event

Sampling without replacement

KEY FORMULAS

Discrete probability distribution:

$$\mu = \Sigma\, xp(x) \qquad \sigma^2 = \Sigma(x - \mu)^2 p(x) = \Sigma\, x^2 p(x) - \mu^2$$

Binomial probability distribution:

$$p(x) = \binom{n}{x}\pi^x(1 - \pi)^{n-x} \qquad x = 0, 1, 2, \ldots, n$$

$$\mu = n\pi \qquad \sigma^2 = n\pi(1 - \pi) \qquad \sigma = \sqrt{n\pi(1 - \pi)}$$

c. Suppose that the number x in the sample of 1,200 who suffer from high blood pressure is equal to 151. If the physician's claim is true, would this represent a rare event? Would you doubt the physician's claim? Explain.

5.67   In recent years, the use of the telephone as a data collection instrument for public opinion polls has been steadily increasing. However, one of the major factors affecting the extent to which the telephone can be used is the refusal rate—that is, the percentage of the eligible subjects actually contacted who refuse to take part in the poll. Suppose that past records indicate a refusal rate of 20% in a large city. If 25 prospective survey subjects are contacted by telephone, what is the probability that:

   a. Five or more refuse to take part in the poll?
   b. Eight or more refuse to take part in the poll?
   c. Fewer than three refuse to take part in the poll?

CASE STUDY 5.1

STATISTICS CAN SCARE YOU TO DEATH

In a special article for the *Tampa Tribune* (Aug. 14, 1988), Doralie Segal discusses the problem of control within our health care and delivery system. Recently, there have been numerous quality control audits conducted at major hospitals and health care organizations in order to determine areas where health care needs to be improved. The results of many of these audits, such as those conducted by the Department of Health and Human Services (HHS), are available to the public. In her article, Segal wonders whether "the average user of these (health) services can really make a better decision about his or her own health care based on the results of (these) quality control audits?"

As an example, Segal notes that several years ago HHS "published lists of hospitals in each state that had significantly more or fewer deaths than were 'expected'.... The public interpreted the figures at face value and were scared to death.... (However), when the data were reviewed in more detail, some interesting facts surfaced. In general, the more a hospital performed a particular operation, the lower the death rates for that procedure. But, the longer a hospital kept patients before discharge, the more likely the death rate was increased. The between-the-lines message, of course, is that if a patient is discharged early to home or a nursing home, the greater the chance that the patient will die there and the hospital won't be charged for the death."

Segal makes two major points in the article. First, "statistics from health care studies," especially those concerning death rates, "need to be interpreted very cautiously." Second, "large numbers (of patients) are necessary for meaningful results" when rates are extremely small, like death rates.

Segal illustrates the last point with the following hypothetical situation. Consider a small hospital with 100 patients, each of which has received a certain operation. Suppose the expected death rate for this operation is 1 per 100 (i.e., .01). Let x be the number of the 100 patients who die from the operation.

a. Verify that x is a binomial random variable.

b. What is the expected value of x?

c. Suppose that the last of the 100 patients is the only one to die from the operation (i.e., $x = 1$). Does the observed death rate match the expected?

d. Now, suppose the second death occurs in the 101st patient. Does the observed death rate match the expected? [*Hint:* Use $n = 101$ and $x = 2$.]

e. Suppose, as Segal describes, that "the next 200 patients go home smiling." How does the observed death rate compare with the expected? [*Hint:* Use $n = 301$ and $x = 2$.]

CASE STUDY 5.2

EXPECTED WINNINGS IN JUNK-MAIL CONTESTS

In an interesting and readable article, James S. Trefil* writes on the role of probabilities and expected values in our everyday lives. Trefil illustrates expected value theory by applying it to junk-mail contests, ploys used by advertisers to interest readers in their products. He writes,

> If you are on a junk-mail list, you probably get regular notices announcing that "You may have already won the $10,000 jackpot!" Is it really worth answering the ad? Well, suppose the mailing went to 100,000 people. Your chances of winning are then 1 in 100,000. Over many contests, therefore, you would expect to win an average of $10,000 $\times (\frac{1}{100,000})$, or ten cents per game. ... You will note that the expected [winnings] in this case is less than the price of the postage stamp you need to enter the contest. In fact, you can expect to lose ten cents every time you play, so the reaction "it's not worth answering this" is correct.

Suppose you randomly select $n = 10$ junk-mail contests to enter, each offering a $10,000 jackpot if you win. Assume that the probability of your winning any one contest is 1 in 100,000 or .00001. Let x be the number of contests that you win.

a. Is this a binomial experiment? Explain.

b. What is the value of π?

c. Find the mean value of x and denote this value μ_x.

d. Now let y be the amount of money you win on any one contest. In Trefil's illustration, either $y = \$10,000$ (i.e., you win the contest) or $y = 0$ (i.e., you lose the contest). In this particular case, it can be shown that the expected value of y is $\mu_y = (\$10,000)\mu_x$. Find your mean (or expected) winnings μ_y.

e. Find μ_y if you enter only a single contest (i.e., $n = 1$). Does this value agree with Trefil's "ten cents" per game?

---

*Trefil, J. S. "Odds are against your breaking that law of averages." *Smithsonian*, Sept. 1984, Vol. 15, No. 6, pp. 66–75.

After reading our discussion of probability theory (Chapter 4) and probability distributions (Chapter 5), you may wonder where and when the notion of probability began. Since probability theory is really the study of chance events, it is not surprising that the idea of probability first occurred to gamblers in their effort to improve the "odds" of winning games of chance.

According to James S. Trefil, "Most historians date the study of the theory of chance occurrences from a day in 1654 when Antoine Gombaud, Chevalier de Méré, Sieur de Baussey, walked into a gaming room in Paris."* Trefil relates the following story about de Méré, which is known to students of probability as "The Chevalier's Dilemma."

In progress was a popular gambling diversion of the time, a game in which the player bet the house that he could roll a die four times in a row without getting a 6. If he did so, he collected his winnings; if a 6 came up, he lost. De Méré and every other gambler in the room knew that this game was stacked in favor of the house. But de Méré got to thinking about a slightly more complicated game: one involving two dice. What sort of game could you set up by betting that you could roll two dice a certain number of times without getting a double 6? The conventional gambling wisdom of the time said the break-even point was 24 throws: fewer than this would put the odds too heavily in favor of the gambler, greater than this would favor the house. De Méré thought the conventional wisdom wrong, and his calculations eventually involved such giants of French mathematics as Blaise Pascal and Pierre de Fermat. One of the first products of modern probability theory was the realization that the real break-even point for the double-die game was 25 throws, not 24.

First, consider the game in which the player bets the house that he can roll a die four times in a row without getting a 6. Let x be the number of 6's observed in four rolls of a balanced die.

a. Is x a binomial random variable? Explain.
b. What are the values of π and n?
c. The player wins if he does not roll a 6 in any of the four rolls of the die, i.e., the player wins if $x = 0$. Find $P(x = 0)$.
d. Use the probability, part c, to explain why "this game was stacked in favor of the house."

The solution to "The Chevalier's Dilemma" involves an application of a discrete probability distribution called a **geometric probability distribution**. The geometric distribution is based on a series of independent trials of the type encountered in a binomial experiment. The difference is that the number of trials is not

*Trefil, J. S. "Odds are against your breaking that law of averages." *Smithsonian*, Sept. 1984, Vol. 15, No. 6, pp. 66–75.

fixed. Instead, the trials continue until the first "success" is observed. Thus, the geometric random variable, x, is the number of trials until the first success is observed. The probability distribution for x is

$$p(x) = \pi(1 - \pi)^{x-1} \quad \text{for } x = 1, 2, 3, \ldots$$

where $\pi = P(\text{success on a single trial})$.

e. To begin to solve "The Chevalier's Dilemma," let x be the number of throws of two dice until the first "double 6" occurs. Define a success for this geometric random variable.

f. What is the value of π?

g. Find $P(x = 25)$.

h. Find $P(x > 24)$, the probability that double 6's will not occur until after the 24th throw. [*Hint:* It can be shown (proof omitted) that for the geometric probability distribution, $P(x > k) = (1 - \pi)^k$.]

i. Find $P(x < 24)$, the probability that double 6's will occur before the 24th throw. [*Hint:* Since $P(x < k) = 1 - P(x > k - 1)$, the hint given in part **h** implies that $P(x < k) = 1 - (1 - \pi)^{k-1}$.]

j. Refer to the probabilities computed in parts **h** and **i**. Check to verify that more than 24 throws put the odds in favor of the gambler, and fewer than 24 throws puts the odds in favor of the house. (This reasoning was used to support the original belief that the break-even point for the double-die game was 24 throws.)

A version of the dice game of interest to de Méré can be formally described as in the table. "The Chevalier's Dilemma" is to find the break-even point B. In the language of probability, the break-even point is the value of B such that a gambler's expected winnings in a long series of dice throws is $0. This expected (or mean) value can be written as

$$E(\text{winnings}) = (1)P(x > B) + (-1)P(x < B) + (0)P(x = B)$$
$$= P(x > B) - P(x < B)$$

Using the hints given in parts **h** and **i**, we have

$$E(\text{winnings}) = (1 - \pi)^B - [1 - (1 - \pi)^{B-1}]$$
$$= (1 - \pi)^B + (1 - \pi)^{B-1} - 1$$

We must find an integer B such that $E(\text{winnings}) = 0$, or, equivalently, such that $(1 - \pi)^B + (1 - \pi)^{B-1} - 1 = 0$.

k. Find $E(\text{winnings})$ for $B = 24$.

l. Find $E(\text{winnings})$ for $B = 25$.

m. Do the results of parts **k** and **l** support the Chevalier de Méré's findings that the break-even point is 25 throws, not 24? [Check to see that $E(\text{winnings})$ is closer to 0 for $B = 25$ than for $B = 24$.]

OUTCOME	WINNINGS (dollars)
Gambler wins: $x > B$	1
Gambler loses: $x < B$	-1^a
Tie: $x = B$	0

^{a}The gambler forfeits $1.

REFERENCES

Feller, W. *An Introduction to Probability Theory and Its Applications*, Vol. I, 3rd ed. New York: Wiley, 1968.

Hogg, R. V. and Craig, A. T. *Introduction to Mathematical Statistics*, 4th ed. New York: Macmillan, 1978.

Mendenhall, W., Scheaffer, R. L., and Wackerly, D. *Mathematical Statistics with Applications*, 4th ed. Boston: Duxbury Press, 1989.

Mood, A. M., Graybill, F. A., and Boes, D. C. *Introduction to the Theory of Statistics*, 3rd ed. New York: McGraw-Hill, 1963.

Mosteller, F., Rourke, R. E. K., and Thomas, G. B. *Probability with Statistical Applications*, 2nd ed. Reading, Mass.: Addison-Wesley, 1970.

Parzen, E. *Modern Probability Theory and Its Applications*. New York: Wiley, 1964.

Parzen, E. *Stochastic Processes*. San Francisco: Holden-Day, 1962.

A CONTINUOUS PROBABILITY DISTRIBUTION: THE NORMAL DISTRIBUTION

CHAPTER 6

According to one theory of stock price behavior, publicized and popularized by B. G. Malkiel's book *A Random Walk Down Wall Street*, stock prices actually change (walk) upward and downward in a random manner and produce a relative frequency distribution of changes in price that has a familiar bell-shaped curve known as a **normal distribution** (see the chapter case in Section 6.4). Does the real world of stock price changes agree with this theory? For reasons that you will subsequently learn, many random variables possess normal relative frequency distributions. Consequently, we shall study the characteristics of normal distributions in this chapter and learn how to identify improbable or rare events. This will help us decide whether the real world of stock prices disagrees with the random walk theory of stock price changes.

CONTENTS

PROBABILITY MODELS FOR CONTINUOUS RANDOM VARIABLES

6.1 Suppose you want to predict the length of time of your wait in a dentist's office, the sale price of a home, the annual amount of rainfall in your city, or an evening's winnings at blackjack. If you do, you will need to know something about continuous random variables.

Recall that continuous random variables are those that can assume (at least theoretically) any of the infinitely large number of values contained in an interval. Thus, we might envision a population of patient waiting times in a dentist's office, the sale prices of houses, the annual amounts of rainfall in your city since 1900, or the gains (or losses) of many evenings of blackjack. Since our ultimate goal is to make inferences about a population based on the measurements contained in a sample, we shall need to know the probability that the sample observations (or sample statistics) assume specific values.

For example, suppose that all University of Florida graduates between fall 1989 and spring 1991 had secured employment at the time of graduation, instead of only the 1,795 graduates recorded in Appendix A. Then the corresponding set of starting salaries for all the graduates would be the population that describes the financial compensations of all graduates of the university. What is the probability that a single graduate selected at random had a starting salary of less than $35,000? To answer this question, we need to know the proportion of all graduates who had starting salaries less than $35,000. This proportion is given by the area under the population relative frequency distribution that lies to the left of $35,000. For example, if the shaded area under the *hypothetical* relative frequency distribution of Figure 6.1 is equal to $\frac{8}{10}$ of the total area under the curve, then the probability that a randomly selected graduate had a starting salary of less than $35,000 is .8.

FIGURE 6.1

Hypothetical Relative Frequency Distribution of Starting Salaries of University of Florida Graduates, Fall 1989 through Spring 1991

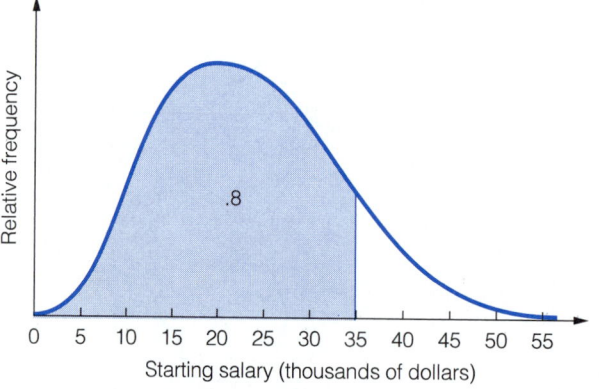

The problem, of course, is that not all of the fall 1989–spring 1991 graduates had secured employment at the time of graduation and responded to the employment questionnaire. Hence, we do not know the exact shape of the population relative frequency distribution sketched in Figure 6.1. Then, as in the case of a coin-tossing experiment, we postulate a model, i.e., we select a smooth curve (similar to the one shown in Figure 6.1) as a **model** for the population relative frequency distribution. To find the probability that a particular observation (say, a starting salary) will fall in a particular interval, we use the model and find the area under the curve that falls over that interval. Of course, in order for this approximate probability to be realistic, we need to be fairly certain that the model and the population relative frequency distribution are very similar. In Chapter 7 we shall show why we believe that the models we use are good approximations to reality.

In the next section, we introduce one of the most important and useful models for population relative frequency distributions and show how it can be used to find probabilities associated with specific sample observations.

6.2 THE NORMAL DISTRIBUTION

6.2 One of the most useful models for a population relative frequency distribution is known as the **normal distribution**. A graph of the normal distribution (often called the **normal curve**) is shown in Figure 6.2.* The normal distribution was proposed by C. F. Gauss (1777–1855) as a model for the relative frequency distribution of *errors*, such as errors of measurement. Amazingly, this curve provides an adequate model for the relative frequency of data collected from many different disciplines.

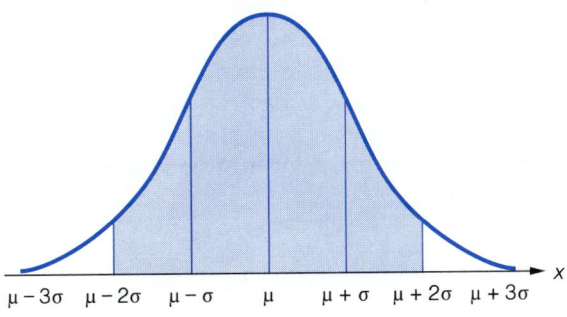

FIGURE 6.2
The Normal Curve

You can see from the figure that the mound-shaped normal curve is symmetric about its mean μ. Furthermore, approximately 68% of the area under a normal curve lies within the interval $\mu \pm \sigma$. Approximately 95% of the area lies within the interval $\mu \pm 2\sigma$ (shaded in Figure 6.2), and almost all (99.7%) lies within the interval $\mu \pm 3\sigma$. Note that these percentages agree with the Empirical Rule of Section 3.6. (This is because the Empirical Rule is based on data that can be modeled by a normal distribution.)

Remember that areas under the normal curve have a probabilistic interpretation. Thus, if a population of measurements has approximately a normal distribution, then the probability that a randomly selected observation falls within the interval $\mu \pm 2\sigma$ is approximately .95.

---

*The formula for the normal curve, denoted $f(x)$ and called the **normal probability density function**, is

$$f(x) = \frac{1}{\sigma\sqrt{2\pi}}e^{-(1/2)[(x-\mu)/\sigma]^2}$$

Although always mound-shaped and symmetric, the exact shape of the normal curve will depend on the specific values of μ and σ. Several different normal curves are shown in Figure 6.3. You can see from Figure 6.3 that the mean μ measures the location of the distribution and the standard deviation σ measures its spread.

FIGURE 6.3

Three Normal Distributions with Different Means and Standard Deviations

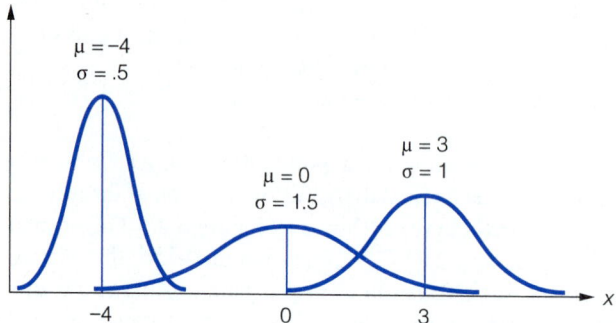

The areas under the normal curve have been computed and are given in Table 4 of Appendix G. Since the normal curve is symmetric, we need give areas on only one side of the mean. Consequently, the entries in Table 4 are areas between the mean and a point x to the right of the mean.

Since the values of μ and σ vary from one normal distribution to another, the easiest way to express a distance from the mean is in terms of a z score (see Section 3.8), i.e., the number of standard deviations between the mean and the point x. Thus,

$$z = \frac{x - \mu}{\sigma}$$

is the distance between x and μ, expressed in units of σ.

EXAMPLE 6.1   Suppose a population relative frequency distribution has mean $\mu = 500$ and standard deviation $\sigma = 100$. Give the z score corresponding to $x = 650$.

Solution   The value $x = 650$ lies 150 units above $\mu = 500$. This distance, expressed in units of σ ($\sigma = 100$), is 1.5. We can get this answer directly by substituting x, μ, and σ into the formula for z:

$$z = \frac{x - \mu}{\sigma} = \frac{650 - 500}{100} = \frac{150}{100} = 1.5$$

A partial reproduction of Table 4 of Appendix G is shown in Table 6.1. The entries in the complete table give the areas to the right of the mean for distances from $z = 0.00$ to $z = 3.09$.

TABLE 6.1

Reproduction of Part of Table 4 of Appendix G

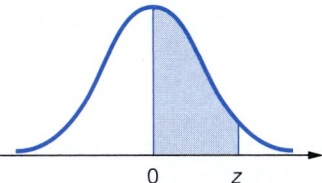

z	.00	.01	.02	.03	.04	.05	.06	.07	.08	.09
.0	.0000	.0040	.0080	.0120	.0160	.0199	.0239	.0279	.0319	.0359
.1	.0398	.0438	.0478	.0517	.0557	.0596	.0636	.0675	.0714	.0753
.2	.0793	.0832	.0871	.0910	.0948	.0987	.1026	.1064	.1103	.1141
.3	.1179	.1217	.1255	.1293	.1331	.1368	.1406	.1443	.1480	.1517
.4	.1554	.1591	.1628	.1664	.1700	.1736	.1772	.1808	.1844	.1879
.5	.1915	.1950	.1985	.2019	.2054	.2088	.2123	.2157	.2190	.2224
.6	.2257	.2291	.2324	.2357	.2389	.2422	.2454	.2486	.2517	.2549
.7	.2580	.2611	.2642	.2673	.2704	.2734	.2764	.2794	.2823	.2852
.8	.2881	.2910	.2939	.2967	.2995	.3023	.3051	.3078	.3106	.3133
.9	.3159	.3186	.3212	.3238	.3264	.3289	.3315	.3340	.3365	.3389
1.0	.3413	.3438	.3461	.3485	.3508	.3531	.3554	.3577	.3599	.3621
1.1	.3643	.3665	.3686	.3708	.3729	.3749	.3770	.3790	.3810	.3830
1.2	.3849	.3869	.3888	.3907	.3925	.3944	.3962	.3980	.3997	.4015
1.3	.4032	.4049	.4066	.4082	.4099	.4115	.4131	.4147	.4162	.4177
1.4	.4192	.4207	.4222	.4236	.4251	.4265	.4279	.4292	.4306	.4319
1.5	.4332	.4345	.4357	.4370	.4382	.4304	.4406	.4418	.4429	.4441

EXAMPLE 6.2 Find the area under a normal curve between the mean and a point $z = 1.26$ standard deviations to the right of the mean, i.e., $P(0 \leq z \leq 1.26)$.

Solution   To locate the proper entry, proceed down the left (z) column of the table to the row corresponding to $z = 1.2$. Then move across the top of the table to the column headed .06. The intersection of the .06 column and the 1.2 row contains the desired area, .3962, as shown in Figure 6.4.

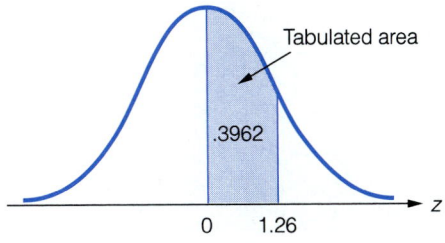

FIGURE 6.4
The Tabulated Area
Corresponding to $z = 1.26$

The normal distribution of the z statistic (as shown in Figure 6.4) is called the **standard normal distribution**. The mean of a standard normal distribution is 0 (since $z = 0$ when $x = \mu$); the standard deviation is equal to 1 (proof omitted). Since the mean is 0, z values to the right of the mean are positive; those to the left are negative.

EXAMPLE 6.3 Find the area beneath the standard normal curve between the mean $z = 0$ and the point $z = -1.26$; i.e., find $P(-1.26 \leq z \leq 0)$.

Solution The best way to solve a problem of this type is to draw a sketch of the distribution (see Figure 6.5). Since $z = -1.26$ is negative, we know that it lies to the left of the mean, and the area that we seek is the shaded area shown.

FIGURE 6.5
Standard Normal Distribution
for Example 6.3

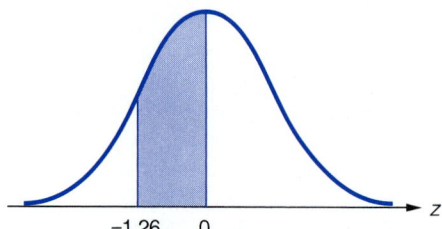

Since the normal curve is symmetric, the area between the mean 0 and $z = -1.26$ is exactly the same as the area between the mean 0 and $z = +1.26$. We found this area in Example 6.2 to be .3962. Therefore, the area between $z = -1.26$ and $z = 0$ is .3962.

EXAMPLE 6.4 Find the probability that a normally distributed random variable will lie within $z = 2$ standard deviations of its mean; that is, find the probability $P(-2 \leq z \leq 2)$.

Solution The probability that we seek is the shaded area shown in Figure 6.6. Since the area between the mean and $z = 2.0$ is exactly the same as the area between the mean and $z = -2.0$, we need find only the area between the mean and $z = 2$ standard deviations to the right of the mean and multiply by 2. This area is given in Table 4 of Appendix G as .4772. Therefore, the probability P that

FIGURE 6.6
Standard Normal Distribution
for Example 6.4

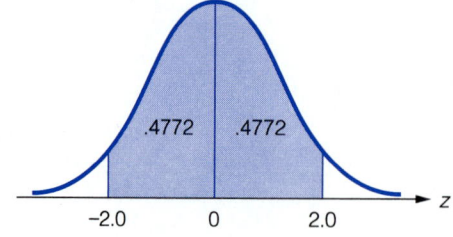

a normally distributed random variable will lie within two standard deviations of its mean is

$$P = 2(.4772) = .9544$$

EXAMPLE 6.5 Find the probability that a normally distributed random variable x will lie more than $z = 2$ standard deviations above its mean; i.e., find $P(z > 2)$.

Solution The probability we seek is the darker shaded area shown in Figure 6.7. The total area under a standard normal curve is 1; half this area lies to the left of the mean, half to the right. Consequently, the probability P that x will lie more than two standard deviations above the mean is equal to .5 less the area A:

$$P = .5 - A$$

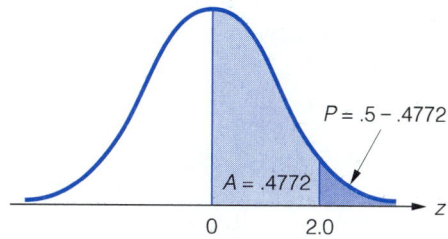

FIGURE 6.7
Standard Normal Distribution for Example 6.5

The area A corresponding to $z = 2.0$ is .4772. Therefore,

$$P = .5 - .4772 = .0228$$

EXAMPLE 6.6 Find the area under the normal curve between $z = 1.2$ and $z = 1.6$; i.e., find $P(1.2 \leq z \leq 1.6)$.

Solution The area A that we seek lies to the right of the mean because both z values are positive. It will appear as the shaded area shown in Figure 6.8. Let A_1 represent the area between $z = 0$ and $z = 1.2$, and A_2 represent the area between $z = 0$ and $z = 1.6$. Then the area A that we desire is

$$A = A_2 - A_1$$

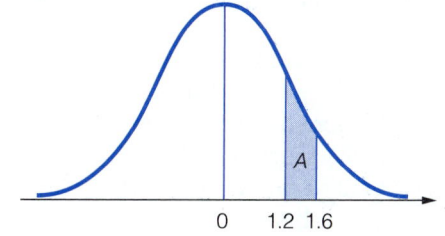

FIGURE 6.8
Standard Normal Distribution for Example 6.6

From Table 4 of Appendix G, we obtain:

$$A_1 = .3849 \qquad \text{and} \qquad A_2 = .4452$$

Then

$$A = A_2 - A_1$$
$$= .4452 - .3849$$
$$= .0603$$

EXAMPLE 6.7 Let $z_{.10}$ denote the value of z such that the area to the right of $z_{.10}$ is .10. Find $z_{.10}$.

Solution   The z value that we seek appears as shown in Figure 6.9. Note that we show an area to the right of z equal to .10. Since the total area to the right of the mean $z = 0$ is equal to .5, the area between the mean 0 and the unknown z value is $.5 - .1 = .4$ (as shown in the figure). Consequently, to find $z_{.10}$, we just look in Table 4 of Appendix G for the z value that corresponds to an area equal to .4.

FIGURE 6.9
Standard Normal Distribution for Example 6.7

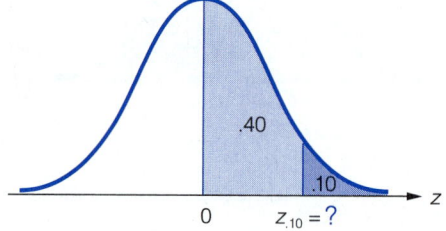

The area .4000 does not appear in Table 4. The closest values are .3997, corresponding to $z = 1.28$, and .4015, corresponding to $z = 1.29$. Since the area .3997 is closer to .4000 than is .4015, we will choose $z = 1.28$ as our answer. That is, $z_{.10} = 1.28$.

Examples 6.2–6.7 demonstrate how to solve the following two types of normal probability problems:

1. Examples 6.2–6.6 use Table 4 of Appendix G to **find areas under the standard normal curve**. These problems may be further classified into one of three types:
   a. Finding the area between the mean $\mu_z = 0$, and some value of z—say, z_0—that is located above or below $\mu_z = 0$ (Examples 6.2 and 6.3)
   b. Finding the area between the values z_1 and z_2, where z_1 or z_2 is not equal to 0   (Examples 6.4 and 6.6)
   c. Finding the area in either the upper or the lower tail of the standard normal z distribution   (Example 6.5)

2. Example 6.7 uses Table 4 of Appendix G to **find the z value, denoted** z_α, **corresponding to an area** α in the upper tail of the standard normal z distribution. A similar procedure may be used to find a z value corresponding to a lower-tail area under the curve.

Many distributions of data that occur in the real world are approximately normal, but few are *standard* normal. However, Examples 6.8–6.12 use what you have learned about the standard normal curve to solve the same two types of problems involving *any* normal distribution:

1. Finding the probability that a normal random variable x falls between the values x_1 and x_2, or the probability that it falls in either the upper or the lower tail of the normal distribution   (Examples 6.8–6.10)
2. Finding the value of x—say, x_0—that places a probability P in the upper (or lower) tail of a normal distribution   (Examples 6.11 and 6.12)

EXAMPLE 6.8   Medical research has linked excessive consumption of salt to hypertension (high blood pressure). Yet, salt is America's second leading food additive (after sugar), in both factory-processed foods and home cooking. The average amount of salt consumed per day by an American is 15 grams (15,000 milligrams), although the actual physiological minimum daily requirement for salt is only 220 milligrams. Suppose that the amount of salt intake per day is approximately normally distributed with a standard deviation of 5 grams. What proportion of all Americans consume between 14 and 22 grams of salt per day?

Solution   The proportion P of Americans who consume between $x = 14$ grams and $x = 22$ grams of salt is the total shaded area in Figure 6.10a. Before we can compute this area, we need to determine the z values that correspond to $x = 14$ and $x = 22$. Substituting $\mu = 15$ and $\sigma = 5$ into the formula for z, we compute the z value for $x = 14$ as

$$z_1 = \frac{x - \mu}{\sigma} = \frac{14 - 15}{5} = \frac{-1}{5} = -.20$$

The corresponding z value for $x = 22$ is

$$z_2 = \frac{x - \mu}{\sigma} = \frac{22 - 15}{5} = \frac{7}{5} = 1.40$$

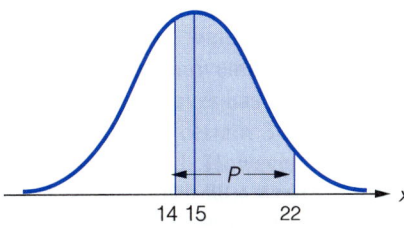

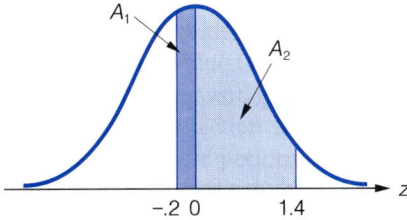

a. Amount of salt intake (in grams) per day   **b.** Standard normal

FIGURE 6.10
Normal Curve Sketches for Example 6.8

EXERCISES
LEARNING THE MECHANICS

6.1 Find the area under the standard normal curve:

a. Between $z = 0$ and $z = 1.2$ b. Between $z = 0$ and $z = 1.49$
c. Between $z = -.48$ and $z = 0$ d. Between $z = -1.37$ and $z = 0$
e. For values of z larger than 1.33

Show the z values and the corresponding area of interest on a sketch of the normal curve for each part of the exercise.

6.2 Find the area under the standard normal curve:

a. Between $z = 1.21$ and $z = 1.94$ b. For values of z larger than 2.33
c. For values of z less than -2.33 d. Between $z = -1.50$ and $z = 1.79$

Show the values of z and corresponding area of interest on a sketch of the normal curve for each part of the exercise.

6.3 Find the z value (to two decimal places) that corresponds to a tabulated area (Table 4 of Appendix G) equal to:

a. .1000 b. .3200 c. .4000 d. .4500 e. .4750

Show the area and corresponding value of z on a sketch of the normal curve for each part of the exercise.

6.4 Find the value of z (to two decimal places) that cuts off an area in the upper tail of the standard normal curve equal to:

a. .025 b. .05 c. .005 d. .01 e. .10

Show the area and corresponding value of z on a sketch of the normal curve for each part of the exercise.

6.5   Suppose that a normal random variable x has mean $\mu = 20.0$ and standard deviation $\sigma = 4.0$. Find the z score corresponding to:

     a. $x = 23.0$ **b.** $x = 16.0$ **c.** $x = 13.5$
     d. $x = 28.0$ **e.** $x = 12.0$

     For each part of the exercise, locate x and μ on a sketch of the normal curve. Check to make sure that the sign and magnitude of your z score agree with your sketch.

6.6   Find the approximate value for z_0 such that the probability that z is larger than z_0 is:

     a. $P = .10$ **b.** $P = .15$ **c.** $P = .20$ **d.** $P = .25$

     Locate z_0 and the corresponding probability P on a sketch of the normal curve for each part of the exercise.

6.7   Find the approximate value for z_0 such that the probability that z is less than z_0 is:

     a. $P = .10$ **b.** $P = .15$ **c.** $P = .30$ **d.** $P = .50$

     Locate z_0 and the corresponding probability P on a sketch of the normal curve for each part of the exercise.

APPLYING THE CONCEPTS

6.8   Birdwatchers, or "birders" as the more committed prefer to be called, visiting a National Wildlife Refuge located near Atlantic City, New Jersey, report a high degree of satisfaction with their visits (*Leisure Sciences*, Vol. 9, 1987). On a standard satisfaction scale that ranged from 1 to 6 (where 1 = poor rating and 6 = perfect rating), the birders have a mean score of 5.05 and a standard deviation of .98. Assume the population of satisfaction scores of birders visiting the National Wildlife Refuge is approximately normal.

     a. Find the probability that a randomly selected birder visiting the refuge will have a satisfaction score of at least 5.
     b. One-fourth of the birders have satisfaction scores below what value?

6.9   Pacemakers are used to control the heartbeat of cardiac patients, with over 120,000 of the devices implanted each year. A single pacemaker is made up of several biomedical components that must be of a high quality for the pacemaker to work. It is vitally important for manufacturers of pacemakers to use parts that meet specifications. One particular plastic part, called a connector module, mounts on the top of the pacemaker. Connector modules are required to have a length between .304 inch and .322 inch to work properly. Any module with length outside these limits are *out-of-spec*. *Quality* (Aug. 1989) reported on one supplier of connector modules that had been shipping out-of-spec parts to the manufacturer for 12 months.

     a. The lengths of the connector modules produced by the supplier were found to follow an approximate normal distribution with mean $\mu = .3015$ inch and standard deviation $\sigma = .0016$ inch. Use this information to find the probability that the supplier produces an out-of-spec part.
     b. Once the problem was detected, the supplier's inspection crew began to employ an automated data-collection system designed to improve product quality. After two months, the process was producing connector modules with mean $\mu = .3146$ inch and standard deviation $\sigma = .0030$ inch. Find the probability that an out-of-spec part will be produced. Compare your answer to part **a**.

6.10 A *chewing cycle* is defined as an upward movement followed by a downward movement of the chin. Clinicians have found that the chewing cycles of "normal" children differ from the chewing cycles of children with eating difficulties. In one study (*The American Journal of Occupational Therapy*, Mar. 1984), the number of chewing cycles required for a "normal" preschool child to swallow a bite of graham cracker was found to have a mean of 15.09 and a standard deviation of 3.4. Suppose a "normal" preschool child is fed a bite of graham cracker, and assume that the number of chewing cycles required to swallow a bite of graham cracker follows an approximate normal distribution.

     a. Find $P(x < 10)$.
     b. Find $P(8 < x < 16)$.

c. Find $P(x \geq 22)$.

d. One preschool child, thought to have eating difficulties, required $x = 22$ cycles to chew and swallow the bite of graham cracker. Is it likely that this child is from the group of "normal" children? Explain.

6.11   The U.S. Department of Agriculture (USDA) has recently patented a process that uses a bacterium for removing bitterness from citrus juices (*Chemical Engineering*, Feb. 3, 1986). In theory, almost all the bitterness could be removed by the process, but for practical purposes the USDA aims at 50% overall removal. Suppose a USDA spokesperson claims that the percentage of bitterness removed from an 8-ounce glass of freshly squeezed citrus juice is normally distributed with mean 50.1 and standard deviation 10.4. To test this claim, the bitterness removal process is applied to a randomly selected 8-ounce glass of citrus juice.

a. Find the probability that the process removes less than 43.7% of the bitterness.

b. Refer to your answer to part **a**. If the test on the single glass of citrus juice yielded a bitterness removal percentage of 43.7%, would you tend to doubt the USDA spokesperson's claim?

6.12   In a laboratory experiment, researchers at Barry University (Miami Shores, Florida) studied the rate at which sea urchins ingested turtle grass (*Florida Scientist*, Summer/Autumn 1991). The urchins, starved for 48 hours, were fed 5-cm. blades of green turtle grass. The mean ingestion time was found to be 2.83 hours and the standard deviation was .79 hour. Assume that green turtle grass ingestion time for the sea urchins has an approximate normal distribution.

a. Find the probability that a sea urchin will require 4 or more hours to ingest a 5-cm. blade of green turtle grass.

b. Find the probability that a sea urchin will require between 2 and 3 hours to ingest a 5-cm. blade of green turtle grass.

6.13   Foresters "cruising" British Columbia's boreal forest have determined that the diameter at breast height of white spruce trees in a particular community is approximately normal, with mean 17 meters and standard deviation 6 meters.*

a. Find the probability that the breast height diameter of a randomly selected white spruce in the forest community is less than 12 meters.

b. Suppose you observe a white spruce with a breast height diameter of 12 meters. Is this an unusual event? Explain.

c. Find the probability that the breast height diameter of a randomly selected white spruce in the forest community will exceed 37 meters.

d. Suppose you observe a tree in the community forest with a breast height diameter of 38 meters. Is this tree likely to be a white spruce? Explain.

6.14   The metropolitan airport commission is considering the establishment of limitations on the extent of noise pollution around a local airport. At the present time the noise level per jet takeoff in one neighborhood near the airport is approximately normally distributed with a mean of 100 decibels and a standard deviation of 6 decibels.

a. What is the probability that a randomly selected jet will generate a noise level greater than 108 decibels in this neighborhood?

b. What is the probability that a randomly selected jet will generate a noise level of exactly 100 decibels?

c. Suppose a regulation is passed that requires jet noises in this neighborhood to be lower than 105 decibels 95% of the time. Assuming the standard deviation of the noise distribution remains the same, how much will the mean noise level have to be lowered to comply with the regulations?

6.15   The Trail Making Test is frequently used by clinical psychologists to test for brain damage. Patients are required to connect consecutively numbered circles on a sheet of paper. It has been determined that the mean length of time required for a patient to perform this task is 32 seconds and the standard deviation is 4 seconds. Assume that the distribution of the lengths of time required to connect the circles is normal.

*Scholz, H. "Fish Creek Community Forest: Exploratory statistical analysis of selected data." Working paper, Northern Lights College, British Columbia, Canada.

a. Find the probability that a randomly selected patient will take longer than 40 seconds to perform the task.

b. Find the probability that a randomly selected patient will take between 24 and 40 seconds to complete the task.

c. A psychologist would like to retest those persons with completion times in the highest 5% of the distribution of times required. What time would a person need to exceed on the Trail Making Test to be considered for retesting?

6.16 Behaviorists have developed an instrument designed to measure the maturity of small groups. The 10-item questionnaire is based on the assumptions that a mature group is able to function independently of its leader, is active, organized, and has an established work history, while an immature group has the opposite attributes. Krayer (1988) divided a class of undergraduate college students into two groups, mature and immature, based on their answers to the 10-item questionnaire. A final project was then assigned and, at the end of the semester, student performances were evaluated. A summary of the grades on the project for the two groups is provided below. Assume these represent population means and standard deviations.

GROUP	MEAN GRADE	STANDARD DEVIATION
Mature	91.50	8.48
Immature	84.20	6.98

Source: Krayer, K. J. "Exploring group maturity in the classroom." *Small Group Behavior*, Vol. 19, No. 2, May 1988, p. 268.

a. Assuming the population of project grades for the mature group is approximately normal, find the probability that a mature student will score below 80 on the final project.

b. Repeat part **a** for the immature group.

c. Why might the assumption of normality in parts **a** and **b** be suspect? [*Hint*: Consider the fact that the highest grade that can be assigned to a project is 100.)

6.17 Each year the top marlin fishermen from around the world compete in the Hawaiian International Bluefish Tournament. In a recent year, one fisherman landed a 987-pound Pacific blue marlin—a world record until a check showed his fishing line was a few pounds over the 80-pound limit. The "80 pounds" refers to the strength of the fishing line, i.e., the weight the line is tested to hold outside water. Suppose the actual strength of manufactured "80-pound-test-line" is normally distributed with a mean of 80 pounds and a standard deviation of .2 pound.

a. What is the probability that an "80-pound-test-line" randomly selected from the production process will have a strength of at least 1 pound over the 80-pound limit?

b. Based on the probability computed in part **a**, is it likely that the fisherman actually used "80-pound-test-line" to catch the 987-pound blue marlin? Explain.

6.18 A television cable company receives numerous phone calls throughout the day from customers reporting service troubles and from would-be subscribers to the cable network. Most of these callers are put "on hold" until a company operator is free to help them. The company has determined that the length of time a caller is on hold is normally distributed with a mean of 3.1 minutes and a standard deviation of .9 minute. Company experts have decided that if as many as 5% of the callers are put on hold for 4.8 minutes or longer, more operators should be hired.

a. What proportion of the company's callers are put on hold for at least 4.8 minutes? Should the company hire more operators? Show the pertinent quantities on a sketch of the normal curve.

b. At this company, 5% of the callers are put on hold for longer than x minutes. Find the value of x and show the pertinent quantities on a sketch of the normal curve.

6.19 Steel used for water pipelines is often coated on the inside with cement mortar to prevent corrosion. In a study of the mortar coatings of the pipeline used in a water transmission project in California (*Transportation Engineering Journal*, Nov. 1979), researchers noted that the mortar thickness was specified to be $\frac{7}{16}$ inch. A very large sample of thickness measurements produced a mean equal to .635 inch and a standard deviation equal to .082 inch. If the thickness measurements were normally distributed, approximately what proportion were less than $\frac{7}{16}$ inch?

6.20 How does the stock market react when a firm announces its stock earnings in *The Wall Street Journal*? To examine this issue, a comprehensive study of 240 stocks was conducted over a 10-year period (*The Accounting Review*, Jan. 1991). One of the variables used to measure market reaction was excess trading volume. Excess trading volume is defined as the difference between the percentage of shares traded on the day of the earnings announcement and the average percentage of shares traded for a 3-day period prior to the announcement. The researcher assumes that excess trading volume, when standardized, has an approximate standard normal distribution. Let x equal the standardized excess trading volume for a particular stock.

a. What is the mean of x?
b. What is the standard deviation of x?
c. Find the median (50th percentile) of x.
d. Find the lower quartile (25th percentile) of x.
e. Find the upper quartile (75th percentile) of x.

6.21 Refer to Exercise 6.20. Using simulation, the researcher found that standardized excess trading volume x is approximately normal, but with mean $\mu = -1.767$ and standard deviation $\sigma = .956$.

a. Find $P(x < 0)$.
b. Find $P(x < Q_L)$, where Q_L is the lower quartile of a standard normal (z) distribution.
c. Find $P(x < Q_U)$, where Q_U is the upper quartile of a standard normal (z) distribution.
d. Based on the results, parts **a–c**, how well does the standard normal (z) distribution approximate the true distribution of x?

DESCRIPTIVE METHODS FOR ASSESSING NORMALITY

6.3 In the chapters that follow, we learn how to make inferences about the population based on information in the sample. Several of these techniques are based on the assumption that the population is approximately normally distributed. Consequently, it will be important to determine whether the sample data come from a normal population before we can properly apply these techniques.

Several descriptive methods can be used to check for normality. In this section, we consider the three methods summarized in the box.

DETERMINING WHETHER THE DATA ARE FROM AN APPROXIMATELY NORMAL DISTRIBUTION

1. Construct either a **relative frequency histogram** or **stem-and-leaf display** for the data. If the data are approximately normal, the shape of the graph will be similar to the normal curve, Figure 6.2 (i.e., mound-shaped and symmetric about the mean).

2. Find the **interquartile range, IQR**, and **standard deviation, s**, for the sample, then calculate the ratio IQR/s. If the data are approximately normal, then IQR/$s \approx 1.3$.

3. Construct a **normal probability plot** for the data. (See the box on page 275.) If the data are approximately normal, the points will fall (approximately) on a straight line.

EXAMPLE 6.12 Consider the data set consisting of the tar contents of the 372 cigarette brands listed in Appendix F. Numerical and graphical descriptive measures for the data are shown on the SAS printouts, Figures 6.14a–c. Determine whether the tar contents have an approximately normal distribution.

Solution As a first check, we examine the horizontal frequency histogram of the data shown in Figure 6.14b. Clearly, the tar contents fall in an approximately mound-shaped, symmetric distribution centered around the mean of 11.60 milligrams. Thus, from Check #1 in the box, the data appear to be approximately normal.

Check #2 in the box requires that we find the interquartile range (i.e., the difference between the 75th and 25th percentiles) and the standard deviation of the data set, and compute the ratio of these two numbers. The ratio IQR/s for a sample from a normal distribution will approximately equal 1.3.* The values

a. SAS Descriptive Statistics for Tar Contents of Appendix F

FIGURE 6.14
SAS Printouts for Example 6.12

UNIVARIATE PROCEDURE

Variable=TAR

Moments

N	372	Sum Wgts	372
Mean	11.60215	Sum	4316
Std Dev	4.966089	Variance	24.66204
Skewness	0.247064	Kurtosis	0.301387
USS	59224.5	CSS	9149.618
CV	42.80318	Std Mean	0.25748
T:Mean=0	45.06044	Prob>|T|	0.0001
Sgn Rank	34689	Prob>|S|	0.0001
Num ^= 0	372		

Quantiles(Def=5)

100% Max	26	99%	25
75% Q3	15	95%	19
50% Med	11	90%	17
25% Q1	9	10%	5
0% Min	0.5	5%	4
		1%	0.5
Range	25.5		
Q3-Q1	6		
Mode	9		

Extremes

Lowest	Obs	Highest	Obs
0.5(	250)	25(	258)
0.5(	249)	25(	276)
0.5(	61)	25(	298)
0.5(	59)	26(	255)
0.5(	57)	26(	271)

Continued

---

*You can see that this property holds for normal distributions by noting that the z values (obtained from Table 4 in Appendix G) corresponding to the 75th and 25th percentiles are .67 and $-.67$, respectively. Since $\sigma = 1$ for a standard normal (z) distribution, IQR/$\sigma = [.67 - (-.67)]/1 = 1.34$.

FIGURE 6.14 Continued **b.** SAS Histogram and Box Plot for Tar Contents of Appendix F

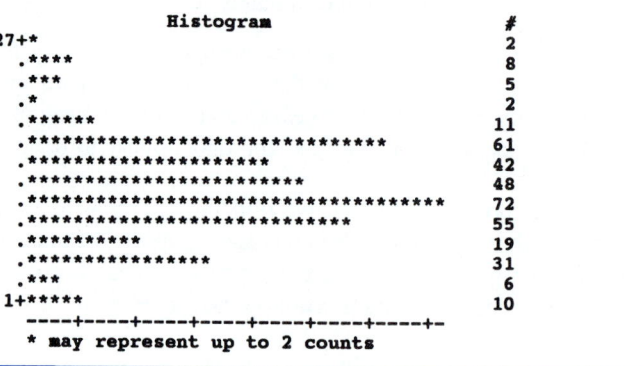

c. SAS Normal Probability Plot for Tar Contents of Appendix F

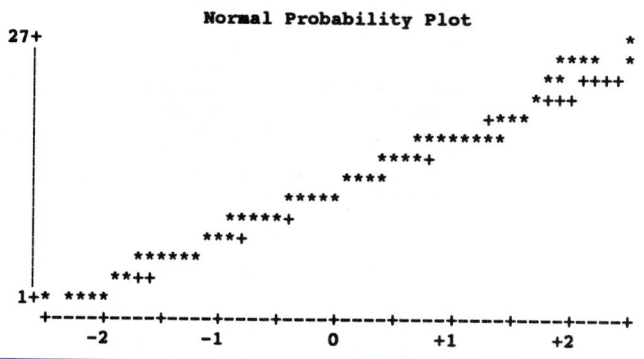

of IQR and s, shaded in Figure 6.14a, are IQR $= Q_U - Q_L = 6$ and $s = 4.966$. Then the ratio is

$$\frac{\text{IQR}}{s} = \frac{6}{4.966} = 1.21$$

Since this value is approximately equal to 1.3, we have further confirmation that the data are approximately normal.

A third descriptive technique for checking normality is a **normal probability plot**. In a normal probability plot, the observations in the data set are ordered and then plotted against the standardized expected values of the observations under the assumption that the data are normally distributed. When the data are, in fact, normally distributed, an observation will approximately equal its expected value. Thus, a linear (straight-line) trend on the normal probability plot suggests that the data are from an approximate normal distribution, while a nonlinear trend indicates that the data are nonnormal.

Normal probability plots can be constructed by hand, as shown in the box. However, it is easier to generate these plots by computer. A SAS normal probability plot for the 372 tar measurements is shown in Figure 6.14c. Notice that the ordered measurements (represented by the plotting symbol "*") fall reasonably close to a straight line (plotting symbol "+"). Thus, check #3 also suggests that the data are likely to be approximately normally distributed.

CONSTRUCTING A NORMAL PROBABILITY PLOT FOR A DATA SET

1. List the observations in the sample data set in ascending order, where x_i represents the ith ordered value.

2. For each observation, calculate the corresponding tail area of the standard normal (z) distribution,

$$A = \frac{i - .375}{n + .25}$$

where n is the sample size.

3. Calculate the estimated expected value of x_i under normality using the following formula:

$$E(x_i) = (s)[Z(A)]$$

where s is the sample standard deviation and $Z(A)$ is the z value that cuts off an area A in the lower tail of the standard normal distribution.

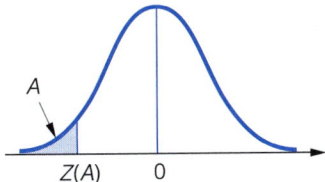

4. Plot the ordered observations x_i on the vertical axis and the corresponding estimated expected values, $E(x_i)$, on the horizontal axis.

The checks for normality given in the box are simple, yet powerful, techniques to apply, but they are only descriptive in nature. It is possible (although unlikely) that the data are nonnormal even when the checks are reasonably satisfied. Thus, we should be careful not to claim that the 372 tar measurements in Appendix F are, in fact, normally distributed. We can only state that it is reasonable to believe that the data are from a normal distribution.*

---

*Statistical tests of normality that provide a measure of reliability for the inference are available. However, these tests tend to be very sensitive to slight departures from normality, i.e., they tend to reject the hypothesis of normality for any distribution that is not perfectly symmetrical and mound-shaped. Consult the references if you want to learn more about these tests.

EXERCISES

LEARNING THE MECHANICS

6.22 The data for Exercises 2.17 and 3.43 are reproduced below. Determine whether distribution of the data is approximately normal.

5.9	5.3	1.6	7.4	8.6	1.2	2.1
4.0	7.3	8.4	8.9	6.7	4.5	6.3
7.6	9.7	3.5	1.1	4.3	3.3	8.4
1.6	8.2	6.5	1.1	5.0	9.4	6.4

APPLYING THE CONCEPTS

6.23 *Meta-analysis* is a procedure to compare the findings of independent studies on the same subject matter. Stock et al. (1985) conducted a meta-analysis of the relation between race and subjective well-being among noninstitutionalized adults based on results obtained from 54 sources. For each source, a measure of the relationship between race and well-being, called "effect size," was calculated. (Effect sizes ranged from -1 to 1, with positive values implying that whites had a higher subjective well-being than blacks and negative values implying the reverse.) A stem-and-leaf display (see Chapter 2) of the 54 effect sizes is reproduced below. The mean and standard deviation of the data were .10 and .09, respectively.

STEM	LEAF
.3	0 1
.2	5 6 7
.2	0 0 0 2 4
.1	5 5 6 9 9 9
.1	0 0 0 1 1 2 2 2 3 4 4 4
.0	5 6 6 6 6 7 7 8 8 8 9
.0	2
−.0	1 0 0 0
−.0	9 9 8
−.1	2 2 1 1 1
−.1	6
−.2	4

Source: Stock, W. A., Okun, M. A., Haring, M. J., & Witter, R. A. "Race and subjective well-being in adulthood: A black-white research synthesis." *Human Development*, Vol. 28, 1985, p. 195. By permission of S. Karger AG, Basel.

a. Examine the stem-and-leaf display. Is the distribution of effect sizes approximately normal?

b. Assuming normality, calculate the probability that a randomly selected study on the relationship between race and subjective well-being will have a negative effect size? (Assume $\mu = .10$ and $\sigma = .09$.)

c. Use the stem-and-leaf display to calculate the actual percentage of the 54 effect sizes that were negative. Compare the result to your answer to part b.

6.24 Refer to the *Small Group Behavior* study of mature and immature groups, Exercise 6.11. A stem-and-leaf display and normal probability plot for each of the two groups are shown in the accompanying SAS printouts. Based on these graphs, assess whether the grade distributions are approximately normal.

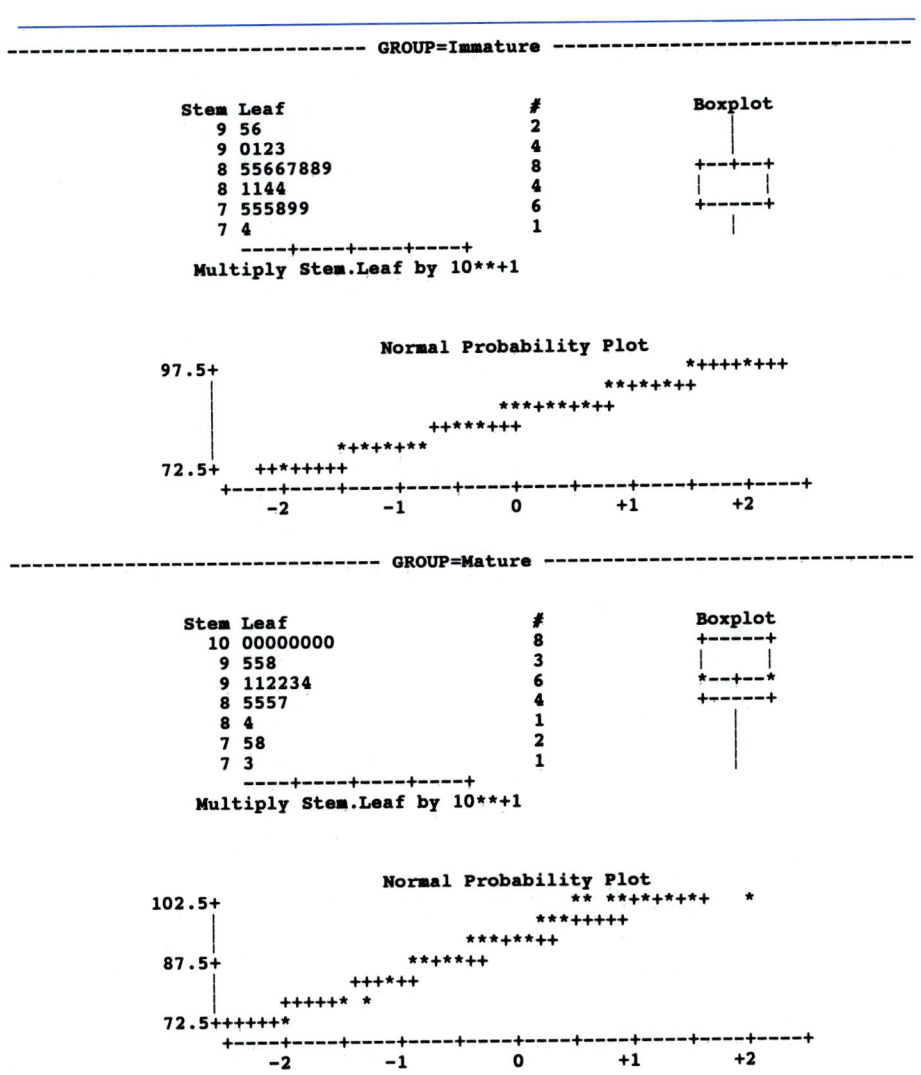

6.25 At the end of each academic semester, the Career Resource Center at the University of Florida mails out questionnaires pertaining to employment status and starting salary of all students who graduated that particular semester. From fall 1989 through spring 1991, 1,795 graduates returned the questionnaire and indicated they had secured a job as of the date of employment. SAS printouts describing the 1,795 starting salaries are displayed on the next page. Use this information to assess whether the data are approximately normal.

SAS Printouts for Exercise 6.25

UNIVARIATE PROCEDURE

Variable=SALARY

Moments

N	1795	Sum Wgts	1795
Mean	28474.83	Sum	51112325
Std Dev	9369.484	Variance	87787229
Skewness	1.093351	Kurtosis	2.301454
USS	1.613E12	CSS	1.575E11
CV	32.90444	Std Mean	221.1482
T:Mean=0	128.759	Prob>|T|	0.0
Sgn Rank	805955	Prob>|S|	0.0
Num ^= 0	1795		
W:Normal	0.934422	Prob<W	0.0

Quantiles(Def=5)

100% Max	80000		99%	60000
75% Q3	33000		95%	46000
50% Med	27000		90%	40000
25% Q1	22000		10%	19000
0% Min	7100		5%	16000
			1%	10000
Range	72900			
Q3-Q1	11000			
Mode	20000			

UNIVARIATE PROCEDURE

Variable=SALARY

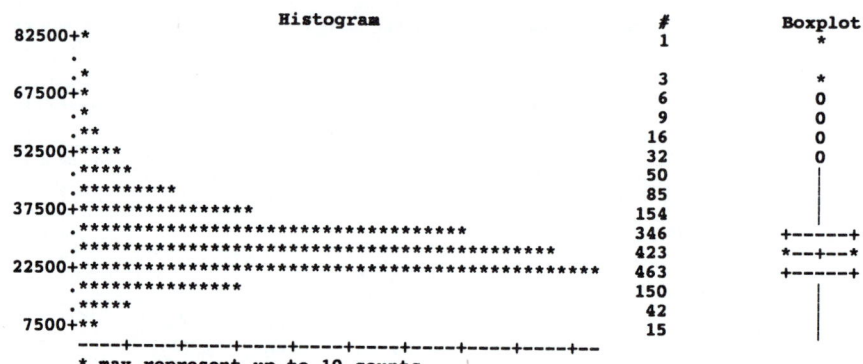

```
                        Histogram                          #        Boxplot
 82500+*                                                    1          *
      .
      .*                                                     3
 67500+*                                                     6          *
      .*                                                     9          0
      .**                                                   16          0
 52500+****                                                 32          0
      .*****                                                50
      .********                                             85
 37500+****************                                    154
      .*****************************************           346      +-----+
      .************************************************    423      *--+--*
 22500+**************************************************  463      +-----+
      .****************                                    150
      .*****                                                42
  7500+**                                                  15
      ----+----+----+----+----+----+----+----+----+--
      * may represent up to 10 counts
```

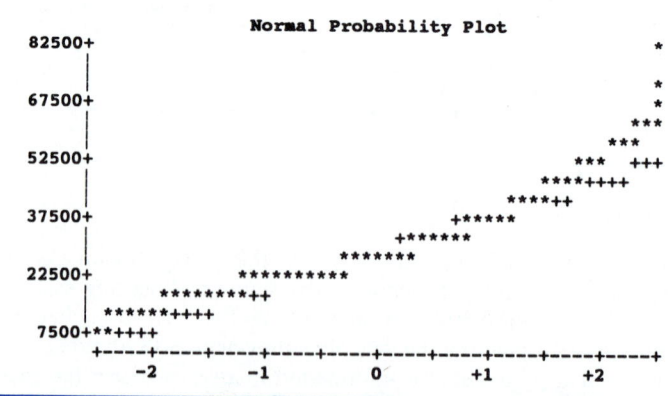

```
                        Normal Probability Plot
 82500+                                                    *
      |
 67500+                                                  *
      |                                               ***
      |                                             ***
 52500+                                      *** +++
      |                                 ****++++
      |                            ****++
 37500+                       +*****
      |                   +******
      |                *******
 22500+          **********
      |     ********++
      | ******+++
  7500+**+++++
      +----+----+----+----+----+----+----+----+----+
          -2        -1         0        +1        +2
```

CHAPTER 6   A CONTINUOUS PROBABILITY DISTRIBUTION: THE NORMAL DISTRIBUTION

6.26 Refer to the study of British Columbia's boreal forest, Exercise 6.13. The diameters at breast height (in meters) for a sample of 28 trembling aspen trees are listed below. Determine whether the sample data are from an approximately normal distribution.

12.4	17.3	27.3	19.1	16.9	16.2	20.0
16.6	16.3	16.3	21.4	25.7	15.0	19.3
12.9	18.6	12.4	15.9	18.8	14.9	12.8
24.8	26.9	13.5	17.9	13.2	23.2	12.7

Source: Scholz, H. "Fish Creek Community Forest: Exploratory statistical analysis of selected data." Working paper, Northern Lights College, British Columbia, Canada.

6.27 It is well-known that children with developmental delays (i.e., mild mental retardation) are slower, cognitively, than normally developing children. Are their social skills also lacking? A study compared the social interactions of the two groups of children in a controlled playground environment (*American Journal on Mental Retardation*, Jan. 1992). One variable of interest was the number of intervals of "no play" by each child. Children with developmental delays had a mean of 2.73 intervals of "no play" and a standard deviation of 2.58 intervals. Based on this information, is it possible for the variable of interest to be normally distributed? Explain.

6.4   CHAPTER CASE: A RANDOM WALK DOWN WALL STREET

In his lively and insightful book, *A Random Walk Down Wall Street*,* Burton G. Malkiel devotes Chapter 6 to "Technical Analysis and the Random-Walk Theory." In brief, this chapter discusses the theory used by some stock market technical analysts to forecast the upward or downward movement of specific stocks or the market as a whole.

Technical analysts called *chartists* believe that "knowledge of a stock's past behavior can help predict its probable future behavior." According to Malkiel, chartists strongly favor stocks that have made an upward move, and they advocate selling stocks that have moved downward in price. Similarly, when the stock market averages have shown strength and moved upward, they forecast further upward movement in the market averages. Malkiel clearly has little faith in the forecasts of chartists. He states, "On close examination, technicians are often seen with holes in their shoes and frayed shirt colors. I, personally, have never known a successful technician, but I have seen wrecks of several successful ones. (This is, of course, in terms of following their own technical advice. Commissions from urging customers to act on their recommendations are very lucrative.)" In brief, Malkiel does not agree that knowledge of a stock's past performance can be used to predict its future behavior.

The antithesis of the theory held by chartists is known as the *random-walk theory*. According to this theory, the price movement of a stock (or the stock market) today is completely independent of its movement in the past; the price will rise or fall today by a random amount. A sequence of these random increases and decreases is known as a **random walk**. Note that this theory does not rule out the possibility of long-term trends—say, a long-term upward trend in stock prices fueled by increased earnings or dividends. It simply states that the *daily* change in price is independent of the changes that have occurred in the past.

---

*Malkiel, B. G. *A Random Walk Down Wall Street*. New York: Norton, 1975.

To support the random-walk theory, Malkiel generated the stock price chart for several fictitious stocks by tossing a coin to decide whether a stock would move up (a head) or down (a tail) by a fixed amount, say $\frac{1}{2}$ dollar, on a given day. This procedure was repeated for a large number of days. The movement of the fictitious stock was plotted, thus revealing its random walk. We performed an equivalent coin-tossing experiment for this case and our fictitious stock chart for 50 days is shown in Figure 6.15. You can see how the price seems to surge upward and downward in this chart (suggesting short-term trends), although the coin tossings were independent.

FIGURE 6.15

Fictitious Stock Price Chart: Random Walk

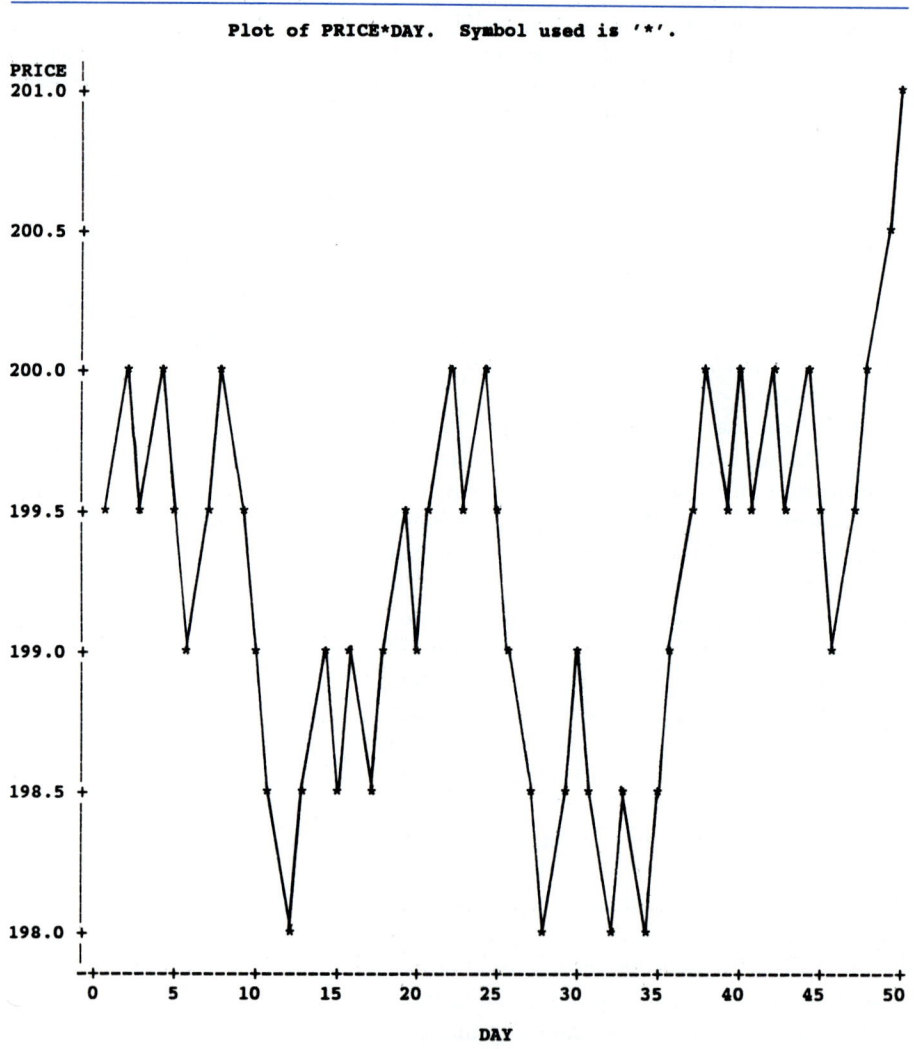

Plot of PRICE*DAY. Symbol used is '*'.

One way that we can refine the random-walk theory is to postulate the probability distribution of the daily price change in a stock (or the daily change in a market average). Although stock prices change by discrete amounts, the most common assumption is that the change has a distribution that is approximately normal. To examine this theory, we recorded the daily change in the Standard & Poor's (S&P) Stock Index for each market day during a recent 5-year period. The relative frequency distribution for these changes, shown in Figure 6.16, has a mean equal to $.095 and a standard deviation equal to $1.527. Does this distribution differ markedly from a normal distribution with $\mu = .095$ and $\sigma = 1.527$?

FIGURE 6.16

Distribution of Changes in Standard and Poor's Index for a Recent 5-Year Period

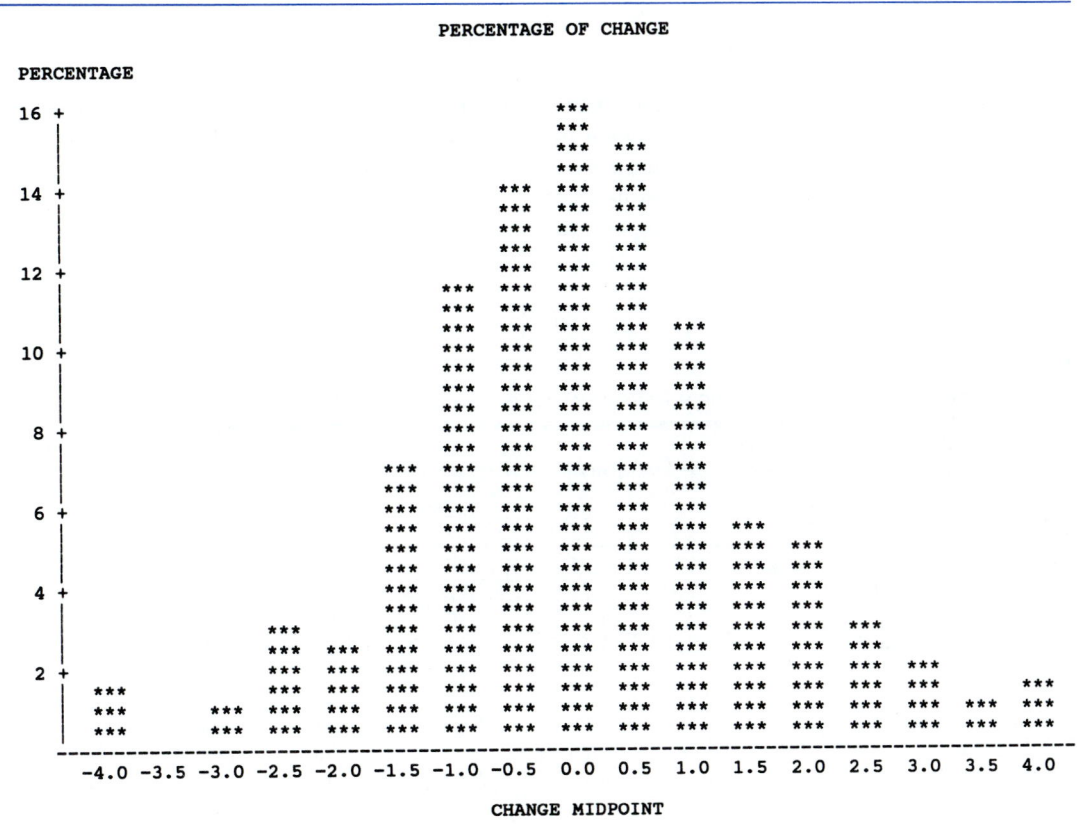

If the daily change in the Standard and Poor's Stock Index does trace a random walk, then a positive daily change is just as likely as a negative change, and the mean change over a long period of time should equal 0. Consequently, we will compare a normal curve with mean $\mu = 0$ and standard deviation $\sigma = 1.50$ with the relative frequency histogram for our data (Figure 6.16). To do this, we will use a normal curve with $\mu = 0$ and $\sigma = 1.50$ to calculate the area under the normal curve over each of the 17 class intervals used to construct Figure 6.16. These areas are shown in Figure 6.17 (page 282). We will then plot these theoretical relative frequencies and sketch in the normal curve with $\mu = 0$ and $\sigma = 1.50$.

FIGURE 6.17

Areas Under Normal Curve with
$\mu = 0$ and $\sigma = 1.50$, Over the
Class Intervals Used to Construct
the Relative Frequency
Distribution in Figure 6.16

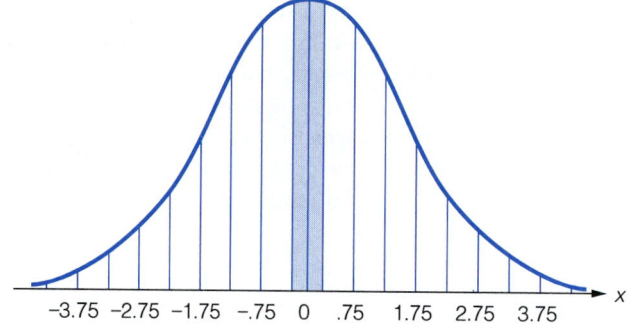

The calculations for determining the normal curve areas over each class interval are shown in Table 6.2. The first column of the table gives the class intervals, beginning with the class located over $\mu = 0$ and proceeding to the right. Class intervals to the left of the mean are not shown because the normal curve is symmetric about its mean. Therefore, the areas over class intervals lying the same distance above and below the mean will be equal.

TABLE 6.2

Calculations for the Class Theoretical Relative Frequencies in Figure 6.16

CLASS	CLASS MIDPOINT	x_L	x_U	$z_L = \dfrac{x_L}{\sigma}$	$z_U = \dfrac{x_U}{\sigma}$	A_L	A_U	CLASS THEORETICAL RELATIVE FREQUENCY
−.25− .25	.00	−.25	.25	−.17	.17	.0675	.0675	.1350
.25− .75	.50	.25	.75	.17	.50	.0675	.1915	.1240
.75−1.25	1.00	.75	1.25	.50	.83	.1915	.2967	.1052
1.25−1.75	1.50	1.25	1.75	.83	1.17	.2967	.3790	.0823
1.75−2.25	2.00	1.75	2.25	1.17	1.50	.3790	.4332	.0542
2.25−2.75	2.50	2.25	2.75	1.50	1.83	.4332	.4664	.0332
2.75−3.25	3.00	2.75	3.25	1.83	2.17	.4664	.4850	.0186
3.25−3.75	3.50	3.25	3.75	2.17	2.50	.4850	.4938	.0088
3.75−4.25	4.00	3.75	4.25	2.50	2.83	.4938	.4977	.0039

The second column of Table 6.2 gives the class midpoint and the third and fourth columns give the upper and lower boundaries, x_L and x_U, respectively, for each class. Columns 5 and 6 give the z values, z_L and z_U, corresponding to x_L and x_U. Assuming $\mu = 0$ and $\sigma = 1.50$,

$$z_L = \frac{x_L - \mu}{\sigma} = \frac{x_L - 0}{1.50} = \frac{x_L}{1.50} \quad \text{and} \quad z_U = \frac{x_U}{1.50}$$

Columns 7 and 8 give the areas A_L and A_U between $z = 0$ and the z values, z_L and z_U.

Figure 6.18 also shows how the theoretical relative frequency is calculated for the class interval from $x_L = .25$ to $x_U = .75$. This interval is the one containing the first bar located to the right of the mean in Figure 6.16. The values of z_L and z_U shown in Figure 6.18 correspond to $x_L = .25$ and $x_U = .75$. Thus,

$$z_L = \frac{x_L}{\sigma} = \frac{.25}{1.50} = .17 \quad \text{and} \quad z_U = \frac{x_U}{\sigma} = \frac{.75}{1.50} = .50$$

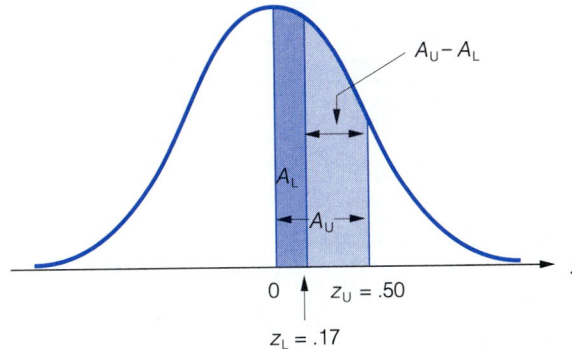

FIGURE 6.18
Sketch Showing How Class
Theoretical Relative Frequencies
are Calculated

The areas $A_L = .0675$ and $A_U = .1915$ are the tabulated areas corresponding to $z_L = .17$ and $z_U = .50$. (These areas are given in Table 4 of Appendix G.) The area over the class interval is the interval's theoretical relative frequency. It is equal to the difference between the areas A_L and A_U:

Theoretical relative frequency for a class interval $= A_U - A_L$
$$= .1915 - .0675 = .1240$$

This is close to the observed relative frequency for this class. You can see from Figure 6.16 that the observed relative frequency is approximately .15.

The areas over the other class intervals, their theoretical relative frequencies, are calculated in exactly the same way. They are shown in Table 6.2. The only exception is the class centered at $\mu = 0$. It will have lower and upper class boundaries $x_L = -.17$ and $x_U = .17$. Thus, half of the area over this interval will lie above $\mu = 0$; half will lie below. This area, .1350, is equal to $A_L + A_U$.

Figure 6.19 (page 284) shows the relative frequency histogram for the observed daily changes in the Standard and Poor's Stock Index. The theoretical relative frequency is plotted for each class interval (see dots) and the approximating normal curve with $\mu = 0$ and $\sigma = 1.50$ is sketched through these points. As you can see, the normal curve provides a reasonably good approximation to the relative frequency distribution of daily changes in our data set.* The observed distribution seems to indicate a slight preponderance of positive daily changes. This probably is a reflection of the long-term upward trends in stock prices, due to inflation, growth in company earnings, and so forth.

---

*Statistical tests are available to detect distributional departures from normality, but they are beyond the scope of this text.

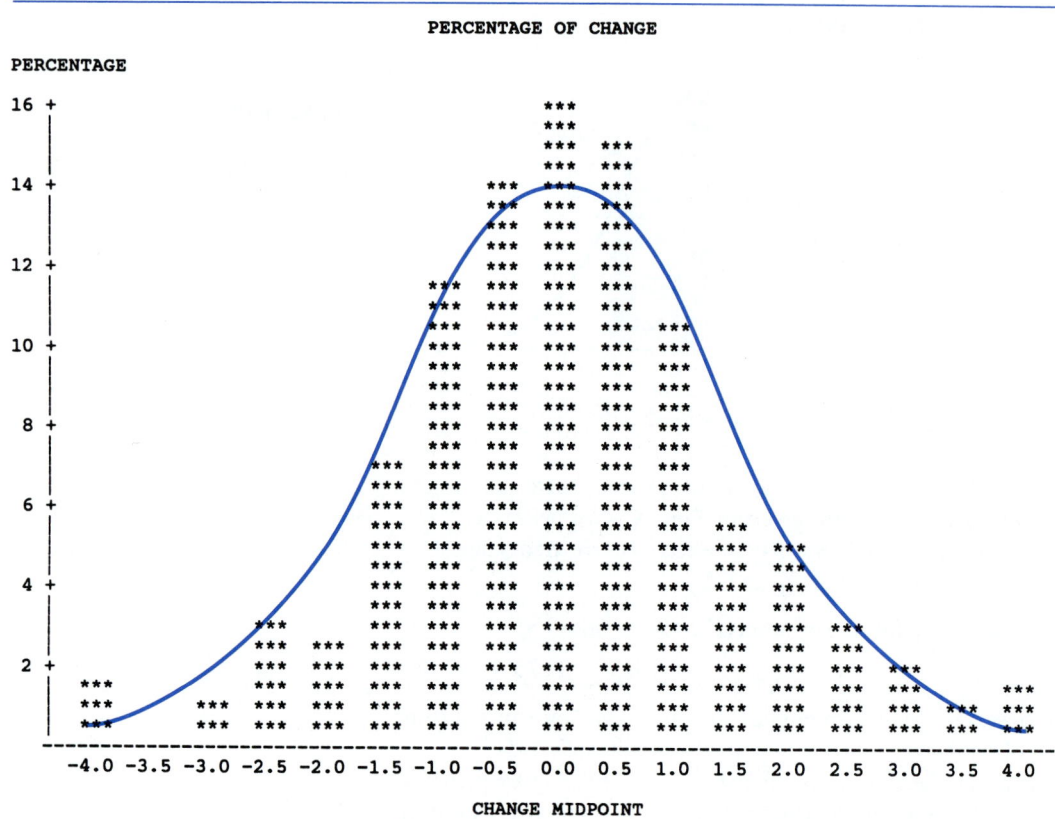

THE NORMAL
APPROXIMATION TO
THE BINOMIAL
DISTRIBUTION

6.5 It can be shown (proof omitted) that the binomial probability distribution of Chapter 5 becomes more nearly normal as the sample size n becomes larger. That is, the histogram for a binomial random variable is shaped nearly like a normal distribution when n is large. Consequently, we can use the normal probability tables to approximate binomial probabilities using the technique of Section 6.2.

The normal approximation to a binomial probability distribution is reasonably good even for small samples—say, n as small as 10—when $\pi = .5$ and the distribution of x is therefore symmetric about its mean $\mu = n\pi$. When π is near 0 (or 1), the binomial probability distribution tends to be skewed to the right (or left), but this skewness disappears as n becomes large. In general, the approximation will be good when n is large enough so that both $n\pi$ and $n(1 - \pi)$ are greater than or equal to 5.

EXAMPLE 6.13   Suppose x has a binomial probability distribution with $n = 10$ and $\pi = .5$.

a. Graph $p(x)$ and superimpose on the graph a normal distribution with

$$\mu = n\pi \qquad \text{and} \qquad \sigma = \sqrt{n\pi(1 - \pi)}.$$

b. Use Table 3 of Appendix G to find $P(x \leq 4)$.

c. Use the normal approximation to the binomial probability distribution to find an approximation to $P(x \leq 4)$.

Solution

a. The graphs of $p(x)$ and a normal distribution with

$$\mu = n\pi = (10)(.5) = 5 \qquad \text{and} \qquad \sigma = \sqrt{n\pi(1 - \pi)}$$
$$= \sqrt{(10)(.5)(.5)} = 1.58$$

are shown in Figure 6.20. Note that $n\pi = n(1 - \pi) = 5$; thus, the normal distribution with $\mu = 5$ and $\sigma = 1.58$ provides a good approximation to $p(x)$.

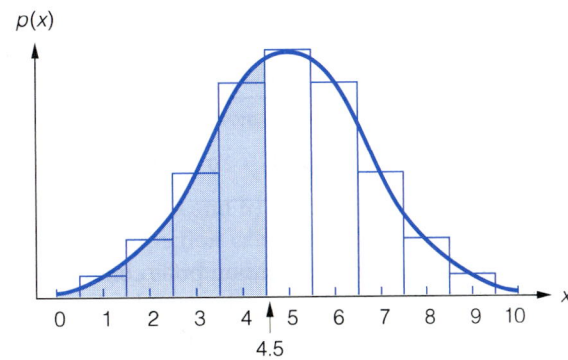

$p(x)$

4.5

FIGURE 6.20
A Binomial Probability
Distribution ($n = 10$, $\pi = .5$)
and the Approximating Normal
Distribution [$\mu = n\pi = 5$ and
$\sigma = \sqrt{n\pi(1 - \pi)} = 1.58$]

b. From Table 3 of Appendix G, we obtain

$$\sum_{x=0}^{4} p(x) = .3770$$

c. By examining Figure 6.20, you can see that $P(x \leq 4)$ is the area under the normal curve to the left of $x = 4.5$. Note that the area to the left of $x = 4$ would *not* be appropriate because it would omit half the probability rectangle corresponding to $x = 4$. We need to add .5 and 4 before calculating the probability in order to correct for the fact that we are using a continuous

6.49   The difference between the actual and the scheduled arrival time for your local commuter train is normally distributed with a mean of 5 minutes (i.e., on the average, it is 5 minutes late) and a standard deviation of 11 minutes. On a randomly selected day, what is the probability that:

   a. The train will be late?
   b. The train will be early?
   c. The train will be more than 5 minutes late?
   d. The train will be at least 10 minutes late?

   Locate μ and the required probability on a sketch of the normal curve for each part of the exercise.

6.50   Refer to Exercise 6.49. The management of the commuter line would like to adjust the scheduled arrival time so that the train appears to be operating more efficiently. To do this, they would like to adjust the scheduled time of arrival so that only 10% of the trains would arrive later than the scheduled time. How many minutes should they add to the current scheduled time in order to accomplish this goal? Show the pertinent quantities on a sketch of the normal curve.

6.51   A standard fluorescent tube has a lifelength that is normally distributed with a mean of 7,000 hours and a standard deviation of 1,000 hours. A competitor has developed a compact fluorescent lighting system that will fit into incandescent sockets. It claims that the new compact tube has a lifelength that is normally distributed with a mean of 7,500 hours and a standard deviation of 1,200 hours.

   a. Which fluorescent tube is more likely to have a lifelength greater than 9,000 hours?
   b. Which tube is more likely to have a lifelength less than 5,000 hours?

6.52   Blending feeders are used to break up tobacco that has been aged in tightly packed hogsheads. One cigarette manufacturer determined that the time between breakdowns for its blending feeders is best represented by a normal distribution with a mean of 100 hours and a standard deviation of 35 hours. Suppose that a particular feeder was just repaired and put back into service.

   a. What is the probability that the feeder will not break down for at least 50 more hours?
   b. What is the probability that the feeder will break down within the next 100 hours? Locate μ and the desired probability on a sketch of the normal curve for each part of the exercise.

6.53   Recall that Appendix B contains the starting salaries of University of Florida bachelor degree graduates in five colleges. For each college, use the computer to determine whether the starting salaries are approximately normal.

THE THEORY OF SIGNAL DETECTION

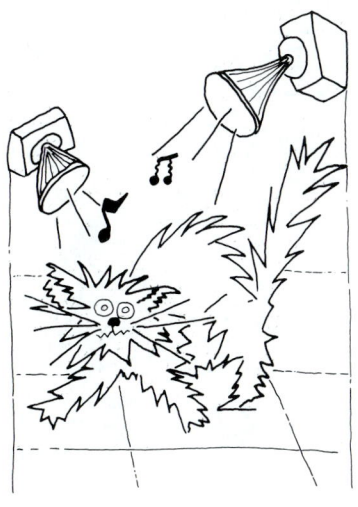

Acoustical engineers have developed a technique to measure the effect of acoustic stimuli in the auditory nerve. Cats (selected because of their keen sense of hearing) are exposed to bursts of noise in a laboratory setting. A sophisticated machine then records the number of spikes per 200 milliseconds of noise burst received by the auditory nerve fibers of each. This number is called the discharge (or response) rate of auditory nerve fibers. The higher the response rate, the more "noise" detected by the auditory nerve.

An empirical study of auditory nerve response rates in cats was reported in the *Journal of the Acoustical Society of America* (Feb. 1986). A key question addressed by the research is whether rate changes (i.e., changes in number of spikes per burst of noise) produced by tones in the presence of background noise are large enough to detect reliably. That is, can the tone be detected reliably when background noise is present?

In the theory of signal detection, the problem involves a comparison of two probability distributions. Let x represent the auditory nerve response rate (i.e., the number of spikes observed) under two conditions: when the stimulus is background noise only (N) and when the stimulus is a tone plus background noise (T). Assume also that the mean response rate under the background-noise-only condition is less than the mean response rate under the tone-plus-noise condition, i.e., $\mu_N < \mu_T$.

Empirical research has found that the probability distribution of x under either condition N or T, can be approximated by a normal distribution. Based on the results of the *Journal of the Acoustical Society of America* study, assume that the two normal distributions have means $\mu_N = 10.1$ spikes per burst and $\mu_T = 13.6$ spikes per burst, respectively, and equal variances $\sigma_N^2 = \sigma_T^2 = 2$.

Given these conditions, an observer sets a threshold C and decides that a tone is present if $x \geq C$ and decides that no tone is present if $x < C$.

a. For a threshold of $C = 11$ spikes per burst, find the probability of detecting the tone given that the tone is present. That is, find $P(x \geq 11)$ under condition T. (This is known as the **detection probability**.)

b. For a threshold of $C = 11$ spikes per burst, find the probability of detecting the tone given that only background noise is present. That is, find $P(x \geq 11)$ under condition N. (This is known as the **probability of false alarm**.)

c. Usually, it is desirable to maximize detection probability while minimizing false alarm probability. Can you find a value of C that will both increase the detection probability (part **a**) and decrease the probability of false alarm (part **b**)? [*Hint:* Sketch the two probability distributions for conditions N and T, side-by-side, allowing some overlap between them. For any value C, shade the two probabilities in parts **a** and **b**. As you move C right or left, what happens to the probabilities?]

INTERPRETING THOSE WONDERFUL EPA MILEAGE ESTIMATES

One common ploy of advertisements for new automobiles is to list the Environmental Protection Agency (EPA) estimated miles per gallon (mpg) for the make of car being advertised. A recent advertisement in a national magazine boasts that the Dodge Aries-K station wagon has the highest mileage of all 6-passenger cars. Its EPA estimated miles per gallon is listed as 28 mpg. However, footnoted in fine print is the following statement:

> Use EPA estimated mpg for comparison only. Your mileage may vary depending on speed, weather, and distance. Actual highway mileage will probably be less.

How should the observant reader, in view of the footnote, interpret the 28 mpg figure? The EPA tests cars under conditions (weather, brand of gasoline, speed, terrain, etc.) ideally suited for maximum mileage performance. Nevertheless, even under identical conditions, it is unreasonable and impractical to assume that all Dodge Aries-K wagons tested will obtain the same gas mileage. If the EPA tested 50 Aries-K wagons, we would expect to observe 50 different mpg's. Conceivably, if all Dodge Aries-K wagons were tested under "ideal conditions," we would obtain a set of numbers representing the population. It is most likely then, that the 28 mpg figure is the average miles per gallon obtained by the sample of wagons tested. The EPA uses this sample average mpg to estimate the population average mpg of all Dodge Aries-K wagons.

If you are an owner of (or are considering buying) a new Dodge Aries-K wagon, you may be interested in the likelihood that the wagon you own (or purchase) will perform as advertised. The answer, of course, requires knowledge of the probability model for the continuous random variable of interest, miles per gallon of Dodge Aries-K wagons.

Let us assume that the EPA estimated mpg for this type of car is accurate, i.e., the true mean mpg obtained under "ideal" conditions for all Dodge Aries-K wagons is, in fact, 28. Also, suppose that the distribution of miles per gallon is approximately normal with a standard deviation of 2.

a. What proportion of all Dodge Aries-K wagons tested under "ideal" conditions will obtain at least 32 mpg?
b. What is the probability that a Dodge Aries-K wagon tested under "ideal" conditions will obtain less than 20 mpg?
c. Fifteen percent of all Dodge Aries-K wagons tested will obtain an mpg rating above a particular value. Find this value.
d. Suppose you test your new Dodge Aries-K wagon under "ideal" conditions and find that your car obtains 20 mpg. Does this result imply that you have bought a "lemon," or is it more likely that the EPA estimated mpg figure of 28 is too high? [*Hint:* Use your answer to part **b**.]

BREAK-EVEN ANALYSIS—WHEN TO MARKET A NEW PRODUCT

The inherent risk involved with marketing a new product is a key consideration for market researchers. How much will it cost to produce? Should the product be marketed? And if marketed, what should be the price of the new product? Will it be profitable? Various statistical models have been developed to aid speculators and businesses in making such decisions. Typically, these decision models are based on the assumption that production costs and selling price are known, and that the quantity produced is determined without knowledge of the demand for the product.

One area of marketing research that uses decision models is called *break-even analysis*. The underlying assumption of break-even analysis is that the demand for a product is normally distributed, with known mean μ and standard deviation σ. Of interest is the relationship between actual demand D and the break-even point BE, where the break-even point is defined as the number of units of the product the company must sell in order to "break even" on the investment.

Shih (1981) developed two practical decision criteria using break-even analysis.* Both decision rules require knowledge of probability and the normal distribution.

DECISION RULE A

Market the new product if the chance is better than 50% that demand D will exceed the break-even point BE—that is, market the product if

$$P(D \geq BE) > .5$$

Under the assumption that demand is normally distributed, decision rule A is equivalent to marketing the product when $\mu \geq BE$. Thus, no knowledge of σ is needed to arrive at a decision.

The probability $P(D \geq BE)$ is often called the *level of risk*. Shih's second decision criterion "allows top management to vary the level of risk it could tolerate [for each new product] in light of its outlook on the uncertainty."

DECISION RULE B

For a specified level of risk $p(0 \leq p \leq 1)$, market the new product if

$$P(D \geq BE) > p$$

---

*Shih, W. "A general decision model for cost-volume-profit analysis under uncertainty: A reply." *The Accounting Review*, Vol. 56, No. 2, 1981, pp. 404–408.

WHY THE METHOD OF SAMPLING IS IMPORTANT

7.1 We now return to the objective of statistics—namely, the use of sample information to infer the nature of a population. Predicting your waiting time in a dentist's office, the annual amount of rainfall in your city, or an evening's winnings at blackjack based upon your previous (successful and unsuccessful) treks to the casino are examples of statistical inferences, and each involves an element of uncertainty. In this chapter we discuss a technique for measuring the uncertainty associated with making inferences.

Suppose we want to make an inference about the mean, or average, starting salary of all recent University of Florida graduates discussed in earlier chapters. To completely characterize these salaries, we would need the actual starting salary for each graduate. This complete listing of starting salaries constitutes the population of interest to us, and we are particularly interested in the parameter μ, the mean of the population. Unfortunately, in this and in most practical situations, the entire population of starting salaries is unavailable to us. However, we do have available (in Appendix A) a subset or *sample* of 1,795 observations from the target population. But for this sample to be used to infer the characteristics of the population, the sample must be *representative* of the population about which inferences are to be made, i.e., the sample must possess characteristics similar to those that would be observed in the entire population, if it were available. In Example 1.8, we explained why this sample of 1,795 starting salaries may not be characteristic of the much larger (and partly conceptual) population of all starting salaries of recent graduates.

EXAMPLE 7.1 Researchers have determined that the time patients spend waiting in physicians' offices plays an important role in an efficiently run practice (see Section 7.4). Suppose an orthodontist is interested in examining the waiting time of his patients over the past year as part of an annual evaluation of his practice. Now, unknown to the orthodontist, suppose the population relative frequency distribution for the waiting time for each of the orthodontist's 2,000 patients last year appears as in Figure 7.1. (We emphasize that this example is for illustration

FIGURE 7.1

Relative Frequency Distribution of Waiting Times for 2,000 Patients

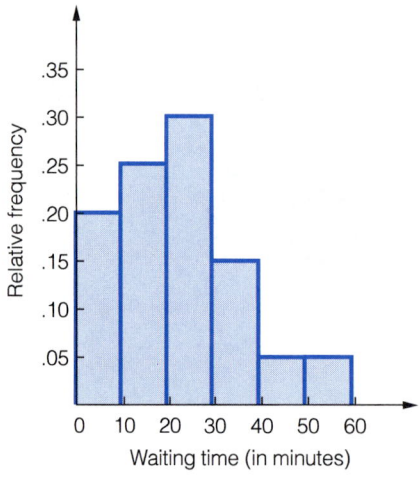

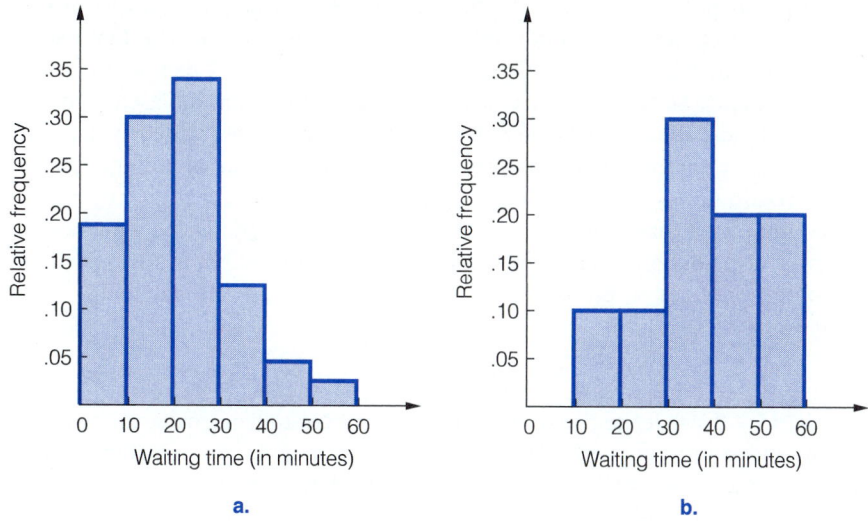

FIGURE 7.2
Relative Frequency Distributions of Waiting Times for Each of Two Samples of 50 Patients Selected from 2,000 Patients

only. In actual practice, the entire population of 2,000 waiting times may not be easily accessible.) Now, assume that a member of his staff provides the orthodontist with the relative frequency distributions of waiting times for each of two samples of 50 patients (Figures 7.2a and 7.2b) selected from the 2,000 patients last year.

Compare the distributions of patient waiting times for the two samples. Which appears to better characterize patient waiting time for the population?

Solution   It is clear that the two samples lead to quite different conclusions about the same population from which they were both selected. From Figure 7.2b, we see that 20% of the sampled patients waited in the orthodontist's office at least 50 minutes before being served, whereas from Figure 7.2a, we see that only 2% of the sampled patients had such a long wait. This may be compared to the relative frequency distribution for the population (shown in Figure 7.1), in which we observe that 5% of all the patients last year waited at least 50 minutes. In addition, note that none of the patients in the second sample (Figure 7.2b) had waits of less than 10 minutes, whereas 18% of the patients in the first sample (Figure 7.2a) had waits of less than 10 minutes. This value from the first sample compares favorably with the 20% of the waiting times for the entire population (Figure 7.1) that were less than 10 minutes.

To rephrase the question posed in the example, we could ask: Which of the two samples is most representative of, or characteristic of, patient waiting time for all 2,000 of the orthodontist's patients last year? Clearly, the information provided by the first sample (Figure 7.2a) gives a better picture of the actual population of waiting times. Its relative frequency distribution is more similar to that for the entire population (Figure 7.1) than is the one provided by the second sample (Figure 7.2b). Thus, if the orthodontist were to rely on information from the second sample only, he may have a distorted, or *biased*, impression of the true situation with respect to patient waiting time last year.

How is it possible that two samples from the same population can provide (apparently) contradictory information about the population? The key issue is the method by which the samples are obtained. Example 7.1 demonstrated that great care must be taken to select a sample that will give an unbiased picture of the population about which inferences are to be made. We learned (Chapter 1) that one way to cope with this problem is to use **random sampling**. You will recall (Definition 1.10) that random sampling is a process that guarantees that every sample of size n has an equal chance of selection. Thus, with random samples, the possibility of sample bias is greatly reduced. In addition, random sampling provides a probabilistic basis for evaluating the reliability of an inference.

EXAMPLE 7.2 Refer to the 1,795 starting salaries of college graduates in Appendix A, and assume that this data set is our target population. In Example 1.10, we used the computer (SAS) to generate a random sample of $n = 50$ starting salaries from Appendix A. These 50 salaries are listed in Table 7.1. Compute the mean of the sample and compare it to the mean μ of the target population. (Recall, from Chapter 3, that the mean of all 1,795 starting salaries is $28,475.)

TABLE 7.1
SAS-Generated Random Sample of 50 Salaries from Appendix A

OBS	GRADUATE	SALARY	OBS	GRADUATE	SALARY	OBS	GRADUATE	SALARY
1	38	$26,500	18	621	$17,000	35	1302	$24,600
2	42	$32,000	19	673	$28,000	36	1332	$18,000
3	62	$33,000	20	684	$27,900	37	1349	$30,000
4	99	$19,000	21	711	$70,000	38	1380	$20,000
5	168	$25,000	22	718	$22,000	39	1387	$31,500
6	208	$18,000	23	774	$26,000	40	1420	$33,800
7	318	$8,500	24	785	$23,000	41	1441	$26,000
8	341	$32,000	25	848	$26,000	42	1450	$27,000
9	354	$27,000	26	966	$30,000	43	1494	$40,000
10	355	$24,000	27	1100	$33,100	44	1531	$21,000
11	374	$31,200	28	1130	$19,000	45	1536	$24,700
12	426	$15,000	29	1138	$20,000	46	1563	$23,900
13	441	$28,000	30	1141	$33,700	47	1602	$24,900
14	451	$27,100	31	1154	$14,000	48	1624	$30,000
15	480	$33,000	32	1254	$60,000	49	1677	$26,000
16	510	$30,000	33	1260	$29,000	50	1685	$18,000
17	613	$23,000	34	1277	$20,000			

Solution A Minitab printout giving descriptive statistics for the sample of 50 starting salaries is displayed in Figure 7.3. The mean of the sample, $\bar{x} = 27,008$, is shaded on the printout. Note that the sample mean is close to $\mu = \$28,475$. Will this always occur with random samples? Not always; but because we used random sampling to generate the sample, we can compute the probability of obtaining a sample mean "close" to the population mean. This information allows us to attach a measure of reliability to any inferences made about the population mean.

	N	MEAN	MEDIAN	TRMEAN	STDEV	SEMEAN
salary	50	27008	26000	25975	9945	1406
	MIN	MAX	Q1	Q3		
salary	8500	70000	20750	30300		

FIGURE 7.3
Minitab Printout: Descriptive Statistics for a Sample of 50 Starting Salaries

In the next two sections, we demonstrate how to judge the performance of a sample mean computed from a random sample.

7.2 SAMPLING DISTRIBUTIONS

In inferential statistics, the ultimate goal is to use information from the sample to make an inference about the nature of the population. In many situations, the objective will be to estimate a numerical characteristic of the population, called a **parameter**, using information in the sample. To illustrate, in Example 7.2 we computed $\bar{x} = \$27,008$, the mean starting salary for a random sample of $n = 50$ observations from the data on the starting salaries recorded in Appendix A. In other words, we used the sample information to compute a **statistic**—namely, the sample mean, $\bar{x}$.

DEFINITION 7.1

A numerical descriptive measure of a population is called a **parameter**.

DEFINITION 7.2

A quantity computed from the observations in a sample is called a **statistic**.

You may have observed that the value of a population parameter (for example, the mean μ) is constant (although it is usually unknown to us); its value does not vary from sample to sample. However, the value of a sample statistic (for example, the sample mean, $\bar{x}$) is highly dependent on the particular sample that is selected. If, in Example 7.2, we had used a different random number generator (e.g., a random number table), we would have obtained a different random sample of 50 observations, and thus a different value of $\bar{x}$.

Since statistics vary from sample to sample, any inferences based on them will necessarily be subject to some uncertainty. How, then, do we judge the reliability of a sample statistic as a tool in making an inference about the corresponding population parameter? Fortunately, the uncertainty of a statistic generally has characteristic properties that are known to us, and that are reflected in its **sampling distribution**. Knowledge of the sampling distribution of a particular statistic provides us with information about its performance over the long run.

The **sampling distribution** of a sample statistic (based on n observations) is the relative frequency distribution of the values of the statistic theoretically generated by taking repeated random samples of size n and computing the value of the statistic for each sample.

The notion of a sampling distribution can be illustrated with the data in Appendix A. Assume that our interest focuses only on the starting salaries of fall 1989–spring 1991 University of Florida graduates *who indicated that they had secured a job* on the CRC questionnaire described in Chapter 1. In particular, we wish to estimate the mean starting salary of all such graduates. Then the target population consists of the 1,795 observations on starting salary contained in Appendix A. (Although the true value of μ, the mean of these 1,795 observations, is already known to us, this example will serve to illustrate the concepts.)

EXAMPLE 7.3   How could we physically generate the sampling distribution $\bar{x}$, the mean of a random sample of $n = 5$ observations from the population of 1,795 starting salaries in Appendix A?

Solution   The sampling distribution for the statistic $\bar{x}$, based on a random sample of $n = 5$ measurements, would be generated in this manner: Select a random sample of five measurements from the population of 1,795 observations on starting salary in Appendix A; compute and record the value of $\bar{x}$ for this sample. Now return these five measurements to the population and repeat the procedure, i.e., draw another random sample of $n = 5$ measurements and record the value of $\bar{x}$ for this sample. Return these measurements and repeat the process. If this sampling procedure could be repeated an infinite number of times, as shown in Figure 7.4, the infinite number of values of $\bar{x}$ obtained would form a new population. The data in this population could be summarized in a relative frequency distribution, called the **sampling distribution of $\bar{x}$**.

FIGURE 7.4
Generating the Theoretical
Sampling Distribution of the
Sample Mean, $\bar{x}$

Select sample of size $n = 5$
salaries from target population   Calculate $\bar{x}$

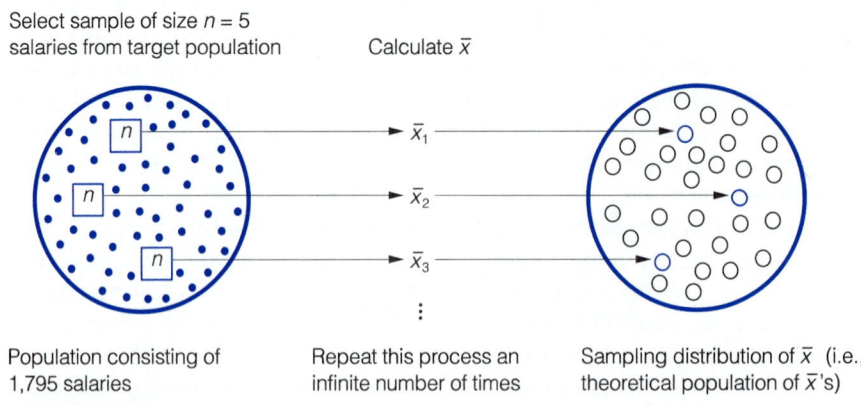

Population consisting of   Repeat this process an   Sampling distribution of $\bar{x}$ (i.e.,
1,795 salaries   infinite number of times   theoretical population of $\bar{x}$'s)

The task described in Example 7.3, which may seem impractical if not impossible, is not performed in actual practice. Instead, the sampling distribution of a statistic is obtained by applying mathematical theory or computer simulation, as illustrated in the next example.

EXAMPLE 7.4 Use computer simulation to find the approximate sampling distribution of $\bar{x}$, the mean of a random sample of $n = 5$ observations from the population of 1,795 starting salaries in Appendix A.

Solution We obtained a large number—namely, 100—of computer-generated random samples of size $n = 5$ from the target population. The first ten samples are presented in Table 7.2.

TABLE 7.2
First 10 Samples of $n = 5$ Starting Salaries from Appendix A

SAMPLE	STARTING SALARIES (DOLLARS)				
1	27,000	28,100	34,500	10,000	25,500
2	23,000	23,600	26,000	23,000	20,000
3	35,000	24,000	30,900	38,000	36,000
4	38,000	35,000	20,000	25,000	33,000
5	41,100	20,000	20,000	24,000	16,000
6	15,000	30,000	24,000	26,000	30,000
7	34,900	26,400	18,000	22,000	26,400
8	26,400	20,000	15,000	26,000	22,500
9	30,900	28,000	12,000	26,400	24,000
10	18,700	35,000	40,000	30,000	16,000

For example, the first computer-generated sample contained the following measurements: 27,000, 28,100, 34,500, 10,000, 25,500. The corresponding value of the sample mean is

$$\bar{x} = \frac{\Sigma x}{n} = \frac{27,000 + 28,100 + 34,500 + 10,000 + 25,500}{5} = \$25,020$$

For each sample of five observations, the sample mean $\bar{x}$ was computed. The 100 values of $\bar{x}$ are summarized in the SAS relative frequency histogram shown in Figure 7.5 (page 304). This distribution approximates the sampling distribution of $\bar{x}$ for a sample of size $n = 5$.

Let us compare the relative frequency histogram for $\bar{x}$ (Figure 7.5) with the relative frequency histogram for the population, shown in Figure 7.6. Note that the values of $\bar{x}$ in Figure 7.5 tend to cluster around the population mean, $\mu = \$28,475$. Also, the values of the sample mean are less spread out (that is, they have less variation) than the population values shown in Figure 7.6 on page 305. These two observations are borne out by comparing the means and standard deviations of the two sets of observations, as shown in Table 7.3.

FIGURE 7.5

Sampling Distribution of $\bar{x}$: 100 Random Samples of $n = 5$ Starting Salaries from Appendix A

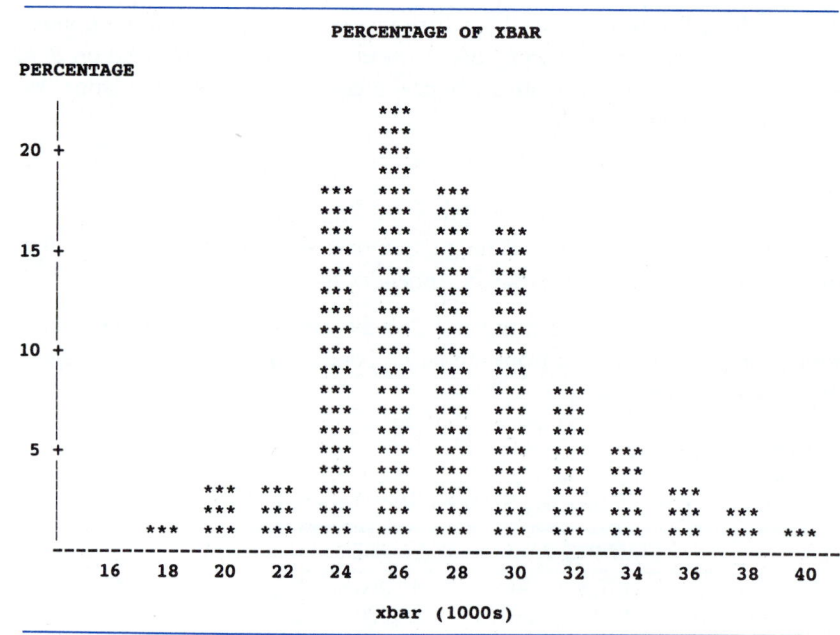

TABLE 7.3

Comparison of the Population Distribution and the Approximate Sampling Distribution of $\bar{x}$, Based on 100 Samples of Size $n = 5$

	MEAN	STANDARD DEVIATION
Population of 1,795 starting salaries (Figure 7.6)	$\mu = \$28,475$	$\sigma = \$9,369$
100 values of $\bar{x}$ based on samples of size $n = 5$ (Figure 7.5)	$\$27,891$	$\$4,152$

EXAMPLE 7.5 Refer to Example 7.4. Simulate the sampling distribution of $\bar{x}$ for samples of size $n = 25$ from the population of 1,795 starting salary observations. Compare the result with the sampling distribution of $\bar{x}$ based on samples of size $n = 5$, obtained in Example 7.4.

Solution We obtained 100 computer-generated random samples of size $n = 25$ from the target population. A SAS relative frequency histogram for the 100 corresponding values of $\bar{x}$ is shown in Figure 7.7 on page 306.

First, note that, as with the sampling distribution based on samples of size $n = 5$, the values of $\bar{x}$ tend to center about the population mean. Second, notice that the variation of the $\bar{x}$ values about their mean in Figure 7.7 is less than the variation in the values of $\bar{x}$ based on samples of size $n = 5$ (Figure 7.5). The mean and standard deviation for these 100 values of $\bar{x}$ are shown in Table 7.4 for comparison with previous results.

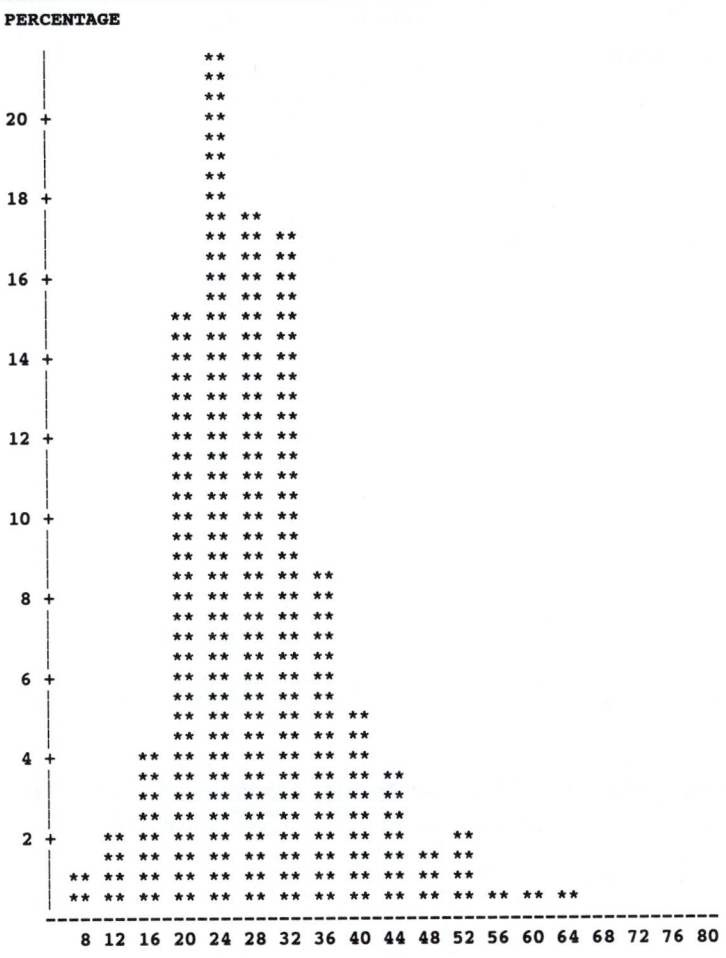

PERCENTAGE

```
                    **
                    **
                    **
  20 +              **
                    **
                    **
                    **
  18 +              **
                    ** **
                    ** ** **
                    ** ** **
  16 +              ** ** **
                    ** ** **
              ** ** ** **
              ** ** ** **
  14 +        ** ** ** **
              ** ** ** **
              ** ** ** **
              ** ** ** **
  12 +        ** ** ** **
              ** ** ** **
              ** ** ** **
              ** ** ** **
  10 +        ** ** ** **
              ** ** ** **
              ** ** ** **
              ** ** ** ** **
   8 +        ** ** ** ** **
              ** ** ** ** **
              ** ** ** ** **
              ** ** ** ** **
   6 +        ** ** ** ** **
              ** ** ** ** **
              ** ** ** ** ** **
              ** ** ** ** ** **
   4 +  ** ** ** ** ** ** **
        ** ** ** ** ** ** ** **
        ** ** ** ** ** ** ** **
        ** ** ** ** ** ** ** **
   2 +  ** ** ** ** ** ** ** ** **       **
        ** ** ** ** ** ** ** ** ** ** **
  ** ** ** ** ** ** ** ** ** ** ** **
  ** ** ** ** ** ** ** ** ** ** ** ** ** ** ** **
  ---------------------------------------------------
    8 12 16 20 24 28 32 36 40 44 48 52 56 60 64 68 72 76 80
```

Salary (thousands) Midpoint

FIGURE 7.6
Relative Frequency Histogram of the Population of 1,795 Starting Salaries in Appendix A

TABLE 7.4
Comparison of the Population Distribution and the Approximate Sampling Distribution of $\bar{x}$, Based on 100 Samples of Size $n = 25$

	MEAN	STANDARD DEVIATION
Population of 1,795 starting salaries (Figure 7.6)	$\mu = \$28,475$	$\sigma = \$9,369$
100 values of $\bar{x}$ based on samples of size $n = 5$ (Figure 7.5)	$27,891	$4,152
100 values of $\bar{x}$ based on samples of size $n = 25$ (Figure 7.7)	$28,349	$1,564

FIGURE 7.7

Sampling Distribution of $\bar{x}$: 100 Random Samples of $n = 25$ Starting Salaries from Appendix A

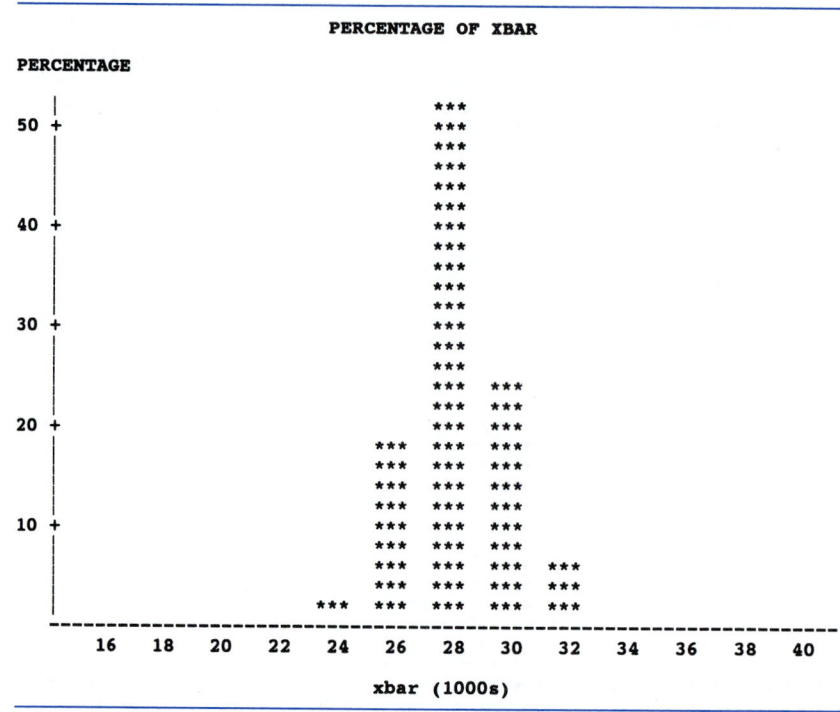

From Table 7.4 we observe that, as the sample size increases, there is less variation in the sampling distribution of $\bar{x}$; that is, the values of $\bar{x}$ cluster more closely about the population mean as n gets larger. This intuitively appealing result will be stated formally in the next section.

EXERCISES

LEARNING THE MECHANICS

7.1 The table contains 50 random samples of $n = 5$ measurements selected from a population with $\mu = 4.5$ and $\sigma^2 = 8.25$.

		SAMPLE		
1, 8, 0, 6, 6	1, 6, 0, 0, 9	3, 6, 4, 2, 0	4, 5, 3, 4, 8	2, 3, 7, 6, 3
2, 1, 7, 2, 9	6, 8, 5, 2, 8	1, 5, 0, 5, 8	5, 6, 7, 8, 2	2, 0, 6, 3, 3
4, 5, 7, 7, 1	2, 4, 9, 4, 6	4, 6, 2, 6, 2	3, 8, 6, 0, 1	1, 9, 0, 3, 2
3, 6, 1, 8, 1	6, 7, 0, 4, 3	1, 8, 8, 2, 1	1, 4, 4, 9, 0	8, 9, 2, 7, 0
9, 8, 6, 2, 9	0, 5, 9, 9, 6	9, 0, 6, 1, 7	7, 7, 9, 8, 1	1, 5, 0, 5, 1
6, 8, 8, 3, 5	4, 4, 7, 5, 6	3, 7, 3, 4, 3	9, 2, 9, 8, 7	7, 8, 7, 7, 6
9, 5, 7, 7, 9	6, 6, 5, 5, 6	4, 5, 2, 6, 6	6, 8, 9, 6, 0	9, 3, 7, 3, 9
7, 6, 4, 4, 7	5, 0, 6, 6, 5	9, 3, 7, 1, 3	3, 4, 6, 7, 0	5, 1, 1, 4, 0
6, 5, 6, 4, 2	3, 0, 4, 9, 6	1, 9, 6, 9, 2	8, 4, 7, 6, 9	2, 5, 7, 7, 9
8, 6, 8, 6, 0	3, 0, 7, 4, 1	5, 1, 2, 3, 4	6, 9, 4, 4, 2	3, 0, 6, 9, 7

a. Calculate $\bar{x}$ for each of the 50 samples.
b. Construct a relative frequency histogram for the 50 sample means. This figure represents an approximation to the sampling distribution of $\bar{x}$ based on samples of size $n = 5$.
c. Compute the mean and standard deviation for the 50 sample means. Locate these values on the histogram of part b. Note how the sample means cluster about $\mu = 4.5$.

7.2 Refer to Exercise 7.1. Combine pairs of samples (moving down the columns of the table) to obtain 25 samples of $n = 10$ measurements.

a. Calculate $\bar{x}$ for each of the 25 samples.
b. Construct a relative frequency histogram for the 25 sample means. This figure represents an approximation to the sampling distribution of $\bar{x}$ based on samples of size $n = 10$. Compare with the figure constructed in Exercise 7.1.
c. Compute the mean and standard deviation for the 25 sample means, and locate them on the relative frequency histogram. Note how the sample means cluster about $\mu = 4.5$.
d. Compare the standard deviations of the two sampling distributions in Exercises 7.1 and 7.2. Which sampling distribution has less variation?

APPLYING THE CONCEPTS

7.3 Use computer simulation or Table 1 of Appendix G to obtain 30 random samples of size $n = 5$ from the "population" of 500 supermarket customer service times in Appendix E. (Alternatively, each class member may generate several random samples, and the results can be pooled.)

a. Calculate $\bar{x}$ for each of the 30 samples. Construct a relative frequency histogram for the 30 sample means. Compare with the population relative frequency histogram shown in Exercise 3.25.
b. Compute the average of the 30 sample means.
c. Compute the standard deviation of the 30 sample means.
d. Locate the average of the 30 sample means, computed in part b, on the relative frequency distribution. This value could be used as an estimate for μ, the mean of the entire population of 500 supermarket customer service times.

7.4 Repeat parts a, b, c, and d of Exercise 7.3, using random samples of size $n = 10$. Compare the relative frequency distribution with that of Exercise 7.1a. Do the values of $\bar{x}$ generated from samples of size $n = 10$ cluster more closely about μ?

7.5 Generate the approximate sampling distribution of $\bar{x}$, the mean of a random sample of $n = 15$ observations from the population of DDT content in fish in Appendix D. [*Hint:* Obtain 50 random samples of size $n = 15$ from the data in Appendix D, compute $\bar{x}$ for each sample, and then construct a relative frequency histogram for the 50 sample means. Again, each class member could generate several random samples, and the results could be pooled.]

7.6 A research cardiologist wants to examine the rate at which a person's heartbeat increases after 15 minutes of vigorous exercise. One method would be to consider the population of ratios of heart rate (in beats per minute) after exercising to heart rate before exercising for all human subjects. The cardiologist has reason to believe that the population relative frequency distribution for ratio of heart rate after exercise to heart rate before exercise would be markedly skewed to the right since no value can be less than 0, most values would presumably be greater than 1, and, occasionally, very large values would be observed. Now, recalling from Section 3.4 that the mean is sensitive to very large observations, one could argue that the population median ratio would provide more information than would the mean about the increase in human heart rate after vigorous exercise.

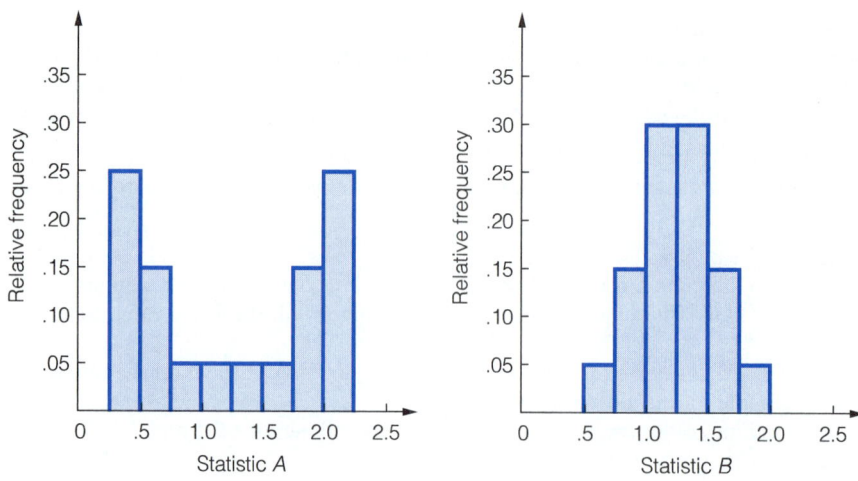

Suppose a medical researcher has proposed two different statistics (call them *A* and *B*) for estimating the population median. To judge which of the statistics is more suitable, you simulated the approximate sampling distributions for each of the statistics, based on random samples of size $n = 10$ human subjects, with the results shown in the accompanying figures. Comment on the two sampling distributions. Which of the statistics, *A* or *B*, would you recommend for use? (In the next section we will discuss desirable properties of a sampling distribution.)

7.7   Generate the sampling distribution of $\bar{x}$, the mean of a random sample of $n = 20$ observations from the "population" of tar contents in cigarettes given in Appendix F.

THE SAMPLING DISTRIBUTION OF $\bar{x}$; THE CENTRAL LIMIT THEOREM

7.3   Estimating the mean starting salary for all graduates of a certain university, or the average increase in heart rate for all human subjects after 15 minutes of vigorous exercise, or the mean yield per acre of farmland, are all examples of practical problems in which the goal is to make an inference about the mean μ, of some target population. In previous sections, we have indicated that the sample mean $\bar{x}$ is often used as a tool for making an inference about the corresponding population parameter μ, and we have shown how to approximate its sampling distribution. The following theorem, of fundamental importance in statistics, provides information about the actual sampling distribution of $\bar{x}$.

> **THE CENTRAL LIMIT THEOREM**
>
> If the sample size is *sufficiently large*, then the mean $\bar{x}$ of a random sample from a population has a sampling distribution that is approximately normal, *regardless of the shape of the relative frequency distribution of the target population*. As the sample size increases, the better will be the normal approximation to the sampling distribution.

The sampling distribution of $\bar{x}$, in addition to being approximately normal, has other known characteristics, which are summarized in the next box.

PROPERTIES OF THE SAMPLING DISTRIBUTION OF $\bar{x}$

If $\bar{x}$ is the mean of a random sample of size n from a population with mean μ and standard deviation σ, then:

1. The sampling distribution of $\bar{x}$ has a mean equal to the mean of the population from which the sample was selected. That is, if we let $\mu_{\bar{x}}$ denote the mean of the sampling distribution of $\bar{x}$, then

$$\mu_{\bar{x}} = \mu$$

2. The sampling distribution of $\bar{x}$ has a standard deviation equal to the standard deviation of the population from which the sample was selected, divided by the square root of the sample size. That is, if we let $\sigma_{\bar{x}}$ denote the standard deviation of the sampling distribution of $\bar{x}$ (also called the **standard error of $\bar{x}$**), then

$$\sigma_{\bar{x}} = \frac{\sigma}{\sqrt{n}}$$

EXAMPLE 7.6 Show that the empirical evidence obtained in Examples 7.4 and 7.5 supports the Central Limit Theorem and the two properties of the sampling distribution of $\bar{x}$. Recall that, in Examples 7.4 and 7.5, we obtained repeated random samples of sizes $n = 5$ and $n = 25$ from the population of starting salaries recorded in Appendix A. For this target population, we know the values of the parameters μ and σ:

Population mean: $\mu = \$28{,}475$

Population standard deviation: $\sigma = \$9{,}369$

Solution   In Figures 7.5 and 7.7, we noted that the values of $\bar{x}$ cluster about the population mean, $\mu = \$28{,}475$. This is guaranteed by property 1, which implies that, in the long run, the average of *all* values of $\bar{x}$ that would be generated in infinite repeated sampling would be equal to μ.

We also observed, from Table 7.4, that the standard deviation of the sampling distribution of $\bar{x}$, called the **standard error of $\bar{x}$**, decreases as the sample size increases from $n = 5$ to $n = 25$. Property 2 quantifies the decrease and relates it to the sample size. As an example, note that, for our approximate (simulated) sampling distribution based on samples of size $n = 5$, we obtained a standard deviation of \$4,152, whereas property 2 tells us that, for the actual sampling distribution of $\bar{x}$, the standard deviation is equal to

$$\sigma_{\bar{x}} = \frac{\sigma}{\sqrt{n}} = \frac{\$9{,}369}{\sqrt{5}} = \$4{,}190$$

Similarly, for samples of size $n = 25$, the sampling distribution of $\bar{x}$ actually has a standard deviation of

$$\sigma_{\bar{x}} = \frac{\sigma}{\sqrt{n}} = \frac{\$9,369}{\sqrt{25}} = \$1,874$$

The value we obtained by simulation was $1,564.

Finally, for sufficiently large samples, the Central Limit Theorem guarantees an approximately normal distribution for $\bar{x}$, regardless of the shape of the original population. In our examples, the population from which the samples were selected is seen in Figure 7.6 to be moderately skewed to the right. Note from Figures 7.5 and 7.7 that, although the sampling distribution of $\bar{x}$ tends to be mound-shaped in each case, the normal approximation improves when the sample size is increased from $n = 5$ (Figure 7.5) to $n = 25$ (Figure 7.7).

EXAMPLE 7.7   Three population relative frequency distributions that provide reasonably accurate probability models for certain real-world phenomena are the **normal distribution** (which we discussed in Chapter 6), the **uniform distribution**, and the **exponential distribution**. Their vastly different shapes are shown in Figure 7.8. Simulate the sampling distributions of $\bar{x}$ by drawing 1,000 samples of $n = 5$ observations from populations that have the relative frequency distributions shown in Figure 7.8. Repeat the procedure for $n = 15, 25, 50,$ and 100. Does the Central Limit Theorem appear to provide adequate information about the shapes of the sampling distributions of $\bar{x}$?

FIGURE 7.8
Three Population Relative
Frequency Distributions

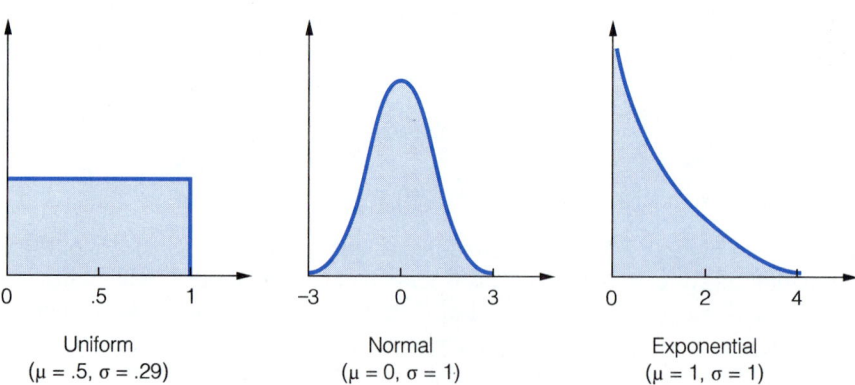

| Uniform | Normal | Exponential |
| ($\mu = .5$, $\sigma = .29$) | ($\mu = 0$, $\sigma = 1$) | ($\mu = 1$, $\sigma = 1$) |

Solution   For each population and each sample size n, we obtained 1,000 computer-generated random samples. The SPSS relative frequency distributions for the 1,000 values of $\bar{x}$ obtained for samples of size $n = 5, 15, 25, 50,$ and 100 from the uniform distribution are displayed in Figure 7.9. Similarly, the simulated (SPSS) sampling distributions of $\bar{x}$ for samples from the normal and exponential distribution are shown in Figures 7.10 and 7.11, respectively.

a. $n = 5$

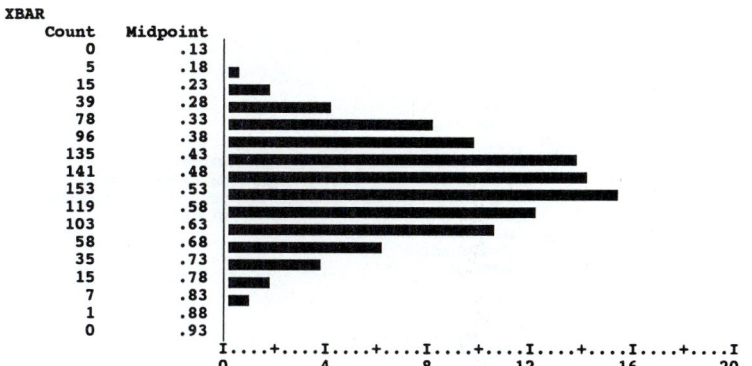

FIGURE 7.9
Sampling Distributions of $\bar{x}$: Uniform Population

b. $n = 15$

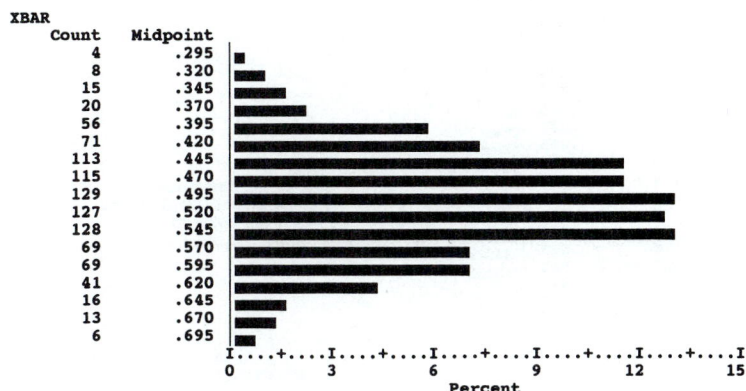

c. $n = 25$

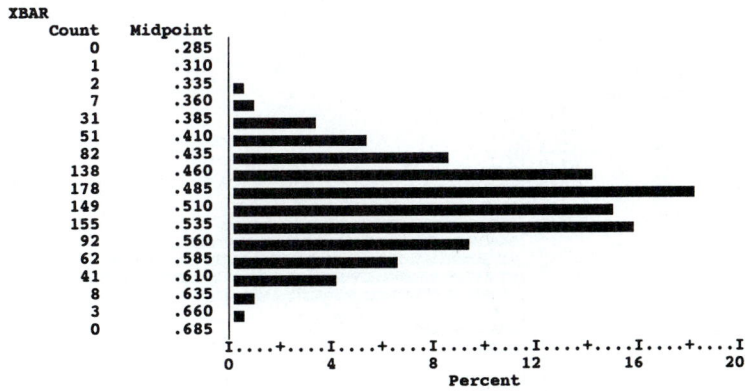

Continued

FIGURE 7.9, Continued **d.** $n = 50$

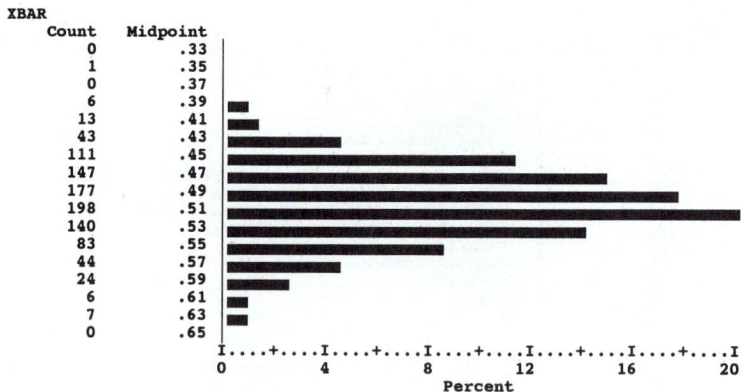

e. $n = 100$

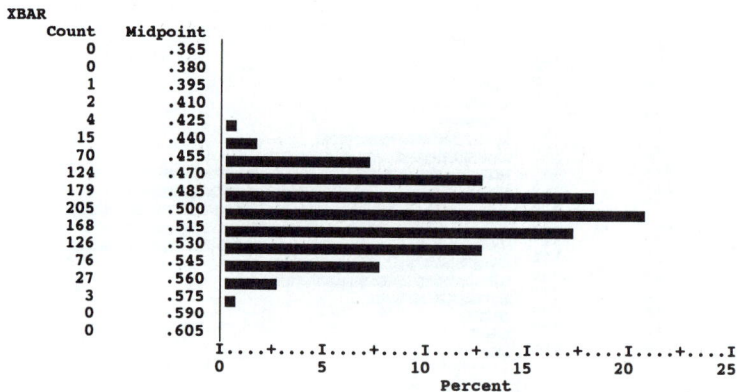

FIGURE 7.10

Sampling Distributions of $\bar{x}$: Normal Population

a. $n = 5$

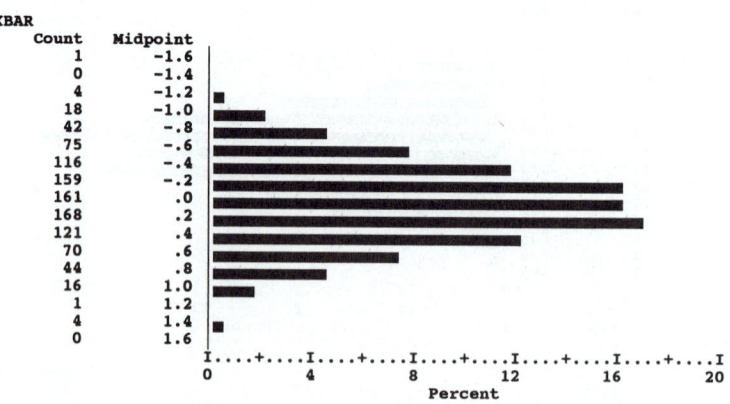

b. $n = 15$

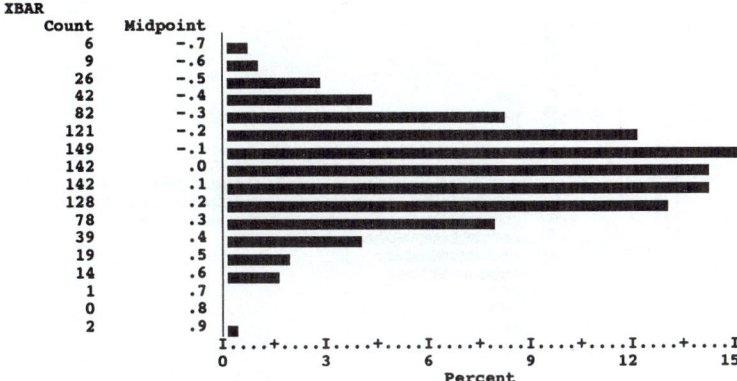

c. $n = 25$

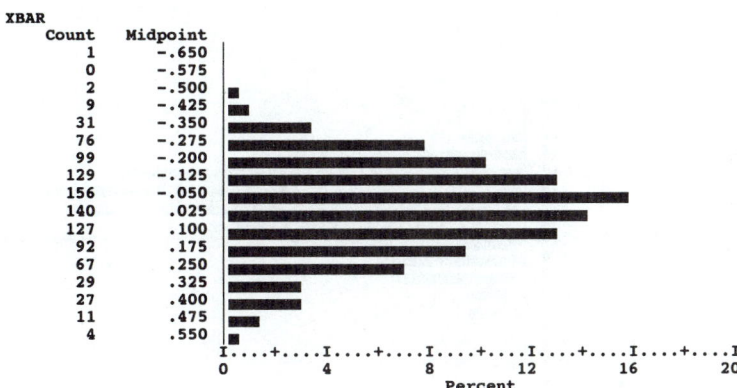

d. $n = 50$

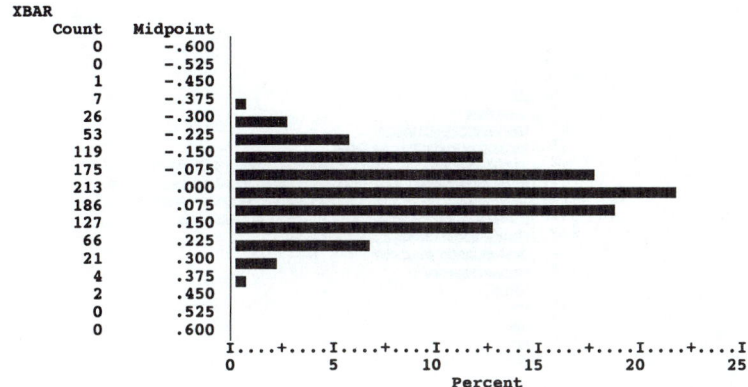

Continued

---

e. $n = 100$

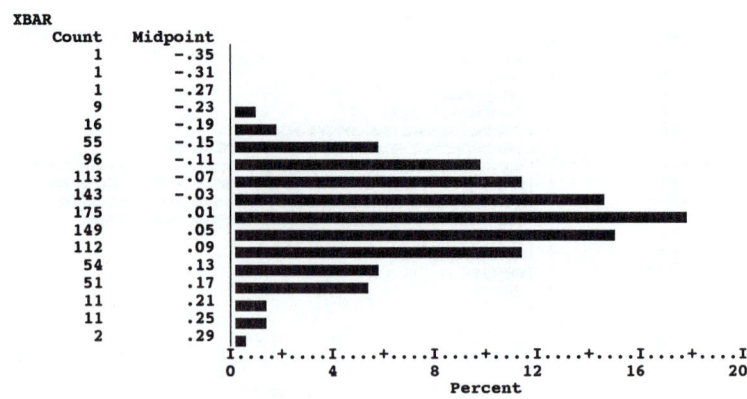

FIGURE 7.11

Sampling Distributions of $\bar{x}$:
Exponential Population

a. $n = 5$

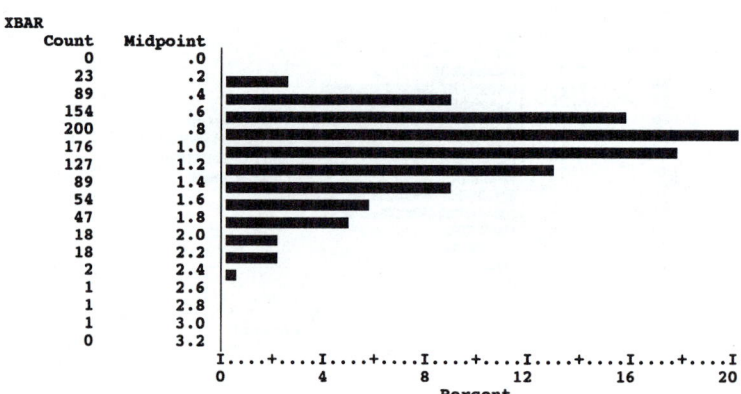

b. $n = 15$

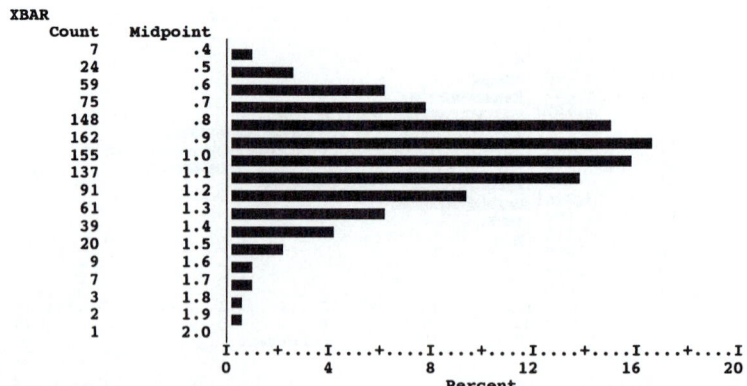

c. $n = 25$

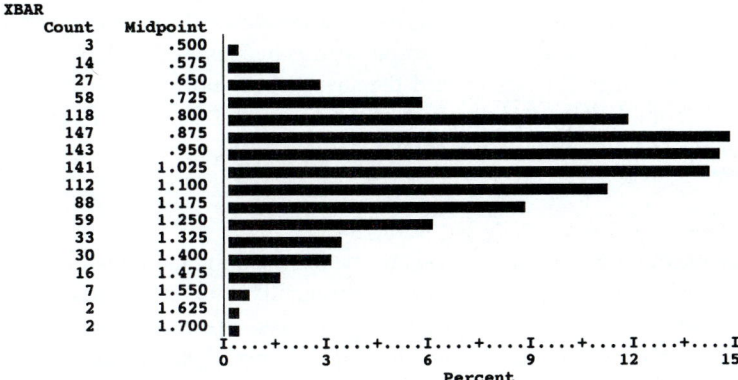

```
XBAR
   Count   Midpoint
      3      .500
     14      .575
     27      .650
     58      .725
    118      .800
    147      .875
    143      .950
    141     1.025
    112     1.100
     88     1.175
     59     1.250
     33     1.325
     30     1.400
     16     1.475
      7     1.550
      2     1.625
      2     1.700
             I....+....I....+....I....+....I....+....I....+....I
             0         3         6         9        12        15
                                   Percent
```

d. $n = 50$

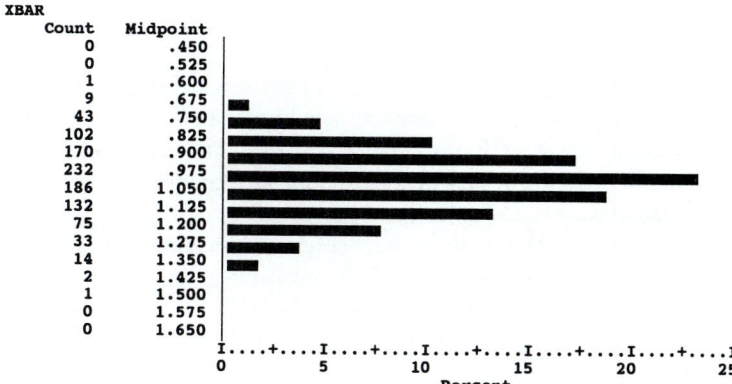

```
XBAR
   Count   Midpoint
      0      .450
      0      .525
      1      .600
      9      .675
     43      .750
    102      .825
    170      .900
    232      .975
    186     1.050
    132     1.125
     75     1.200
     33     1.275
     14     1.350
      2     1.425
      1     1.500
      0     1.575
      0     1.650
             I....+....I....+....I....+....I....+....I....+....I
             0         5        10        15        20        25
                                   Percent
```

e. $n = 100$

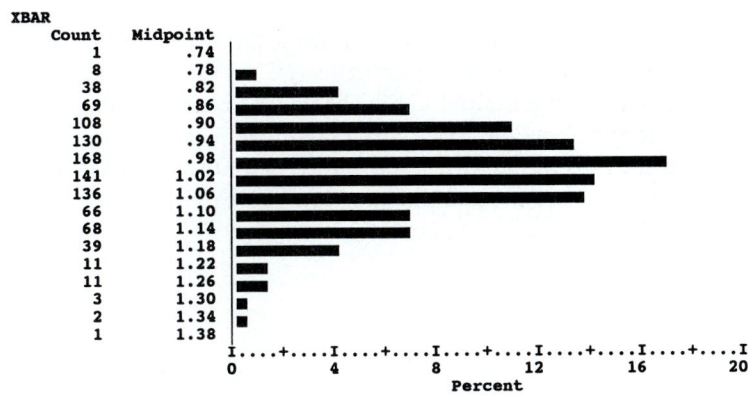

```
XBAR
   Count   Midpoint
      1      .74
      8      .78
     38      .82
     69      .86
    108      .90
    130      .94
    168      .98
    141     1.02
    136     1.06
     66     1.10
     68     1.14
     39     1.18
     11     1.22
     11     1.26
      3     1.30
      2     1.34
      1     1.38
             I....+....I....+....I....+....I....+....I....+....I
             0         4         8        12        16        20
                                   Percent
```

Examine each of the figures; three patterns emerge. First, you can see that the values of $\bar{x}$ cluster about the mean of the probability distribution from which the samples were taken. Second, as n increases, there is less variation in the sampling distribution. Third, as the sample size n increases, the shape of the sampling distribution of $\bar{x}$ tends toward the shape of the normal distribution (symmetric and mound-shaped), regardless of the shape of the relative frequency histogram of the sampled population shown in Figure 7.8.

EXAMPLE 7.8   Engineers responsible for the design and maintenance of aircraft pavements traditionally use pavement-quality concrete. A study was conducted at Luton Airport (United Kingdom) to assess the suitability of concrete blocks as a surface for aircraft pavements (*Proceedings of the Institute of Civil Engineers*, Apr. 1986). The original pavement-quality concrete of the western end of the runway was overlaid with 80-mm-thick concrete blocks. A series of plate-bearing tests was carried out to determine the load classification number (LCN)—a measure of breaking strength—of the surface. Let $\bar{x}$ represent the mean LCN of a sample of 25 concrete block sections on the western end of the runway.

a. Prior to resurfacing, the mean LCN of the original pavement-quality concrete of the western end of the runway was known to be $\mu = 60$, and the standard deviation was $\sigma = 10$. If the mean strength of the new concrete block surface is no different from that of the original surface, describe the sampling distribution of $\bar{x}$.

b. If the mean strength of the new concrete block surface is no different from that of the original surface, find the probability that $\bar{x}$, the sample mean LCN of the 25 concrete block sections, exceeds 65.

c. The plate-bearing tests on the new concrete block surface resulted in $\bar{x} = 73$. Based on this result, what can you infer about the true mean LCN of the new surface?

Solution

a. Although we have no information about the shape of the relative frequency distribution of the breaking strengths (LCNs) for sections of the new surface, we can apply the Central Limit Theorem to conclude that the sampling distribution of $\bar{x}$, the mean LCN of the sample, is approximately normally distributed. In addition, if $\mu = 60$ and $\sigma = 10$, the mean $\mu_{\bar{x}}$, and the standard deviation, $\sigma_{\bar{x}}$, of the sampling distribution are given by

$$\mu_{\bar{x}} = \mu = 60$$

and

$$\sigma_{\bar{x}} = \frac{\sigma}{\sqrt{n}} = \frac{10}{\sqrt{25}} = 2$$

b. If the two surfaces are of equal strength, then $P(\bar{x} \geq 65)$, the probability of observing a mean LCN of 65 or more in the sample of 25 concrete block sections, is equal to the shaded area shown in Figure 7.12.

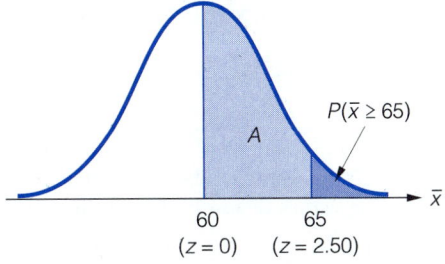

FIGURE 7.12
Sampling Distribution of $\bar{x}$ in Example 7.8

Since the sampling distribution is approximately normal, with mean and standard deviation as obtained in part **a**, we can compute the desired area by obtaining the z score for $\bar{x} = 65$:

$$z = \frac{\bar{x} - \mu_{\bar{x}}}{\sigma_{\bar{x}}} = \frac{65 - 60}{2} = 2.50$$

Thus, $P(\bar{x} \geq 65) = P(z \geq 2.50)$, and this probability (area) may be found using Table 4 of Appendix G and the methods of Chapter 6:

$$P(\bar{x} \geq 65) = P(z \geq 2.50)$$
$$= .5 - A \qquad \text{(see Figure 7.12)}$$
$$= .5 - .4938 = .0062$$

c. If there is no difference between the true mean strengths of the new and original surfaces (i.e., $\mu = 60$ for both surfaces), the probability that we would obtain a sample mean LCN for concrete block of 65 or greater is only .0062. Observing $\bar{x} = 73$ provides strong evidence that the true mean breaking strength of the new surface exceeds $\mu = 60$. Our reasoning stems from the rare event philosophy of Chapter 4, which states that such a large sample mean ($\bar{x} = 73$) is very unlikely to occur if $\mu = 60$.

In practical terms, the Central Limit Theorem and the two properties of the sampling distribution of $\bar{x}$ assure us that the sample mean $\bar{x}$ is a reasonable statistic to use in making inferences about the population mean μ, and they allow us to compute a measure of the reliability of inferences made about μ. (This topic will be treated more thoroughly in Chapter 8.)

As we noted earlier, we will not be required to obtain sampling distributions by simulation or by mathematical arguments. Rather, for all the statistics to be used in this course, the sampling distribution and its properties (which are a matter of record) will be presented as the need arises.

EXERCISES
LEARNING THE MECHANICS

7.8   Suppose a random sample of n measurements is selected from a population with mean $\mu = 60$ and variance $\sigma^2 = 100$. For each of the following values of n, give the mean and standard deviation of the sampling distribution of the sample mean, $\bar{x}$:

  a. $n = 10$ **b.** $n = 25$ **c.** $n = 50$
  d. $n = 75$ **e.** $n = 100$ **f.** $n = 500$
  g. $n = 1,000$

7.9   Suppose a random sample of $n = 100$ measurements is selected from a population with mean μ and standard deviation σ. For each of the following values of μ and σ, give the values of $\mu_{\bar{x}}$ and $\sigma_{\bar{x}}$:

  a. $\mu = 10$, $\sigma = 20$ **b.** $\mu = 20$, $\sigma = 10$
  c. $\mu = 50$, $\sigma = 300$ **d.** $\mu = 100$, $\sigma = 200$

7.10   A random sample of $n = 50$ observations is selected from a population with $\mu = 21$ and $\sigma = 6$. Calculate each of the following probabilities:

  a. $P(\bar{x} < 23.1)$ **b.** $P(\bar{x} > 21.7)$ **c.** $P(22.8 < \bar{x} < 23.6)$

7.11   A random sample of $n = 225$ observations is selected from a population with $\mu = 70$ and $\sigma = 30$. Calculate each of the following probabilities:

  a. $P(\bar{x} > 72.5)$ **b.** $P(\bar{x} < 73.6)$
  c. $P(69.1 < \bar{x} < 74.0)$ **d.** $P(\bar{x} < 65.5)$

APPLYING THE CONCEPTS

7.12   The National Institute for Occupational Safety and Health (NIOSH) recently completed a study to evaluate the level of exposure of workers to the chemical dioxin, 2,3,7,8-TCDD. The distribution of TCDD levels in parts per trillion (ppt) of production workers at a Newark, New Jersey, chemical plant had a mean of 293 ppt and a standard deviation of 847 ppt (*Chemosphere*, Vol. 20, 1990). A graph of the distribution is shown here.

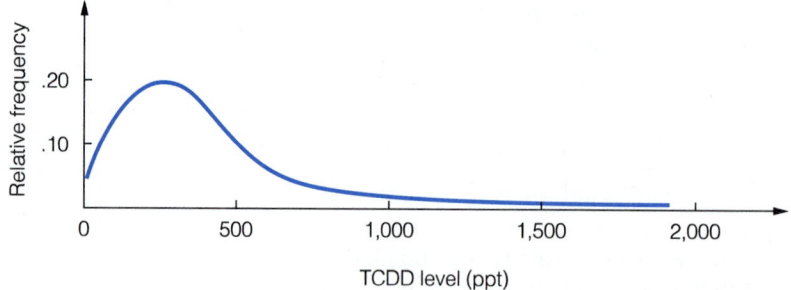

In a random sample of $n = 50$ workers selected at the New Jersey plant, let $\bar{x}$ represent the sample mean TCDD level.

  a. Find the mean and standard deviation of the sampling distribution of $\bar{x}$.
  b. Draw a sketch of the sampling distribution of $\bar{x}$. Locate the mean on the graph.
  c. Find the probability that $\bar{x}$ exceeds 550 ppt.

7.13 Many species of terrestrial frogs that hibernate at or near the ground surface can survive prolonged exposure to low winter temperatures. In freezing conditions, the frog's body temperature, called its *supercooling temperature*, remains relatively higher because of an accumulation of glycerol in its body fluids. Studies have shown that the supercooling temperature of terrestrial frogs frozen at $-6°C$ has a relative frequency distribution with a mean of $-2.18°C$ and a standard deviation of $.32°C$ (*Science*, May 1983). Consider the mean supercooling temperature, $\bar{x}$, of a random sample of $n = 42$ terrestrial frogs frozen at $-6°C$.

a. Find the probability that $\bar{x}$ exceeds $-2.05°C$.
b. Find the probability that $\bar{x}$ falls between $-2.20°C$ and $-2.10°C$.

7.14 According to researchers, "One of the primary reasons relationships sour is that people stop listening to one another" (*USA Today*, Aug. 14, 1985). In a 10-year study of the listening habits of 150 couples, the researchers discovered that couples in conflict listen to each other for a maximum of 14 seconds. Assume that the relative frequency distribution for the length of time couples in conflict listen to each other has a mean of $\mu = 8$ seconds and a standard deviation of $\sigma = 5$ seconds.

a. In a sample of $n = 37$ couples in conflict, let $\bar{x}$ represent the mean length of time the couples listen to each other. Describe the sampling distribution of $\bar{x}$.
b. Find the probability that $\bar{x}$ exceeds 10 seconds.
c. Within what limits would you expect $\bar{x}$ to fall?

7.15 The rash of recent incidents of unethical business tactics has organizations searching for ways to dissuade unethical behavior. But how much will a company's stated concern for ethical conduct influence the behavior of its decision makers? To answer this question, researchers at Marquette University presented MBA students (believed to be representative of entry-level managers) with decision-making situations that were clearly unethical in nature (*Journal of Business Ethics*, Vol. 6, 1987). The subject's decisions were then rated on a scale of 1 ("definitely unethical") to 5 ("definitely ethical"). When no references to ethical concern by the "company" were explicitly stated, the ratings had a mean of 3.00 and a standard deviation of 1.03. Assume that these values represent the population mean and standard deviation, respectively, under the condition "no reference to ethical concern."

a. Suppose we present a random sample of 30 entry-level managers with a similar situation and record the ratings of each. Find the probability that $\bar{x}$, the sample mean rating, is greater than 3.40.
b. Refer to part **a**. Prior to making their decisions, the 30 entry-level managers were all read a statement from the president of the "company" concerning the company's code of business ethics. The code advocates socially responsible behavior by all employees. The researchers theorize that the population mean rating of the managers under this condition will be larger than for the "no reference to ethical concern" condition. (A higher mean indicates a more ethical response.) If the sample mean, $\bar{x}$, is 3.55, what can you infer about the population mean under the "stated concern" condition?

7.16 *Cost estimation* is the term used to describe the process by which engineers estimate the cost of work contracts (e.g., road construction, building construction) that are to be awarded to the lowest bidder. The engineers' estimate is the baseline against which the low (winning) bid is compared. A recent study investigated the factors that affect the accuracy of engineers' estimates (*Cost Engineering*, Oct. 1988), where accuracy is measured as the percentage difference between the low bid and the engineers' estimate. One of the most important factors is number of bidders—the more bidders on the contract, the more likely the engineers are to overestimate the cost. For building contracts with five bidders, the mean percentage error was -7.02 and the standard deviation was 24.66. Consider a sample of 50 building contracts, each with five bidders.

a. Describe the sampling distribution of $\bar{x}$, the mean percentage difference between the low bid and the engineers' estimate, for the 50 contracts.
b. Find $P(\bar{x} < 0)$. (This is the probability of an overestimate.)
c. Suppose you observe $\bar{x} = -17.83$ for a sample of 50 building contracts. Based on the information given here, are all these contracts likely to have five bidders? Explain.

7.17 Many firms are using research and development limited partnerships (R&D LPs) as innovative fund-raising vehicles. According to the Securities and Exchange Commission (SEC), funds raised through an R&D LP should be reported as debt on the firm's balance sheet; most firms, however, violate this policy. To gain more insight into this problem, the *Accounting Review* (Jan. 1991) investigated the financial statements of firms with R&D LPs. The mean and standard deviation of the population, consisting of present values of all R&D LPs, were estimated to be $\mu = \$28.5$ million and $\sigma = \$51.8$ million. Consider a random sample of $n = 75$ R&D LPs selected from the population.

a. Describe the sampling distribution of $\bar{x}$, the mean present value of the sample of 75 R&D LPs.
b. What is the probability that $\bar{x}$ falls between \$25.2 and \$36.6 million?
c. What is the probability that $\bar{x}$ is less than \$30 million?

7.18 Water availability is of prime importance in the life cycle of most reptiles. To determine the rate of evaporative water loss of a certain species of lizard at a particular desert site, 34 such lizards were randomly collected, weighed, and placed under the appropriate experimental conditions. After 24 hours, each lizard was removed, weighed, and its total water loss was calculated by subtracting its body weight after treatment from its initial body weight. Previous studies have shown that the relative frequency distribution of water loss for this species of lizard has a mean of 3.1 grams and a standard deviation of .8 gram.

a. Compute the probability that the 34 lizards will have a mean water loss of less than 2.7 grams.
b. Suppose the sample mean water loss for the lizards in the experiment is computed to be 2.58 grams. Based on the probability computed in part **a**, do you believe that the mean and standard deviation of the relative frequency distribution of water loss for this species of lizard may have changed since the previous studies? Explain.

7.19 Medical researchers theorize that the low death rate from coronary heart disease among the Greenland Eskimos is due to their high fish consumption. The average amount of fish consumed by the Eskimos is estimated to be 400 grams per day (*New England Journal of Medicine*, May 9, 1985). Assume the standard deviation is 50 grams per day.

a. Describe the sampling distribution of $\bar{x}$, the mean amount of fish consumed per day for a sample of 25 Greenland Eskimos.
b. Find the probability that $\bar{x}$ is less than 390 grams.
c. Find the probability that $\bar{x}$ falls between 405 and 425 grams.

7.20 By definition, an entrepreneur is "one who undertakes to start and conduct an enterprise or business, assuming full control and risks" (Funk and Wagnall's *Standard Dictionary*). Thus, a distinguishing characteristic of entrepreneurs is their propensity for taking risks. R. H. Brockhaus used a choice dilemma questionnaire (CDQ) to measure the risk-taking propensities of successful entrepreneurs (*Academy of Management Journal*, Sept. 1980). He found that the CDQ scores of entrepreneurs had a mean of 71 and a standard deviation of 12. (Lower scores are associated with a greater propensity for taking risks.) Let $\bar{x}$ be the mean CDQ score for a random sample of $n = 50$ entrepreneurs.

a. Describe the sampling distribution of $\bar{x}$.
b. Find $P(69 \leq \bar{x} \leq 72)$.
c. Find $P(\bar{x} \leq 67)$.
d. Would you expect to observe a sample mean CDQ score of 67 or lower? Explain.

7.21 The determination of the percent canopy closure of a forest is essential for wildlife habitat assessment, watershed runoff estimation, erosion control, and other forest management activities. One way in which geoscientists estimate the percent forest canopy closure is through the use of a satellite sensor called the Landsat Thematic Mapper. A study of the percent canopy closure in the San Juan National Forest (Colorado) was conducted by examining Thematic Mapper Simulator (TMS) data collected by aircraft at various forest sites (*IEEE Transactions on Geoscience and Remote Sensing*, Jan. 1986). The mean and standard deviation of the readings obtained from TMS Channel 5 were found to be 121.74 and 27.52, respectively.

a. Let $\bar{x}$ be the mean TMS reading for a sample of 32 forest sites. Assuming the figures given are population values, describe the sampling distribution of $\bar{x}$.

b. Use the sampling distribution of part **a** to find the probability that $\bar{x}$ falls between 118 and 130.

7.4

F. A. Sloan and J. H. Lorant report on the advantages and disadvantages of the time patients spend waiting in physicians' offices before they receive health care services.* They write:

> Past studies by economists have considered waiting time to be fully unproductive in the provision of a particular service; their conceptual discussions have emphasized the dead-weight loss associated with a queue [i.e., a line waiting for services]. By contrast, operations researchers, especially in health care applications, have assessed productive aspects of waiting. The latter type of study is based on the premise that increasing patient waiting time is likely to reduce idle time of doctors and their staffs. The queue in the office serves at least three roles. First, with patients waiting, the pace of the physicians' practice is less likely to be disturbed by late patient arrivals. If a patient arrives late for his appointment, the physician can draw from the queue of waiting patients. Second, given unanticipated variability in visit lengths (and visit complexity), the physician may use waiting patients to fill up unexpected idle moments in his schedule and that of staff when other patients are receiving X-rays and the like. Finally, even if all patients were punctual and there were no variability in visit lengths, the patient may use waiting time to complete forms and/or undress prior to the medical examination. Were patient waiting to be reduced to an absolute minimum, the physician and/or his staff might have to wait for these tasks to be completed.

A result of maintaining queues, say Sloan and Lorant, is that patient demand for the physicians' services is reduced because of the higher patient (opportunity) time price. (For example, some patients will be reluctant to give up their opportunity to earn income during this possibly long waiting-time period.) It is up to physicians, then, to "determine the optimal mean wait in their practices by balancing the efficiency of their operations against patient demand considerations."

As part of a study to determine the relationship between the length of time patients wait in the physician's office and certain demand and cost factors, Sloan and Lorant obtained data on the typical patient waiting times for 4,500 physicians in the five largest specialities—general practice, general surgery, internal medicine, obstetrics/gynecology, and pediatrics. They reported a mean waiting time of 24.7 minutes and a standard deviation of 19.3 minutes. (Sloan and Lorant note that the 4,500 observations in the data set represent estimates of waiting time based on recall by the physicians and warn that these "estimates may be biased downward because physicians may tend to underestimate the amount of time their patients wait.")

*Sloan, F. A., and Lorant, J. H. "The role of patient waiting time: Evidence from physicians' practices." *Journal of Business*, Oct. 1977, Vol. 50, pp. 486–507. Reprinted by permission of The University of Chicago Press. © 1977 The University of Chicago.

One important aspect in the study of waiting times is the effect of policy changes on the waiting time distribution. For example, suppose the 4,500 typical waiting times were actually collected over a period of time at a health clinic and that the clinic wants to change its operating procedure to increase the number of patients treated by a physician in a given period of time. What effect will this change in operating procedure have on the mean length of time that a patient will have to wait for treatment?

To answer this question, the clinic places the new procedure in operation and then collects a sample of, say, $n = 100$ patient waiting times. We know that the distribution of waiting times using the old operating procedure possessed a mean $\mu = 24.7$ minutes and a standard deviation $\sigma = 19.3$ minutes. Suppose the mean of the sample of 100 waiting times is $\bar{x} = 19.8$, a value smaller than the population mean waiting time, $\mu = 24.7$. Is it likely that the sample mean $\bar{x}$ could be as small as 19.8 if, in fact, the population mean is still $\mu = 24.7$? Or is the population mean waiting time for the new operating procedure less than 24.7?

To determine whether the value $\bar{x} = 19.8$ is improbably small, if in fact $\mu = 24.7$, we must find the sampling distribution of $\bar{x}$. Applying the Central Limit Theorem, we know that the sampling distribution of $\bar{x}$ has a shape that is approximately normal. Also, the mean of the sampling distribution of $\bar{x}$ is

$$\mu_{\bar{x}} = \mu = 24.7 \text{ minutes}$$

and the standard deviation of the sampling distribution of $\bar{x}$ (standard error) is

$$\sigma_{\bar{x}} = \frac{\sigma}{\sqrt{n}} = \frac{19.3}{\sqrt{100}} = \frac{19.3}{10} = 1.93 \text{ minutes}$$

This distribution is shown in Figure 7.13.

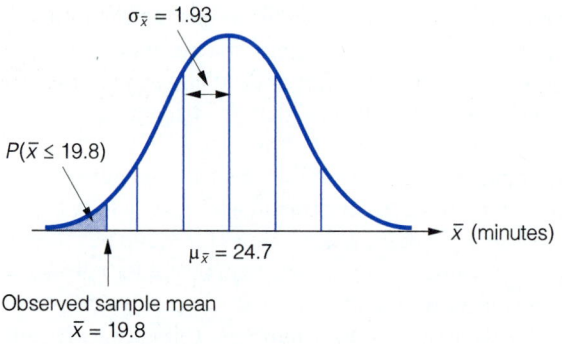

The probability that $\bar{x}$ could be as small as or smaller than the observed sample mean, $\bar{x} = 19.8$, is the shaded area in the lower tail of the sampling distribution shown in Figure 7.13. To find this probability (area), we calculate the z value corresponding to 19.8:

$$z = \frac{\bar{x} - \mu_{\bar{x}}}{\sigma_{\bar{x}}} = \frac{19.8 - 24.7}{1.93} = -2.53$$

From Table 4 of Appendix G, we find the area A between $z = 0$ and $z = -2.53$ is .4943 (see Figure 7.14). Therefore, the probability that $\bar{x}$ is less than or equal to 19.8 is only $.5 - .4943 = .0057$. Thus, we see that a sample mean as small as 19.8 is highly improbable if, in fact, the population mean waiting time is as large as 24.7.

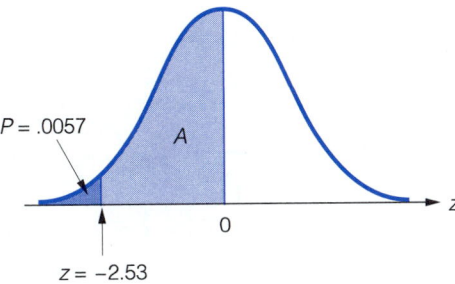

This would lead us to conclude that the new operating procedure has decreased the mean waiting time at the clinic.

The logic employed in arriving at this inference about the mean of the new population of waiting times is discussed in detail in Chapters 9 and 10. At this point, it is important only to note the key role that the sampling distribution plays in the process.

7.5 OTHER TYPES OF SAMPLES: STRATIFIED, CLUSTER, AND SYSTEMATIC SAMPLES

Selecting a random sample can be difficult and costly. For example, suppose we want to sample the opinion of all residents in a certain community age 18 or over, on the issue of mandatory testing for AIDS in the workplace. The first obstacle to the sample selection is acquiring a **frame**, a complete listing of all the experimental units in the population. In this example, the frame is all residents age 18 or over in the community. The second obstacle is contacting the residents who appear in the sample to obtain their opinions.

DEFINITION 7.4

A **frame** is a list of all sampling (or experimental) units in the population.

To reduce the difficulty and costs associated with acquiring a frame and selecting the sample, and to increase the precision of the sample information, experts trained in the art of **survey sampling** have devised some modifications to the simple random sampling procedure. In this section we discuss three alternative methods of sampling: **stratified random sampling, cluster sampling,** and **systematic sampling.**

EXAMPLE 7.9 Suppose we wish to sample the opinions of all heads of household in a state on some issue (e.g., mandatory AIDS testing) and further suppose that the state contains 10 counties (see Figure 7.15). It might be difficult to obtain a frame listing all households within the state, but suppose we know that each county possesses a frame—namely, the households listed on its tax roll. How could we proceed?

FIGURE 7.15

Map of a Fictitious State with 10 Counties

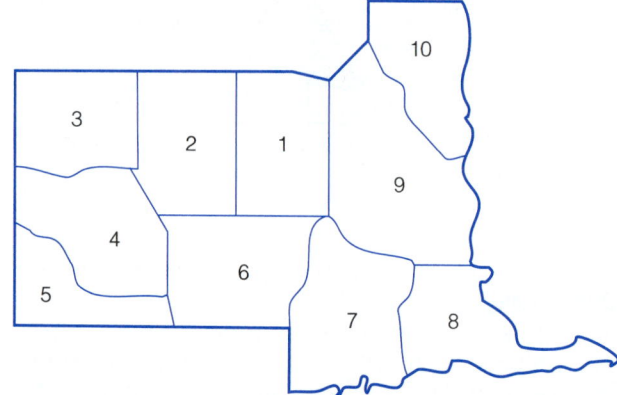

Solution   Instead of combining the 10 frames and selecting a random sample from the whole state, it would be easier and less costly to select a random sample of heads of household from each county. This would enable us to obtain sample opinions not only for each county (so that particular counties could be compared) but also for the entire state. This method of sampling is called **stratified random sampling** because the population is partitioned into a number of **strata** (in this example, counties), and random samples are then selected from among the elements in each **stratum**.

Stratification is advantageous because it allows you to acquire sample information on the individual strata (e.g., counties), as well as on the entire population (e.g., state). It is often less costly to select the sample and, because the variability of the responses within a stratum (county) is usually less than the variability in responses between strata, stratification may provide more accurate estimates of strata and population parameters.

DEFINITION 7.5

A **stratified random sample** is obtained by partitioning the sampling units in the population into nonoverlapping subpopulations called **strata**. Random samples are then selected from each stratum.

EXAMPLE 7.10   Suppose we want to sample the opinions of the heads of household in a city but we know that many households, for one reason or another, are not listed on the tax roll. How could we proceed?

Solution   Since the tax roll cannot be used as a frame, we could construct a frame by numbering on a map each of the city blocks. The list of all blocks in the city would then provide a frame for selecting a random sample of city blocks, each representing a **cluster** of households. After the random sample of clusters is selected, interviewers are sent out to contact and interview all heads of household within each cluster (block) that appears in the sample. The information contained in the clusters is ultimately combined to make inferences about the population of heads of household within the city. This method of sampling is commonly known as **cluster sampling**. The advantages of cluster sampling are clear: It is often easy to form a frame consisting of clusters, and it is less costly to conduct interviews within clusters than it is to interview individual elements selected at random from the population.

DEFINITION 7.6

A **cluster sample** is a random sample in which the sampling units consist of clusters of elements.

EXAMPLE 7.11   Suppose our objective is to sample the opinions of all persons who use the downtown area of the city concerning a prospective increase in bus fares. Since it would be difficult, if not impossible, to construct a frame consisting of all people who use the downtown area, how should we proceed?

Solution   The sample could be collected by *systematically* questioning every fifth person (or every 10th person or, in general, every *k*th person) encountered at a particular street location. This technique, known as **systematic sampling** is very valuable when choosing large samples. Instead of generating many random numbers, you can order the sampling frame and select every *n*th element.

Although this type of sampling is easy and inexpensive, the disadvantage is that it is difficult to obtain the sampling distribution of a statistic computed from the sample data. This makes it difficult to assess the accuracy of estimates based on a systematic sample.*

---

*One alternative is to treat the systematic sample the same as a simple random sample. This approach is dangerous, however, if a periodic or cyclical pattern exists in the data. (For example, if prices are sampled every 30th day at a retail outlet, the prices will tend to be too low because of end-of-the-month clearance sales.) Another alternative is to select multiple systematic samples (e.g., one sample for each of several street locations in Example 7.11) and treat them as clusters in a cluster sample. Whatever method is employed, the standard errors of the estimate will only be approximate.

> **DEFINITION 7.7**
>
> A **systematic sample** is obtained by systematically selecting every kth element in the population when the elements are ordered from 1 to N.

Other sampling methods have been devised to reduce the difficulties and costs of sampling. To use a particular method, we must be able to determine the sampling distributions of the statistics that we plan to use to estimate the parameters of the sampled population. A discussion of these statistics and their sampling distributions is beyond the scope of this text. However, the use of the sampling distribution is the same as described in this chapter for random sampling. If you would like to learn more about these sampling methods, we refer you to a text that describes sample survey methodology. Several of these are listed in the references at the end of the chapter.

PROBLEMS OF NONRESPONSE AND INVALID RESPONSES

7.6 There are other problems, in addition to constructing a sampling frame and selecting a suitable sampling design, associated with conducting a sample survey. One of these is the problem of **nonresponse**, the inability of an interviewer to contact one or more of the persons (or elements) listed in the sample or the refusal of a sampled person to respond. Nonresponse is an important problem because the exclusion of the nonrespondents may produce a serious **bias** in the resulting sample. For example, it is quite common for television stations to request viewers to call in their opinions regarding some political candidate or issue (see Case Study 7.1). This type of "sampling" is likely to include people who want to produce a strong showing for one side or the other and exclude people who are less ardent in expressing their political opinions. Nonresponse often occurs in mailed surveys because busy people, a particular and unique social class, are too busy to complete the survey's questionnaire. Methods for coping with nonresponse (based primarily on resampling the nonresponders) are available and are discussed in the literature.

Another problem encountered in some surveys is the difficulty in eliciting valid responses from persons included in the sample. For example, suppose that one question in a survey of employees asks whether the employee wrongly took sick leave during the year and used it for vacation. An employee who habitually misused sick leave might want to hide this fact by falsifying his or her response to the question. Sampling techniques using a **randomized response model** are available for coping with this problem. With this method, each person is presented two questions, one sensitive and one of a trivial nature. One of these questions is randomly selected by some mechanism and the person interviewed responds "yes" or "no." The interviewer does not know which question has been answered, but the technique enables us to obtain an estimate of the proportion of employees who misuse sick leave.

To deal with the numerous problems encountered in conducting business surveys, consumer preference polls, and so forth, a particular area of statistics known as **survey sampling** has been developed. Consult the references at the end of this chapter for details on this methodology.

SUMMARY

The objective of most statistical investigations is to make an inference about a *population parameter*. Since we often base inferences upon information contained in a sample from the target population, it is essential that the sample be properly selected. For most practical problems, selecting a *random sample* will usually yield a "representative," nonbiased sample.

After the sample has been selected, we compute a statistic that contains information about the target parameter. The *sampling distribution* of the statistic characterizes the relative frequency distribution of values of the statistic over an infinitely large number of samples.

The *Central Limit Theorem* provides information about the sampling distribution of the sample mean, $\bar{x}$. In particular, if you have used random sampling, the sampling distribution of $\bar{x}$ will be approximately normal if the sample size is sufficiently large. Other types of sampling include *stratified random sampling*, *cluster sampling*, and *systematic sampling*. These procedures reduce the difficulty and costs associated with selecting the sample, and they can increase the precision of the sample information.

KEY WORDS

Biased sample	Random sample
Central Limit Theorem	Randomized response model
Cluster sample	Sampling distribution
Frame	Statistic
Nonresponse	Stratified random sample
Parameter	Survey sampling
Population	Systematic sample

KEY FORMULAS

Sampling distribution of $\bar{x}$:

Mean: $\quad \mu_{\bar{x}} = \mu$

Standard deviation: $\quad \sigma_{\bar{x}} = \dfrac{\sigma}{\sqrt{n}}$

SUPPLEMENTARY EXERCISES

7.22   Suppose you work for a market research firm and you are assigned the task of surveying a random sample of 10 companies from the *Fortune 500* list of corporations. Describe how you would select the sample.

7.23   Let $\bar{x}_{25}$ represent the mean of a random sample of size 25 obtained from a population with mean $\mu = 17$ and standard deviation $\sigma = 10$. Similarly, let $\bar{x}_{100}$ represent the mean of a random sample of size 100 selected from the same population.

   a. Describe the sampling distribution of $\bar{x}_{25}$.
   b. Describe the sampling distribution of $\bar{x}_{100}$.
   c. Which of the probabilities, $P(15 < \bar{x}_{25} < 19)$ or $P(15 < \bar{x}_{100} < 19)$, would you expect to be the larger?
   d. Calculate the two probabilities in part c. Was your answer to part c correct?

7.24   Dioxin, often described as the most toxic chemical known, is created as a by-product in the manufacture of herbicides such as Agent Orange. Scientists have found that .000005 gram (five millionths of a gram) of dioxin—a dot barely visible to the human eye—will kill more than half the guinea pigs used in experiments to test toxic products, making dioxin 2,000 times more toxic than strychnine. Assume that the amount of dioxin required to kill a guinea pig has a relative frequency distribution with mean $\mu = .000005$ gram and a standard deviation of .000002 gram. Consider an experiment in which the amount of dioxin required to kill each of $n = 50$ guinea pigs is measured, and the sample mean $\bar{x}$ is computed.

   a. Calculate $\mu_{\bar{x}}$ and $\sigma_{\bar{x}}$.
   b. Find the probability that the mean amount of dioxin required to kill the 50 guinea pigs is larger than .0000053 gram.

7.25   The manufacturer of a new instant-picture camera claims that its product has "the world's fastest-developing color film by far." Extensive laboratory testing has shown that the relative frequency distribution for the time it takes the new instant camera to begin to reveal the image after shooting has a mean of 9.8 seconds and a standard deviation of .55 second. Suppose 50 of these cameras are randomly selected from the production line and tested. The time until the image is first revealed, $\bar{x}$, is recorded for each.

   a. Describe the sampling distribution of $\bar{x}$, the mean time it takes the sample of 50 cameras to begin to reveal the image.
   b. Find the probability that the mean time until the image is first revealed for the 50 sampled cameras is greater than 9.70 seconds.
   c. If the mean and standard deviation of the population relative frequency distribution for the times until the cameras begin to reveal the image are correct, would you expect to observe a value of $\bar{x}$ less than 9.55 seconds? Explain.

7.26   Refer to Exercise 7.25. Describe the changes in the sampling distribution of $\bar{x}$ if the sample size were:

   a. Decreased from $n = 50$ to $n = 20$
   b. Increased from $n = 50$ to $n = 100$

7.27   The administrator of a hospital for mentally ill patients reports that, for patients who have received treatment at the hospital, the mean score on a test to measure rehabilitation potential is 90, and the standard deviation is 4. A clinical psychologist randomly samples 60 recent patient records and notes the score on the rehabilitation test for each patient. Of special interest is the sample mean of the 60 test scores.

   a. Assuming the administrator's claim is true, describe the sampling distribution of the mean test score for the sample of 60 patients.
   b. Assuming the administrator's claim is true, what is the probability that the sample mean test score will be 88 or less?
   c. Suppose the clinical psychologist obtains a sample mean test score of 85. Is this sufficient evidence to contradict the administrator's claim? Explain.

7.28   Prior to the occurrence of a minor forest fire, the number of trees per square mile in a certain national forest had a relative frequency distribution with a mean of 7,500 trees and a standard deviation of 80 trees. To determine the extent of the fire's destruction, a forest ranger "cruised" (i.e., counted) the number of trees in four randomly selected square-mile sectors and calculated the mean $\bar{x}$ of these four measurements.

   a. If the relative frequency distribution of the number of trees per square mile in the forest after the fire is the same as before the fire, describe the sampling distribution of $\bar{x}$. Do you think $\bar{x}$ will have a sampling distribution that is approximately normal? Explain.
   b. Answer part **a** assuming the forest ranger cruised 100 randomly selected square-mile sectors and calculated $\bar{x}$, the mean number of trees per square mile.

7.29   Every 10 years the U.S. population census provides essential information about our nation and its people. The 1990 census included questions on age, sex, race, marital status, family relationship, and income; this census was mailed to every household in the country. In some cities, however, a series of questions was added for a 5% random sample of the city's households. That is, each of a random sample of the city's households was mailed a census form that included additional questions. Suppose a particular city has 100,000 households and, of these, 5,000 are to be selected and mailed the longer census form.

   a. If you worked for the Bureau of the Census and were assigned the task of selecting a random sample of 5,000 of the city's households, describe how you would proceed.
   b. Suppose one of the additional questions on the long form of the census concerned energy consumption. The city planned to use this sample information to project the average energy consumption for the city's 100,000 households. Explain why it is important that the sample of 5,000 households be random.

7.30   This past year, an elementary school began using a new method to teach arithmetic to first graders. A standardized test, administered at the end of the year, was used to measure the effectiveness of the new method. The relative frequency distribution of test scores in past years (before implementation of the new teaching method) had a mean of 75 and a standard deviation of 10. Consider the standardized test scores for a random sample of 36 first graders taught by the new method.

   a. If the relative frequency distribution of test scores for first graders taught by the new method is no different from that of the old method, describe the sampling distribution of $\bar{x}$, the mean test score for a random sample of 36 first graders.
   b. If the sample mean test score was computed to be $\bar{x} = 79$, what would you conclude about the effectiveness of the new method of teaching arithmetic? Explain.

7.31   Electric power plants that use water for cooling their condensers sometimes discharge heated water into rivers, lakes, or oceans. It is known that water heated above certain temperatures has a detrimental effect on plant and animal life in the water. Suppose it is known that the increase in the temperature of the heated water discharged by a certain power plant on any given day has a relative frequency distribution with a mean of 5.2°C and a standard deviation of 3.6°C.

   a. An ecologist investigating the plant and animal life of the area recorded the increase in the temperature of the heated water discharged by the power plant for a total of 25 randomly selected days. What is the probability that $\bar{x}$, the mean increase in water temperature for the sample of 25 days, will fall between 5.0°C and 6.5°C? (Assume the sample size is large enough to apply the Central Limit Theorem.)
   b. What is the probability that $\bar{x}$, the mean increase in water temperature for the sample of 25 days, will exceed 6.0°C?
   c. Within what limits should the ecologist expect $\bar{x}$ to fall?

7.32   Suppose you are in charge of student ticket sales for a major college football team. From past experience, you know that the number of tickets purchased by a student standing in line at the ticket window has a relative frequency distribution with a mean of 2.4 and a standard deviation of 2.0. For today's game, there are 100 eager students standing in line to purchase tickets. If only 250 tickets remain, what is the probability that all 100 students will be able to purchase the tickets they desire?

A DECISION PROBLEM FOR FINANCIAL MANAGERS: WHEN TO INVESTIGATE COST VARIANCES

Financial managers are faced daily with the job of controlling costs, and several useful management techniques for implementing that control are available. In this case study, we focus on a method that has been received favorably in the health care field—**cost variance analysis.**

In variance analysis, actual performance (usually measured as either cost or level of activity) is compared against some standard of expected performance. The difference between the two measures of performance is called a *variance.** These differences, or variances, can direct managers to potential problem areas or situations where costs are out of control. The question is, which variance should the manager investigate? That is, when is the deviation between the expected and the actual results large enough to warrant an expensive investigation?

W. A. Robbins and F. A. Jacobs[†] discuss the relevant issues surrounding the cost variance investigation decision problem. They illustrate the problem as follows:

> Suppose a laboratory technician is able to perform a given test in 45 minutes, "on average," and management observes the test performed during the previous week took 60 minutes, on average, to perform. An investigation of this 15-minute unfavorable variance might reveal the technicians are being poorly supervised, resulting in an inefficient use of their time. On the other hand, management might find newly installed laboratory testing equipment is more sophisticated and requires a longer set-up time than originally anticipated, perhaps an additional 15 minutes on average. The investigation in this latter situation may result in management changing their expectation (standard) of efficient performance to 60 minutes or to some other new time dictated by the new technology.
>
> In either instance, the variance has directed attention to a problem that can be corrected upon investigation.

Robbins and Jacobs developed a statistical model for the cost variance investigation decision problem. "The model," they write, "is based on the concept that a standard [usually the mean or average of an activity distribution] is best described by a band or area of acceptability, rather than a single point. The UCL and LCL [upper and lower control limits] form the bounds on the area of acceptability. Any variations falling within this area are considered to be due to random causes and do not require investigation. If a variance falls outside these limits, it is deemed to have a controllable source (nonrandom) and should be considered for possible investigation." The decision model is illustrated in Figure 7.16.

The control limits of the decision model can be either arbitrarily selected by management or mathematically calculated. For the purposes of this case study,

---

*The word *variance* is used loosely here. Students should not confuse its meaning with the statistical definition of variance given in Chapter 3.

[†]Robbins, W. A. and Jacobs, F. A. *Healthcare Financial Management*, Sept. 1985, pp. 36–41.

FIGURE 7.16
Decision Model for Case
Study 7.2

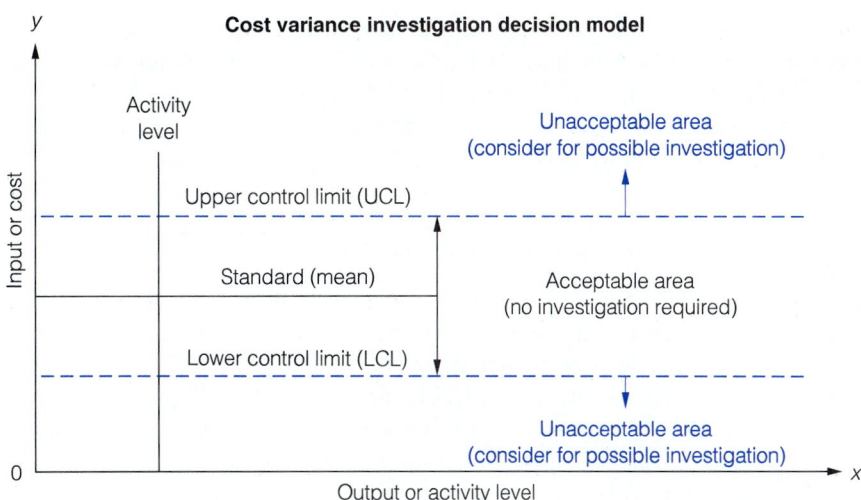

consider the following three decision rules, each with a different method of calculating LCL and UCL:

RULE 1 $LCL = \mu - p\mu$ and $UCL = \mu + p\mu$

where

μ = Mean (or expected) level of activity

p = Percentage of mean $(0 < p < 1)$

$\bar{x}$ = Actual mean activity level of a sample of n observations

Acceptable area: $(\mu - p\mu) < \bar{x} < (\mu + p\mu)$ (Do not investigate)

RULE 2 $LCL = \mu - 2\sigma$ and $UCL = \mu + 2\sigma$

where

μ = Mean (or expected) level of activity

σ = Standard deviation of distribution of expected activity levels

$\bar{x}$ = Actual mean activity level of a sample of n observations

Acceptable area: $(\mu - 2\sigma) < \bar{x} < (\mu + 2\sigma)$ (Do not investigate)

RULE 3 $LCL = \mu - \dfrac{2\sigma}{\sqrt{n}}$ and $UCL = \mu + \dfrac{2\sigma}{\sqrt{n}}$

where

μ = Mean (or expected) level of activity

σ = Standard deviation of expected activity

$\bar{x}$ = Actual mean activity level of a sample of n observations

Acceptable area: $\left(\mu - \dfrac{2\sigma}{\sqrt{n}}\right) < \bar{x} < \left(\mu + \dfrac{2\sigma}{\sqrt{n}}\right)$ (Do not investigate)

To illustrate the differences among the three rules, we use a variation of a problem proposed by Robbins and Jacobs. Suppose the time required to fill a certain type of prescription in a hospital pharmacy is expected to have a probability distribution with $\mu = 24$ minutes and $\sigma = 2.7$ minutes. As a check on the prescription-filling process, the manager of the pharmacy sampled $n = 30$ prescriptions of this type and recorded the time it took to fill each prescription. If the sample mean, $\bar{x}$, falls outside the control limits, the manager will conduct a costly investigation of the prescription-filling process.

a. Suppose the process is out of control, i.e., the mean time required to fill the prescription is greater than the expected mean of $\mu = 24$. Assuming the actual mean is $\mu = 26$, describe the sampling distribution of the sample mean, $\bar{x}$.

b. Using the expected mean ($\mu = 24$), calculate LCL and UCL for rule 1 when $p = .10$. (This is a situation in which management believes a time variance of 10%, greater than or less than the mean, is significant enough to warrant an investigation.) Assuming the process is out of control (i.e., $\mu = 26$), what is the probability that $\bar{x}$ falls outside the control limits? This can be viewed as the probability that the manager will proceed with an investigation when, in fact, the process is out of control. [*Hint*: Use the sampling distribution from part **a** to calculate the probability.]

c. Using the expected mean ($\mu = 24$), calculate LCL and UCL for rule 2. Assuming the process is out of control (i.e., $\mu = 26$), what is the probability that $\bar{x}$ falls outside the control limits?

d. Using the expected mean ($\mu = 24$), calculate LCL and UCL for rule 3. Assuming the process is out of control (i.e., $\mu = 26$), what is the probability that $\bar{x}$ falls outside the control limits?

e. Based on the three probabilities computed in parts **b–d**, which decision rule would you recommend? Explain.

INCREASING SURVEY RESPONSE RATES: THE FOOT AND THE FACE TECHNIQUES

As discussed in Section 7.6, sample surveys often suffer from a lack of a suitable number of respondents; any inferences derived from surveys with low response rates could very well be biased. Many strategies have been devised for the purpose of increasing survey response rates. Although these compliance-gaining tactics originated in the nonbusiness behavioral sciences (social psychology, personality, etc.), much attention has recently been given to them in business and marketing literature.

Mowen and Cialdini* give brief descriptions of various manipulative strategies. The most popular of these among business and marketing researchers is the "foot-in-the-door" or, more simply, the "foot" principle. Mowen and Cialdini wrote: "In using this compliance-gaining tactic, a requester first makes a request so small that nearly anyone would comply, in effect getting a 'foot in the door.' After compliance with the first request occurs, a second, larger request is made—actually the one desired from the outset." For example, Hansen and Robinson† conducted an experiment in which a random group of subjects were contacted by phone and initially asked whether they had purchased a new car within the last 3 years. If they had, they were asked some basic questions on general perceptions toward automobile dealers, such as, "All car dealers overcharge on their repair work; do you agree or disagree?" After the brief (no longer than 5 minutes) "foot-in-the-door" interview, the subject was asked if he or she would be willing to participate in the mail portion of the survey (the desired, larger request). This "foot" technique has been shown to increase response rates in a number of business settings, typical of the one described above. The key to the success of the "foot" principle, say Hansen and Robinson, is that it allows the respondent to become involved in the subject area, which eventually leads to a greater degree of participation in the subsequent larger request.

A second strategy discussed by Mowen and Cialdini is labeled the "door-in-the-face" principle. In the "face" approach, "the [person administering the survey] begins with an initial request so large that nearly everyone refuses (i.e., the door is slammed in his face). [After the first refusal,] the requester then retreats to a smaller favor—actually the one desired from the outset." The "face" principle is based on the social rule of reciprocation that states, "One should make concessions to those who make concessions to oneself." Mowen and Cialdini explain: "The requester's movement from the initial, extreme favor to the second, more moderate one is seen by the [potential respondent] as a concession. To reciprocate this concession, the [respondent] must move from his or her initial position of

---

*Mowen, J. C. and Cialdini, R. B. "On implementing the door-in-the-face compliance technique in a business context." *Journal of Marketing Research*, May 1980, Vol. 17, pp. 253–258.

†Hansen, R. A. and Robinson, L. M. "Testing the effectiveness of alternative foot-in-the-door manipulations." *Journal of Market Research*, Aug. 1980, Vol. 17, pp. 359–363.

noncompliance with the large request to a position of compliance with the smaller request." The key to the successful "face" approach is that the respondent perceive the original request as being legitimate, and that a concession was clearly made in the movement from the large to the small request.

An example of the "door-in-the-face" technique is given by Mowen and Cialdini. Subjects were approached by experimenters representing a fictitious corporation, the California Mutual Insurance Company. The experimenters' initial request went as follows:

Hello, I'm doing a survey for the California Mutual Insurance Company. For each of the last twelve years, we have been on campus to gather survey information on safety in the home or dorm. The survey takes about one hour to administer. Would you be willing to take an hour, right now, to answer the questions?

After the subject declined to participate, the experimenter would make the second smaller request:

Oh, . . . well, look, one part of the survey is particularly important and is fairly short. It will take only fifteen minutes to administer. If you take fifteen minutes right now to complete this short survey, it would really help us out.

The "foot-in-the-door" and "door-in-the-face" strategies present an interesting contrast in sample survey designs. The "foot" approach uses an initial, small request to enhance the likelihood of compliance with a second, larger (desired) request; the "face" approach uses an initial, large request to increase the response rate on a second, smaller (desired) request.

a. Think of a survey that might be best implemented by using the "foot" approach.
b. Repeat part **a** for the "face" approach.

REFERENCES Cochran, W. G. *Sampling Techniques*, 2nd ed. New York: Wiley & Sons, Inc., 1963.

Deming, W. E. *Sample Design in Business Research*. New York: Wiley & Sons, Inc., 1960.

Hansen, M. H., Hurwitz, W. N., and Madow, W. G. *Sample Survey Methods and Theory*, Vol. 1. New York: Wiley & Sons, Inc., 1953.

Hogg, R. V. and Craig, A. T. *Introduction to Mathematical Statistics*, 4th ed. New York: Macmillan, 1978.

Kish, L. *Survey Sampling*. New York: Wiley & Sons, Inc., 1965.

Mendenhall, W. and Sincich, T. *Statistics for Engineering and the Sciences*, 3rd ed. San Francisco: Dellen, 1992.

Scheaffer, R., Mendenhall, W., and Ott, R. L. *Elementary Survey Sampling*, 2nd ed. Boston: Duxbury, 1979.

Warner, S. L. "Randomized response: A survey technique for eliminating evasive answer bias." *Journal of the American Statistical Association*, Vol. 60, 1965, pp. 63–69.

Yamane, T. *Elementary Sampling Theory*, 3rd ed. Englewood Cliffs, N.J.: Prentice-Hall, 1967.

ESTIMATION OF POPULATION PARAMETERS: CONFIDENCE INTERVALS

CHAPTER 8

According to a magazine advertisement, the new twin-blade Daisy shaver from Gillette shaves legs smoother, closer, and safer than any single-blade shaver. And, says the ad, "The Crazy Daisy Shave costs less than 25 cents. That's really crazy!"

To substantiate the advertisement's claim, a consumer affairs newspaper columnist conducted a survey to determine the proportion of all women who shave their legs and prefer the Gillette Daisy razor over its major competitor. How should we calculate the estimate and how accurate will it be? How does the sample size affect the accuracy of the estimate? These and other questions will be answered in this chapter. You will learn how sample size and other factors affect the behavior of sample statistics and, in the chapter case in Section 8.9, we will reexamine the consumer survey.

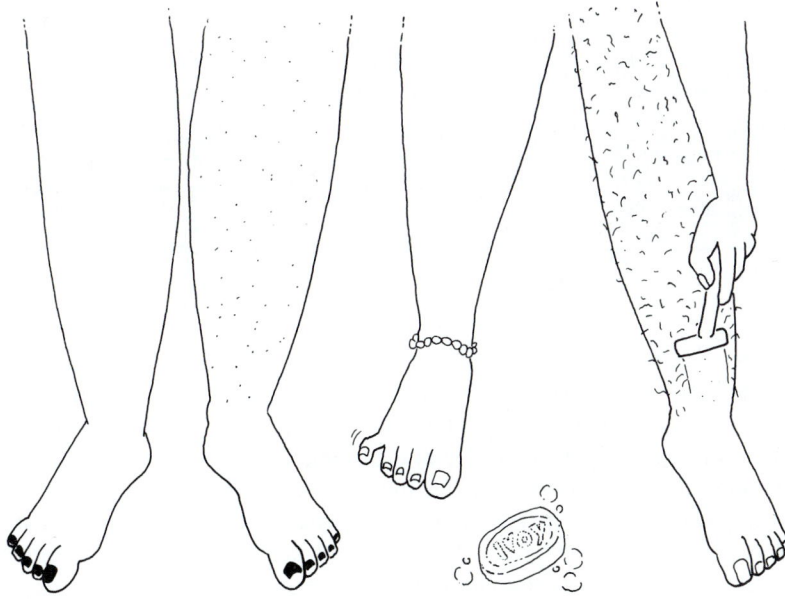

CONTENTS

8.1

In preceding chapters we learned that populations are characterized by numerical descriptive measures (parameters), and that inferences about parameter values are based on statistics computed from the information in a sample selected from the population of interest. In this chapter, we will demonstrate how to estimate population parameters and assess the reliability of our estimates, based on knowledge of the sampling distributions of the statistics being used.

EXAMPLE 8.1   Suppose we are interested in estimating the average starting salary of all graduates of the University of Florida (during the period Fall 1989– Spring 1991) who indicated they had secured employment on the CRC questionnaire described in Chapter 1. (Recall that the target population consists of the 1,795 observations on starting salary in Appendix A. Although we already know the value of the population mean, this example will be continued to illustrate the concepts involved in estimation.) How could one estimate the parameter of interest in this situation?

Solution   An intuitively appealing estimate of a population mean, μ, is the sample mean, $\bar{x}$, computed from a random sample of n observations from the target population. Assume, for example, that we obtain a random sample of size $n = 100$ from the starting salary measurements in Appendix A, and then compute the value of the sample mean to be $\bar{x} = \$27,800$. This value of $\bar{x}$ provides a **point estimate** of the population mean.

DEFINITION 8.1

A **point estimate** of a parameter is a statistic, a single value computed from the observations in a sample, that is used to estimate the value of the target parameter.

How reliable is a point estimate for a parameter? To be truly practical and meaningful, an inference concerning a parameter (in this case, estimation of the value of μ) must consist not only of a point estimate, but also must be accompanied by a measure of the reliability of the estimate; that is, we need to be able to state how close our estimate is likely to be to the true value of the population parameter. This can be done by using the characteristics of the sampling distribution of the statistic that was used to obtain the point estimate; the procedure will be illustrated in the next section.

8.2

Recall from Section 7.3 that, for sufficiently large sample sizes, the sampling distribution of the sample mean, $\bar{x}$, is approximately normal, as indicated in Figure 8.1.

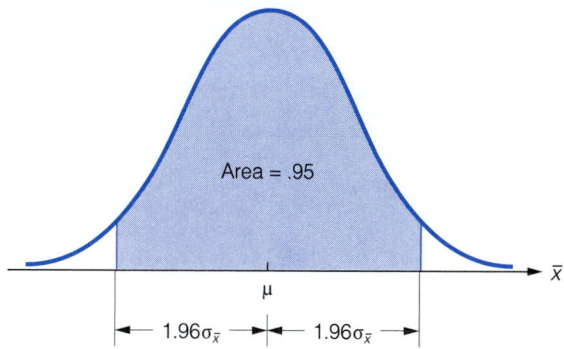

Area = .95

μ

$\leftarrow 1.96\sigma_{\bar{x}} \rightarrow|\leftarrow 1.96\sigma_{\bar{x}} \rightarrow$

$\bar{x}$

FIGURE 8.1
Sampling Distribution of $\bar{x}$

EXAMPLE 8.2 Refer to Example 8.1. Recall that our target population is the data set of Appendix A, and we are interested in estimating μ, the mean starting salary of recent University of Florida graduates. Suppose we plan to take a sample of $n = 100$ measurements from the population of starting salaries in Appendix A and construct the interval

$$\bar{x} \pm 1.96\sigma_{\bar{x}} = \bar{x} \pm \left(\frac{\sigma}{\sqrt{n}}\right)$$

where σ is the population standard deviation of the 1,795 starting salary values and $\sigma_{\bar{x}} = \sigma/\sqrt{n}$ is the **standard error** of $\bar{x}$. In other words, we will construct an interval 1.96 standard deviations around the sample mean, $\bar{x}$. What can we say about how likely it is that this interval will contain the true value of the population mean, μ?

Solution We arrive at a solution by the following three-step process:

STEP 1 First note that the area beneath the sampling distribution of $\bar{x}$ between $\mu - 1.96\sigma_{\bar{x}}$ and $\mu + 1.96\sigma_{\bar{x}}$ is approximately .95. (This area, shaded in Figure 8.1, is obtained from Table 4 of Appendix G.) This implies that before the sample of measurements is drawn, the probability that $\bar{x}$ will fall within the interval $\mu \pm 1.96\sigma_{\bar{x}}$ is .95.

STEP 2 If, in fact, the sample yields a value of $\bar{x}$ that falls within the interval $\mu \pm 1.96\sigma_{\bar{x}}$, then it is also true that the interval $\bar{x} \pm 1.96\sigma_{\bar{x}}$ will contain μ. This concept is illustrated in Figure 8.2. For a particular value of $\bar{x}$ (shown with a vertical arrow) that falls within the interval $\mu \pm 1.96\sigma_{\bar{x}}$, a distance of $1.96\sigma_{\bar{x}}$ is marked off both to the left and to the right of $\bar{x}$. You can see that the value of μ must fall within $\bar{x} \pm 1.96\sigma_{\bar{x}}$.

STEP 3 Steps 1 and 2 combined imply that, before the sample is drawn, the probability that the interval $\bar{x} \pm 1.96\sigma_{\bar{x}}$ will enclose μ is approximately .95.

FIGURE 8.2
Sampling Distribution of $\bar{x}$ in Example 8.2

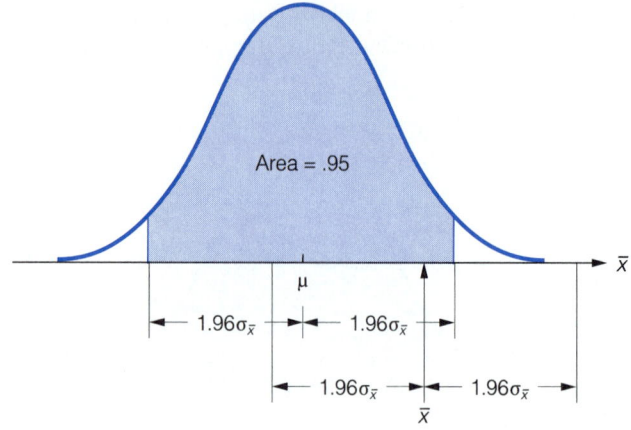

The interval $\bar{x} + 1.96\sigma_{\bar{x}}$ in Example 8.2 is called a large-sample 95% **confidence interval** for the population mean μ. The term *large-sample* refers to the sample being of a sufficiently large size that we can apply the Central Limit Theorem to determine the form of the sampling distribution of $\bar{x}$. Although it is arbitrary, the conventional rule of thumb is that a sample size of $n \geq 30$ is required to employ large-sample confidence interval procedures.

DEFINITION 8.2

A **confidence interval** for a parameter is an interval of numbers within which we expect the true value of the population parameter to be contained. The endpoints of the interval are computed based on sample information.

EXAMPLE 8.3 Refer to Example 8.2. Suppose a random sample of 100 observations from the population of starting salaries yielded a sample mean of $\bar{x} =$ $27,800. Construct a 95% confidence interval for μ, the population mean starting salary, based on this sample information. Recall that the standard deviation of the population of starting salaries is $\sigma =$ $9,369.

Solution   A 95% confidence interval for μ, based on a sample of size $n = 100$, is given by

$$\bar{x} + 1.96\sigma_{\bar{x}} = \bar{x} + 1.96\left(\frac{\sigma}{\sqrt{n}}\right)$$

$$= 27,800 \pm 1.96\left(\frac{9,369}{\sqrt{100}}\right)$$

$$= 27,800 \pm 1,836$$

or (25,964, 29,636). Hence, we estimate that the population mean starting salary falls within the interval from $25,964 to $29,636.

How much confidence do we have that μ, the true population mean starting salary, lies within the interval ($25,964, $29,636)? Although we cannot be certain whether the sample interval contains μ (unless we calculate the true value of μ for all 1,795 observations), we can be reasonably sure that it does. This confidence is based on the interpretation of the confidence interval procedure: If we were to select repeated random samples of size $n = 100$ starting salaries and form a 1.96 standard deviation interval around $\bar{x}$ for each sample, then approximately 95% of the intervals constructed in this manner would contain μ. Thus, we are 95% confident that the particular interval ($25,964, $29,636) contains μ; this is our measure of the reliability of the point estimate $\bar{x}$.

EXAMPLE 8.4   To illustrate the classical interpretation of a confidence interval, we generated 40 random samples, each of size $n = 100$, from the population of starting salaries in Appendix A. For each sample, the sample mean was calculated and used to construct a 95% confidence interval for μ as in Example 8.3. Interpret the results, which are shown in Table 8.1.

TABLE 8.1
95% Confidence Intervals for μ for 40 Random Samples of 100 Starting Salaries from Appendix A

SAMPLE	$\bar{x}$	LOWER CONFIDENCE LIMIT	UPPER CONFIDENCE LIMIT	SAMPLE	$\bar{x}$	LOWER CONFIDENCE LIMIT	UPPER CONFIDENCE LIMIT
1	28965	27129	30801	21	29551	27715	31387
2	28692	26856	30528	22	29210	27374	31046
3	27816	25980	29652	23	28203	26367	30039
4	29760	27924	31596	24	27844	26008	29680
5	27469	25633	29305	25	29280	27444	31116
6	28334	26498	30170	26	29365	27529	31201
7	30234	28398	32070	27	29044	27208	30880
8	29592	27756	31428	28	29336	27500	31172
9	29194	27358	31030	29	28769	26933	30605
10	29436	27600	31272	30	26581	24745	28417*
11	27962	26126	29798	31	28498	26662	30334
12	28299	26463	30135	32	28521	26685	30357
13	29210	27374	31046	33	27085	25249	28921
14	30355	28519	32191*	34	27151	25315	28987
15	29304	27468	31140	35	29749	27913	31585
16	28538	26702	30374	36	28567	26731	30403
17	29068	27232	30904	37	28120	26284	29956
18	26705	24869	28541	38	28356	26520	30192
19	27721	25885	29557	39	27918	26082	29754
20	28161	26325	29997	40	29512	27676	31348

Note: Asterisks (*) identify the intervals that do not contain $\mu = 28,475$.

Solution For the target population of 1,795 starting salaries, we have previously obtained the population mean value, $\mu = \$28,475$ (Section 3.4). In the 40 repetitions of the confidence interval procedure described here, note that only two of the intervals (those based on samples 14 and 30, indicated by asterisks in Table 8.1) do not contain the value of μ, whereas the remaining 38 of the 40 intervals (or 95% of the 40 intervals) do contain the true value of μ.

Keep in mind that, in actual practice, you would not know the true value of μ and you would not perform this repeated sampling; rather you would select a single random sample and construct the associated 95% confidence interval. The one confidence interval you form may or may not contain μ, but you can be fairly sure it does because of your *confidence in the statistical procedure*, the basis for which was illustrated in this example.

Suppose you want to construct an interval that you believe will contain μ with some degree of confidence other than 95%; in other words, you want to choose a **confidence coefficient** other than .95.

DEFINITION 8.3

The **confidence coefficient** is the proportion of times that a confidence interval encloses the true value of the population parameter if the confidence interval procedure is used repeatedly a very large number of times.

The first step in constructing a confidence interval with any desired confidence coefficient is to notice from Figure 8.1 that, for a 95% confidence interval, the confidence coefficient of .95 is equal to the total area under the sampling distribution (1.00), less .05 of the area, which is divided equally between the two tails of the distribution. Thus, each tail has an area of .025. Second, consider that the tabulated value of z (from Table 4 of Appendix G) that cuts off an area of .025 in the right tail of the standard normal distribution is 1.96 (see Figure 8.3). The value $z = 1.96$ is also the distance, in terms of standard deviations, that $\bar{x}$ is from each endpoint of the 95% confidence interval. By assigning a confidence coefficient other than .95 to a confidence interval, we change the area under the sampling distribution between the endpoints of the interval, which in turn changes the tail area associated with z. Thus, this z value provides the key to

FIGURE 8.3
Tabulated z Value Corresponding to a Tail Area of .025

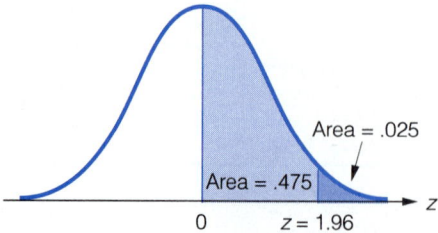

Area = .025

Area = .475

0    $z = 1.96$ z

constructing a confidence interval with any desired confidence coefficient. In our subsequent discussion, we will use the notation defined in the next box.

DEFINITION 8.4

We define $z_{\alpha/2}$ to be the z value such that an area of $\alpha/2$ lies to its right (see Figure 8.4).

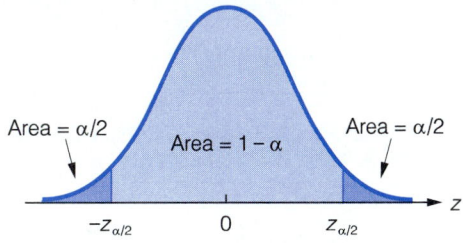

Area = $\alpha/2$

Area = $1 - \alpha$

Area = $\alpha/2$

$-z_{\alpha/2}$ 0 $z_{\alpha/2}$

FIGURE 8.4
Locating $z_{\alpha/2}$ on the Standard Normal Curve

Now, if an area of $\alpha/2$ lies beyond $z_{\alpha/2}$ in the right tail of the standard normal (z) distribution, then an area of $\alpha/2$ lies to the left of $-z_{\alpha/2}$ in the left tail (Figure 8.4) because of the symmetry of the distribution. The remaining area, $(1 - \alpha)$, is equal to the confidence coefficient—that is, the probability that $\bar{x}$ falls within $z_{\alpha/2}$ standard deviations of μ is $(1 - \alpha)$. Thus, a large-sample confidence interval for μ, with confidence coefficient equal to $(1 - \alpha)$, is given by

$$\bar{x} \pm z_{\alpha/2}\sigma_{\bar{x}}$$

EXAMPLE 8.5   In statistical problems using confidence interval techniques, a very common confidence coefficient is .90. Determine the value of $z_{\alpha/2}$ that would be used in constructing a 90% confidence interval for a population mean based on a large sample.

Solution   For a confidence coefficient of .90, we have

$$1 - \alpha = .90$$
$$\alpha = .10$$
$$\frac{\alpha}{2} = .05$$

and we need to obtain the value $z_{\alpha/2} = z_{.05}$ that locates an area of .05 in the upper tail of the standard normal distribution. Since the total area to the right of 0 is .50, $z_{.05}$ is the value such that the area between 0 and $z_{.05}$ is .50 − .05 = .45. From the body of Table 4 in Appendix G, we find $z_{.05} = 1.645$ (see Figure 8.5). We conclude that a large-sample 90% confidence interval for a population mean is given by

$$\bar{x} \pm 1.645\sigma_{\bar{x}}$$

FIGURE 8.5
Location of $z_{\alpha/2}$ for Example 8.5

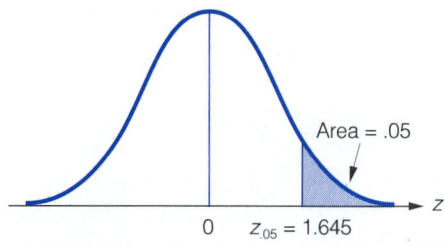

Area = .05

$0 \quad z_{.05} = 1.645$

TABLE 8.2
Commonly Used Confidence
Coefficients and Their
Corresponding z Values

CONFIDENCE COEFFICIENT		
$1 - \alpha$	$\alpha/2$	$z_{\alpha/2}$
.90	.05	1.645
.95	.025	1.96
.98	.01	2.33
.99	.005	2.575

In Table 8.2 we present the values of $z_{\alpha/2}$ for the most commonly used confidence coefficients.

A summary of the large-sample confidence interval procedure for estimating a population mean appears in the box.

LARGE-SAMPLE $(1 - \alpha)$100% CONFIDENCE INTERVAL FOR A POPULATION MEAN, μ

$$\bar{x} \pm z_{\alpha/2}\sigma_{\bar{x}} = \bar{x} + z_{\alpha/2}\left(\frac{\sigma}{\sqrt{n}}\right)$$

where $z_{\alpha/2}$ is the z value that locates an area of $\alpha/2$ to its right, σ is the standard deviation of the population from which the sample was selected, n is the sample size, and $\bar{x}$ is the value of the sample mean.

Assumption: A large random sample (usually, $n \geq 30$) is selected.

[*Note:* When the value of σ is unknown (as will usually be the case), the sample standard deviation s may be used to approximate σ in the formula for the confidence interval. The approximation is generally quite satisfactory for large samples.]

EXAMPLE 8.6 A fact long known but little understood is that twins, in their early years, tend to have lower intelligence quotients and pick up language more slowly than nontwins. Recently, psychologists have speculated that the slower intellectual growth of twins may be caused by benign parental neglect. Suppose we want to investigate this phenomenon. A random sample of $n = 50$ sets of $2\frac{1}{2}$-year-old twin boys is selected, and the total parental attention time given to each pair during 1 week is recorded. The data (in hours) are listed in Table 8.3.

TABLE 8.3
Attention Time for Random Sample of $n = 50$ Sets of Twins

20.7	14.0	16.7	20.7	22.5	48.2	12.1	7.7	2.9	22.2
23.5	20.3	6.4	34.0	1.3	44.5	39.6	23.8	35.6	20.0
10.9	43.1	7.1	14.3	46.0	21.9	23.4	17.5	29.4	9.6
44.1	36.4	13.8	0.8	24.3	1.1	9.3	19.3	3.4	14.6
15.7	32.5	46.6	19.1	10.6	36.9	6.7	27.9	5.4	14.0

Estimate μ, the mean attention time given to all $2\frac{1}{2}$-year-old twin boys by their parents, using a 99% confidence interval. Interpret the interval in terms of the problem.

Solution To compute the interval, we require the sample mean $\bar{x}$ and the population standard deviation σ. In most practical applications, however, the value of σ will be unknown. For large samples, the fact that σ is unknown poses only a minor problem since the sample standard deviation s provides a good approximation to σ. Consequently, we may substitute s for σ in the confidence interval formula given in the box.*

A SAS printout showing descriptive statistics for the sample of $n = 50$ attention times is displayed in Figure 8.6. The values of $\bar{x}$ and s, shaded on the printout, are $\bar{x} = 20.85$ and $s = 13.41$. Because n is large, we substitute s for σ in the formula to obtain the approximate 99% confidence interval:

$$20.85 \pm 2.575 \left(\frac{13.41}{\sqrt{50}} \right)$$

or (15.97, 25.73). We can be 99% confident that the interval (15.97, 25.73) encloses the true mean weekly attention time given to $2\frac{1}{2}$-year-old twin boys by their parents. Since all the values in the interval fall below 28 hours, we conclude that there is a general tendency for $2\frac{1}{2}$-year-old twin boys to receive less than 4 hours of parental attention time per day, on the average. Further investigation would be required to relate this phenomenon to the intellectual growth of the twins.

```
Analysis Variable : ATIME

N Obs   N       Minimum       Maximum          Mean      Std Dev
-----------------------------------------------------------------
   50  50     0.8000000    48.2000000    20.8480000   13.4138253
-----------------------------------------------------------------
```

FIGURE 8.6
SAS Descriptive Statistics for $n = 50$ Sample Attention Times

EXAMPLE 8.7 Refer to Example 8.6.

a. Using the sample information provided in Example 8.6, construct a 95% confidence interval for the mean weekly attention time given to all $2\frac{1}{2}$-year-old twin boys by their parents.
b. For a fixed sample size, how is the width of the confidence interval related to the confidence coefficient?

---

*We discuss an alternative form of the confidence interval when σ is unknown in Section 8.3.

Solution

a. The form of a large-sample 95% confidence interval for a population mean μ is

$$\bar{x} \pm 1.96\left(\frac{\sigma}{\sqrt{n}}\right) \approx \bar{x} \pm 1.96\left(\frac{s}{\sqrt{n}}\right)$$

$$= 20.85 \pm 1.96\left(\frac{13.41}{\sqrt{50}}\right)$$

$$= 20.85 \pm 3.72$$

or (17.13, 24.57).

b. The 99% confidence interval for μ was determined in Example 8.6 to be (15.97, 25.73). The 95% confidence interval, obtained in part **a** and based on the same sample information, is narrower than the 99% confidence interval. This relationship holds in general, as noted in the accompanying box.

> **RELATIONSHIP BETWEEN WIDTH OF CONFIDENCE INTERVAL AND CONFIDENCE COEFFICIENT**
>
> For a given sample size, the width of the confidence interval for a parameter increases as the confidence coefficient increases. Intuitively, the interval must become wider for us to have greater confidence that it contains the true parameter value.

EXAMPLE 8.8 Refer to Example 8.6.

a. Assume that the given values of the statistics $\bar{x}$ and s were based on a sample of size $n = 100$ instead of a sample of size $n = 50$. Construct a 99% confidence interval for μ, the population mean weekly attention time given to $2\frac{1}{2}$-year-old twin boys by their parents.

b. For a fixed confidence coefficient, how is the width of the confidence interval related to the sample size?

Solution

a. Substituting the values of the sample statistics into the general formula for a 99% confidence interval for μ yields

$$\bar{x} \pm 2.575\left(\frac{\sigma}{\sqrt{n}}\right) = 20.85 \pm 2.575\left(\frac{13.41}{\sqrt{100}}\right)$$

$$= 20.85 \pm 3.46$$

or (17.40, 24.30).

b. The 99% confidence interval based on a sample of size $n = 100$, constructed in part **a**, is narrower than the 99% confidence interval based on a sample of size $n = 50$, constructed in Example 8.6. This will also hold true in general, as noted in the box.

RELATIONSHIP BETWEEN WIDTH OF CONFIDENCE INTERVAL AND SAMPLE SIZE

For a fixed confidence coefficient, the width of the confidence interval decreases as the sample size increases. In other words, larger samples generally provide more information about the target population than do smaller samples.

In this section, we have introduced the concepts of point and interval estimation of the population mean μ, based on large samples. The general theory appropriate for the estimation of μ also carries over to the estimation of other population parameters. Hence, in subsequent sections we will present only the point estimate, its sampling distribution, the general form of a confidence interval for the parameter of interest, and any assumptions required for the validity of the procedure.

EXERCISES
LEARNING THE MECHANICS

8.1 In a large-sample confidence interval for a population mean, what does the confidence coefficient represent?

8.2 Use Table 4 of Appendix G to determine the value of $z_{\alpha/2}$ that would be used to construct a large-sample confidence interval for μ, for each of the following confidence coefficients:

a. .85   b. .95   c. .975

8.3 Suppose a random sample of size $n = 100$ produces a mean of $\bar{x} = 81$ and a standard deviation of $s = 12$.

a. Construct a 90% confidence interval for μ.
b. Construct a 95% confidence interval for μ.
c. Construct a 99% confidence interval for μ.

8.4 A random sample of size n is selected from a population with unknown mean μ and standard deviation σ. Calculate a 95% confidence interval for μ for each of the following situations:

a. $n = 35$ $\bar{x} = 26$ $s^2 = 228.2$
b. $n = 70$ $\bar{x} = 24.1$ $s^2 = 198.4$
c. $n = 105$ $\bar{x} = 24.2$ $s^2 = 216.9$

8.5   A random sample of size 400 is taken from an unknown population with mean μ and standard deviation σ. The following values are computed:

$$\sum x = 2,280 \qquad \sum x^2 = 38,532$$

  a. Find a 90% confidence interval for μ.
  b. Find a 99% confidence interval for μ.

8.6   The mean and standard deviation of a random sample of n measurements are equal to 22 and 16, respectively.

  a. Construct a 95% confidence interval for μ if $n = 100$.
  b. Construct a 95% confidence interval for μ if $n = 500$.

APPLYING THE CONCEPTS

8.7   Give a precise interpretation of the statement, "We are 95% confident that the interval estimate contains μ."

8.8   Tropical swarm-founding wasps, like ants and bees, rely on workers to raise their offspring. Interestingly, the workers of this species of wasp are mostly female, capable of producing offspring of their own. Instead, they rear the young of others in the brood. One possible explanation for this strange behavior is inbreeding, which increases relatedness among the wasps and makes it easier for the workers to pick out and aid their closest relatives. To test this theory, 197 swarm-founding wasps were captured in Venezuela, frozen at $-70°C$, and then subjected to a series of genetic tests (*Science*, Nov. 1988). The data were used to generate an inbreeding coefficient x, for each wasp specimen, with the following results: $\bar{x} = .044$ and $s = .884$.

  a. Construct a 90% confidence interval for the mean inbreeding coefficient of this species of wasp.
  b. A coefficient of 0 implies that the wasp has no tendency to inbreed. Use the confidence interval, part **a**, to make an inference about the tendency for this species of wasp to inbreed.

8.9   When a university professor attempts to publish a research article in a professional journal, the manuscript goes through a rigorous review process. Usually, anywhere from three to five reviewers read and critique the article, then pass judgment on whether the article should be published. Recently, a study was undertaken to seek information on how reviewers for research journals pursue their activities (*Academy of Management Journal*, Mar. 1989). A sample of 73 reviewers for the Academy of Management's *Journal* (AMJ) and *Review* (AMR) were asked how many hours they spent per paper for a typical complete review process. The sample mean and standard deviation were computed to be $\bar{x} = 5.4$ hours and $s = 3.6$ hours.

  a. Find a point estimate for μ, the true mean number of hours spent by a reviewer in conducting a complete review of a paper submitted to AMJ or AMR.
  b. Compute a 99% confidence interval for μ.
  c. Interpret the interval, part **b**.

8.10   Refer to the data on physician costs given in Appendix C. The president of the health maintenance organization (HMO) wants to estimate the average per-member per-month cost of the physicians in the HMO.

  a. Use one of the methods of Chapter 7 to select a random sample of 30 physician costs.
  b. Use the data from part **a** to construct a 97% confidence interval for the true average physician cost of this HMO.
  c. What is the confidence coefficient for the interval of part **b**? Interpret this value.
  d. Based on your interval obtained in part **b**, would you expect the true mean physician cost to exceed $200 per member per month?

8.11   A study was conducted to investigate the perceived unit effectiveness of purchasing companies (*Journal of Applied Behavioral Science*, Vol. 22, 1986). The researchers define unit effectiveness as "the relative ability of the members of a unit (e.g., office) to mobilize their centers of power to produce, adapt, and handle temporarily unpredicted overloads of work." A sample of 115 purchasing agents participated in the study by responding to questionnaires on the organizational effectiveness of his or her office. Each agent rated each of eight "effectiveness" items on a scale of 1 to 5 (where 1 = not effective and 5 = very effective). The sum of the eight values was used as a measure of perceived unit effectiveness. A summary of the results follows:

$$\bar{x} = 29.07 \qquad s = 4.68$$

Construct a 95% confidence interval for the true mean "perceived unit effectiveness" rating of all purchasing agents. Interpret the interval.

8.12 Adult students are enrolling in colleges and universities in ever-increasing numbers, and many are majoring in marketing. Recently, a study was conducted to determine marketing faculty attitudes toward the adult students in their classes (*Journal of Marketing Education*, Summer, 1987). A sample of 290 faculty, drawn at random from the American Marketing Association's membership directory, responded to a series of attitudinal statements, the first of which was: "Adult students (i.e., undergraduates 24 years or older) participate more actively in classroom discussions than do younger students." Attitudes were measured using a 5-point Likert scale (1 = strongly agree, 2 = agree, 3 = no opinion, 4 = disagree, and 5 = strongly disagree). For the participation statement, the mean attitudinal score for the sample was 1.94 and the standard deviation was .92.

a. Estimate the true mean attitudinal score of marketing faculty with regard to classroom participation of adult students using a 98% confidence interval. Interpret the result.

b. How could you reduce the width of the confidence interval from part a?

8.13 The Beck Depression Inventory (BDI) is a widely used psychological test designed to measure depressive symptoms in humans. One study was conducted to assess the effect of a "life event" on the BDI scores of college undergraduates (*Journal of Human Stress*, Mar. 1983). In this experiment, the "life event" was "doing very poorly on an important exam." Thirty-three students enrolled in an introductory psychology class were identified as students who felt they performed poorly on a midterm exam. Following the exam, the students were administered the BDI. The post-exam BDI scores are summarized as follows (higher scores indicate higher levels of depression):

$$\bar{x} = 10.18 \qquad s = 5.26$$

Estimate the mean postexam BDI score of introductory psychology students using a 99% confidence interval.

8.14 Unusual rocks at "The Seven Islands," located along the lower St. Lawrence River in Canada, have attracted geologists to the area for over a century. A major geological survey of "The Seven Islands" was recently completed for the purpose of developing a three-dimensional gravity model of the area (*Canadian Journal of Earth Sciences*, Vol. 27, 1990). One of the keys to an objective model is obtaining an accurate estimate of the rock densities. Based on samples of several varieties of rock, the following information on rock density (grams per cubic centimeter) was obtained.

TYPE OF ROCK	SAMPLE SIZE	MEAN DENSITY	STANDARD DEVIATION
Late gabbro	36	3.04	.13
Massive gabbro	148	2.83	.11
Cumberlandite	135	3.05	.31

Source: Loncarevic, B. D., Feninger, T., and Lefebvre, D. "The Sept-Iles layered mafic intrusion: Geophysical expression." *Canadian Journal of Earth Sciences*, Vol. 27, Aug. 1990, p. 505.

a. For each rock type, estimate the mean density with a 90% confidence interval.

b. Interpret the intervals, part a.

8.15 Refer to the data on student-loan default rates for 66 Florida colleges, Exercise 2.28. A Minitab printout showing descriptive statistics for the sample data and a 95% confidence interval for the mean student-loan default rate is presented on the next page.

TABLE 8.4

Reproduction of a Portion of Table 5 of Appendix G

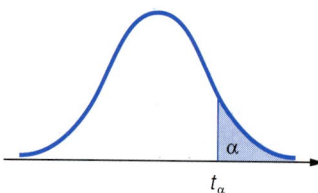

DEGREES OF FREEDOM	$t_{.100}$	$t_{.050}$	$t_{.025}$	$t_{.010}$	$t_{.005}$	$t_{.001}$	$t_{.0005}$
1	3.078	6.314	12.706	31.821	63.657	318.31	636.62
2	1.886	2.920	4.303	6.965	9.925	22.326	31.598
3	1.638	2.353	3.182	4.541	5.841	10.213	12.924
4	1.533	2.132	2.776	3.747	4.604	7.173	8.610
5	1.476	2.015	2.571	3.365	4.032	5.893	6.869
6	1.440	1.943	2.447	3.143	3.707	5.208	5.959
7	1.415	1.895	2.365	2.998	3.499	4.785	5.408
8	1.397	1.860	2.306	2.896	3.355	4.501	5.041
9	1.383	1.833	2.262	2.821	3.250	4.297	4.781
10	1.372	1.812	2.228	2.764	3.169	4.144	4.587
11	1.363	1.796	2.201	2.718	3.106	4.025	4.437
12	1.356	1.782	2.179	2.681	3.055	3.930	4.318
13	1.350	1.771	2.160	2.650	3.102	3.852	4.221
14	1.345	1.761	2.145	2.624	2.977	3.787	4.140
15	1.341	1.753	2.131	2.602	2.947	3.733	4.073

EXAMPLE 8.9 Use Table 5 of Appendix G to determine the t value that would be used in constructing a 95% confidence interval for μ based on a sample of size $n = 14$.

Solution For a confidence coefficient of .95, we have

$$1 - \alpha = .95$$

$$\alpha = .05$$

$$\frac{\alpha}{2} = .025$$

We thus require the value of $t_{.025}$ for a t distribution based on $(n - 1) = (14 - 1) = 13$ degrees of freedom. Now in Table 5, at the intersection of the column labeled $t_{.025}$ and the row corresponding to df $= 13$, we find the entry 2.160 (see Figure 8.8). Hence, a 95% confidence interval for μ, based on a sample of $n = 14$ observations, would be given by

$$\bar{x} \pm 2.160 \left(\frac{s}{\sqrt{14}} \right)$$

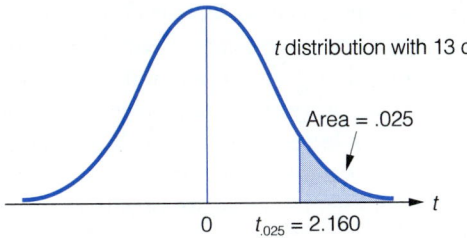

FIGURE 8.8
Location of $t_{.025}$ for Example 8.9

t distribution with 13 df

Area = .025

$0 \quad t_{.025} = 2.160$

EXAMPLE 8.10 Recall that when estimating μ in Section 8.2, we used the arbitrary cutoff point of $n = 30$ for distinguishing between large and small samples. Explain why $n \geq 30$ is selected as the arbitrary cutoff.

Solution   Turn to Appendix G and observe that the values in the last row of Table 5 (corresponding to df $= \infty$) are the values from the standard normal z distribution. This phenomenon occurs because, as the sample size increases, the t distribution becomes more and more like the z distribution (recall Figure 8.7). By the time n reaches 30, i.e., df $= 29$, there is very little difference between tabulated values of t and z. Of course, the values $n = 40$, $n = 60$, and even $n = 120$ could also have been selected as the cutoff for defining a "large" sample. However, $n = 30$ seems to be the smallest value of n for which the t values reasonably approximate the corresponding z values.

EXAMPLE 8.11   The Geothermal Loop Experimental Facility, located in the Salton Sea in southern California, is a U.S. Department of Energy operation for studying the feasibility of generating electricity from the hot, highly saline water of the Salton Sea. Operating experience has shown that these brines leave silica scale deposits on metallic plant piping, causing excessive plant outages. Jacobsen et al. (*Journal of Testing and Evaluation*, Vol. 9, No. 2, Mar. 1981, pp. 82–92) have found that scaling can be reduced somewhat by adding chemical solutions to the brine. In one screening experiment, each of five antiscalants was added to an aliquot of brine, and the solutions were filtered. A silica determination (parts per million of silicon dioxide) was made on each filtered sample after a holding time of 24 hours, with the following results:

229   255   280   203   229

Estimate the mean amount of silicon dioxide present in the five antiscalant solutions. Use a 99% confidence interval.

Solution   The first step in constructing the confidence interval is to compute the mean, $\bar{x}$, and standard deviation, s, of the sample of five silicon dioxide amounts. These values, $\bar{x} = 239.2$ and $s = 29.3$, are provided in Figure 8.9.

	N	MEAN	MEDIAN	TRMEAN	STDEV	SEMEAN
ppm	5	239.2	229.0	239.2	29.3	13.1

	MIN	MAX	Q1	Q3
ppm	203.0	280.0	216.0	267.5

FIGURE 8.9
Minitab Descriptive Statistics for Example 8.11

For a confidence coefficient of $1 - \alpha = .99$, we have $\alpha = .01$ and $\alpha/2 = .005$. Since the sample size is small ($n = 5$), our estimation technique requires the assumption that the amount of silicon dioxide present in an antiscalant solution has an approximately normal distribution (i.e., the sample of five silicon amounts is selected from a normal population).

Substituting the values for $\bar{x}$, s, and n into the formula for a small-sample confidence interval for μ, we obtain

$$\bar{x} \pm t_{\alpha/2}\left(\frac{s}{\sqrt{n}}\right) = \bar{x} \pm t_{.005}\left(\frac{s}{\sqrt{n}}\right)$$

$$= 239.2 \pm t_{.005}\left(\frac{29.3}{\sqrt{5}}\right)$$

where $t_{.005}$ is the value corresponding to an upper-tail area of .005 in the Student's t distribution based on $(n - 1) = 4$ degrees of freedom. From Table 5 of Appendix G, the required t value is $t_{.005} = 4.604$. Substituting this value yields

$$239.2 \pm t_{.005}\left(\frac{29.3}{\sqrt{5}}\right) = 239.2 \pm (4.604)\left(\frac{29.3}{\sqrt{5}}\right)$$

$$= 239.2 \pm 60.3$$

or, 178.9 to 299.5 ppm. Thus, if the distribution of silicon dioxide amounts is approximately normal, then we can be 99% confident that the interval (178.9, 299.5) encloses μ, the true mean amount of silicon dioxide present in an antiscalant solution.

EXAMPLE 8.12   Refer to Example 8.6 and the problem of estimating μ, the mean ratio of attention time given to $2\frac{1}{2}$-year-old twin boys by their parents. Recall that a 99% confidence interval was computed based on data collected for a random sample of $n = 50$ sets of twins. A Minitab printout showing a 99% confidence interval for μ is displayed in Figure 8.10. Compare the results to the interval calculated in Example 8.6.

FIGURE 8.10
Minitab Printout for Example 8.12

	N	MEAN	STDEV	SE MEAN	99.0 PERCENT C.I.	
atentime	50	20.85	13.41	1.90	(15.76,	25.93)

Solution   The 99% confidence interval, shaded on the Minitab printout, is (15.76, 25.93). The interval calculated in Example 8.6 is (15.97, 25.73). The differences in the end points of the interval, although relatively minor, are because σ is unknown for the target population. As do most statistical software packages, Minitab computes the confidence interval using the t statistic, i.e.,

$$\bar{x} \pm t_{.005}\left(\frac{s}{\sqrt{n}}\right)$$

where $t_{.005} \approx 2.68$ (based on $n - 1 = 49$ df). The confidence interval in Example 8.6, you will recall, was calculated using the z statistic, i.e.,

$$\bar{x} \pm z_{.005}\left(\frac{s}{\sqrt{n}}\right)$$

where $z_{.005} = 2.575$. Theoretically, the Minitab confidence interval is the correct one, since $\bar{x}$ has a t distribution when σ is unknown. The confidence interval in Example 8.6 is approximate. But you can see the approximation is good when the sample size n is large.

Before concluding this section, we will comment on the assumption that the sampled population is normally distributed. In the real world, we rarely know whether a sampled population has an exactly normal distribution. However, empirical studies indicate that moderate departures from this assumption do not seriously affect the confidence coefficients for small-sample confidence intervals. For example, if the population of silicon dioxide amounts for the antiscalant solutions of Example 8.11 has a distribution that is mound-shaped but nonnormal, it is likely that the actual confidence coefficient for the 99% confidence interval will be close to .99—at least close enough to be of practical use. As a consequence, the small-sample confidence interval given in the box is frequently used by experimenters when estimating the population mean of a nonnormal distribution as long as the distribution is mound-shaped and only moderately skewed.

For populations that depart greatly from normality, other estimation techniques (such as robust estimation) or methods that are distribution-free (called **nonparametrics**) are recommended. Nonparametric statistics are the topic of Chapter 15.

EXERCISES
LEARNING THE MECHANICS

8.16 Use Table 5 of Appendix G to determine the values of $t_{\alpha/2}$ that would be used to construct a confidence interval for a population mean for each of the following combinations of confidence coefficient and sample size:

a. Confidence coefficient .99, $n = 18$ **b.** Confidence coefficient .95, $n = 10$
c. Confidence coefficient .90, $n = 15$

8.17 Give two reasons why the interval estimation procedure of Section 8.2 may not be applicable when the sample size is small, i.e., when $n < 30$.

8.18 A random sample of $n = 10$ measurements from a normally distributed population yielded $\bar{x} = 9.4$ and $s = 1.8$.

a. Calculate a 90% confidence interval for μ.
b. Calculate a 95% confidence interval for μ.
c. Calculate a 99% confidence interval for μ.

8.19 The following data represent a random sample of five measurements from a normally distributed population:

7   4   2   5   7

a. Find a 90% confidence interval for μ.
b. Find a 99% confidence interval for μ.

8.20   The mean and standard deviation of n measurements randomly sampled from a normally distributed population are 33 and 4, respectively. Construct a 95% confidence interval for μ when:

  a. $n = 5$ b. $n = 15$ c. $n = 25$

8.21   How are the t distribution and the z distribution similar? How are they different?

APPLYING THE CONCEPTS

8.22   Refer to the *Risk Management* study on fires in compartmented fire-resistant buildings, Exercise 3.13. The data shown in the table give the number of victims who died attempting to evacuate for a sample of 14 recent fires.

FIRE	NUMBER OF VICTIMS
Las Vegas Hilton (Las Vegas)	5
Inn on the Park (Toronto)	5
Westchase Hilton (Houston)	8
Holiday Inn (Cambridge, Ohio)	10
Conrad Hilton (Chicago)	4
Providence College (Providence)	8
Baptist Towers (Atlanta)	7
Howard Johnson (New Orleans)	5
Cornell University (Ithaca, New York)	9
Westport Central Apartments (Kansas City, Missouri)	4
Orrington Hotel (Evanston, Illinois)	0
Hartford Hospital (Hartford, Connecticut)	16
Milford Plaza (New York)	0
MGM Grand (Las Vegas)	36

Source: Macdonald, J. N. "Is evacuation a fatal flaw in fire fighting philosophy?" *Risk Management*, Vol. 33, No. 2, Feb. 1986, p. 37.

  a. State the assumption, in terms of the problem, that is required for a small-sample confidence interval technique to be valid.
  b. Construct a 98% confidence interval for the true mean number of victims per fire who die attempting to evacuate compartmented fire-resistant buildings.
  c. Interpret the interval constructed in part **b**.

8.23   An evaluation of trace metal chemistry and cycling in an acidic Adirondack lake was reported in *Environmental Science & Technology* (Dec. 1985). Twenty-four water samples were collected from Darts Lake, New York, and analyzed for concentrations of both lead and aluminum particulates.

  a. The lead concentration measurements had a mean of 9.9 nmol/l and a standard deviation of 8.4 nmol/l. Calculate a 99% confidence interval for the true mean lead concentration in water samples collected from Darts Lake.
  b. The aluminum concentration measurements had a mean of 6.7 nmol/l and a standard deviation of 10.8 nmol/l. Calculate a 99% confidence interval for the true mean aluminum concentration in water samples collected from Darts Lake.
  c. What assumptions are necessary for the intervals of parts **a** and **b** to be valid?

8.24   In one recent study, 16 handicapped children enrolled in special education preschool in the state of Washington had a mean IQ of 80.2 and a standard deviation of 20.7 (*Exceptional Children*, Sept. 1984). All 16 children met Washington's criteria for classification as developmentally delayed (i.e., performing at least 25% below chronological age in development). Estimate the mean IQ of handicapped children classified as developmentally delayed in the state of Washington using a 95% confidence interval. Interpret the interval.

8.25 Many Vietnam veterans have dangerously high levels of the dioxin 2,3,7,8-TCDD in blood and fat tissue as a result of their exposure to the defoliant Agent Orange. A study published in *Chemosphere* (Vol. 20, 1990) reported on the TCDD levels of 20 Massachusetts Vietnam veterans who were possibly exposed to Agent Orange. The amounts of TCDD (measured in parts per trillion) in blood plasma drawn from each veteran are shown in the table.

VETERAN	TCDD LEVELS IN PLASMA	VETERAN	TCDD LEVELS IN PLASMA
1	2.5	11	6.9
2	3.1	12	3.3
3	2.1	13	4.6
4	3.5	14	1.6
5	3.1	15	7.2
6	1.8	16	1.8
7	6.0	17	20.0
8	3.0	18	2.0
9	36.0	19	2.5
10	4.7	20	4.1

Source: Schecter, A. et al. "Partitioning of 2,3,7,8-chlorinated dibenzo-*p*-dioxins and dibenzofurans between adipose tissue and plasma lipid of 20 Massachusetts Vietnam veterans." *Chemosphere*, Vol. 20, Nos. 7–9, 1990, pp. 954–955 (Table I).

a. Construct a 90% confidence interval for the true mean TCDD level in plasma of all Vietnam veterans exposed to Agent Orange.
b. Interpret the interval, part **a**.
c. What assumption is required for the interval estimation procedure to be valid?
d. Use one of the methods of Section 6.3 to determine whether the assumption, part **c**, is approximately satisfied.

8.26 Refer to the *Journal of Marketing Research* study of the effect of brand name and store name on product quality, Exercise 2.23. The stem-and-leaf displays showing the distribution of effect size (an index ranging from 0 to 1) for the 15 brand name studies and 17 store name studies are reproduced here.

Brand Name (15 studies)   Store Name (17 studies)

STEM	LEAF		STEM	LEAF
.6	0		.6	
.5	7		.5	
.4			.4	3 4
.3	4		.3	
.2	5 5		.2	
.1	0 1 1 2 4		.1	2
.0	3 3 5 5 7		.0	0 0 0 1 1 2 2 3 3 4 6 7 8 8

Source: Rao, A. R. and Monroe, K. B. "The effect of price, brand name, and store name on buyers' perceptions of product quality: An integrative review." *Journal of Marketing Research*, Vol. 26, Aug. 1989, p. 354 (Table 2).

a. Use the method of this section to find a 95% confidence interval for the mean effect size for all brand name studies. Interpret the result.
b. Repeat part **a** for store name studies.
c. Why might the validity of the confidence intervals, parts **a** and **b**, be suspect?

8.27 Refer to the data on DDT concentrations in fish inhabiting the Tennessee River and its tributary creeks, given in Appendix D. Suppose you want to estimate the mean DDT content of all fish in the river and its tributary creeks.

a. Select a random sample of size $n = 5$ from the DDT measurements listed in Appendix D.
b. Compute $\bar{x}$ and s for this sample.
c. Using the summary statistics you computed in part b, construct a 99% confidence interval for the true mean DDT content of all fish in the river.
d. Now select a random sample of size $n = 30$ from the DDT measurements in Appendix D. Repeat parts b and c.
e. Compare the widths of the two intervals. Which interval gives a more reliable estimate of the true mean DDT content? Explain.

8.28 During the budget preparation process at a hospital, forecasts of inpatient utilization (measured in patient-days) for the coming fiscal year play a pivotal role. Planners at Sisters of St. Joseph of Peace Health and Hospital Services, Bellevue, Washington, have developed a budget early warning technique (BEWT) for forecasting future fiscal year utilizations. The method involves collecting past data on monthly utilization and calculating, for each month, the ratio of monthly utilization to utilization for the entire fiscal year preceding the month. Confidence intervals established on this utilization ratio enable planners to establish budget forecasts. At one of the smaller (50 beds) Sisters of St. Joseph's hospitals, the April utilization ratios for each of the past 20 years were calculated and found to have a sample mean of .0817 and a sample standard deviation of .0069 (*Hospitals & Health Services Administration*, Jan./Feb. 1986). Establish a 95% confidence interval for μ, the true mean April utilization ratio. Interpret the interval.

ESTIMATION OF A POPULATION PROPORTION

8.4 We will now consider the method for estimating the binomial proportion of successes—that is, the **proportion** of elements in a population that have a certain characteristic. For example, a sociologist may be interested in the proportion of urban New York City residents who are black; a pollster may be interested in the proportion of Americans who favor the president's policy of limited government spending; or a supplier of heating oil may be interested in the proportion of homes in its service area that are heated by natural gas. How would you estimate a binomial proportion π (e.g., the proportion of crimes related to firearms), based on information contained in a sample from the population?

EXAMPLE 8.13 The United States Commission on Crime is interested in estimating the proportion of crimes related to firearms in an area with one of the highest crime rates in the country. The commission selects a random sample of 300 files of recently committed crimes in the area and determines that a firearm was reportedly used in 180 of them. Estimate the true proportion π of all crimes committed in the area in which some type of firearm was reportedly used.

Solution A logical candidate for a point estimate of the population proportion π is the proportion of observations in the sample that have the characteristic of interest (called a "success"); we will call this sample proportion p. In this example, the sample proportion of crimes related to firearms is given by

$$p = \frac{\text{Number of crimes in sample in which a firearm was reportedly used}}{\text{Total number of crimes in sample}}$$

$$= \frac{180}{300} = .60$$

That is, 60% of the crimes in the sample were related to firearms; the value $p = .60$ serves as our point estimate of the population proportion π.

To assess the reliability of the point estimate p, we need to know its sampling distribution. This information may be derived by an application of the Central Limit Theorem (details are omitted here). Properties of the sampling distribution of p (illustrated in Figure 8.11) are given in the next box.

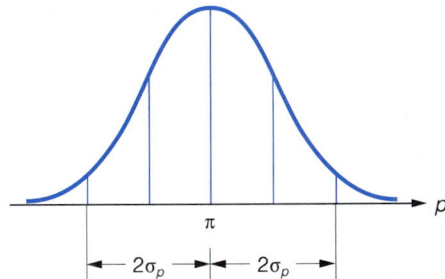

FIGURE 8.11
Sampling Distribution of p

SAMPLING DISTRIBUTION OF THE SAMPLE PROPORTION, p

For sufficiently large samples, the sampling distribution of $p = x/n$ is approximately normal, with

   Mean $\mu_p = \pi$

and

   Standard deviation $\sigma_p = \sqrt{\dfrac{\pi(1 - \pi)}{n}}$

where π = True population proportion of successes
         x = Number of successes in the sample
         n = Sample size

A large-sample confidence interval for π may be constructed by using a procedure analogous to that used for estimating a population mean. We begin with the point estimator p, then add and subtract a certain number of standard deviations of p to obtain the desired level of confidence. The details are given in the next box (page 360).

To construct the interval, note that we must substitute p and q into the formula for $\sigma_p = \sqrt{\pi(1 - \pi)/n}$. This approximation will be valid as long as the sample size n is sufficiently large. Many researchers adopt the rule of thumb that n is "sufficiently large" if the interval $p \pm 2\sqrt{pq/n}$ does not contain 0 or 1.

EXAMPLE 8.14   Refer to Example 8.13. Construct a 95% confidence interval for π, the population proportion of crimes committed in the area in which some type of firearm is reportedly used.

Solution   For a confidence coefficient of .95, we have $1 - \alpha = .95$; $\alpha = .05$; $\alpha/2 = .025$; and the required z value is $z_{.025} = 1.96$. In Example 8.13, we obtained $p = 180/300 = .60$. Thus, $q = 1 - .60 = .40$. Substituting these values into the formula for an approximate confidence interval for π yields

$$p \pm z_{\alpha/2}\sqrt{\frac{pq}{n}} = .60 \pm 1.96\sqrt{\frac{(.60)(.40)}{300}}$$

$$= 60 \pm .06$$

or (.54, .66). Note that the approximation is valid since the interval does not contain 0 or 1.

EXAMPLE 8.15   Potential advertisers value television's well-known Nielsen ratings as a barometer of a TV show's popularity among viewers. The Nielsen rating of a certain TV program is an estimate of the proportion of viewers, expressed as a percentage, who tune their sets to the program on a given night. In a random sample of 165 families who regularly watch television, a Nielsen survey indicated that 101 of the families were tuned to NBC's "The Cosby Show" on the night of its premiere. Estimate π, the true proportion of all TV-viewing families who watched the premiere of "The Cosby Show," using a 90% confidence interval. Interpret the interval.

Solution   In this problem the variable of interest is the response (yes or no) to the question, "Did you watch 'The Cosby Show' on the night of its premiere?"

The sample proportion of families that watched the premiere of "The Cosby Show" is

$$p = \frac{x}{n} = \frac{\text{Number of families in sample that watched the premiere}}{\text{Number of families in sample}}$$

$$= \frac{101}{165} = .612$$

Thus $q = 1 - .612 = .388$. Using the formula in the box, the approximate 90% confidence interval is

$$p \pm z_{.05}\sqrt{\frac{pq}{n}} = .612 \pm 1.645\sqrt{\frac{(.612)(.388)}{165}}$$

$$= .612 \pm .062$$

or (.550, .674).

We are 90% confident that the interval from .550 to .674 encloses the true proportion of TV-viewing families that watched the premiere of "The Cosby Show." If we repeatedly selected random samples of $n = 165$ families and constructed a 90% confidence interval based on each sample, then we would expect 90% of the confidence intervals constructed to contain π.

Although small-sample procedures are available to estimate a population proportion π, they are beyond the scope of this introductory text. However, most surveys of binomial populations (e.g., opinion polls) conducted in actual practice use samples that are large enough to employ the procedure of this section.

EXERCISES
LEARNING THE MECHANICS

8.29 Random samples of n measurements are selected from a population with unknown proportion of successes π. Compute an estimate of σ_p for each of the following combinations of sample size n and sample proportion of successes p.

    a. $n = 250, p = .4$ **b.** $n = 500, p = .85$ **c.** $n = 100, p = .25$

8.30 Random samples of n measurements are selected from a population with unknown proportion of successes π. Calculate a 95% confidence interval for π for each of the following combinations of sample size n and sample proportion of successes p.

    a. $n = 500, p = .38$ **b.** $n = 100, p = .45$ **c.** $n = 1,000, p = .43$

8.31 The proportion of successes in a random sample of size n is $p = .20$.

    a. Find a 95% confidence interval for π if $n = 100$.
    b. Find a 95% confidence interval for π if $n = 500$.

8.32 A random sample of size 150 is selected from a population and the number of successes is 60.

    a. Find p, the sample proportion of successes.
    b. Construct a 90% confidence interval for π.
    c. Construct a 95% confidence interval for π.
    d. Construct a 99% confidence interval for π.

8.33 A University of Minnesota survey of brand names (e.g., Levi's, Lee, and Calvin Klein), as well as private labels manufactured for retail chains, found a high percentage of jeans with incorrect waist and/or inseam measurements on the label. The study found that only 18 of 240 pairs of men's five-pocket, prewashed jeans sold in Minneapolis stores came within a half inch of all their label measurements (*Tampa Tribune*, May 20, 1991). Let π represent the true proportion of men's five-pocket, prewashed jeans sold in Minneapolis that have inseam and waist measurements that fall within .5 inch of the labeled measurements.

    a. Find a point estimate of π.
    b. Find an interval estimate of π. Use a confidence coefficient of .90.
    c. Interpret the interval, part **b**.

8.34 Astronauts often report episodes of disorientation as they move around the zero-gravity spacecraft. To compensate, crew members rely heavily on visual information to establish a top-down orientation. An empirical study was conducted to assess the potential of using color brightness as a body orientation clue (*Human Factors*, Dec. 1988). Ninety college students, reclining on their backs in the dark, were disoriented when positioned on a rotating platform under a slowly rotating disk that filled their entire field of vision. Half the disk was painted with a brighter level of color than the other half. The students were asked to say "stop" when they believed they were right-side up, and the brightness level of the disk was recorded. Of the 90 students, 58 selected the brighter color level.

    a. Use this information to estimate the true proportion of subjects who use the bright color level as a cue to being right-side up. Construct a 95% confidence interval for the true proportion.
    b. Can you infer from the result, part **a**, that a majority of subjects would select bright color levels over dark color levels as a cue to being right-side up? Explain.

8.35 *Agoraphobia* is the psychiatric term for an abnormal fear of being in an open space. In an experiment reported in *Behavioral Research Theory* (Vol. 26, 1988), 51 agoraphobic patients were administered the Behavioral Avoidance Test (BAT). The BAT consisted of three tasks, one of which required the patient to walk unaccompanied from the hospital into a crowded urban center and to cross a wide and congested avenue. Even though most of the 51 patients experienced at least one panic attack, only seven were unable to complete the task. Construct a 90% confidence interval for the proportion of all agoraphobic patients who cannot complete the BAT task. Interpret the interval.

8.36 "Are today's undergraduates more willing to cheat in order to get good grades?" This was the question posed to a national sample of 5,000 college professors by the Carnegie Foundation for the Advancement of Teaching. Given the "make it at all costs" mentality of the past decade, it is not surprising that 43% of the professors responded "yes" (*Tampa Tribune*, Mar. 7, 1990). Based on this survey, estimate the proportion of all college professors who feel their undergraduate students are more willing to cheat to get good grades. Use a confidence coefficient of .90.

8.37 According to an Internal Revenue Service (IRS) study, most IRS officials believe that taxpayers are unethical when filing their taxes. Of a sample of 800 IRS executives and managers from across the country, only 144 rated the ethics of the average taxpayer as good or excellent (*Arizona Republic*, Mar. 23, 1991). Estimate the true proportion of IRS officials who rate the ethics of the average taxpayer as good or excellent with a 99% confidence interval. Interpret the result.

8.38 The U.S. Food and Drug Administration (FDA) recently approved the marketing of a new chemical solution, Caridex, which dissolves cavities. In a study conducted by dental researchers at Northwestern University, 21 of 35 patients with cavities preferred treatment with Caridex to drilling (*Gainesville Sun*, Feb. 11, 1988). Estimate the true proportion of dental patients who prefer having their cavities dissolved with Caridex rather than drilled. Use a 99% confidence interval and interpret the result.

8.39 In 1963, the Rev. Martin Luther King, Jr., gave his inspirational "I have a dream" speech, which established civil rights on the forefront of the nation's social agenda. Since that time, the United States has made strides toward racial equality. However, a Media General–Associated Press poll conducted 25 years later found that a majority (55%) of Americans still believe the American society is racist overall (*Tampa Tribune*, Aug. 8, 1988). The results were based on telephone interviews with a randomly selected sample of 1,223 adults across the nation. Use this information to find a 90% confidence interval for π, the proportion of all Americans who believe that American society is racist overall.

8.40 Do most state lottery winners who win big payoffs quit their jobs? Not according to a study conducted by sociologist and professor H. Roy Kaplan (*Journal of the Institute for Socioeconomic Studies*, Sept. 1985). Kaplan mailed questionnaires to over 2,000 lottery winners who won at least $50,000 in the past 10 years. Of the 576 who responded, only 11% had quit their jobs during the first year after striking it rich.

    a. Use a 95% confidence interval to estimate π, the proportion of all state lottery winners (at least $50,000) who quit their jobs during the first year after striking it rich.

    b. Do you think the 576 lottery winners who returned the questionnaire represent a random sample of all state lottery winners? Explain how a nonrandom sample could bias the results.

8.5   ESTIMATION OF THE DIFFERENCE BETWEEN TWO POPULATION MEANS: INDEPENDENT SAMPLES

In Sections 8.2 and 8.3, we learned how to estimate the mean μ of a single population. We now proceed to a technique for using the information in two samples to estimate the difference between two population means. For example, we may want to compare the mean starting salaries of college graduates with engineering and journalism degrees; or the mean gasoline consumptions that may be expected this year for drivers in two areas of the country; or the mean reaction times of men and women to a visual stimulus. The technique to be presented is a straightforward extension of that used for estimation of a single population mean.

LARGE-SAMPLE ESTIMATION

EXAMPLE 8.16 We wish to estimate the difference between the mean starting salaries for all bachelor's degree graduates of the University of Florida in the colleges of Engineering and Journalism. The following information is extracted from Appendix B.

1. A random sample of 40 starting salaries for College of Engineering graduates produced a sample mean of $30,653 and a standard deviation of $4,172.
2. A random sample of 30 starting salaries for College of Journalism graduates produced a sample mean of $21,291 and a standard deviation of $3,864.

Calculate a point estimate for the difference between starting salaries for graduates of the two colleges.

Solution We will let the subscript "1" refer to the College of Engineering and the subscript "2" to the College of Journalism. We will also define the following notation:

μ_1 = Population mean starting salary of all bachelor's degree graduates of the College of Engineering

μ_2 = Population mean starting salary of all bachelor's degree graduates of the College of Journalism

Similarly, let $\bar{x}_1$ and $\bar{x}_2$ denote the respective sample means; s_1 and s_2, the respective sample standard deviations; and n_1 and n_2, the respective sample sizes. The given information may be summarized as in Table 8.5.

TABLE 8.5
Summary of Information for Example 8.16

	COLLEGE OF ENGINEERING	COLLEGE OF JOURNALISM
Sample size	$n_1 = 40$	$n_2 = 30$
Sample mean	$\bar{x}_1 = \$30,653$	$\bar{x}_2 = \$21,291$
Sample standard deviation	$s_1 = \$4,172$	$s_2 = \$3,864$

Now, to estimate $(\mu_1 - \mu_2)$, it seems logical to use the difference between the sample means

$$(\bar{x}_1 - \bar{x}_2) = (\$30,653 - \$21,291) = \$9,362$$

as our point estimate of the difference between the population means.

Confidence intervals for $(\mu_1 - \mu_2)$ will be based on the sampling distribution of the point estimate $(\bar{x}_1 - \bar{x}_2)$. The properties of the sampling distribution of $(\bar{x}_1 - \bar{x}_2)$ are shown in the next box (see also Figure 8.12).

FIGURE 8.12
Sampling Distribution of $(\bar{x}_1 - \bar{x}_2)$

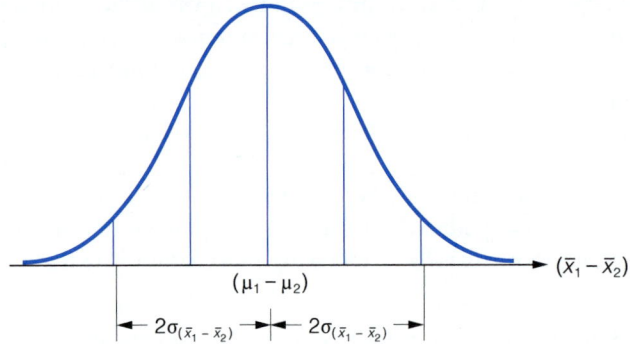

SAMPLING DISTRIBUTION OF $(\bar{x}_1 - \bar{x}_2)$

For sufficiently large sample sizes (say, $n_1 \geq 30$ and $n_2 \geq 30$), the sampling distributions of $(\bar{x}_1 - \bar{x}_2)$, based on independent random samples from two populations, is approximately normal with

Mean: $\mu_{(\bar{x}_1 - \bar{x}_2)} = (\mu_1 - \mu_2)$

Standard deviation: $\sigma_{(\bar{x}_1 - \bar{x}_2)} = \sqrt{\dfrac{\sigma_1^2}{n_1} + \dfrac{\sigma_2^2}{n_2}}$

where σ_1^2 and σ_2^2 are the variances of the two populations from which the samples were selected.

As was the case with large-sample estimation of a single population mean, the requirements of "large" sample sizes enables us to apply the Central Limit Theorem to obtain the sampling distribution of $(\bar{x}_1 - \bar{x}_2)$; it also justifies the use of s_1^2 and s_2^2 as approximations to the respective population variances, σ_1^2 and σ_2^2.

The procedure for forming a large-sample confidence interval for $(\mu_1 - \mu_2)$ appears in the accompanying box.

LARGE-SAMPLE $(1 - \alpha)100\%$ CONFIDENCE INTERVAL FOR $(\mu_1 - \mu_2)$

$$(\bar{x}_1 - \bar{x}_2) \pm z_{\alpha/2}\sigma_{(\bar{x}_1 - \bar{x}_2)} = (\bar{x}_1 - \bar{x}_2) \pm z_{\alpha/2}\sqrt{\frac{\sigma_1^2}{n_1} + \frac{\sigma_2^2}{n_2}}$$

$$\approx (\bar{x}_1 - \bar{x}_2) \pm z_{\alpha/2}\sqrt{\frac{s_1^2}{n_1} + \frac{s_2^2}{n_2}}$$

[*Note:* We have used the sample variances s_1^2 and s_2^2 as approximations to the corresponding population parameters.]

Assumptions:

1. The two random samples are selected in an independent manner from the target populations. That is, the choice of elements in one sample does not affect, and is not affected by, the choice of elements in the other sample.

2. The sample sizes n_1 and n_2 are sufficiently large. (We recommend $n_1 \geq 30$ and $n_2 \geq 30$.)

EXAMPLE 8.17   Refer to Example 8.16. Construct a 95% confidence interval for $(\mu_1 - \mu_2)$, the difference between the mean starting salaries of all bachelor's degree graduates of the colleges of Engineering and Journalism. Interpret the interval.

Solution   The general form of a 95% confidence interval for $(\mu_1 - \mu_2)$, based on large samples from the target populations, is given by

$$(\bar{x}_1 - \bar{x}_2) + z_{.025}\sqrt{\frac{\sigma_1^2}{n_1} + \frac{\sigma_2^2}{n_2}}$$

Recall that $z_{.025} = 1.96$ and use the information in Table 8.5 to make the following substitutions to obtain the desired confidence interval:

$$(30{,}653 - 21{,}291) \pm 1.96\sqrt{\frac{\sigma_1^2}{40} + \frac{\sigma_2^2}{30}}$$

$$\approx (30{,}653 - 21{,}291) \pm 1.96\sqrt{\frac{(4{,}172)^2}{40} + \frac{(3{,}864)^2}{30}}$$

$$= 9{,}362 \pm 1{,}893$$

or ($7,469, $11,255).

Using this method of estimation produces confidence intervals that will enclose $(\mu_1 - \mu_2)$, the difference between population means, 95% of the time in repeated sampling. Since all the values in our interval are positive, we can be reasonably confident that the mean starting salary of College of Engineering bachelor's degree graduates was between $7,469 and $11,255 higher than the mean starting salary of College of Journalism bachelor's degree graduates during the period Fall 1989–Spring 1991. (The actual difference between the two population means, calculated by computer, is $10,822. Note that this value falls within the 95% confidence interval.)

EXAMPLE 8.18 The personnel manager for a large steel company suspects there is a difference between the mean amounts of work time lost to sickness for blue-collar and white-collar workers at the plant. She randomly samples the records of 45 blue-collar workers and 38 white-collar workers, and records the number of days lost to sickness within the past year. The data were entered into a computer; Minitab was used to estimate $(\mu_1 - \mu_2)$, the difference between the population mean times lost to sickness for blue-collar and white-collar workers at the steel company last year, with a 90% confidence interval. The Minitab printout is shown in Figure 8.13. Interpret the results.

FIGURE 8.13
Minitab Printout for the Data in
Example 8.18

```
TWOSAMPLE T FOR sickdays
worker    N        MEAN       STDEV     SE MEAN
1         45       11.5       10.2         1.5
2         38       9.00       5.58        0.91

90 PCT CI FOR MU 1 - MU 2:  (-0.4, 5.47)

TTEST MU 1 = MU 2 (VS NE):  T= 1.42   P=0.16   DF=  70
```

Solution The 90% confidence interval for $\mu_1 - \mu_2$, shaded on the printout, is $(-.40, 5.47)$. Thus, the personnel manager is 90% confident that $(\mu_1 - \mu_2)$, the difference between the mean days lost to sickness for the two groups of workers, falls between $-.40$ and 5.47. In other words, the manager estimates that μ_2, the mean days lost to sickness for white-collar workers, could be *larger* than μ_1, the mean days lost to sickness for blue-collar workers, by as much as .40 day, or it could be *less* than μ_1 by as much as 5.47 days. Since the interval contains the value 0, she is unable to conclude that there is a real difference between the mean numbers of sick days lost by the two groups. If, in fact, such a difference exists, she would have to increase the sample sizes to be able to detect it. This would reduce the width of the confidence interval and provide more information about the phenomenon under investigation.

SMALL-SAMPLE ESTIMATION When estimating the difference between two population means, based on small samples from each population, we must make specific assumptions about the relative frequency distributions of the two populations, as indicated in the box.

ASSUMPTIONS REQUIRED FOR SMALL-SAMPLE ESTIMATION OF $(\mu_1 - \mu_2)$

1. Both of the populations from which the samples are selected have relative frequency distributions that are approximately normal.

2. The variances σ_1^2 and σ_2^2 of the two populations are equal.

3. The random samples are selected in an independent manner from the two populations.

Figure 8.14 illustrates the form of the population distributions implied by assumptions 1 and 2. Observe that both populations have relative frequency distributions that are approximately normal. Although the means of the two populations may differ, we require the variances σ_1^2 and σ_2^2, which measure the spread of the two distributions, to be equal. When these assumptions are satisfied, we may use the Student's t distribution (specified in the next box) to construct a confidence interval for $(\mu_1 - \mu_2)$, based on small samples (say, $n_1 < 30$ or $n_2 < 30$) from the respective populations. Since we assume that the two populations have equal variances (i.e., $\sigma_1^2 = \sigma_2^2 = \sigma^2$), we construct an estimate of σ^2 based on the information contained in *both* samples. This **pooled estimate** is denoted s_p^2 and is computed as shown in the box.

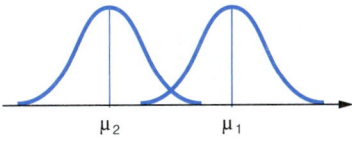

FIGURE 8.14
Assumptions Required for Small-Sample Estimation of $(\mu_1 - \mu_2)$: Normal Distributions with Equal Variances

SMALL-SAMPLE $(1 - \alpha)100\%$ CONFIDENCE INTERVAL FOR $(\mu_1 - \mu_2)$

$$(\bar{x}_1 - \bar{x}_2) \pm t_{\alpha/2}\sqrt{s_p^2\left(\frac{1}{n_1} + \frac{1}{n_2}\right)}$$

where

$$s_p^2 = \frac{(n_1 - 1)s_1^2 + (n_2 - 1)s_2^2}{n_1 + n_2 - 2}$$

and the value of $t_{\alpha/2}$ is based on $(n_1 + n_2 - 2)$ degrees of freedom.

EXAMPLE 8.19 To study the effectiveness of a new type of dental anesthetic, a dentist conducted an experiment with 10 randomly selected patients. Five patients were randomly assigned to receive the standard anesthetic (Novocain), whereas the remaining five patients received the proposed new anesthetic. While being treated, each patient was asked to give a measure of his or her discomfort, on a scale from 0 to 100. (Higher scores indicate greater discomfort.) The discomfort scores for the 10 patients are shown in Table 8.6. Suppose the dentist wants to estimate $(\mu_1 - \mu_2)$, the true difference between the mean discomfort levels of the two groups of patients.

TABLE 8.6

Discomfort Scores for Example 8.19

NOVOCAIN	62	71	44	50	42
NEW ANESTHETIC	38	45	27	42	30

a. Construct a 95% confidence interval for the difference between the mean discomfort levels for dental patients who receive Novocain and those who receive the new anesthetic.

b. If the new anesthetic is more effective, then its associated mean discomfort level will be less than the mean discomfort level for Novocain. Use the confidence interval, part **a**, to make an inference about the effectiveness of the new anesthetic.

Solution

a. Let μ_1 and μ_2 represent the true mean discomfort levels of patients receiving Novocain and the new anesthetic, respectively. Since the samples selected for the study are small ($n_1 = n_2 = 5$), the following assumptions are required:

1. The populations of discomfort levels of dental patients receiving Novocain and treated with the new anesthetic both have approximately normal distributions.

2. The variance, σ^2, in the discomfort levels is the same for both groups of patients.

3. The samples were independently and randomly selected from the two target populations.

To compute the confidence interval, we need to find $\bar{x}_1$, $\bar{x}_2$, s_1, and s_2. These summary statistics are given in the Minitab printout, Figure 8.15, as $\bar{x}_1 = 53.8$, $\bar{x}_2 = 36.4$, $s_1 = 12.4$, and $s_2 = 7.7$. Since we have assumed that the two populations have equal variances (i.e., $\sigma_1^2 = \sigma_2^2 = \sigma^2$), the next step is to compute an estimate of this common variance. Our pooled estimate is given by

$$s_p^2 = \frac{(n_1 - 1)s_1^2 + (n_2 - 1)s_2^2}{n_1 + n_2 - 2} = \frac{(5 - 1)(12.4)^2 + (5 - 1)(7.7)^2}{5 + 5 - 2}$$

$$= \frac{852.2}{8} = 106.53$$

FIGURE 8.15

Minitab Printout for Example 8.19

```
TWOSAMPLE T FOR Novacain VS New
              N       MEAN       STDEV     SE MEAN
Novacain      5       53.8       12.4        5.5
New           5       36.40      7.70        3.4

95 PCT CI FOR MU Novacain - MU New: (2.4, 32.4)

TTEST MU Novacain = MU New (VS NE): T= 2.67   P=0.028   DF=  8

POOLED STDEV =        10.3
```

The 95% confidence interval for $(\mu_1 - \mu_2)$ will be based on the value of $t_{.025}$, where t has $(n_1 + n_2 - 2) = 8$ degrees of freedom. From Table 5 of Appendix G, we obtain $t_{.025} = 2.306$. We now substitute the appropriate quantities into the general formula:

$$(\bar{x}_1 - \bar{x}_2) \pm t_{.025} \sqrt{s_P^2\left(\frac{1}{n_1} + \frac{1}{n_2}\right)}$$

$$= (53.8 - 36.4) \pm 2.306 \sqrt{106.53\left(\frac{1}{5} + \frac{1}{5}\right)}$$

$$= 17.4 \pm 15.05$$

or (2.35, 32.45). We estimate, with 95% confidence, that the difference $(\mu_1 - \mu_2)$ falls in the interval from 2.35 to 32.45. Note that this interval is also given (shaded) on the Minitab printout, Figure 8.15.

b. From the 95% confidence interval, part a, we can infer that μ_1, the mean discomfort level for patients receiving Novocain, exceeds μ_2, the mean discomfort level for patients treated with the new anesthetic, by an amount that lies in the interval from 2.35 to 32.45. Thus, it appears that the new anesthetic is more effective than the standard one.

As with the one-sample case, the assumptions required for estimating $(\mu_1 - \mu_2)$ with small samples do not have to be satisfied exactly for the interval estimate to be useful in practice. Slight departures from these assumptions do not seriously affect the level of confidence in the procedure. For example, when the variances σ_1^2 and σ_2^2 of the sampled populations are unequal, researchers have found that the small-sample confidence interval for $(\mu_1 - \mu_2)$ given in the box will still yield valid results in practice as long as the two populations are normal and the sample sizes are equal, i.e., $n_1 = n_2$. In the case where $\sigma_1^2 \neq \sigma_2^2$ and $n_1 \neq n_2$, an approximate confidence interval for $(\mu_1 - \mu_2)$ can be constructed by modifying the degrees of freedom associated with the t distribution.

In the box on the next page, we give the approximate small-sample confidence intervals for $(\mu_1 - \mu_2)$ for two situations when the assumption of equal variances is violated: $n_1 = n_2$ and $n_1 \neq n_2$.

EXAMPLE 8.20 A key aspect of organizational buying is negotiation. S. W. Clopton investigated several issues pertaining to buyer-seller negotiations (*Journal of Marketing Research*, Feb. 1984). One aspect of the analysis involved a comparison of two types of bargaining strategies—competitive bargaining and coordinative bargaining. A *competitive strategy* is characterized by inflexible behavior aimed at forcing concessions, whereas a *coordinative strategy* involves a problem-solving orientation to negotiations with a high degree of trust and cooperation. A sample of organizational buyers were recruited to participate in

APPROXIMATE SMALL-SAMPLE $(1 - \alpha)100\%$ CONFIDENCE INTERVAL FOR $(\mu_1 - \mu_2)$ WHEN $\sigma_1^2 \neq \sigma_2^2$

1. $n_1 = n_2 = n$: $\quad (\bar{x}_1 - \bar{x}_2) \pm t_{\alpha/2} \sqrt{\dfrac{s_1^2}{n} + \dfrac{s_2^2}{n}}$

where the distribution of t depends on $\nu = n_1 + n_2 - 2 = 2(n - 1)$ degrees of freedom.

2. $n_1 \neq n_2$: $\quad (\bar{x}_1 - \bar{x}_2) \pm t_{\alpha/2} \sqrt{\dfrac{s_1^2}{n_1} + \dfrac{s_2^2}{n_2}}$

where the distribution of t has degrees of freedom equal to

$$\nu = \frac{\left(\dfrac{s_1^2}{n_1} + \dfrac{s_2^2}{n_2}\right)^2}{\dfrac{\left(\dfrac{s_1^2}{n_1}\right)^2}{n_1 - 1} + \dfrac{\left(\dfrac{s_2^2}{n_2}\right)^2}{n_2 - 1}}$$

[*Note:* In the case of $n_1 \neq n_2$, the value of ν will not generally be an integer. Round ν down to the nearest integer to use the t table (Table 5 of Appendix G).

Assumptions: 1. Both of the populations from which the samples are selected have relative frequency distributions that are approximately normal.
2. The random samples are selected in an independent manner from the two populations.

a particular negotiation experiment. In one negotiation setting where the maximum profit was fixed, eight buyers used the competitive bargaining strategy and eight buyers used the coordinative bargaining strategy.

Data on the individual savings for the two groups of buyers are provided in Table 8.7. A SAS printout giving descriptive statistics for the two samples follows (Figure 8.16).

TABLE 8.7
Data for Example 8.20

COMPETITIVE BARGAINING		COORDINATIVE BARGAINING	
$1,857	1,566	$1,544	2,137
1,700	663	2,640	2,327
1,829	1,712	1,645	2,152
2,644	1,679	2,275	2,130

Source: Clopton, S. W. "Seller and buying firm factors affecting industrial buyers' negotiation behavior and outcomes." *Journal of Marketing*, Feb. 1984, pp. 39–53, published by the American Marketing Association.

Solution

a. Let μ_1 and μ_2 represent the true mean savings of buyers using the competitive and coordinative bargaining strategies, respectively. Since the samples selected for the study are small ($n_1 = n_2 = 8$), one of the assumptions required is that $\sigma_1^2 = \sigma_2^2$. In Figure 8.16, note that $s_1^2 = 289,432$ and $s_2^2 = 127,797$. These values are so dissimilar that extreme doubt is cast on the validity of the equal variances assumption. (We demonstrate how to test this assumption inferentially in Chapter 10.) Therefore, we will use the procedure in the preceding box to compute the small-sample confidence interval.

FIGURE 8.16
SAS Printout for Example 8.20

```
         Analysis Variable : SAVINGS

-------------------------- STRATEGY=Compet --------------------------

    N Obs        Mean         Variance      Std Dev
    -----------------------------------------------------
      8        1706.25       289431.93    537.9887811
    -----------------------------------------------------


-------------------------- STRATEGY=Coordin -------------------------

    N Obs        Mean         Variance      Std Dev
    -----------------------------------------------------
      8        2106.25       127796.50    357.4863634
    -----------------------------------------------------
```

Since the sample sizes are equal ($n = 8$), the appropriate formula for a 95% confidence interval is:

$$(\bar{x}_1 - \bar{x}_2) \pm t_{.025} \sqrt{\frac{s_1^2}{n} + \frac{s_2^2}{n}}$$

where $t_{.025}$ depends on $\nu = 2(n - 1) = 2(8 - 1) = 14$ degrees of freedom. From the SAS printout, $\bar{x}_1 = 1,706$, $\bar{x}_2 = 2,106$, $s_1 = 538$, and $s_2 = 357$. Also, from Table 5 of Appendix G, $t_{.025} = 2.145$. Substituting, we obtain

$$(1,706 - 2,106) \pm 2.145 \sqrt{\frac{289,432}{8} + \frac{127,796}{8}} = -400 \pm 490$$

or $(-890, 90)$.

We estimate, with 95% confidence, that the difference $(\mu_1 - \mu_2)$ falls in the interval from -890 to 90.

b. From the 95% confidence interval, part **a**, we can infer that μ_1, the mean buyer savings for the competitive bargaining strategy, is anywhere from $890 less to $90 more than μ_2, the corresponding mean for coordinative bargaining. Since 0 is included in the interval, there is no evidence to support the theory that μ_1 is less than μ_2.

EXERCISES

LEARNING THE MECHANICS

8.41 Independent random samples are selected from two populations with means μ_1 and μ_2, respectively. Determine approximate values of $\mu_{(\bar{x}_1 - \bar{x}_2)}$ and $\sigma_{(\bar{x}_1 - \bar{x}_2)}$ for each of the following situations. (Assume $\sigma_1^2 = \sigma_2^2$ in each case.)

    a. $\bar{x}_1 = 150$, $s_1^2 = 36$; $\bar{x}_2 = 140$, $s_2^2 = 24$; $n_1 = n_2 = 35$
    b. $\bar{x}_1 = 125$, $s_1^2 = 225$, $n_1 = 90$; $\bar{x}_2 = 112$, $s_2^2 = 90$, $n_2 = 60$

8.42 Consider two independent random samples with 30 observations selected from population 1 and 40 from population 2. The resulting sample means and variances are shown in the accompanying table. (Assume $\sigma_1^2 \neq \sigma_2^2$.)

SAMPLE FROM POPULATION 1	SAMPLE FROM POPULATION 2
$\bar{x}_1 = 15$	$\bar{x}_2 = 23$
$s_1^2 = 16$	$s_2^2 = 100$
$n_1 = 30$	$n_2 = 40$

    a. Construct a 90% confidence interval for $(\mu_1 - \mu_2)$.
    b. Construct a 95% confidence interval for $(\mu_1 - \mu_2)$.
    c. Construct a 99% confidence interval for $(\mu_1 - \mu_2)$.

8.43 To use the t statistic in a small-sample confidence interval for the difference between the means of two populations, what assumptions must be made about the two populations? About the two samples?

8.44 The following tables show summary statistics for random samples selected from two normal populations that are assumed to have the same variance. In each case, find s_p^2, the pooled estimate of the common variance.

a.	SAMPLE FROM POPULATION 1	SAMPLE FROM POPULATION 2	b.	SAMPLE FROM POPULATION 1	SAMPLE FROM POPULATION 2
	$\bar{x}_1 = 552$	$\bar{x}_2 = 369$		$\bar{x}_1 = 10.8$	$\bar{x}_2 = 8.4$
	$s_1^2 = 4{,}400$	$s_2^2 = 7{,}481$		$s_1^2 = .313$	$s_2^2 = .499$
	$n_1 = 6$	$n_2 = 7$		$n_1 = 8$	$n_2 = 8$

8.45 Independent random samples from two normal populations with equal variances produced the sample means and sample variances listed in the accompanying table.

SAMPLE FROM POPULATION 1	SAMPLE FROM POPULATION 2
$n_1 = 14$	$n_2 = 7$
$\bar{x}_1 = 53.2$	$\bar{x}_2 = 43.4$
$s_1^2 = 96.8$	$s_2^2 = 102.0$

    a. Find a 90% confidence interval for $(\mu_1 - \mu_2)$.
    b. Find a 95% confidence interval for $(\mu_1 - \mu_2)$.
    c. Find a 99% confidence interval for $(\mu_1 - \mu_2)$.

8.46 Two independent random samples selected from normal populations produced the summary statistics shown in the table.

SAMPLE 1	SAMPLE 2
$\bar{x}_1 = 25$	$\bar{x}_2 = 23$
$s_1^2 = 20$	$s_2^2 = 20$

  a. Find a 95% confidence interval for $(\mu_1 - \mu_2)$ if $n_1 = n_2 = 100$.
  b. Find a 95% confidence interval for $(\mu_1 - \mu_2)$ if $n_1 = n_2 = 10$ and $\sigma_1^2 = \sigma_2^2$.
  c. Find a 95% confidence interval for $(\mu_1 - \mu_2)$ if $n_1 = n_2 = 10$ and $\sigma_1^2 \neq \sigma_2^2$.
  d. Find a 95% confidence interval for $(\mu_1 - \mu_2)$ if $n_1 = 10$, $n_2 = 20$, and $\sigma_1^2 = \sigma_2^2$.
  e. Find a 95% confidence interval for $(\mu_1 - \mu_2)$ if $n_1 = 10$, $n_2 = 20$, and $\sigma_1^2 \neq \sigma_2^2$.

APPLYING THE CONCEPTS

8.47 Students enrolled in music classes at the University of Texas (Austin) participated in a study to compare the observations and teacher evaluations of music education majors and nonmusic majors (*Journal of Research in Music Education*, Winter 1991). Independent random samples of 100 music majors and 100 nonmajors rated the overall performance of their teacher using a 6-point scale, where 1 = lowest rating and 6 = highest rating. Use a 95% confidence interval to compare the mean teacher ratings of the two groups of music students. Interpret the result.

	MUSIC MAJORS	NONMUSIC MAJORS
Sample size	100	100
Mean "overall" rating	4.26	4.59
Standard deviation	.81	.78

Source: Duke, R. A. and Blackman, M. D. "The relationship between observers' recorded teacher behavior and evaluation of music instruction." *Journal of Research in Music Education*, Vol. 39, No. 4, Winter 1991 (Table 2).

8.48 The neuropsychological consequences of long-term alcoholism are well known and include changes in memory, learning, and problem-solving ability. How do these memory impairments affect the treatment of hospitalized alcoholics? In a recent study of the memory performance of alcoholics during the early stage of recovery, 17 inpatient alcoholics who had been abstinent for less than 1 week and 10 nonalcoholic patients from an orthopedic surgery ward ("controls") were all shown a film on the disease of alcoholism. Following the film, subjects were given a test of recognition memory (i.e., multiple choice) in which they were asked to identify 25 items that had been presented in the film. The scores on the multiple-choice exam (percentage correct) are summarized in the table.

	INPATIENT ALCOHOLICS	CONTROL (ORTHOPEDIC SURGERY) PATIENTS
n	17	10
$\bar{x}$	42.2	62.1
s	17.4	12.6

Source: Becker, J. T. and Jaffe, J. H. "Impaired memory for treatment-relevant information in inpatient men alcoholics." Reprinted by permission from *Journal of Studies on Alcohol*, Vol. 45, No. 4, 1984, pp. 339–343. Copyright by Alcohol Research Documentation, Inc., Rutgers Center of Alcohol Studies, New Brunswick, NJ 08903.

a. Estimate the difference between the mean memory recognition scores of the two groups of patients with a 90% confidence interval.

b. Interpret the interval obtained in part **a**.

c. What assumptions are necessary for the validity of the interval estimation procedure used in part **a**?

8.49 The label "Machiavellian" was derived from the sixteenth century Florentine writer Niccolo Machiavelli, who wrote on ways of manipulating others to accomplish one's objective. Critics often accuse marketers of being manipulative and unethical, or "Machiavellian" in nature. S. D. Hunt and L. B. Chanko explored the question of whether "marketers are more Machiavellian than others" (*Journal of Marketing*, Summer 1984). The Machiavellian scores (measured by the Mach IV scale) for a sample of marketing professionals were recorded and compared to the Machiavellian scores for other groups of people, including a sample of college students in an earlier study. The results are summarized in the accompanying table. (Higher scores are associated with Machiavellian attitudes.)

	MARKETING PROFESSIONALS	COLLEGE STUDENTS
Sample size	1,076	1,782
Mean score	85.7	90.7
Standard deviation	13.2	14.3

a. Construct a 99% confidence interval for the mean difference in Machiavellian scores between marketing professionals and college students.

b. Interpret the interval constructed in part **a**.

8.50 Many business decisions are made because of offered incentives that are intended to make the decision maker "feel good." How does such a positive affect influence the risk preference of decision makers? This question was the subject of a study conducted at Ohio State University. Before the experiment began, subjects were divided randomly into two groups: the "positive affect" group and the control group. The positive affect group were given a bag of candies as a token of appreciation for participating in the study, whereas the control group subjects were not given the gift. All subjects were then given 10 gambling chips (worth credit for participation) as a reward for participating in the study, and presented with a choice of (1) betting five chips on any one of the bets available in roulette, or (2) not betting. After a short explanation of the probabilities associated with the different roulette bets, the subjects were instructed to indicate on a scale of .00 to 1.00, marked at intervals of .10, the riskiest bet they were willing to make (i.e., what the probability of winning would have to be for them to bet). A summary of the results (winning probabilities) follows.

	POSITIVE AFFECT GROUP	CONTROL GROUP
Number of subjects	11	13
Mean probability of winning	.65	.52
Standard deviation	.18	.15

Source: Isen, A. M. and Geva, N. "The influence of positive affect on acceptable level of risk: The person with the large canoe has a large worry." *Organizational Behavior and Human Decision Processes*, Vol. 39, 1987, p. 149.

a. Construct a 95% confidence interval for the difference between the mean probabilities of winning indicated by the two groups of subjects.

b. Interpret the interval obtained in part **a**. Is there evidence of a difference between the mean probabilities selected by subjects in the two groups?

c. What assumptions are required for the inference made in part **b** to be valid?

8.51 Laboratory tests were conducted to investigate the stability and permeability of open-graded asphalt concrete (*Journal of Testing & Evaluation*, July 1981). In one part of the experiment, four concrete specimens were prepared for asphalt contents of 3% and 7% by total weight of mix. The water permeability of each concrete specimen was determined by flowing de-aired water across the specimen and measuring the amount of water loss. The permeability measurements (recorded in inches per hour) for the eight concrete specimens are shown in the table, followed by a SAS printout giving descriptive statistics for the two samples.

ASPHALT CONTENT			
3%		7%	
1,189	1,020	853	733
840	980	900	785

Source: Woelfl, G., Wei, I., Faulstich, C., and Litwack, H. "Laboratory testing of asphalt concrete for porous pavements." *Journal of Testing and Evaluation*, Vol. 9, No. 4, July 1981, pp. 175–181. Copyright 1981, ASTM, 1916 Race Street, Philadelphia, PA 19103. Reprinted with permission.

```
N Obs   Variable   N        Mean        Variance      Std Dev
-------------------------------------------------------------
   4    ASPH3PCT   4      1007.25       20636.92     143.6555487
        ASPH7PCT   4    817.7500000      5420.92      73.6268746
-------------------------------------------------------------
```

a. Find a 90% confidence interval for the difference between the mean permeabilities of concrete made with asphalt contents of 3% and 7%.

b. List any assumptions required for the interval of part a to be valid.

8.52 The *Florida Scientist* (Summer/Autumn 1991) reported on a study of the feeding habits of sea urchins. A sample of 20 urchins were captured from Biscayne Bay (Miami), placed in marine aquaria, then starved for 48 hours. Each sea urchin was then fed a 5-cm blade of turtle grass. Ten of the urchins received only green blades, whereas the other half received only decayed blades. (Assume that the two samples of sea urchins—10 urchins per sample—were randomly and independently selected.) The ingestion time, measured from the time the blade first made contact with the urchin's teeth to the time the urchin had finished eating the blade, was recorded. A summary of the results is provided in the table.

	GREEN BLADES	DECAYED BLADES
Number of sea urchins	10	10
Mean ingestion time (hours)	3.35	2.36
Standard deviation (hours)	.79	.47

Source: Montague, J. R., et al. "Laboratory measurement of ingestion rate for the sea urchin *Lytechinus variegatus*." *Florida Scientist*, Vol. 54, Nos. 3/4, Summer/Autumn, 1991 (Table 1).

a. Construct a 90% confidence interval for the difference between the mean ingestion times of sea urchins feeding on green and decayed turtle grass.

b. According to the researchers, "the difference in rates at which the urchins ingested the blades suggest that green, unblemished turtle grass may not be a particularly palatable food compared with decayed turtle grass. If so, urchins in the field may find it more profitable to selectively graze on decayed portions of the leaves." Does the result, part a, support this conclusion?

8.53 As a college student, you may wonder how different variables affect your performance on a test. One such variable studied by *Educational and Psychological Measurement* (Summer 1987) was the format of answer sheets. One hundred college students were asked to complete the DAT-Clerical Speed and Accuracy Test on one of two differently formatted answer sheets, a new Westinghouse answer sheet and the test's published answer sheet. Half the students were randomly assigned to the Westinghouse answer sheet and half to the published sheet, then the test scores were recorded. A summary of the results follows:

	WESTINGHOUSE ANSWER SHEET	PUBLISHED ANSWER SHEET
Sample size	50	50
Mean	49.12	52.94
Standard deviation	9.26	11.29

Source: Hodgkinson, G. P. "The effect of variations in answer sheet format on performance on the DAT clerical speed and accuracy test." *Educational and Psychological Measurements*, Vol. 47, No. 2, Summer 1987, pp. 473–475.

a. Form a 95% confidence interval for the difference between the mean test scores associated with the two different answer sheets.
b. The article concludes that the "Westinghouse answer sheet may be used routinely as an alternative to the published answer sheet." Do you agree? Explain.

8.54 Epidemiologists have theorized that the risk of coronary heart disease can be reduced by an increased consumption of fish. One study, begun in 1960, monitored the diet and health of a random sample of middle-age Dutchmen. The men were divided into groups according to the number of grams of fish consumed per day. Twenty years later, the level of dietary cholesterol (one of the risk factors for coronary disease) present in each was recorded. The results for two groups of subjects, the "no fish consumption" group (0 grams per day) and the "high fish consumption" group (greater than 45 grams per day), are summarized in the table. (Dietary cholesterol is measured in milligrams per 1,000 calories.)

	NO FISH CONSUMPTION (0 GRAMS/DAY)	HIGH FISH CONSUMPTION (> 45 GRAMS/DAY)
Sample size	159	79
Mean	146	158
Standard deviation	66	75

Source: Kromhout, D., Bosschieter, E. B., and Coulander, C. L. "The inverse relationship between fish consumption and 20-year mortality from coronary heart disease." *New England Journal of Medicine*, May 9, 1985, Vol. 312, No. 19, pp. 1205–1209. Reprinted by permission.

a. Calculate a 99% confidence interval for the difference between the mean levels of dietary cholesterol present in the two groups.
b. Based on the interval constructed in part a, what can you infer about the true difference? Explain.

ESTIMATION OF THE DIFFERENCE BETWEEN TWO POPULATION MEANS: MATCHED PAIRS

8.6 The procedures for estimating the difference between two population means presented in Section 8.5 were based on the assumption that the samples were randomly and independently selected from the target populations. Sometimes we can obtain more information about the difference between population means, $(\mu_1 - \mu_2)$, by selecting **paired observations**.

EXAMPLE 8.21 An elementary education teacher wants to compare two methods for teaching reading skills to first graders. One way to design the experiment is to randomly select 20 students from all available first graders, and then randomly assign 10 students to method 1 and 10 students to method 2 (see Figure 8.17). The reading achievement test scores obtained after completion of the experiment would represent independent random samples of scores attained by students taught reading skills by the two different methods. The confidence interval procedure described in Section 8.5 could be used to estimate $(\mu_1 - \mu_2)$, the difference between the mean achievement test scores of the two methods.

	Method 1	Method 2
Student Identification Number	1	2
	5	3
	6	4
	8	7
	11	9
	14	10
	15	12
	17	13
	19	16
	20	18

FIGURE 8.17
Independent Random Samples of First Graders, Example 8.21

a. Comment on the potential drawbacks of using independent random samples to estimate $(\mu_1 - \mu_2)$.
b. Propose a better method of sampling, one that will yield more information on the parameter of interest.

Solution

a. Assume that method 1 is truly more effective than method 2 in teaching reading skills to first graders. A potential drawback to the independent sampling plan shown in Figure 8.17 is that the differences in the reading skills of first graders because of IQ, learning ability, socioeconomic status, and other factors are not taken into account. For example, by chance the sampling plan may assign the 10 "worst" students to method 1 and the 10 "best" students to method 2. This unbalanced assignment may mask the fact that method 1 is more effective than method 2, i.e., the resulting confidence interval on $(\mu_1 - \mu_2)$ may fail to show that μ_1 exceeds μ_2.
b. A better method of sampling is one that attempts to remove the variation in achievement test scores that result from extraneous factors such as IQ, learning ability, and socioeconomic status. One way to do this is to match the first graders in pairs, where the students in each pair have similar IQ, socioeconomic status, etc. From each pair, one member would be randomly selected to be taught by method 1; the other member would be assigned to the class taught by method 2 (see Figure 8.18). The differences between the **matched pairs** of achievement test scores should provide a clearer picture of the true difference in achievement for the two reading methods because the matching would cancel the effects of the extraneous factors that formed the basis of the matching.

First-Grader Pair	Assignment	
	Method 1	Method 2
1	A	B
2	B	A
$\vdots$	$\vdots$	$\vdots$
10	A	B

The sampling plan shown in Figure 8.18 is commonly known as a **matched-pairs experiment**. In the box we give the procedure for estimating the difference between two population means based on matched-pairs data. You can see that once the differences in the paired observations are obtained, the analysis proceeds as a one-sample problem. That is, a confidence interval on a single mean (the mean of the difference, μ_d) is computed.

**$(1 - \alpha)100\%$ CONFIDENCE INTERVAL FOR
$\mu_d = (\mu_1 - \mu_2)$ MATCHED PAIRS**

Let $d_1, d_2, \ldots, d_n$ represent the differences between the pairwise observations in a *random sample* of n matched pairs, $\bar{d}$ = mean of the n sample differences, and s_d = standard deviation of the n sample differences.

LARGE SAMPLE

$$\bar{d} \pm z_{\alpha/2}\left(\frac{\sigma_d}{\sqrt{n}}\right)$$

where σ_d is the population standard deviation of differences.

Assumption: $n \geq 30$

[*Note:* When σ_d is unknown (as is usually the case), use s_d to approximate σ_d.]

SMALL SAMPLE

$$\bar{d} \pm t_{\alpha/2}\left(\frac{s_d}{\sqrt{n}}\right)$$

where $t_{\alpha/2}$ is based on $(n - 1)$ degrees of freedom.

Assumption: The population of paired differences is normally distributed.

EXAMPLE 8.22 In the comparison of two methods for teaching reading discussed in Example 8.21, suppose that the $n = 10$ pairs of achievement test scores were as shown in Table 8.8. Find a 95% confidence interval for the difference in mean achievement, $\mu_d = (\mu_1 - \mu_2)$.

Solution The differences between the $n = 10$ matched pairs of reading achievement test scores are computed as

$$d = (\text{Method 1 score}) - (\text{Method 2 score})$$

and are shown in the third row of Table 8.8.

TABLE 8.8
Reading Achievement Test Scores for Example 8.22

	STUDENT PAIR									
	1	2	3	4	5	6	7	8	9	10
Method 1 score	78	63	72	89	91	49	68	76	85	55
Method 2 score	71	44	61	84	74	51	55	60	77	39
Pair difference	7	19	11	5	17	−2	13	16	8	16

To proceed with this small-sample estimation, we must assume that these differences are from an approximately normal population. The mean and standard deviation of these sample differences are shown (shaded) on the SPSS printout, Figure 8.19. From the printout, $\bar{d} = 11.0$ and $s_d = 6.53$.

FIGURE 8.19
SPSS Printout for Example 8.22

```
-----------------------------------------------------------------
Number of Valid Observations (Listwise) =        10.00

Variable      Mean     Std Dev    Minimum   Maximum    N   Label

SCORDIFF     11.00      6.53       -2.00     19.00     10
-----------------------------------------------------------------
```

The value of $t_{.025}$, based on $(n - 1) = (10 - 1) = 9$ degrees of freedom is given in Table 5 of Appendix G as $t_{.025} = 2.262$. Substituting these values into the formula for the confidence interval, we obtain

$$\bar{d} \pm t_{.025}\left(\frac{s_d}{\sqrt{n}}\right) = 11.0 \pm 2.262\left(\frac{6.53}{\sqrt{10}}\right)$$

$$= 11.0 \pm 4.7$$

or (6.3, 15.7).

We estimate, with 95% confidence, that the difference between mean reading achievement test scores for methods 1 and 2 falls within the interval 6.3 to 15.7. Since all the values within the interval are positive, method 1 seems to produce a mean achievement test score that is statistically higher than the mean score for method 2.

Often, the pairs of observations in a matched-pairs experiment arise naturally by recording two measurements on the same experimental unit at two different points in time. For example, in Case Study 1.2, we discussed a study that investigated the effectiveness of a prescription drug in improving the SAT scores of nervous test takers. The experimental units (the objects upon which the measurements are taken) were 22 high school juniors who took the SAT twice, once without and once with the drug. Thus, two measurements were taken on each junior: (1) SAT score with no drug and (2) SAT score with the drug. These two observations, taken for all students in the study, formed the "matched pairs" of the experiment.

In an analysis of matched-pair observations, it is important to stress that the pairing of the experimental units must be performed *before* the data are collected. Recall that the objective is to compare two methods of "treating" the experimental units. By using matched pairs of units that have similar characteristics, we are able to cancel out the effects of the variables used to match the pairs.

EXERCISES
LEARNING THE MECHANICS

8.55 The data for a random sample of four paired observations are shown in the table.

PAIR	OBSERVATION FROM POPULATION A	OBSERVATION FROM POPULATION B
1	2	0
2	5	7
3	10	6
4	8	5

a. Calculate the difference within each pair, subtracting observation B from observation A. Use the differences to calculate $\bar{d}$ and s_d.
b. If μ_1 and μ_2 are the means of populations A and B, respectively, express μ_d in terms of μ_1 and μ_2.
c. Construct a 95% confidence interval for μ_d.

8.56 A random sample of 10 paired observations yielded the following summary information:

$$\bar{d} = 2.3 \qquad s_d = 2.67$$

a. Find a 90% confidence interval for μ_d.
b. Find a 95% confidence interval for μ_d.
c. Find a 99% confidence interval for μ_d.

8.57 A random sample of $n = 50$ paired observations yielded the following summary statistics: $\bar{d} = 19.3$, $s_d = 5.2$. Construct a 95% confidence interval for μ_d.

8.58 The data for a random sample of seven paired observations are shown in the accompanying table. Find a 90% confidence interval for $\mu_d = (\mu_A - \mu_B)$.

PAIR	OBSERVATION FROM POPULATION A	OBSERVATION FROM POPULATION B
1	48	54
2	50	56
3	47	50
4	50	55
5	63	64
6	65	65
7	55	61

8.59 List the assumptions required to construct a valid confidence interval for μ_d based on matched-pairs data.

APPLYING THE CONCEPTS

APPLYING THE CONCEPTS

8.60 A recent supermarket advertisement states: "Winn-Dixie offers you the lowest total food bill! Here's the proof!" The "proof" (shown here) is a side-by-side listing of the prices of 60 grocery items purchased at Winn-Dixie and at Publix on the same day.

ITEM	WINN-DIXIE	PUBLIX	ITEM	WINN-DIXIE	PUBLIX
Big Thirst Towel	1.21	1.49	Keb Graham Crust	.79	1.29
Camp Crm/Broccoli	.55	.67	Spiffits Glass	1.98	2.19
Royal Oak Charcoal	2.99	3.59	Prog Lentil Soup	.79	1.13
Combo Chdr/Chz Snk	1.29	1.29	Lipton Tea Bags	2.07	2.17
Sure Sak Trash Bag	1.29	1.79	Carnation Hot Coco	1.59	1.89
Dow Handi Wrap	1.59	2.39	Crystal Hot Sauce	.70	.87
White Rain Shampoo	.96	.97	C/F/N Coffee Bag	1.17	1.15
Post Golden Crisp	2.78	2.99	Soup Start Bf Veg	1.39	2.03
Surf Detergent	2.29	1.89	Camp Pork & Beans	.44	.49
Sacramento T/Juice	.79	.89	Sunsweet Pit Prune	.98	1.33
SS Prune Juice	1.36	1.61	DM Vgcls Grdn Duet	1.07	1.13
V-8 Cocktail	1.18	1.29	Argo Corn Starch	.69	.89
Rodd Kosher Dill	1.39	1.79	Sno Drop Bowl Clnr	.53	1.15
Bisquick	2.09	2.19	Cadbury Milk Choc	.79	1.29
Kraft Italian Drs	.99	1.19	Andes Crm/De Ment	1.09	1.30
BC Hamburger Helper	1.46	1.75	Combat Ant & Roach	2.33	2.39
Comstock Chrry Pie	1.29	1.69	Joan/Arc Kid Bean	.45	.56
Dawn Liquid King	2.59	2.29	La Vic Salsa Pican	1.22	1.75
DelMonte Ketchup	1.05	1.25	Moist N Beef/Chz	2.39	3.19
Silver Floss Kraut	.77	.81	Ortega Taco Shells	1.08	1.33
Trop Twist Beverag	1.74	2.15	Fresh Step Cat Lit	3.58	3.79
Purina Kitten Chow	1.09	1.05	Field Trial Dg/Fd	3.49	3.79
Niag Spray Starch	.89	.99	Tylenol Tablets	5.98	5.29
Soft Soap Country	.97	1.19	Rolaids Tablets	1.88	2.20
Northwood Syrup	1.13	1.37	Plax Rinse	2.88	3.14
Bumble Bee Tuna	.58	.65	Correctol Laxative	3.44	3.98
Mueller Elbow/Mac	2.09	2.69	Tch Scnt Potpourri	1.50	1.89
Kell Nut Honey Crn	2.95	3.25	Chld Enema 2.250	.98	1.15
Cutter Spray	3.09	3.95	Gillette Atra Plus	5.00	5.24
Lawry Season Salt	2.28	2.97	Colgate Shave	.94	1.10

Source: Advertisement in *Tampa Tribune*, June 2, 1991.

a. Explain why the data should be analyzed as matched pairs.

b. A Minitab printout showing a 95% confidence interval for $(\mu_{\text{Winn}} - \mu_{\text{Publix}})$, the difference between the mean prices of grocery items purchased at the two supermarkets, is shown here. Interpret the result.

```
           N     MEAN    STDEV   SE MEAN    95.0 PERCENT C.I.
Diff      60   -0.2540   0.2741   0.0354   ( -0.3248, -0.1832)
```

8.61 Research reported in the *American Journal of Sociology* indicates that publicized stories of suicide may trigger an increase in the number of suicides and that some of these suicides may be disguised as motor vehicle fatalities a short time after the publicized suicide stories appear. The researchers compiled a list of all suicide stories on the front pages of both the *Detroit News* and the *Detroit Free Press* from 1973 to 1976. The number of motor vehicle fatalities in Detroit 4 days before and 3 days after the publicized suicide story were also recorded, as shown in the accompanying table.

PUBLICIZED SUICIDE	MOTOR VEHICLE FATALITIES	
	4 Days Before	3 Days After
Murder suspect (Mar. 8, 1973)	7	8
Ex-POW (June 3, 1973)	6	8
President of Chile (Sept. 11, 1973)	10	14
Detroit woman (Dec. 14, 1973)	2	6
Newscaster (July 15, 1974)	3	2
Confessed murderer (Sept. 29, 1974)	4	4
Spy (Feb. 29, 1976)	2	4
Murder suspect (Apr. 6, 1976)	1	1
Deputy police chief (Sept. 29, 1976)	8	11

Source: Bollen, K. A. and Phillips, D. P. "Suicide motor vehicle fatalities in Detroit: A replication." *American Journal of Sociology*, Vol. 87, No. 2, Sept. 1981, pp. 404–412. Reprinted by permission of the University of Chicago Press. © 1981 The University of Chicago.

a. Find a 95% confidence interval of the difference in the mean number of Detroit motor vehicle fatalities 4 days before and 3 days after a publicized suicide.

b. Does it appear that Detroit motor vehicle fatalities increase significantly 3 days after a suicide is published?

8.62 In the past, many bodily functions were thought to be beyond conscious control. However, recent experimentation suggests that it may be possible for a person to control certain body functions if he or she is trained in a program of *biofeedback* exercises. An experiment is conducted to determine whether blood pressure levels can be consciously reduced in people trained in this program. The blood pressure measurements (in millimeters of mercury) listed in the table represent readings before and after the biofeedback training of six subjects.

SUBJECT	BEFORE	AFTER
1	136.9	130.2
2	201.4	180.7
3	166.8	149.6
4	150.0	153.2
5	173.2	162.6
6	169.3	160.1

a. Construct a 90% confidence interval for the difference in mean blood pressure measurements before and after the biofeedback training.

b. Interpret the interval obtained in part a.

8.63 Medical researchers believe that exposure to dust from cotton bract induces respiratory disease in susceptible field workers. An experiment was conducted to determine the effect of air-dried green cotton bract extract (GBE) on the cells of non–dust-exposed mill workers (*Environmental Research*, Feb. 1986). Blood samples taken on eight workers were incubated with varying concentrations of GBE. After a short period of time, the cyclic AMP level (a measure of cell activity expressed in picomoles per million cells) of each blood sample was measured. The data for two GBE concentrations, 0 mg/ml (salt buffer, control solution) and .2 mg/ml, are reproduced in the next table. [Note that one blood sample was taken from each worker, with one aliquot exposed to the salt buffer solution and the other of the GBE.]

a. Find a 95% confidence interval for the mean difference between the cyclic AMP levels of blood samples exposed to the two concentrations of GBE.

b. Based on the interval obtained in part a, is there evidence that exposure to GBE blocks cell activity?

WORKER	GBE CONCENTRATION (mg/ml)	
	0	.2
A	8.8	4.4
B	13.0	5.7
C	9.2	4.4
D	6.5	4.1
F	9.1	4.4
H	17.0	7.9

Source: Butcher, B. T., Reed, M. A., and O'Neil, C. E. "Biochemical and immunologic characterization of cotton bract extract and its effect on *in vitro* cyclic AMP production." *Environmental Research*, Vol. 39, No. 1, Feb. 1986, p. 119.

8.64 Refer to the *Chemosphere* study of Vietnam veterans exposure to Agent Orange, Exercise 8.25. In addition to the amount of TCDD (measured in parts per million) in blood plasma, the TCDD in fat tissue drawn from 20 exposed Vietnam veterans was recorded. The data are shown in the table. Construct a confidence interval that will allow you to compare the mean TCDD level in plasma to the mean TCDD level in fat tissue for Vietnam veterans exposed to Agent Orange. Interpret the result.

VETERAN	TCDD LEVELS IN PLASMA	TCDD LEVELS IN FAT TISSUE	VETERAN	TCDD LEVELS IN PLASMA	TCDD LEVELS IN FAT TISSUE
1	2.5	4.9	11	6.9	7.0
2	3.1	5.9	12	3.3	2.9
3	2.1	4.4	13	4.6	4.6
4	3.5	6.9	14	1.6	1.4
5	3.1	7.0	15	7.2	7.7
6	1.8	4.2	16	1.8	1.1
7	6.0	10.0	17	20.0	11.0
8	3.0	5.5	18	2.0	2.5
9	36.0	41.0	19	2.5	2.3
10	4.7	4.4	20	4.1	2.5

Source: Schecter, A. et al. "Partitioning of 2,3,7,8-chlorinated dibenzo-*p*-dioxins and dibenzofurans between adipose tissue and plasma lipid of 20 Massachusetts Vietnam veterans." *Chemosphere*, Vol. 20, Nos. 7–9, 1990, pp. 954–955 (Tables I & II).

8.65 One of the keys to occupational therapy is patient motivation. A study was conducted to determine whether *purposeful activity* (defined as tasks that are goal-directed) provides intrinsic motivation to exercise performance (*American Journal of Occupational Therapy*, Mar. 1984). Twenty-six females were recruited to take part in the study. Each female subject was instructed to perform two similar exercises, jumping rope (the purposeful activity) and jumping without a rope (the nonpurposeful activity), until their perceived exertion level reached 17 on the RPE scale (i.e., until they felt they had worked their bodies "very hard"). The increase in heart rate was then recorded for each of the two exercises and the differences

$$d_i = (\text{Heart rate increase with rope}) - (\text{Heart rate increase without rope})$$

were calculated. The mean of the differences for the 26 females was 11.15 beats per minute and the standard deviation of the differences was 16.84 beats per minute. The data are summarized in the table.

a. Since each female subject performed both the purposeful and nonpurposeful activities, the data were collected as matched pairs. Give one advantage of using a matched-pairs experiment instead of one based on independent random samples.

	HEART RATE INCREASES	
	With rope	Without rope
Sample size	26	26
Mean (beats/minute)	103.19	92.04
Standard deviation (beats/minute)	24.33	27.83

Source: Kircher, M. A. "Motivation as a factor of perceived exertion in purposeful versus nonpurposeful activity." *American Journal of Occupational Therapy,* Mar. 1984, Vol. 38, No. 3, pp. 165–170. Copyright 1984 by the American Occupational Therapy Association, Inc. Reprinted with permission.

b. Calculate a 95% confidence interval for the difference between mean heart rate increases of the purposeful (jumping with rope) and nonpurposeful (jumping without rope) activities. Interpret your result.

c. Use the given information and the formula for independent samples to construct a 95% confidence interval for the mean difference.

d. Compare the intervals obtained in parts b and c. Note that the interval from part c is wider than the interval from part b. How do you explain this result?

8.66 One desirable characteristic of water pipes is that the quality of the water they deliver is equal to or near the quality of the water entering the system at the water treatment plant. A type of ductile iron pipe has provided an excellent water delivery system for the St. Louis County Water Company. The chlorine level of water emerging from the south water treatment plant and at the fire station (Fenton Zone 13) was measured over a 12-month period, with the results shown in the accompanying table. Estimate the mean difference in monthly chlorine content between the two locations using a 90% confidence interval.

		MONTH											
		Jan.	Feb.	Mar.	Apr.	May	June	July	Aug.	Sept.	Oct.	Nov.	Dec.
LOCATION	South Plant	2.0	2.0	2.1	1.9	1.7	1.8	1.7	1.9	2.0	2.0	2.1	2.2
	Fire Station	2.2	2.2	2.1	2.0	1.9	1.9	1.8	1.7	1.9	1.9	1.8	2.0

Source: "St. Louis County standardizes pipes and procedures for reliability." Staff Report, Water and Sewage Works, Dec. 1980.

ESTIMATION OF THE DIFFERENCE BETWEEN TWO POPULATION PROPORTIONS

8.7 This section extends the method of Section 8.4 to the case in which we want to estimate the difference between two population proportions. For example, one may be interested in comparing the proportions of married and unmarried persons who are overweight, or the proportions of homes in two states that are heated by natural gas, or the proportions of adults today and 20 years ago who smoke, etc.

EXAMPLE 8.23 Over the years, one of the issues studied in depth by organizational behavior theorists involves the area of ethical management decision making. Prior to 1980, these studies focused on the male manager because the management field was male-dominated. Today, with women entering management careers in record numbers, researchers are studying the differences in ethical perceptions between male and female managers. In one study, 48 of 50 female managers responded that concealing one's on-the-job errors was very unethical, whereas only 30 of 50 male managers responded in like manner (*Journal of Business Ethics*, Aug. 1987). Construct a point estimate for the difference between the proportions of female and male managers who believe that concealing one's errors is very unethical.

Solution   For this example, define

π_1 = Population proportion of female managers who believe that concealing on-the-job errors is unethical

π_2 = Population proportion of male managers who believe that concealing on-the-job errors is unethical

x_1 = Number of females in the sample who believe that concealing on-the-job errors is unethical

x_2 = Number of males in the sample who believe that concealing on-the-job errors is unethical

As a point estimate of $(\pi_1 - \pi_2)$, we will use the difference between the corresponding sample proportions $(p_1 - p_2)$, where

$$p_1 = \frac{x_1}{n_1} = \frac{48}{50} = .96$$

and

$$p_2 = \frac{x_2}{n_2} = \frac{30}{50} = .60$$

Thus, the point estimate of $(\pi_1 - \pi_2)$ is

$$(p_1 - p_2) = .96 - .60 = .36$$

To judge the reliability of the point estimate $(p_1 - p_2)$, we need to know the characteristics of its performance in repeated independent sampling from two binomial populations. This information is provided by the sampling distribution of $(p_1 - p_2)$, illustrated in Figure 8.20 and described in the accompanying box.

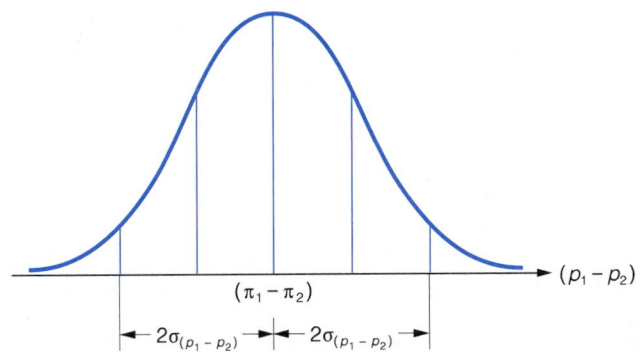

FIGURE 8.20
Sampling Distribution of $(p_1 - p_2)$

SAMPLING DISTRIBUTION OF $(p_1 - p_2)$

For sufficiently large sample sizes, n_1 and n_2, the sampling distribution of $(p_1 - p_2)$, based on independent random samples from two binomial populations, is approximately normal with

Mean: $\quad \mu_{(p_1-p_2)} = (\pi_1 - \pi_2)$

and

Standard deviation: $\quad \sigma_{(p_1-p_2)} = \sqrt{\dfrac{\pi_1(1 - \pi_1)}{n_1} + \dfrac{\pi_2(1 - \pi_2)}{n_2}}$

It follows that a large-sample confidence interval for $(\pi_1 - \pi_2)$ may be obtained as shown in the next box.

LARGE-SAMPLE $(1 - \alpha)$ 100% CONFIDENCE INTERVAL FOR $(\pi_1 - \pi_2)$

$$(p_1 - p_2) \pm z_{\alpha/2}\, \sigma_{(p_1-p_2)} \approx (p_1 - p_2) \pm z_{\alpha/2} \sqrt{\dfrac{p_1 q_1}{n_1} + \dfrac{p_2 q_2}{n_2}}$$

where p_1 and p_2 are the sample proportions of observations with the characteristic of interest, $q_1 = 1 - p_1$, and $q_2 = 1 - p_2$.

[*Note:* We have followed the usual procedure of substituting the sample values p_1, q_1, p_2, and q_2 for the corresponding population values required for $\sigma_{(p_1-p_2)}$].

Assumptions: 1. Independent random samples are collected.
2. The samples are sufficiently large so that the approximation is valid. As a general rule of thumb we will require that the intervals

$$p_1 \pm 2\sqrt{\dfrac{p_1 q_1}{n_1}} \quad \text{and} \quad p_2 \pm 2\sqrt{\dfrac{p_2 q_2}{n_2}}$$

do not contain 0 or 1.

EXAMPLE 8.24   Refer to Example 8.23. Estimate the difference between the proportions of female and male managers who believe that concealing one's errors is very unethical, using a 95% confidence interval. Interpret your results.

Solution   For a confidence coefficient of .95, we will use $z_{.025} = 1.96$ in constructing the confidence interval. From Example 8.23, we have $n_1 = 50$, $n_2 =$

50, $p_1 = .96$ and $p_2 = .60$. Thus, $q_1 = 1 - .96 = .04$, $q_2 = 1 - .60 = .40$, and the 95% confidence interval for $(\pi_1 - \pi_2)$ is given by

$$(p_1 - p_2) \pm z_{.025} \sqrt{\frac{p_1 q_1}{n_1} + \frac{p_2 q_2}{n_2}}$$

$$= (.96 - .60) \pm 1.96 \sqrt{\frac{(.96)(.04)}{50} + \frac{(.60)(.40)}{50}}$$

$$= .36 + .146$$

or (.214, .506). Thus, we estimate that the interval (.214, .506) encloses the difference $(\pi_1 - \pi_2)$ with 95% confidence. It appears that there are between 21% and 51% more female managers than male managers who believe that concealing on-the-job errors is very unethical.

Small-sample estimation procedures for $(\pi_1 - \pi_2)$ will not be discussed here for the reasons outlined at the end of Section 8.4.

EXERCISES

LEARNING THE MECHANICS

8.67 Suppose you want to estimate $\pi_1 - \pi_2$, the difference between the proportions of two populations. Based on independent random samples, you can compute the sample proportions of successes, p_1 and p_2, respectively. Estimate $\mu_{(p_1 - p_2)}$ and $\sigma_{(p_1 - p_2)}$ for each of the following results:

a. $n_1 = 150$, $p_1 = .3$; $n_2 = 130$, $p_2 = .4$
b. $n_1 = 100$, $p_1 = .10$; $n_2 = 100$, $p_2 = .05$
c. $n_1 = 200$, $p_1 = .76$; $n_2 = 200$, $p_2 = .96$

8.68 Independent random samples are taken from two populations. The accompanying table shows the sample sizes and the sample proportions of observations with the characteristic of interest.

SAMPLE FROM POPULATION 1	SAMPLE FROM POPULATION 2
$n_1 = 400$	$n_2 = 350$
$p_1 = .50$	$p_2 = .60$

a. Find a 90% confidence interval for $(\pi_1 - \pi_2)$.
b. Find a 95% confidence interval for $(\pi_1 - \pi_2)$.
c. Find a 99% confidence interval for $(\pi_1 - \pi_2)$.

8.69 Independent random samples from populations 1 and 2 produced sample proportions $p_1 = .44$ and $p_2 = .52$.

a. Find a 95% confidence interval for $(\pi_1 - \pi_2)$ if $n_1 = n_2 = 500$.
b. Find a 99% confidence interval for $(\pi_1 - \pi_2)$ if $n_1 = n_2 = 500$.

8.70 Independent random samples of 250 observations each are selected from two populations. The samples from populations 1 and 2 produced, respectively, 100 and 75 observations possessing the characteristic of interest.

a. Construct a 90% confidence interval for $(\pi_1 - \pi_2)$.
b. Construct a 99% confidence interval for $(\pi_1 - \pi_2)$.

8.71 According to an American Heart Association (AHA) researcher, "people who are hostile and mistrustful are more likely to die young or develop life-threatening heart disease than those with more 'trusting hearts'" (*Tampa Tribune*, Jan. 17, 1989). A sample of 118 male doctors, lawyers, and workers in a large industrial firm in Chicago were divided into two groups based on a standard psychological test designed to measure hostility. Of the 35 men who scored high in hostility, 7 died at a relatively early age (i.e., between the ages of 25 and 50). In contrast, only 4 of the 83 men whose hostility rating was low died at an early age.

  a. Estimate the true difference between the proportion of men with high hostility scores who die at an early age and the corresponding proportion of men with low hostility scores. Use a 95% confidence interval.
  b. Are the sample sizes large enough for the interval, part a, to be valid?
  c. Interpret the interval, part a. Do you agree with the AHA researcher?

8.72 The *Journal of Fish Biology* (Aug. 1990) reported on a study to compare the prevalence of parasites (tapeworms) found in species of Mediterranean and Atlantic fish. In the Mediterranean Sea, 588 brill were captured and dissected; 211 were found to be infected by the parasite. In the Atlantic Ocean, 123 brill were captured and dissected; 26 were found to be infected. Compare the proportions of infected brill at the two capture sites using a 90% confidence interval. Interpret the interval.

8.73 A research team at the University of South Florida conducted a study of former cigarette smokers. One phase of the study compared the percentage of black former smokers who quit smoking on their own to the corresponding percentage of white former smokers (*USF Magazine*, Summer 1992). Of the 153 black former smokers sampled, 150 quit on their own; of the 381 white former smokers sampled, 324 quit without assistance.

  a. Compare the percentages of the two groups of former smokers with a 90% confidence interval.
  b. Based on the interval, part a, can you infer that one group of former smokers had more difficulty quitting on their own than the other group? Explain.

8.74 Americans shop for food, clothing, housewares, furniture, and other necessities and luxuries week after week. But is shopping considered a pleasant or unpleasant experience? According to R. H. Bruskin's "Update on America" study, about one in three individuals feels shopping is an unpleasant experience (*Journal of Marketing Research*, Feb./Mar. 1984). A national sample of 2,025 male and female adults were surveyed to determine each respondent's opinion on the pleasantness of shopping. The survey produced the results shown in the table.

	MALES	FEMALES
Sample size	1,012	1,013
Number who feel shopping is an unpleasant experience	425	283

Reprinted from the *Journal of Marketing Research*, published by the American Marketing Association.

  a. Compute the proportion of males in the sample who feel that shopping is an unpleasant experience.
  b. Compute the proportion of females in the sample who feel that shopping is an unpleasant experience.
  c. Construct a 98% confidence interval for the difference between the true proportions of males and females who feel shopping is an unpleasant experience.
  d. Which group, males or females, appears to dislike shopping more?

8.75 A study was conducted to compare the attitudes of American and Russian teenagers on nuclear war (*New England Journal of Medicine*, Aug. 18, 1988). A team of American and Russian researchers surveyed 3,370 public school students in Maryland and 2,148 students in central Russia. One question asked whether the students believe a nuclear war will occur in their lifetime. Forty-two percent of the Maryland students and 9% of the Russian students responded affirmatively.

a. Calculate a 99% confidence interval for the difference between the proportions of Maryland and Russian students who believe that a nuclear war will occur in their lifetime. Interpret the interval.

b. How could the width of the interval of part **a** be reduced?

c. Although Maryland students were recruited randomly for the study, there is speculation that the Russian students were selected much more carefully. How could the nonrandom Russian sample bias the results obtained in part **a**?

8.76   The incidence of acquired immune deficiency syndrome (AIDS) has been increasing dramatically. Proven to be of viral origin, AIDS is characterized by the development of Kaposi's sarcoma (a form of cancer), pneumonia and other microorganism infections. In the early 1980s, medical research found that AIDS patients also have increased rates of exposure to human T-cell leukemia virus (HTLV). In one study reported in *Science* (May 20, 1983), serum samples from 75 AIDS patients and 336 control subjects were submitted to a well-known immunofluorescence technique and were judged to be positive if at least 50% of the cells were infected with the leukemia virus. Results of the study revealed that 19 of the AIDS patients and 2 of the control subjects were exposed to HTLV.

a. Construct a 90% confidence interval for the difference between the population rates of exposure to HTLV for AIDS patients and control subjects.

b. Interpret the interval obtained in part **a**.

c. Are the sample sizes sufficiently large for the interval to be valid?

8.8   CHOOSING THE SAMPLE SIZE

In the preceding sections we have overlooked a problem that usually must be faced in the initial stages of an experiment. Before constructing a confidence interval for a parameter of interest, we will have to decide on the number n of observations to be included in a sample. Should we sample $n = 10$ observations, $n = 20$, or $n = 100$? To answer this question we need to decide how wide a confidence interval we are willing to tolerate and the measure of confidence—that is, the confidence coefficient—that we wish to place in it. The following example will illustrate the method for determining the appropriate sample size for estimating a population mean.

EXAMPLE 8.25   A mail-order house wants to estimate the mean length of time between shipment of an order and receipt by the customer. The management plans to randomly sample n orders and determine, by telephone, the number of days between shipment and receipt for each order. If the management wants to estimate the mean shipping time correct to within .5 day with probability equal to .95, how many orders should be sampled?

Solution   We will use $\bar{x}$, the sample mean of the n measurements, to estimate μ, the mean shipping time. Its sampling distribution will be approximately normal and the probability that $\bar{x}$ will lie within

$$1.96\sigma_{\bar{x}} = 1.96\left(\frac{\sigma}{\sqrt{n}}\right)$$

of the mean shipping time, μ, is approximately .95 (see Figure 8.21). Therefore, we want to choose the sample size n so that $1.96\sigma/\sqrt{n}$ equals .5 day:

$$1.96\left(\frac{\sigma}{\sqrt{n}}\right) = .5$$

FIGURE 8.21
Sampling Distribution of the Sample Mean, $\bar{x}$

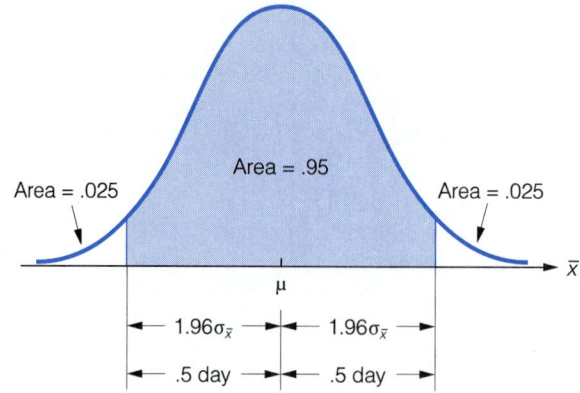

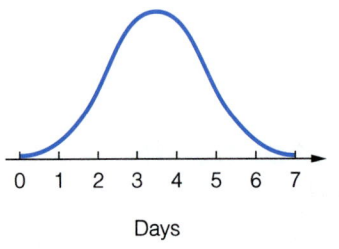

To solve the equation $1.96\sigma/\sqrt{n} = .5$, we need to know the value of σ, a measure of variation of the population of all shipping times. Since σ is unknown (as will usually be the case in practical applications), we must approximate its value using the standard deviation of some previous sample data or deduce an approximate value from other knowledge about the population. Suppose, for example, that we know almost all shipments will be delivered within 7 days. Then the population of shipping times might appear as shown in Figure 8.22.

Figure 8.22 provides the information we need to find an approximation for σ. Since the Empirical Rule tells us that almost all the observations in a data set will fall within the interval $\mu \pm 3\sigma$, it follows that the range of a population is approximately 6σ. If the range of the population of shipping times is 7 days, then

$$6\sigma = 7 \text{ days}$$

and σ is approximately equal to $\frac{7}{6}$ or 1.17 days.

The final step in determining the sample size is to substitute this approximate value of σ into the equation obtained previously and solve for n. Thus, we have

$$1.96\left(\frac{1.17}{\sqrt{n}}\right) = .5$$

or

$$\sqrt{n} = \frac{1.96(1.17)}{.5} = 4.59$$

Squaring both sides of this equation yields

$$n = 21.07$$

We will follow the usual convention of rounding the calculated sample size upward. Therefore, the mail-order house needs to sample approximately $n = 22$ shipping times to estimate the mean shipping time correct to within .5 day with probability equal to .95.

In Example 8.25, we wanted our sample estimate to lie within .5 day of the true mean shipping time, μ, with probability .95, where .95 represents the confidence coefficient. We could calculate the sample size for a confidence coefficient other than .95 by changing the z value in the equation. In general, if we want $\bar{x}$ to lie within a distance d of μ with probability $(1 - \alpha)$, we would solve for n in the equation

$$z_{\alpha/2}\left(\frac{\sigma}{\sqrt{n}}\right) = d$$

where the value of $z_{\alpha/2}$ is obtained from Table 4 of Appendix G. The solution is given by

$$n = \left(\frac{z_{\alpha/2}\sigma}{d}\right)^2$$

For example, for a confidence coefficient of .90, we would require a sample size of

$$n = \left(\frac{1.645\sigma}{d}\right)^2$$

CHOOSING THE SAMPLE SIZE FOR ESTIMATING A POPULATION MEAN μ TO WITHIN d UNITS WITH CONFIDENCE COEFFICIENT $(1 - \alpha)$

$$n = \left(\frac{z_{\alpha/2}\sigma}{d}\right)^2$$

[*Note:* The population standard deviation σ will usually have to be approximated.]

The procedures for determining the sample sizes needed to estimate a population proportion, the difference between two population means, or the difference between two population proportions are analogous to the procedure for determining the sample size for estimating a population mean. In the remainder of this section, we will present the appropriate formulas and illustrate their use with examples. We first consider the estimation of a population proportion.

CHOOSING THE SAMPLE SIZE FOR ESTIMATING A POPULATION PROPORTION π TO WITHIN d UNITS WITH CONFIDENCE COEFFICIENT $(1 - \alpha)$

$$n = \left(\frac{z_{\alpha/2}}{d}\right)^2 \pi(1 - \pi)$$

where π is the value of the population proportion that you are attempting to estimate.

[*Note:* This technique requires a previous estimate of π. If none is available, use $\pi = .5$ for a conservative choice of n.]

EXAMPLE 8.26   In Example 8.13, the United States Commission on Crime sampled recently committed crimes to estimate the proportion in which firearms were used. Suppose the Commission wants to acquire an estimate of π that is correct to within .02 with 90% confidence. How many cases would have to be included in the Commission's sample?

Solution   For a confidence coefficient of $(1 - \alpha) = .90$, we have $\alpha = .10$. Thus, to calculate n using the equation given in the box, we must find the value of $z_{\alpha/2} = z_{.05}$ and find an approximation for the unknown population proportion π.

From Table 4 of Appendix G, the z value corresponding to an area of $\alpha/2 = .05$ in the upper tail of the standard normal distribution is $z_{.05} = 1.645$. As an approximation to π, we will use the sample estimate, $p = .60$, obtained for the sample of 300 cases in Example 8.13.

Substituting the value of .6 for π and $z_{.05} = 1.645$ into the equation for n, we have

$$n = \left(\frac{z_{\alpha/2}}{d}\right)^2 \pi(1 - \pi) = \left(\frac{1.645}{.02}\right)^2 (.6)(.4) = 1,623.6$$

Therefore, to estimate π to within .02 with 90% confidence, the Commission will have to sample approximately $n = 1,624$ cases.

In Example 8.26, we used a prior estimate of π in computing the required sample size. If such prior information were not available, we could approximate π in the sample size equation using $\pi = .5$. The nearer the substituted value of π is to .5, the larger will be the sample size obtained from the formula. Hence, if you take $\pi = .5$ as the approximation to π, you will always obtain a sample size that is at least as large as required.

The procedure for determining the sample sizes for estimating the difference between a pair of population means is summarized in the box.

CHOOSING THE SAMPLE SIZES FOR ESTIMATING THE DIFFERENCE $(\mu_1 - \mu_2)$ BETWEEN A PAIR OF POPULATION MEANS CORRECT TO WITHIN d UNITS WITH CONFIDENCE COEFFICIENT $(1 - \alpha)$

$$n_1 = n_2 = \left(\frac{z_{\alpha/2}}{d}\right)^2 (\sigma_1^2 + \sigma_2^2)$$

where n_1 and n_2 are the numbers of observations sampled from each of the two populations, and σ_1^2 and σ_2^2 are the variances of the two populations.

EXAMPLE 8.27   Refer to Example 8.25. Suppose the mail-order house wanted to estimate the difference in mean shipping times for two different express services. If the company specifies that the estimated difference in means is to be correct to within $d = .5$ day with 99% confidence, how many shipping times would have to be included in each sample?

Solution We will assume that the population standard deviations are approximately equal and will estimate them using the value $\sigma \approx 1.17$ obtained in Example 8.25. Consulting Table 4 of Appendix G, we find that for $\alpha = .01$, $z_{\alpha/2} = z_{.005} \approx 2.575$. Substituting these values into the equation for the required sample sizes, we obtain

$$n_1 = n_2 = \left(\frac{z_{\alpha/2}}{d}\right)^2 (\sigma_1^2 + \sigma_2^2) = \left(\frac{2.575}{.5}\right)^2 [2(1.17)^2] = 72.6$$

Therefore, to estimate the difference in mean shipping times correct to within .5 day with 99% confidence, the mail-order house will have to sample approximately 73 shipments for each express service.

In the next box we give the equation for determining the sample sizes required to estimate the difference between two population proportions.

CHOOSING THE SAMPLE SIZES FOR ESTIMATING THE DIFFERENCE $(\pi_1 - \pi_2)$ BETWEEN TWO POPULATION PROPORTIONS TO WITHIN d UNITS WITH CONFIDENCE COEFFICIENT $(1 - \alpha)$

$$n_1 = n_2 = \left(\frac{z_{\alpha/2}}{d}\right)^2 [\pi_1(1 - \pi_1) + \pi_2(1 - \pi_2)]$$

where π_1 and π_2 are the proportions for populations 1 and 2, respectively, and n_1 and n_2 are the numbers of observations to be sampled from each population.

EXAMPLE 8.28 A soft-drink bottler wants to assess the effect of an advertising campaign designed to increase customer recognition of a new cola drink. Random samples of consumers are to be selected from the marketing area both before and after the advertising campaign and asked whether they have heard of the new cola drink. Suppose the bottler wants to estimate the difference in the proportions of consumers who recognize the brand name of the cola drink correct to within .05 with confidence coefficient equal to .95. How many people should be included in each sample? (Assume that samples of equal size are to be selected before and after the campaign.)

Solution We have no prior information on the values of π_1 and π_2. Therefore, to be certain that the sample sizes are large enough to estimate $(\pi_1 - \pi_2)$ to within .05 with confidence coefficient .95, we will approximate both π_1 and π_2 with the value .5. The z value corresponding to $(1 - \alpha) = .95$ is $z_{\alpha/2} = z_{.025} = 1.96$. Substituting these values into the formula for n_1 and n_2, we obtain

$$n_1 = n_2 = \left(\frac{z_{\alpha/2}}{d}\right)^2 [\pi_1(1 - \pi_1) + \pi_2(1 - \pi_2)]$$

$$= \left(\frac{1.96}{.05}\right)^2 [(.5)(.5) + (.5)(.5)] = 768.3$$

Therefore, the bottler should include approximately 769 consumers in each of the two samples.

The formulas given in this section are appropriate when the sample size n is small relative to the population size N. For situations in which n may be large relative to N, adjustments to these formulas must be made. Sample size determination for this special case (called *survey sampling*) is beyond the scope of this text. Consult the references if you want to learn more about this particular application.

EXERCISES
LEARNING THE MECHANICS

8.77 Determine the sample size needed to estimate μ for each of the following situations:

a. $d = 3$ $\sigma = 40$ $(1 - \alpha) = .95$
b. $d = 5$ $\sigma = 40$ $(1 - \alpha) = .95$
c. $d = 5$ $\sigma = 40$ $(1 - \alpha) = .99$

8.78 Find the sample size needed to estimate π for each of the following situations:

a. $d = .04$ $\pi \approx .9$ $(1 - \alpha) = .90$
b. $d = .04$ $\pi \approx .5$ $(1 - \alpha) = .90$
c. $d = .01$ $\pi \approx .5$ $(1 - \alpha) = .90$

8.79 Find the appropriate value of $n_1 = n_2$ needed to estimate $(\mu_1 - \mu_2)$ to within:

a. 5 units with confidence coefficient .95 (Assume $\sigma_1 \approx 12$ and $\sigma_2 \approx 15$.)
b. 5 units with confidence coefficient .99 (Assume $\sigma_1 \approx 12$ and $\sigma_2 \approx 15$.)
c. 1 unit with confidence coefficient .90 (Assume $\sigma_1^2 \approx 100$ and $\sigma_2^2 \approx 120$.)

8.80 Assuming that $n_1 = n_2$, find the appropriate sample sizes needed to estimate $(\pi_1 - \pi_2)$ for each of the following situations:

a. $d = .01$ $(1 - \alpha) = .99$ $\pi_1 \approx .3$ $\pi_2 \approx .6$
b. $d = .05$ $(1 - \alpha) = .95$ $\pi_1 \approx .2$ $\pi_2 \approx .08$
c. $d = .05$ $(1 - \alpha) = .90$ $\pi_1 \approx .5$ $\pi_2 \approx .5$

APPLYING THE CONCEPTS

8.81 A child psychologist wants to estimate the mean age at which a child learns to walk. How many children must be sampled if the psychologist desires an estimate that is correct to within 1 month of the true mean with 99% confidence? Assume the psychologist knows only that the age at which a child begins to walk ranges from 8 to 26 months.

8.82 Parent battering is one of the problems confronting family-care professionals today. The National Institute of Mental Health (NIMH) estimates that 3% of all American children age 3–18 commit severe acts of physical aggression against their parents. Suppose you want to estimate to within .005 with 90% confidence the true proportion of American children age 3–18 who commit acts of physical aggression against their parents.

    a. How many families should be included in the sample if no prior information is available about the value of the proportion?

    b. What sample size is required if the NIMH proportion is used as an estimate?

8.83 Some power plants are located near rivers or oceans so that the available water can be used for cooling the condensers. As part of an environmental impact study, suppose a power company wants to estimate the difference in mean water temperature between the discharge of its plant and the offshore waters. How many sample measurements must be taken at each site to estimate the true difference between means to within .2°C with 95% confidence? Assume the range in readings will be about 4°C at each site and the same number of readings will be taken at each site.

8.84 *Cost Engineering* (Oct. 1988) reports on a study of the percentage difference between the low bid and the engineer's estimate of the cost for building contracts (see Exercise 7.16). For contracts with four bidders, the mean percentage error is $\mu = -7.02$ and the standard deviation is $\sigma = 24.66$. Suppose you want to estimate the mean percentage error for building contracts with five bidders. How many five-bidder contracts must be sampled to estimate with 90% confidence the mean to within 5 percentage points of its true value? Assume that the standard deviation for five-bidder contracts is approximately equal to the standard deviation for four-bidder contracts.

8.85 Rat damage creates a large financial loss in the production of sugar cane. One aspect of the problem that has been investigated by the U.S. Department of Agriculture concerns the optimal place to locate rat poison. To be more effective in reducing rat damage, should the poison be located in the middle of the field or on the outer perimeter? One way to answer this question is to determine where the greater amount of damage occurs. If damage is measured by the proportion of cane stalks that have been damaged by rats, how many stalks from each section of the field should be sampled to estimate the true difference between the proportions of stalks damaged in the two sections to within .02 with probability .95? (Assume that samples of equal size are to be selected from each section.)

8.86 Refer to the U.S. Department of Energy study of brine antiscalants, Example 8.11. How many brine solutions must be treated with antiscalants and tested to estimate the true mean amount of silicon present in an aliquot of brine to within 10 ppm with 90% confidence? Use the sample standard deviation calculated in Example 8.11 as an estimate of σ.

8.87 Refer to the *Human Factors* study on the use of color brightness as a body orientation clue, Exercise 8.34. How many subjects are required for a similar experiment to estimate the true proportion who use a bright color level as a cue to being right-side up to within .05 with 95% confidence? Use the sample proportion calculated in Exercise 8.34 as an estimate of π.

8.9

According to consumer affairs reporter Judy Hill,

CHAPTER CASE: THE CRAZY DAISY SHAVE

Many adolescent girls can't wait until they are old enough to shave their legs: They consider it a sign of growing up. But by the time they are adults—and have been shaving a while—the thrill is gone. Shaving is a pain. There's the soap or the cream all over the place, there's the stubble, the nicks, the cuts—and the cost. But a [disposable shaver] promises to end at least some of those problems.*

---

*Gainesville Sun, Apr. 26, 1981.

Hill refers to an advertisement for the twin-blade Daisy shaver from Gillette that claims, "[it] shaves legs smoother, closer, and safer than any single-blade shaver."

To substantiate the advertisement's claim, Hill conducted her own survey. Using a test similar to Gillette's, she asked 13 women ranging in age from 19 to 50 to shave one leg with a Daisy and the other leg with a Lady Bic (a single-blade disposable razor for women). After shaving according to their own routine, they each gave their opinion as to which of the two shavers gave a "smoother, closer, and safer" shave. Hill reports the results as follows: "Nine of the women chose the Daisy as being superior in all three categories; two said they could tell no difference; one chose the Lady Bic as being superior in all three categories, and one said the Bic gave a close shave while the Daisy gave a smoother shave."

How accurate is the estimate,

$$p = \frac{9}{13} = .69$$

of the proportion π of all women shavers who would find the Daisy superior to the Lady Bic? Most people would intuitively say that a sample containing only 13 women is too small to obtain an accurate estimate. After studying Section 8.4, we would agree.

Even if we could assume that the normal distribution would provide an adequate approximation to the sampling distribution of p (which it does not), a 95% confidence interval for π would be

$$p \pm z_{\alpha/2} \sqrt{\frac{pq}{n}}$$

$$.69 \pm 1.96 \sqrt{\frac{(.69)(.31)}{13}}$$

$$.69 \pm .25$$

You can see that the resulting interval, .44 to .94, is very wide. Our interval estimate indicates that a value of π as low as .44 or as high as .94 could have produced Hill's sample results.

How large a survey would be required to obtain an accurate estimate of the proportion of women who favor the Daisy razor over the Lady Bic? To answer this question, we need to know how accurate we want our estimate of π to be and what level of confidence we wish to place in our conclusions. Suppose we want to estimate π to within .03 with 95% confidence. Then, from Section 8.8, we would find the required sample size using the formula

$$n = \left(\frac{z_{\alpha/2}}{d}\right)^2 \pi(1 - \pi)$$

where π is the population proportion that we are attempting to estimate and d is the distance from π within which we want our estimate to lie. For this example, $\alpha = .05$, $z_{\alpha/2} = 1.96$, and $d = .03$. Since our estimate of π is very unreliable, we will use .5 to approximate the value of π in the equation. This will ensure that the resulting sample size will be adequate. Substituting into the formula for n, we obtain

$$n = \left(\frac{z_{\alpha/2}}{d}\right)^2 \pi(1 - \pi)$$
$$= \left(\frac{1.96}{.03}\right)^2 (.5)(.5)$$
$$= 1,067.1$$

Thus, a random sample of 1,068 women will provide an estimate of π that will lie within .03 of the proportion of all women shavers who prefer the Daisy razor over the Lady Bic with 95% confidence.

8.10 In the previous sections, we considered interval estimates for population means or proportions. In this optional section, we discuss a confidence interval for a population variance, σ^2.

ESTIMATION OF A POPULATION VARIANCE (OPTIONAL)

EXAMPLE 8.29 Refer to the U.S. Army Corps of Engineers study of contaminated fish in the Tennessee River, Alabama (Case Study 1.1). It is important for the Corps of Engineers to know how stable the weights of the contaminated fish are. That is, how large is the variation in the fish weights?

a. Identify the parameter of interest to the U.S. Army Corps of Engineers.
b. Explain how to build a confidence interval for the parameter, part **a**.

Solution

a. The Corps of Engineers is interested in the *variation* of the fish weights. Consequently, the target population parameter is σ^2, the variance of the weights of all contaminated fish inhabiting the Tennessee River.
b. Intuitively, it seems reasonable to use the sample variance s^2 to estimate σ^2 and to construct our confidence interval around this value. However, unlike sample means and sample proportions, the sampling distribution of the sample variance s^2 does not possess a normal (z) distribution or a t distribution.

Rather, when certain assumptions are satisfied (we discuss these later), the sampling distribution of s^2 possesses approximately a **chi-square (χ^2) distribution**.* The chi-square probability distribution, like the t distribution, is characterized by a quantity called the *degrees of freedom* associated with the distribution. Several chi-square probability distributions with different degrees of freedom are shown in Figure 8.23. Unlike z and t distributions, the chi-square distribution is not symmetric about 0.

*Throughout this section (and this text), we will use the words *chi-square* and the Greek symbol χ^2 interchangeably.

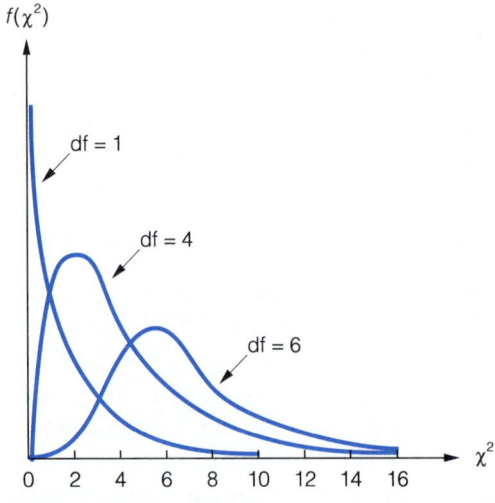

FIGURE 8.23
Several Chi-Square Probability Distributions

$f(\chi^2)$

df = 1

df = 4

df = 6

0   2   4   6   8   10 12 14 16 χ^2

EXAMPLE 8.30 Tabulated values of the χ^2 distribution are given in Table 6 of Appendix G; a partial reproduction of this table is shown in Table 8.9. Entries in the table give an upper-tail value of χ^2, call it χ^2_α, such that $P(\chi^2 > \chi^2_\alpha) = \alpha$. Find the tabulated value of χ^2 corresponding to 9 degrees of freedom that cuts off an upper-tail area of .05.

Solution The value of χ^2 that we seek appears (shaded) in the partial reproduction of Table 6 of Appendix G given in Table 8.9. The columns of the table identify the value of α associated with the tabulated value of χ^2_α and the rows correspond to the degrees of freedom. For this example, we have df = 9 and α = .05. Thus, the tabulated value of χ^2 corresponding to 9 degrees of freedom is

$$\chi^2_{.05} = 16.9190$$

We use the tabulated values of χ^2 to construct a confidence interval for σ^2, as the next example illustrates.

EXAMPLE 8.31 Refer to Example 8.29. The 144 samples of fish in the U.S. Army Corps of Engineers study produced the following summary statistics: $\bar{x}$ = 1,049.7 grams, s = 376.6 grams. Use this information to construct a 95% confidence interval for the true variation in weights of contaminated fish in the Tennessee River.

Solution A $(1 - \alpha)100\%$ confidence interval for σ^2 depends on the quantities s^2, $(n - 1)$, and critical values of χ^2 as shown in the box. Note that $(n - 1)$ represents the degrees of freedom associated with the χ^2 distribution. To construct the interval, we first locate the critical values $\chi^2_{(1-\alpha/2)}$ and $\chi^2_{\alpha/2}$. These are the

TABLE 8.9
Reproduction of Part of Table 6 of Appendix G

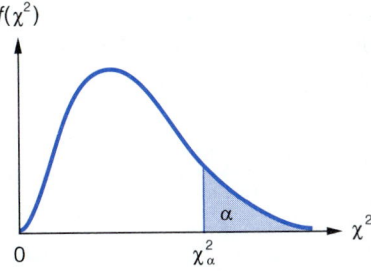

$f(\chi^2)$

DEGREES OF FREEDOM	$\chi^2_{.100}$	$\chi^2_{.050}$	$\chi^2_{.025}$	$\chi^2_{.010}$	$\chi^2_{.005}$
1	2.70554	3.84146	5.02389	6.63490	7.87944
2	4.60517	5.99147	7.37776	9.21034	10.5966
3	6.25139	7.81473	9.34840	11.3449	12.8381
4	7.77944	9.48773	11.1433	13.2767	14.8602
5	9.23635	11.0705	12.8325	15.0863	16.7496
6	10.6446	12.5916	14.4494	16.8119	18.5476
7	12.0170	14.0671	16.0128	18.4753	20.2777
8	13.3616	15.5073	17.5346	20.0902	21.9550
9	14.6837	16.9190	19.0228	21.6660	23.5893
10	15.9871	18.3070	20.4831	23.2093	25.1882
11	17.2750	19.6751	21.9200	24.7250	26.7569
12	18.5494	21.0261	23.3367	26.2170	28.2995
13	19.8119	22.3621	24.7356	27.6883	29.8194
14	21.0642	23.6848	26.1190	29.1413	31.3193
15	22.3072	24.9958	27.4884	30.5779	32.8013
16	23.5418	26.2962	28.8454	31.9999	34.2672
17	24.7690	27.5871	30.1910	33.4087	35.7185
18	25.9894	28.8693	31.5264	34.8053	37.1564
19	27.2036	30.1435	32.8523	36.1908	38.5822

A $(1 - \alpha)$100% CONFIDENCE INTERVAL FOR A POPULATION VARIANCE, σ^2

$$\frac{(n - 1)s^2}{\chi^2_{\alpha/2}} \le \sigma^2 \le \frac{(n - 1)s^2}{\chi^2_{(1-\alpha/2)}}$$

where $\chi^2_{\alpha/2}$ and $\chi^2_{(1-\alpha/2)}$ are values of χ^2 that locate an area of $\alpha/2$ to the right and $\alpha/2$ to the left, respectively, of a chi-square distribution based on $(n - 1)$ degrees of freedom.

Assumption: The population from which the random sample is selected has an approximate normal distribution.

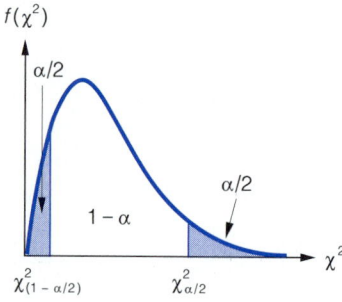

$f(\chi^2)$

$\alpha/2$

$1 - \alpha$

$\alpha/2$

$\chi^2_{(1-\alpha/2)}$ $\chi^2_{\alpha/2}$ χ^2

FIGURE 8.24
The Location of $\chi^2_{(1-\alpha/2)}$ and $\chi^2_{\alpha/2}$ for a Chi-Square Distribution

values of χ^2 that cut off an area of $\alpha/2$ in the lower and upper tails, respectively, of the chi-square distribution (see Figure 8.24).

For a 95% confidence interval, $(1 - \alpha) = .95$ and $\alpha/2 = .05/2 = .025$. Therefore, we need the tabulated values $\chi^2_{.025}$ and $\chi^2_{.975}$ for $(n - 1) = 143$ df. Looking in the df = 150 row of Table 6, Appendix G (the row with the df values closest to 143), we find $\chi^2_{.025} = 185.800$ and $\chi^2_{.975} = 117.985$. Substituting into the formula given in the box, we obtain

$$\frac{(144 - 1)(376.6)^2}{185.800} \le \sigma^2 \le \frac{(144 - 1)(376.6)^2}{117.985}$$

$$109{,}156.8 \le \sigma^2 \le 171{,}898.4$$

We are 95% confident that the true variance in weights of contaminated fish in the Tennessee River falls between 109,156.8 and 171,898.4. The Army Corps of Engineers could use this interval to determine whether the weights of the fish are stable enough to allow further testing for DDT contamination.

EXAMPLE 8.32   Refer to Example 8.31. Find a 95% confidence interval for σ, the true standard deviation of the fish weights.

Solution   A confidence interval for σ is obtained by taking the square roots of the lower and upper endpoints of a confidence interval for σ^2. Thus, the 95% confidence interval is

$$\sqrt{109{,}156.8} \le \sigma \le \sqrt{171{,}898.4}$$

$$330.4 \le \sigma \le 414.6$$

Thus, we are 95% confident that the true standard deviation of the fish weights is between 330.4 grams and 414.6 grams.

Note that the procedure for calculating a confidence interval for σ^2 in the previous examples requires an assumption regardless of whether the sample size n is large or small (see box on previous page). We must assume that the population from which the sample is selected has an approximately normal distribution. It is reasonable to expect this assumption to be satisfied in Examples 8.31 and 8.32 since the histogram of the 144 fish weights in the sample, shown in Figure 3.12, is approximately normal.

EXERCISES
LEARNING THE MECHANICS

8.88   For each of the following combinations of α and degrees of freedom (df), use Table 6, Appendix G to find the values of $\chi^2_{\alpha/2}$ and $\chi^2_{(1-\alpha/2)}$ that would be used to form a confidence interval for σ^2.

a. $\alpha = .05$, df = 7   b. $\alpha = .10$, df = 16
c. $\alpha = .01$, df = 10   d. $\alpha = .05$, df = 20

8.89 Given the following values of $\bar{x}$, s, and n, calculate a 90% confidence interval for σ^2

    a. $\bar{x} = 21$, $s = 2.5$, $n = 50$
    b. $\bar{x} = 1.3$, $s = .02$, $n = 15$
    c. $\bar{x} = 167$, $s = 31.6$, $n = 22$
    d. $\bar{x} = 9.4$, $s = 1.5$, $n = 5$

What assumption about the population must be satisfied for the confidence interval to be valid?

8.90 Refer to Exercise 8.89. For each part **a–d**, calculate a 90% confidence interval for σ.

8.91 A random sample of $n = 6$ observations from a normal distribution resulted in the following measurements: 8, 2, 3, 7, 11, 6. Find a 95% confidence interval for σ^2.

APPLYING THE CONCEPTS

8.92 A machine used to fill beer cans must operate so that the amount of beer actually dispensed varies very little. If too much beer is released, the cans will overflow, causing waste. If too little beer is released, the cans will not contain enough beer, causing complaints from customers. A random sample of the fills for 20 cans yielded a standard deviation of .07 ounce. Estimate the true variance of the fills using a 95% confidence interval.

8.93 Refer to the *Science* (Nov. 1988) study of swarm-founding wasps in Exercise 8.8. Recall that an inbreeding coefficient, x, was genetically determined for each of 197 wasps, resulting in a sample mean of $\bar{x} = .044$ and a sample standard deviation of $s = .884$. Use this information to construct a 90% confidence interval for σ, the true standard deviation of the inbreeding coefficients of the wasps. Interpret the interval.

8.94 Refer to the *Exceptional Children* (Sept. 1984) study of developmentally delayed children in Exercise 8.24. Recall that the sample of 16 children had a mean IQ of 80.2 and a standard deviation of 20.7. Find an estimate of the variance in IQ of developmentally delayed children in the state using a 99% confidence interval.

8.95 Geologists analyze fluid inclusions in rock to infer the compositions of fluids present when the rocks crystallized. A new technique, called laser Raman microprobe (LRM) spectroscopy, has been developed for this purpose. An experiment was conducted to estimate the precision of the LRM technique (*Applied Spectroscopy*, Feb. 1986). A chip of natural Brazilian quartz with several artificially produced fluid inclusions was subjected to LRM spectroscopy. The amount of liquid carbon dioxide (CO_2) present in the inclusion was recorded for the same inclusion on four different days. The data (in mole percentage) are:

    86.6   84.6   85.5   85.9

    a. Obtain an estimate of the precision of the LRM technique by constructing a 99% confidence interval for the variation in the CO_2 concentration measurements.
    b. What assumption is required for the interval estimate to be valid?

8.96 Snyder and Chrissis (1990) presented a hybrid algorithm for solving a polynomial zero–one mathematical programming problem. The algorithm incorporates a mixture of pseudo-Boolean concepts and time-proven implicit enumeration procedures. Fifty-two random problems were solved using the hybrid algorithm; the times to solution (CPU time in seconds) are listed in the accompanying table.

.045	3.985	.506	.145	1.267	.049	.333	.379	.091	.036	.336	.219	.209
1.070	.130	.579	.045	.118	1.894	.136	1.639	.064	.258	.412	.209	.070
8.788	3.046	.179	.136	3.888	.242	.227	.182	.136	.600	.394	.258	.327
.445	1.055	.670	.088	4.170	.567	.079	.554	.912	.194	.182	.361	.258

Source: Snyder, W. S. and Chrissis, J. W. "A hybrid algorithm for solving zero–one mathematical programming problems." *IEEE Transactions*, Vol. 22, No. 2, June 1990, p. 166 (Table 1).

A SAS printout giving descriptive statistics for the sample of 52 solution times follows. Use this information to compute a 95% confidence interval for the variance of the solution times. Interpret the result.

```
Analysis Variable : CPU


N Obs   N          Mean       Variance       Std Dev
-----------------------------------------------------
   52  52      0.8121923     2.2643035      1.5047603
-----------------------------------------------------
```

SUMMARY

This chapter presented the technique of *estimation*—that is, using sample information to make an inference about the value of a population parameter, or the difference between two population parameters. In each instance, we presented the point estimate of the parameter of interest, its sampling distribution, the general form of a *confidence interval*, and any assumptions required for the validity of the procedure. These results are summarized in Tables 8.10a and 8.10b. In addition, we provided techniques for determining the sample size necessary to estimate each of these parameters.

KEY WORDS

Chi-square (χ^2) distribution*
Confidence coefficient
Confidence interval
Degrees of freedom
Independent samples
Matched pairs

Point estimate
Pooled estimate of variance
Proportion
Standard error
Student's *t* distribution

KEY FORMULAS

Large-sample confidence interval for means or proportions:

$$\text{Point estimator} \pm (z_{\alpha/2})(\text{Standard error})$$

Small-sample confidence interval for means or proportions:

$$\text{Point estimator} \pm (t_{\alpha/2})(\text{Standard error})$$

[*Note:* The respective point estimator and standard error for each parameter discussed in this chapter are provided in Table 8.10.]

*Confidence interval for variances:**

$$\frac{(n-1)s^2}{\chi^2_{\alpha/2}} \le \sigma^2 \le \frac{(n-1)s^2}{\chi^2_{(1-\alpha/2)}}$$

---

*From the optional section in this chapter.

TABLE 8.10

a. Summary of Estimation Procedures: One-Population Cases

PARAMETER	POINT ESTIMATE	STANDARD ERROR	$(1-\alpha)100\%$ CONFIDENCE INTERVAL	ASSUMPTIONS/CONDITIONS
μ Population mean	$\bar{x}$ Sample mean	$\sigma/\sqrt{n}$ where σ is standard deviation of sample population	$\bar{x} \pm z_{\alpha/2}(\sigma/\sqrt{n})$ $\approx \bar{x} \pm z_{\alpha/2}(s/\sqrt{n})$	$n \geq 30$ (large sample) Random sample
μ Population mean	$\bar{x}$ Sample mean	(see above)	$\bar{x} \pm t_{\alpha/2}(s/\sqrt{n})$ where $t_{\alpha/2}$ is based on $(n-1)$ degrees of freedom	$n < 30$ (small sample) Random sample Relative frequency distribution of population is approximately normal.
μ_d Population mean difference, matched pairs	$\bar{d}$ Sample mean difference	$\sigma_d/\sqrt{n}$ where σ_d is standard deviation of population of differences	$\bar{d} \pm t_{\alpha/2}(s_d/\sqrt{n})$ where $t_{\alpha/2}$ is based on $(n-1)$ degrees of freedom	$n < 30$ (small sample) Random sample Relative frequency distribution of population of differences is approximately normal.
π Proportion of population with specified characteristic(s)	Sample proportion with specified characteristic(s): $p = \dfrac{\text{Number in sample with characteristic}}{n}$ where n is number of observations sampled	$\sqrt{\dfrac{\pi(1-\pi)}{n}}$	$p \pm z_{\alpha/2}\sqrt{\dfrac{\pi(1-\pi)}{n}}$ $\approx p \pm z_{\alpha/2}\sqrt{\dfrac{pq}{n}}$ where $q = 1 - p$	The interval $p \pm 2\sqrt{\dfrac{pq}{n}}$ does not contain 0 or 1 (large sample). Random sample
σ^2 Population variance (optional)	s^2 Sample variance	(not required)	$\dfrac{(n-1)s^2}{\chi^2_{\alpha/2}} \leq \sigma^2 \leq \dfrac{(n-1)s^2}{\chi^2_{(1-\alpha/2)}}$ where the values $\chi^2_{\alpha/2}$ and $\chi^2_{(1-\alpha/2)}$ are based on $(n-1)$ degrees of freedom	Relative frequency distribution of population is approximately normal. Random sample

Continued

TABLE 8.10, continued

b. Summary of Estimation Procedures: Two-Population Cases

PARAMETER	POINT ESTIMATE	STANDARD ERROR	$(1 - \alpha)100\%$ CONFIDENCE INTERVAL	ASSUMPTIONS/CONDITIONS
$(\mu_1 - \mu_2)$ Difference between population means, independent samples	$(\bar{x}_1 - \bar{x}_2)$ Difference between sample means	$\sqrt{\dfrac{\sigma_1^2}{n_1} + \dfrac{\sigma_2^2}{n_2}}$ where σ_1^2 and σ_2^2 are the variances of the sampled populations	$(\bar{x}_1 - \bar{x}_2) \pm z_{\alpha/2} \sqrt{\dfrac{\sigma_1^2}{n_1} + \dfrac{\sigma_2^2}{n_2}}$ $\approx (\bar{x}_1 - \bar{x}_2) \pm z_{\alpha/2} \sqrt{\dfrac{s_1^2}{n_1} + \dfrac{s_2^2}{n_2}}$	$n_1 \geq 30$ and $n_2 \geq 30$ (large samples) Samples are randomly and independently selected from the two populations.
$(\mu_1 - \mu_2)$ Difference between population means, independent samples	$(\bar{x}_1 - \bar{x}_2)$ Difference between sample means	(see above)	$(\bar{x}_1 - \bar{x}_2) \pm t_{\alpha/2} \sqrt{s_p^2 \left(\dfrac{1}{n_1} + \dfrac{1}{n_2} \right)}$ where $s_p^2 = \dfrac{(n_1 - 1)s_1^2 + (n_2 - 1)s_2^2}{n_1 + n_2 - 2}$ and $t_{\alpha/2}$ is based on $(n_1 + n_2 - 2)$ degrees of freedom	$n_1 < 30$ or $n_2 < 30$ (small samples) 1. Relative frequency distributions of both populations are approximately normal. 2. Variances of both populations are equal. 3. Samples are randomly and independently selected from the two populations.
$\mu_d = (\mu_1 - \mu_2)$ Difference between population means, matched pairs			(see Table 8.10a)	
$(\pi_1 - \pi_2)$ Difference between population proportions	$(p_1 - p_2)$ Difference between sample proportions	$\sqrt{\dfrac{\pi_1(1 - \pi_1)}{n_1} + \dfrac{\pi_2(1 - \pi_2)}{n_2}}$	$(p_1 - p_2) \pm z_{\alpha/2} \sqrt{\dfrac{\pi_1(1 - \pi_1)}{n_1} + \dfrac{\pi_2(1 - \pi_2)}{n_2}}$ $\approx (p_1 - p_2) \pm z_{\alpha/2} \sqrt{\dfrac{p_1 q_1}{n_1} + \dfrac{p_2 q_2}{n_2}}$ where $q_1 = 1 - p_1$ and $q_2 = 1 - p_2$	The intervals $p_1 \pm 2\sqrt{\dfrac{p_1 q_1}{n_1}}$ and $p_2 \pm 2\sqrt{\dfrac{p_2 q_2}{n_2}}$ do not contain 0 or 1 (large samples). Samples are randomly and independently selected from the two populations.

SUPPLEMENTARY EXERCISES

[*Note:* List the assumptions necessary to ensure the validity of the interval estimation procedures you use to solve these exercises. Exercises marked with an asterisk (*) are from the optional section in this chapter.]

8.97 The "Black Hole" survey, sponsored by the Professional Employment Research Council, reports on the toughest jobs to fill on recruiters' lists. In the most recent survey, 95 of 285 recruiters listed engineering positions as the "toughest to fill" (*Industrial Engineering*, Aug. 1990).

   a. Estimate the true percentage of recruiters who find it toughest to fill engineering positions. Use a 99% confidence interval.

   b. Give a precise interpretation of the phrase, "We are 99% confident that the interval encloses the true value of π."

8.98 The theoretical relationship between heat flux and temperature gradient for homogeneous materials is well known and described by a Fourier equation. However, this relationship does not hold for nonhomogeneous materials such as porous-capillary bodies, cellular systems, suspensions, and pastes. An experiment was conducted to estimate the mean thermal relaxation time (defined as the mean time needed for accumulating the thermal energy required for propagative transfer of heat) for several nonhomogeneous materials (*Journal of Heat Transfer*, Aug. 1990). A 95% confidence interval for the mean thermal relaxation time of sand was found to be 20.0 ± 6.4 seconds.

   a. Give a practical interpretation of the 95% confidence interval.

   b. Give a theoretical interpretation of the 95% confidence interval.

8.99 In a recent clinical study, it was learned that the use of aspirin to alleviate the symptoms of viral infections in children may lead to serious complications (Reyes' syndrome). A random sample of 500 children with viral infections received no aspirin to alleviate symptoms, and 12 developed Reyes' syndrome. In a random sample of 450 children with viral infections who were given aspirin, 23 developed Reyes' syndrome. Construct a 95% confidence interval for the difference in the proportions of children who develop Reyes' syndrome between those who receive no aspirin and those who receive aspirin during a viral infection. Interpret the interval.

8.100 What do college recruiters believe are the most important topics to be covered in a job interview? To answer this and other questions, a recent study elicited the opinions of recruiters interviewing at a small midwestern college and a large midwestern university (*Journal of Occupational Psychology*, Vol. 57, 1984). Recruiters were asked to rate on a 105-point scale the importance of each in a list of 25 interview topics [where 0 = least important (i.e., can often be omitted without hurting the interview), 52.5 = average importance (i.e., can sometimes be omitted without hurting the interview), and 105 = most important (i.e., can never be omitted without hurting the interview)]. The topic concerning "applicant's skill in communicating ideas to others" received the highest ratings of the $n = 58$ college recruiters who returned the questionnaire. The sample mean rating and sample standard deviation for this topic were $\bar{x} = 84.84$ and $s = 15.67$, respectively.

   a. Give a point estimate for the true mean rating of "applicant's skill in communicating ideas to others" by all college recruiters.

   b. Use the sample information to construct a 95% confidence interval for the true mean rating.

   c. What is the confidence coefficient for the interval of part **b**? Interpret this value.

8.101 To investigate the possible link between fluoride content of drinking water and cancer, researchers have examined cancer death rates (number of deaths per 100,000 population) from 1952–1969 in 20 selected U.S. cities—the 10 largest fluoridated cities and the 10 largest cities not fluoridated by 1969. For each city, the annual rate of increase in cancer death rate over this 18-year period is calculated for each of four age groups: under 25, 25–44, 45–64, and 65 or older. The data for the 45–64 age group are reproduced in the accompanying table.

FLUORIDATED		NONFLUORIDATED	
City	Annual Increase in Cancer Death Rate	City	Annual Increase in Cancer Death Rate
Chicago	1.0640	Los Angeles	.8875
Philadelphia	1.4118	Boston	1.7358
Baltimore	2.1115	New Orleans	1.0165
Cleveland	1.9401	Seattle	.4923
Washington	3.8772	Cincinnati	4.0155
Milwaukee	−.4561	Atlanta	−1.1744
St. Louis	4.8359	Kansas City	2.8132
San Francisco	1.8875	Columbus	1.7451
Pittsburgh	4.4964	Newark	−.5676
Buffalo	1.4045	Portland	2.4471

Source: Maritz, J. S. and Jarrett, R. G. "The use of statistics to examine the association between fluoride in drinking water and cancer death rates." *Applied Statistics*, Vol. 32, No. 2, 1983, pp. 97–101.

a. Construct a 95% confidence interval for the difference between the mean annual increases in cancer death rates for fluoridated and nonfluoridated cities.

b. Interpret the interval obtained in part a.

c. What assumptions are necessary for the validity of the interval estimation procedure and any inferences derived from it? Do you think these assumptions are satisfied?

8.102 Wall and Peterson (1986) developed a heat transfer model for predicting winter heat loss in wastewater treatment clarifiers. Part of their analysis involved a comparison of clear-sky solar irradiation for horizontal surfaces at different sites in the midwest. The day-long solar irradiation levels (in BTU/sq. ft.) at two midwestern locations of different latitudes (St. Joseph, Missouri, and Iowa Great Lakes) were recorded on each of seven clear-sky winter days. The data are given in the table. Find a 95% confidence interval for the mean difference between the day-long, clear-sky solar irradiation levels at the two sites.

DATE	ST. JOSEPH, MO	IOWA GREAT LAKES
December 21	782	593
January 6	965	672
January 21	948	750
February 6	1,181	988
February 21	1,414	1,226
March 7	1,633	1,462
March 21	1,852	1,698

Source: Wall, D. J. and Peterson, G. "Model for winter heat loss in uncovered clarifiers." *Journal of Environmental Engineering*, Vol. 112, No. 1, Feb. 1986, p. 128.

8.103 An experiment was conducted to determine whether individuals could be taught how to make decisions rationally and if such training would improve the quality of their career decisions (*Journal of Vocational Behavior*, Aug. 1986). A sample of 69 California community college students, all classified as rational decision makers, was randomly divided into two groups. The experimental group (34 students) received instruction in rational decision making, whereas the control group (35 students) did not. At the end of the instruction period, all subjects completed a multiple-choice test designed to assess the extent to which an individual knows how to apply rational principles in job-decision situations. The results are summarized in the table. (Higher scores indicate greater adherence to the rational style of decision making). Construct a 90% confidence interval for the difference between the mean test scores of the two groups of students. Interpret the interval.

	EXPERIMENTAL (TRAINING) GROUP	CONTROL (NO TRAINING) GROUP
Sample size	34	35
Mean	50.26	47.34
Standard deviation	6.67	11.52

Source: Krumboltz, J. D. et al. "Teaching a rational approach to career decision making: Who benefits most?" *Journal of Vocational Behavior*, Vol. 29, No. 8, Aug. 1986, pp. 1–6 (Table 1).

8.104 Operation Kidsafe is a nationwide safety program designed for children between the ages of 3 and 7. One of the goals of Operation Kidsafe is to help children learn vital information (e.g., name, address, and phone number) in case of emergency. One study, commissioned by the sponsor of Operation Kidsafe, found that 340 of 500 children between the ages of 3 and 7 do not know their home phone number (*USA Today*, Aug. 14, 1985). Estimate the true percentage of children between the ages of 3 and 7 who do not know their home phone number. Use a 95% confidence interval.

8.105 The pesticide Temik is used for controlling insects that feed on potatoes, oranges, and other crops. According to federal standards, drinking water wells with levels of Temik above 1 part per billion are considered contaminated. The accompanying table lists the results of tests for Temik contamination conducted in five states over the past decade.

STATE	NUMBER OF WELLS TESTED	NUMBER OF CONTAMINATED WELLS
New York	10,500	2,750
Wisconsin	700	105
Maine	124	82
Florida	825	4
Virginia	76	17

Source: *Orlando Sentinel Star*, July 4, 1983.

a. For each state, construct a 95% confidence interval for the true proportion of wells contaminated with Temik.
b. Find a 90% confidence interval for the difference between the proportions of wells in Wisconsin and Maine that are contaminated with Temik.

8.106 The growth of "off-price" retail stores in the 1980s has been phenomenal. Off-price stores are specialty stores that claim to sell brand-name and designer women's career clothes, casual wear, and active wear for less than traditional retail stores. A study was conducted to investigate apparel price variations in off-price and department stores. The average prices of 20 fall women's apparel items at both off-price stores and department stores in Montgomery County, Maryland, were recorded for each of 13 weeks. The data are shown in the accompanying table. Estimate the mean difference between the average retail prices of the items at the two types of stores. Use a 99% confidence interval.

WEEK	OFF-PRICE STORE AVERAGE PRICE	DEPARTMENT STORE AVERAGE PRICE	WEEK	OFF-PRICE STORE AVERAGE PRICE	DEPARTMENT STORE AVERAGE PRICE
1	$55.63	$81.18	8	55.16	74.98
2	55.63	81.80	9	54.11	71.66
3	55.32	79.36	10	54.65	68.64
4	54.11	79.18	11	53.01	68.56
5	54.79	79.55	12	42.61	67.39
6	54.36	78.21	13	50.66	67.01
7	55.33	77.70			

Source: Kirby, G. M. and Dardis, R. "Research note: A pricing study of women's apparel in off-price and department stores." *Journal of Retailing*, Vol. 62, No. 3, Fall 1986, p. 325.

8.107 The methodology for conducting a stress analysis of newly designed timber structures is well known. However, few data are available on the actual or allowable stress for repairing damaged structures. Consequently, design engineers often propose a repair scheme (e.g., gluing) without any knowledge of its structural effectiveness. To partially fill this void, a stress analysis was conducted on epoxy-repaired truss joints (*Journal of Structural Engineering*, Feb. 1986). Tests were conducted on epoxy-bonded truss joints made of various species of wood to determine actual glue-line shear stress (recorded in pounds per square inch). Summary information for independent random samples of southern pine and ponderosa pine truss joints is given in the accompanying table. Estimate the difference between the mean shear strengths of epoxy-repaired truss joints for the two species of wood with a 90% confidence interval.

	SOUTHERN PINE	PONDEROSA PINE
Sample size	100	47
Mean shear stress (psi)	1,312	1,352
Standard deviation	422	271

Source: Avent, R. R. "Design criteria for epoxy repair of timber structures." *Journal of Structural Engineering*, Vol. 112, No. 2, Feb. 1986, p. 232.

8.108 Do colleges provide good value for the dollar? A majority of Americans do not think so, according to a 1987 Media General—Associated Press telephone poll. Of the 1,348 adult Americans who participated in the nationwide poll, 805 believe tuition at most private colleges and universities is too high for the quality of education provided (*Gainesville Sun*, Sept. 1, 1987).

a. Calculate a point estimate of π, the true proportion of adult Americans who think private colleges are too expensive, given the quality of education provided.
b. Construct a 95% confidence interval for π.
c. Interpret the interval obtained in part **b**.
d. How would the width of the confidence interval in part **b** change if the confidence coefficient were increased from .95 to .99?
e. How many adult Americans must be surveyed to estimate the true proportion who believe college tuition is too high to within .02 of its true value, with 95% confidence? Use the sample proportion calculated in part **a**, as an estimate of π.

*8.109 An experiment was conducted to investigate the precision of measurements of a saturated solution of iodine after an extended period of continuous stirring. The data shown in the table represent $n = 10$ iodine concentration measurements on the same solution. The population variance σ^2 measures the variability—i.e., the precision— of a measurement. Use the data to find a 95% confidence interval for σ^2.

RUN	CONCENTRATION	RUN	CONCENTRATION
1	5.507	6	5.527
2	5.506	7	5.504
3	5.500	8	5.490
4	5.497	9	5.500
5	5.506	10	5.497

8.110 What do managers stress most on the job? In a survey prepared for Towers, Perrin, Forster, and Crosby (an international management consultant firm), 462 senior human resource and compensation executives in private industry, government, and nonprofit organizations were asked this question (*Personnel Journal*, Oct. 1984). The most frequent response, given by 226 of 462 managers, was "pay employees for performance." Construct a 95% confidence interval for the true proportion of managers who stress pay for performance on the job.

8.111   As part of a cooperative research agreement between Japan and the United States, a full-scale reinforced concrete building was designed and tested under simulated earthquake loading conditions in Japan (*Journal of Structural Engineering*, Jan. 1986). For one part of the study, several U.S. design engineers were asked to evaluate the new design. Of the 48 engineers surveyed, 36 believed the shear wall of the structure to be too lightly reinforced. Find a 95% confidence interval for the true proportion of U.S. design engineers who consider the shear wall of the building too lightly reinforced.

*8.112   A regional data center serving a major university and surrounding city wants to estimate the variability in the number of computer jobs it processes per day. A random sample of 18 days during the past 4 months was selected and the number of jobs processed each day was recorded. The following sample statistics were computed:

$$\bar{x} = 1,266 \qquad s = 175$$

Estimate the true variability in the number of jobs processed per day using a 99% confidence interval. What assumptions are necessary for the validity of the estimation technique?

CASE STUDY 8.1

PUBLIC OPINION POLLS: HOW ACCURATE ARE THEY?

[In the 1984 presidential election, then incumbent] Ronald Reagan captured 59% of the nation's popular votes, and all of the electoral votes save those in challenger Walter Mondale's home state. Published preelection polls generally picked Reagan as the likely winner. Yet, even late in the campaign, quite discrepant estimates of the victory margin were appearing. At the extremes, a Gordan Black survey conducted for *USA Today* gave Reagan a lead over Mondale by 60% to 35%, with 5% undecided, while a Roper poll for the Public Broadcasting Station (PBS) showed Reagan ahead 52.5% to 42.5% , also with 5% undecided. Earlier polls, even when simultaneous, had diverged more widely.

This is the opening paragraph of an article published in a 1986 issue of *Science* magazine.* The authors, P. E. Converse and M. W. Traugott, both directors at the Institute for Social Research, University of Michigan, are writing on the accuracy of public opinion polls. "Discrepancies as glaring as [the two polls referenced in the previous paragraph] are not common for reputable sample surveys of the same population at the same time," they state, "but they do occur. And while polls reported in the national media tend to state error margins, usually plus or minus three percentage points, a reader diligent enough to compare competing polls . . . is likely to conclude that error margins must somehow exceed this three percentage point value by an appreciable amount."

The "three-percentage-point" error margin the authors are referring to arises from the sampling error associated with estimating a population proportion π. From Section 8.4, we know that this sampling error is proportional to the standard error of the sample proportion, σ_p, where

$$\sigma_p = \sqrt{\frac{\pi(1 - \pi)}{n}}$$

---

*Converse, P. E. and Traugott, M. W. "Assessing the accuracy of polls and surveys." *Science*, Vol. 234, 1986, pp. 1094–1098. Copyright 1986 by the AAAS.

Since most conventional opinion polls utilize a sample size of $n = 1,500$, a conservative estimate of σ_p is

$$\sigma_p \approx \sqrt{\frac{(.5)(.5)}{1,500}} = .0129$$

(Note that we substituted $\pi = .5$ into the equation to obtain this value. Recall from Section 8.8 that $\pi = .5$ is a conservative estimate of π because it is the value that maximizes σ_p.) For a 95% confidence interval for π, the bound on the error of estimation is approximately

$$2\sigma_p \approx 2(.0129) = .0258$$

Rounding up, we obtain $2\sigma_p \approx .03$, or 3%. In other words, the estimate of π will be within about 3% of the true value at probability level .95 when a sample of size $n = 1,500$ is employed.

a. Assuming that the *USA Today* poll previously referenced included a sample of 1,500 voters, find a 95% confidence interval for π, the true proportion of voters who favored Reagan just prior to the 1984 presidential election.
b. Repeat part **a** for the PBS poll.
c. Compare the two intervals obtained in parts **a** and **b**. If the two samples were selected from the same population (i.e., the population of eligible voters) at the same time, would you expect to see such disparate results? Explain.

Converse and Traugott note that the plus or minus three percent "refers only to the most obvious source of variability, the error arising because the population at issue has not been fully enumerated, but merely sampled." In addition to sampling error, they identify several other reasons why such discrepancies in public opinion polls exist.

1. The target populations for the polls are often not the same. For example, one poll may sample the population of eligible voters, whereas another may sample the population consisting of those citizens who actually do vote.
2. Most national surveys, although they claim to cover the entire adult population of the United States, miss a small margin of the population. "Few surveys include Alaska or Hawaii, or institutionalized members of the population in hospitals, barracks, dormitories, and jails." Also, telephone surveys miss about "8% of the household population [that] remains inaccessible by residential phone."
3. Every survey is faced with the problem of nonresponse by eligible sample members, either by a refusal of the polled person to answer questions or by failure of the polling service to contact the selected member. This leads to biased survey results, even when repeated attempts to contact the person are made. For example, during the 1984 campaign the authors found that "Democratic partisans were most accessible at early calls than Republican ones. A trial heat gave Reagan a mere three-percentage-point margin over Mondale among those interviewed at one call; . . . after up to 30 callbacks for the most difficult-to-locate respondents, the lead had advanced to 13 percentage points."

4. The way the written questionnaire or verbal questions are constructed and asked has a significant impact on the survey results. For example, researchers have found that "both black and white respondents report positions on race-related issues that are less supportive of blacks when talking to white interviewers than when talking to black ones."

CASE STUDY 8.2

AN IQ COMPARISON OF IDENTICAL TWINS REARED APART

How much of our personality, our likes and dislikes, our individuality, is predetermined by our genes? And which of our traits are shaped and changed by our environment? Twins, because they share an identical genotype, make ideal subjects for investigating the degree to which various environmental conditions may instigate change. The classical method of studying this phenomenon and the subject of an interesting book by Susan Farber (1981) is the study of identical twins separated early in life and reared apart.*

Identical twins, genetically called *monozygotic (MZ) twins*, are formed when a single egg, fertilized by a single sperm, splits into two parts and each develops into a separate embryo. In contrast, fraternal twins, or *dizygotic (DZ) twins*, develop when two eggs are released from one or both ovaries and each is fertilized by a different sperm. Although they have been the subjects in many studies of twins, DZ twins are no more genetically similar than two siblings. Therefore, a study of DZ twins would leave unanswered the question of whether a given individual trait is due to hereditary or environmental differences. Likewise, MZ twins reared in the same family share almost identical environments and make it difficult to separate the two factors. Thus, claims Farber, "in theory at least, the clearest demarcation of heredity and environment is found when identical twins have been separated early in life and reared apart in different homes, by different parents, and often in widely varying socioeconomic and geographic circumstances."

Over the years, several studies of MZ twins have been conducted. Farber's book contains a chronicle and reanalysis of 95 pairs of identical twins reared apart. Much of her discussion focuses on a comparison of IQ scores; in this case study we will apply the matched-pairs technique of Section 8.6 to her data.

The question of concern is, "Are there significant differences between the IQ scores of identical twins, where one member of the pair is reared by the natural parents and the other member of the pair is not?" The data for this analysis (extracted from Table E6 of Farber's book) appear in Table 8.11. One member (A) of each of the $n = 32$ pairs of twins was reared by a natural parent, whereas the other member (B) was reared by a relative or some other person.

*Adapted from *Identical Twins Reared Apart*, by Susan L. Farber. Copyright © 1981 by Basic Books, Inc. Reprinted by permission of Basic Books, Inc., Publishers, New York.

TABLE 8.11
IQ Scores of Identical Twins Reared Apart

PAIR ID	TWIN A	TWIN B	PAIR ID	TWIN A	TWIN B
112	113	109	228	100	88
114	94	100	232	100	104
126	99	86	236	93	84
132	77	80	306	99	95
136	81	95	308	109	98
148	91	106	312	95	100
170	111	117	314	75	86
172	104	107	324	104	103
174	85	85	328	73	78
180	66	84	330	88	99
184	111	125	338	92	111
186	51	66	342	108	110
202	109	108	344	88	83
216	122	121	350	90	82
218	97	98	352	79	76
220	82	94	416	97	98

a. Explain why the data should be analyzed as matched pairs.
b. Compute the difference between IQ scores within each of the $n = 32$ pairs by subtracting the score for twin B from the score for twin A.
c. Construct a 95% confidence interval for the mean difference in IQ scores, $\mu_d = (\mu_A - \mu_B)$.
d. Based on the interval constructed in part c, is there evidence of a difference in mean IQ scores between twins reared by a natural parent and twins reared by a relative or some other person? Explain.
e. What assumptions are necessary for the validity of the confidence interval technique?

AN ASPIRIN A DAY KEEPS THE HEART DOCTOR AWAY?

According to the National Center for Health Statistics, heart disease is the leading cause of death in the United States; heart attacks and strokes account for about 40% of all deaths.* With this statistic in mind, it is no surprise that the following headline appeared across the front page of nearly every U.S. daily newspaper in late January 1988: "Aspirin cuts the risk of heart attack."

The exciting news was based on a nationwide study of 22,071 U.S. physicians, the results of which were reported in the *New England Journal of Medicine* (Jan. 27, 1988). The U.S. Physicians' Health Study, as it is known, involved a randomized clinical trial in which about half (11,037) of the physicians were assigned at random to receive one Bufferin brand aspirin tablet every other day. The other half (11,034) received a placebo, a harmless and ineffective substitute. The study was designed so that neither the participants (i.e., the physicians) nor the medical scientists who were conducting the research knew which tablet, the Bufferin or the placebo, was being administered. (This is commonly known as a **double-blind** study.) After 5 years, the researchers found that the incidence of fatal heart attacks among the "placebo" group was over three times greater than that for the "aspirin" group. (See Table 8.12a.) Based on these findings, the study was halted so that those physicians who were taking the placebo could immediately switch to aspirin and receive its "extreme beneficial effects on fatal heart attacks."

The good news of the U.S. Physicians' Health Study was followed by an avalanche of television commercials for aspirin, promoting its new-found status as the "heart attack prevention drug." Unfortunately, the excitement over the "cure" for heart disease was short-lived "and turned to confusion, when," according to *Chance* (Fall 1988), "three days later a headline in the *New York Times* read, 'Value of Daily Aspirin Disputed in British Study of Heart Attacks.' It seemed that a similar study conducted in England did not show that aspirin had any beneficial effect in reducing the risk of heart attack."

TABLE 8.12

Results of Two Studies on the Use of Aspirin in the Prevention of Heart Attacks

a. U.S. PHYSICIANS' HEALTH STUDY	ASPIRIN GROUP	PLACEBO GROUP
Sample size	11,037	11,034
Number of fatal heart attacks	5	18
b. BRITISH STUDY	ASPIRIN GROUP	CONTROL GROUP
Sample size	3,429	1,710
Number of fatal heart attacks	89	47

*Greenhouse, J. B. and Greenhouse, S. M. "An aspirin a day . . . ?" *Chance: New Directions for Statistics and Computing*, Vol. 1, No. 4, Fall 1988, pp. 24–31, New York, Springer-Verlag.

The 6-year British study involved 5,139 doctors in which two-thirds (3,429) were randomly chosen to take daily aspirin. The remaining physicians (1,710) were not given a placebo, but instead were instructed "to avoid aspirin and products containing aspirin unless some specific indication for aspirin was thought to have developed." The results, reported in the *British Medical Journal* (Jan. 1988), showed that the fatal heart attack rate was essentially the same in both groups. (See Table 8.12b.)

a. Consider the results of the U.S. Physicians' Health Study. Construct a 95% confidence interval for the true difference between the fatal heart attack rates of the aspirin group and the placebo group. Interpret the interval.

b. Now consider the results of the British study. Construct a 95% confidence interval for the true difference between the heart attack rates of the aspirin group and the control group. Interpret the interval.

c. Refer to parts **a** and **b**. Does your inference about the beneficial effect of aspirin in the prevention of heart attacks depend on which study you consider?

d. Why might the two studies yield contrasting results? [*Hint:* Consider one or more of the following issues: sample size; the fact that the U.S. study used physicians who had extraordinarily low cardiovascular mortality rates; double-blind study versus unblinded study; placebo versus "no aspirin."]

COMPUTER LAB
CONFIDENCE INTERVALS

Of the three commercially available statistical software packages discussed in this text, only Minitab has routines for producing confidence intervals for population parameters based on simple random sampling.* Minitab can produce confidence intervals for population means, but not for population proportions or variances.

a. **Confidence Interval for μ: Large sample, σ known** (data from Example 8.6)   **MINITAB**

```
Command
  Line
    1    SET C1                              } Data entry instruction
    2    20.7    16.7    22.5    12.1    2.9  ⎫
    .     .       .       .       .      .   ⎪  Input data values
    .     .       .       .       .      .   ⎬  (5 observations per line)
    6    32.5    19.1    36.9    27.9    14.0 ⎭
    7    NAME C1='ATENTIME'
    8    ZINTERVAL 99 13.41 C1  } Confidence interval command
```

*To generate confidence intervals in SAS and SPSS, you will need to obtain a printout of the descriptive statistics for the sample and substitute the appropriate numbers (e.g., n, $\bar{x}$, and s) into the formula for the parameter of interest. Both SAS and SPSS, however, will directly produce confidence intervals for means in more sophisticated designed experiments. We discuss these sampling designs in Chapter 13.

COMMAND 8 The ZINTERVAL command produces a confidence interval for the mean of the data stored in C1. The confidence level (99%) is specified after ZINTERVAL followed by the known value of σ. (The default confidence level is 95%.)

b. Confidence Interval for μ: Small sample, σ unknown (data from Example 8.11).

Command Line		
1	SET SILICON PPM IN C1	} Data entry instruction
2	229 255 280 203 229	} Input data values
3	NAME C1 = 'PPM'	
4	TINTERVAL 99 C1	} Confidence interval command

COMMAND 4 The TINTERVAL command produces a confidence interval for the mean of the data stored in C1. The confidence level (in this case, 99%) is specified following TINTERVAL. (The default is a 95% confidence interval.)

NOTE When σ is unknown, as is usually the case, TINTERVAL uses the appropriate value from the t distribution to calculate the interval regardless of the size of the sample. For large samples, recall that $t_{\alpha/2} \approx z_{\alpha/2}$.

c. Confidence Interval for $\mu_1 - \mu_2$: Independent samples (data from Example 8.19)

Command Line		
1	SET NOVACAIN IN C1	} Data entry instruction #1
2	62 71 44 50 42	} Input data values
3	SET NEW IN C2	} Data entry instruction #2
4	38 45 27 42 30	} Input data values
5	NAME C1 = 'NOVACAIN' C2 = 'NEW'	
6	TWOSAMPLE 95 C1 C2;	} Confidence interval commands
7	POOLED.	

COMMAND 6 TWOSAMPLE produces a confidence interval on the difference between the mean of the data in C1 and the mean of the data in C2. By default, a 95% confidence interval is computed. To change the confidence level, specify 99, 90, etc., following the TWOSAMPLE command.

COMMAND 7 The POOLED subcommand instructs Minitab to use s_p^2 in the calculation of a small-sample confidence interval. Omit the POOLED subcommand if you want Minitab to compute a large sample confidence interval for $\mu_1 - \mu_2$.

NOTE TWOSAMPLE uses the appropriate value from the t distribution to compute the confidence interval regardless of the sample size. For large samples, recall that $t_{\alpha/2} \approx z_{\alpha/2}$.

d. **Confidence Interval for $\mu_d = (\mu_1 - \mu_2)$: Matched pairs** (data from Example 8.22)

Command Line		
1	`READ C1 C2` } Data entry instruction	
2	`78  71`	
3	`63  44`	
.	.	Input data values
.	.	(2 observations per line)
.	.	
11	`55  39`	
12	`SUBTRACT C2 FROM C1, PUT IN C3 = 'DIFF'`	} Create DIFF variable
13	`NAME C1 = 'SCORE1' C2 = 'SCORE2' C3 = 'DIFF'`	
14	`TINTERVAL 95 C3`	} Confidence interval command

COMMAND 12 Use the SUBTRACT command to calculate the differences for the paired observations in C1 and C2.

COMMAND 14 Use the TINTERVAL command to compute a 95% confidence interval for the mean of the differences in C3.

COMPUTER ACTIVITIES

Consider the starting salary data recorded in Appendix A.

a. Obtain independent random samples of size $n_1 = 10$ male graduates and $n_2 = 10$ female graduates from the data in Appendix A. Enter the data into the computer.

b. Use a statistical software package to construct a 95% confidence interval for the difference between the mean starting salaries of male and female graduates. Interpret the interval.

c. Compute the mean starting salary of *all* male graduates and the mean starting salary of *all* female graduates in the listing of 1,795 graduates in Appendix A. Does the difference in mean starting salaries of the two populations fall within the interval you constructed in part b?

REFERENCES Freedman, D., Pisani, R., and Purves, R. *Statistics*. New York: Norton, 1978.

Mendenhall, W., Wackerly, D., and Scheaffer, R. *Mathematical Statistics with Applications*, 4th ed. Boston: Duxbury, 1988.

Snedecor, G. W. and Cochran, W. G. *Statistical Methods*, 7th ed. Ames, Iowa: Iowa State University Press, 1980.

COLLECTING EVIDENCE TO SUPPORT A THEORY: GENERAL CONCEPTS OF HYPOTHESIS TESTING

CHAPTER 9

The famous Schlitz versus Budweiser confrontation (Mug to Mug) was viewed by sports enthusiasts across the country during halftime of the 1980 NFL wildcard football game between the Houston Oilers and the Oakland (now Los Angeles) Raiders. According to Schlitz, 100 "loyal" Budweiser drinkers were selected to taste each of the two unmarked mugs of beer, one containing Budweiser and the other containing Schlitz. On live television, 46 of the Budweiser drinkers chose the mug containing Schlitz. Ignoring the favorable publicity obtained from this live taste test, do the test results, 46 out of 100, suggest that Schlitz might carve a larger share of the market for itself? If all confirmed Budweiser drinkers were to conduct a similar taste test, would as many as 40% prefer Schlitz? In this chapter we will learn how sample data can be used to make decisions about population parameters, and we will examine the Mug-to-Mug confrontation in greater detail in the chapter case presented in Section 9.5.

CONTENTS

9.1

As stated in Chapter 8, there are two general methods available for making inferences about population parameters. We can estimate their values using confidence intervals (the subject of Chapter 8) or we can make decisions about them. Making decisions about specific values of the population parameters—**testing hypotheses** about these values—is the topic of this chapter.

Confidence intervals and hypothesis tests are related, and either can be used to make decisions about parameters. For example, suppose an investigator for the Environmental Protection Agency (EPA) wants to determine whether the mean level μ of a certain type of pollutant released into the atmosphere by a chemical company meets the EPA guidelines. If 3 parts per million is the upper limit allowed by the EPA, the investigator would want to use sample data (daily pollution measurements) to decide whether the company is violating the law, i.e., to decide whether $\mu > 3$. If, say, a 99% confidence interval for μ contained only numbers greater than 3, then the EPA would be confident that the mean exceeds the established limit.

As a second example, consider a manufacturer who purchases terminal fuses in lots of 10,000. Suppose that the supplier of the fuses guarantees that no more than 1% of the fuses in any given lot are defective. Since the manufacturer cannot test each of the 10,000 fuses in a lot, he or she must decide whether to accept or reject a lot based on an examination of a sample of fuses selected from the lot. If the number x of defective fuses in a sample of, say, $n = 100$, is large, the lot will be rejected and sent back to the supplier. Thus, the manufacturer wants to decide whether the proportion π of defectives in the lot exceeds .01, based on information contained in a sample. If a confidence interval for π falls below .01, then the manufacturer will accept the lot and be confident that the proportion of defectives is less than 1%; otherwise, it will be rejected.

The examples in the preceding paragraphs illustrate how a confidence interval can be used to make a decision about a parameter. Note that both applications are one-directional: The EPA wants to determine whether $\mu > 3$, and the manufacturer wants to know if $\pi > .01$. (In contrast, if the manufacturer is interested in determining whether $\pi > .01$ or $\pi < .01$, the inference would be two-directional.)

Recall from Chapter 8 that to find the value of z (or t) used in a $(1 - \alpha)100\%$ confidence interval, the value of α is divided in half and $\alpha/2$ is placed in both the upper and lower tails of the z (or t) distribution. Consequently, confidence intervals are designed to be two-directional. Use of a two-directional technique in a situation where a one-directional method is desired will lead the researcher (e.g., the EPA or the manufacturer) to understate the level of confidence associated with the method. As we will explain in this chapter, hypothesis tests are appropriate for either one- or two-directional decisions about a population parameter.

This chapter will treat the general concepts involved in hypothesis testing; specific applications will be demonstrated in Chapter 10.

FORMULATION OF HYPOTHESES

9.2

When a researcher in any field sets out to test a new theory, he or she first formulates a **hypothesis**, or claim, that he or she believes to be true. For example, a college recruiter may claim that the mean starting salary of graduates of the College of Liberal Arts and Sciences is less than the mean starting salary of graduates of the College of Engineering. In statistical terms, the hypothesis that the researcher tries to establish is called the **alternative hypothesis**, or **research hypothesis**. To be paired with the alternative hypothesis is the **null hypothesis**, which is the "opposite" of the alternative hypothesis. In this way, the null and alternative hypotheses, both stated in terms of the appropriate population parameters, describe two possible states of nature that cannot simultaneously be true. When the researcher begins to collect information about the phenomenon of interest, he or she generally tries to present evidence that lends support to the alternative hypothesis. As you will subsequently learn, we take an indirect approach to obtaining support for the alternative hypothesis: Instead of trying to show that the alternative hypothesis is true, we attempt to produce evidence to show that the null hypothesis (which may often be interpreted as "no change from the status quo") is false.

DEFINITION 9.1

A statistical **hypothesis** is a statement about the value of a population parameter.

DEFINITION 9.2

The **alternative** (or **research**) **hypothesis**, denoted by H_a, is usually the hypothesis for which the researcher wants to gather supporting evidence.

DEFINITION 9.3

The **null hypothesis**, denoted H_0, is usually the hypothesis that the researcher wants to gather evidence against.

EXAMPLE 9.1   A metal lathe is checked periodically by quality control inspectors to determine whether it is producing machine bearings with a mean diameter of .5 inch. If the mean diameter of the bearings is larger or smaller than .5 inch, then the process is out of control and needs to be adjusted. Formulate the null and alternative hypotheses that could be used to test whether the bearing production process is out of control.

Solution   The hypotheses must be stated in terms of a population parameter. Thus, we define

μ = True mean diameter (in inches) of all bearings produced by the lathe

---

If either $\mu > .5$ or $\mu < .5$, then the metal lathe's production process is out of control. Since we wish to be able to detect either possibility, the null and alternative hypotheses would be

H_0: $\mu = .5$ (i.e., the process is in control)

H_a: $\mu \neq .5$ (i.e., the process is out of control)

EXAMPLE 9.2   Since 1970, cigarette advertisements have been required by law to carry the following statement: "Warning: The surgeon general has determined that cigarette smoking is dangerous to your health." However, this warning is often located in inconspicuous corners of the advertisements and printed in small type. Consequently, a spokesperson for the Federal Trade Commission (FTC) believes that over 80% of those who read cigarette advertisements fail to see the warning. Specify the null and alternative hypotheses that would be used in testing the spokesperson's theory.

Solution   The FTC spokesperson wants to make an inference about π, the true proportion of all readers of cigarette advertisements who fail to see the surgeon general's warning. In particular, the FTC spokesperson wishes to collect evidence to support the claim that π is greater than .80; thus, the null and alternative hypotheses are

H_0: $\pi \leq .80$

H_a: $\pi > .80$

The sign in H_0 is "$\leq$" because we want to cover all situations for which H_a *does not* occur. In other words, the event that H_0 occurs is the complement of the event that H_a occurs.

An accepted convention in hypothesis testing is to write H_0 with an equality sign ($=$). Consequently, in Example 9.2 we may also write the null hypothesis as H_0: $\pi = .80$. The reasoning is as follows: Since the alternative of interest is that $\pi > .80$, then any evidence that would cause you to reject the null hypothesis H_0: $\pi = .80$ in favor of H_a: $\pi > .80$ would also cause you to reject H_0: $\pi = \pi'$, for any value of π' that is *less than* .80. In other words, H_0: $\pi = .80$ represents the worst possible case, from the researcher's point of view, if the alternative hypothesis is *not* correct. Thus, for mathematical ease, we combine all possible situations for describing the opposite of H_a into one statement involving an equality.

An alternative hypothesis may hypothesize a change from H_0 in a particular direction, or it may merely hypothesize a change without specifying a direction. In Example 9.2, the researcher is interested in detecting departure from H_0 in a particular direction; interest focuses on whether the proportion of cigarette advertisement readers who fail to see the surgeon general's warning is *greater than* .80. This test is called a **one-tailed** (or **one-sided**) **test**. In contrast, Example 9.1 illustrates a **two-tailed** (or **two-sided**) **test** in which we are interested in whether the mean diameter of the machine bearings differs in either direction from .5 inch, i.e., whether the process is out of control.

DEFINITION 9.4

A **one-tailed test** of hypothesis is one in which the alternative hypothesis is directional, and includes either the "<" symbol or the ">" symbol.

DEFINITION 9.5

A **two-tailed test** of hypothesis is one in which the alternative hypothesis does not specify departure from H_0 in a particular direction; such an alternative will be written with the "$\neq$" symbol.

EXAMPLE 9.3 The incidence of tuberculosis among Dade County (Miami) residents is known to be no more than .0002 (i.e., no more than 2 cases per 10,000 people). After conducting medical checks of Haitian refugees arriving in Miami from their Caribbean island homeland, a medical researcher believes that these Haitians have a much higher incidence of tuberculosis. To check this belief, the researcher will test the null hypothesis

$$H_0: \quad \pi = .0002$$

where π is the true proportion of Haitians living in Miami who contract tuberculosis. Formulate the appropriate alternative hypothesis for the researcher.

Solution The medical researcher is interested in detecting whether the true incidence of tuberculosis in the Haitian population living in Miami is larger than .0002. Thus, the alternative hypothesis of interest to the researcher is

$$H_a: \quad \pi > .0002$$

Note that the null hypothesis

$$H_0: \quad \pi = .0002$$

actually represents all possible situations for which the incidence of tuberculosis among the Haitian refugees is no more than 2 cases per 10,000 people, i.e., $\pi \leq .0002$. Since the alternative is directional—that is, since the researcher is interested in detecting a departure from H_0 in the direction of values of π larger than .0002—a one-tailed test is to be performed.

EXAMPLE 9.4 A study was conducted to determine the impact of a multifunction workstation (MFWS) on the way managers work (*Datamation*, Feb. 15, 1986). Two groups of managers at a St. Louis–based defense agency took part in the study: a group of managers who currently use MFWS software and a control group of non-MFWS users. To determine whether the proportion of MFWS users who rely on the computer as their major information source differs from the corresponding proportion of non-MFWS users, the researcher tested the null hypothesis

$$H_0: \quad (\pi_1 - \pi_2) = 0$$

where π_1 and π_2 represent the true proportions of MFWS users and non-MFWS users, respectively, who rely on the computer as their major information source. Specify the appropriate alternative hypothesis for this test.

Solution   The researcher is interested only in detecting whether there is a difference between the proportions of MFWS and non-MFWS users who rely on the computer as their major information source. If there is a difference, then $\pi_1 \neq \pi_2$, or equivalently, the difference between proportions $(\pi_1 - \pi_2)$ differs from 0. Thus, the alternative hypothesis of interest to the researcher is the two-tailed alternative

$$H_a: \quad (\pi_1 - \pi_2) \neq 0$$

EXERCISES

APPLYING THE CONCEPTS

9.1   Explain the difference between an alternative hypothesis and a null hypothesis.

9.2   Formulate the appropriate null and alternative hypotheses for each of the following problems.

   a. Kimberly-Clark Corporation, the makers of Kleenex, periodically conducts market surveys to determine the average number of tissues used by people when they have a cold. Currently, the company puts 60 tissues in a box. Suppose marketing experts at the company want to test whether the mean number of tissues used by people with colds exceed 60.
   b. Cannibalism among chickens is common when the birds are confined in small areas. A breeder and seller of live chickens wants to test whether the mortality rate resulting from cannibalism is less than .04 for a certain breed of chickens.
   c. To determine whether car ownership has an effect on academic achievement, a university investigator wishes to test whether there is a difference between the mean grade point averages of students who own cars and students who do not own cars.
   d. A medical researcher would like to determine whether the proportion of males admitted to a hospital because of heart disease differs from the corresponding proportion of females.
   e. Federal scientists tracking storms in Florida must forecast whether the storms will become hurricanes. Suppose we want to test whether the accuracy rate of the forecasts (i.e., the proportion of times the scientists correctly forecast the outcome of the storm) is greater than .90.
   f. Clinical researchers speculate that children who drink milk fortified with calcium tend to develop stronger and denser bones as adults and, consequently, are less likely to suffer from osteoporosis (a crippling bone disease). One study, conducted at the University of Pittsburgh Health Center (*American Journal of Clinical Nutrition*, Aug. 1985), was designed to test whether the mean density of bones of women who drank milk with each meal as children is greater than the mean density of bones of women who drank milk less frequently.

9.3   State whether each of the tests in Exercise 9.2 is one-tailed or two-tailed.

9.3 The goal of any hypothesis test is to make a decision; in particular, we will decide whether to reject the null hypothesis, H_0, in favor of the alternative hypothesis, H_a. Although we would like to be able to make a correct decision always, we must remember that the decision will be based on sample information. Thus we are subject to make one of two types of error, as the following examples illustrate.

EXAMPLE 9.5 Identify the two types of errors that can be made in a hypothesis test.

Solution The null hypothesis can be either true or false; further, we will make a decision either to reject or not to reject the null hypothesis. Thus, four possible situations may arise in testing a hypothesis, as summarized in Table 9.1. Note that an error occurs in two of the four situations. We can reject H_0 when H_0 is true (a **Type I error**), or we can accept H_0 when H_0 is false (a **Type II error**).

TABLE 9.1
Conclusions and Consequences for Testing a Hypothesis

		TRUE STATE OF NATURE	
		H_0 true (H_a false)	H_0 false (H_a true)
DECISION	Accept H_0	Correct decision	Type II error
	Reject H_0	Type I error	Correct decision

Note that we risk a Type I error only if the null hypothesis is rejected, and we risk a Type II error only if the null hypothesis is accepted. Thus, we may make no error, or we may make either a Type I error or a Type II error, but not both. Notationally, let α represent the probability of a Type I error and let β represent the probability of a Type II error. There is an intuitively appealing relationship between the probabilities for the two types of error: *As α increases, β decreases; similarly, as β increases, α decreases. The only way to reduce α and β simultaneously is to increase the amount of information available in the sample, i.e., to increase the sample size.*

DEFINITION 9.6

A **Type I error** occurs if we reject a null hypothesis when it is true. The probability of committing a Type I error is usually denoted by α.

DEFINITION 9.7

A **Type II error** occurs if we accept a null hypothesis when it is false. The probability of making a Type II error is usually denoted by β.

EXAMPLE 9.6 Refer to Example 9.1. Specify what Type I and Type II errors would represent, in terms of the problem.

Solution   From Definition 9.6, a Type I error results from incorrectly rejecting the null hypothesis. In our example, this would occur if we conclude that the process is out of control when, in fact, the process is in control, i.e., if we conclude that the mean bearing diameter is different from .5 inch, when the mean is equal to .5 inch. The consequence of making such an error would be that unnecessary time and effort would be expended to repair a metal lathe that is operating properly.

From Definition 9.7, a Type II error results from incorrectly accepting the null hypothesis. This occurs if we conclude that the mean bearing diameter is equal to .5 inch when, in fact, the mean differs from .5 inch. The practical significance of making a Type II error is that the metal lathe would not be repaired, when, in fact, the process is out of control.

Subsequently, we will see that the probability of making a Type I error is controlled by the researcher (Section 9.4); thus, it is often used as a measure of the reliability of the conclusion and is called the **significance level** of the test.

DEFINITION 9.8

The probability, α, of making a Type I error is called the **level of significance** (or **significance level**) for a hypothesis test.

In practice, we will carefully avoid stating a decision in terms of "accept the null hypothesis H_0." Instead, if the sample does not provide enough evidence to support the alternative hypothesis H_a, we prefer a decision "fail to reject H_0," or "insufficient evidence to reject H_0." This is because, if we were to "accept H_0," the reliability of the conclusion would be measured by β, the probability of a Type II error. Unfortunately, the value of β is not constant, but depends on the specific alternative value of the parameter and is difficult to compute in most testing situations. (Guidelines on how to calculate an estimate of β are given in optional Section 9.7.)

In summary, we recommend the procedure outlined in the box on the next page for formulating hypotheses and stating conclusions.

EXAMPLE 9.7 The logic used in hypothesis testing has often been likened to that used in the courtroom in which a defendant is on trial for committing a crime. Assume that the judge has issued the standard instruction to the jury: The defendant should be acquitted unless evidence of guilt is beyond a "reasonable doubt."

a. Formulate appropriate null and alternative hypotheses for judging the guilt or innocence of the defendant.
b. Interpret the Type I and Type II errors in this context.
c. If you were the defendant, would you want α to be small or large? Explain.

Solution

a. Under our judicial system, a defendant is "innocent until proven guilty." That is, the burden of proof is *not* on the defendant to prove his or her innocence; rather, the court must collect sufficient evidence to support the claim that the defendant is guilty "beyond a reasonable doubt." Thus, the null and alternative hypotheses would be

H_0: Defendant is innocent

H_a: Defendant is guilty

b. The four possible outcomes are shown in Table 9.2. A Type I error would be to conclude that the defendant is guilty, when, in fact, he or she is innocent; a Type II error would be to conclude that the defendant is innocent, when he or she is guilty.

TABLE 9.2
Conclusions and Consequences in Example 9.7

		TRUE STATE OF NATURE	
		Defendant Is Innocent	Defendant Is Guilty
DECISION OF COURT	Defendant Is Innocent	Correct decision	Type II error
	Defendant Is Guilty	Type I error	Correct decision

c. Most would probably agree that the Type I error in this situation is by far the more serious. Thus, we would want α, the probability of committing a Type I error, to be very small indeed.

A convention that is generally observed when formulating the null and alternative hypotheses of any statistical test is to *state H_0 so that the possible error of incorrectly rejecting H_0 (Type I error) is considered more serious than the possible error of incorrectly accepting H_0 (Type II error)*. In many cases, the decision as to which type of error is more serious is admittedly not as clear-cut as that of Example 9.7; a little experience will help to minimize this potential difficulty.

EXERCISES
APPLYING THE CONCEPTS

9.4   Refer to Exercise 9.2. Interpret the Type I and Type II errors in the context of each problem.

9.5   Explain why each of the following statements is incorrect:
  a. The probability that the null hypothesis is correct is equal to α.
  b. If the null hypothesis is rejected, then the test proves that the alternative hypothesis is correct.
  c. In all statistical tests of hypothesis, $\alpha + \beta = 1$.

9.6   Why do we avoid stating a decision in terms of "accept the null hypothesis H_0"?

9.7   Last month, a large supermarket chain received many consumer complaints about the quantity of chips in 16-ounce bags of a particular brand of potato chips. Suspecting that the complaints were merely the result of the potato chips settling to the bottom of the bags during shipping, but wanting to be able to assure its customers they were getting their money's worth, the chain decided to test the following hypotheses concerning μ, the mean weight (in ounces) of a bag of potato chips in the next shipment of chips received from their largest supplier:

$$H_0: \quad \mu = 16$$
$$H_a: \quad \mu < 16$$

If there is evidence that $\mu < 16$, then the shipment would be refused and a complaint registered with the supplier.
  a. What is a Type I error, in terms of the problem?
  b. What is a Type II error, in terms of the problem?
  c. Which type of error would the chain's customers view as more serious? Which type of error would the chain's supplier view as more serious?

TEST STATISTICS AND REJECTION REGIONS

9.4   In this section we will describe how to arrive at a decision when testing hypotheses. Recall that when making any type of statistical inference (of which hypothesis testing is a special case), we collect information by obtaining a random sample from the population(s) of interest. In all our applications, we will assume that the appropriate sampling process has already been carried out.

EXAMPLE 9.8 Suppose we want to test the hypotheses

H_0: $\mu = 72$

H_a: $\mu > 72$

What is the general format for carrying out a statistical test of hypothesis?

Solution   Once we have specified H_0 and H_a (step 1), the second step is to obtain a random sample from the population of interest. The information provided by this sample, in the form of a sample statistic, will help us decide whether to reject the null hypothesis. The sample statistic upon which we base our decision is called the **test statistic**.

The third step is to determine a test statistic that is reasonable in the context of a given hypothesis test. For this example, we are hypothesizing about the value of the population mean μ. Since our best guess about the value of μ is the sample mean $\bar{x}$ (see Section 8.2), it seems reasonable to use $\bar{x}$ as a test statistic. We will learn how to choose the test statistic for other hypothesis-testing situations in the examples that follow.

The fourth step is to specify the range of possible computed values of the test statistic for which the null hypothesis will be rejected. That is, what specific values of the test statistic will lead us to reject the null hypothesis in favor of the alternative hypothesis? These specific values are known collectively as the **rejection region** for the test. For this example, we would need to specify the values of $\bar{x}$ that would lead us to believe that H_a is true, i.e., that μ is greater than 72. We will learn how to find an appropriate rejection region in later examples.

DEFINITION 9.9

The **test statistic** is a sample statistic, computed from the information provided by the sample, upon which the decision concerning the null and alternative hypotheses is based.

DEFINITION 9.10

The **rejection region** is the set of possible computed values of the test statistic for which the null hypothesis will be rejected.

Finally, in the fifth step, we make our decision by observing whether the computed value of the test statistic lies within the rejection region. If the computed value falls within the rejection region, we will reject the null hypothesis; otherwise, we do not reject the null hypothesis.

An outline of the hypothesis-testing procedure developed in Example 9.8 is given in the box. Each step in this approach will be explained in greater detail as we proceed.

OUTLINE FOR TESTING A HYPOTHESIS

1. Specify the **null** and **alternative hypotheses**, H_0 and H_a, and the **significance level**, α.
2. Obtain a **random sample** from the population(s) of interest.
3. Determine an appropriate **test statistic** and compute its value using the sample data.
4. Specify the **rejection region**. (This will depend on the value of α selected.)
5. Make the appropriate **conclusion** by observing whether the computed value of the test statistic lies within the rejection region. If so, reject the null hypothesis; otherwise, do not reject the null hypothesis.

Recall that the null and alternative hypotheses (step 1) will be stated in terms of specific population parameters. Thus, in step 3 we decide on a test statistic that will provide information about the target parameter.

EXAMPLE 9.9   Refer to Example 9.2. The spokesperson for the FTC wants to test

H_0: $\pi = .80$

H_a: $\pi > .80$

where π is the proportion of all readers of cigarette advertisements who fail to notice the surgeon general's warning. Suggest a test statistic that may be useful in deciding whether to reject H_0.

Solution   Since the target parameter is a population proportion π, it would be logical to use the sample proportion p as a tool in the decision-making process. Recall from Section 8.4 that p is the point estimate of π used in the interval estimation procedure.

EXAMPLE 9.10   A college recruiter believes that the mean starting salary of graduates with nursing degrees is less than the mean starting salary of graduates with journalism degrees. Consequently, the recruiter wants to test:

H_0: $(\mu_1 - \mu_2) = 0$

H_a: $(\mu_1 - \mu_2) < 0$

where μ_1 and μ_2 are the population mean starting salaries of all nursing graduates and journalism graduates, respectively. Suggest an appropriate test statistic in the context of this problem.

Solution   The parameter of interest is $(\mu_1 - \mu_2)$, the difference between the two population means. Therefore, we will use $(\bar{x}_1 - \bar{x}_2)$, the difference between the corresponding sample means, as a basis for deciding whether to reject H_0. If the difference between the sample means, $(\bar{x}_1 - \bar{x}_2)$, falls greatly below the hypothesized value of $(\mu_1 - \mu_2) = 0$, then we have evidence that disagrees with the null hypothesis. In fact, it would support the alternative hypothesis that $(\mu_1 - \mu_2) < 0$. Again, we are using the point estimate of the target parameter as the test statistic in the hypothesis-testing approach.

> ### GUIDELINE FOR STEP 3 OF HYPOTHESIS TESTING
>
> In general, when the hypothesis test involves a specific population parameter, **the test statistic to be used is based on the conventional point estimate of that parameter**.

In step 4, we divide all possible values of the test statistic (or a standardized version of it) into two sets: the **rejection region** and its complement. If the computed value of the test statistic falls within the rejection region, we reject the null hypothesis. If the computed value of the test statistic does not fall within the rejection region, we do not reject the null hypothesis.

EXAMPLE 9.11   Refer to Example 9.8. For the hypothesis test

H_0: $\mu = 72$
H_a: $\mu > 72$

indicate which decision you may make for each of the following values of the test statistic:

a. $\bar{x} = 110$ b. $\bar{x} = 59$ c. $\bar{x} = 73$

Solution

a. If $\bar{x} = 110$, then much doubt is cast upon the null hypothesis. In other words, *if the null hypothesis were true* (i.e., if μ is equal to 72), then it is very unlikely that we would observe a sample mean $\bar{x}$ as large as 110. We would thus reject the null hypothesis on the basis of information contained in this sample.
b. Since the alternative of interest is $\mu > 72$, this value of the sample mean, $\bar{x} = 59$, provides no support for H_a. Thus, we would *not* reject H_0 in favor of H_a: $\mu > 72$, based on this sample.

c. Does a sample value of $\bar{x} = 73$ cast sufficient doubt on the null hypothesis to warrant its rejection? Although the sample mean $\bar{x} = 73$ is larger than the null hypothesized value of $\mu = 72$, is this due to chance variation, or does it provide strong enough evidence to conclude in favor of H_a? We think you will agree that the decision is not as clear-cut as in parts **a** and **b**, and that we need a more formal mechanism for deciding what to do in this situation.

We now illustrate how to determine a rejection region that takes into account such factors as the sample size and the maximum probability of a Type I error that you are willing to tolerate.

EXAMPLE 9.12 Refer to Example 9.11. Specify completely the form of the rejection region for a test of

H_0: $\mu = 72$

H_a: $\mu > 72$

at a significance level of $\alpha = .05$.

Solution We are interested in detecting a directional departure from H_0; in particular, we are interested in the alternative that μ is *greater than* 72. Now, what values of the sample mean $\bar{x}$ would cause us to reject H_0 in favor of H_a? Clearly, values of $\bar{x}$ which are "sufficiently greater" than 72 would cast doubt on the null hypothesis. But how do we decide whether a value—say, $\bar{x} = 73$—is "sufficiently greater" than 72 to reject H_0? A convenient measure of the distance between $\bar{x}$ and 72 is the z score, which "standardizes" the value of the test statistic $\bar{x}$:

$$z = \frac{\bar{x} - \mu_{\bar{x}}}{\sigma_{\bar{x}}} = \frac{\bar{x} - 72}{\sigma/\sqrt{n}} \approx \frac{\bar{x} - 72}{s/\sqrt{n}}$$

(The z score is obtained by using the values of $\mu_{\bar{x}}$ and $\sigma_{\bar{x}}$ that would be valid if the null hypothesis were true, i.e., if $\mu = 72$.) *The z score then gives us a measure of how many standard deviations the observed $\bar{x}$ is from what we would expect to observe if H_0 were true.*

Now examine Figure 9.1a and observe that the chance of obtaining a value of $\bar{x}$ more than 1.645 standard deviations above 72 is only .05, *when the true value of μ is 72.* (We are assuming that the sample size is large enough to ensure that the sampling distribution of $\bar{x}$ is approximately normal.) Thus, if we observe a sample mean located more than 1.645 standard deviations above 72, then either H_0 is true and a relatively rare (with probability .05 or less) event has occurred, *or* H_a is true and the population mean exceeds 72. We would favor the latter explanation for obtaining such a large value of $\bar{x}$, and would then reject H_0.

In summary, our rejection region for this example consists of all values of z that are greater than 1.645 (i.e., all values of $\bar{x}$ that are more than 1.645 standard deviations above 72). The **critical value** 1.645 is shown in Figure 9.1b. In this situation, the probability of a Type I error—that is, deciding in favor of H_a if in fact H_0 is true—is equal to $\alpha = .05$.

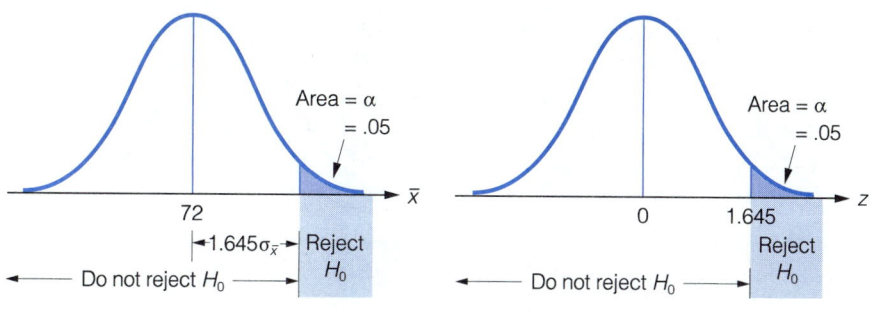

FIGURE 9.1

Location of Rejection Region for Example 9.12

a. Rejection region in terms of $\bar{x}$

b. Rejection region in terms of z

DEFINITION 9.11

In specifying the rejection region for a particular test of hypothesis, the value at the boundary of the rejection region is called the **critical value**.

EXAMPLE 9.13   Specify the form of the rejection region for a test of

H_0: $\mu = 72$

H_a: $\mu < 72$

at significance level $\alpha = .01$.

Solution   Here, we want to be able to detect the directional alternative that μ is *less than* 72; in this case, it is "sufficiently small" values of the test statistic $\bar{x}$ that would cast doubt on the null hypothesis. As in Example 9.12, we will standardize the value of the test statistic to obtain a measure of the distance between $\bar{x}$ and the null hypothesized value of 72:

$$z = \frac{\bar{x} - \mu_{\bar{x}}}{\sigma_{\bar{x}}} = \frac{\bar{x} - 72}{\sigma/\sqrt{n}} \approx \frac{\bar{x} - 72}{s/\sqrt{n}}$$

This z value tells us how many standard deviations the observed $\bar{x}$ is from what would be expected if H_0 *were true*. (Again, we have assumed that the sample size n is large so that the sampling distribution of $\bar{x}$ will be approximately normal. The appropriate modifications for small samples will be discussed in Chapter 10.)

Figure 9.2a (page 432) shows us that, *when the true value of μ is 72*, the chance of observing a value of $\bar{x}$ more than 2.33 standard deviations below 72 is only .01. Thus, at significance level (probability of Type I error) equal to .01, we would reject the null hypothesis for all values of z that are less than -2.33 (see Figure 9.2b), i.e., for all values of $\bar{x}$ that lie more than 2.33 standard deviations below 72.

FIGURE 9.2
Location of Rejection Region for
Example 9.13

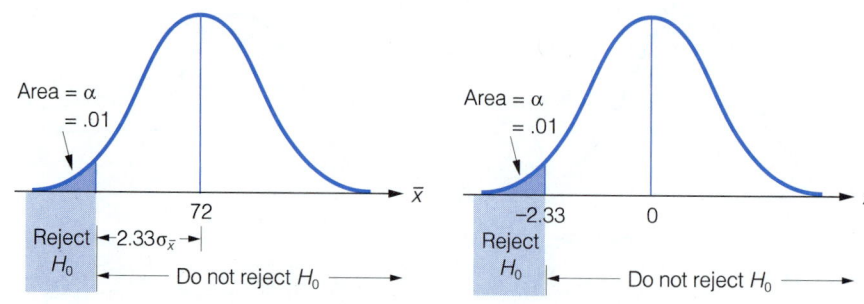

a. Rejection region in terms of $\bar{x}$ **b.** Rejection region in terms of z

EXAMPLE 9.14   Specify the form of the rejection region for a test of

$$H_0: \quad \mu = 72$$
$$H_a: \quad \mu \neq 72$$

where we are willing to tolerate a .05 chance of making a Type I error.

Solution   For this two-sided (nondirectional) alternative, we would reject the null hypothesis for "sufficiently small" *or* "sufficiently large" values of the standardized test statistic

$$z = \frac{\bar{x} - \mu_{\bar{x}}}{\sigma_{\bar{x}}} = \frac{\bar{x} - 72}{\sigma/\sqrt{n}} \approx \frac{\bar{x} - 72}{s/\sqrt{n}}$$

Now, from Figure 9.3a, we note that the chance of observing a sample mean $\bar{x}$ more than 1.96 standard deviations below 72 *or* more than 1.96 standard deviations above 72, *when H_0 is true*, is only $\alpha = .05$. Thus, the rejection region consists of two sets of values: We will reject H_0 if z is either less than -1.96 or greater than 1.96 (see Figure 9.3b). For this rejection rule, the probability of a Type I error is .05.

FIGURE 9.3
Location of Rejection Region for
Example 9.14

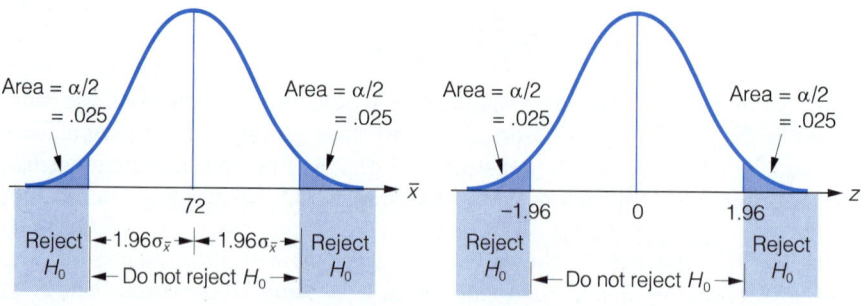

a. Rejection region in terms of $\bar{x}$ **b.** Rejection region in terms of z

The three previous examples all exhibit certain common characteristics regarding the rejection region, as indicated in the box.

GUIDELINES FOR STEP 4 OF HYPOTHESIS TESTING: SPECIFYING THE REJECTION REGION

1. The value of α, the probability of a Type I error, is specified in advance by the researcher. It can be made as small or as large as desired; typical values are $\alpha = .01, .02, .05$, and $.10$. For a fixed sample size, the size of the rejection region decreases as the value of α decreases (see Figure 9.4 on page 434). That is, for smaller values of α, more extreme departures of the test statistic from the null hypothesized parameter value are required to permit rejection of H_0.

2. The test statistic (i.e., the point estimate of the target parameter) is standardized to provide a measure of how great its departure is from the null hypothesized value of the parameter. The standardization is based on the sampling distribution of the point estimate, assuming H_0 is true. (It is through the standardization that the rejection rule takes into account the sample sizes.) For means and proportions, the general formula is:

$$\text{Standardized test statistic} = \frac{\text{Point estimate} - \text{Hypothesized value}}{\text{Standard deviation of point estimate}}$$

3. The location of the rejection region depends on whether the test is one-tailed or two-tailed, and on the prespecified significance level, α.

   a. For a one-tailed test in which the symbol ">" occurs in H_a, the rejection region consists of values in the upper tail of the sampling distribution of the standardized test statistic. The critical value is selected so that the area to its right is equal to α. (See Figure 9.1b.)

   b. For a one-tailed test in which the symbol "<" occurs in H_a, the rejection region consists of values in the lower tail of the sampling distribution of the standardized test statistic. The critical value is selected so that the area to its left is equal to α. (See Figure 9.2b.)

   c. For a two-tailed test, in which the symbol "$\neq$" occurs in H_a, the rejection region consists of two sets of values. The critical values are selected so that the area in each tail of the sampling distribution of the standardized test statistic is equal to $\alpha/2$. (See Figure 9.3b.)

Once we have chosen a test statistic from the sample information and computed its value (step 3), we determine if its standardized value lies within the rejection region so we can decide whether to reject the null hypothesis (step 5).

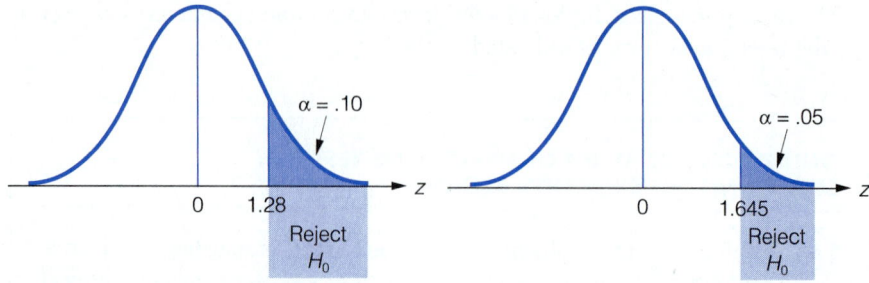

EXAMPLE 9.15 Refer to Example 9.12. Suppose the following statistics were calculated based on a random sample of $n = 30$ measurements: $\bar{x} = 73$, $s = 13$. Perform a test of

H_0: $\mu = 72$

H_a: $\mu > 72$

Solution In Example 9.12, we determined the following rejection rule for the given value of α and the alternative hypothesis of interest:

Reject H_0 if $z > 1.645$.

The standardized test statistic, computed assuming H_0 is true, is given by

$$z = \frac{\bar{x} - \mu_{\bar{x}}}{\sigma_{\bar{x}}} = \frac{\bar{x} - 72}{\sigma/\sqrt{n}} \approx \frac{\bar{x} - 72}{s/\sqrt{n}} = \frac{73 - 72}{13/\sqrt{30}} = .42$$

Since this value does not lie within the rejection region (shown in Figure 9.5), we fail to reject H_0 and conclude there is insufficient evidence to support the alternative hypothesis, H_a: $\mu > 72$. (Note that we do *not* conclude that H_0 is true; rather, we state that we have insufficient evidence to reject H_0.)

FIGURE 9.5

Location of Rejection Region and Test Statistic for Example 9.15

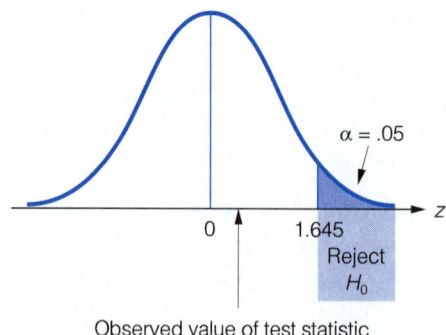

EXAMPLE 9.16 Recall from Example 9.1 that a metal lathe that produces machine bearings is out of control if the mean diameter of the bearings differs

from .5 inch. Suppose that a random sample of $n = 50$ bearings produced a mean diameter of $\bar{x} = .46$ inch and a standard deviation of $s = .075$ inch. Perform a test, at significance level .05, of

H_0: $\mu = .5$

H_a: $\mu \neq .5$

where μ is the true mean diameter of all machine bearings being produced by the lathe.

Solution The standardized test statistic is computed as follows:

$$z = \frac{\bar{x} - \mu_{\bar{x}}}{\sigma_{\bar{x}}} = \frac{\bar{x} - .5}{\sigma/\sqrt{50}} \approx \frac{\bar{x} - .5}{s/\sqrt{50}} = \frac{.46 - .5}{.075/\sqrt{50}} = -3.77$$

This value lies within the rejection region shown in Figure 9.6; therefore, we conclude that the mean diameter of the bearings is not equal to .5 inch. It appears that the production process is out of control. We acknowledge that we may be making a Type I error, with probability $\alpha = .05$.

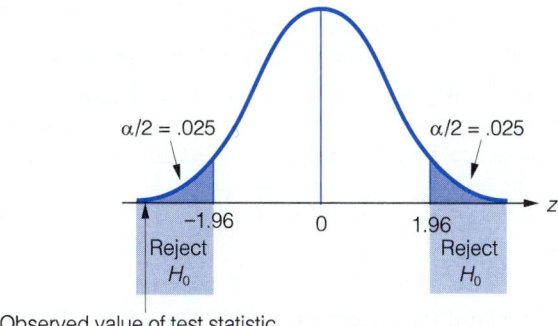

FIGURE 9.6
Location of Rejection Region and Test Statistic for Example 9.16

In the subsequent discussions on hypothesis testing, we will not differentiate between the test statistic (point estimate) and its standardized value. We will employ the common usage, in which *test statistic* refers to the standardized value of the point estimate for the target parameter. Thus, in Example 9.16, the value of the *test statistic* was computed to be $z = -3.77$.

EXERCISES
LEARNING THE MECHANICS

9.8 Suppose it is desired to test H_0: $\mu = 65$. Specify the form of the rejection region for each of the following (assume that the sample size will be sufficient to guarantee the approximate normality of the sampling distribution of $\bar{x}$):

a. H_a: $\mu \neq 65$, $\alpha = .02$ b. H_a: $\mu > 65$, $\alpha = .05$
c. H_a: $\mu < 65$, $\alpha = .01$ d. H_a: $\mu < 65$, $\alpha = .10$

9.9 Refer to Exercise 9.8. Calculate the test statistic for each of the following sample results:

a. $n = 100$, $\bar{x} = 60$, $s = 15$ b. $n = 50$, $\bar{x} = 60$, $s = 15$
c. $n = 50$, $\bar{x} = 70$, $s = 15$ d. $n = 50$, $\bar{x} = 70$, $s = 30$

9.10 Refer to Exercises 9.8 and 9.9. Give the appropriate conclusions for each of the tests, parts **a**, **b**, **c**, and **d**.

9.11 For each of the following rejection regions, determine the value of α, the probability of a Type I error:

a. $z < -1.96$ b. $z > 1.645$ c. $z < -2.58$ or $z > 2.58$

APPLYING THE CONCEPTS

9.12 Refer to Exercise 9.7. The supermarket chain randomly samples $n = 50$ bags of potato chips from the shipment and measures the weight of the chips in each. The mean weight of the sample was determined to be $\bar{x} = 15.7$ ounces and the sample standard deviation was $s = .8$ ounce.

a. Calculate the appropriate (standardized) test statistic for this test.
b. Specify the form of the rejection region if the level of significance is $\alpha = .01$. Locate the rejection region, α, and the critical value on a sketch of the standard normal curve.
c. Use the results of parts **a** and **b** to make the proper conclusion in terms of the problem.

9.13 A certain species of beetle produces offspring with either blue eyes or black eyes. Suppose a biologist wants to determine which, if either, of the two eye colors is dominant for this species of beetle. Let π represent the true proportion of offspring that possess blue eyes. If the beetles produce blue-eyed and black-eyed offspring at an equal rate, then $\pi = .5$. Thus, the biologist desires to test

H_0: $\pi = .5$

H_a: $\pi \neq .5$

Give the form of the rejection region if the biologist is willing to tolerate a Type I error probability of $\alpha = .05$. Locate the rejection region, α, and the critical value(s) on a sketch of the normal curve. (Assume that the sample size n is large.)

9.14 Each year, hundreds of thousands of investors hope to strike it rich by playing in a controversial U.S. government lottery. For $75, one can compete for a federal oil and gas lease that the Interior Department assumes is practically worthless. All the land in the lottery is supposed to have been checked by the government for oil and gas. However, the *Wall Street Journal* (Mar. 29, 1984) reports on evidence suggesting that the Interior Department is including valuable oil leases in the lottery, some worth millions of dollars. Of 328 winners in a July 1980 lottery of Wyoming lands, 184 (or 56%) were able to sell their leases to an oil company for a substantial profit. Yet the Interior Department claims that at most 10% of the leases in the lottery are salable to an oil company. Let π represent the true proportion of lottery leases that are salable to an oil company. Then we want to test

H_0: $\pi = .10$

H_a: $\pi > .10$

a. Give the form of the rejection region for $\alpha = .05$. Locate the rejection region, α, and the critical value on a sketch of the normal curve.
b. Calculate the value of the test statistic. [*Hint:* The standard deviation of the point estimate p is (from Section 8.4) $\sqrt{\pi(1 - \pi)/n}$, where π is the hypothesized value in H_0.]
c. In terms of the problem, what is the proper conclusion?

9.5 In 1980, the Schlitz Brewing Company reported that its beer sales had decreased 50% over a 5-year period. To revive depressed sales, CEO Frank Sellinger announced that Schlitz would broadcast a taste test featuring 100 beer drinkers on live television during halftime of the December 28, 1980, National Football League AFC wildcard playoff game between the Houston Oilers and the Oakland (now Los Angeles) Raiders.*

During the live broadcast, Schlitz claimed that the 100 beer drinkers selected for the taste test were "loyal" drinkers of Budweiser, the industry's best-selling beer. Each of the participants was served two beers, one Schlitz and one Budweiser, in unlabeled ceramic mugs. Tasters were then told to make a choice by pulling an electronic switch left or right in the direction of the beer they preferred. (Before the test, the tasters were informed that one of the mugs contained their regular beer, Budweiser, and the other contained Schlitz, but the ordering was not revealed.) The percentage of the 100 "loyal" Budweiser drinkers who preferred Schlitz was then tabulated live, in front of millions of football fans.

One beer industry observer was quoted as calling the test "a giant roll of the dice." However, CEO Sellinger disagreed that the move was a gamble: "Some people thought it was risky to do live TV taste tests. But it didn't take nerve, it just took confidence."

The results of the live TV taste test showed that 46 of the 100 "loyal" Budweiser beer drinkers preferred Schlitz. Schlitz, of course, labeled the outcome "an impressive showing" in a magazine advertisement following the test. For the purposes of this case, suppose that market experts hired by Schlitz informed the company that the taste test would be successful in boosting sales if more than 40 of the 100 Budweiser drinkers selected Schlitz as their favorite. Since 46 tasters pulled the switch in the direction of Schlitz, the brewer called the outcome "impressive," and anxiously awaited sales of Schlitz beer to increase. However, do these sample results indicate that the proportion of "loyal" Budweiser drinkers who prefer Schlitz is larger than 40%?

To answer this question, we employ a test of hypothesis about the proportion π of all "loyal" Budweiser drinkers who prefer Schlitz. If Schlitz's claim is correct, then we want to detect a value of π larger than .40. Thus, we want to test the null hypothesis that π equals .40 against the alternative hypothesis that π is larger than .40:

H_0: $\pi = .40$

H_a: $\pi > .40$

It would seem natural to base our decision on the value of the point estimator p of the population proportion π. This estimator, like the sample mean $\bar{x}$, is approximately normally distributed when the sample size n is large. Therefore, we will reject H_0 if the difference $(p - .4)$ is too large. "Too large" means too many standard deviations away from the hypothesized value of π, $\pi = .4$. Just as in the test of hypothesis about a population mean μ (Example 9.15), we

---

*Reported in the Orlando *Sentinel Star*, Dec. 11, 1980.

measure this distance, $p - .4$, in units of standard deviation of p and use the standardized test statistic

$$z = \frac{p - .4}{\sqrt{\dfrac{\pi(1 - \pi)}{n}}}$$

where the value of π used in the denominator of z is the value given in H_0 (i.e., $\pi = .40$). The rejection region for the test for $\alpha = .05$ will be values of z larger than 1.645 (see Figure 9.7).

FIGURE 9.7

Location of the Rejection Region and the Test Statistic for the Schlitz Mug-to-Mug Test

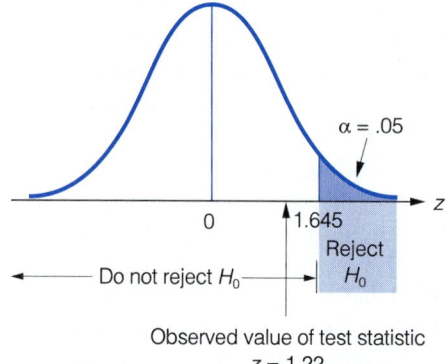

$\alpha = .05$

0   1.645

Reject H_0

Do not reject H_0

Observed value of test statistic
$z = 1.22$

For the Schlitz Mug-to-Mug test, 46 out of the 100 "loyal" Budweiser drinkers preferred Schlitz to Budweiser. Therefore, $p = \frac{46}{100} = .46$ and the test statistic is

$$z = \frac{p - .40}{\sqrt{\pi(1 - \pi)/n}} = \frac{.46 - .40}{\sqrt{(.40)(.60)/100}} = \frac{.06}{.049} = 1.22$$

This tells us that $p = .46$ lies 1.22 standard deviations above $\pi = .40$. The value of the test statistic, $z = 1.22$, does not fall in the rejection region. Therefore, the Schlitz Mug-to-Mug test does not provide sufficient evidence to indicate that more than 40% of "loyal" Budweiser drinkers prefer Schlitz.

This chapter case demonstrates the similarity of tests about population means and proportions. We will learn more about both of these tests in Chapter 10.

REPORTING TEST RESULTS: p-VALUES

9.6 The statistical hypothesis-testing technique that we have developed in Section 9.4 requires us to choose the significance level α (i.e., the maximum probability of a Type I error that we are willing to tolerate) before obtaining the data and computing the test statistic. By choosing α a priori, we in effect fix the rejection region for the test. Thus, no matter how large or how small the observed value of the test statistic, our decision regarding H_0 is clear-cut: Reject H_0 (i.e., conclude that the test results are statistically significant) if the observed value of the test statistic falls into the rejection region, and do not reject H_0 (i.e., conclude that the test results are insignificant) otherwise. This **"fixed" significance level**, α, then serves as a measure of the reliability of our inference. However, there is one drawback to a test conducted in this manner—

namely, a measure of the *degree* of significance of the test results is not readily available. That is, if the value of the test statistic falls into the rejection region, we have no measure of the extent to which the data disagree with the null hypothesis.

EXAMPLE 9.17 A large-sample test of H_0: $\mu = 72$ against H_a: $\mu > 72$ is to be conducted at a fixed significance level of $\alpha = .05$. Consider the following possible values of the computed test statistic:

$$z = 1.82 \quad \text{and} \quad z = 5.66$$

a. Which of these values of the test statistic provides stronger evidence for the rejection of H_0?

b. How can we measure the extent of disagreement between the sample data and H_0 for each of the computed values?

Solution

a. From Example 9.15, the appropriate rejection region for this test, at $\alpha = .05$, is given by

$$z > z_{.05} = 1.645$$

Clearly, for either of the previous test statistic values, $z = 1.82$ or $z = 5.66$, we will reject H_0; hence, the result in each case is statistically significant. Recall, however, that the appropriate test statistic for a large-sample test concerning μ is simply the z score for the observed sample mean $\bar{x}$, calculated by using the hypothesized value of μ in H_0 (in this case, $\mu = 72$). The larger the z score, the greater the distance (in units of standard deviations) that $\bar{x}$ is from the hypothesized value of $\mu = 72$. Thus, a z score of 5.66 would present stronger evidence that the true mean is larger than 72 than would a z score of 1.82. This reasoning stems from our knowledge of the sampling distribution of $\bar{x}$; if, in fact, $\mu = 72$, we would certainly not expect to observe an $\bar{x}$ with a score as large as 5.66.

b. One way of measuring the amount of disagreement between the observed data and the value of μ in the null hypothesis is to calculate the probability of observing a value of the test statistic equal to or greater than the actual computed value, if H_0 were true. That is, if z_c is the computed value of the test statistic, calculate

$$P(z \geq z_c)$$

assuming that the null hypothesis is true. This "disagreement" probability, or *p*-value, is calculated here for each of the computed test statistics, $z = 1.82$ and $z = 5.66$ using Table 4 of Appendix G:

$$P(z \geq 1.82) = .5 - .4656 = .0344$$
$$P(z \geq 5.66) \approx .5 - .5 = 0$$

From the discussion in part a, you can see that the smaller the *p*-value, the greater the extent of disagreement between the data and the null hypothesis— that is, the more significant the result.

In general, p-values for tests based on large samples are computed as shown in the box below. (p-values for small-sample tests are discussed in Chapter 10.)

MEASURING THE DISAGREEMENT BETWEEN THE DATA AND H_0: p-VALUES

Upper-tailed test: p-value $= P(z \geq z_c)$

Lower-tailed test: p-value $= P(z \leq z_c)$

Two-tailed test: p-value $= 2P(z \geq |z_c|)$

where z_c is the computed value of the test statistic and $|z_c|$ denotes the absolute value of z_c (which will always be positive).

Notice that the p-value for a two-tailed test is twice the probability for the one-tailed test. This is because the disagreement between the data and H_0 can be in two directions.

When publishing the results of a statistical test of hypothesis in journals, case studies, reports, etc., many researchers use p-values. Instead of selecting α a priori and then conducting a test as outlined in Section 9.4, the researcher will compute and report the value of the appropriate test statistic and its associated p-value. It is left to the reader of the report to judge the significance of the result—that is, the reader must determine whether to reject the null hypothesis in favor of the alternative, based on the reported p-value. This p-value is often referred to as the **observed significance level** of the test. Usually, the null hypothesis will be rejected if the observed significance level is *less* than the fixed significance level, α, chosen by the reader. There are two inherent advantages of reporting test results in this manner: (1) Readers are permitted to select the maximum value of α that they would be willing to tolerate in carrying out a standard test of hypothesis in the manner outlined in this chapter; and (2) It is an easy way to present the results of test calculations performed by a computer. Most statistical software packages perform the calculations for a test, give the value of the test statistic, and report the observed significance level (p-value) of the test; this makes it easy for the user to decide whether to reject H_0.

REPORTING TEST RESULTS AS p-VALUES: HOW TO DECIDE WHETHER TO REJECT H_0

---

1. Choose the maximum value of α that you are willing to tolerate.
2. If the observed significance level (p-value) of the test is less than the maximum value of α, then reject the null hypothesis.

EXAMPLE 9.18   Refer to Example 9.15 and the test of H_0: $\mu = 72$ versus H_a: $\mu > 72$. Compute the observed significance level of the test and interpret its value.

Solution In this large-sample test concerning a population mean μ, the computed value of the test statistic was $z_c = .42$. Since the test is upper-tailed, the associated p-value is given by

$$P(z \geq z_c) = P(z \geq .42) = .5 - .1628 = .3372$$

Thus, the observed significance level of the test is .3372. To reject the null hypothesis $H_0: \mu = 72$, we would have to be willing to risk a Type I error probability, α, of at least .3372. Most researchers would not be willing to take this risk and would deem the result insignificant (i.e., conclude that there is insufficient evidence to reject H_0).

EXAMPLE 9.19 Refer to Example 9.16 and the test of $H_0: \mu = .5$ versus H_a: $\mu \neq .5$.

a. Compute the observed significance level of the test.
b. Make the appropriate conclusion if you are willing to tolerate a Type I error probability of $\alpha = .01$.

Solution

a. The computed test statistic for this large-sample test about μ was given as $z_c = -3.77$. Since the test is two-tailed, the associated p-value is

$$2P(z \geq |z_c|) = 2P(z \geq |-3.77|) = 2P(z \geq 3.77)$$

Since $P(z \geq 3.77)$ is very near 0, the observed significance level of the test is approximately 0.

b. Since the approximate p-value of 0 is less than the maximum tolerable Type I error probability of $\alpha = .01$, we will reject H_0 and conclude that the mean diameter of the bearings is significantly different from .5 inch. In fact, we could choose an even smaller Type I error probability (e.g., $\alpha = .001$) and still have sufficient evidence to reject H_0. Thus, the result is highly significant.

Whether we conduct a test using p-values or the rejection region approach, our choice of a maximum tolerable Type I error probability becomes critical to the decision concerning H_0 and should not be hastily made. In either case, care should be taken to weigh the seriousness of committing a Type I error in the context of the problem.

EXERCISES

LEARNING THE MECHANICS

9.15   For a large-sample test of

$$H_0: (\mu_1 - \mu_2) = 0$$
$$H_a: (\mu_1 - \mu_2) > 0$$

compute the p-value associated with each of the following computed test statistic values:

a. $z_c = 1.96$ b. $z_c = 1.645$ c. $z_c = 2.67$ d. $z_c = 1.25$

9.16   For a large-sample test of

$$H_0: (\pi_1 - \pi_2) = 0$$
$$H_a: (\pi_1 - \pi_2) \neq 0$$

compute the p-value associated with each of the following computed test statistic values:

a. $z_c = -1.01$ b. $z_c = -2.37$ c. $z_c = 4.66$ d. $z_c = -1.45$

9.17   Give the approximate observed significance level of the test $H_0: \mu = 16$ for each of the following combinations of test statistic value and H_a:

a. $z_c = 3.05$, $H_a: \mu \neq 16$ b. $z_c = -1.58$, $H_a: \mu < 16$
c. $z_c = 2.20$, $H_a: \mu > 16$ d. $z_c = -2.97$, $H_a: \mu \neq 16$

APPLYING THE CONCEPTS

9.18   Refer to Exercises 9.7 and 9.12. Compute the observed significance level (p-value) of the test. Interpret the result.

9.19   Refer to Exercise 9.14. Compute the approximate observed significance level (p-value) of the test. What is your decision regarding H_0 if you are willing to risk a maximum Type I error probability of only $\alpha = .01$?

9.20   Refer to Exercise 9.2. For each problem, the p-value and the value of α for the appropriate test is given here. State the conclusions in the words of each problem.

a. p-value $= .217$, $\alpha = .10$ b. p-value $= .033$, $\alpha = .05$ c. p-value $= .001$, $\alpha = .05$
d. p-value $= .866$, $\alpha = .01$ e. p-value $= .025$, $\alpha = .01$

CALCULATING THE PROBABILITY OF A TYPE II ERROR AND THE POWER OF A TEST (OPTIONAL)

9.7   In Section 9.3 we discussed the problem with making the decision "accept H_0." The chance that we incorrectly accept H_0, i.e., accept H_0 when H_0 is in fact false, is β, the probability of a Type II error. Therefore, β serves as a measure of reliability for the decision "accept H_0." Unlike α, the value of β is typically not controlled by the researcher, and, in most testing situations, is very difficult to calculate. Consequently, when the test statistic does not fall in the rejection region, we state our conclusion as "do not reject H_0" rather than risk making a Type II error with unknown probability of occurrence.

In certain situations, however, *it is possible to calculate β for a specified value of the parameter in H_a*, as the following example illustrates.

EXAMPLE 9.20   Suppose we want to conduct the test of hypothesis of Example 9.15,

$$H_0: \quad \mu = 72$$
$$H_a: \quad \mu > 72$$

at significance level $\alpha = .05$. Recall that the test was based on a random sample of $n = 30$ measurements, with $\bar{x} = 73$ and $s = 13$. Calculate the value of β if the value of μ in H_a is $\mu_a = 75$.

Solution

STEP 1 The first step is to find the value of $\bar{x}$ on the border between the rejection region and the acceptance region. From Example 9.15, the rejection region is $z > 1.645$. Since the test statistic is calculated as

$$z = \frac{\bar{x} - 72}{\sigma/\sqrt{n}}$$

we can write the rejection region in terms of $\bar{x}$:

$$\frac{\bar{x} - 72}{\sigma/\sqrt{n}} > 1.645$$

or,

$$\bar{x} > 72 + 1.645 \left(\frac{\sigma}{\sqrt{n}}\right)$$

Substituting $n = 30$ and $s = 13$ (as an approximation for σ) into the expression for the rejection region, we obtain

$$\bar{x} > 72 + 1.645 \left(\frac{13}{\sqrt{30}}\right) = 75.90$$

Thus, at $\alpha = .05$, we will reject H_0 if the sample mean $\bar{x}$ exceeds 75.90. Or equivalently, we will accept H_0 if $\bar{x} < 75.90$. These rejection and acceptance regions are illustrated in Figure 9.8.

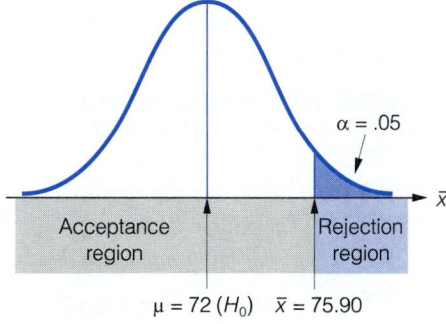

FIGURE 9.8
Rejection Region for Example 9.20, in Terms of $\bar{x}$

STEP 2 The next step is to write β as a probability statement involving $\bar{x}$:

$$\beta = P(\text{Type II error})$$
$$= P(\text{Accept } H_0 \text{ when } H_0 \text{ is false})$$
$$= P(\bar{x} < 75.90 \text{ when } \mu_a = 75)$$

[Note that we have substituted $\mu_a = 75$ for "H_0 is false," because this is the value of μ specified in H_a.] This probability is shown in Figure 9.9 (page 444).

FIGURE 9.9
Value of β When $\mu_a = 75$

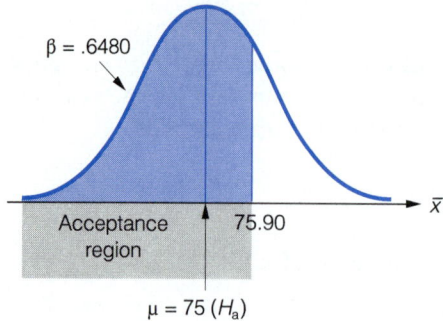

STEP 3   The final step is to calculate β using the standard normal (z) table (Table 4 of Appendix G):

$$\beta = P(\bar{x} < 75.90 \text{ when } \mu = 75)$$

$$= P\left(\frac{\bar{x} - \mu}{\sigma/\sqrt{n}} < \frac{75.90 - 75}{\sigma/\sqrt{n}}\right)$$

$$\approx P\left(z < \frac{75.90 - 75}{13/\sqrt{30}}\right)$$

$$= P(z < .38) = .5 + .1480 = .6480$$

Thus, the probability that the test procedure will lead us to incorrectly accept H_0: $\mu = 72$ when, in fact, $\mu = 75$ is approximately .65. The fact that this probability is so large implies that it will be difficult to detect departures from H_0 for values of μ close, but not equal, to 72.

GUIDELINES FOR CALCULATING β FOR A LARGE-SAMPLE TEST ABOUT μ

Consider a large-sample test of H_0: $\mu = \mu_0$ at significance level α. The value of β for a specific value of the alternative, $\mu = \mu_a$, is calculated as follows:

STEP 1   For one-tailed tests, find the value of $\bar{x}$ corresponding to the border of the rejection region. The calculation of this value, $\bar{x}_0$, depends on whether the test is upper-tailed or lower-tailed:

$$\textit{Upper-tailed test:} \quad \bar{x}_0 = \mu_0 + z_\alpha\left(\frac{\sigma}{\sqrt{n}}\right) \approx \mu_0 + z_\alpha\left(\frac{s}{\sqrt{n}}\right)$$

$$\textit{Lower-tailed test:} \quad \bar{x}_0 = \mu_0 - z_\alpha\left(\frac{\sigma}{\sqrt{n}}\right) \approx \mu_0 - z_\alpha\left(\frac{s}{\sqrt{n}}\right)$$

For two-tailed tests, two border values exist—one in the upper tail of the distribution ($\bar{x}_{0,U}$) and one in the lower tail ($\bar{x}_{0,L}$):

Two-tailed test: $\bar{x}_{0,U} \approx \mu_0 + z_{\alpha/2}\left(\dfrac{s}{\sqrt{n}}\right)$

$$\bar{x}_{0,L} \approx \mu_0 - z_{\alpha/2}\left(\dfrac{s}{\sqrt{n}}\right)$$

STEP 2 Write β as a probability involving the border value(s) of $\bar{x}$ and the alternative value of μ:

Upper-tailed test: $\beta = P(\bar{x} < \bar{x}_0 \text{ when } \mu = \mu_a)$

Lower-tailed test: $\beta = P(\bar{x} > \bar{x}_0 \text{ when } \mu = \mu_a)$

Two-tailed test: $\beta = P(\bar{x}_{0,L} < \bar{x} < \bar{x}_{0,U} \text{ when } \mu = \mu_a)$

STEP 3 Convert the border value(s) of $\bar{x}$ to z values in the probability statement of step 2. Then find the probability using Table 4 of Appendix G.

Upper-tailed test: $\beta \approx P\left(z < \dfrac{\bar{x}_0 - \mu_a}{s/\sqrt{n}}\right)$

Lower-tailed test: $\beta \approx P\left(z > \dfrac{\bar{x}_0 - \mu_a}{s/\sqrt{n}}\right)$

Two-tailed test: $\beta \approx P\left(\dfrac{\bar{x}_{0,L} - \mu_a}{s/\sqrt{n}} < z < \dfrac{\bar{x}_{0,U} - \mu_a}{s/\sqrt{n}}\right)$

EXAMPLE 9.21 Refer to Example 9.20. Calculate the value of β when the value of μ in H_a is $\mu_a = 80$.

Solution From step 1 of Example 9.20 we know that the "acceptance region" for the test with $\alpha = .05$ is $\bar{x} < 75.90$. Then for $\mu_a = 80$, we have

$$\beta = P(\text{Accept } H_0 \text{ when } H_0 \text{ is false})$$
$$= P(\bar{x} < 75.90 \text{ when } \mu_a = 80)$$
$$\approx P\left(\dfrac{\bar{x} - \mu_a}{\sigma/\sqrt{n}} < \dfrac{75.90 - 80}{13/\sqrt{30}}\right)$$
$$= P(z < -1.73)$$

From Table 4 of Appendix G, this probability is

$$\beta = .5 - .4582 = .0418 \qquad \text{(see Figure 9.10 on page 446)}$$

Examples 9.20 and 9.21 illustrate an important property of the statistical test of hypothesis: The value of β *decreases* as the true value of μ departs from the value hypothesized in H_0. In other words, the farther away the true value of μ lies from μ_0, the less likely you will be to incorrectly accept H_0 or, equivalently,

FIGURE 9.10
Value of β When $\mu_a = 80$

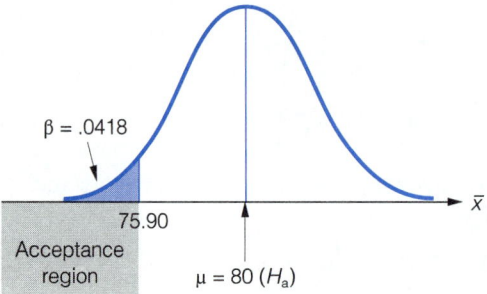

the more likely you will be to correctly reject H_0. Also, it can be easily shown that β decreases as α increases for fixed n, or as n increases for fixed α. These properties of β are summarized in the box.

PROPERTIES OF β FOR A TEST OF HYPOTHESIS ABOUT μ

1. β decreases as the distance between μ_0 and μ_a increases.
2. For fixed sample size n, β decreases as α increases.
3. For fixed significance level α, β decreases as n increases.

In practice, it is useful to interpret the value of $(1 - \beta)$, which is known as the **power of the test**. Since β and $(1 - \beta)$ are complementary probabilities, the power represents the probability that you will reject H_0 when H_0 is false, i.e., **the power of the test is the probability that you detect a departure from H_0 for a specific value of μ in H_a.** For example, the power of the test for $\mu_a = 80$ in Example 9.21 is

$$\text{Power} = 1 - \beta = 1 - .0418 = .9582$$

Thus, the test is very likely (probability of about .96) to lead to a rejection of H_0: $\mu = 72$ in favor of the alternative H_a: $\mu > 72$ when the true value of μ is $\mu_a = 80$. The greater the value of $(1 - \beta)$, the more powerful the test is for the particular alternative value of μ.

DEFINITION 9.12

The **power** of a test is the probability that the test will lead to a rejection of the null hypothesis H_0 when, in fact, the alternative hypothesis H_a is true. For a particular alternative, the power is equal to $(1 - \beta)$.

EXAMPLE 9.22 Refer to Example 9.21. Calculate the power of the test for the following values of μ_a: 73, 74, 75, . . . , 85. Plot these values on a graph. What pattern do you observe?

Solution For this upper-tailed $\alpha = .05$ test, the value of β for some alternative μ_a is given by

$$\beta = P\left(z < \frac{\bar{x}_0 - \mu_a}{s/\sqrt{n}}\right) = P\left(z < \frac{75.90 - \mu_a}{13/\sqrt{30}}\right)$$

Therefore, the power of the test is

$$\text{Power} = 1 - \beta \approx 1 - P\left(z < \frac{75.90 - \mu_a}{13/\sqrt{30}}\right)$$

Substituting $\mu_a = 73, 74, 75, \dots, 85$ into this equation and finding the respective probabilities from Table 4, Appendix G, we obtain the powers shown in Table 9.3.

A plot of powers in Table 9.3 against μ_a is shown in Figure 9.11. This graph is known as a **power curve** for the large-sample test.

Notice that the power of the test increases the farther μ_a deviates from the hypothesized value, $\mu_0 = 72$. That is, the greater the true value of μ deviates from the value hypothesized in H_0, the higher the probability of correctly rejecting H_0. Also, you can see that for alternative values of μ exceeding 83, the power of the test is approximately 1.

TABLE 9.3

Power of the Test H_0: $\mu = 72$ Versus H_a: $\mu > 72$ for Several Different Values of μ_a

μ_a	POWER
73	.1112
74	.2119
75	.3520
76	.5160
77	.6772
78	.8106
79	.9049
80	.9582
81	.9842
82	.9949
83	.9986
84	.9997
85	≈ 1.0000

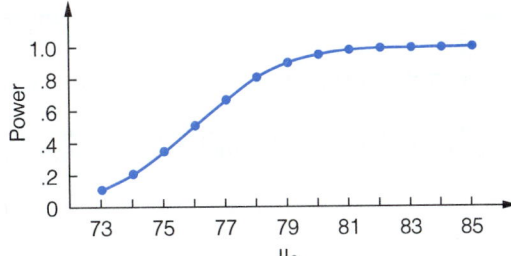

FIGURE 9.11
Power Curve for Example 9.22

Since the power equals $(1 - \beta)$, the properties of the power are similar to those of β. These properties are summarized in the box.

PROPERTIES OF POWER $= (1 - \beta)$ FOR A TEST OF HYPOTHESIS ABOUT μ

1. The power increases as the distance between μ_0 and μ_a increases.
2. For fixed sample size n, the power increases as α increases.
3. For fixed significance level α, the power increases as n increases.

When testing a hypothesis about a population parameter, we want to use the procedure with the highest power. In Chapter 10, we present testing procedures that have been shown to be most powerful for a general set of alternatives.

EXERCISES

LEARNING THE MECHANICS

9.21 Explain how you can increase the power of a test.

9.22 How are β and the power of a test for a fixed alternative related?

9.23 Suppose you want to test $H_0: \mu = \mu_0$ against $H_a: \mu > \mu_0$. Assuming $\sigma = 42$, find the value of $\bar{x}$ on the border of the rejection region if:

a. $\mu_0 = 1,000,\quad n = 36,\quad \alpha = .01$ b. $\mu_0 = 68,\quad n = 100,\quad \alpha = .05$
c. $\mu_0 = 7.8,\qquad n = 85,\quad \alpha = .10$

9.24 Suppose you want to test $H_0: \mu = 400$ against $H_a: \mu < 400$ at significance level $\alpha = .05$. Find the value of β for the following sets of values:

a. $n = 100,\quad s = 20,\quad \mu_a = 395$ b. $n = 50,\quad s = 20,\quad \mu_a = 395$
c. $n = 50,\quad s = 20,\quad \mu_a = 390$

9.25 Suppose you want to test $H_0: \mu = 25$ against $H_a: \mu \neq 25$. A sample size of $n = 100$ yielded a value of $s = 12$. Find the power of the test for the alternative $\mu_a = 23$ for each of the following values of α:

a. .10   b. .05   c. .01

APPLYING THE CONCEPTS

9.26 Refer to Examples 9.1 and 9.16. Calculate the power of the test if the true mean diameter of the bearings is .495 inch. Interpret your results.

9.27 Refer to Exercises 9.7 and 9.12. Calculate the power of the test if the true mean weight of the chips is 15 ounces. Interpret your result.

SUMMARY

In this chapter, we have introduced the logic and general concepts involved in the statistical procedure of *hypothesis testing*. The techniques will be illustrated more fully with practical applications in Chapter 10.

KEY WORDS

Alternative hypothesis	Power curve*
Critical value	Power of the test*
Fixed significance level	Rejection region
Hypothesis testing	Test statistic
Null hypothesis	Two-tailed test
Observed significance level	Type I error
One-tailed test	Type II error
p-value	

_____

*From the optional section in this chapter.

H_0: Null hypothesis

H_a: Alternative hypothesis

$\alpha = P(\text{Type I error}) = P(\text{Reject } H_0 \text{ when } H_0 \text{ is true})$

$\beta = P(\text{Type II error}) = P(\text{Accept } H_0 \text{ when } H_0 \text{ is false})$

Power $= 1 - \beta$*

SUPPLEMENTARY EXERCISES

[*Note:* Exercises marked with an asterisk (*) are from the optional section in this chapter.]

9.28 Explain the difference between the hypothesis and the alternative hypothesis in a statistical test.

9.29 In a test of hypothesis, is the size of the rejection region increased or decreased when the significance level α is reduced?

9.30 What are the two possible conclusions in a statistical test of hypothesis?

9.31 If the calculated value of the test statistic falls in the rejection region, we reject H_0 in favor of H_a. Does this prove that H_a is correct? Explain.

9.32 When do you risk making a Type I error? A Type II error?

9.33 Define each of the following:
a. Type I error   b. Type II error   c. α
d. β e. Critical value   f. Level of significance
g. One-tailed test   h. Two-tailed test   *i. Power

9.34 What happens to the power of the test as α is decreased?

9.35 Specify the form of the rejection region for a two-tailed test of hypothesis conducted at each of the following significance levels:
a. $\alpha = .01$ b. $\alpha = .02$ c. $\alpha = .04$
Locate the rejection region, α, and the critical values on a sketch of the standard normal curve for each part of the exercise. (Assume that the sampling distribution of the test statistic is approximately normal.)

9.36 For each of the following rejection regions, determine the value of α, the probability of a Type I error:
a. $z > 2.58$ b. $z < -1.29$ c. $z < -1.645$ or $z > 1.645$
Locate the rejection region, α, and the critical value(s) on a sketch of the standard normal curve.

9.37 Formulate the appropriate null and alternative hypotheses for each problem. Define all notation used.

a. As part of a study of the relationship between birth order and college success, an investigator wants to determine whether there is a difference in the proportions of college graduates and nongraduates who were firstborn or only children.

b. A manufacturer of fishing line wants to show that the mean breaking strength of a competitor's 22-pound line is really less than 22 pounds.

c. A craps player who has experienced a long run of bad luck at the craps table wants to test whether the casino dice are "loaded," i.e., whether the proportion of "sevens" occurring in many tosses of the two dice is different from $\frac{1}{6}$. (If the dice are fair, the probability of tossing a "seven" is $\frac{1}{6}$.)

---

*From the optional section in this chapter.

a. Write the null and alternative hypotheses for testing whether the mean IQ of high-lead children differs from the mean IQ of low-lead children.

b. Write the null and alternative hypotheses for testing whether the true proportion of children between the ages of 6 months and 5 years with a high level of lead in their blood exceeds .04.

c. Write the hypotheses for testing whether the mean level of lead in rural white children is less than the mean level of lead in inner-city black children.

d. For each test in parts a–c, what is a Type I error? A Type II error?

e. For each test in parts a–c, give the rejection region if $\alpha = .05$.

CASE STUDY 9.2

DRUG SCREENING: A STATISTICAL DECISION PROBLEM

Pharmaceutical companies are continually searching for new drugs. Charles W. Dunnett, in his essay,* "Drug Screening: The Never-Ending Search for New and Better Drugs," writes that

> Research chemists often know what types of chemical structures to look for to treat a particular disease, and the chemists can set about synthesizing compounds of the desired type. Sometimes, however, their knowledge may be vague, resulting in such a wide range of possibilities that many, many compounds have to be made and tested. In such a case, the search is very lengthy and requires years of effort by many people to develop a useful new drug.

Testing these thousands of compounds for the few that might be effective is known in the pharmaceutical industry as *drug screening*. Because of the obvious impact on human health, drug screening requires highly organized, efficient testing methods, and "anything that improves the efficiency of the testing procedure," writes Dunnett, "increases the chance of discovering a new cure."

Drug-screening techniques have improved tremendously over the years, and one of the major contributors of this continual improvement is the discipline of statistics. In fact, Dunnett views the drug-screening procedure in its preliminary stage in terms of a statistical decision problem: "In drug screening, two actions are possible: (1) to 'reject' the drug, meaning to conclude that the tested drug has little or no effect, in which case it will be set aside and a new drug selected for screening; and (2) to 'accept' the drug provisionally, in which case it will be subjected to further, more refined experimentation." Since it is the goal of the researcher to find a drug that effects a cure, the null and alternative hypotheses in a statistical test would take the following form:

H_0: Drug is ineffective in treating a particular disease.

H_a: Drug is effective in treating a particular disease.

---

*From Tanur, J. M., et al., eds. *Statistics: A Guide to the Unknown.* San Francisco: Holden-Day, 1978.

Dunnett comments on the possible errors associated with the drug-screening procedure: "To abandon a drug when in fact it is a useful one (a *false negative*) is clearly undesirable, yet there is always some risk in that. On the other hand, to go ahead with further, more expensive testing of a drug that is in fact useless (a *false positive*) wastes time and money that could have been spent on testing other compounds." Thus, to a statistician, a false positive result corresponds to a Type I error (i.e., to reject H_0 when in fact H_0 is true), and a false negative result corresponds to a Type II error (i.e., to fail to reject H_0 when in fact H_0 is false).

For this case study, we will consider the following hypothetical drug-screening experiment. A drug developed by a pharmaceutical company for possible treatment of cancerous tumors is to be screened. An investigator implants cancer cells in 100 laboratory mice. From this group, 50 mice are randomly selected and treated with the drug. The remaining 50 are left untreated, and comprise what is known as the *control group*. After a fixed length of time, the actual tumor weights of all the mice in the experiment are measured. If μ_1, the population mean tumor weight of all mice who could be treated with the drug, is significantly less than μ_2, the population mean tumor weight of all untreated mice, then the drug will be provisionally accepted and subjected to further testing; otherwise, the drug will be rejected.

a. Give the appropriate null and alternative hypotheses for the drug-screening test.
b. What are the Type I and Type II errors for this test? (Explain in terms of false positive and false negative results.) Which error is more serious?
c. Using a significance level of $\alpha = .01$, set up the rejection region for the test.
d. The experimental results are summarized in the accompanying table. Compute the difference between sample means, $(\bar{x}_1 - \bar{x}_2)$. Explain why we should not base our decision on this value alone.

TREATED GROUP	CONTROL GROUP
$\bar{x}_1 = 1.23$ grams	$\bar{x}_2 = 1.37$ grams
$s_1 = .55$ gram	$s_2 = .21$ gram

e. Use the results given in the table to calculate the required test statistic. [*Hint:* The standard deviation of the point estimate is

$$\sqrt{\frac{\sigma_1^2}{n_1} + \frac{\sigma_2^2}{n_2}} \approx \sqrt{\frac{s_1^2}{n_1} + \frac{s_2^2}{n_2}}$$

from Section 8.5.]
f. Should the pharmaceutical company provisionally accept the drug and subject it to further testing?

CONSUMER PRODUCT SAFETY WARNINGS—HOW EFFECTIVE ARE THEY?

According to the Consumer Product Safety Commission, millions of people are injured each year in accidents related to manufactured products. To minimize injury from product usage, the federal government now strictly enforces guidelines that require specific safety warnings to be placed on all potentially dangerous products. Although these regulations have led to a multitude of court cases, little research has been done to determine the impact of safety warnings on consumers.

To fill this void, Michael Ursic conducted a study* at the University of Florida involving 91 business undergraduates. Each student was presented with display boards containing information on price, smell, ease of application, and safety for two hypothetical brands of bug killers.† One of the two brands was randomly assigned a safety warning (such as "Danger: Do Not Inhale"), whereas the other had no safety message. Also, each nonsafety attribute (e.g., price, smell, ease of application) for each brand was randomly assigned a label ranging from very good to very poor. An example of a display board is given in Figure 9.12. The students were then instructed to rate each brand regarding its effectiveness and safety on a scale of 1 to 25 (where 1 = very poor and 25 = very good).

An analysis of the data led Ursic to conclude the following:

The brands of bug killers . . . with the safety warnings were perceived as significantly safer than the brands without the safety warnings. Therefore, it seems that a safety warning, instead of causing a person to perceive a product as more dangerous, may cause an individual to believe that the product is safer.

The bug killer brands that had the safety warning were also perceived as being significantly more effective than the other brands. Thus, people may use the safety warning as a surrogate indicator of the product's strength or quality.

FIGURE 9.12
Sample Display Board for Bug Killers

ATTRIBUTE	BRAND A	BRAND B
Price	Fair	Poor
Smell	Very good	Poor
Ease of application	Very poor	Very good
Safety message	None	Danger: Do not inhale

---

*Ursic, M. "The impact of safety warnings on perception and memory." *Human Factors*, Vol. 26, No. 6, Dec. 1984, pp. 677–682.

†The experiment actually included a third hypothetical brand on the display board.

Ursic based his inferences on the results of a statistical test of hypothesis. The parameter of interest is $\mu_d = \mu_1 - \mu_2$ where

$$\mu_1 = \text{Mean safety/effectiveness rating of brand A}$$
$$\text{(no safety warning)}$$

and

$$\mu_2 = \text{Mean safety/effectiveness rating of brand B}$$
$$\text{(safety warning)}$$

Since each student rated each of the two brands, the data were analyzed as matched pairs. That is, the differences

$$d_i = \text{(Rating for brand A)} - \text{(Rating for brand B)}$$

were calculated for each of the 91 students in the sample and then the test was conducted on the differences. In Chapter 11, we will give the formula for the test statistic when the data are collected as matched pairs and the sample size (i.e., number of pairs) is small. Since the sample size is large ($n = 91$ students) in this experiment, the test statistic for testing H_0: $\mu_d = 0$ is approximated by

$$z = \frac{\bar{d} - 0}{s_d}$$

where

$\bar{d} = $ Sample mean of the 91 differences

$s_d = $ Standard deviation of the 91 differences

a. Suppose we want to detect whether the brand of bug killer with the safety warning (brand B) is perceived as being significantly safer (i.e., has a higher mean safety rating) than the brand with no safety warning (brand A). Write the appropriate null and alternative hypotheses in terms of the parameter of interest, μ_d.

b. Give the rejection region for the test described in part a, for $\alpha = .01$.

c. The exact value of the test statistic was not reported in the article. However, from the information given we know the value falls below $z = -2.33$. In terms of the problem, what is the appropriate conclusion?

REFERENCES

McClave, J. T. and Dietrich, F. H. *Statistics*, 5th ed. San Francisco: Dellen, 1991.

Mendenhall, W. *Introduction to Probability and Statistics*, 7th ed. Boston: Duxbury, 1987.

Mendenhall, W., Wackerly, D. D., and Scheaffer, R. L. *Mathematical Statistics with Applications*, 4th ed. Boston: PWS-Kent, 1989.

Snedecor, G. W. and Cochran, W. G. *Statistical Methods*, 7th ed. Ames, Iowa: Iowa State University Press, 1980.

Tanur, J. M., Mosteller, F., Kruskal, W. H., Link, R. F., Pieters, R. S., and Rising, G. R., eds. *Statistics: A Guide to the Unknown*. San Francisco: Holden-Day, 1978.

HYPOTHESIS TESTING: APPLICATIONS

CHAPTER 10

What actually caused the downfall of the Roman Empire? One popular theory is that lead poisoning, contracted by drinking wine brewed in lead pots, was a major contributor to the Empire's collapse. The recently discovered skeletal remains of 55 Romans killed by the eruption of Mount Vesuvius in A.D. 79 were found to have a sample mean level of lead of 84 parts per million. This is in contrast to a sample mean level of lead of 35 parts per million found in the average modern American. Does this imply a difference between the mean levels of lead found in the early Romans and modern Americans? We will learn how to answer this and similar questions in this chapter, and we will examine the lead data findings more closely in the chapter case in Section 10.5.

**DIAGNOSING A
HYPOTHESIS TEST:
DETERMINING THE
TARGET PARAMETER**

10.1 In this chapter, we will present applications of the hypothesis-testing logic developed in Chapter 9. Among the population parameters to be considered are those for which we developed estimation procedures in Chapter 8: μ, $(\mu_1 - \mu_2)$, π, and $(\pi_1 - \pi_2)$.

The concepts of a hypothesis test are the same for all these parameters; the null and alternative hypotheses, test statistic, and rejection region all have the same general form (see Chapter 9).* However, the manner in which the test statistic is actually computed depends on the parameter of interest. For example, in Chapter 9 we saw that the large-sample test statistic for testing a hypothesis about a population mean μ is given by

$$z = \frac{\bar{x} - \mu_0}{\sigma/\sqrt{n}} \quad \text{(see Example 9.12)}$$

whereas the test statistic for testing a hypothesis about the parameter π is

$$z = \frac{p - \pi}{\sqrt{\pi_0(1 - \pi_0)/n}} \quad \text{(see Section 9.5)}$$

The key, then, to correctly diagnosing a hypothesis test is first to determine the parameter of interest—a task that can sometimes present difficulties for the introductory statistics student, especially when the parameter is stated in words rather than symbols.

Those who are routinely successful in diagnosing a hypothesis test generally follow a three-step process. First, identify the experimental unit (i.e., the objects upon which the measurements are taken). Second, identify the type of variable, quantitative or qualitative, measured on each experimental unit. Third, determine the target parameter based on the phenomenon of interest and the variable measured. For quantitative data, the target parameter will be either a population mean or a variance; for qualitative data the parameter will be a population proportion.

Often, there are one or more key words in the statement of the problem that indicate the appropriate population parameter. In this section, we will present several examples illustrating how to determine the parameter of interest. First, we state in the box the key words to look for when conducting a hypothesis test about a population parameter.

EXAMPLE 10.1 The "Pepsi Challenge" was a marketing strategy recently used by Pepsi-Cola. A consumer is presented with two cups of cola and asked to select the one that tastes best. Unknown to the consumer, one cup is filled with Pepsi, the other with Coke. Marketers of Pepsi claim that the true fraction of consumers who select their product will exceed .50. What is the parameter of interest to the Pepsi marketers?

---

*In optional Sections 10.8 and 10.9, we consider the population parameters σ^2 and σ_1^2/σ_2^2, respectively. For these cases, the form of the test statistic and rejection region will be different.

DIAGNOSING A HYPOTHESIS TEST:
DETERMINING THE TARGET PARAMETER

PARAMETER	KEY WORDS OR PHRASES
μ	Mean; average
$(\mu_1 - \mu_2)$	Difference in means or averages; mean difference, comparison of means or averages
π	Proportion; percentage; fraction; rate
$(\pi_1 - \pi_2)$	Difference in proportions, percentages, fractions, or rates; comparison of proportions, percentages, fractions, or rates
σ^2 (Optional)	Variance; variation; spread; precision
$\dfrac{\sigma_1^2}{\sigma_2^2}$ (Optional)	Ratio of variances; difference in variation; comparison of variances

---

Solution   In this problem, the experimental units are the consumers and the variable measured is *qualitative*—the consumer chooses either Pepsi or Coke. The key word in the statement of the problem is "fraction." Thus, the parameter of interest is π, where

π = True fraction of consumers who favor Pepsi over Coke in the taste test

To test the hypothesis, the marketers will need to present a sample of consumers with the Pepsi Challenge and determine the number in the sample who select Pepsi over Coke.

EXAMPLE 10.2   A dietitian wants to determine whether a new liquid protein diet is effective in reducing the weight of obese persons. The dietitian will conclude that the diet is effective if the mean weight of obese persons after 1 month of dieting is significantly less than their mean initial weight. What is the parameter of interest to the dietitian?

Solution   In this problem, the experimental units are obese persons and the variables measured are initial weight and weight after dieting—both *quantitative* in nature. The key word in the statement of this problem is "mean" and, although the word "difference" or "comparison" does not appear, it is clear that there are two means to be compared:

μ_1 = Mean weight before dieting

and

μ_2 = Mean weight 1 month after dieting

Thus, the parameter of interest is $(\mu_1 - \mu_2)$. Furthermore, the sample data that the dietitian will collect will be *paired* since two measurements—initial weight and weight after dieting—will be recorded for each obese person in the study. Therefore, a hypothesis test for matched pairs is appropriate.

EXAMPLE 10.3 A quality control engineer wants to determine whether a difference exists in the proportions of defective bolts produced by two different assembly lines. What is the parameter of interest?

Solution For the quality control engineer, the experimental units are the bolts produced by the assembly lines, and the variable measured is *qualitative*—either the bolt is defective or it is not. The two key words in the statement of the problem are "difference" and "proportions." Therefore, the parameter of interest is $(\pi_1 - \pi_2)$, where

π_1 = Proportion of bolts produced by assembly line 1 that are defective

and

π_2 = Proportion of bolts produced by assembly line 2 that are defective

In the following sections we present a summary of the hypothesis-testing procedures for each of the parameters listed in the previous box. In the exercises that follow each section, the target parameter can easily be identified by simply noting the title of the section. However, to properly diagnose a hypothesis test, it is essential to search for the key words in the statement of the problem. For practice, we strongly recommend that you read through all the Supplementary Exercises at the end of the chapter (where the specific parameter is not known a priori) and determine the parameter of interest before attempting to analyze the data.

TESTING A POPULATION MEAN

10.2 When testing a hypothesis about a population mean, μ, the procedure that we use will depend on whether the sample size n is large (say, $n \geq 30$) or small. The accompanying box contains the elements of a large-sample hypothesis test for μ based on the z statistic. Note that for this case, the only assumption required for the validity of the procedure is that the random sample is, in fact, large so that the sampling distribution of $\bar{x}$ is normal. Technically, the true population standard deviation σ must be known to use the z statistic, and this is rarely, if ever, the case. However, we established in Chapter 8 that when n is large, the sample standard deviation s provides a good approximation to σ and the z statistic can be approximated as shown in the box.

LARGE SAMPLES

EXAMPLE 10.4 Prior to the institution of a new safety program, the average number of on-the-job accidents per day at a factory was 4.5. To determine whether the safety program has been effective in reducing the average number of accidents per day, a random sample of 120 days is taken after the institution of the new safety program, and the number of accidents per day is recorded. The sample mean and standard deviation were computed as follows:

$$\bar{x} = 3.7 \qquad s = 2.6$$

a. Is there sufficient evidence to conclude (at significance level .01) that the average number of on-the-job accidents per day at the factory has decreased since the institution of the safety program?
b. What is the practical interpretation of the test statistic computed in part **a**?

Solution

a. To determine whether the safety program was effective, we will conduct a test of

> H_0: $\mu = 4.5$ (i.e., no change in average number of on-the-job accidents per day)
>
> H_a: $\mu < 4.5$ (i.e., average number of on-the-job accidents per day has decreased)

where μ represents the mean number of on-the-job accidents per day at the factory after institution of the new safety program. Note that the sample size $n = 120$ is sufficiently large so that the sampling distribution of $\bar{x}$ is approximately normal and that s provides a good approximation to σ. Since the required assumption is satisfied, we may proceed with a large-sample test about μ.

this section. Therefore, just like z, the computed value of t indicates the direction and approximate distance (in units of standard deviations) that the sample mean, $\bar{x}$, is from the hypothesized population mean, μ_0.

EXAMPLE 10.7 The building specifications in a certain city require that the sewer pipe used in residential areas have a mean breaking strength of more than 2,500 pounds per lineal foot. A manufacturer who would like to supply the city with sewer pipe has submitted a bid and provided the following additional information: An independent contractor randomly selected seven sections of the manufacturer's pipe and tested each for breaking strength. The results (pounds per lineal foot) are shown here:

$$2{,}610 \quad 2{,}750 \quad 2{,}420 \quad 2{,}510 \quad 2{,}540 \quad 2{,}490 \quad 2{,}680$$

a. Compute $\bar{x}$ and s for the sample.
b. Is there sufficient evidence to conclude that the manufacturer's sewer pipe meets the required specifications? Use a significance level of $\alpha = .10$.
c. Compute the p-value of the test, part b. Interpret the result.

Solution

a. The data were entered into a computer, and descriptive statistics were generated using SPSS. The SPSS printout is shown in Figure 10.4. The values of the sample mean breaking strength and standard deviation of breaking strengths (shaded) are $\bar{x} = 2{,}571.4$ and $s = 115.1$.

FIGURE 10.4

SPSS Printout for Example 10.7

Paired samples t-test: STRENGTH
                         MU

Variable	Number of Cases	Mean	Standard Deviation	Standard Error
STRENGTH	7	2571.4286	115.098	43.503
MU	7	2500.0000	.000	.000

(Difference) Mean	Standard Deviation	Standard Error	2-Tail Corr. Prob.	t Value	Degrees of Freedom	2-Tail Prob.
71.4286	115.098	43.503	. .	1.64	6	.152

b. The relevant hypothesis test has the following elements:

H_0: $\mu = 2{,}500$ (i.e., the manufacturer's pipe does not meet the city's specifications)

H_a: $\mu > 2{,}500$ (i.e., the pipe meets the specifications)

where μ represents the true mean breaking strength (in pounds per lineal foot) for all sewer pipe produced by this manufacturer.

This small-sample ($n = 7$) test requires the assumption that the relative frequency distribution of the population values of breaking strength for the manufacturer's pipe is approximately normal. Then the test will be based on a t distribution with $(n - 1) = 6$ degrees of freedom. We will thus reject H_0 if

$$t > t_{.10} = 1.440 \quad \text{(see Figure 10.5)}$$

Substituting the values $\bar{x} = 2{,}571.4$ and $s = 115.1$ yields the test statistic

$$t = \frac{\bar{x} - \mu_0}{s/\sqrt{n}} = \frac{2{,}571.4 - 2{,}500}{115.1/\sqrt{7}} = 1.64$$

This value is also shaded on the SPSS printout of the test results, Figure 10.4. Since this value of t is larger than the critical value of 1.440, we reject H_0. There is sufficient evidence (at significance level $\alpha = .10$) that the manufacturer's pipe meets the city's building specifications.

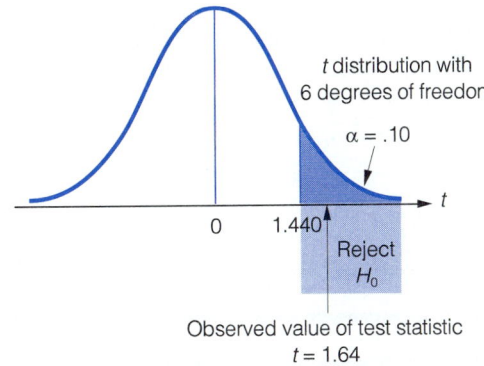

FIGURE 10.5

Rejection Region for Example 10.7

c. p-values for small-sample tests are computed in the same way as those for large-sample tests, except that we use the t distribution rather than the z distribution. For this upper-tailed test, the computed value of the test statistic is $t_c = 1.64$, and the associated p-value is

$$P(t \geq t_c) = P(t \geq 1.64)$$

where the distribution of t is based on $(n - 1) = 6$ degrees of freedom. To find the p-value from the table of critical t values provided in Table 5 of Appendix G, search for the value 1.64 in the row corresponding to 6 df. You can see that 1.64 does not appear in this row but falls between the values 1.943 (in the $t_{.05}$ column) and 1.440 (in the $t_{.10}$ column). The p-value associated with 1.943 is .05, and the p-value associated with 1.440 is .10. Thus, the p-value associated with $t_c = 1.64$ is somewhere between .05 and .10. Since the exact p-value is unknown, we take the conservative approach and report the approximate p-value as the larger of the two endpoints—namely, .10. This (approximate) p-value indicates that the null hypothesis $H_0: \mu = 2{,}500$ will be rejected in favor of $H_a: \mu > 2{,}500$ for any fixed significance level α larger than or equal to .10.

Although we can approximate the p-value using Table 5 of Appendix G, usually we will rely on the computer to calculate its exact value. The two-tailed p-value for the test of part **b** is shaded on the SPSS printout, Figure

10.4. Recall (Section 9.6) that the *p*-value for a one-tailed test is half that for a two-tailed test; consequently, the exact *p*-value $= .152/2 = .076$.

Remember that the small-sample test of Example 10.7 requires the assumption that the sampled population has a relative frequency distribution that is approximately normal. If you know that the population is highly skewed, then any inferences derived from the *t* test are suspect. In this case, we do not perform the *t* test but use one of the nonparametric statistical methods discussed in Chapter 15.

WARNING

When the sampled population is decidedly nonnormal (e.g., highly skewed), any inferences derived from the small-sample *t* test for μ are suspect. In this case, one alternative is to use the nonparametric sign test discussed in Section 15.2.

EXERCISES
LEARNING THE MECHANICS

10.1 Compute the value of the test statistic z for each of the following situations:

a. H_0: $\mu = 9.8$, H_a: $\mu > 9.8$; $\bar{x} = 10.0$, $s = 4.3$, $n = 50$
b. H_0: $\mu = 80$, H_a: $\mu < 80$; $\bar{x} = 75$, $s^2 = 19$, $n = 86$
c. H_0: $\mu = 8.3$, H_a: $\mu \neq 8.3$; $\bar{x} = 8.2$, $s = .79$, $n = 175$

10.2 A random sample of n observations is selected from a population with unknown mean μ and variance σ^2. For each of the following situations, specify the test statistic and rejection region.

a. H_0: $\mu = 50$, H_a: $\mu > 50$; $n = 36$, $\bar{x} = 60$, $s^2 = 64$; $\alpha = .05$
b. H_0: $\mu = 140$, H_a: $\mu \neq 140$; $n = 40$, $\bar{x} = 143.2$, $s = 9.4$; $\alpha = .01$
c. H_0: $\mu = 10$, H_a: $\mu < 10$; $n = 50$, $\bar{x} = 9.5$, $s = .35$; $\alpha = .10$

10.3 To test the null hypothesis H_0: $\mu = 10$, a random sample of n observations is selected from a normal population. Specify the rejection region for each of the following combinations of H_a, n, and α:

a. H_a: $\mu \neq 10$, $n = 15$, $\alpha = .05$
b. H_a: $\mu \neq 10$, $n = 15$, $\alpha = .01$
c. H_a: $\mu < 10$, $n = 15$, $\alpha = .05$
d. H_a: $\mu > 10$, $n = 5$, $\alpha = .10$
e. H_a: $\mu > 10$, $n = 25$, $\alpha = .10$

10.4 A random sample of n observations is selected from a normal population. For each of the following situations, specify the rejection region, test statistic, and conclusion.

a. H_0: $\mu = 3,000$ H_a: $\mu \neq 3,000$, $\bar{x} = 2,958$, $s = 39$, $n = 8$, $\alpha = .05$
b. H_0: $\mu = 6$, H_a: $\mu > 6$, $\bar{x} = 6.3$, $s = .3$, $n = 7$, $\alpha = .01$
c. H_0: $\mu = 22$, H_a: $\mu < 22$, $\bar{x} = 13.0$, $s = 6$, $n = 17$, $\alpha = .05$

10.5 A random sample of 49 measurements produced the following sums:

$$\sum x = 50.3 \qquad \sum x^2 = 68$$

    a. Test the null hypothesis that $\mu = 1.18$ against the alternative that $\mu < 1.18$. Use $\alpha = .01$.
    b. Test the null hypothesis that $\mu = 1.18$ against the alternative that $\mu < 1.18$. Use $\alpha = .10$.
    c. Find the p-value of the test, part **a**.
    d. Find the p-value of the test, part **b**.

10.6 A random sample of five measurements from a normally distributed population yielded the following data:

    12 4 3 5 5

    a. Test the null hypothesis that $\mu = 4$, against the alternative hypothesis that $\mu \ne 4$. Use $\alpha = .01$.
    b. Test the null hypothesis that $\mu = 4$, against the alternative hypothesis that $\mu > 4$. Use $\alpha = .01$.
    c. Find the p-value of the test, part **a**.
    d. Find the p-value of the test, part **b**.

APPLYING THE CONCEPTS

[*Note:* In all the Applying the Concepts exercises for this chapter, you should carefully define any notation used, perform all steps of the relevant hypothesis test, state a conclusion in terms of the problem, and specify any assumptions required for the validity of the procedure.]

10.7 Refer to the *Science* (Nov. 1988) study of inbreeding in tropical swarm-founding wasps, Exercise 8.8. A sample of 197 wasps, captured, frozen, and subjected to a series of genetic tests, yielded a sample mean inbreeding coefficient of $\bar{x} = .044$ with a standard deviation of $s = .884$. Recall that if the wasp has no tendency to inbreed, the true mean inbreeding coefficient μ for the species will equal 0.

    a. Test the hypothesis that the true mean inbreeding coefficient μ for this species of wasp exceeds 0. Use $\alpha = .05$.
    b. Compare the inference, part **a**, to the inference obtained in Exercise 8.8 using a confidence interval. Do the inferences agree? Explain.

10.8 The effect of machine breakdowns on the performance of a manufacturing system was investigated using computer simulation (*Industrial Engineering*, Aug. 1990). The simulation study focused on a single machine tool system with several characteristics, including a mean interarrival time of 1.25 minutes, a constant processing time of 1 minute, and a machine that breaks down 10% of the time. After $n = 5$ independent simulation runs of length 160 hours, the mean throughput per 40-hour week was $\bar{x} = 1,908.8$ parts. For a system with no breakdowns, the mean throughput for a 40-hour week will be equal to 1,920 parts. Assuming the standard deviation of the five sample runs was $s = 18$ parts per 40-hour week, test the hypothesis that the true mean throughput per 40-hour week for the system is less than 1,920 parts. Test using $\alpha = .05$.

10.9 How do the makers of Kleenex know how many tissues to put in a box? According to the *Wall Street Journal* (Sept. 21, 1984), the marketing experts at Kimberly-Clark Corporation have "little doubt that the company should put 60 tissues in each pack." The researchers determined that 60 is "the average number of times people blow their nose during a cold" by asking hundreds of customers to keep count of their Kleenex use in diaries. Suppose a random sample of 250 Kleenex users yielded the following summary statistics on the number of times they blew their nose when they had a cold:

    $\bar{x} = 57$ $s = 26$

Is this sufficient evidence to dispute the researcher's claim? Test at $\alpha = .05$.

LARGE-SAMPLE TEST OF HYPOTHESIS ABOUT $(\mu_1 - \mu_2)$

ONE-TAILED TEST

H_0: $(\mu_1 - \mu_2) = D_0$
H_a: $(\mu_1 - \mu_2) > D_0$
   [or H_a: $(\mu_1 - \mu_2) < D_0$]

TWO-TAILED TEST

H_0: $(\mu_1 - \mu_2) = D_0$
H_a: $(\mu_1 - \mu_2) \neq D_0$

Test statistic:

$$z = \frac{(\bar{x}_1 - \bar{x}_2) - D_0}{\sigma_{(\bar{x}_1 - \bar{x}_2)}} \approx \frac{(\bar{x}_1 - \bar{x}_2) - D_0}{\sqrt{\dfrac{s_1^2}{n_1} + \dfrac{s_2^2}{n_2}}}$$

Rejection region:

$z > z_\alpha$ (or $z < -z_\alpha$)

Rejection region:

$|z| > z_{\alpha/2}$

[Note: D_0 is our symbol for the particular numerical value specified for $(\mu_1 - \mu_2)$ in the null hypothesis. In many practical applications, we wish to hypothesize that there is no difference between the population means; in such cases, $D_0 = 0$.]

Assumptions: 1. The sample sizes n_1 and n_2 are sufficiently large—say, $n_1 \geq 30$ and $n_2 \geq 30$.
2. The samples are selected randomly and independently from the target populations.

Solution

a. For this problem, let μ_1 represent the mean initial performance rating of stayers and μ_2 represent the mean initial performance rating of leavers. Then the researcher wants to test the hypotheses

H_0: $(\mu_1 - \mu_2) = 0$ (i.e., no difference between mean initial performance ratings of stayers and leavers)

H_a: $(\mu_1 - \mu_2) \neq 0$ (i.e., mean initial performance ratings of stayers and
leavers differ)

This two-tailed, large-sample (since both n_1 and n_2 exceed 30) test is based on a z statistic. Thus, we will reject H_0: if $|z| > z_{\alpha/2} = z_{.005}$. Since $z_{.005} = 2.575$, the rejection region is given by

$z > 2.575$ or $z < -2.575$ (see Figure 10.7)

We compute the test statistic as follows:

$$z \approx \frac{(\bar{x}_1 - \bar{x}_2) - D_0}{\sqrt{\dfrac{s_1^2}{n_1} + \dfrac{s_2^2}{n_2}}} = \frac{(3.51 - 3.24) - 0}{\sqrt{\dfrac{(.51)^2}{174} + \dfrac{(.52)^2}{355}}} = 5.68$$

FIGURE 10.7

Rejection Region for Example 10.10

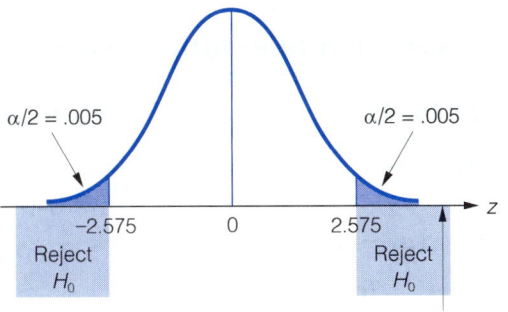

Observed value of test statistic
$z = 5.68$

Since this computed value of $z = 5.68$ lies in the rejection region, there is sufficient evidence (at $\alpha = .01$) to conclude that the mean initial performance rating of stayers is significantly different from the mean initial performance rating of leavers. The probability of our having committed a Type I error is $\alpha = .01$.

b. As we discovered in Section 10.2, the information provided by a two-tailed test of hypothesis is limited. In part **a**, we determined that the mean initial peformance ratings of the two groups of employees differ; but we have only a point estimate of the magnitude of the difference, namely, $(\bar{x}_1 - \bar{x}_2) = (3.51 - 3.24) = .27$. To find an interval estimate of the difference, we construct a 99% confidence interval for $(\mu_1 - \mu_2)$:

$$(\bar{x}_1 - \bar{x}_2) \pm z_{.005} \sqrt{\frac{\sigma_1^2}{n_1} + \frac{\sigma_2^2}{n_2}} \approx (3.51 - 3.24) \pm 2.575 \sqrt{\frac{(.51)^2}{174} + \frac{(.52)^2}{355}}$$

$$= .27 \pm .12$$

or $(.15, .39)$. Note that the interval includes only positive numbers. This implies that the difference $(\mu_1 - \mu_2)$ is positive. That is, the mean initial performance rating of stayers (μ_1) exceeds the mean initial performance rating of leavers (μ_2). This difference, although statistically significant, may be too small for the researcher to attach any practical significance to the result.

SMALL SAMPLES

When the sample sizes n_1 and n_2 are inadequate to permit use of the large-sample procedure of Example 10.10, modifications may be made to perform a small-sample test of hypothesis about the difference between two population means. The test procedure is based on assumptions that are, again, more restrictive than in the large-sample case. The elements of the hypothesis test and required assumptions are listed in the box on page 478.

EXAMPLE 10.11 The Federal Trade Commission (FTC) conducts periodic tests on the tar and nicotine content and carbon monoxide emission of each brand of cigarette sold in the United States (see Case Study 2.3). Suppose an investigator for the FTC suspects there is a significant difference between the mean amounts

a. Give the null and alternative hypotheses appropriate for testing whether the true mean satisfaction score for the more knowledgeable birders is less than the corresponding mean for the less knowledgeable birders.

b. Explain why it is dangerous to conclude that the true means differ based on the observed difference between the sample means.

c. The test statistic for part a (reported in *Leisure Sciences*) is $z = -2.97$. What is the appropriate conclusion? Test using $\alpha = .01$.

d. *Leisure Sciences* also compared the mean times spent on the tour for the two groups. A test of the difference between the mean tour time of the more knowledgeable group ($\bar{x}_1 = 142$ minutes) and the less knowledgeable group ($\bar{x}_2 = 88$ minutes) resulted in a test statistic of $z = 4.05$. According to *Leisure Sciences*, the more knowledgeable birders are "doing it more but enjoying it less." Do you agree? Test using $\alpha = .01$.

10.35   Refer to the study on the differences in the ethical perceptions of male and female managers, described in Exercise 8.66. Another objective of the research was to determine how each gender perceived their counterparts would respond to the same set of ethical decision situations (*Journal of Business Ethics*, Aug. 1987). For example, the 50 male executives in the survey had an average self-rating of 2.44 on the question of whether using company materials and supplies for personal use is unethical (1 = very unethical, 5 = not at all unethical). When the 50 female executives were asked to estimate how the men would respond to such a question, the average rating was 3.06. Assuming that the variances in the ratings of the two samples are both equal to 2.2, conduct a test for a difference between the mean ethics ratings of the two groups using $\alpha = .05$. Interpret the result in the context of the problem.

10.36   Recent research in nursing education has been focused on teaching strategies that link scientific theory and practice. One study compared a traditional approach to teaching basic nursing skills with an innovative approach (*Journal of Nursing Education*, Jan. 1992). The innovative approach utilizes two strategies (Vee heuristics and concept maps) that consciously link theory with practice. Forty-two students enrolled in an upper-division nursing course participated in the study. Half (21) were randomly assigned to labs that utilized the innovative approach. After completing the course, all students were given short-answer questions about scientific principles underlying each of 10 nursing skills. The objective of the research is to compare the mean scores of the two groups of students.

a. What is the appropriate test to use to compare the two groups?

b. Are any assumptions required for the test?

c. One question dealt with the use of clean/sterile gloves. The mean scores for this question were 3.28 (traditional) and 3.40 (innovative). Is there sufficient information to perform the test?

d. Refer to part c. The p-value for the test was reported as $p = .79$. Interpret this result.

e. Another question concerned the choice of a stethoscope. The mean scores of the two groups were 2.55 (traditional) and 3.60 (innovative) with an associated p-value of .02. Interpret these results.

10.37   In psychiatry, a *life event* is defined as a stressful event that some persons believe will have a major impact on their lives. One study was conducted to investigate how the life event "doing poorly on an important exam" affected the level of depression of college students (*Journal of Human Stress*, Mar. 1983). One hundred and sixty-five introductory psychology students were asked how satisfied they were with their grade on a just-completed midterm exam. Of these, 33 indicated they were not at all satisfied with their grade. These students were considered to have experienced the life event. All students were then administered the Depressive Adjective Checklist (DACL) to measure symptoms of depression. The DACL scores for the two groups of students are summarized in the table. (Higher scores indicate higher levels of depression.) Conduct a test to determine whether the mean depression level of students who experience the life event exceeds the mean depression level of students who do not experience the life event. Test using $\alpha = .05$.

	LIFE EVENT EXPERIENCED	LIFE EVENT NOT EXPERIENCED
Sample size	33	132
Mean	16.30	9.02
Standard deviation	4.94	5.42

10.38 A recent study of the impact of mass media violence on aggression investigated the effect of a heavyweight championship prize fight on the number of homocides in the United States 3 days after the fight. One theory is that prize fights may trigger homicides through some type of modeling of aggression. If so, then prize fights that receive much publicity (such as those discussed on the network evening news) should be followed by a greater mean increase in the number of homicides than prize fights that receive less publicity. For all heavyweight championship fights in the period from 1973 to 1978, the accompanying table records the observed increase in U.S. homicides from the "norm" 3 days after each fight and whether the fight was publicized (mentioned on the network evening news). The objective of the study is to determine whether publicized fights yield a greater increase in mean number of homicides than unpublicized fights. A SAS printout of the analysis follows. Interpret the test results.

PUBLICIZED FIGHTS	INCREASE IN HOMICIDES	UNPUBLICIZED FIGHTS	INCREASE IN HOMICIDES
Foreman/Frazier	12.90	Foreman/Roman	−3.43
Ali/Foreman	19.99	Foreman/Norton	.67
Ali/Wepner	−2.78	Ali/Bugner	23.07
Ali/Lyle	6.97	Ali/Coopman	8.98
Ali/Frazier	26.31	Ali/Young	−2.62
Ali/Dunn	8.53	ali/Evangelista	−6.11
Ali/Norton	11.43	Ali/Shavers	−.86
Spinks/Ali	10.04	Holmes/Norton	4.03
Ali/Spinks	6.75	Holmes/Evangelista	1.76

Source: Phillips, D. P. "The impact of mass media violence on U.S. homicides." *American Sociological Review*, Vol. 48, Aug. 1983, pp. 560–567.

```
                            TTEST PROCEDURE

Variable: HOMICIDE

PUBLICTY      N              Mean           Std Dev        Std Error
-----------------------------------------------------------------------
PUB           9         11.12666667       8.29217251      2.76405750
UNPUB         9          2.83222222       8.78322517      2.92774172

Variances       T       DF     Prob>|T|
-----------------------------------------------
Unequal      2.0600    15.9     0.0561
Equal        2.0600    16.0     0.0560

For HO: Variances are equal, F' = 1.12     DF = (8,8)     Prob>F' = 0.8747
```

10.5

CHAPTER CASE: THE DOWNFALL OF THE ROMAN EMPIRE

Over the years, historians have debated the reasons for the downfall of the Roman Empire. One popular theory is that lead poisoning, which can cause mental retardation and erratic behavior, was a major contributor to the Empire's collapse. Historical records, in fact, reveal that the food and wine of the early Romans were heavily contaminated with lead. Recently, a *New England Journal of Medicine* article suggested "a link between the legendary Roman appetite for reinforced wines, laced with a syrup brewed in lead pots, and the bizarre behavior of the Roman aristocrats, particulary such emperors as Nero, Claudius, and Caligula."*

*Reported in the *Orlando Sentinel Star*, June 1, 1983.

**SMALL-SAMPLE TEST OF HYPOTHESIS
ABOUT $(\mu_1 - \mu_2)$: MATCHED PAIRS**

ONE-TAILED TEST

H_0: $(\mu_1 - \mu_2) = D_0$

H_a: $(\mu_1 - \mu_2) > D_0$

  [or H_a: $(\mu_1 - \mu_2) < D_0$]

TWO-TAILED TEST

H_0: $(\mu_1 - \mu_2) = D_0$

H_a: $(\mu_1 - \mu_2) \neq D_0$

Test statistic:

$$t = \frac{\bar{d} - D_0}{s_d/\sqrt{n}}$$

Rejection region:

  $t > t_\alpha$ (or $t < -t_\alpha$)

Rejection region:

  $|t| > t_{\alpha/2}$

[*Note:* D_0 is our symbol for the particular numerical value specified for $(\mu_1 - \mu_2)$ in the null hypothesis. In many pracitical applications, we want to hypothesize that there is no difference between the population means; in such cases, $D_0 = 0$.]

Assumptions: 1. The relative frequency distribution of the population of differences is approximately normal.

2. The paired differences are randomly selected from the population of differences.

TABLE 10.5

Weeks of Jogging for Example 10.12

	JOGGER									
	1	2	3	4	5	6	7	8	9	10
Shoe type A	27	35	19	39	34	32	15	26	18	17
Shoe type B	23	28	16	31	38	30	17	22	15	16
d	4	7	3	8	−4	2	−2	4	3	1

Solution   Let μ_1 and μ_2 represent the mean number of weeks of durable wear for shoe types A and B, respectively. To determine whether μ_1 exceeds μ_2, we want to test the hypothesis:

H_0: $\mu_1 - \mu_2 = 0$

H_a: $\mu_1 - \mu_2 > 0$

The test statistic will have a t distribution based on $(n - 1) = (10 - 1) = 9$ degrees of freedom. We will reject the null hypothesis if

$$t > t_{.05} = 1.833 \quad \text{(see Figure 10.11)}$$

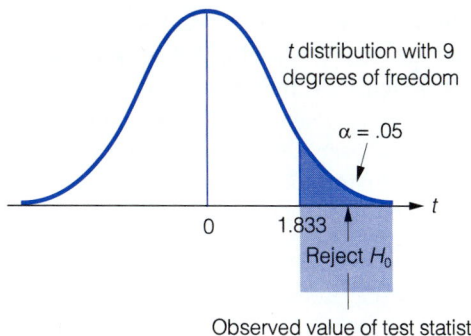

FIGURE 10.11
Rejection Region for Example
10.12

t distribution with 9
degrees of freedom

$\alpha = .05$

0   1.833

Reject H_0

Observed value of test statistic
$t = 2.25$

To conduct the test, we must first calculate the difference d in length of use of the two shoe types for each jogger. These differences (where the observation for shoe type B is subtracted from the observation for shoe type A within each pair) are shown in the last row of Table 10.5. Next, we would calculate the mean $\bar{d}$ and standard deviation s_d for this sample of $n = 10$ differences to obtain the test statistic.

Rather than perform these calculations, we will rely on the output of a computer. A Minitab printout for the analysis is shown in Figure 10.12.

TEST OF MU = 0.000 VS MU N.E. 0.000

	N	MEAN	STDEV	SE MEAN	T	P VALUE
DiffA_B	10	2.600	3.658	1.157	2.25	0.051

FIGURE 10.12
Minitab Printout for Example
10.12

Substituting the values of $\bar{d} = 2.6$ and $s_d = 3.66$ (both shown on the printout) into the formula for the test statistic, we have

$$t = \frac{\bar{d} - D_0}{s_d/\sqrt{n}} = \frac{2.6 - 0}{3.66/\sqrt{10}} = 2.25$$

Since this value of the test statistic (shaded on the printout) exceeds the critical value, $t_{.05} = 1.833$, there is sufficient evidence to indicate that the mean number of weeks of durable service for shoe type A exceeds the mean for shoe type B.

Note that we can derive the same inference by examining the *p*-value of the test. The two-tailed *p*-value, shaded in Figure 10.12, is .051. Thus, the *p*-value for our one-tailed test is $\frac{.051}{2} = .0255$. Since this value is less than our preselected value of $\alpha = .05$, we have sufficient evidence to reject H_0.

EXAMPLE 10.13   Refer to the matched pairs experiment, Example 10.12. Why did we collect the data in matched pairs rather than using independent random samples of joggers, with some assigned to use only one shoe type A and others to use only shoe type B?

Solution Clearly, we expect some joggers to wear their shoes out more rapidly than others. To cancel out this variation from jogger to jogger, the experiment was designed so that each jogger would rate both shoe types. Then each shoe type within a given pair (i.e., for a given jogger) would be subjected to the same pattern of wear. By comparing usable service time *within* each jogger, we were able to obtain more information on the difference in mean usable service time than we could have obtained by independent random sampling.

As in the previous section, we close this section with a warning.

WARNING

It is inappropriate to apply the small-sample matched-pairs t test when the population of differences is decidedly nonnormal (e.g., highly skewed): In this case, an alternative procedure is the nonparametric Wilcoxon signed ranks test of Section 15.4.

EXERCISES

LEARNING THE MECHANICS

10.39 A matched-pairs experiment is used to test the hypothesis $H_0: (\mu_1 - \mu_2) = 0$. For each of the following situations, specify the rejection region, test statistic value, and conclusion.

   a. $H_a: (\mu_1 - \mu_2) \neq 0$, $\bar{d} = 400$, $s_d = 435$, $n = 100$, $\alpha = .01$
   b. $H_a: (\mu_1 - \mu_2) > 0$, $\bar{d} = .48$, $s_d = .08$, $n = 5$, $\alpha = .05$
   c. $H_a: (\mu_1 - \mu_2) < 0$, $\bar{d} = -1.3$, $s_d^2 = .95$, $n = 6$, $\alpha = .10$

10.40 The data for a random sample of six paired observations are shown in the table.

PAIR	SAMPLE 1	SAMPLE 2
1	6	4
2	2	1
3	5	8
4	10	7
5	8	6
6	4	2

   a. Test the null hypothesis $H_0: (\mu_1 - \mu_2) = 0$ against the alternative hypothesis $H_a: (\mu_1 - \mu_2) \neq 0$. Use $\alpha = .05$.
   b. What assumptions must be made when performing a test of hypothesis based on matched-pairs data?

10.41 Consider the following summary statistics for a matched-pairs experiment:

$$\bar{d} = 10.5 \qquad s_d = 10$$

   a. Suppose $n = 10$. Test $H_0: (\mu_1 - \mu_2) = 0$ against $H_a: (\mu_1 - \mu_2) > 0$ at $\alpha = .05$.
   b. Suppose $n = 4$. Test $H_0: (\mu_1 - \mu_2) = 0$ against $H_a: (\mu_1 - \mu_2) > 0$ at $\alpha = .05$.

10.42   *USA Today* recently published the results of the *NCAA Graduation-Rates Report*. The report makes public, for the first time, the graduation rates of student-athletes at individual universities. The data in the table represents a random sample of schools selected from the NCAA report. The percentage of students and the percentage of men's basketball players from the 1983–84 and 1984–85 freshmen classes who graduated within six years are presented in the table. Use a test of hypothesis to compare the mean graduation rates of students and male basketball players. Is there evidence (at $\alpha = .05$) of a difference between the two groups?

SCHOOL	GRADUATION RATES Students	GRADUATION RATES Men's Basketball
Duquesne	70	86
Florida State	52	40
Seton Hall	58	60
Oklahoma	42	13
Michigan	81	50
UNLV	26	25
Virginia Tech	71	33
Drexel	68	70
Arizona	45	36
Georgia	60	13
San Diego State	36	0
Vanderbilt	77	83

Source: *USA Today*, Aug. 13, 1992.

10.43   Identical twins, separated at birth and reared apart, provide researchers with a natural "controlled" experiment in which the interplay between heredity and environment can be examined (refer to Case Study 8.2). One researcher suggests that differences in education may influence differences in IQ scores. In particular, the researcher believes that twins with a higher level of education tend to have higher verbal IQ scores. The table shows the verbal IQ scores for four pairs of identical twins reared apart, where one member of the pair (twin A) completed a higher level of education than the other member (twin B). Is there evidence that the mean verbal IQ score of twin members with a higher level of education is larger than the mean IQ score of their siblings? Test using $\alpha = .01$.

PAIR	VERBAL IQ SCORE Twin A	VERBAL IQ SCORE Twin B
1	104	96
2	113	109
3	88	87
4	96	92

10.44   For the perception of speech, profoundly deaf persons rely mainly on speechreading, i.e., they perceive spoken language by observing the articulatory movements, facial expressions, and gestures of the speaker. Can speech perception be improved by supplementing the speechreader with auditorily presented information about the prosody of the speech signal? To investigate this phenomenon, 10 normal-hearing subjects participated in an experiment in which they were asked to verbally reproduce sentences spoken but not heard on a video monitor (*Journal of the Acoustical Society of America*, Feb. 1986). The sentences were presented to the subjects under each of two conditions: (1) speechreading with information about the frequency and amplitude of the speech signal (denoted S + F + A), and (2) speechreading only (denoted S). For each of the 10 subjects, the difference between the percentage of correctly reproduced syllables under condition S + F + A and under condition S was calculated. The mean and standard deviation of the differences are provided here:

$$\bar{d} = 20.4 \qquad s_d = 17.44$$

Test the hypothesis that the mean percentage of correct syllables under condition S + F + A exceeds the corresponding mean under condition S. Use $\alpha = .05$.

10.45   Refer to the paired comparison of grocery items at Winn-Dixie and Publix supermarkets, Exercise 8.60. A SAS printout for testing the hypothesis of no difference between the mean prices of grocery items purchased at the two supermarkets is displayed here.

```
Analysis Variable : DIFF (Winn minus Publix)

N Obs          Mean        Std Dev       Std Error          T  Prob>|T|
----------------------------------------------------------------------
  60     -0.2540000     0.2741223      0.0353890    -7.1773633    0.0001
----------------------------------------------------------------------
```

a. Locate the test statistic on the SAS printout. Interpret its value.
b. Locate the p-value of the test statistic on the SAS printout. Interpret its value.

10.46   Merck Research Labs conducted an experiment to evaluate the effect of a new drug using the Single-T Swim maze. Nineteen impregnated dam rats were captured and allocated a dosage of 12.5 milligrams of the drug. One male and one female pup were randomly selected from each resulting litter to perform in the swim maze. Each rat pup is placed in the water at one end of the maze and allowed to swim until it successfully escapes at the opposite end. If the rat pup fails to escape after a certain period of time, it is placed at the beginning of the maze and given another attempt to escape. The experiment is repeated until three successful escapes are accomplished by each rat pup. The number of swims required by each pup to perform three successful escapes is reported in the accompanying table. Is there sufficient evidence of a difference between the mean number of swims required by male and female rat pups? Test using $\alpha = .10$.

LITTER	MALE	FEMALE	LITTER	MALE	FEMALE
1	8	5	11	6	5
2	8	4	12	6	3
3	6	7	13	12	5
4	6	3	14	3	8
5	6	5	15	3	4
6	6	3	16	8	12
7	3	8	17	3	6
8	5	10	18	6	4
9	4	4	19	9	5
10	4	4			

Source: Thomas E. Bradstreet, Merck Research Labs, BL 3-2, West Point, Penn. 19486.

10.47   Are learning preferences of college students affected by program instruction, or do they remain stable over time? To answer this question, researchers administered the Learning Preference Inventory (LPI) exam to 37 junior students at the University of Illinois during the first week the students were on campus (*The American Journal of Occupational Therapy*, Oct. 1984). The LPI measures preference for learning in a well-organized, teacher-directed class, with expectations, assignments, and goals clearly defined. The higher the LPI score, the greater the preference. Following four quarters of academic course work, which included independent study methods and small group tutorials as well as the traditional classroom lectures, the students were again administered the LPI. The differences between the LPI scores (initial score minus second score) for the sample of 37 students had a mean of 4.11 and a standard deviation of 15.82.

  a. Give the null and alternative hypotheses appropriate for testing whether the mean score on the LPI exam given at the beginning of the term differs from the mean score on the LPI exam given after four quarters of academic course work.
  b. Note that the data were collected as matched pairs. For this application, what is the advantage of using a matched-pairs experiment rather than independent samples?
  c. Conduct the test of part **a**, using $\alpha = .05$. Can you infer that students' preferences for learning in a teacher-structured atmosphere (as measured by the LPI) changed after four quarters of academic course work? Explain.

10.48   Tetrachlorodibenzo-p-dioxin (TCDD) is a highly toxic substance found in industrial wastes. A study was conducted to determine the amount of TCDD present in the tissues of bullfrogs inhabiting the Rocky Branch Creek in central Arkansas, an area known to be contaminated by TCDD (*Chemosphere*, Feb. 1986). The level of TCDD (in parts per trillion) was measured in several specific tissues of four female bull frogs and the ratio of TCDD in the tissue to TCDD in the leg muscle of the frog was recorded for each. The relative ratios of contaminant for two tissues, the liver and the ovaries, are given for each of the four frogs in the accompanying table. According to the researchers, "the data set suggests that the [mean] relative level of TCDD in the ovaries of female frogs is higher than the [mean] level in the liver of the frogs." Test this claim using $\alpha = .05$.

FROG	A	B	C	D
Liver	11.0	14.6	14.3	12.2
Ovaries	34.2	41.2	32.5	26.2

Source: Korfmacher, W. A., Hansen, E. B. Jr., and Rowland, K. L. "Tissue distribution of 2,3,7,8-TCDD in bullfrogs obtained from a 2,3,7,8-TCDD-contaminated area." *Chemosphere*, Vol. 15, No. 2, Feb. 1986, p. 125. Reprinted with permission. Copyright 1986, Pergamon Press, Ltd.

10.7   TESTING THE DIFFERENCE BETWEEN TWO POPULATION PROPORTIONS

Suppose we are interested in comparing two population proportions. Then the target parameter about which we will test a hypothesis is $(\pi_1 - \pi_2)$. Recall that π_1 and π_2 also represent the probabilities of success for two binomial experiments. The method for performing a large-sample test of hypothesis about $(\pi_1 - \pi_2)$, the difference between two binomial proportions, is outlined in the accompanying box (page 494).

When testing the null hypothesis that $(\pi_1 - \pi_2)$ equals some specified difference—say, D_0—we make a distinction between the cases $D_0 = 0$ and $D_0 \neq 0$. For the special case $D_0 = 0$, i.e., when we are testing $H_0: (\pi_1 - \pi_2) = 0$ or, equivalently, $H_0: \pi_1 = \pi_2$, the best estimate of $\pi_1 = \pi_2 = \pi$ is found by dividing the total number of successes in the combined samples by the total number of observations in the two samples. That is, if x_1 is the number of successes in sample 1 and x_2 is the number of successes in sample 2, then

$$p = \frac{x_1 + x_2}{n_1 + n_2}$$

LARGE-SAMPLE TEST OF HYPOTHESIS ABOUT $(\pi_1 - \pi_2)$

ONE-TAILED TEST

H_0: $(\pi_1 - \pi_2) = D_0$

H_a: $(\pi_1 - \pi_2) > D_0$

(or H_a: $(\pi_1 - \pi_2) < D_0$)

TWO-TAILED TEST

H_0: $(\pi_1 - \pi_2) = D_0$

H_a: $(\pi_1 - \pi_2) \neq D_0$

Test statistic: $z = \dfrac{(p_1 - p_2) - D_0}{\sigma_{(p_1-p_2)}} = \dfrac{(p_1 - p_2) - D_0}{\sqrt{\dfrac{\pi_1(1 - \pi_1)}{n_1} + \dfrac{\pi_2(1 - \pi_2)}{n_2}}}$

Rejection region:

$z > z_\alpha$ (or $z < -z_\alpha$)

Rejection region:

$|z| > z_{\alpha/2}$

where

$$p_1 = \frac{x_1}{n_1} = \frac{\text{Number of successes in sample 1}}{n_1}$$

$$p_2 = \frac{x_2}{n_2} = \frac{\text{Number of successes in sample 2}}{n_2}$$

When $D_0 \neq 0$, calculate $\sigma_{(p_1-p_2)}$ using p_1 and p_2:

$$\sigma_{(p_1-p_2)} \approx \sqrt{\frac{p_1 q_1}{n_1} + \frac{p_2 q_2}{n_2}}$$

where $q_1 = 1 - p_1$ and $q_2 = 1 - p_2$.

For the special case where $D_0 = 0$, calculate

$$\sigma_{(p_1-p_2)} \approx \sqrt{pq\left(\frac{1}{n_1} + \frac{1}{n_2}\right)}$$

where the total number of successes in the combined samples is $(x_1 + x_2)$ and

$$p_1 = p_2 = p = \frac{x_1 + x_2}{n_1 + n_2}$$

Assumptions: 1. Independent random samples are selected from the two populations.
2. The sample sizes n_1 and n_2 are large; this will usually be satisfied if the intervals

$$p_1 \pm 2\sqrt{\frac{p_1 q_1}{n_1}} \quad \text{and} \quad p_2 \pm 2\sqrt{\frac{p_2 q_2}{n_2}}$$

do not contain 0 or 1.

In this case, the best estimate of the standard deviation of the sampling distribution of $(p_1 - p_2)$ is found by substituting p for both π_1 and π_2:

$$\sigma_{(p_1 - p_2)} = \sqrt{\frac{\pi_1(1 - \pi_1)}{n_1} + \frac{\pi_2(1 - \pi_2)}{n_2}}$$

$$\approx \sqrt{\frac{pq}{n_1} + \frac{pq}{n_2}} = \sqrt{pq\left(\frac{1}{n_1} + \frac{1}{n_2}\right)}$$

For all cases in which $D_0 \neq 0$ [for example, when testing H_0: $(\pi_1 - \pi_2) = .2$], we use p_1 and p_2 in the formula for $\sigma_{(p_1 - p_2)}$. However, in most practical situations, we will want to test for a difference in proportions—that is, we will want to test H_0: $(\pi_1 - \pi_2) = 0$.

The sample sizes n_1 and n_2 must be sufficiently large to ensure that the sampling distributions of p_1 and p_2, and hence of the difference $(p_1 - p_2)$, are approximately normal. The rule of thumb given in the box may be used to determine whether the sample sizes are "sufficiently large."

EXAMPLE 10.14 In recent years there has been a trend toward both parents working outside the home. Do working mothers experience the same burdens and family pressures as their spouses? A popular belief is that the proportion of working mothers who feel they have enough spare time for themselves is significantly less than the corresponding proportion of working fathers. To test this claim, independent random samples of 100 working mothers and 100 working fathers were selected. Working parents' views on spare time for themselves were recorded. A summary of the data is given in Table 10.6. (Assume that the spouses of all individuals sampled were also working outside the home.) Does the sample information support the belief that the proportion of working mothers who feel they have enough spare time for themselves is less than the corresponding proportion of working fathers? Test at significance level $\alpha = .01$.

TABLE 10.6
Data on Working Parents for Example 10.14

	WORKING MOTHERS	WORKING FATHERS
Number sampled	100	100
Number in sample who feel they have enough spare time for themselves	37	56

Solution   We wish to perform a test of

H_0: $(\pi_1 - \pi_2) = 0$

H_a: $(\pi_1 - \pi_2) < 0$

where

$\pi_1 = $ Proportion of all working mothers who feel they have enough spare time for themselves

π_2 = Proportion of all working fathers who feel they have enough spare time for themselves

For this large-sample, one-tailed test, the null hypothesis will be rejected if

$$z < -z_{.01} = -2.33 \quad \text{(see Figure 10.13)}$$

(see Figure 10.13)

FIGURE 10.13
Rejection Region for Example
10.14

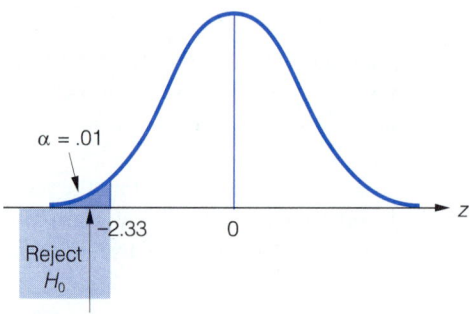

$\alpha = .01$

-2.33

0

z

Reject
H_0

Observed value of test statistic
$z = -2.69$

The sample proportions p_1 and p_2 are computed for substitution into the formula for the test statistic:

p_1 = Sample proportion of working mothers who feel they have enough spare time for themselves

$$= \frac{37}{100} = .37$$

p_2 = Sample proportion of working fathers who feel they have enough spare time for themselves

$$= \frac{56}{100} = .56$$

Hence,

$$q_1 = 1 - p_1 = 1 - .37 = .63$$
$$q_2 = 1 - p_2 = 1 - .56 = .44$$

Since $D_0 = 0$ for this test of hypothesis, the test statistic is given by

$$z = \frac{(p_1 - p_2) - D_0}{\sqrt{pq\left(\dfrac{1}{n_1} + \dfrac{1}{n_2}\right)}}$$

where

$$p = \frac{\left(\begin{array}{c}\text{Total number of sampled mothers and fathers who} \\ \text{feel they have enough spare time for themselves}\end{array}\right)}{\text{Total number of mothers and fathers sampled}}$$

$$= \frac{37 + 56}{100 + 100} = .465$$

and $q = 1 - p = .535$. Then we have

$$z = \frac{(.37 - .56) - 0}{\sqrt{(.465)(.535)\left(\dfrac{1}{100} + \dfrac{1}{100}\right)}} = -2.69$$

This value falls below the critical value of -2.33. Thus, at $\alpha = .01$, we reject the null hypothesis; there is sufficient evidence to conclude that the proportion of working mothers who feel they have enough spare time for themselves is significantly less than the corresponding proportion of working fathers; i.e., $\pi_1 < \pi_2$.

The inference derived from the test in Example 10.14 is valid only if the sample sizes, n_1 and n_2, are sufficiently large to guarantee that the intervals

$$p_1 \pm 2\sqrt{\frac{p_1 q_1}{n_1}} \quad \text{and} \quad p_2 \pm 2\sqrt{\frac{p_2 q_2}{n_2}}$$

do not contain 0 or 1. This requirement is satisfied for Example 10.14, as the following calculations show:

$$p_1 \pm 2\sqrt{\frac{p_1 q_1}{n_1}} = .37 \pm 2\sqrt{\frac{(.37)(.63)}{100}}$$

$$= .37 \pm .097 \quad \text{or} \quad (.273, .467)$$

$$p_2 \pm 2\sqrt{\frac{p_2 q_2}{n_2}} = .56 \pm 2\sqrt{\frac{(.56)(.44)}{100}}$$

$$= .56 \pm .099 \quad \text{or} \quad (.461, .659)$$

When the samples are not "sufficiently large," we must resort to another statistical technique.

WARNING

The z test for comparing π_1 and π_2 is inappropriate when the sample sizes are not "sufficiently large" (see the rule of thumb given in the previous box). In this case, π_1 and π_2 can be compared using a statistical technique to be discussed in Section 14.4.

EXERCISES

LEARNING THE MECHANICS

10.49 Independent random samples of size n_1 and n_2 are selected from two binomial populations to test the null hypothesis $H_0: (\pi_1 - \pi_2) = 0$. For each of the following situations, specify the rejection region, test statistic value, and conclusion.

a. $H_a: (\pi_1 - \pi_2) \neq 0$, $p_1 = .4$, $p_2 = .3$, $p = .343$, $n_1 = 1,500$, $n_2 = 2,000$, $\alpha = .01$

b. $H_a: (\pi_1 - \pi_2) > 0$, $p_1 = .05$, $p_2 = .02$, $p = .035$, $n_1 = n_2 = 1,000$, $\alpha = .05$

c. $H_a: (\pi_1 - \pi_2) < 0$, $p_1 = .60$, $p_2 = .65$, $p = .625$, $n_1 = n_2 = 120$, $\alpha = .01$

10.50   Independent random samples selected from two binomial populations produced the results given in the accompanying table.

	SAMPLE 1	SAMPLE 2
Number of successes	82	76
Sample size	100	100

a. Test $H_0: (\pi_1 - \pi_2) = 0$ against $H_a: (\pi_1 - \pi_2) > 0$ at $\alpha = .05$.

b. Suppose $n_1 = n_2 = 1,000$, but the sample estimates p_1, p_2, and p remain the same as in part **a**. Test H_0: $(\pi_1 - \pi_2) = 0$ against $H_a: (\pi_1 - \pi_2) > 0$ at $\alpha = .05$.

10.51   Independent random samples were selected from two binomial populations, 100 observations from population 1 and 150 from population 2. The numbers of successes were 78 and 87, respectively.

a. At $\alpha = .05$, test $H_0: (\pi_1 - \pi_2) = 0$ against $H_a: (\pi_1 - \pi_2) > 0$.

b. What assumptions must be made to conduct a large-sample test of hypothesis about $(\pi_1 - \pi_2)$?

APPLYING THE CONCEPTS

10.52   Calcium blockers are among several classes of medicines commonly prescribed to relieve high blood pressure. A study in Denmark has found that calcium blockers may also be effective in reducing the risk of heart attacks (*Tampa Tribune*, Mar. 23, 1990). A total of 897 Danish patients, each recovering from a heart attack, were given a daily dose of the drug Verapamil, a calcium blocker. After 18 months of follow-up, 146 of these patients had recurring heart attacks. In a control group of 878 people—each of whom took placebos—180 had a heart attack. Do the data provide sufficient evidence to infer that calcium blockers are effective in reducing the risk of heart attacks? Test using $\alpha = .01$.

10.53   In the travel industry, destination-specific travel literature (DSTL) refers to booklets, brochures, and pamphlets that describe a destination in detail (e.g., information on activities, facilities, and prices). DSTL is made available to travelers free of charge upon request. A study was undertaken to investigate the differences between information seekers (i.e., those who request DSTL) and nonseekers on a variety of consumer travel dimensions. Independent random samples of 288 seekers and 367 nonseekers were asked several questions about their "most recent pleasure trip or vacation of two or more days away from home." One interesting question asked whether the vacation was "active" (i.e., involved mainly challenging events or educational activities) or "passive" (i.e., involved mainly rest and relaxation). The number of "passive" vacations in each group is given in the table. Do the data provide sufficient evidence to indicate that information seekers are less likely to have a "passive" vacation than nonseekers? Test using $\alpha = .10$.

	SEEKERS	NONSEEKERS
Number surveyed	288	367
Number who experienced a "passive" vacation	197	301

Source: Etzel, M. J. and Wahlers, R. G. "The use of requested promotional material by pleasure travelers." *Journal of Travel Research*, Vol. 23, No. 4, Spring 1985, pp. 2–6.

10.54 According to new research, inositol—a sugar alcohol nutrient found in breast milk—has been found to reduce the risk of lung and eye damage in premature infants. The study, published in the *New England Journal of Medicine* (May 7, 1992), involved 220 infants born prematurely. The infants were randomly divided into two groups; half (110) of the infants received an intravenous feeding of inositol, whereas the other half received standard nutrition. The researchers found that 14 of the inositol-fed infants suffered retinopathy of prematurity, an eye injury that results from high oxygen levels needed to compensate for poorly developed lungs. In contrast, 29 of the infants on the standard diet developed this disease. Analyze the results with a test of hypothesis. Use $\alpha = .01$.

10.55 The nuclear accident on Three Mile Island near Harrisburg, Pennsylvania, on March 28, 1979, forced many local residents to evacuate their homes—some temporarily, others permanently. To assess the impact of the accident on the area population, a questionnaire was designed and mailed to a sample of 150 households within 2 weeks after the accident occurred. Residents were asked how they felt both before and after the accident about having some of their electricity generated from nuclear power. The summary results are provided in the table. Is there sufficient evidence to indicate a difference in the proportions of Three Mile Island residents who favor nuclear power before and after the accident? Use $\alpha = .01$.

	ATTITUDE TOWARD NUCLEAR POWER			TOTALS
	Favor	Oppose	Indifferent	
Before accident	62	35	53	150
After accident	52	72	26	150

Source: Brown, A., et al. *Final report on a survey of Three Mile Island area residents.* Department of Geography, Michigan State University, Aug. 1979.

10.56 Are black teachers in Florida's public universities earning tenure at the same rate as their white colleagues? A Florida university system study found that of the 20 black professors hired in 1983, only one had received tenure by their seventh year. Comparatively, 60 of the 150 white professors hired in 1983 earned tenure by their seventh year (*Tampa Tribune*, Sept. 16, 1990).

  a. Is there sufficient evidence to indicate that black professors in Florida have a lower tenure rate than white professors? Test using $\alpha = .05$.
  b. Find the *p*-value of the test, part a. Interpret the result.

10.57 A study was conducted to determine the impact of a multifunction workstation (MFWS) on the way managers work (*Datamation*, Feb. 15, 1986). Two groups of managers at a St. Louis-based defense agency took part in the survey: a test group consisting of 12 managers who currently use MFWS software and a control group of 25 non-MFWS users. One question on the survey concerned the information sources of the managers. In the test group (MFWS users), four of the 12 managers reported that their major source of information is the computer, whereas two of the 25 managers in the control group (non-MFWS users) rely on the computer as their major source of information.

  a. Is there evidence of a difference between the proportions of MFWS users and non-MFWS users who rely on the computer as their major information source? Test using $\alpha = .10$.
  b. Refer to the test in part a. Are the samples sufficiently large to properly apply the test? How does this affect the validity of the inference derived in part a?

10.58 According to a Gallup study for the American Society of Quality Control, "men focus on product performance and durability when assessing quality [of American-made goods] while women also consider the availability of service, whether a product can be repaired, and type of warranty" (*American Demographics*, June 1986). Overall, 58% of the women surveyed rated American products "high" in quality, compared to 43% of the men. Assuming that the sample survey included 500 men and 500 women, conduct a test to determine whether the true percentage of women who rate American products "high" in quality exceeds the true percentage of men. Use $\alpha = .01$.

10.59   Researchers at Mount Sinai Medical Center in New York believe that "Lou Gehrig's disease" (a neurological disorder that slowly paralyzes and kills its victims, one of the first of whom was New York Yankee's baseball slugger Lou Gehrig in 1941), may be linked to household pets, especially small dogs. A five-member medical team has found that 72% of the afflicted patients studied had small household dogs at least 20 years before contracting Lou Gehrig's disease. In contrast, only 33% of a healthy control group had pet dogs early in life. (Assume that 100 afflicted patients and 100 healthy controls were studied.) Is there sufficient evidence to indicate that the percentage of all patients with Lou Gehrig's disease who had pet dogs early in life is larger than the corresponding percentage for nonafflicted people? Test using $\alpha = .01$.

TESTING A POPULATION VARIANCE (OPTIONAL)

10.8

Hypothesis tests about a population variance σ^2 are conducted using the chi-square (χ^2) distribution introduced in optional Section 8.10. The test is outlined in the box. Note that the assumption of a normal population is required regardless of whether the sample size n is large or small.

TEST OF HYPOTHESIS ABOUT A POPULATION VARIANCE σ^2

ONE-TAILED TEST

H_0: $\sigma^2 = \sigma_0^2$

H_a: $\sigma^2 > \sigma_0^2$)

  (or H_a: $\sigma^2 < \sigma_0^2$)

TWO-TAILED TEST

H_0: $\sigma^2 = \sigma_0^2$

H_a: $\sigma^2 \neq \sigma_0^2$

Test statistic:

$$\chi^2 = \frac{(n-1)s^2}{\sigma_0^2}$$

Rejection region:

$\chi^2 > \chi_\alpha^2$ (or $\chi^2 < \chi_{(1-\alpha)}^2$)

Rejection region:

$\chi^2 < \chi_{(1-\alpha/2)}^2$ or $\chi^2 > \chi_{\alpha/2}^2$

where χ_α^2 and $\chi_{(1-\alpha)}^2$ are values of χ^2 that locate an area of α to the right and α to the left, respectively, of a chi-square distribution based on $(n-1)$ degrees of freedom.

[*Note:* σ_0^2 is our symbol for the particular numerical value specified for σ^2 in the null hypothesis.]

Assumption: The population from which the random sample is selected has an approximate normal distribution.

EXAMPLE 10.15   A quality control supervisor in a cannery knows that the exact amount each can contains will vary, since there are certain uncontrollable factors that affect the amount of fill. The mean fill per can is important, but equally important is the variation σ^2 of the amount of fill. If σ^2 is large, some cans will contain too little and others too much. Suppose regulatory agencies specify that

the standard deviation of the amount of fill for 8-ounce cans should be less than
.1 ounce. The quality control supervisor sampled $n = 10$ 8-ounce cans and
measured the amount of fill in each. The data (in ounces) are shown here.

  7.96 7.90 7.98 8.01 7.97 7.96 8.03 8.02 8.04 8.02

Is there sufficient evidence to indicate that the standard deviation, σ, of the fill
measurements is less than .1 ounce?

Solution   Since the null and alternative hypotheses must be stated in terms of
σ^2 (rather than σ), we will want to test the null hypothesis that $\sigma^2 = .01$ against
the alternative that $\sigma^2 < .01$. Therefore, the elements of the test are

- H_0: $\sigma^2 = .01$ (i.e., $\sigma = .1$)
  H_a: $\sigma^2 < .01$ (i.e., $\sigma < .1$)
- Assumption: The population of amounts of fill of the cans is approximately
  normal.
- Test statistic: $\chi^2 = \dfrac{(n-1)s^2}{\sigma_0^2}$
- Rejection region: The smaller the value of s^2 we observe, the stronger the
  evidence in favor of H_a. Thus, we reject H_0 for "small values" of the test statistic.
  With $\alpha = .05$ and $(n-1) = 9$ df, the critical χ^2 value is found in Table 6 of
  Appendix G and illustrated in Figure 10.14. We will reject H_0 if $\chi^2 < 3.32511$.
  (Remember that the area given in Table 6 is the area to the *right* of the numerical
  value in the table. Thus, to determine the lower-tail value that has $\alpha = .05$ to
  its *left*, we use the $\chi^2_{.95}$ column in Table 6.)
    To compute the test statistic, we need to find the sample standard deviation,
  s. Numerical descriptive statistics for the sample data are provided in the SAS
  printout, Figure 10.15 (page 502). The value of s, shaded in Figure 10.15 is
  $s = .043$. Substituting $s = .043$, $n = 10$, and $\sigma_0^2 = .01$ into the formula for
  the test statistic, we obtain

$$\chi^2 = \frac{(10-1)(.043)^2}{.01} = 1.66$$

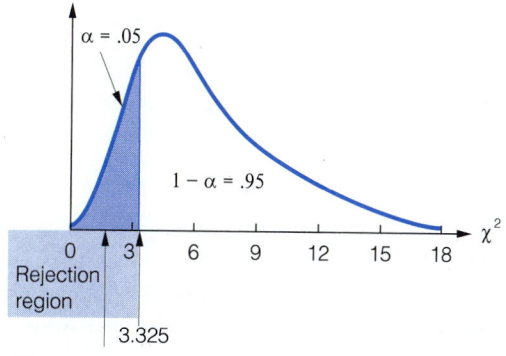

FIGURE 10.14
Rejection Region for Example
10.15

FIGURE 10.15

SAS Printout: Descriptive
Statistics for Example 10.15

```
Variable=FILL

                      Moments

N                 10    Sum Wgts            10
Mean           7.989    Sum              79.89
Std Dev     0.043063    Variance      0.001854
Skewness     -0.8538    Kurtosis      0.479371
USS         638.2579    CSS            0.01669
CV          0.539032    Std Mean      0.013618
T:Mean=0    586.6587    Prob>|T|        0.0001
Sgn Rank       27.5     Prob>|S|        0.0020
Num ^= 0          10

                 Quantiles(Def=5)

    100% Max     8.04        99%      8.04
    75% Q3       8.02        95%      8.04
    50% Med     7.995        90%     8.035
    25% Q1       7.96        10%      7.93
     0% Min       7.9         5%       7.9
                              1%       7.9

    Range        0.14
    Q3-Q1        0.06
    Mode         7.96
```

■ *Conclusion:* Since the test statistic, $\chi^2 = 1.66$, is less than 3.32511, the supervisor can conclude that the variance of the population of all amounts of fill is less than .01 ($\sigma < .1$) with 95% confidence. As usual, the confidence is in the procedure used—the χ^2 test. If this procedure is repeatedly used, it will incorrectly reject H_0 only 5% of the time. Thus, the quality control supervisor is confident in the decision that the cannery is operating within the desired limits of variability.

EXERCISES

LEARNING THE MECHANICS

10.60 Calculate the test statistic for testing H_0: $\sigma^2 = \sigma_0^2$ in a normal population for each of the following:

a. $n = 25$, $\bar{x} = 17$, $s^2 = 8$, $\sigma_0^2 = 10$
b. $n = 12$, $\bar{x} = 6.2$, $s^2 = 1.7$, $\sigma_0^2 = 1$
c. $n = 50$, $\bar{x} = 106$, $s^2 = 31$, $\sigma_0^2 = 50$
d. $n = 100$, $\bar{x} = .35$, $s^2 = .04$, $\sigma_0^2 = .01$

10.61 A random sample of n observations, selected from a normal population, is used to test the null hypothesis H_0: $\sigma^2 = 9$. Specify the appropriate rejection region for each of the following:

a. H_a: $\sigma^2 > 9$, $n = 20$, $\alpha = .01$
b. H_a: $\sigma^2 \neq 9$, $n = 20$, $\alpha = .01$
c. H_a: $\sigma^2 < 9$, $n = 12$, $\alpha = .05$
d. H_a: $\sigma^2 < 9$, $n = 12$, $\alpha = .10$

10.62 A random sample of $n = 10$ observations yielded $\bar{x} = 231.7$ and $s^2 = 15.5$. Test the null hypothesis H_0: $\sigma^2 = 20$ against the alternative hypothesis H_a: $\sigma^2 < 20$. Use $\alpha = .05$. What assumptions are necessary for the test to be valid?

10.63 The following measurements represent a random sample of $n = 5$ observations from a normal population: 11, 7, 2, 9, 13. Is this sufficient evidence to conclude that $\sigma^2 \neq 2$? Test using $\alpha = .10$.

10.64 Recording electrical activity of the brain is important in clinical problems as well as neurophysiological research. To improve the signal-to-noise ratio (SNR) in the electrical activity, it is necessary to repeatedly stimulate subjects and average the responses—a procedure that assumes that single responses are homogeneous. A study was conducted to test the homogeneous signal theory (*IEEE Engineering in Medicine and Biology Magazine*, Mar. 1990). The null hypothesis is that the variance of the SNR readings of subjects equals the "expected" level under the homogeneous signal theory. For this study, the "expected" level was assumed to be .54. If the SNR variance exceeds this level, the researchers will conclude that the signals are nonhomogeneous.

   a. Set up the null and alternative hypotheses for the researchers.
   b. SNRs recorded for a sample of 41 normal children ranged from .03 to 3.0. Use this information to obtain an estimate of the sample standard deviation. [*Hint:* Assume that the distribution of SNRs is normal, and that most of the SNRs in the population will fall within $\mu \pm 2\sigma$, i.e., from $\mu - 2\sigma$ to $\mu + 2\sigma$. Note that the range of the interval equals 4σ.]
   c. Use the estimate of s in part **b** to conduct the test of part **a**. Test using $\alpha = .10$.

10.65 Polychlorinated biphenyls (PCBs), used in the manufacture of large electrical transformers and capacitors, are extremely hazardous contaminants when released into the environment. The Environmental Protection Agency (EPA) is experimenting with a new device for measuring PCB concentration in fish. To check the precision of the new instrument, seven PCB readings were taken on the same fish sample. The data are recorded here (in parts per million):

   6.2   5.8   5.7   6.3   5.9   5.8   6.0

   Suppose the EPA requires an instrument that yields PCB readings with a variance of less than .1. Does the new instrument meet the EPA's specifications? Test using $\alpha = .05$.

10.66 The most common method of disinfecting water for potable use is free residual chlorination. Recently, preammoniation (i.e., the addition of ammonia to the water before applying free chlorine) has received considerable attention as an alternative treatment. In one study, 44 water specimens treated with preammoniation were found to have a mean effluent turbidity of 1.8 and a standard deviation of .16 (*American Water Works Journal*, Jan. 1986). Is there sufficient evidence to indicate that the variance of the effluent turbidity in water specimens disinfected by the preammoniation method exceeds .0016? (The value .0016 represents the known effluent turbidity variance of water specimens treated with free chlorine.) Test using $\alpha = .01$.

10.67 In any canning process, a manufacturer will lose money if the cans contain either significantly more or significantly less than is claimed on the label. Accordingly, canners pay close attention to the amount of their product being dispensed by the can-filling machines. Consider a company that produces a fast-drying rubber cement in 32-ounce aluminum cans. A quality control inspector is interested in testing whether the variance of the amount of rubber cement dispensed into the cans is more than .3. If so, the dispensing machine needs adjustment. Since inspection of the canning process requires that the dispensing machines be shut down, and shutdowns for any lengthy period of time cost the company thousands of dollars in lost revenue, the inspector is able to obtain a random sample of only 10 cans for testing. After measuring the weights of their contents, the inspector computes the following summary statistics:

$$\bar{x} = 31.55 \text{ ounces} \qquad s = .48 \text{ ounce}$$

   a. Does the sample evidence indicate that the dispensing machines are in need of adjustment? Test at significance level $\alpha = .05$.
   b. What assumption is necessary for the hypothesis test of part **a** to be valid?

10.68 Refer to the study of sleep-deprived college students in Exercise 10.10. Some sleep experts believe that although most people lose their ability to think creatively when deprived of sleep, others are not nearly as affected. Consequently, they theorize that the variation in test scores for sleep-deprived subjects is larger than 225, the

TABLE 10.8

Data on ETO Levels, Example 10.17

	TASK 1	TASK 2
Sample size	19	11
Mean	5.90	5.60
Standard deviation	1.93	4.10

Solution   Let

σ_1^2 = Population variance of ETO levels in subjects assigned task 1

σ_2^2 = Population variance of ETO levels in subjects assigned task 2

For this test to yield valid results, we must assume that both samples of ETO levels come from normal populations and that the samples are independent.

Now, following the guidelines in the box, the elements of this two-tailed hypothesis test follow:

- H_0: $\sigma_1^2/\sigma_2^2 = 1$ $(\sigma_1^2 = \sigma_2^2)$
  H_a: $\sigma_1^2/\sigma_2^2 \neq 1$ $(\sigma_1^2 \neq \sigma_2^2)$

- *Test statistic:* $F = \dfrac{\text{Larger } s^2}{\text{Smaller } s^2} = \dfrac{s_2^2}{s_1^2} = \dfrac{(4.10)^2}{(1.93)^2} = 4.51$

- *Rejection region:*   For this two-tailed test, $\alpha = .10$ and $\alpha/2 = .05$. Thus, the rejection region is

$$F > F_{.05} = 1.98 \quad \text{(from Table 8, Appendix G)}$$

where the distribution of F is based on $\nu_1 = (n_2 - 1) = 10$ and $\nu_2 = (n_1 - 1) = 18$ degrees of freedom.

- *Conclusion:*   Since the test statistic, $F = 4.51$, falls in the rejection region, we reject H_0. Therefore, the data provide sufficient evidence to indicate that the population variances differ. It appears that hospital personnel involved with opening the sterilization package (task 1) have less variable ETO levels than those involved with opening and unloading the sterilizer gun (task 2).

Example 10.17 illustrates the technique for calculating the test statistic and rejection region for a two-tailed F test. The reason we place the larger sample variance in the numerator of the test statistic is that only upper-tail values of F are shown in the F tables of Appendix G—no lower-tail values are given. By placing the larger sample variance in the numerator, we make certain that only the upper tail of the rejection region is used. The fact that the upper-tail area is $\alpha/2$ reminds us that the test is two-tailed.

The problem of not being able to locate an F value in the lower tail of the F distribution is easily avoided in a one-tailed test because we can control how we specify the ratio of the population variances in H_0 and H_a. That is, we can always make a one-tailed test an *upper-tailed* test. For example, if we want to test whether σ_1^2 is greater than σ_2^2, then we write the alternative hypothesis as

H_a: $\dfrac{\sigma_1^2}{\sigma_2^2} > 1$ (i.e., $\sigma_1^2 > \sigma_2^2$)

and the appropriate test statistic is $F = s_1^2/s_2^2$. Conversely, if we want to test whether σ_1^2 is less than σ_2^2 (i.e., whether σ_2^2 is greater than σ_1^2), we write

H_a: $\dfrac{\sigma_2^2}{\sigma_1^2} > 1$ (i.e., $\sigma_2^2 > \sigma_1^2$)

and the corresponding test statistic is $F = s_2^2/s_1^2$.

EXERCISES

LEARNING THE MECHANICS

10.69   Find F_α for an F distribution with 15 numerator df and 12 denominator df for the following values of α:

a. $\alpha = .025$ b. $\alpha = .05$ c. $\alpha = .10$

10.70   Find $F_{.05}$ for an F distribution with:

a. Numerator df $= 7$,   denominator df $= 25$ b. Numerator df $= 10$,   denominator df $= 8$
c. Numerator df $= 30$,   denominator df $= 60$ d. Numerator df $= 15$,   denominator df $= 4$

10.71   Under what conditions does the sampling distribution of s_1^2/s_2^2 have an F distribution?

10.72   Calculate the value of the test statistic for testing H_0: $\sigma_1^2/\sigma_2^2 = 1$ in each of the following cases:

a. H_a: $\sigma_1^2/\sigma_2^2 > 1$; $s_1^2 = 1.75$, $s_2^2 = 1.23$
b. H_a: $\sigma_1^2/\sigma_2^2 < 1$; $s_1^2 = 1.52$, $s_2^2 = 5.90$
c. H_a: $\sigma_1^2/\sigma_2^2 \neq 1$; $s_1^2 = 2{,}264$, $s_2^2 = 4{,}009$

APPLYING THE CONCEPTS

10.73   Wet samplers are standard devices used to measure the chemical composition of precipitation. The accuracy of the wet deposition readings, however, may depend on the number of samplers stationed in the field. Experimenters in The Netherlands collected wet deposition measurements using anywhere from one to eight identical wet samplers (*Atmospheric Environment*, Vol, 24A, 1990). For each sampler (or sampler combination) data were collected every 24 hours for an entire year; thus, 365 readings were collected per sampler (or sampler combination). When one wet sampler was used, the standard deviation of the hydrogen readings (measured as percentage relative to the average reading from all eight samplers) was 6.3%. When three wet samplers were used, the standard deviation of the hydrogen readings (measured as percentage relative to the average reading from all eight samplers) was 2.6%. Conduct a test to compare the variation in hydrogen readings for the two sampling schemes (i.e., one wet sampler versus three wet samplers). Test using $\alpha = .05$.

10.74   An experiment was conducted in England to examine the diet metabolizable energy (ME) content of commercial cat foods. The researchers monitored the diets of 57 adult domestic short-haired cats; 28 cats were fed a diet of commercial canned cat food, whereas 29 cats were fed a diet of dry cat food over a 3-week period. At the end of the trial, the ME content was determined for each cat, with the results shown in the table. Conduct a test to determine whether the variation in ME content of cats fed canned food differs from the variation in ME content of cats fed dry food. Use $\alpha = .10$.

	CANNED FOOD	DRY FOOD
Sample size	28	29
Mean ME content	.96	3.70
Standard deviation	.26	.48

Source: Kendall, P. T., Burger, I. N., and Smith, P. M. "Methods of estimation of the metabolizable energy content of cat foods." *Feline Practice*, Vol. 15, No. 2, Feb. 1986, pp. 38–44.

10.75 An experiment was conducted to study the effect of reinforced flanges on the torsional capacity of reinforced concrete T-beams (*Journal of the American Concrete Institute*, Jan.–Feb. 1986). Several different types of T-beams were used in the experiment, each type having a different flange width. The beams were tested under combined torsion and bending until failure (cracking). One variable of interest is the cracking torsion moment at the top of the flange of the T-beam. Cracking torsion moments for eight beams with 70-cm slab widths and eight beams with 100-cm slab widths are recorded here:

70-cm slab width: 6.00, 7.20, 10.20, 13.20, 11.40, 13.60, 9.20, 11.20

100-cm slab width: 6.80, 9.20, 8.80, 13.20, 11.20, 14.90, 10.20, 11.80

a. Is there evidence of a difference in the variation in the cracking torsion moments of the two types of T-beams? Use $\alpha = .10$.
b. What assumptions are required for the test to be valid?

10.76 Refer to the speechreading study introduced in Exercise 10.44. A second experiment was conducted to compare the variability in the sentence perception of normal-hearing individuals with no prior experience in speechreading to those with experience in speechreading. The sample consisted of 24 inexperienced and 12 experienced subjects. All subjects were asked to verbally reproduce sentences under several conditions, one of which was speechreading supplemented with sound-pressure information. A summary of the results (percentage of correct syllables) for the two groups is given in the table. Conduct a test to determine whether the variance in the percentage of correctly reproduced syllables differs between the two groups of speechreaders. Test using $\alpha = .10$.

INEXPERIENCED SPEECHREADERS	EXPERIENCED SPEECHREADERS
$n_1 = 24$	$n_2 = 12$
$\bar{x}_1 = 87.1$	$\bar{x}_2 = 86.1$
$s_1 = 8.7$	$s_2 = 12.4$

Source: Breeuwer, M. and Plomp, R. "Speechreading supplemented with auditorily presented speech parameters." *Journal of the Acoustical Society of America*, Vol. 79, No. 2, Feb. 1986, p. 487.

10.77 General trace organic monitoring describes the process in which water engineers analyze water samples for various types of organic material (e.g., contaminants). The total organic carbon (TOC) level was measured in water samples collected at two sewage treatment sites in England. The accompanying table gives the summary information on the TOC levels (measured in mg/l) found in the rivers adjacent to the two sewage facilities. Since the river at the Foxcote sewage treatment works was subject to periodic spillovers, not far upstream of the plant's intake, it is believed that the TOC levels found at Foxcote will have greater variation than the levels at Bedford. Does the sample information support this hypothesis? Test using $\alpha = .05$.

BEDFORD	FOXCOTE
$n_1 = 61$	$n_2 = 52$
$\bar{x}_1 = 5.35$	$\bar{x}_2 = 4.27$
$s_1 = .96$	$s_2 = 1.27$

Source: Pinchin, M. J. "A study of the trace organics profiles of raw and potable water systems." *Journal of the Institute of Water Engineers & Scientists*, Vol. 40, No. 1, Feb. 1986, p. 87.

SUMMARY

In this chapter, we have summarized the procedures for testing hypotheses about various population parameters. As we noted with the estimation techniques of Chapter 8, fewer assumptions about the sampled populations are required when the sample sizes are large. We also wish to emphasize that *statistical* significance differs from *practical* significance, and the two must not be confused. A reasonable approach to hypothesis testing blends a valid application of the formal statistical procedures with the researcher's knowledge of the subject matter.

KEY WORDS

Practical significance
Statistical significance

KEY FORMULAS

Large-sample test statistic: $\quad z = \dfrac{\text{Estimator} - \text{Hypothesized } (H_0) \text{ value}}{\text{Standard error}}$

Small-sample test statistic: $\quad t = \dfrac{\text{Estimator} - \text{Hypothesized } (H_0) \text{ value}}{\text{Standard error}}$

Note: The respective estimators and standard errors for the population parameters, μ, $(\mu_1 - \mu_2)$, π, and $(\pi_1 - \pi_2)$, are provided in Table 8.10 of Chapter 8. [The test statistic for the parameters σ^2 and σ_1^2/σ_2^2 from the optional sections are given here.]

Test statistic for σ^2:* $\quad \chi^2 = \dfrac{(n-1)s^2}{\sigma_0^2}$

Test statistic for σ_1^2/σ_2^2:* $\quad F = \dfrac{s_1^2}{s_2^2}$

*From the optional sections of this chapter.

Note: Starred (*) exercises are from the optional sections of this chapter.

10.78 Refer to the matched-pairs study of purposeful and nonpurposeful activity described in Exercise 8.65 (*Journal of Occupational Therapy*, Mar. 1984). Recall that a sample of 26 female subjects were instructed to jump both with and without a rope, each time jumping until they felt they had worked their bodies "very hard." In addition to heart rate increase, the length of time that each subject jumped was recorded for both exercises. The 26 differences

$$d_i = (\text{Length of jumping time with rope}) - (\text{Length of jumping time without rope})$$

are calculated and summarized as follows:

$$\bar{d} = 41.84 \text{ seconds} \qquad s_d = 110.28 \text{ seconds}$$

One theory held by occupational therapists is that those performing a purposeful activity are more motivated and, hence, tend to fatigue less easily. Test the hypothesis that the mean exercise time for the purposeful activity (jumping with a rope) exceeds the mean exercise time for the nonpurposeful activity (jumping without a rope). Use $\alpha = .05$.

10.79 A field experiment was conducted to ascertain the impact of desert granivores (seedeaters) on the density and distribution of seeds in the soil. Since some desert rodents are known to hoard seeds in surface caches, the study was specifically designed to determine if these caches eventually produce more seedlings, on the average, than an adjacent control area. Forty small areas excavated by rodents were located and covered with plastic cages to prevent rodents from reusing the caches. A caged control area was set up adjacent to each of the caged caches. The numbers of seedlings germinating from the caches and from the control areas were then observed. A summary of the data is provided in the following table. Is there sufficient evidence (at $\alpha = .05$) to indicate that the average number of seedlings germinating from the seed caches of desert rodents is significantly higher than the corresponding average for the control areas?

CACHES	CONTROL AREAS
$n_1 = 40$	$n_2 = 40$
$\bar{x}_1 = 5.3$	$\bar{x}_2 = 2.7$
$s_1 = 1.3$	$s_2 = 0.7$

Source: Reichman, O. J. "Desert granivore foraging and its impact on seed densities and distributions." *Ecology*, Dec. 1979, Vol. 60, pp. 1085–1092. Copyright 1979, the Ecological Society of America. Reprinted by permission.

*10.80 The application of adrenaline is the prevailing treatment to reduce eye pressure in glaucoma patients. Theoretically, a new synthetic drug will cause the same standard deviation in blood pressure drop (1.2 units) without the side effects caused by adrenaline. The new drug is given to $n = 50$ glaucoma patients, and the reduction in pressure for each patient is measured. The results are summarized as follows:

$$\bar{x} = 4.68 \qquad s = .82$$

Is this sufficient evidence (at significance level $\alpha = .05$) to conclude that the standard deviation in blood pressure reduction resulting from the new drug is different from that produced by adrenaline?

*10.81   The testing department of a tire and rubber company schedules truck and passenger tires for durability tests. Currently, tires are scheduled twice weekly on flexible processors (machines that can handle either truck or passenger tires) using the shortest processing time (SPT) approach. Under SPT, the tire with the shortest processing time is scheduled first. Company researchers have developed a new scheduling rule that they believe will reduce the variation in flow time (i.e., the variation in the completion time of a test) and lead to a reduction in the variation in tardiness of a scheduled test. To compare the two scheduling rules, 64 tires were randomly selected and divided into two groups of equal size. One set of tires was scheduled using SPT, the other using the proposed rule. A summary of the flow times and tardiness (in hours) of the tire tests is provided in the table.

	FLOW TIME		TARDINESS	
	Mean	Variance	Mean	Variance
SPT	158.28	8,532.80	5.26	452.09
Proposed rule	117.07	5,208.53	4.52	319.41

a. Is there sufficient evidence at $\alpha = .05$ to conclude that the variance in flow time is less under the proposed scheduling rule than under the SPT approach?
b. Is there sufficient evidence at $\alpha = .05$ to conclude that the proposed scheduling rule will lead to a reduction in the variation in tardiness of tire tests?

10.82   To what extent, if any, can we influence local weather conditions? Some Texas farmers have hired a meteorologist to investigate the effectiveness of cloud seeding in the artificial production of rainfall. Two farming areas in Texas with similar past meteorological records were selected for the experiment. One is seeded regularly throughout the year, whereas the other is left unseeded. Data on the monthly precipitation (in inches) at the farms for the first 6 months of the year are recorded in the table. A SAS printout for a test to compare the mean monthly precipitation in the seeded and unseeded farm areas follows. Interpret the results of the test.

MONTH	SEEDED FARM AREA	UNSEEDED FARM AREA
1	1.75	1.62
2	2.12	1.83
3	1.53	1.40
4	1.10	0.75
5	1.70	1.71
6	2.42	2.33

```
Analysis Variable : DIFF    (SEED minus UNSEED)

 N Obs         Mean         Std Dev          T   Prob>|T|
 --------------------------------------------------------
    6      0.1633333      0.1330664    3.0066442    0.0299
 --------------------------------------------------------
```

10.83 Marketers spend billions of dollars annually to develop humorous advertising. But is humor in advertising an effective means of marketing a product? To determine the views of successful advertising practitioners concerning humor in advertising, T. J. Madden and M. G. Weinberger surveyed two groups of advertising personnel: (1) vice-presidents/directors of research and (2) vice-presidents/directors of creative services (*Journal of Advertising Research*, Aug./Sept. 1984). The two groups of executives, research and creative, were selected because historically they have conflicting views on advertising techniques and objectives. One hundred thirty-seven research directors and 145 creative directors took part in the study. The table shows the numbers of directors in each group who agreed with the stated communication objectives of humorous advertising. (Numbers are based on percentages given in the original article.) For each of the three communication objectives, conduct a test to determine if the percentage who agree with the objective differs for the two groups of directors. Use $\alpha = .01$.

COMMUNICATION OBJECTIVE	RESEARCH SAMPLE $n_1 = 137$	CREATIVE SAMPLE $n_2 = 145$
1. Humor helps gain awareness of new products	81	120
2. Humor harms comprehension more than nonhumor	88	57
3. Humor increases persuasion more than nonhumor	34	39

Source: Adapted from the *Journal of Advertising Research*, © Copyright 1984 by the Advertising Research Foundation.

10.84 Studies have shown that in a nonbusiness (e.g., academic) setting, those who have job mobility are predominantly better performers. To examine the performance turnover relationship in a business setting, G. G. Dreher examined the personnel records of a large national oil company (*Academy of Management Journal*, Mar. 1982.). Dreher's sample consisted of 174 employees who were classified as "stayers" (those who stayed with the company from 1964 through 1979) and 355 former employees who were classified as "leavers" (those who left the company at varying points during the 15-year period). The company's annual performance appraisals corresponding to the initial years of service were used to form an initial performance rating for each employee. Summary statistics on initial performance for the two groups of employees are provided in the table. Is there evidence of a difference between the mean initial performance ratings of stayers and leavers? Test using $\alpha = .01$.

STAYERS	LEAVERS
$n_1 = 174$	$n_2 = 355$
$\bar{x}_1 = 3.51$	$\bar{x}_2 = 3.24$
$s_1 = .51$	$s_2 = .52$

10.85 A school official wants to compare a new method of teaching reading to "slow learners" with the current standard method. A random sample of 10 slow learners is selected and each is matched with another slow learner of the same age, IQ, and standardized prereading comprehension test score. One member of each pair is taught by the new method and the other member is taught by the standard method. All 20 children are taught by qualified instructors under similar conditions for a 6-month period. The results of the reading test given at the end of this period are recorded in the table. Does the sample evidence indicate (at significance level $\alpha = .01$) that the mean reading test score of children taught by the new method exceeds the mean reading test score of children taught by the standard method?

PAIR	NEW METHOD	STANDARD METHOD
1	70	69
2	79	71
3	73	60
4	82	75
5	66	62
6	73	70
7	77	72
8	68	69
9	75	71
10	80	68

10.86 The ion balance of our atmosphere has a significant effect on human health. A high concentration of positive ions in a room can induce fatigue, stress, and respiratory problems in the room's occupants. However, research has shown that introducing additional negative ions into the room's atmosphere (through a negative ion generator), in combination with constant ventilation, regains the natural balance of ions that is conducive to human health. One experiment was conducted as follows. Two hundred employees of a large factory were randomly selected and divided into two groups of 100 each. Both groups were told that they would be working in an atmosphere with an ion balance controlled through negative ion generators. However, unknown to the employees, the generators were switched on only in the experimental group's work area. At the end of the day, eight employees in the experimental group (ion generators on) and 24 employees in the control group (ion generators off) reported migraine, nausea, fatigue, faintness, or some other physical discomfort. Is the proportion of employees in the experimental group who experience some type of physical discomfort at the end of the day significantly less than the corresponding proportion for the control group? Test using $\alpha = .05$.

10.87 To significantly reduce soil erosion in our country, the Conservation Title of the 1985 farm bill requires that conservation compliance be implemented by 1995. Despite controversy over the bill, the U.S. Soil Conservation Service (SCS) claims that 80% of farmers who already have a soil conservation plan feel their plan is reasonable and practical (*Prairie Farmer*, Mar. 20, 1990). An independent survey conducted by *Prairie Farmer* magazine found that of 144 Indiana farmers who have a conservation plan, only 78 believe their plan is realistic.

a. Does this survey refute or support the SCS claim? Test using $\alpha = .01$.
b. Compute the *p*-value of the test. Interpret the result.

10.88 A psychologist is studying the effects of lack of sleep on the performance of various perceptual-motor tasks. After a given period of sleep deprivation, a measurement of reaction time to an auditory stimulus was taken for each of six adult male subjects. The reaction times (in seconds) are summarized as follows:

$$\bar{x} = 1.82 \qquad s = .22$$

Previous psychological studies show that the true mean reaction time for non–sleep-deprived male subjects is 1.70 seconds. Does the sample evidence indicate (at significance level $\alpha = .10$) that the mean reaction time for sleep-deprived male subjects is longer than 1.70 seconds? State any assumptions that are necessary for the hypothesis test to be valid.

10.89 As part of the evaluation for an environmental impact statement of proposed hydroelectric design on the Stikine River in British Columbia, researchers conducted preliminary investigations of the effects of human-induced disturbances on the behavior of the resident mountain goat population (*Environmental Management*, Mar. 1983).

Goat responses to exploration activities, inlcuding close-flying helicopters, fixed-wing aircraft, human bipedal movement, and loud blasts from geological drilling activities, were recorded for $n = 804$ goats. The researchers observed that 265 goats displayed a severe flight response to local rock or plant cover. Test the hypothesis that over 30% of the resident mountain goats will elicit a severe response to human-induced disturbances. Use $\alpha = .05$.

10.90   Does competition between separate research and development (R&D) teams in the U.S. Department of Defense, working independently on the same project, improve performance? To answer this question, performance ratings were assigned to each of 58 multisource (competitive) and 63 sole-source R&D contracts (*IEEE Transactions on Engineering Management*, Feb. 1990). With respect to quality of reports and products, the competitive contracts had a mean performance rating of 7.62, whereas the sole-source contracts had a mean of 6.95.

a. Set up the null and alternative hypotheses for determining whether the mean quality performance rating of competitive R&D contracts exceeds the mean for sole-source contracts.
b. Find the rejection region for the test using $\alpha = .05$.
c. The p-value for the test was reported to be between .02 and .03. What is the appropriate conclusion?

10.91   To examine potential gender differences in the industrial sales force, a sample of 244 males and a sample of 153 females were administered a questionnaire (*Journal of Personal Selling & Sales Management*, Summer 1990). All respondents were either sales managers or salespeople at one of 16 industrial firms located in the southeastern United States. One of the variables measured in the study was months of experience in sales. Summary statistics for this variable are given in the table.

	MALES	FEMALES
Sample size	244	155
Mean number of months experience	64.99	27.69
Standard deviation	79.59	68.88

Source: Schul, P. L., Remington, S., and Berl, R. L. "Assessing gender differences in relationships between supervisory behaviors and job-related outcomes in the industrial sales force." *Journal of Personal Selling & Sales*, Summer 1990, Vol. X, p. 7 (Table 2).

a. Is there evidence of a difference between the mean number of months of experience in sales of males and females? Test using $\alpha = .01$.
b. The p-value of the test, part **a**, was reported as $p < .001$. Interpret this result.
c. The researchers also compared the male and female samples with respect to the variables job satisfaction, pay satisfaction, motivation, and organizational commitment. Why might the results of these tests be misleading indicators of gender differences? [*Hint:*   Consider the possibility that differences in experience may explain why the two groups differ with respect to job satisfaction, pay satisfaction, etc.]

THE SAT PILL REVISITED

Refer to the experiment on using beta blockers to reduce anxiety in students taking the Scholastic Aptitude Test (SAT) in Case Study 1.2. Recall that 22 high school juniors who had not performed as well as expected on the SAT were administered the "SAT pill" (i.e., a beta blocker) 1 hour prior to retaking the test in their senior year. The sample mean increase in SAT scores for these 22 students was $\bar{x}_d = 120$ points, compared to the national average increase of $\mu_d = 38$ points (reported in *Newsweek*, Nov. 16, 1987).

a. Assuming that the standard deviation of the difference in SAT scores for the sample of 22 students was $s_d = 125$, test the hypothesis that the true mean increase in SAT scores for those students who take an "SAT pill" prior to the exam exceeds the national average increase of 38. Use $\alpha = .05$.

b. Repeat part **a** assuming that the standard deviation of the differences in SAT scores for the sample was $s_d = 250$.

c. The *Newsweek* article did not provide the value of the sample standard deviation, s_d, of the matched pairs experiment. Given this omission, why should one interpret the results of the experiment with caution?

STRESS IN NURSING STUDENTS: MYTH OR MYSTERY?

Studies indicate that nursing students generally experience much more undue stress, discomfort, and dissatisfaction than their peers in college. Researchers have theorized that these differences are because nursing students are expected to demonstrate a high level of responsibility on the job, are often unable to avoid identification with their patient's illnesses, are usually perceived at the bottom of the health care hierarchical structure, and consistently work in an environment that uses drugs as a solution to many problems.

However, Dr. Elizabeth W. Carter, a professor in the School of Nursing at Columbia University, questions "whether the distress observed in nursing students is related to the educational/clinical experiences or in fact is experienced in a similar manner by women in other undergraduate programs of study."* Based on her work with senior nursing students over the past 20 years and the marked change in women's values and expectations during this same period, Dr. Carter believes that the observed stress in nursing students is not a situational crisis, as has been assumed in the past, but a developmental crisis—one that affects most college women.

---

*Carter, E. W. "Stress in nursing students: Dispelling some of the myth." *Nursing Outlook*, Vol. 30, No. 4, Apr. 1982, pp. 248–252. Copyright by the American Journal of Nursing Company. Reprinted from *Nursing Outlook*.

at in the TV series, but they are not seen as particularly desirable in real-life situations.

[Radar O'Reilly] exemplifies the qualities students see as desirable in someone they would like to supervise . . . [He] is seen as being dependable, competent, sensitive, cooperative and loyal . . . [Hawkeye] is a noncomformist in dress, speech, and action in relating to superiors, and students can see that such a person would be much more difficult to supervise than Radar would be.

Hawkeye, however, is the overwhelming choice as a peer or colleague. Students respond positively to his sense of humor and feel that a Hawkeye would keep life interesting in an organization. Pierce is also seen as being competent, intelligent, and broadminded—qualities admired in peers.

Another interesting result of the study is that students with a lower grade point average (GPA) seem more likely to choose Hawkeye as a desirable superior. Of the 411 students with GPA below 2.99, 222 selected Hawkeye as their first or second choice; of the 76 students with GPA above 3.75, only 23 selected Hawkeye as their first or second choice.

a. Is there sufficient evidence to indicate that only 50% of all college students would prefer a Colonel Potter type to be their superior in a real-world organization? Test using $\alpha = .01$.
b. Is there sufficient evidence to indicate that over 50% of all college students would prefer a Hawkeye Pierce type to be their peer in a real-world organization? Test using $\alpha = .01$.
c. Is there sufficient evidence to indicate that over 50% of all college students would prefer a Radar O'Reilly type to be their subordinate in a real-world organization? Test using $\alpha = .01$.
d. Is there sufficient evidence to indicate that college students with lower GPAs are more likely to choose Hawkeye as a desirable superior than college students with higher GPAs? Test using $\alpha = .01$.

COMPUTER LAB

In this section we present the computer commands for conducting tests of hypotheses. All three commercially available packages, SAS, SPSS, and Minitab, can perform t tests about μ and $(\mu_1 - \mu_2)$ for independent samples, and $(\mu_1 - \mu_2)$ for matched pairs. The computer output consists of the test statistic value and observed significance level (p-value) of the test. (Remember, for large samples, the t and z statistics are nearly equivalent.) Tests about variances and proportions are not available in SAS, SPSS, or Minitab.

TESTS OF HYPOTHESES

SAS a. One-sample t test (e.g., H_0: $\mu = 50$)

```
Command
  Line
   1    DATA ONESAMP;        ⎫
   2    INPUT X @@;          ⎬  Data entry instructions
   3    NEWX = X-50;         ⎪
   4    CARDS;               ⎭

        [Input data values (multiple observations per line)]

   5    PROC MEANS T PRT;    ⎫  Student's t test
   6    VAR NEWX;            ⎭
```

COMMAND 3 The transformed variable NEWX is computed by subtracting the hypothesized mean ($\mu = 50$) from each value of x.

COMMANDS 5–6 The PROC MEANS statement commands SAS to conduct a t test on the values of the variable NEWX (specified in line 6). SAS will test the null hypothesis H_0: $\mu_{NEWX} = 0$, which is equivalent to testing H_0: $\mu_X = 50$.

GENERAL The p-value reported in SAS is a *two-tailed* observed significance level. Divide this reported value in half to obtain the p-value for a one-tailed test.

b. Two-sample t test, independent samples (e.g., H_0: $\mu_1 - \mu_2 = 0$)

```
Command
  Line
   1    DATA TWOSAMP;        ⎫
   2    INPUT X SAMPLE;      ⎬  Data entry instructions
   3    CARDS;               ⎭

        [Input data values (one observation per line)]

   4    PROC TTEST;          ⎫  Student's t test
   5       CLASS SAMPLE; VAR X;  ⎭
```

COMMAND 2 X is the variable of interest. SAMPLE is a grouping variable that takes on two values (e.g., 1 and 2).

COMMANDS 4–5 The TTEST procedure conducts a t test on the difference in means of the variable X for the two groups identified by SAMPLE.

OUTPUT SAS calculates the t value for both the equal population variances case and the unequal variances case.

c. Two-sample t test, matched pairs (e.g., H_0: $\mu_d = \mu_1 - \mu_2 = 0$)

```
Command
   Line
     1    DATA MATCHSAM;     ⎫
     2    INPUT X1 X2;       ⎬  Data entry instructions
     3    DIFF=X1-X2;        ⎭
     4    CARDS;

          [Input data values (one observation per line)]

     5    PROC MEANS T PRT;  ⎫  Student's t test
     6    VAR DIFF;          ⎭
```

COMMANDS 2–3   The variables X1 and X2 contain the measurements for each member of the matched pair. The difference, DIFF, is computed in line 3.

a. One-sample t test (e.g., H_0: $\mu = 50$)

SPSS

```
Command
   Line
     1    DATA LIST FREE / X.  }  Data entry instruction
     2    BEGIN DATA.

          [Input data values (multiple observations per line)]

     3    END DATA.
     4    COMPUTE MU=50.
     5    T-TEST PAIRS=X MU.   }  Student's t test
```

COMMAND 4   The value of the hypothesized mean (or median) is specified.

COMMAND 5   The T-TEST procedure tests the difference between the mean of X and the constant MU. (The t value and corresponding two-tailed p-value will appear on the printout under the heading POOLED VARIANCE ESTIMATE.)

GENERAL   Remember to omit the period at the end of each command when using SPSS on a mainframe computer.

b. Two-sample t test, independent samples (e.g., H_0: $\mu_1 - \mu_2 = 0$)

```
Command
   Line
     1    DATA LIST FREE / X SAMPLE.  }  Data entry instruction
     2    BEGIN DATA.

          [Input data values (multiple observations per line)]

     3    END DATA.
     4    T-TEST GROUPS=SAMPLE(1,2)/VARIABLES=X.  }  Student's t test
```

COMMAND 1   X is the variable of interest. SAMPLE is a grouping variable name that takes on the value 1 if the observation (X) is from the first sample, and the value 2 if the observation (X) is from the second sample.

COMMAND 4   The T-TEST procedure tests the difference between the X means of the two groups identified by SAMPLE.

GENERAL   The t value and corresponding two-tailed p-value will appear on the printout under the heading POOLED VARIANCE ESTIMATE. (Remember to omit the period at the end of each command when using SPSS on a mainframe computer.)

c. **Two-sample t test, matched pairs** (e.g., $H_0: \mu_d = \mu_1 - \mu_2 = 0$)

Command Line		
1	DATA LIST FREE /X1, X2, }	Data entry instruction
2	BEGIN DATA.	
	[Input data values (multiple observations per line)]	
3	END DATA.	
4	T-TEST PAIRS=X1 X2, }	Student's t test

COMMAND 1   The variables X1 and X2 contain the measurements for each member of the matched pair.

COMMAND 4   The SPSS t test is performed on the difference between the pairs of observations, X1 and X2.

MINITAB   a. **One-sample t test** (e.g., $H_0: \mu = 50$)

Command Line		
1	SET DATA IN C1 }	Data entry instruction
	[Input data values (multiple observations per line)]	
2	TTEST OF MU=50 ON C1; }	Student's t test
3	ALTERNATIVE=+1. }	

COMMANDS 2–3   The TTEST procedure performs a t test on the difference between the mean of the variable read in C1 and the hypothesized value specified in the MU= subcommand (line 2). The subcommand ALTERNATIVE=+1 (line 3) requests that a one-tailed, upper-tailed test be performed. Use ALTERNATIVE=−1 for a lower-tailed test. If the subcommand is not used, a two-tailed test is performed.

b. **Two-sample test, independent samples** (e.g., $H_0: \mu_1 - \mu_2 = 0$)

Command
Line

  1   SET DATA FOR 1ST SAMPLE IN C1 } Data entry instruction

      [Input data values]

  2   SET DATA FOR 2ND SAMPLE IN C2 } Data entry instruction

      [Input data values]

  3   TWOSAMPLE T ON DATA IN C1 C2;   } Student's *t* test
  4   POOLED.

COMMANDS 3–4 TWOSAMPLE performs a *t* test on the difference between the means of the data in C1 and C2. The subcommand POOLED (line 4) requests that a pooled sample variance be used. (This is appropriate when the population variances are equal.) If you want Minitab to adjust the *t* statistic and degrees of freedom for the unequal variances case, omit the POOLED subcommand.

GENERAL Use the ALTERNATIVE subcommand to obtain a one-tailed test.

c. **Two-sample test, paired samples** (e.g., $H_0: \mu_d = \mu_1 - \mu_2 = 0$)

Command
Line

  1   READ C1 C2 } Data entry instruction

      [Input data values (one observation per line)]

  2   SUBTRACT C2 FROM C1, PUT IN C3   } Student's *t* test
  3   TTEST OF MU=0 ON DATA IN C3

COMMANDS 1–2 The data in columns C1 and C2 contain the measurements for each member of the matched pair. C3 contains the difference between the measurements.

COMMAND 3 TTEST performs a *t* test on the mean of the difference in C3.

GENERAL Use the ALTERNATIVE subcommand to obtain a one-tailed test.

COMPUTER ACTIVITIES

Select one or more of the data sets in the Appendix, enter the data (or a portion of the data) into the computer, then use a statistical software package to test the following hypotheses:

Appendix A The mean starting salary of all recent UF graduates exceeds $25,000.

Appendix B The mean starting salary of College of Engineering bachelor degree graduates exceeds the mean starting salary of College of Business Administration bachelor degree graduates.

Appendix C The mean per-member per-month cost of all Tampa physicians is less than $100.

Appendix D The mean DDT level found in channel catfish differs from the mean DDT level found in smallmouth buffalo.

Appendix E The mean tar content of cigarettes differs from the mean carbon monoxide content.

Appendix F The mean checkout time of all supermarket customers exceeds 1 minute.

REFERENCES

Freedman, D., Pisani, R., and Purves, R. *Statistics*. New York: Norton, 1978.

McClave, J. T. and Benson, P. G. *Statistics for Business and Economics*, 5th ed. San Francisco: Dellen, 1991.

Mendenhall, W. *Introduction to Probability and Statistics*, 8th ed. Boston: PWS-Kent, 1989.

Mendenhall, W., Wackerly, D., and Scheaffer, R. *Mathematical Statistics with Applications*, 4th ed. Boston: Duxbury, 1988.

Snedecor, G. W. and Cochran, W. G. *Statistical Methods*, 7th ed. Ames, Iowa: Iowa State University Press, 1980.

SIMPLE LINEAR REGRESSION AND CORRELATION

CHAPTER 11

A t major colleges and universities, administrators (e.g., deans, chairpersons, provosts, presidents) are among the highest paid state employees. Is there a link between the raises administrators receive and their performance on the job? This and other questions concerning the relationship between two variables will be discussed in this chapter. You will learn more about the raise–performance relationship of college administrators in the chapter case, Section 11.11.

CONTENTS

INTRODUCTION TO BIVARIATE RELATIONSHIPS

11.1 The procedures discussed in the previous four chapters are most useful in cases where we are interested in testing hypotheses about or estimating the values of one or more population parameters based on random sampling. However, a more important concern may be the relationship between two different random variables, x and y, known as a **bivariate relationship**. For example, a farmer may be interested in the relationship between the level of fertilizer, x, and the yield of potatoes, y; or a psychologist may be interested in the bivariate relationship between a child's creativity score, x, and flexibility score, y; or a medical researcher may be interested in the bivariate relationship between a patient's blood pressure, x, and heart rate, y; and so on. In each case, the object of this interest is not merely academic. The farmer wants to know the level of fertilizer that gives the maximum yield of potatoes; the psychologist would like to know if a child's creativity score is a reliable predictor of his or her flexibility score; and the medical researcher wishes to determine whether blood pressure is a good indicator of a patient's heart rate.

How can we determine if one variable, x, is a reliable predictor of another variable, y? To answer this question, we must be able to *model* the bivariate relationship—that is, describe how the two variables, x and y, are related using a mathematical equation. In this chapter we present a method useful for modeling the (straight-line) relationship between two variables—a method called **simple linear regression analysis**.

STRAIGHT-LINE PROBABILISTIC MODELS

11.2 Consider the Federal Trade Commission's tar, nicotine, and carbon monoxide rankings of domestic cigarettes given in Appendix F. Suppose you want to model the relationship between the carbon monoxide (CO) ranking and the nicotine content of the cigarette brands. Do you believe that an *exact* mathematical relationship exists between the two variables? That is, would it be possible to state or calculate the exact CO ranking of a cigarette brand if you knew the nicotine content of the brand? In reality, the answer is a very definite no! The amount of carbon monoxide in the smoke of any given cigarette will depend not only on the nicotine content of the brand, but also on such variables as the tar content, length, filter type, light type, and menthol flavor of the cigarette. You can probably think of additional variables that play an important role in determining the FTC's carbon monoxide ranking of a cigarette. How can we construct a model, then, for two variables for which no exact relationship exists? We illustrate with an example.

EXAMPLE 11.1 Table 11.1 is a list of the nicotine and CO measurements, extracted from Appendix F, for a sample of 20 different cigarette brands. Hypothesize a reasonable model for the relationship between CO ranking and nicotine content.

Solution   Let y represent the carbon monoxide ranking of a cigarette brand and let x represent the brand's nicotine content (both measured in milligrams). We can gain insight into the bivariate relationship by constructing a **scattergram** for the sample data. A scattergram is constructed by plotting the pairs of sample

TABLE 11.1

Carbon Monoxide–Nicotine Data for Example 11.1

OBS #	CO	NICOTINE	OBS #	CO	NICOTINE
5	15	0.9	191	6	0.5
60	6	0.4	220	22	1.2
71	13	1.3	227	15	0.8
73	12	0.8	260	17	1.3
84	12	1.2	270	13	1.0
120	9	0.7	282	11	0.7
132	13	1.0	286	14	1.1
146	16	1.1	292	14	1.4
154	13	1.1	328	12	0.9
175	12	0.9	347	8	0.4

observations in Table 11.1 on a two-dimensional plot (x values on the horizontal axis and y values on the vertical axis).

Figure 11.1 shows a scattergram for the sample data in Table 11.1. You can see that, on average, CO ranking (y) increases linearly as nicotine content (x) increases. Thus, we could select a model that proposes a straight-line relationship between y and x. Such a **deterministic model**—one that attempts to predict y exactly for a given x value—might be adequate if all the points in Figure 11.1 fell exactly on a straight line. However, you can see that this idealistic situation will not occur for the data in Table 11.1. No matter how you draw a line through

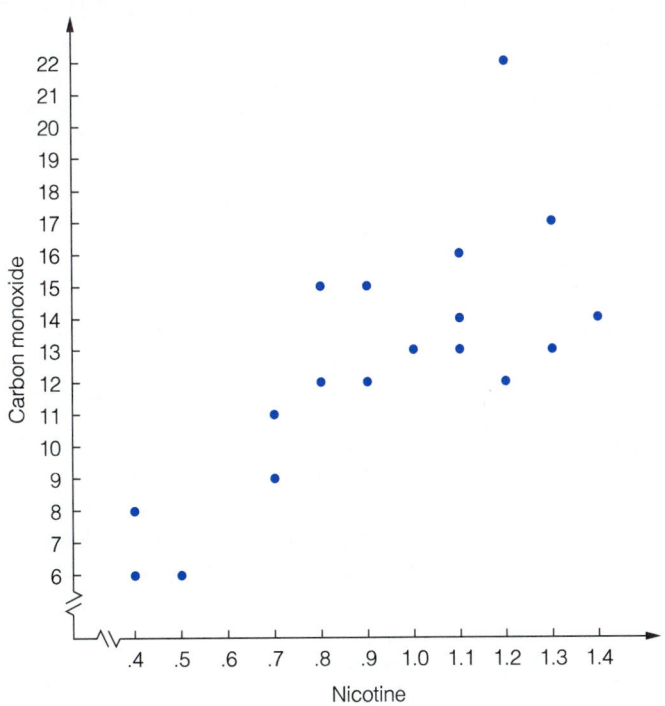

FIGURE 11.1

Scattergram of CO Ranking Versus Nicotine Content for Example 11.1

the points of Figure 11.1, at least some (if not most) of the points will deviate from the line.

A more reasonable model is one that will allow for unexplained variation in CO ranking caused by important variables not included in the model (such as those discussed earlier) or simply by random phenomena that cannot be modeled or explained. Models that account for this **random error** are called **probabilistic models**, as shown in the box.

GENERAL FORM OF A PROBABILISTIC MODEL

y = Deterministic component + Random error

where y = Variable to be predicted

Assumption: The mean value of the random error equals 0. This is equivalent to assuming that the mean value of y, $E(y)$, equals the deterministic component of the model, i.e.,

$E(y)$ = Deterministic component

Note that probabilistic models include two components: a **deterministic component** and a **random error component**. In regression we assume that the deterministic component represents the mean value of y, denoted $E(y)$. The deterministic component will always be a function of the x variables in the model. The random error component, typically denoted by the Greek symbol ε, allows for random fluctuation of the y values about their mean, $E(y)$. Consequently, the probabilistic model relating CO ranking (y) to nicotine content (x) will be of the form

$$y = E(y) + \varepsilon$$

where $E(y)$ will be a straight-line function of x.

EXAMPLE 11.2   Refer to Example 11.1 and the probabilistic model $y = E(y) + \varepsilon$. Write the straight-line equation relating $E(y)$ to nicotine content (x).

Solution   A straight-line equation involves two parameters, the y-intercept and slope of the line. We will use the Greek symbols β_0 and β_1 to represent the y-intercept and slope, respectively, since they are population parameters that will be known only if we have access to the entire population of (x, y) measurements. Therefore, the equation of the deterministic portion of the model is

$$E(y) = \beta_0 + \beta_1 x$$

A graph of this model is shown in Figure 11.2. Note that β_0 (the y-intercept) is the point at which the line cuts through the y-axis, and β_1 (the slope) is the amount change in y for every 1-unit increase in x.

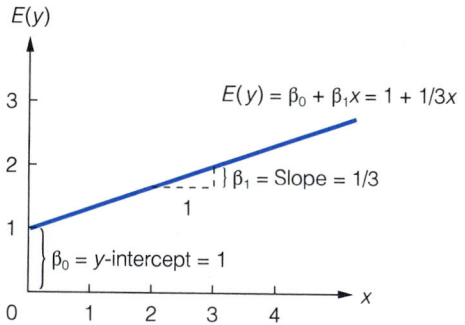

FIGURE 11.2
The Straight-Line Model

In this chapter, we consider only the simplest of probabilistic models—the **straight-line model**—which derives its name from the fact that the deterministic portion of the model graphs as a straight line. The elements of the straight-line model are summarized in the box.

THE STRAIGHT-LINE PROBABILISTIC MODEL

$$y = \beta_0 + \beta_1 x + \varepsilon$$

where

y = Variable to be predicted, called the **dependent** (or **response**) **variable**

x = Variable to be used as a predictor of y, called the **independent variable**

$E(y) = \beta_0 + \beta_1 x$ is the deterministic portion of the model (the equation of a straight line)

β_0 (beta zero) = **y-intercept of the line**, i.e., the point at which the line intercepts or cuts through the y-axis (see Figure 11.2)

β_1 (beta one) = **Slope of the line**, i.e., amount of increase (or decrease) in the deterministic component of y for every 1-unit increase in x (see Figure 11.2)

ε = Random error

In the sections that follow, we will use the sample data to estimate the slope (β_1) and the y-intercept (β_0) of the deterministic portion of the straight-line model.

EXERCISES
LEARNING THE MECHANICS

11.1   Suppose that y is exactly related to x by the equation $y = 1.5 + 2x$.

  a. Find the value of y that corresponds to $x = 1$.
  b. Find the value of y that corresponds to $x = 2$.
  c. Plot the two (x, y) points found in parts **a** and **b** on graph paper, and draw a line through the points. This line corresponds to the equation $y = 1.5 + 2x$.
  d. Find the value of y that corresponds to $x = 1.5$. Plot this point on the graph of part **c**, and confirm that it falls on the line that passes through the points found in parts **a** and **b**.
  e. Part **d** illustrates an important relationship between graphs and equations: All the points that satisfy the equation $y = 1.5 + 2x$ have a common property. What is it?

11.2   Refer to Exercise 11.1.

  a. Find the y-intercept for the line and interpret its value.
  b. Find the slope of the line and interpret its value.
  c. If you increase x by 1 unit, how much will y increase or decrease?
  d. If you decrease x by 1 unit, how much will y increase or decrease?
  e. What is the value of y when $x = 0$?

11.3   Answer the questions posed in Exercise 11.1 using the line $y = 1.5 - 2x$.

11.4   Refer to Exercise 11.3.

  a. Find the y-intercept for the line and interpret its value.
  b. Find the slope of the line and interpret its value.
  c. If you increase x by 1 unit, how much will y increase or decrease?
  d. What is the value of y when $x = 0$?
  e. What do the two lines in Exercises 11.1 and 11.3 have in common? How do they differ?

11.5   Graph the lines corresponding to each of the following equations:

  a. $y = 1 + 3x$ b. $y = 1 - 3x$ c. $y = -1 + \frac{1}{2}x$ d. $y = -1 - 3x$
  e. $y = 2 - \frac{1}{2}x$ f. $y = -1.5 + x$ g. $y = 3x$ h. $y = -2x$

11.6   Give the values of β_0 and β_1 corresponding to each of the lines of Exercise 11.5.

ESTIMATING THE MODEL PARAMETERS: THE METHOD OF LEAST SQUARES

11.3

The following example illustrates the technique we will use to fit the straight-line model to the data, i.e., to estimate the slope and y-intercept of the line using information provided by the sample data.

EXAMPLE 11.3   Suppose a psychologist wants to model the relationship between the creativity score y and the flexibility score x of a mentally retarded child. Based on practical experience, the psychologist hypothesizes the deterministic component of the probabilistic model as

$$E(y) = \beta_0 + \beta_1 x$$

If the psychologist were able to obtain the flexibility and creativity scores of *all* mentally retarded children, i.e., the entire population of (x, y) measurements, then the values of the population parameters β_0 and β_1 could be determined exactly. Collecting this mass of data would, of course, be impossible. The problem, then, is to estimate the unknown population parameters based on the information contained in a sample of (x, y) measurements. Suppose the psychologist tests five randomly selected mentally retarded children. The creativity and flexibility scores (measured on a scale of 1 to 20) are given in Table 11.2.* How can we best use the sample information to estimate the unknown y-intercept β_0 and the slope β_1?

TABLE 11.2
Creativity–Flexibility Scores for Example 11.3

CHILD	FLEXIBILITY SCORE x	CREATIVITY SCORE y
1	2	2
2	3	5
3	4	7
4	5	10
5	6	11

Solution   Estimates of the unknown parameters β_0 and β_1 are obtained by finding the best-fitting straight line through the sample data points of Table 11.2. (These points are plotted in Figure 11.3.) We will denote the estimates as $\hat{\beta}_0$ and $\hat{\beta}_1$, respectively. The procedure we will use to find the best fit is known as the **method of least squares**, and the best-fitting line, called the **least squares line**, is written

$$\hat{y} = \hat{\beta}_0 + \hat{\beta}_1 x$$

where $\hat{y}$ is the predicted value of creativity score, y.

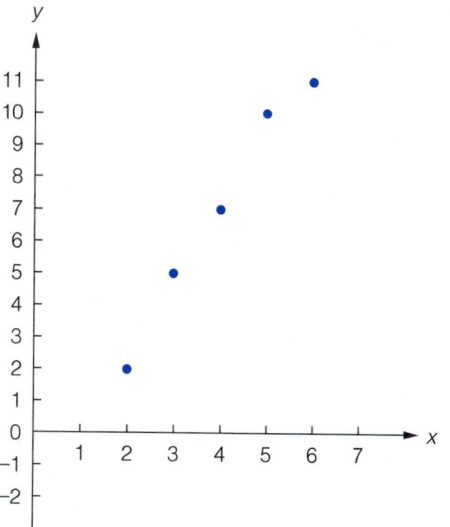

FIGURE 11.3
Scattergram of Creativity–Flexibility Score Data of Table 11.2

*The sample size ($n = 5$) is unrealistically small to demonstrate the calculations involved in a simple linear regression.

The first step in finding the least squares line is to construct a sums of squares table to find the sums of the x^2 values (Σx^2), and the cross-products of the corresponding x and y values (Σxy). The sums of squares table for the creativity–flexibility score data is given in Table 11.3.

TABLE 11.3
Sums of Squares for Data of Table 11.2

	x	y	x^2	xy
	2	2	4	4
	3	5	9	15
	4	7	16	28
	5	10	25	50
	6	11	36	66
TOTALS	$\Sigma x = 20$	$\Sigma y = 35$	$\Sigma x^2 = 90$	$\Sigma xy = 163$

The second step is to substitute the values of Σx, Σy, Σx^2, and Σxy into the formulas for SS_{xy} and SS_{xx} given in the accompanying box:

$$SS_{xy} = \Sigma xy - \frac{(\Sigma x)(\Sigma y)}{n} = 163 - \frac{(20)(35)}{5}$$

$$= 163 - 140 = 23$$

$$SS_{xx} = \Sigma x^2 - \frac{(\Sigma x)^2}{n} = 90 - \frac{(20)^2}{5}$$

$$= 90 - 80 = 10$$

Next, use these values of SS_{xy} and SS_{xx} to compute the estimate $\hat{\beta}_1$, as shown in the box:

SLOPE OF THE LEAST SQUARES LINE

$$\hat{\beta}_1 = \frac{SS_{xy}}{SS_{xx}}$$

where $SS_{xx} = \Sigma x^2 - \dfrac{(\Sigma x)^2}{n}$

$$SS_{xy} = \Sigma xy - \frac{(\Sigma x)(\Sigma y)}{n}$$

Substituting, we find $\hat{\beta}_1$, the slope of the least squares line, to be

$$\hat{\beta}_1 = \frac{SS_{xy}}{SS_{xx}} = \frac{23}{10} = 2.3$$

Finally, calculate $\hat{\beta}_0$, the y-intercept of the least squares line, as follows:

---

y-INTERCEPT OF THE LEAST SQUARES LINE

$$\hat{\beta}_0 = \bar{y} - \hat{\beta}_1 \bar{x}$$

For our example, we obtain

$$\hat{\beta}_0 = \bar{y} - \hat{\beta}_1 \bar{x} = \frac{\sum y}{5} - \hat{\beta}_1 \left(\frac{\sum x}{5}\right)$$

$$= \frac{35}{5} - (2.3)\left(\frac{20}{5}\right)$$

$$= 7 - (2.3)(4) = 7 - 9.2 = -2.2$$

Therefore, $\hat{\beta}_0 = -2.2$, $\hat{\beta}_1 = 2.3$, and the least squares line is

$$\hat{y} = -2.2 + 2.3x$$

A graph of this line is shown in Figure 11.4.

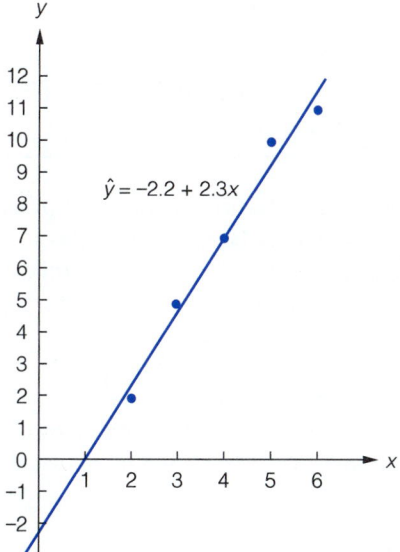

FIGURE 11.4
Least Squares Line for
Example 11.3

This four-step least squares procedure is summarized in the box on page 536.

EXAMPLE 11.4 Refer to Example 11.3. In what sense is the least squares line the "best-fitting" straight line to the data of Table 11.3?

Solution In deciding whether a line provides a good fit to a set of data, we examine the vertical distances, or **deviations**, between the data points and the fitted line. (Since we are attempting to predict y, a measure of fit will involve the difference between the observed value y and the predicted value $\hat{y}$—a quantity that is represented by the *vertical* deviation between the data point and the fitted line.) The deviations for the least squares line

$$\hat{y} = -2.2 + 2.3x$$

are shown in Figure 11.5a.

Let us compare the deviations of the least squares line with the deviations of another fitted line (one fitted visually), given by the equation

$$\hat{y} = -1 + 2x$$

The deviations of the visually fitted line are shown in Figure 11.5b. Notice first that some of the deviations are positive, some are negative, and that even though three of the five data points fall exactly on the visually fitted line, the individual deviations tend to be smaller for the least squares line than for the visually fitted line. Second, note that the sum of squares of deviations, $SSE = \Sigma(y - \hat{y})^2$, is smaller for the least squares line. (The values of SSE for the least squares and visually fitted lines are given at the bottom of Figures 11.5a and 11.5b, respectively.) In fact, it can be shown that **there is one and only one line that will minimize the sum of squares of deviations of the points about the fitted line. It is the least squares line.**

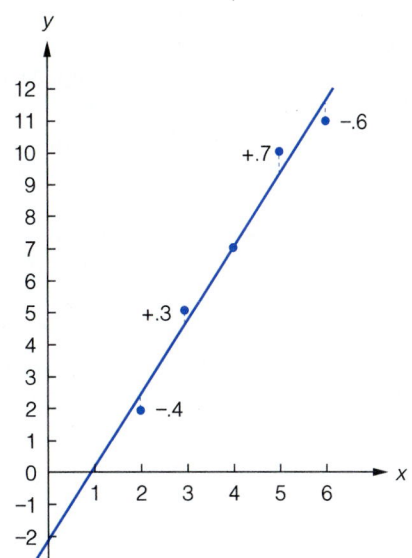

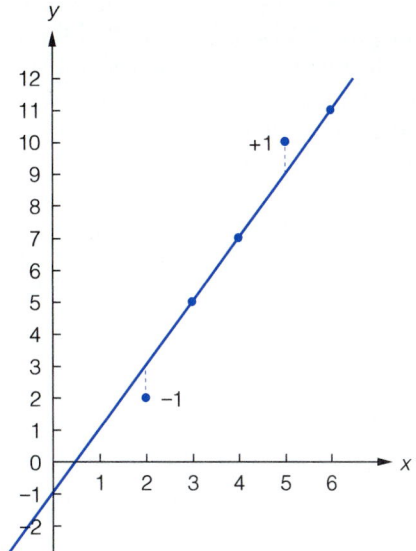

$$\text{SSE} = \Sigma(y - \hat{y})^2$$
$$= (-.4)^2 + (.3)^2 + (0)^2 + (.7)^2 + (-.6)^2$$
$$= 1.10$$

a. Least squares line: $\hat{y} = -2.2 + 2.3x$

$$\text{SSE} = \Sigma(y - \hat{y})^2$$
$$= (-1)^2 + (0)^2 + (0)^2 + (1)^2 + (0)^2$$
$$= 2$$

b. Visually fitted line: $\hat{y} = -1 + 2x$

FIGURE 11.5
Deviations About the Least Squares and Visually Fitted Lines of Example 11.4

LEAST SQUARES CRITERION FOR FINDING THE "BEST-FITTING" LINE

Choose the line that minimizes the sum of squared deviations,

$$\text{SSE} = \sum(y - \hat{y})^2$$

This is called the **least squares line**, or the **least squares prediction equation**.

The fact that the least squares line is the one that minimizes the sum of squared deviations does not guarantee that it is the "best" line to fit the data. However, intuitively, it would seem that this is a desirable property for a good-fitting line.* A second advantage of the method of least squares is that we know the sampling distributions of the estimates of β_0 and β_1—something that would be unknown for lines fitted intuitively or visually. A third desirable property is that, under certain conditions, the sampling distributions of the least squares estimators of β_0 and β_1 will have smaller standard deviations than other types of estimators.

---

*The sum of the deviations from the least squares line will also always equal 0. Since there are many other fitted lines that also have this property, we do not use this as the only criterion for choosing the "best-fitting" line.

EXAMPLE 11.5   One aspect of the preliminary evaluation of a new food product is determination of the nutritive quality of the product. This is often accomplished by feeding the food product to animals (e.g., rats) whose metabolic processes are very similar to our own. General Foods Corporation used this technique to evaluate the protein efficiency of two forms of a product (known by the pseudonym H)—one solid and the other liquid.* Thirty male rats, all newly weaned, were used in the experiment. Ten rats were randomly assigned a diet of solid H, 10 a diet of liquid H, and 10 a standard (control) diet. During the feeding period, each rat was permitted to eat as much as it wished. At the end of 28 days, the total protein intake x (in grams) and the weight gain y (in grams) were recorded for each of the 30 rats. General Foods fit a straight line to the 10 (x, y) data points for each diet by the method of least squares. The three least squares prediction equations follow:

Liquid H $\hat{y} = 109.3 + 3.72x$

Solid H $\hat{y} = 106.7 + 3.66x$

Control $\hat{y} = 50.6 + 2.91x$

a. Interpret the estimated y-intercept and slope for each diet.
b. Predict the weight gain of a rat on liquid H if the rat's 28-day protein intake is 40 grams.

Solution   **a.** First, we will interpret the least squares slopes. Recall that the slope is the change in y for every 1-unit increase in x. Therefore, $\hat{\beta}_1$ represents the estimated change in weight gain (y) for every 1-gram increase in total protein intake (x) for rats on that particular diet. Since all the least squares slopes are *positive*, we are estimating a positive change (i.e., an *increase*) in y.

Liquid H   For every 1-gram increase in total protein intake (x), we estimate weight gain (y) to increase $\hat{\beta}_1 = 3.72$ grams.

Solid H   For every 1-gram increase in total protein intake (x), we estimate weight gain (y) to increase $\hat{\beta}_1 = 3.66$ grams.

Control   For every 1-gram increase in total protein intake (x), we estimate weight gain (y) to increase $\hat{\beta}_1 = 2.91$ grams.

You can see that liquid H yields the greatest estimated increase in weight gain for every 1-gram increase in protein intake. For this reason, General Foods concentrated on developing the liquid version of the new food product.

The y-intercept of the least squares line is the point at which the line crosses the y-axis. This point has an x value of 0. Consequently, the y-intercept can be interpreted as the predicted weight gain (y) for rats with a total protein

*Source: Street, E. and Carroll, M. D. "Preliminary evaluation of a new food product." In *Statistics: A Guide to the Unknown*, Tanur, J. M., et al., eds. San Francisco: Holden-Day, 1978, pp. 269–278.

intake of $x = 0$ grams. In this experiment, the estimated y-intercept is meaningless since it is impractical for rats on any diet to have a total protein intake of 0 grams after 28 days.

In simple linear regression, the estimated y-intercept will usually not have a practical interpretation. It will, however, be practical if the value $x = 0$ is meaningful and is within the range of the sample x-values.

b. To predict the weight gain y of a rat with a protein intake of 40 grams, we substitute $x = 40$ into the least squares prediction equation for liquid H:

$$\hat{y} = 109.3 + 3.72(40) = 258.1 \text{ grams}$$

Thus, we estimate that the rat will gain 258.1 grams in weight if it gets 40 grams of protein while on the 28-day liquid H diet. [*Note:* We will obtain a measure of reliability for a prediction such as this in Section 11.9.]

EXERCISES
LEARNING THE MECHANICS

11.7 Consider the data listed in the table.

x	−1	0	1	2	3
y	−1	1	2	4	5

a. Calculate SS_{xy}.   b. Calculate SS_{xx}.   c. Calculate $\bar{y}$.
d. Calculate $\bar{x}$.   e. Find $\hat{\beta}_1$.   f. Find $\hat{\beta}_0$.

11.8 Consider the five data points:

x	−1	0	1	2	3
y	−1	1	1	2.5	3.5

a. Construct a scattergram for the data.
b. Find the least squares prediction equation.
c. Graph the least squares line on the scattergram and visually confirm that it provides a good fit to the data points.

11.9 Consider the four data points shown in the table.

x	1	1.5	1.9	2.5
y	3.1	2.2	1.0	.3

a. Construct a scattergram for the data.
b. Find the least squares prediction equation.
c. Graph the least squares line on the scattergram and visually confirm that it provides a good fit to the data points.

11.10 Consider the following seven data points:

x	−5	−3	−1	0	1	3	5
y	.8	1.1	2.5	3.1	5.0	4.7	6.2

a. Construct a scattergram for the data.
b. Find the least squares prediction equation.
c. Graph the least squares line on the scattergram and visually confirm that it provides a good fit to the data points.

11.11 Consider the four data points in the accompanying table.

x	−3.0	2.4	−1.1	2.0
y	2.7	.4	1.3	.5

a. Construct a scattergram for the data.
b. Find the least squares prediction equation.
c. Graph the least squares line on the scattergram and visually confirm that it provides a good fit to the data points.

APPLYING THE CONCEPTS

11.12 Two processes for hydraulic drilling of rock are dry drilling and wet drilling. In a dry hole, compressed air is forced down the drill rods to flush the cuttings and drive the hammer; in a wet hole, water is forced down. An experiment was conducted to determine whether the time y it takes to dry drill a distance of 5 feet in rock increases with depth x (*The American Statistician*, Feb. 1991). The results for one portion of the experiment are shown in the accompanying table.

DEPTH AT WHICH DRILLING BEGINS x, feet	TIME TO DRILL 5 FEET y, minutes
0	4.90
25	7.41
50	6.19
75	5.57
100	5.17
125	6.89
150	7.05
175	7.11
200	6.19
225	8.28
250	4.84
275	8.29
300	8.91
325	8.54
350	11.79
375	12.12
395	11.02

Source: Penner, R. and Watts, D. G. "Mining information." *The American Statistician*, Vol. 45, No. 1, Feb. 1991, p. 6 (Table 1).

a. Construct a scattergram for the data.
b. Find the least squares prediction equation.
c. Graph the least squares line on the scattergram.
d. Interpret the values of $\hat{\beta}_0$ and $\hat{\beta}_1$.

11.13   The electroencephalogram (EEG) is a device used to measure brain waves. Neurologists have found that the peak EEG frequency in normal children increases with age. In one study (reported in *Science*, 1982), 287 normal children from 2 to 16 years old were instructed to hold a 65-gram weight in the palm of their outstretched hand for a brief but unspecified time. The peak EEG frequency (measured in hertz) was then recorded for each child. The children were grouped according to age, and the average peak frequency for each age group was recorded. The data appear in the accompanying table.

AGE x, years	AVERAGE PEAK EEG FREQUENCY y, hertz	AGE x, years	AVERAGE PEAK EEG FREQUENCY y, hertz
2	5.33	10	7.28
3	5.75	11	7.06
4	5.80	12	7.60
5	5.60	13	7.45
6	6.00	14	8.23
7	5.78	15	8.50
8	5.90	16	9.38
9	6.23		

Source: Tryon, W. W. "Development equation for postural tremor." *Science*, Vol. 215, No. 2, 1982, pp. 300–301. Copyright 1982 by the AAAS.

a. Construct a scattergram for the data.
b. Find the least squares prediction equation.
c. Graph the least squares line on the scattergram.
d. Use the least squares prediction equation to predict the average peak EEG frequency y of x = 7-year-old normal children. [*Note:*   We will find a measure of the reliability of this prediction in Section 11.9.]

11.14   Each year, *Fortune* ranks the top American cities according to their ability to provide high-quality, low-cost labor for companies that are relocating. One important measure used to form the rankings is the *labor market stress index* (y), which indicates the availability of workers in the city. (The higher the index, the tighter the labor market.) A second important variable is the *unemployment rate* (x). The values of these two variables for each of the top 10 cities in 1990 are listed in the table.

RANK	CITY	LABOR MARKET STRESS INDEX y	UNEMPLOYMENT RATE $x\%$
1	Salt Lake City	107	4.5
2	Minneapolis–St. Paul	107	3.8
3	Atlanta	100	5.1
4	Sacramento	100	4.9
5	Austin (Texas)	80	5.4
6	Columbus (Ohio)	100	4.8
7	Dallas/Fort Worth	100	5.5
8	Phoenix	93	4.3
9	Jacksonville (Florida)	87	5.7
10	Oklahoma City	80	4.6

Source: *Fortune*, Oct. 22, 1990, pp. 58–63.

a. Construct a scattergram for the data.
b. Find the least squares prediction equation.
c. Graph the least squares line on the scattergram.
d. Interpret the values of $\hat{\beta}_0$ and $\hat{\beta}_1$.

11.15 A study was conducted to model the thermal performance of integral-fin tubes used in the refrigeration and process industries (*Journal of Heat Transfer*, Aug. 1990). Twenty-four specially manufactured integral-fin tubes having rectangular-shaped fins made of copper were used in the experiment. Vapor was released downward into each tube, and the vapor-side heat transfer coefficient (based upon the outside surface area of the tube) was measured. The dependent variable for the study is the heat transfer enhancement ratio, y, defined as the ratio of the vapor-side coefficient of the fin tube to the vapor-side coefficient of a smooth tube evaluated at the same temperature. Theoretically, heat transfer will be related to the area at the top of the tube that is "unflooded" by condensation of the vapor. The data in the table are the unflooded area ratio (x) and heat transfer enhancement (y) values recorded for the 24 integral-fin tubes.

UNFLOODED AREA RATIO	HEAT TRANSFER ENHANCEMENT	UNFLOODED AREA RATIO	HEAT TRANSFER ENHANCEMENT
x	y	x	y
1.93	4.4	2.00	5.2
1.95	5.3	1.77	4.7
1.78	4.5	1.62	4.2
1.64	4.5	2.77	6.0
1.54	3.7	2.47	5.8
1.32	2.8	2.24	5.2
2.12	6.1	1.32	3.5
1.88	4.9	1.26	3.2
1.70	4.9	1.21	2.9
1.58	4.1	2.26	5.3
2.47	7.0	2.04	5.1
2.37	6.7	1.88	4.6

Source: Marto, P. J., et al. "An experimental study of R-113 film condensation on horizontal integral-fin tubes." *Journal of Heat Transfer*, Vol. 112, Aug. 1990, p. 763 (Table 2).

a. Find the least squares line relating heat transfer enhancement y to unflooded area ratio x.
b. Plot the data points and graph the least squares line as a check on your calculations.
c. Interpret the values of $\hat{\beta}_0$ and $\hat{\beta}_1$.

11.16 An automated system for marking large numbers of student computer programs, called AUTOMARK, has been used successfully at McMaster University in Ontario, Canada. AUTOMARK takes into account both program correctness and program style when marking student assignments. To evaluate the effectiveness of the automated system, AUTOMARK was used to grade the FORTRAN77 assignments of a class of 33 students. These grades were then compared to the grades assigned by the instructor. The results are shown in the table on the next page.

a. Construct a scattergram for the data.
b. Find the least squares prediction equation.
c. Graph the least squares line on the scattergram.
d. Interpret the values of $\hat{\beta}_0$ and $\hat{\beta}_1$.

11.17 Civil engineers often use the straight-line equation $E(y) = \hat{\beta}_0 + \hat{\beta}_1 x$ to model the relationship between the mean shear strength $E(y)$ of masonry joints and precompression stress x. To test this theory, a series of stress tests was performed on solid bricks arranged in triplets and joined with mortar (*Proceedings of the Institute of Civil Engineers*, Mar. 1990). The precompression stress was varied for each triplet; the ultimate shear load just before failure (called the shear strength) was recorded. The stress results for seven triplets (measured in N/mm^2) is shown in the accompanying table.

TRIPLET TEST	1	2	3	4	5	6	7
Shear Strength, y	1.00	2.18	2.24	2.41	2.59	2.82	3.06
Precompression Stress, x	0	.60	1.20	1.33	1.43	1.75	1.75

Source: Riddington, J. R. and Ghazali, M. Z. "Hypothesis for shear failure in masonry joints." *Proceedings of the Institute of Civil Engineers, Part 2,* Mar. 1990, Vol. 89, p. 96 (Figure 7).

a. Plot the seven data points in a scattergram. Does the relationship between shear strength and precompression stress appear to be linear?
b. Use the method of least squares to estimate the parameters of the linear model.
c. Interpret the values of $\hat{\beta}_0$ and $\hat{\beta}_1$.

Data for Exercise 11.16

AUTOMARK GRADE	INSTRUCTOR GRADE	AUTOMARK GRADE	INSTRUCTOR GRADE	AUTOMARK GRADE	INSTRUCTOR GRADE
x	y	x	y	x	y
12.2	10	18.2	15	19.0	17
10.6	11	15.1	16	19.3	17
15.1	12	17.2	16	19.5	17
16.2	12	17.5	16	19.7	17
16.6	12	18.6	16	18.6	18
16.6	13	18.8	16	19.0	18
17.2	14	17.8	17	19.2	18
17.6	14	18.0	17	19.4	18
18.2	14	18.2	17	19.6	18
16.5	15	18.4	17	20.1	18
17.2	15	18.6	17	19.2	19

Source: Redish, K. A. and Smyth, W. F. "Program style analysis: A natural by-product of program compilation." *Communications of the Association for Computing Machinery,* Vol. 29, No. 2, Feb. 1986, p. 132, Figure 4.

11.4

In the remainder of this chapter, we describe the statistical methods (e.g., tests of hypotheses and confidence intervals) appropriate for making inferences from a simple linear regression analysis. As with most statistical procedures, the validity of the inferences depends on certain assumptions being satisfied. These assumptions, made about the random error term in the straight-line probabilistic model, are summarized in the box on page 544.

Figure 11.6 (page 544) shows a pictorial representation of the assumptions given in the box. For each value of x shown in the figure, the relative frequency distribution of the errors is normal with mean 0 and with a constant variance (all the distributions shown have the same amount of spread or variability) equal to σ^2.

Statistical techniques are available for detecting when one or more of the assumptions are grossly violated. These methods are based on an analysis of the least squares errors of prediction, or **residuals**. (We present an analysis of residuals in Chapter 12.) In practice, however, you will never know whether the data satisfy exactly the four assumptions. Fortunately, the estimators and test statistics

ASSUMPTIONS REQUIRED FOR A LINEAR REGRESSION ANALYSIS

FIGURE 11.6

The Probability Distribution of the Random Error Component, ε

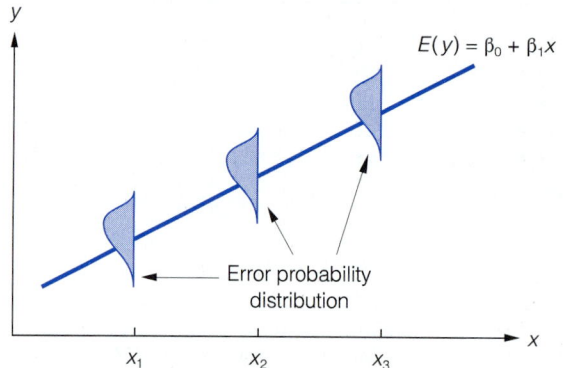

used in a simple linear regression have sampling distributions that remain relatively stable for minor departures from the assumptions.

There is an additional assumption that is implied in a regression analysis, but often forgotten—namely, the assumption that the relationship between the mean value of y, $E(y)$, and the independent variable x is correctly modeled by a straight line. In a real application, the relationship between $E(y)$ and x probably possesses some curvature. Therefore, when we conduct a simple linear regression analysis, we are assuming that this curvature is minimal over the set of values for which x is measured. The implications of this assumption can be seen in Figure 11.7. If x is measured over the interval between two points, say, x_L and x_U, a simple linear regression analysis may produce a very good prediction equation for estimating $E(y)$ or predicting y for values of x between x_L and x_U, but very poor estimates and predictions for values of x outside this range. The rule, then, is

---

*The assumption that $E(\varepsilon) = 0$ is *not* guaranteed by the method of least squares. The method of least squares yields a *sample* mean error of 0, i.e., $[\Sigma(y - \hat{y})]/n = 0$, for all models regardless of whether the model is the correct one. The assumption that $E(\varepsilon) = 0$ in simple linear regression is equivalent to assuming that the form of the correct model is, in fact, a straight line.

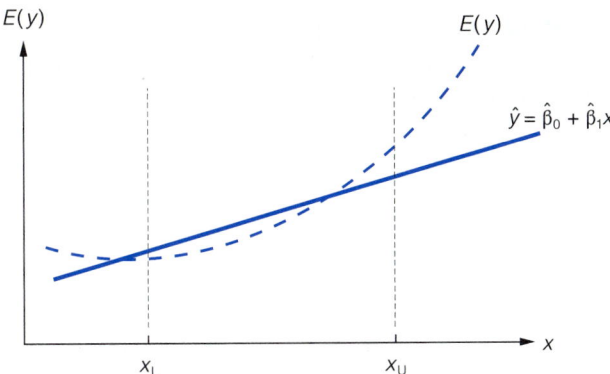

FIGURE 11.7
Hypothetical Comparison of the
True Relationship Between $E(y)$
and x with the Simple Linear
Regression Model

never to attempt to predict values of y for values of x outside the range of values used in the regression analysis.

11.5 Is the creativity score y of Example 11.3 really related to the flexibility score x, or is the linear relation that we seem to see a result of chance? That is, could it be the case that x and y are completely unrelated, and that the apparent linear configuration of the data points in the scattergram of Figure 11.3 is due to random variation? The statistical method that will answer this question requires that we know how much y will vary for a given value of x. That is, we need to know the value of the quantity, called σ^2, that measures the variability of the y values about the least squares line. This value is the variance σ^2 of the random error identified in Section 11.4. Since the variance σ^2 is rarely known, we estimate its value using the sum of the squared deviations (sum of the squared errors, SSE). The procedure is shown in the next box.

ESTIMATING σ^2

ESTIMATION OF σ^2, A MEASURE OF THE VARIABILITY OF THE y VALUES ABOUT THE LEAST SQUARES LINE

An estimate of σ^2 is given by

$$s^2 = \frac{\text{SSE}}{n-2}$$

where $\text{SSE} = \sum(y - \hat{y})^2 = \text{SS}_{yy} - \hat{\beta}_1 \text{SS}_{xy}$

$$\text{SS}_{yy} = \sum(y - \bar{y})^2 = \sum y^2 - \frac{\left(\sum y\right)^2}{n}$$

Warning: When performing these calculations, you may be tempted to round the calculated values of SS_{yy}, $\hat{\beta}_1$, and SS_{xy}. We recommend carrying at least six significant figures for each of these quantities to avoid substantial rounding errors in the calculation of SSE.

[*Note:* The denominator of s^2 is termed the **number of degrees of freedom for error variance estimation**.]

APPLYING THE CONCEPTS

11.21 The data for Exercise 11.12 are reproduced in the table.

DEPTH AT WHICH DRILLING BEGINS x, feet	TIME TO DRILL 5 FEET y, minutes
0	4.90
25	7.41
50	6.19
75	5.57
100	5.17
125	6.89
150	7.05
175	7.11
200	6.19
225	8.28
250	4.84
275	8.29
300	8.91
325	8.54
350	11.79
375	12.12
395	11.02

Source: Penner, R. and Watts, D. G. "Mining information." *The American Statistician*, Vol. 45, No. 1, Feb. 1991, p. 6 (Table 1).

a. Find SSE and s^2 for the data.
b. Calculate s, the estimate of σ.
c. Interpret the value of s obtained in part **b**.

11.22 The data for Exercise 11.13 are reproduced in the table.

AGE x, years	AVERAGE PEAK EEG FREQUENCY y, hertz	AGE x, years	AVERAGE PEAK EEG FREQUENCY y, hertz
2	5.33	10	7.28
3	5.75	11	7.06
4	5.80	12	7.60
5	5.60	13	7.45
6	6.00	14	8.23
7	5.78	15	8.50
8	5.90	16	9.38
9	6.23		

Source: Tryon, W. W. "Developmental equations for postural tremor." *Science*, Vol. 215, No. 2, 1982, pp. 300–301. Copyright 1982 by the AAAS.

a. Find SSE and s^2 for the data.
b. Calculate s, the estimate of σ.
c. Interpret the value of s obtained in part **b**.

11.23 The data for Exercise 11.14 are reproduced in the accompanying table.

RANK	CITY	LABOR MARKET STRESS INDEX y	UNEMPLOYMENT RATE $x\%$
1	Salt Lake City	107	4.5
2	Minneapolis–St. Paul	107	3.8
3	Atlanta	100	5.1
4	Sacramento	100	4.9
5	Austin (Texas)	80	5.4
6	Columbus (Ohio)	100	4.8
7	Dallas/Fort Worth	100	5.5
8	Phoenix	93	4.3
9	Jacksonville (Florida)	87	5.7
10	Oklahoma City	80	4.6

Source: *Fortune*, Oct. 22, 1990, pp. 58–63.

a. Find SSE and s^2 for the data.
b. Calculate s, the estimate of σ.
c. Interpret the value of s.

11.24 The data for Exercise 11.16 is reproduced in the table.

AUTOMARK GRADE x	INSTRUCTOR GRADE y	AUTOMARK GRADE x	INSTRUCTOR GRADE y	AUTOMARK GRADE x	INSTRUCTOR GRADE y
12.2	10	18.2	15	19.0	17
10.6	11	15.1	16	19.3	17
15.1	12	17.2	16	19.5	17
16.2	12	17.5	16	19.7	17
16.6	12	18.6	16	18.6	18
16.6	13	18.8	16	19.0	18
17.2	14	17.8	17	19.2	18
17.6	14	18.0	17	19.4	18
18.2	14	18.2	17	19.6	18
16.5	15	18.4	17	20.1	18
17.2	15	18.6	17	19.2	19

Source: Redish, K. A. and Smyth, W. F. "Program style analysis: A natural by-product of program compilation." *Communications of the Association for Computing Machinery*, Vol. 29, No. 2, Feb. 1986, p. 132, Figure 4.

a. Find SSE and s^2 for the data.
b. Calculate s, the estimate of σ.
c. Interpret the value of s.

11.6   MAKING INFERENCES ABOUT THE SLOPE β_1

After fitting the model to the data and computing an estimate of σ^2, we can statistically check the usefulness of the model. That is, we can use a statistical procedure (a test of hypothesis or confidence interval) to determine whether the least squares straight-line (linear) model is a reliable tool for predicting y for a given value of x.

EXAMPLE 11.7 Consider the probabilistic model

$$y = \beta_0 + \beta_1 x + \varepsilon$$

How do we determine statistically whether this model is useful for prediction? In other words, how could we test whether x provides useful information for the prediction of y?

Solution Suppose x is *completely unrelated* to y. What could we say about the values of β_0 and β_1 in the probabilistic model, if x contributes no information for the prediction of y? For y to be independent of x, the true slope of the line, β_1, must be equal to 0. Therefore, to test the null hypothesis that x contributes no information for the prediction of y against the alternative that these variables are linearly related with a slope differing from 0, we test

$$H_0: \quad \beta_1 = 0$$
$$H_a: \quad \beta_1 \neq 0$$

If the data support the alternative hypothesis, we will conclude that x does contribute information for the prediction of y using the straight-line model, although the true relationship between $E(y)$ and x could be more complex than a straight line.

Using the hypothesis-testing techniques developed in Chapters 9 and 10, we set up the test for the predictive ability of the model as shown in the next box.

TEST OF HYPOTHESIS FOR DETERMINING WHETHER THE STRAIGHT-LINE MODEL IS USEFUL FOR PREDICTING y FROM x

ONE-TAILED TEST

$$H_0: \quad \beta_1 = 0$$
$$H_a: \quad \beta_1 > 0$$
$$(\text{or } H_a: \quad \beta_1 < 0)$$

TWO-TAILED TEST

$$H_0: \quad \beta_1 = 0$$
$$H_a: \quad \beta_1 \neq 0$$

Test statistic: $t = \dfrac{\hat{\beta}_1}{s/\sqrt{SS_{xx}}}$

Rejection region:

$$t > t_\alpha \quad (\text{or } t < -t_\alpha)$$

Rejection region:

$$|t| > t_{\alpha/2}$$

where the distribution of t is based on $(n - 2)$ degrees of freedom, t_α is the t value such that $P(t > t_\alpha) = \alpha$, and $t_{\alpha/2}$ is the t value such that $P(t > t_{\alpha/2}) = \alpha/2$.

Assumptions: See Section 11.4.

[*Note:* The test statistic is derived from the sampling distribution of the least squares estimator of the slope, $\hat{\beta}_1$.]

Inferences based on this hypothesis test require the standard least squares assumptions about the random error term listed in the box in Section 11.4. However, the test statistic has a sampling distribution that remains relatively stable for minor departures from the assumptions. That is, our inferences remain valid for practical cases in which the assumptions are nearly, but not completely, satisfied.

EXAMPLE 11.8 Refer to the creativity–flexibility scores of Example 11.3. At significance level $\alpha = .05$, test the hypothesis that flexibility score x contributes useful information for the prediction of creativity score y, i.e., test the predictive ability of the least squares straight-line model

$$\hat{y} = -2.2 + 2.3x$$

Solution Testing the usefulness of the model requires testing the hypotheses

H_0: $\beta_1 = 0$

H_a: $\beta_1 \neq 0$

With $n = 5$ and $\alpha = .05$, the critical value based on $(5 - 2) = 3$ df is obtained from Table 5 of Appendix G:

$$t_{\alpha/2} = t_{.025} = 3.182$$

Thus, we will reject H_0 if $|t| > 3.182$, i.e., if $t < -3.182$ or $t > 3.182$.

To compute the test statistic, we need the values of $\hat{\beta}_1$, s, and SS_{xx}. In previous examples, we computed $\hat{\beta}_1 = 2.3$, $s = .6055$, and $SS_{xx} = 10$. Hence, our test statistic is

$$t = \frac{\hat{\beta}_1}{s/\sqrt{SS_{xx}}} = \frac{2.3}{.6055/\sqrt{10}} = 12.01$$

Since this calculated t value falls in the upper tail of the rejection region (see Figure 11.9), we reject the null hypothesis and conclude that the slope β_1 is not 0.

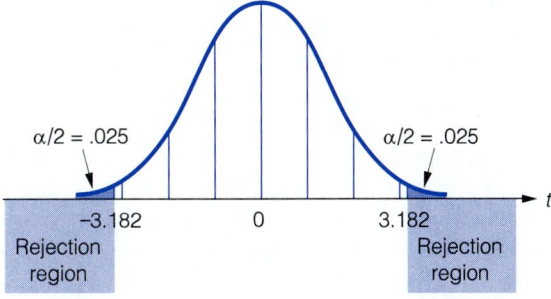

FIGURE 11.9
Rejection Region for Example 11.8

At the $\alpha = .05$ level of significance then, the sample data provide sufficient evidence to conclude that flexibility score *does* contribute useful information for the prediction of creativity score using the linear model.

If the test statistic in Example 11.8 had not fallen in the rejection region, would we have concluded that $\beta_1 = 0$? The answer to this question is "no" (recall the discussion on Type II errors in Chpater 9). Rather, we acknowledge that additional data might indicate that β_1 differs from 0, or that a more complex relationship (other than a straight line) may exist between y and x. We may also want to examine the observed significance level of the test, as illustrated in the next example.

EXAMPLE 11.9 A consumer investigator obtained the following least squares straight-line model (based on a sample of $n = 100$ families) relating the yearly food cost y for a family of four to annual income x:

$$\hat{y} = 467 + .26x$$

In addition, the investigator computed the quantities $s = 1.1$ and $SS_{xx} = 26$. Compute the observed significance level (p-value) for a test to determine whether mean yearly food cost y increases as annual income x increases, i.e., whether the slope of the line, β_1, is positive. Interpret this value.

Solution The consumer investigator wishes to test

$$H_0: \quad \beta_1 = 0$$
$$H_a: \quad \beta_1 > 0$$

To compute the observed significance level of the test we must first find the calculated value of the test statistic, t_c. Since $\hat{\beta}_1 = .26$, $s = 1.1$, and $SS_{xx} = 26$, we have

$$t_c = \frac{\hat{\beta}_1}{s/\sqrt{SS_{xx}}} = \frac{.26}{1.1/\sqrt{26}} = 1.21$$

From Section 9.6, the observed significance level or p-value is given by

$$P(t > t_c) = P(t > 1.21)$$

where the distribution of t is based on $(n - 2) = (100 - 2) = 98$ degrees of freedom. Since df is greater than 30, we can approximate the t distribution with the z distribution. Thus,

$$p\text{-value} = P(t > 1.21) \approx P(z > 1.21) = .5 - .3869 = .1131$$

To conclude that the mean yearly food cost increases as annual income increases (i.e., $\beta_1 > 0$), the investigator would have to be willing to tolerate a Type I error probability, α, of .1131 or larger. Since it is very doubtful that the investigator would be willing to take such a risk, we consider the sample result to be statistically insignificant.

In addition to testing whether the slope β_1 is 0, we may also be interested in estimating its value with a confidence interval. The procedure is illustrated in the following example.

EXAMPLE 11.10   Using the information supplied in Example 11.8, construct a 95% confidence interval for the slope β_1 in the straight-line model relating creativity score to flexibility score.

Solution   The methods of Chapter 8 can be used to construct a confidence interval for β_1. The interval, derived from the sampling distribution of $\hat{\beta}_1$, is given in the following box.

A $(1 - \alpha)$100% CONFIDENCE INTERVAL FOR THE SLOPE β_1

$$\hat{\beta}_1 \pm t_{\alpha/2}\left(\frac{s}{\sqrt{SS_{xx}}}\right)$$

where the distribution of t is based on $(n - 2)$ degrees of freedom and $t_{\alpha/2}$ is the value of t such that $P(t > t_{\alpha/2}) = \alpha/2$.

For a 95% confidence interval, $\alpha = .05$. Therefore, we need to find the value of $t_{.025}$ based on $(n - 2) = (5 - 2) = 3$ df. In Example 11.8, we found that $t_{.025} = 3.182$. Also we have $\hat{\beta}_1 = 2.3$, $s = .6055$, and $SS_{xx} = 10$. Thus, a 95% confidence interval for the slope in the model relating creativity score to flexibility score is

$$\hat{\beta}_1 \pm t_{.025}\left(\frac{s}{\sqrt{SS_{xx}}}\right) = 2.3 \pm 3.182\left(\frac{.6055}{\sqrt{10}}\right) = 2.3 \pm .61$$

Our interval estimate of the slope parameter β_1 is then 1.69 to 2.91.

EXAMPLE 11.11   Interpret the interval estimate of β_1 derived in Example 11.10.

Solution   Since all the values in the interval $(1.69, 2.91)$ are positive, we say that we are 95% confident that the slope β_1 is positive. That is, we are 95% confident that the mean creativity score, $E(y)$, increases as flexibility score, x, increases. In addition, we can say that for every 1-point increase in flexibility score x, the increase in creativity score $E(y)$ could be as small as 1.69 or as large as 2.91. However, the rather large width of the interval reflects the small number of data points (and, consequently, a lack of information) in the experiment. We could expect a narrower interval if the sample size were increased.

EXERCISES
LEARNING THE MECHANICS

11.25   Suppose you want to test H_0: $\beta_1 = 0$ versus H_a: $\beta_1 \neq 0$ in a simple linear regression model. Give the degrees of freedom associated with the value of the test statistic for each of the following sample sizes.

    a. $n = 6$ **b.** $n = 10$ **c.** $n = 25$ **d.** $n = 50$

11.26 For each of the following combinations of H_a, $\hat{\beta}_1$, s^2, SS_{xx}, n, and α, specify the rejection region, test statistic value, and your conclusion for testing the null hypothesis H_0: $\beta_1 = 0$.

a. H_a: $\beta_1 \neq 0$, $\hat{\beta}_1 = .7$, $s^2 = .36$, $SS_{xx} = 10$, $n = 5$, $\alpha = .05$

b. H_a: $\beta_1 > 0$, $\hat{\beta}_1 = 1.5$, $s^2 = .10$, $SS_{xx} = 10$, $n = 5$, $\alpha = .10$

c. H_a: $\beta_1 < 0$, $\hat{\beta}_1 = -.011$, $s^2 = 3.880$, $SS_{xx} = 21,752$, $n = 7$, $\alpha = .01$

11.27 Refer to the data given in Exercises 11.8 and 11.19.

a. Test the null hypothesis that the slope β_1 of the line equals 0 against the alternative hypothesis that β_1 is not equal to 0. Use $\alpha = .10$.

b. Compute the approximate observed significance level of the test.

c. Find a 90% confidence interval for the slope β_1.

11.28 Refer to the data given in Exercises 11.10 and 11.20.

a. Test the null hypothesis that the slope β_1 of the line equals 0 against the alternative hypothesis that β_1 is not equal to 0. Use $\alpha = .10$.

b. Compute the approximate observed significance level of the test.

c. Find a 90% confidence interval for the slope β_1.

APPLYING THE CONCEPTS

11.29 Theophylline is a drug used to control asthma in children. To be effective, the blood concentration level of the drug must remain between 4 and 20 picograms per milliliter. Thus, frequent monitoring of theophylline concentrations is necessary for successful therapy. A study was conducted to compare the Ames Seralyzer assay system and the enzyme-multiplied immunoassay technique (EMIT), two methods of determining theophylline concentration (*American Journal of Hospital Pharmacy*, July 1986). A total of 102 blood serum samples were obtained from pediatric intensive-care unit patients who were receiving theophylline intravenously. Each sample was analyzed for theophylline concentration using both methods. The data (recorded in picograms per milliliter) were used to fit the model $y = \beta_0 + \beta_1 x + \varepsilon$, where

y = Theophylline concentration as determined by Ames Seralyzer

x = Theophylline concentration as determined by EMIT

A summary of the results follows:

$$\hat{y} = 1.737 + .953x \qquad r = .97 \qquad s_{\hat{\beta}_1} = .025 \qquad s = 1.9$$

a. Construct a 95% confidence interval for β_1.

b. Interpret the interval, part **a**, and explain what it tells you about the relationship between the theophylline measurements of the two methods.

11.30 Refer to the *The American Statistician* investigation of dry drilling in rock, Exercises 11.12 and 11.21. Is there evidence to indicate that dry drill time y increases with depth x? Test using $\alpha = .10$.

11.31 Refer to the *Journal of Heat Transfer* study of the straight-line relationship between heat transfer enhancement (y) and unflooded area ratio (x), Exercise 11.15. Construct a 95% confidence interval for β_1, the slope of the line. Interpret the result.

11.32 Refer to the *Science* (Vol. 215, 1982) study discussed previously in Exercises 11.13 and 11.22. Do the data provide sufficient evidence to indicate that age, x, contributes information for the prediction of average peak frequency, y? Test using $\alpha = .05$.

11.33 Refer to the McMaster University (Canada) study of an automated system for grading student assignments, Exercises 11.16 and 11.24. Is there evidence of a positive linear relationship between AUTOMARK grade (x) and instructor grade (y)? Use $\alpha = .01$.

11.34 The data in the accompanying table consist of the average life expectancy and the population–physician ratio (i.e., the population divided by the number of physicians) for a random sample of 10 developing African countries.

COUNTRY	AVERAGE LIFE EXPECTANCY y, years	POPULATION–PHYSICIAN RATIO x
1	63.00	1,907
2	32.00	47,889
3	48.30	26,447
4	52.70	815
5	53.50	6,411
6	49.05	10,136
7	38.30	7,306
8	50.00	22,291
9	47.35	18,657
10	52.50	7,378

Source: United Nations Statistical Yearbook.

a. Find the least squares prediction equation $\hat{y} = \hat{\beta}_0 + \hat{\beta}_1 x$.
b. Find SSE and s^2.
c. Is there sufficient evidence to indicate that average life expectancy y of developing African countries is linearly related to population–physician ratio x? Test using $\alpha = .10$.

11.35 Refer to the *Chemosphere* (Vol. 20, 1990) study of Vietnam veterans exposed to Agent Orange (and the dioxin 2,3,7,8-TCDD), Exercise 8.64. The table, reproduced here, gives the amounts of 2,3,7,8-TCDD (measured in parts per million) in both blood plasma and fat tissue drawn from each of the 20 veterans studied.

VETERAN	TCDD LEVELS IN PLASMA	TCDD LEVELS IN FAT TISSUE	VETERAN	TCDD LEVELS IN PLASMA	TCDD LEVELS IN FAT TISSUE
1	2.5	4.9	11	6.9	7.0
2	3.1	5.9	12	3.3	2.9
3	2.1	4.4	13	4.6	4.6
4	3.5	6.9	14	1.6	1.4
5	3.1	7.0	15	7.2	7.7
6	1.8	4.2	16	1.8	1.1
7	6.0	10.0	17	20.0	11.0
8	3.0	5.5	18	2.0	2.5
9	36.0	41.0	19	2.5	2.3
10	4.7	4.4	20	4.1	2.5

Source: Schecter, A., et al. "Partitioning of 2,3,7,8-chlorinated dibenzo-p-dioxins and dibenzofurans between adipose tissue and plasma lipid of 20 Massachusetts Vietnam veterans." Chemosphere, Vol. 20, Nos. 7–9, 1990, pp. 954–955 (Table I & II).

One goal of the researchers is to determine the degree of linear association between the level of dioxin found in blood plasma and fat tissue. If a linear association between the two variables can be established, the researchers want to build models to (1) predict blood plasma level of 2,3,7,8-TCDD from the observed level of 2,3,7,8-TCDD in fat tissue and (2) predict fat tissue level from the observed blood plasma level.

a. Find the prediction equations for the researchers. Interpret the results.
b. Test the hypothesis that fat tissue level (x) is a useful linear predictor of blood plasma level (y). Use $\alpha = .05$.

c. Test the hypothesis that blood plasma level (x) is a useful linear predictor of fat tissue level (y). Use $\alpha = .05$.

d. Intuitively, why must the results of the tests, parts b and c, agree?

11.36 Refer to Exercise 11.35b. The blood plasma and fat tissue levels of several other types of dioxin (called cogeners) were also measured for each of the 20 Vietnam veterans. For each cogener, a simple linear regression analysis was conducted to predict (1) fat tissue level from blood plasma level and (2) blood plasma level from fat tissue level. The results for three of these cogeners are shown in the table.

COGENER	y = FAT TISSUE LEVEL x = BLOOD PLASMA LEVEL	y = BLOOD PLASMA LEVEL x = FAT TISSUE LEVEL	t-VALUE FOR TESTING β_1
2,3,4,7,8-P_n CDF	$\hat{y} = .8109 + .9713x$	$\hat{y} = .9855 + .7605x$	7.13
H_x CDD	$\hat{y} = 18.1565 + .7377x$	$\hat{y} = 5.2009 + .9018x$	5.98
OCDD	$\hat{y} = 118.6057 + .3679x$	$\hat{y} = 167.723 + 1.5752x$	4.98

Source: Schecter, A., et al. "Partitioning of 2,3,7,8-chlorinated dibenzo-p-dioxins and dibenzofurans between adipose tissue and plasma lipid of 20 Massachusetts Vietnam veterans." *Chemosphere*, Vol. 20, Nos. 7–9, 1990, pp. 954–955 (Table III).

a. For the cogener 2,3,4,7,8-P_n CDF, are the two regression models statistically adequate for predicting y? Test both using $\alpha = .05$.

b. Repeat part **a** for the cogener H_x CDD.

c. Repeat part **a** for the cogener OCDD.

d. Use the regression results to predict the level of 2,3,4,7,8-P_n CDF in blood plasma for a veteran with a fat tissue level of 8.0 ppm.

e. Use the regression results to predict the level of H_x CDD in fat tissue for a veteran with a blood plasma level of 24.0 ppm.

f. Use the regression results to predict the level of OCDD in blood plasma for a veteran with a fat tissue level of 776 ppm.

THE COEFFICIENT OF CORRELATION

11.7 In Section 11.6 we discovered that the least squares slope, $\hat{\beta}_1$, provides useful information on the linear relationship between two variables y and x. Another way to measure association is to compute the **Pearson product moment correlation coefficient r**. The correlation coefficient, defined in the box, provides a quantitative measure of the strength of the linear relationship between x and y in the sample, just as does the least squares slope $\hat{\beta}_1$. However, unlike the slope, the correlation coefficient r is *scaleless*. The value of r is always between -1 and $+1$, no matter what the units of x and y are.

DEFINITION 11.1

The **Pearson product moment coefficient of correlation r** is a measure of the strength of the linear relationship between two variables x and y in the sample. It is computed (for a sample of n measurements on x and y) as follows:

$$r = \frac{SS_{xy}}{\sqrt{SS_{xx}SS_{yy}}}$$

Since both r and $\hat{\beta}_1$ provide information about the utility of the model, it is not surprising that there is a similarity in their computational formulas. In particular, note that SS_{xy} appears in the numerators of both expressions and, since both denominators are always positive, r and $\hat{\beta}_1$ will always be of the same sign (either both positive or both negative). A value of r near or equal to 0 implies little or no linear relationship between y and x. And, if $r = 1$ or $r = -1$, all the points fall exactly on the least squares line. Positive values of r imply that y increases as x increases; negative values imply that y decreases as x increases. See Figure 11.10.

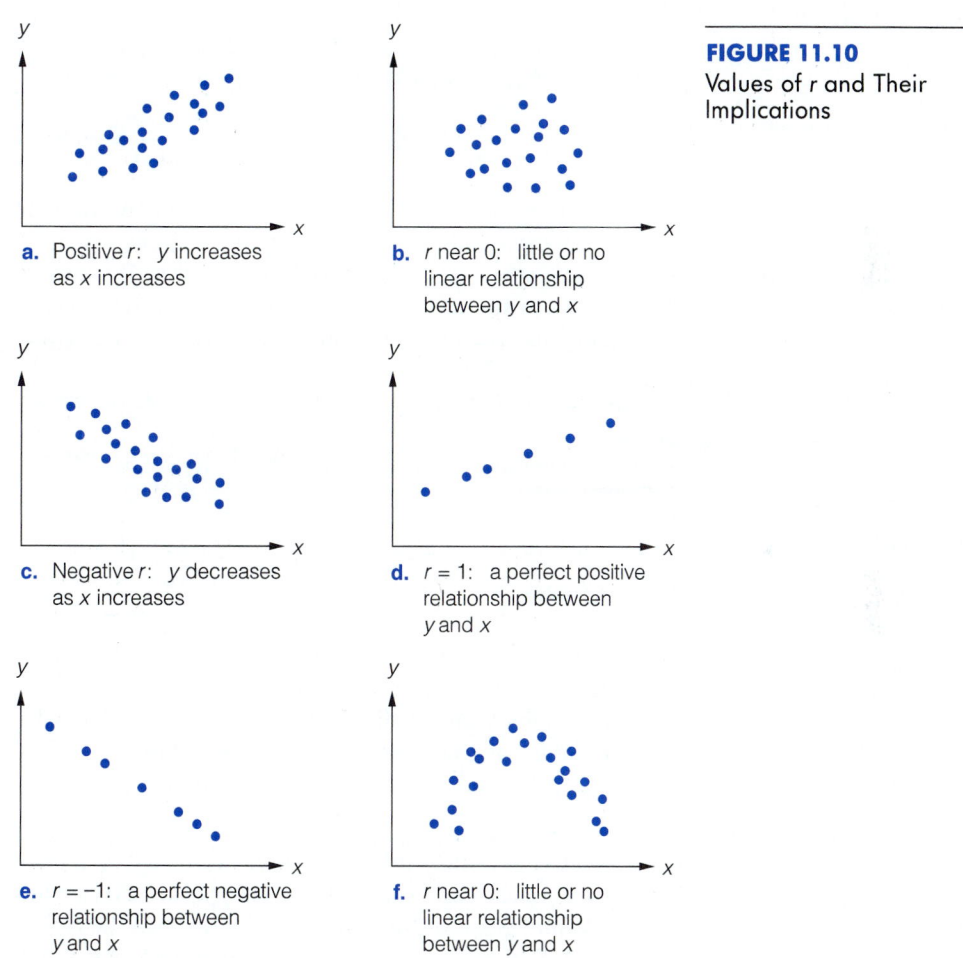

FIGURE 11.10
Values of r and Their Implications

a. Positive r: y increases as x increases

b. r near 0: little or no linear relationship between y and x

c. Negative r: y decreases as x increases

d. $r = 1$: a perfect positive relationship between y and x

e. $r = -1$: a perfect negative relationship between y and x

f. r near 0: little or no linear relationship between y and x

EXAMPLE 11.12   The data for Example 11.3 is reproduced in Table 11.4 (page 558). Calculate the coefficient of correlation r between flexibility score and creativity score y. Interpret the result.

TABLE 11.4
Creativity–Flexibility Score Data for Example 11.12

CHILD	FLEXIBILITY SCORE	CREATIVITY SCORE
1	2	2
2	3	5
3	4	7
4	5	10
5	6	11

Solution From previous calculations (see Examples 11.3 and 11.6), we found $SS_{xy} = 23$, $SS_{xx} = 10$, and $SS_{yy} = 54$. Then, the coefficient of correlation is

$$r = \frac{SS_{xy}}{\sqrt{SS_{xx}SS_{yy}}} = \frac{23}{\sqrt{(10)(54)}} = .9898$$

Thus, creativity score and flexibility score are highly correlated—at least for this sample of five mentally retarded children. The implication is that a strong positive linear relationship exists between these variables. We must be careful, however, not to jump to any unwarranted conclusions. For instance, the psychologist may be tempted to conclude that a high flexibility score will *always* result in a higher creativity score. The implication of such a conclusion is that there is a **causal** relationship between the two variables. However, **high correlation does not imply causality**. Many other factors, such as severity of illness, parental neglect, and IQ may contribute to the change in creativity score.

WARNING

High correlation does not imply causality. If a large positive or negative value of the sample correlation coefficient is observed, it is incorrect to conclude that a change in x causes a change in y. The only valid conclusion is that a linear trend *may* exist between x and y.

Keep in mind that the correlation coefficient r measures the correlation between x values and y values in the sample, and that a similar linear coefficient of correlation exists for the population from which the data points were selected. The **population correlation coefficient** is denoted by the symbol ρ (rho). As you might expect, ρ is estimated by the corresponding sample statistic, r. Or, rather than estimating ρ, we might want to test the hypothesis H_0: $\rho = 0$ against H_a: $\rho \neq 0$, i.e., test the hypothesis that x contributes no information for the prediction of y using the straight-line model against the alternative that the two variables are at least linearly related. However, we have already performed this identical test in Section 11.6 when we tested H_0: $\beta_1 = 0$ against H_a: $\beta_1 \neq 0$.

It is easy to show that $r = \hat{\beta}_1 \sqrt{SS_{xx}/SS_{yy}}$. Thus, $\hat{\beta}_1 = 0$ implies $r = 0$, and vice versa. Consequently, the null hypothesis $H_0: \rho = 0$ is equivalent to the hypothesis $H_0: \beta_1 = 0$. When we tested the null hypothesis $H_0: \beta_1 = 0$ in connection with the creativity–flexibility score example, the data led to a rejection of the hypothesis for $\alpha = .05$. This implies that the null hypothesis of a zero linear correlation between the two variables (creativity score and flexibility score) can also be rejected at $\alpha = .05$. The only real difference between the least squares slope $\hat{\beta}_1$ and the coefficient of correlation r is the measurement scale. Therefore, the information they provide about the utility of the least squares model is to some extent redundant. Furthermore, the slope β_1 gives us additional information on the amount increase (or decrease) in y for every 1-unit increase in x. For this reason, we recommend using the slope to make inferences about the existence of a positive or negative linear relationship between two variables. For those who prefer to test for a linear relationship between two variables using the coefficient of correlation r, we outline the procedure in the box.

TEST OF HYPOTHESIS FOR LINEAR CORRELATION

ONE-TAILED TEST

$H_0: \quad \rho = 0$

$H_a: \quad \rho > 0$

(or $H_a: \quad \rho < 0$)

TWO-TAILED TEST

$H_0: \quad \rho = 0$

$H_a: \quad \rho \neq 0$

$$\text{Test statistic:} \quad t = \frac{r\sqrt{n-2}}{\sqrt{1-r^2}}$$

Rejection region:

$t > t_\alpha \quad (\text{or } t < -t_\alpha)$

Rejection region:

$|t| > t_{\alpha/2}$

where the distribution of t depends on $(n-2)$ df, and t_α and $t_{\alpha/2}$ are the critical values obtained from Table 5 of Appendix G.

Assumptions: The sample of (x, y) values is randomly selected from a (bivariate) normal population.*

The next example illustrates how the correlation coefficient r may be a misleading measure of the strength of the association between x and y in situations where the true relationship is nonlinear.

EXAMPLE 11.13 Underinflated or overinflated tires can increase tire wear and decrease gas mileage. A manufacturer of a new tire tested the tire for wear at different pressures with the results shown in Table 11.5 (page 560). Calculate the coefficient of correlation r for the data. Interpret the result.

---

*A bivariate normal population will result if the probability distributions of both x and y are normal.

TABLE 11.5
Data for Example 11.13

PRESSURE x, pounds per sq. inch	MILEAGE y, thousands	PRESSURE x, pounds per sq. inch	MILEAGE y, thousands
30	29.5	33	37.6
30	30.2	34	37.7
31	32.1	34	36.1
31	34.5	35	33.6
32	36.3	35	34.2
32	35.0	36	26.8
33	38.2	36	27.4

Solution   Rather than perform the calculations by hand, we will use a computer to find the value of r. A SAS printout of the correlation analysis is shown in Figure 11.11. The value of r, shaded on the printout, is $r = -.114$. This relatively small value for r describes a weak linear relationship between pressure (x) and mileage (y). The manufacturer, however, would be remiss in concluding that tire pressure has little or no impact on wear of the tire. On the contrary, the relationship between pressure and wear is fairly strong, as the scattergram in Figure 11.12 illustrates. Note that the relationship is not linear, but curvilinear; the underinflated tires (low pressure values) and overinflated tires (high pressure values) both lead to low mileages.

FIGURE 11.11
SAS Printout of Correlation Analysis of Data in Table 11.5

```
Pearson Correlation Coefficients / Prob > |R|  under Ho: Rho=0 / N = 14

                                    X                    Y

              X               1.00000              -0.11371
                              0.0                   0.6987

              Y              -0.11371               1.00000
                              0.6987                0.0
```

FIGURE 11.12
Scattergram of Data in Table 11.5

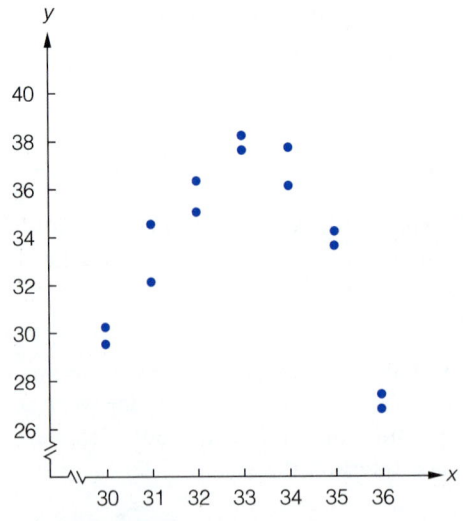

Example 11.13 points out the danger of using r to determine how well x predicts y: The correlation coefficient r describes only the *linear* relationship between x and y. For nonlinear relationships, the value of r may be misleading and we must resort to other methods for describing and testing such a relationship. Regression models for curvilinear relationships are presented in Chapter 12.

EXERCISES

LEARNING THE MECHANICS

11.37 The five data points for Exercise 11.8 are reproduced here. Find the correlation coefficient r and interpret its value.

x	−1	0	1	2	3
y	−1	1	1	2.5	3.5

11.38 The seven data points for Exercise 11.10 are reproduced here. Find the correlation coefficient r and interpret its value.

x	−5	−3	−1	0	1	3	5
y	.8	1.1	2.5	3.1	5.0	4.7	6.2

11.39 What value does r assume if all the sample points fall on the same straight line and if the line has:

a. A positive slope? **b.** A negative slope?

11.40 For each of the following situations, specify the rejection region and state your conclusion for testing the null hypothesis H_0: There is no linear correlation between x and y:

a. H_a: The variables x and y are positively correlated;
$r = .68$, $n = 10$, $\alpha = .01$
b. H_a: The variables x and y are negatively correlated;
$r = -.68$, $n = 52$, $\alpha = .01$
c. H_a: The variables x and y are linearly correlated;
$r = -.84$, $n = 10$, $\alpha = .10$

APPLYING THE CONCEPTS

11.41 Do you believe the grade point average of a college student is correlated with the student's intelligence quotient (IQ)? If so, will the correlation be positive or negative? Explain.

11.42 Research by law enforcement agencies has shown that the crime rate is correlated with the U.S. population. Would you expect the correlation to be positive or negative? Explain.

11.43 An investigation sought to determine whether certain collective behaviors, affective reactions, or performance outcomes are associated with the maturity level of small groups (*Small Group Behavior*, May 1988). Fifty-eight undergraduate students enrolled in MIS or communications courses at a medium-size university participated in the experiment. A 10-item questionnaire (see Exercise 6.16) was used to measure the maturity level, y, of the students on a scale of 0–100, where more mature students received higher scores. One of several other variables

which is the familiar quantity SS_{yy}. The magnitude of SS_{yy} is thus an indicator of how well $\bar{y}$ behaves as a predictor of y.

EXAMPLE 11.15   Refer to Example 11.14. Suppose now that you use the information on flexibility score, x, to predict creativity score, y. How do we measure the additional information provided by using the value of x in the least squares prediction equation, rather than $\bar{y}$, to predict y?

Solution   If we use the information on x to predict y, then the sum of squares of the deviations of the y values about the predicted values obtained from the least squares equation $\hat{y} = \hat{\beta}_0 + \hat{\beta}_1 x$ is

$$SSE = \sum(y - \hat{y})^2$$

A convenient way of measuring how well the least squares equation performs as a predictor of y is to compute the reduction in the sum of squares of deviations that can be attributed to x, expressed as a proportion of SS_{yy}. This quantity, called the **coefficient of determination**, is

$$\frac{SS_{yy} - SSE}{SS_{yy}}$$

It can be shown that this proportion is equal to the square of the simple linear coefficient of correlation r.

> **DEFINITION 11.2**
>
> The **coefficient of determination** is
>
> $$r^2 = \frac{SS_{yy} - SSE}{SS_{yy}} = 1 - \frac{SSE}{SS_{yy}}$$
>
> It represents the proportion of the sum of squares of deviations of the y values about their mean that can be attributed to a linear relationship between y and x. (It may also be computed as the square of the coefficient of correlation.)

Note that r^2 is always between 0 and 1, since r is between -1 and $+1$. Thus, $r^2 = .75$ means that 75% of the sum of squares of deviations of the y values about their mean is attributable to the linear relationship between y and x. In other words, the error of prediction can be reduced by 75% when the least squares equation, rather than $\bar{y}$, is used to predict y.

EXAMPLE 11.16   Calculate the coefficient of determination for the flexibility–creativity score data of Example 11.3, and interpret its value. The data are repeated in Table 11.6 for convenience.

TABLE 11.6

Creativity–Flexibility Score Data for Example 11.16

CHILD	FLEXIBILITY SCORE	CREATIVITY SCORE
1	2	2
2	3	5
3	4	7
4	5	10
5	6	11

Solution We will use the formula given in Definition 11.2 to compute r^2. From previous calculations, we have $SS_{yy} = 54$ and $SSE = 1.1$. Therefore,

$$r^2 = \frac{SS_{yy} - SSE}{SS_{yy}} = \frac{54 - 1.1}{54} = \frac{52.9}{54} = .9796$$

We interpret this value as follows: The use of flexibility score, x, to predict creativity score, y, with the least squares line

$$\hat{y} = -2.2 + 2.3x$$

accounts for approximately 98% of the total sum of squares of deviations of the five sample creativity scores about their mean. That is, we can reduce the total sum of squares of our prediction errors by nearly 98% by using the least squares equation $\hat{y} = -2.2 + 2.3x$, instead of $\bar{y}$, to predict y.

PRACTICAL INTERPRETATION OF THE COEFFICIENT OF DETERMINATION, r^2

About $100(r^2)\%$ of the total sum of squares of deviations of the sample y values about their mean $\bar{y}$ can be explained by (or attributed to) using x to predict y in the straight-line model.

Since the two numerical descriptive measures r and r^2 are very closely related, there may be some confusion as to when each should be used. Our recommendations are as follows: If you are interested only in measuring the strength of the linear relationship between two variables x and y, use the coefficient of correlation r. However, if you want to determine how well the least squares straight-line model fits the data, use the coefficient of determination r^2.

EXERCISES

LEARNING THE MECHANICS

11.51 For a set of $n = 20$ data points, $SS_{yy} = 210$ and $SSE = 31$. Find r^2.

11.52 For a set of $n = 30$ data points, $SS_{xx} = 39$, $SS_{yy} = 12$, and $SS_{xy} = 19$. Find r^2.

11.53 Consider the data listed in the accompanying table.

x	1	2	3	4	5
y	1	1	2	2	4

a. Calculate SS_{xx}. b. Calculate SS_{xy}. c. Calculate SS_{yy}.
d. Calculate $\hat{\beta}_1$. e. Calculate SSE. f. Calculate r^2.

11.54 Refer to the data of Exercises 11.8, 11.19, and 11.27. Calculate the coefficient of determination r^2 and interpret its value.

11.55 Refer to the data of Exercises 11.10, 11.20, and 11.28. Calculate the coefficient of determination r^2 and interpret its value.

APPLYING THE CONCEPTS

11.56 Refer to the data on drill time and depth of penetration provided in Exercise 11.12. Compute r^2 and interpret its value.

11.57 Refer to the data on peak EEG frequency and age of children provided in Exercise 11.13. Calculate the coefficient of determination r^2 and interpret its value.

11.58 A major portion of the effort expended in developing commercial computer software is associated with program testing. A study was undertaken to assess the potential usefulness of various product- and process-related variables in identifying error-prone software (*IEEE Transactions on Software Engineering.* Apr. 1985). A straight-line model relating the number y of module defects to the number x of unique operands in the module was fit to the data collected for a sample of software modules. The coefficient of determination for this analysis was $r^2 = .74$.

a. Interpret the value of r^2.
b. Based on this value, would you infer that the straight-line model is a useful predictor of number y of module defects? Explain.

11.59 Refer to the study of homicide rates in Florida counties, Exercise 11.48. Calculate the coefficient of determination r^2 and interpret its value.

11.60 Refer to the data on AUTOMARK grade and instructor grade provided in Exercises 11.16, 11.24, and 11.50. Find r^2 and interpret its value.

11.61 Refer to the study on theophylline concentration in Exercise 11.29. Find the coefficient of determination r^2 and interpret its value.

11.62 Refer to the data on average life expectancy and population–physician ratio provided in Exercise 11.34. Find the coefficient of determination r^2 and interpret its value.

USING THE MODEL FOR ESTIMATION AND PREDICTION

11.9 After we have statistically checked the usefulness of our straight-line model and are satisfied that x contributes information for the prediction of y, we are ready to accomplish our original objective—using the model for prediction and estimation.

The most common uses of a probabilistic model for making inferences can be divided into two categories, which are listed in the box.

1. Use the model for estimating the mean value of y, $E(y)$, for a specific value of x.
2. Use the model for predicting a particular y value for a given value of x.

In the first case, we want to estimate the mean value of y for a very large number of experiments at a given x value. For example, the psychologist may want to estimate the mean creativity score for all mentally retarded children with flexibility scores of 3. In the second case, we wish to predict the outcome of a single experiment (predict an individual value of y) at the given x value. For example, the psychologist may want to predict the creativity score of a particular mentally retarded child who scored 3 on the flexibility test.

We will use the least squares model

$$\hat{y} = \hat{\beta}_0 + \hat{\beta}_1 x$$

both to estimate the mean value of y, $E(y)$, and to predict a particular value of y for a given x.

EXAMPLE 11.17 Refer to Example 11.3. We found the least squares model relating creativity score, y, to flexibility score, x, to be

$$\hat{y} = -2.2 + 2.3x$$

Give a point estimate for the mean creativity score of all mentally retarded children that have a flexibility score of 3.

Solution We need to find an estimate of $E(y)$. On the basis of the least squares model, our estimate is simply $\hat{y}$. Then, when $x = 3$, we have

$$\hat{y} = -2.2 + (2.3)(3) = -2.2 + 6.9 = 4.7$$

Thus, the estimated mean creativity score for all mentally retarded children with flexibility score 3 is 4.7.

EXAMPLE 11.18 Refer to Example 11.17. Use the least squares model to predict the creativity score of a particular retarded child whose flexibility score is 3.

Solution Just as we use $\hat{y}$ from the least squares model to estimate $E(y)$, we also use $\hat{y}$ to predict a particular value of y for a given value of x. Again, when $x = 3$, we obtain $\hat{y} = 4.7$. Thus, we predict that a retarded child with a flexibility score of 3 would have a creativity score of 4.7.

Since the least squares model is used to obtain both the estimator of $E(y)$ and the predictor of y, how do the two model uses differ? The difference lies in the

a. Estimate the mean value of y when x = −1, using a 90% confidence interval. Interpret the interval.
b. Suppose you plan to observe the value of y for a particular experimental unit with x = −1. Find a 90% prediction interval for the value of y that you will observe. Interpret the interval.
c. Which of the two intervals is wider?

APPLYING THE CONCEPTS

11.67   Refer to Exercise 11.12. Find a 95% prediction interval for drill time y when drilling begins at a depth of 300 feet. Interpret your result.

11.68   In Exercise 11.13 you found the least squares prediction equation relating average peak EEG frequency y to age x, and used it to predict peak EEG when x = 7.

   a. Find a 95% confidence interval for E(y) when x = 7 and interpret it.
   b. Find a 95% prediction interval for y when x = 7 and interpret it.

11.69   In forestry, the diameter of a tree at breast height (which is fairly easy to measure) is used to predict the height of the tree (a difficult measurement to obtain). Silviculturists working in British Columbia's boreal forest conducted a series of spacing trials to predict the heights of several species of trees. The data in the accompanying table are the breast height diameters (in centimeters) and heights (in meters) for a sample of 36 white spruce trees.

BREAST HEIGHT DIAMETER	HEIGHT	BREAST HEIGHT DIAMETER	HEIGHT
x, cm	y, m	x, cm	y, m
18.9	20.0	16.6	18.8
15.5	16.8	15.5	16.9
19.4	20.2	13.7	16.3
20.0	20.0	27.5	21.4
29.8	20.2	20.3	19.2
19.8	18.0	22.9	19.8
20.3	17.8	14.1	18.5
20.0	19.2	10.1	12.1
22.0	22.3	5.8	8.0
23.6	18.9	20.7	17.4
14.8	13.3	17.8	18.4
22.7	20.6	11.4	17.3
18.5	19.0	14.4	16.6
21.5	19.2	13.4	12.9
14.8	16.1	17.8	17.5
17.7	19.9	20.7	19.4
21.0	20.4	13.3	15.5
15.9	17.6	22.9	19.2

Source: Scholz, H., Northern Lights College, British Columbia.

   a. Construct a scattergram for the data.
   b. Assuming the relationship between the variables is best described by a straight line, use the method of least squares to estimate the y-intercept and slope of the line.
   c. Plot the least squares line on your scattergram.
   d. Do the data provide sufficient evidence to indicate that the breast height diameter x contributes information for the prediction of tree height y? Test using $\alpha = .05$.
   e. Use your least squares line to find a 90% confidence interval for the average height of white spruce trees with a breast height diameter of 20 cm. Interpret the interval.

11.70 The Federal Communications Commission (FCC) specifies that radiated electromagnetic emissions from digital devices are to be measured in an open-field test site. To verify test-site acceptability, the site attenuation (i.e., the transmission loss from the input of one half-wave dipole to the output of another when both dipoles are positioned over the ground plane) must be evaluated. A study conducted at a test site in Fort Collins, Colorado, yielded the accompanying data on site attenuation (in decibels or dBL) and transmission frequency (in megahertz or MHz) for dipoles at a distance of 3 meters.

TRANSMISSION FREQUENCY x, MHz	SITE ATTENUATION y, dBL
50	11.5
100	15.8
200	18.2
300	22.6
400	26.2
500	27.1
600	29.5
700	30.7
800	31.3
900	32.6
1,000	34.9

Source: Bennett, W. S. "An error analysis of the FCC site-attenuation approximation." *IEEE Transactions on Electromagnetic Compatibility*, Vol. EMC-27, No. 3, Aug. 1985, p. 113 (Table IV). © 1985 IEEE.

a. Construct a scattergram for the data. Does it appear that x and y are linearly related?
b. Find the least squares line relating site attenuation y to transmission frequency x.
c. Plot the least squares line on your scattergram as a check on your calculations.
d. Test the hypothesis that site attenuation y is linearly related to transmission frequency x. Use $\alpha = .10$.
e. Find a 90% confidence interval for the site attenuation when transmission frequency is $x = 850$ MHz. Interpret the interval.

11.71 Refer to Exercises 11.16, 11.24, 11.50, and 11.60. Find a 90% prediction interval for the instructor assigned grade of a FORTRAN77 assignment that received an AUTOMARK score of 17.5. Interpret the interval.

11.72 Refer to Exercise 11.31. Find a 95% confidence interval for the mean heat transfer coefficient of all tubes with an unflooded area ratio of 1.95. Interpret the results.

11.10 In the previous sections we have presented the basic elements necessary to fit and use the straight-line regression model $E(y) = \beta_0 + \beta_1 x$. Throughout, we have illustrated the numerical techniques through an example relating creativity score to flexibility score for mentally retarded children. Even with a small number of measurements (five data points), the required computations, if performed without the aid of a pocket calculator, can become tedious and cumbersome. There are many statistical computer software packages available that fit a straight-line regression model by the method of least squares. These packages enable the user to greatly decrease the burden of calculation.

In this section, we discuss and interpret the elements of a simple linear regression on a computer printout generated by SAS. Since the SAS linear regression

SIMPLE LINEAR REGRESSION: AN EXAMPLE OF A COMPUTER PRINTOUT

output is similar to that of most other package regression programs (such as Minitab and SPSS), you should have little trouble interpreting output from other packages. We relate all the examples in this section to the SAS output for the creativity–flexibility score data given in Figure 11.15. Again, we let y be the creativity score and x the flexibility score of a mentally retarded child.

FIGURE 11.15

Portion of the SAS Printout for the Creativity–Flexibility Data

Analysis of Variance

Source	DF	Sum of Squares	Mean Square	F Value	Prob>F
Model	1	52.90000	52.90000	144.273	0.0012
Error	3	1.10000	0.36667		
C Total	4	54.00000			

Root MSE	0.60553	R-square	0.9796
Dep Mean	7.00000	Adj R-sq	0.9728
C.V.	8.65043		

Parameter Estimates

Variable	DF	Parameter Estimate	Standard Error	T for H0: Parameter=0	Prob > |T|
INTERCEP	1	-2.200000	0.81240384	-2.708	0.0733
X	1	2.300000	0.19148542	12.011	0.0012

EXAMPLE 11.21 On the SAS printout in Figure 11.15, locate the least squares estimates of the y-intercept β_0 and the slope β_1 for the straight-line model relating creativity score to flexibility score.

Solution The least squares estimates of the y-intercept and slope are located in the column titled **Parameter Estimate** in Figure 11.15. Note that the estimate of the y-intercept ($\hat{\beta}_0 = -2.200000$) and the estimate of the slope ($\hat{\beta}_1 = 2.300000$) given in the printout agree with our previous calculations made by hand. The least squares model relating creativity score to flexibility score can thus be written

$$\hat{y} = -2.2 + 2.3x$$

EXAMPLE 11.22 Locate the value of SSE on the SAS printout in Figure 11.15. Also, find s^2 and s, the estimates of σ^2 and σ, respectively.

Solution The value of SSE is found by locating the entry under the column headed **Sum of Squares** in the row labeled **Error**. This quantity, in Figure 11.15, is SSE = 1.10000. A check of our previous calculations confirms its validity. The estimate of σ^2 is given in the figure as **Mean Square** for **Error** and is located immediately to the right of SSE. Thus, we have $s^2 = 0.36667$. The least squares estimate of σ (found in Figure 11.15 next to the heading **Root MSE**) is 0.60553. Except for rounding errors, our computed values of s^2 and s agree with the figures given in the printout.

EXAMPLE 11.23 Use the SAS output to test the null hypothesis that x contributes no information for the prediction of y against the alternative hypothesis that x and y are linearly related.

Solution   We desire a test of the hypotheses

H_0: $\beta_1 = 0$

H_a: $\beta_1 \neq 0$

The value of the test statistic for this test is found in Figure 11.15 under the column heading **T for HO: Parameter $= 0$** in the lower portion of the printout. The value shown here is $t = 12.011$, which agrees with our computed test statistic. To determine whether this value falls within the rejection region, check the quantity to the immediate right under **Prob $> |T|$**. This quantity is the observed significance level or p-value of the test. Generally, if the p-value is smaller than .05, there is evidence to reject the null hypothesis that the slope is 0 in favor of the alternative hypothesis that the slope differs from 0. In this example, an observed significance level of .0012 indicates that flexibility score x and creativity score y are linearly related (a result consistent with the test conducted in Example 11.8).

EXAMPLE 11.24 To determine how well the least squares straight-line model fits the data, find the value of the coefficient of determination r^2.

Solution   The value of r^2, located in Figure 11.15, is given as **R-Square** $= 0.9796$. The result agrees with the value that we computed previously. We say, then, that by using the least squares equation (instead of $\bar{y}$) to predict y, we can reduce the total sum of squares of our prediction errors by approximately 98%. Notice that the coefficient of correlation r is not given in the SAS printout.

EXAMPLE 11.25 Construct a 95% confidence interval for $E(y)$, the mean creativity score for all mentally retarded children that have a flexibility score of 3.

Solution   A portion of the SAS printout not previously shown is given in Figure 11.16 on page 576. For $x = 3$ (i.e., a flexibility score of 3), we see that the least squares predicted value of y, located in the **Predict Value** column, is $\hat{y} = 4.7000$. The endpoints of a 95% confidence interval for the mean, $E(y)$, for each value of x in the sample data are given in the columns labeled **Lower95% Mean** and **Upper95% Mean**. For $x = 3$, the 95% confidence interval ranges from 3.6445 to 5.7555. Thus, the mean creativity score for mentally retarded children with a flexibility score of 3 falls between 3.6445 and 5.7555, with 95% confidence. These results agree with those in Example 11.19.

EXAMPLE 11.26 Find a 95% prediction interval for the creativity score for an individual mentally retarded child with a flexibility score of 3.

Obs	X	Dep Var Y	Predict Value	Std Err Predict	Lower95% Mean	Upper95% Mean	Residual
1	2	2.0000	2.4000	0.469	0.9073	3.8927	-0.4000
2	3	5.0000	4.7000	0.332	3.6445	5.7555	0.3000
3	4	7.0000	7.0000	0.271	6.1382	7.8618	0
4	5	10.0000	9.3000	0.332	8.2445	10.3555	0.7000
5	6	11.0000	11.6000	0.469	10.1073	13.0927	-0.6000

Sum of Residuals		-8.88178E-16
Sum of Squared Residuals		1.1000
Predicted Resid SS (Press)		4.4337

Solution The endpoints of a 95% prediction interval for an individual value of y are given for each value of x in the printout in Figure 11.17, under the columns **Lower95% Predict** and **Upper95% Predict**. You can see that the 95% prediction interval for y with $x = 3$ is (2.5028, 6.8972). We predict that the creativity score for a particular mentally retarded child with a flexibility score of 3 will fall between 2.5028 and 6.8972. The difference between this result and the interval we computed in Example 11.20, (2.503 to 6.897), is due to rounding error.

Figure 11.17

Portion of the SAS Printout with 95% Prediction Limits for y

Obs	X	Dep Var Y	Predict Value	Std Err Predict	Lower95% Predict	Upper95% Predict	Residual
1	2	2.0000	2.4000	0.469	-0.0376	4.8376	-0.4000
2	3	5.0000	4.7000	0.332	2.5028	6.8972	0.3000
3	4	7.0000	7.0000	0.271	4.8890	9.1110	0
4	5	10.0000	9.3000	0.332	7.1028	11.4972	0.7000
5	6	11.0000	11.6000	0.469	9.1624	14.0376	-0.6000

Sum of Residuals		-8.88178E-16
Sum of Squared Residuals		1.1000
Predicted Resid SS (Press)		4.4337

In Chapter 12 we will discuss the interpretation of those portions of the SAS printout that were not mentioned here. However, the important elements of a linear regression analysis have been located. You should be able to use this example as a guide to interpreting the linear regression printouts of other computer software packages since they will appear very similar to that of SAS. For completeness, we present the SPSS and Minitab linear regression outputs for the creativity–flexibility score in Figures 11.18a–b (page 577), respectively.

CHAPTER CASE: THE S.O.B. EFFECT AMONG COLLEGE ADMINISTRATORS

11.11 Now that we have presented the basic elements of a simple linear regression analysis, we will apply our knowledge in a practical example.

At major colleges and universities, administrators (e.g., deans, chairpersons, provosts, vice presidents, and presidents) are among the highest-paid state employees. Is there a relationship between the raises administrators receive and their performance on the job? This was the question of interest to a group of faculty union members at the University of South Florida called the United Faculty of Florida (UFF).

a. SPSS

```
* * * *   M U L T I P L E   R E G R E S S I O N   * * * *

Equation Number 1    Dependent Variable..   Y


Variable(s) Entered on Step Number
   1..    X


Multiple R              .98976
R Square                .97963
Adjusted R Square       .97284
Standard Error          .60553

Analysis of Variance
                    DF      Sum of Squares      Mean Square
Regression          1          52.90000          52.90000
Residual            3           1.10000            .36667

F =     144.27273      Signif F =   .0012

------------------ Variables in the Equation ------------------

Variable           B          SE B        Beta        T    Sig T

X            2.300000      .191485     .989762     12.011   .0012
(Constant)  -2.200000      .812404                 -2.708   .0733


End Block Number   1    All requested variables entered.
```

---

b. Minitab

```
The regression equation is
y = - 2.20 + 2.30 x

Predictor        Coef       Stdev     t-ratio          p
Constant      -2.2000      0.8124       -2.71      0.073
x              2.3000      0.1915       12.01      0.001

s = 0.6055      R-sq = 98.0%      R-sq(adj) = 97.3%

Analysis of Variance

SOURCE        DF          SS          MS          F          p
Regression     1      52.900      52.900     144.27      0.001
Error          3       1.100       0.367
Total          4      54.000
```

The UFF compared the April 1990 ratings of 15 University of South Florida administrators (as determined by faculty in a survey) to their subsequent raises in August 1990. The data for the analysis are listed in Table 11.7 on page 578. [*Note:* Ratings are measured on a 5-point scale, where 1 = very poor and 5 = very good.] According to the UFF, the "relationship is inverse, i.e., the lower the rating by the faculty, the greater the raise. Apparently, bad administrators are more valuable than good administrators."* (With tongue in cheek, the UFF refers to this phenomenon as "the S.O.B. effect.") The UFF based its conclusions on a simple linear regression analysis of the data in Table 11.7, where y = adminis-

---

UFF Faculty Forum, University of South Florida Chapter, Vol. 3, No. 5, May 1991.

trator's raise and x = average rating of administrator. A plot of the data in Table 11.7 appears in Figure 11.19.

TABLE 11.7

Raises and Ratings of University of South Florida Administrators

ADMINISTRATOR	RAISE[a]	AVERAGE RATING (5-pt scale)[b]
1	$18,000	2.76
2	16,700	1.52
3	15,787	4.40
4	10,608	3.10
5	10,268	3.83
6	9,795	2.84
7	9,513	2.10
8	8,459	2.38
9	6,099	3.59
10	4,557	4.11
11	3,751	3.14
12	3,718	3.64
13	3,652	3.36
14	3,227	2.92
15	2,808	3.00

Sources: [a]Faculty and A&P Salary Report, University of South Florida, Resource Analysis and Planning, 1990.

[b]Administrative Compensation Survey, *Chronicle of Higher Education*, Jan. 1991.

FIGURE 11.19

Plot of the Data in Table 11.7

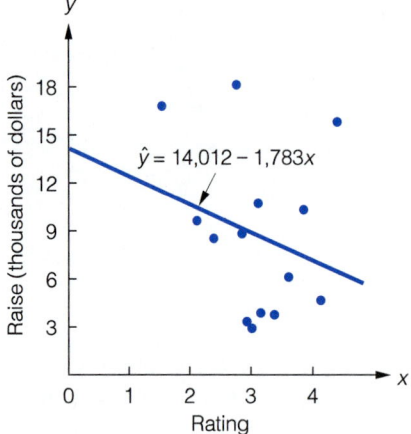

Initially, the UFF conducted the analysis using all 15 data points in Table 11.7. Figure 11.20 is a Minitab printout of the simple linear regression. The least squares line, $\hat{y} = 14{,}012 - 1{,}783x$, is drawn on the scattergram, Figure 11.19.

Note that the t value for testing H_0: $\beta_1 = 0$ is nonsignificant (p-value = .344). Consequently, the estimated straight-line model based on all 15 data points is not a useful predictor of an administrator's raise. This result certainly does not support the UFF's claim of an "inverse relationship" between raises and ratings.

```
The regression equation is
Raise = 14012 - 1783 Rating

Predictor       Coef      Stdev     t-ratio        p
Constant       14012       5800        2.42    0.031
Rating         -1783       1814       -0.98    0.344

s = 5156         R-sq = 6.9%       R-sq(adj) = 0.0%

Analysis of Variance

SOURCE         DF         SS          MS         F        p
Regression      1   25688270    25688270      0.97    0.344
Error          13  345594784    26584214
Total          14  371283040
```

FIGURE 11.20
Minitab Printout of Raise–Rating Regression Analysis

However, the UFF performed a second simple linear regression with only 14 of the data points in Table 11.7. The (x, y) value for administrator #3 was eliminated based on the fact that he was promoted to dean in the middle of the 1989–1990 academic year. No other reason was given for removing this data point from the analysis, although we suspect it may have been eliminated to improve the fit of the straight-line model. You can see in Figure 11.19 that this point appears to be an **outlier*** because it deviates greatly from the least squares line.

When the data point for administrator #3 is removed, the simple linear regression yields the results shown in Figure 11.21.

The p-value for a test of H_0: $\beta_1 = 0$ versus H_a: $\beta_1 < 0$ is $\frac{.041}{2} = .0205$; thus, there is sufficient evidence (at $\alpha = .05$) of a negative slope for the line relating raise (y) to rating (x).

Which regression printout, Figure 11.20 or Figure 11.21, should be used to test the UFF's claim? We will learn in Chapter 12 that it is extremely dangerous to remove a data point from the analysis simply to improve the fit of the model.

```
The regression equation is
Raise = 19680 - 3887 Rating

Predictor       Coef      Stdev     t-ratio        p
Constant       19680       5258        3.74    0.003
Rating         -3887       1699       -2.29    0.041

s = 4268         R-sq = 30.4%      R-sq(adj) = 24.6%

Analysis of Variance

SOURCE         DF         SS          MS         F        p
Regression      1   95266936    95266936      5.23    0.041
Error          12  218540496    18211708
Total          13  313807424

Unusual Observations
Obs.  Rating     Raise     Fit Stdev.Fit  Residual   St.Resid
  1     2.76     18000    8953      1224      9047      2.21R

R denotes an obs. with a large st. resid.

Rating    Fit  Stdev.Fit      95% C.I.         95% P.I.
 2.00   11907       2076   (  7383,  16431)  ( 1564,  22249)
```

FIGURE 11.21
Minitab Printout of Raise–Rating Regression Analysis: Outlier Removed

---

*We formally define an outlier in regression in Chapter 12.

In fact, the outlying data point may be an indicator of a "bad" model. By eliminating this point, we might be throwing away information that could help us develop a better model.

Consequently, good regression analysts carefully check the outlier to determine its cause before eliminating it from the analysis. For example, the UFF might argue that because administrator #3 was promoted to an administration position (dean) in midyear, his salary and/or rating is not indicative of administrators in general, but of professors who have different responsibilities (e.g., teaching and research). On that basis, data point #3 should be removed from the analysis. On the other hand, if the administrator's raise and rating were determined after the promotion to dean, then data point #3 is "typical" of administrators and should be kept in the sample data. The choice of which printout to use depends on which of the two arguments previously stated is more supportable.

Assuming we can support the UFF's removal of data point #3, the least squares prediction equation from Figure 11.21 can be used to predict an administrator's raise. Based on the results of the regression, the UFF computed estimated raises for selected faculty ratings of administrators. These are shown in Table 11.8.

TABLE 11.8

Estimated Raises for Selected Ratings

RATINGS		RAISE
Very poor	1.00	$15,939
	1.50	13,960
Poor	2.00	11,980
	2.50	10,001
Average	3.00	8,021
	3.50	6,042
Good	4.00	4,062
	4.50	2,083
Very good	5.00	103

Note that the table includes ratings ranging from 1.00 to 5.00. Scanning the last column of Table 11.7, you will see that the range of ratings in the sample data is 1.52 to 4.40. The ratings values of 1.00, 1.50, 4.50, and 5.00 shown in Table 11.8 fall outside this range. Consequently, any predictions or estimates of raises made for these four values of x are suspect since a straight-line model may not be appropriate outside the range of the sample data. Also, the UFF did not include a measure of reliability with any of the estimates in Table 11.8. The bottom of the Minitab printout, Figure 11.21, shows a 95% prediction interval for the raise of an administrator with a faculty rating of "poor" ($x = 2.00$). Note that the interval, ($1,564, $22,249), is very wide. Thus, even for ratings within the range of the sample data, the model's predictions are not very practical.

To summarize, we applied our knowledge of simple linear regression to a practical problem: predicting an administrator's raise based on faculty ratings. Two important caveats were addressed. First, although it is tempting to eliminate an outlier to improve the fit of the model, one should investigate fully the data point in question to be sure that it is not the hypothesized model that is suspect. Second, extrapolating the least squares line outside the range of the x-values in the sample can lead to nonsensical and invalid predictions of y.

SUMMARY

In this chapter we introduced *bivariate relationships* and an extremely useful tool—the *method of least squares*—for fitting a straight-line model to a set of data. This procedure, along with associated statistical tests and estimations, is called a **regression analysis**.

After hypothesizing the *straight-line probabilistic model*

$$y = \beta_0 + \beta_1 x + \varepsilon$$

perform the steps outlined in the next box.

STEPS IN A SIMPLE LINEAR REGRESSION ANALYSIS

STEP 1   Use the method of least squares to estimate the unknown parameters in the deterministic component, $\beta_0 + \beta_1 x$. You may obtain the estimates by applying the computational formulas given in this chapter or, if you have access to a computer package, you may obtain them from a computer printout. The least squares estimates will yield a model $\hat{y} = \hat{\beta}_0 + \hat{\beta}_1 x$ with a sum of squared errors (SSE) that is smaller than that produced by any other straight-line model.

STEP 2   Check that the assumptions about the random error ε component (outlined in the box in Section 11.4) are satisfied. (We will explain how to do this in Chapter 12.) You should also determine s^2 (either by hand calculation or from a computer printout), an estimate of σ^2, the variance of the random error component.

STEP 3   Assess the usefulness of the hypothesized model—that is, determine how well x performs as a predictor of y. Included here are making inferences about the slope β_1 and computing the coefficient of determination, r^2.

STEP 4   Finally, if you are satisfied with the model, use it to estimate the mean y value, $E(y)$, for a given x value or to predict an individual y value for a specific x.

Bivariate relationships
Coefficient of correlation, r
Coefficient of determination, r^2
Deterministic component
Least squares line
(prediction equation)

Linear regression analysis
Method of least squares
Probabilistic models
Random error component
Slope of line, β_1
y-intercept of line, β_0

KEY FORMULAS

Least squares estimates of β's: $\hat{\beta}_1 = \dfrac{SS_{xy}}{SS_{xx}}$ $\hat{\beta}_0 = \bar{y} - \hat{\beta}_1 \bar{x}$

$$\text{where } SS_{xy} = \sum xy - \frac{(\sum x)(\sum y)}{n}$$

$$SS_{xx} = \sum x^2 - \frac{(\sum x)^2}{n}$$

$$SS_{yy} = \sum y^2 - \frac{(\sum y)^2}{n}$$

Sum of squared errors: $SSE = SS_{yy} - \hat{\beta}_1 SS_{xy}$

Estimated variance of ε: $s^2 = \dfrac{SSE}{n-2}$

Standard error of $\hat{\beta}_1$: $\sigma_{\hat{\beta}_1} = \sigma/\sqrt{SS_{xx}}$ where $\sigma \approx s$

Coefficient of correlation: $r = \dfrac{SS_{xy}}{\sqrt{SS_{xx}SS_{yy}}}$

Coefficient of determination: $r^2 = 1 - \dfrac{SSE}{SS_{yy}}$

Standard of predicting y: $\sigma_{\hat{y}} = \sigma\sqrt{\dfrac{1}{n} + \dfrac{(x - \bar{x})^2}{SS_{xx}}}$ where $\sigma \approx s$

Standard error of estimating $E(y)$: $\sigma_{(y-\hat{y})} = \sigma\sqrt{1 + \dfrac{1}{n} + \dfrac{(x - \bar{x})^2}{SS_{xx}}}$

where $\sigma \approx s$

SUPPLEMENTARY EXERCISES

11.73   In New York state, common maize rust is a serious disease of sweet corn. Although fungicides are effective in controlling maize rust, the timing of the application is critical. Researchers have developed an action threshold for initiation of fungicide applications based on a regression equation relating maize rust incidence to severity of the disease (*Phytopathology*, Vol. 80, 1990). In one particular field, data were collected on over 100 plants of the sweet corn hybrid Jubilee. For each plant, incidence was measured as the percentage of leaves infected (x) and severity as the log (base 10) of the average number of infections per leaf (y). A simple linear regression analysis of the data produced the following results:

$$\hat{y} = -.939 + .020x \qquad r^2 = .816 \qquad s = .288$$

a. Interpret the value of $\hat{\beta}_1$.
b. Interpret the value of r^2.
c. Interpret the value of s.
d. Calculate the value of r and interpret its value.
e. Use the result, part d, to test the utility of the model. Use $\alpha = .05$.
f. Predict the severity of the disease when the incidence of maize rust for a plant is 80%. [*Note:* Take the antilog (base 10) of $\hat{y}$ to obtain the predicted average number of infections per leaf.]

11.74   A medical item used to administer to a hospital patient is called a *factor*. For example, factors can be intravenous (I.V.) tubing, I.V. fluid, needles, shave kits, bedpans, diapers, dressings, medications, and even code carts. The coronary care unit at Bayonet Point Hospital (St. Petersburg, Florida) investigated the relationship between the number of factors per patient, x, and the patient's length of stay (in days), y. The data for a random sample of 50 coronary care patients are given in the table, followed by a SAS printout of the simple linear regression analysis.

NUMBER OF FACTORS x	LENGTH OF STAY y, days	NUMBER OF FACTORS x	LENGTH OF STAY y, days
231	9	354	11
323	7	142	7
113	8	286	9
208	5	341	10
162	4	201	5
117	4	158	11
159	6	243	6
169	9	156	6
55	6	184	7
77	3	115	4
103	4	202	6
147	6	206	5
230	6	360	6
78	3	84	3
525	9	331	9
121	7	302	7
248	5	60	2
233	8	110	2
260	4	131	5
224	7	364	4
472	12	180	7
220	8	134	6
383	6	401	15
301	9	155	4
262	7	338	8

Source: Bayonet Point Hospital, Coronary Care Unit.

a. Construct a scattergram of the data.
b. Find the least squares line for the data and plot it on your scattergram.
c. Define β_1 in the context of this problem.

```
Model: MODEL1
Dependent Variable: Y

                          Analysis of Variance

                          Sum of         Mean
        Source       DF    Squares       Square    F Value    Prob>F

        Model         1   126.58393    126.58393    28.683    0.0001
        Error        48   211.83607      4.41325
        C Total      49   338.42000

            Root MSE        2.10077    R-square     0.3740
            Dep Mean        6.54000    Adj R-sq     0.3610
            C.V.           32.12193

                          Parameter Estimates

                   Parameter      Standard    T for H0:
     Variable  DF   Estimate        Error    Parameter=0   Prob > |T|

     INTERCEP   1   3.306032     0.67297426       4.913      0.0001
     X          1   0.014755     0.00275502       5.356      0.0001


                Dep Var   Predict   Std Err  Lower95%  Upper95%
    Obs    X       Y       Value    Predict   Predict   Predict   Residual

     1    200       .      6.2570    0.302    1.9898    10.5242      .

Sum of Residuals              9.769963E-15
Sum of Squared Residuals       211.8361
Predicted Resid SS (Press)     234.7934
```

d. Test the hypothesis that the number of factors per patient (x) contributes no information for the prediction of the patient's length of stay (y) when a linear model is used (use $\alpha = .05$). Draw the appropriate conclusions.

e. Find a 90% confidence interval for β_1. Interpret your result.

f. Find the coefficient of correlation for the data. Interpret your result.

g. Find the coefficient of determination for the linear model you constructed in part **b**. Interpret your result.

h. Find a 95% prediction interval for the length of stay of a coronary care patient who is administered a total of $x = 200$ factors.

i. Explain why the prediction interval obtained in part **h** is so wide. How could you reduce the width of the interval?

11.75   The accompanying table shows a portion of the experimental data obtained in a study of the radial tension strength of concrete pipe. The concrete pipe used for the experiment had an inside diameter of 84 inches and a wall thickness of approximately 8.75 inches. In addition, it was reinforced with cold drawn wire. The variable y is the load (in pounds per foot) until the first crack in a pipe specimen was observed. The variable x is the age of the specimen (in days) at the time of the test.

y	x	y	x
11,450	20	10,540	25
10,420	20	9,470	31
11,142	20	9,190	31
10,840	25	9,540	31
11,170	25		

Source: Heger, F. J. and McGrath, T. J. "Radial tension strength of pipe and other curved flexural members." *Journal of the American Concrete Institute,* Vol. 80., No. 1, 1983, pp. 33–39.

a. Construct a scattergram for the data. After examining the scattergram, do you think that x and y are correlated? If correlation is present, is it positive or negative?

b. Find the correlation coefficient r and interpret its value.

c. Do the data provide sufficient evidence to indicate that x and y are linearly correlated? Test using $\alpha = .05$.

d. Find the least squares prediction equation relating load y (in pounds per foot) until the first crack in a pipe specimen and age x (in days) of the specimen.

e. Test the hypothesis $H_0: \beta_1 = 0$ (at $\alpha = .05$) and show that the result agrees with your answer to part **c**.

f. Find a 95% prediction interval for the crack load of a 35-day-old concrete specimen.

g. Why might the prediction interval of part **f** be less reliable than expected? Explain.

11.76 The Consumer Attitude Survey, performed by the University of Florida Bureau of Economic and Business Research (BEBR), is conducted using random-digit telephone dialings of Florida households. The reliability of a telephone survey such as this depends on the *refusal rate*, i.e., the percentage of dialed households that refuse to take part in the study. One factor thought to be related to the refusal rate is personal income. The accompanying table gives the refusal rate, y, and personal income per capita, x, for 12 randomly selected Florida counties from a recent BEBR survey.

COUNTY	REFUSAL RATE y	PER CAPITA INCOME x	COUNTY	REFUSAL RATE y	PER CAPITA INCOME x
1	.296	$ 7,737	7	.429	$11,466
2	.498	12,330	8	.422	10,000
3	.386	12,058	9	.441	10,052
4	.327	9,927	10	.191	8,636
5	.500	6,904	11	.526	7,445
6	.333	9,463	12	.405	9,059

Source: Bureau of Economic and Business Research, University of Florida.

a. Estimate the coefficient of correlation between refusal rate and per capita income.

b. Do the data provide sufficient evidence to indicate a correlation between refusal rate, y, and per capita income, x? Use $\alpha = .05$.

11.77 Spending on research and development (R&D) is essential for a company to maintain a competitive edge in the marketplace. To determine the optimum level for R&D spending and its effect on a company's value, a simple linear regression analysis was performed (*Research Management*, Sept./Oct. 1986). Data collected for the largest R&D spenders (based on 1981–1982 averages) were used to fit a straight-line model relating y to x, where

$$y = \text{Price/earnings (P/E) ratio}$$

$$x = \text{R\&D expenditures/sales (R/S) ratio}$$

The data for 20 of the companies used in the study are provided in the table. An SPSS printout of the simple linear regression follows. Interpret the results.

COMPANY	P/E RATIO y	R/S RATIO x	COMPANY	P/E RATIO y	R/S RATIO x
1	5.6	.003	11	8.4	.058
2	7.2	.004	12	11.1	.058
3	8.1	.009	13	11.1	.067
4	9.9	.021	14	13.2	.080
5	6.0	.023	15	13.4	.080
6	8.2	.030	16	11.5	.083
7	6.3	.035	17	9.8	.091
8	10.0	.037	18	16.1	.092
9	8.5	.044	19	7.0	.064
10	13.2	.051	20	5.9	.028

Source: Wallin, C. C. and Gilman, J. J. "Determining the optimum level for R&D spending." *Research Management*, Vol. 14, No. 5, Sept./Oct. 1986, pp. 19–24 (adapted from Figure 1, p. 20).

```
Equation Number 1     Dependent Variable..   PERATIO

Variable(s) Entered on Step Number
   1..    RSRATIO


Multiple R            .72625
R Square              .52744
Adjusted R Square     .50118
Standard Error        2.07383

Analysis of Variance
                    DF       Sum of Squares       Mean Square
Regression           1            86.40357          86.40357
Residual            18            77.41393           4.30077

F =      20.09024       Signif F =   .0003

------------------ Variables in the Equation ------------------

Variable              B          SE B        Beta        T    Sig T

RSRATIO        74.067607    16.524784     .726249     4.482   .0003
(Constant)      5.977162      .917371                 6.516   .0000
```

11.78 As part of an investigation of medicine use among the rural elderly, 300 individuals age 60 or older were asked a series of questions relating to their health and use of both prescription and nonprescription drugs (*Journal of Health and Social Behavior*, Vol. 26, 1985). The objective of the study was to identify the determinants of the number of prescribed medicines and the number of nonprescribed medicines reportedly used in the previous 2-week period. One variable thought to be related to number of medicines used is perceived morbidity, measured on a scale from 0 (healthy) to 3 (serious or chronic illness). The correlation between perceived morbidity and number of prescribed medicines was measured as $r = .47$, whereas the correlation between perceived morbidity and number of nonprescribed medicines was $r = .29$.

a. Do the data provide sufficient evidence at $\alpha = .05$ to indicate a positive correlation between perceived morbidity and number of prescribed medicines used?

b. Do the data provide evidence (at $\alpha = .05$) to indicate a positive correlation between perceived morbidity and number of nonprescribed medicines used?

11.79 At temperatures approaching absolute zero (273 degrees below zero Celsius), helium exhibits traits that defy many laws of conventional physics. An experiment has been conducted with helium in solid form at various temperatures near absolute zero. The solid helium is placed in a dilution refrigerator along with a solid impure substance, and the proportion (by weight) of the impurity passing through the solid helium is recorded. (This phenomenon of solids passing directly through solids is known as *quantum tunneling*.) The data are given in the table.

PROPORTION OF IMPURITY PASSING THROUGH HELIUM y	TEMPERATURE x, °C
.315	−262
.202	−265
.204	−256
.620	−267
.715	−270
.935	−272
.957	−272
.906	−272
.985	−273
.987	−273

a. Fit a least squares line to the data.

b. Plot the data and graph the line as a check on your calculations.

c. Calculate r and r^2. Interpret these values.

d. Estimate the mean proportion of impurity passing through helium when the temperature is set at $-270°C$. Use a 99% confidence interval.

e. Predict the proportion of impurity passing through helium at a temperature of $-270°C$ using a 99% prediction interval.

11.80   One common symptom of persons who suffer from hepatitis or other diseases of the biliary system (e.g, pancreas or gall bladder) is enlargement of the extrahepatic bile duct. To aid physicians in diagnosing problems of this type, an experiment was conducted to determine the effects of aging on the normal size of the bile duct (*Journal of Clinical Ultrasound*, Vol. 12, 1984). A sample of 256 healthy subjects were examined by ultrasound to locate the widest point in the extrahepatic bile duct. The inner diameter of the bile duct at the widest point was then measured (in millimeters) on the sonogram. A simple linear regression analysis relating maximal inner diameter y to age x (in years) was conducted with the following results:

$$\hat{y} = 2.72 + .06x \qquad r = .60 \qquad n = 256$$

a. Is there sufficient evidence of positive correlation between maximal inner diameter of the bile duct and age? Test using $\alpha = .01$.

b. Predict the maximal inner diameter of a 50-year-old healthy person. Would you expect all normal 50-year-olds to have maximal inner diameters equal to this value? Explain.

11.81   Is the maximal oxygen uptake, a measure often used by physiologists to indicate an individual's cardiovascular fitness, related to the performance of distance runners? Six long-distance runners submitted to treadmill tests for determination of their maximal oxygen uptake. These results, along with each runner's best time for the mile run, are shown in the table.

ATHLETE	MAXIMAL OXYGEN UPTAKE y, milliliters/kilogram	MILE TIME x, seconds
1	63.3	241.5
2	60.1	249.8
3	53.6	246.1
4	58.8	232.4
5	67.5	237.2
6	62.6	238.4

a. Construct a scattergram for the data. After examining the scattergram, do you think that x and y are linearly related? If so, is the relationship positive or negative?

b. The accompanying Minitab printout shows the simple linear regression analysis of the data. Do the data provide sufficient evidence to indicate that x and y are linearly related?

c. Find the coefficient of determination and interpret it.

```
The regression equation is
Y = 127 - 0.274 X

Predictor       Coef       Stdev      t-ratio        p
Constant      126.98       83.55         1.52    0.203
X            -0.2739       0.3467        -0.79    0.474

s = 4.888      R-sq = 13.5%      R-sq(adj) = 0.0%

Analysis of Variance

SOURCE        DF         SS          MS        F        p
Regression     1       14.92       14.92     0.62    0.474
Error          4       95.59       23.90
Total          5      110.51
```

11.82 "In the analysis of urban transportation systems it is important to be able to estimate expected travel time between locations." Cook and Russell (1980) collected data in the city of Tulsa on the urban travel times and distances between locations for two types of vehicles—large hoist compactor trucks and passenger cars. A simple linear regression analysis was conducted for each set of data (y = urban travel time in minutes, x = distance between locations in miles) with the results summarized in the accompanying table.

PASSENGER CARS	TRUCKS
$\hat{y} = 2.50 + 1.93x$	$\hat{y} = 1.85 + 3.86x$
$r^2 = .676$; p-value $< .05$	$r^2 = .758$; p-value $< .01$

Source: Cook, T. M. and Russell, R. A. "Estimating urban travel times: A comparative study." *Transportation Research*, June 1980, Vol. 14A, pp.173–175. Reprinted with permission. Copyright 1980, Pergamon Press, Ltd.

  a. Is there sufficient evidence to indicate that distance between locations is linearly related to urban travel time for passenger cars? Test at $\alpha = .05$.
  b. Is there sufficient evidence to indicate that distance between locations is linearly related to urban travel time for trucks? Test at $\alpha = .01$.
  c. Interpret the values of r^2 for the two prediction equations.
  d. Estimate the mean urban travel time for all passenger cars traveling a distance of 3 miles on Tulsa's highways.
  e. Predict the urban travel time for a particular truck traveling a distance of 5 miles on Tulsa's highways.
  f. How could we attach a measure of reliability to the inferences in parts **d** and **e**?

11.83 The importance of islands as sampling units for flora and fauna population studies has been widely recognized by biogeographers and evolutionists; the theory of equilibrium island biology states that larger islands should have more species than smaller islands. *The American Naturalist* (Jan. 1981) investigated whether such a relationship exists among the species of flora found in the vernal pools (pools of water formed in low-lying areas) of the Central Valley of California. At each of six California sites, 10 to 20 pools were surveyed for species richness (i.e., the number of different species of flora inhabiting the pool) and pool surface area. A linear model relating species richness y and surface area x (in square feet) of the pools was fit to the data using the method of least squares with the following result:

$$\hat{y} = 18.4 + .04x$$

  a. Give the null and alternative hypotheses for a test to determine whether surface area x is useful for predicting species richness y in a linear model.
  b. The reported p-value for the test of part **a** is greater than .05. Interpret this result in terms of the problem.
  c. The coefficient of determination for the simple linear regression is $r^2 = .06$. Interpret this value.

11.84 Concern has been expressed that Americans are becoming a nation of pill takers. Many critics contend that this behavior is learned in childhood and that it is encouraged by TV advertising for proprietary (advertised non-prescription or "over-the-counter") drugs. A study of the relationship between children's attitudes toward proprietary drugs and TV advertising included a sample of children drawn from inner-city and suburban areas of Philadelphia (*Public Opinion Quarterly*, Fall 1980). Two of the many variables measured were drug use intention (measured in response to the question, "How often do you want to take medicine when you have a cold?") and drug advertising exposure (measured as the number of proprietary drug commercials viewed per TV program). The data were analyzed for each of two subsamples: (1) $n = 132$ children, from educationally "disadvantaged" families, who receive little parental instruction about proprietary drugs and the way they are advertised; and (2) $n = 55$ children who are never allowed to take proprietary drugs without parental supervision.

  a. The coefficient of correlation between drug use intention and drug advertising exposure for the low parent education ($n = 132$) children was $r = .44$. Interpret this value.
  b. The coefficient of correlation between drug use intention and drug advertising exposure for the parentally supervised children ($n = 55$) children was $r = .23$. Interpret this value.

THE LONELY HEARTS CLUB

Is there a link between the loneliness of parents and their offspring? Psychologists Lobdell and Perlman examined this question in a recent article (*Journal of Marriage and the Family*, Aug. 1986).

Several previous studies on loneliness had found that positive parent–child relations serve as a buffer against loneliness, whereas those children who lack warm, positive early relationships with their parents will subsequently be prone to loneliness. Past studies on transmission of traits from parents to children (called *intergenerational transmission*) focused on such phenomena as attitudes, values, and political affiliation. However, none of these studies addressed directly the issue of intergenerational transmission of loneliness. Lobdell and Perlman sought to fill this gap in the social science research by answering the following question: Is there a correlation between the loneliness of parents and the loneliness of their offspring?

The participants in the study were 130 female college undergraduates and their parents. (As incentive to participate, students were awarded a 3-hour credit in an introductory psychology course.) Each triad of daughter, mother, and father completed the UCLA Loneliness Scale, a 20-item questionnaire designed to assess loneliness and several variables theoretically related to loneliness, such as social accessibility to others, difficulty in making friends, and depression.

Pearson product moment correlations relating daughter's loneliness to parent's loneliness score as well as the other variables were calculated. The results are summarized in Table 11.9.

TABLE 11.9
Correlations Between Student's Loneliness and Parent-Derived Variables

VARIABLE	CORRELATION (r) BETWEEN DAUGHTER'S LONELINESS AND PARENTAL VARIABLES	
	Mother	Father
Loneliness	.26	.19
Depression	.11	.06
Self-esteem	−.14	−.06
Assertiveness	−.05	.01
Number of friends	−.21	−.10
Quality of friendships	−.17	.01

Source: Lobdell, J. and Perlman, D. "The intergenerational transmission of loneliness: A study of college females and their parents." *Journal of Marriage and the Family*, Vol. 48, No. 8, Aug. 1986, p. 592. Copyright 1986 by the National Council on Family Relations, 3989 Central Ave., N.E., Suite #550, Minneapolis, MN 55421.

a. Lobdell and Perlman conclude that "mother and daughter loneliness scores were (positively) significantly correlated at $\alpha = .01$." Do you agree?

b. Determine which, if any, of the other sample correlations are large enough to indicate (at $\alpha = .01$) that linear correlation exists between daughter's loneliness score and the variable measured.

c. Explain why it would be dangerous to conclude that a causal relationship exists between mother's loneliness and daughter's loneliness.

d. Explain why it would be dangerous to conclude that the variables with non-significant correlations in Table 11.9 are unrelated.

CASE STUDY 11.2
MENTAL IMAGERY AND THE THIRD DIMENSION

How do people form a mental image of an object in three-dimensional space? The question is an interesting one when we consider that our visual senses lack a three-dimensional structure. When we view a scene—for example, a person water-skiing—our eyesight does not process the three-dimensional layout of the scene directly. The scene is reconstructed in our brain from the two-dimensional projections of the scene onto the retina of our eye. Consequently, certain perspective effects result. For example, as the water-skier recedes from our view, he appears to shrink in size or if a large sailboat interrupts our line of sight, the scene of the water-skier is blocked. However, suppose we could explore the scene with our sense of touch. A more distant object (the water-skier) would not feel smaller, nor would the sailboat prevent us from locating the water-skier with our touch. The three-dimensional structure would remain free of the perspective effects that exist in vision.

Psychologists have just begun to research this phenomenon of mental imagery in three dimensions. In this case study, we focus on one of a series of experiments conducted by Steven Pinker (1980)* of Harvard University. Pinker wanted to answer the following question: "Do mental images preserve interval information concerning the distances between objects in three dimensions?" In this experiment, the time that subjects require to scan from one object in a three-dimensional image to another is used as a measure of the "distance" between those objects in the image. Pinker hypothesizes that if three-dimensional distances are preserved in the image, the scanning times should be highly positively linearly correlated with these distances—that is, scanning times should increase linearly with increasing distances in three dimensions.

Ten volunteers for the study, all affiliated with Harvard University, were asked to view a gray box, open at the top and front, in which five small toys (a hat, an apple, a teddy bear, a tire, and a sea shell) were suspended by clear nylon

---

*Pinker, S. "Mental imagery and the third dimension." *Journal of Experimental Psychology: General*, Sept. 1980, Vol. 109, pp. 354–371. Copyright 1980 by the American Psychological Association. Adapted by permission of the publisher and author.

thread. The objects' positions were chosen so that the interobject distances in three dimensions correlated poorly with the corresponding distances in the two-dimensional plane. Each subject was asked to form a mental image of the box and its contents, making sure each object was imagined at its proper location. After the layout was memorized (this was verified through extensive testing), the front of the box was covered with an opaque screen. The subject was then asked to shut his or her eyes, and, upon hearing the name of one of the objects in the box, to mentally focus on that object. Four seconds later another object was named, and the subject was to "scan" to it by moving in a straight line as quickly as possible from the first to the second object. When the subject "arrived" at the destination object, he or she was to press a button indicating the end of the trial. Trials continued until all possible pairs of objects were scanned. The response times (in milliseconds) for scanning between the members of each pair of objects and the three-dimensional distance (in centimeters) between the objects were recorded for each pair of objects for each subject.

a. Pinker plotted the 10 response time means (one for each pair of stimulus objects, averaged over subjects) against the distance between objects. The result is reproduced in Figure 11.22. Does it appear that response time is linearly related to distance between objects?

FIGURE 11.22

Mean Response Times for Scanning Mentally in Three Dimensions

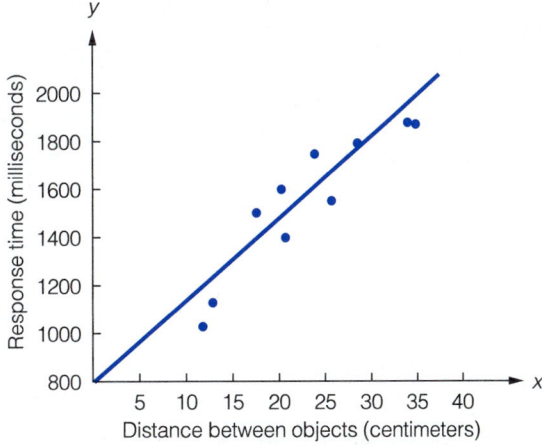

b. The correlation between response (scanning) time and distance between objects is $r = .92$. Interpret this value.

c. The least squares line relating response time y to distance x between objects was found to be $\hat{y} = 833 + 33.8x$. Set up the null and alternative hypotheses for testing whether response time increases linearly as the distance between objects increases.

d. The test statistic for testing the hypothesis of part **c** was found to be $t = 8.28$, corresponding to $df = 81$. Is there sufficient evidence (at $\alpha = .01$) to indicate that response time increases linearly as distance between objects increases?

e. What assumptions are necessary for the validity of the test in part **d**?

TOP CORPORATE EXECUTIVES AND THEIR PAY—ANOTHER LOOK

Each year *Business Week* compiles its Executive Compensation Scoreboard based on a survey of executives at the 1,000 highest-ranked U.S. companies. In Case Study 3.1 we examined the 1990 survey, reported in the May 7, 1990, issue of *Business Week*. In addition to total cash compensation (salary plus bonus, plus long-term compensation), *Business Week* reports the total shareholder return of the executive's company, measured by the dollar value of a $100 investment in the company made 3 years earlier. *Business Week* uses the ratio of total shareholder return to total executive cash compensation (in thousands of dollars) as a useful benchmark in determining which executives are worth their pay. Another approach is to compare the measures of total pay and shareholder return using the methods of this chapter.

Recall that the CEOs in the 1990 Scoreboard are categorized by industry group, of which there are nine. The objective of this case study, as in Case Study 3.1, is to make pay-for-performance comparisons of executives within industry groups. To accomplish this, consider the straight-line model $E(y) = \beta_0 + \beta_1 x$, where y = total cash compensation and x = total shareholder return.

A SAS printout of the simple linear regression for CEOs in the industrial–high-tech industry group is displayed in Figure 11.23. Interpret the key elements of the printout. Would you recommend using the model to predict total executive pay (y)?

FIGURE 11.23
SAS Printout for Case Study 11.3

```
------------------------------ INDUSTRY=IHT ------------------------------
Model: MODEL1
Dependent Variable: TOTCOMP

                            Analysis of Variance

                            Sum of          Mean
       Source       DF      Squares         Square      F Value    Prob>F

       Model         1  871205.46382   871205.46382      0.090     0.7662
       Error        34  329590569.76   9693840.2870
       C Total      35  330461775.22

            Root MSE      3113.49326     R-square      0.0026
            Dep Mean      4835.27778     Adj R-sq     -0.0267
            C.V.            64.39120

                          Parameter Estimates

                    Parameter      Standard     T for H0:
    Variable   DF    Estimate        Error     Parameter=0    Prob > |T|

    INTERCEP    1   5014.627445   791.95042334     6.332        0.0001
    RETURN      1     -1.079697     3.60155065     -0.300        0.7662


            Dep Var   Predict   Std Err  Lower95%  Upper95%
 Obs  RETURN  TOTCOMP    Value   Predict   Predict   Predict   Residual

   1    300       .     4690.7   708.377   -1798.3   11179.8       .

Sum of Residuals              -9.09495E-13
Sum of Squared Residuals       329590569.76
Predicted Resid SS (Press)     353756735.76
```

SIMPLE LINEAR REGRESSION AND CORRELATION

The computer commands and instructions given in this section generate the simple linear regression printouts for the creativity–flexibility score data shown in Figure 11.15 (SAS), Figure 11.18a (SPSS), and Figure 11.18b (Minitab). In addition, we give the commands for calculating the coefficient of correlation r, because it is not given on the regression printouts.

SAS

```
Command
Line
  1    DATA SALES;      ⎫  Data entry instructions
  2    INPUT X Y;       ⎬
  3    CARDS;           ⎭
  4    2  2             ⎫
  5    3  5             ⎪
  6    4  7             ⎬  Input data values
  7    5 10             ⎪  (1 observation per line)
  8    6 11             ⎭
  9    PROC REG;        ⎫  Regression analysis/
 10    MODEL Y = X/P CLI;  ⎬  prediction intervals
 11    ID X;            ⎭
 12    PROC CORR;       ⎫  Correlation analysis
 13    VAR X Y;         ⎬
```

COMMAND 9   The REG procedure performs a complete linear regression analysis on the data.

COMMAND 10   In the MODEL statement, the dependent variable is listed to the left of the equals sign and the independent variable to the right. The option P (following the slash) prints predicted values and residuals, and the option CLI prints corresponding lower and upper 95% prediction limits for all observations in the data set. Specify CLM to obtain 95% confidence intervals for $E(y)$. [*Note:*   To predict y for a value of x that is not included in the data set (e.g., $x = 3.0$), you must include an "extra" observation in the data set. This observation has the specified value of x (e.g., 3.0), but a missing value for y (i.e., a single decimal point).]

COMMAND 11   The optional ID statement identifies the value of x for each 95% prediction (or confidence) interval.

COMMAND 12–13   The CORR procedure calculates the correlation coefficient between the variables specified in the VAR statement. The two-tailed p-value for testing H_0: $\rho = 0$ is also produced.

```
Command
  Line
    1    DATA LIST FREE/X Y.  ⎫  Data entry instructions
    2    BEGIN DATA.          ⎭
    3    2  2                 ⎫
    4    3  5                 ⎪
    5    4  7                 ⎬  Input data values
    6    5 10                 ⎪  (1 observation per line)
    7    6 11                 ⎭
    8    END DATA.
    9    REGRESSION VARIABLES = Y, X/  ⎫
   10              DEPENDENT = Y/      ⎬  Regression analysis
   11              METHOD = ENTER X.   ⎭
   12    CORRELATION VARIABLES = Y, X.    Correlation analysis
```

COMMAND 9–11   The REGRESSION procedure in SPSS performs a complete linear regression analysis on the data. The dependent and independent variables are specified in the DEPENDENT and METHOD subcommands, respectively.

COMMAND 12   The CORRELATION procedure in SPSS calculates the correlation coefficient between the specified variables and gives the one-tailed p-value for testing $H_0: \rho = 0$.

GENERAL   Confidence intervals for $E(y)$ and prediction intervals for y are not available in the SPSS REGRESSSION procedure.

```
Command
  Line
    1    READ X IN C1, Y IN C2    Data entry instruction
    2    2  2                 ⎫
    3    3  5                 ⎪
    4    4  7                 ⎬  Input data values
    5    5 10                 ⎪  (1 observation per line)
    6    6 11                 ⎭
    7    REGRESS C2 ON 1 PREDICTOR IN C1;  ⎫  Regression analysis
    8               PREDICT 3.             ⎭
    9    CORRELATION C1 C2                    Correlation analysis
```

COMMAND 7   The REGRESS procedure in Minitab performs a complete linear regression analysis on the data. The column in which the values of the dependent variables appear must be specified first (e.g., C2), followed by the number of predictors (independent variables) in the model (e.g., 1), and the column(s) in which the values of the predictor(s) appears (e.g., C1).

COMMAND 8   PREDICT is a subcommand of the main REGRESS command that produces a 95% confidence interval for $E(y)$ and a 95% prediction interval for y for the value of the independent variable specified (e.g., 3).

COMMAND 9   The CORRELATION procedure in Minitab calculates the coefficient of correlation between the variables (columns) specified.

GENERAL   When a subcommand is used in Minitab, the main command (e.g., REGRESS in line 7) must end in a semicolon, followed by the subcommand (e.g., PREDICT in line 8), which ends in a period.

COMPUTER ACTIVITIES

Refer to the length, weight, and DDT measurements of the 144 fish in Appendix D. Using a statistical software package, perform the following:

a. Compute and interpret the pairwise correlations between the three variables, length, weight, and DDT level.

b. Select one of the three variables to be the dependent variable and another to be the independent variable. Fit the straight-line model $E(y) = \beta_0 + \beta_1 x$. Interpret the results.

REFERENCES

Draper, N. and Smith, H. *Applied Regression Analysis*. New York: Wiley, 1966.

Mendenhall, W. and Sincich, T. *A Second Course in Business Statistics: Regression Analysis*, 3rd ed. San Francisco: Dellen, 1989.

Montgomery, D. C. and Peck, E. A. *Introduction to Linear Regression Analysis*. New York: Wiley, 1982.

Neter, J., Wasserman, W., and Kutner, K. *Applied Linear Statistical Models*, 3rd ed. Homewood, Ill.: Richard Irwin, 1990.

MULTIPLE REGRESSION AND MODEL BUILDING

CHAPTER 12

Despite the popular oldies tune "It Never Rains in California," meteorologists in the state continually monitor precipitation and the variables (e.g., altitude and latitude) that affect the amount of rainfall at any location. In this chapter, we extend the ideas of Chapter 11 and learn how to build a regression model relating a variable y to two or more independent variables. The question of how precipitation is related to altitude, latitude, and other meteorological independent variables is addressed in the chapter case, Section 12.14.

CONTENTS

Case Study 12.1

The Salary Race: Males Versus Females

Case Study 12.2

The Stock Market Effects of Airline Deregulation

Case Study 12.3

Factors Affecting the Sale Price of Condominium Units

Computer Lab

Multiple Regression and Residual Analysis

Computer Activities

FIGURE 12.1

SAS Output for Sale Price
Model, Example 12.2

Analysis of Variance

Source	DF	Sum of Squares	Mean Square	F Value	Prob>F
Model	3	8779676740.6	2926558913.5	46.662	0.0001
Error	16	1003491259.4	62718203.714		
C Total	19	9783168000.0			

Root MSE	7919.48254	R-Square	0.8974	
Dep Mean	56660.00000	Adj R-Sq	0.8782	
C.V.	13.97720			

Parameter Estimates

Variable	DF	Parameter Estimate	Standard Error	T for H0: Parameter=0	Prob > |T|
INTERCEP	1	1470.275919	5746.3245832	0.256	0.8013
X1	1	0.814490	0.51221871	1.590	0.1314
X2	1	0.820445	0.21118494	3.885	0.0013
X3	1	13.528650	6.58568006	2.054	0.0567

variables are **quantitative** (e.g., numeric variables), the β parameters in the multiple regression model of the form specified in Example 12.2 have similar interpretations. The difference is that when we interpret the β that multiplies one of the variables (e.g., x_1), we must be certain to hold the values of the remaining independent variables (e.g., x_2, x_3) fixed. The following example illustrates the point.

EXAMPLE 12.3   Suppose that the mean value $E(y)$ of a response y is related to two quantitative independent variables, x_1 and x_2, by the model

$$E(y) = \beta_0 + \beta_1 x_1 + \beta_2 x_2$$

where $\beta_0 = 1$, $\beta_1 = 2$, and $\beta_2 = 1$. In other words,

$$E(y) = 1 + 2x_1 + x_2$$

a. Graph the relationship between $E(y)$ and x_1 for $x_2 = 0$, 1, and 2. Interpret the graph.
b. Graph the relationship between $E(y)$, x_1, and x_2 in three dimensions. Interpret the graph.

Solution

a. When $x_2 = 0$, the relationship between $E(y)$ and x_1 is given by

$$E(y) = 1 + 2x_1 + (0) = 1 + 2x_1$$

A graph of this relationship (a straight line) is shown in Figure 12.2. Similar graphs of the relationship between $E(y)$ and x_1 for $x_2 = 1$,

$$E(y) = 1 + 2x_1 + (1) = 2 + 2x_1$$

and for $x_2 = 2$,

$$E(y) = 1 + 2x_1 + (2) = 3 + 2x_1$$

are shown in Figure 12.2. Note that the slopes of the three lines are all equal to $\beta_1 = 2$, the coefficient that multiplies x_1.

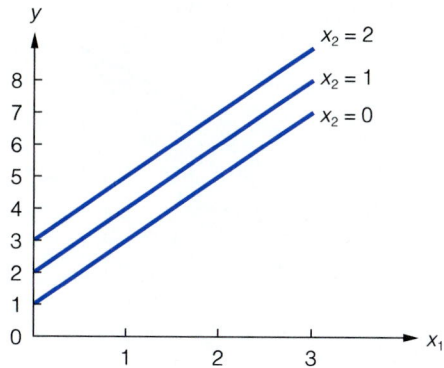

FIGURE 12.2
Graphs of $E(y) = 1 + 2x_1 + x_2$ for $x_2 = 0, 1, 2$

The model $E(y) = 1 + 2x_1 + x_2$ is an example of a **first-order linear model** in two quantitative independent variables, x_1 and x_2. Likewise, the model in Example 12.1 is an example of a first-order model in three quantitative independent variables. A first-order linear model in five quantitative independent variables is shown in the box.

A FIRST-ORDER LINEAR MODEL RELATING TO $E(y)$ TO $x_1, x_2, \ldots, x_5$

$$E(y) = \beta_0 + \beta_1 x_1 + \beta_2 x_2 + \beta_3 x_3 + \beta_4 x_4 + \beta_5 x_5$$

where β_i represents the slope of the line relating y to x_i when all other x's are held fixed (i.e., β_i measures the change in y for every 1-unit increase in x_i, holding all other x's fixed).

Figure 12.2 exhibits a characteristic of all first-order models: If you graph $E(y)$ versus any one variable—say, x_1—for fixed values of the other variables, the response curve will always be a *straight line* with slope equal to β_1. If you repeat the process for other values of the fixed independent variables, you will obtain a set of *parallel* straight lines. This indicates that the effect of the independent variable x_i on $E(y)$ is independent of all the other independent variables in the model, and this effect is measured by the slope β_i. (See the previous box.)

b. A three-dimensional graph of the model $E(y) = 1 + 2x_1 + x_2$ is shown in Figure 12.3 (page 604). Note that the model graphs as a plane. If you slice the plane at a particular value of x_2 (say, $x_2 = 0$), you obtain a straight line relating $E(y)$ to x_1 (e.g., $E(y) = 1 + 2x_1$). Similarly, if you slice the plane at a particular value of x_1, you obtain a straight line relating $E(y)$ to x_2. Since it is more difficult to visualize three-dimensional and, in general, k-dimensional surfaces, we will graph all the models presented in this chapter in two dimensions. The key to obtaining these graphs is to hold fixed all but one of the independent variables in the model.

FIGURE 12.3
The Plane $E(y) = 1 + 2x_1 + x_2$

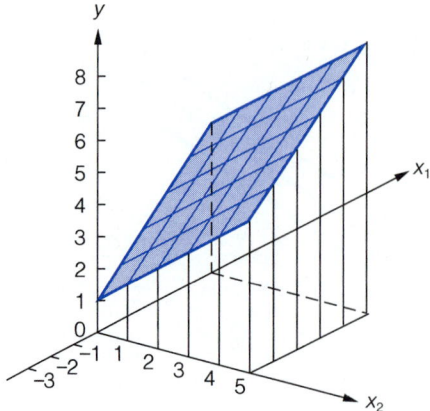

EXAMPLE 12.4 Refer to the first-order model for sale price y considered in Example 12.2. Interpret the estimates of the β parameters in the model.

Solution The least squares prediction equation, given in Example 12.2, is $\hat{y} = 1,470.28 + .8145x_1 + .8204x_2 + 13.53x_3$. From Example 12.3, we know that with first-order models β_1 represents the slope of the y versus x_1 line for fixed x_2 and x_3. That is, β_1 measures the change in $E(y)$ for every 1-unit increase in x_1 when all other independent variables in the model are held fixed. Similar statements can be made about β_2 and β_3, e.g., β_2 measures the change $E(y)$ for every 1-unit increase in x_2 when all other x's in the model are held fixed. Consequently, we obtain the following interpretations:

$\hat{\beta}_1 = .8145$: We estimate the mean sale price of a property, $E(y)$, to increase .8145 dollar for every \$1 increase in appraised land value (x_1) when both appraised improvements (x_2) and area (x_3) are held fixed.

$\hat{\beta}_2 = .8204$: We estimate the mean sale price of a property, $E(y)$, to increase .8204 dollar for every \$1 increase in appraised improvements (x_2) when both appraised land value (x_1) and area (x_3) are held fixed.

$\hat{\beta}_3 = 13.53$: We estimate the mean sale price of a property, $E(y)$, to increase \$13.53 for each additional square foot of living area (x_3) when both appraised land value (x_1) and appraised improvements (x_2) are held fixed.

The value $\hat{\beta}_0 = 1,470.28$ does not have a meaningful interpretation in this example. To see this, note that $\hat{y} = \hat{\beta}_0$ when $x_1 = x_2 = x_3 = 0$. Thus, $\hat{\beta}_0 = 1,470.28$ represents the estimated mean sale price when the values of all independent variables are set equal to 0. Since a residential property with these characteristics—appraised land value of \$0, appraised improvements of \$0, and 0 square feet of living area—is not practical, the value of $\hat{\beta}_0$ has no meaningful interpretation. In general, β_0 will not have a practical interpretation unless it makes sense to set the values of the x's simultaneously equal to 0.

WARNING

The interpretation of the β parameters in the general linear model will depend on the terms specified in the model. The previous interpretations are for a first-order linear model only. In practice, you should be sure that a first-order model is the correct model for $E(y)$ before making these β interpretations. [We discuss alternative models for $E(y)$ in Sections 12.9–12.11.]

12.4 As in simple linear regression, the variance σ^2 of the random error term in a multiple regression model plays a key role in determining the utility of the model. In this section, we show how to obtain an estimate of σ^2 and give a practical interpretation of its value.

EXAMPLE 12.5 Refer to Example 12.2, where we fit the first-order linear model

$$E(y) = \beta_0 + \beta_1 x_1 + \beta_2 x_2 + \beta_3 x_3$$

Locate the minimum value of SSE on the SAS printout reproduced in Figure 12.4. Use this value to obtain estimates of σ^2 and σ, the variance and standard deviation, respectively, of the random error term in the probabilistic model.

Analysis of Variance

Source	DF	Sum of Squares	Mean Square	F Value	Prob>F
Model	3	8779676740.6	2926558913.5	46.662	0.0001
Error	16	1003491259.4	62718203.714		
C Total	19	9783168000.0			

Root MSE	7919.48254	R-Square	0.8974	
Dep Mean	56660.00000	Adj R-Sq	0.8782	
C.V.	13.97720			

Parameter Estimates

Variable	DF	Parameter Estimate	Standard Error	T for H0: Parameter=0	Prob > |T|
INTERCEP	1	1470.275919	5746.3245832	0.256	0.8013
X1	1	0.814490	0.51221871	1.590	0.1314
X2	1	0.820445	0.21118494	3.885	0.0013
X3	1	13.528650	6.58568006	2.054	0.0567

FIGURE 12.4
SAS Printout for Sale Price Model, Example 12.5

Solution The minimum value of SSE, 1,003,491,259.4, is shaded in the row labeled **Error** under the column labeled **Sum of Squares** in the printout shown in Figure 12.4. Recall from Section 11.4 that we can use this quantity to estimate

We use the symbol $s_{\hat{\beta}_2}$ to represent the estimated standard deviation of $\hat{\beta}_2$. Since the formula for $s_{\hat{\beta}_2}$ is very complex, we will not present it here. However, this will not cause difficulty because the printouts from most statistical computer software packages list the estimated standard deviation $s_{\hat{\beta}_i}$ for each of the estimated model coefficients $\hat{\beta}_i$ in the linear model as well as the corresponding calculated t values.

To test the null hypothesis that $\beta_2 = 0$, we again consult the SAS printout for the sale price model. From Figure 12.5, we see that the computed value of the test statistic corresponding to the test of H_0: $\beta_2 = 0$ (shaded under the column headed **T for HO: Parameter=0**) is $t = 3.885$. The appropriate rejection region is obtained from Table 5 of Appendix G. For $\alpha = .05$ and $(n - 4) = 16$ degrees of freedom, we have $t_{\alpha/2} = t_{.025} = 2.120$. Note that the critical t value used to specify the rejection region depends on $(n - 4)$ degrees of freedom because the first-order linear model contains four parameters $(\beta_0, \beta_1, \beta_2, \beta_3)$. Then the rejection region (shown in Figure 12.6) is

$$|t| > 2.120$$

Since $t = 3.885$ falls into the upper tail of the rejection region, we conclude that appraised improvements (x_2) makes an important contribution to the prediction model for the sale price of residential properties in the midsize city.

FIGURE 12.5

SAS Printout for the Sale Price Model, Example 12.7

Analysis of Variance

Source	DF	Sum of Squares	Mean Square	F Value	Prob>F
Model	3	8779676740.6	2926558913.5	46.662	0.0001
Error	16	1003491259.4	62718203.714		
C Total	19	9783168000.0			

Root MSE	7919.48254	R-Square	0.8974	
Dep Mean	56660.00000	Adj R-Sq	0.8782	
C.V.	13.97720			

Parameter Estimates

Variable	DF	Parameter Estimate	Standard Error	T for HO: Parameter=0	Prob > |T|
INTERCEP	1	1470.275919	5746.3245832	0.256	0.8013
X1	1	0.814490	0.51221871	1.590	0.1314
X2	1	0.820445	0.21118494	3.885	0.0013
X3	1	13.528650	6.58568006	2.054	0.0567

This result could be obtained directly from the SAS printout, which lists the two-tailed observed significance level (p-value) for each t value under the column headed **Prob > |T|**. [*Note:* One-tailed observed significance levels are obtained by dividing the two-tailed p-values (shown on the SAS printout) in half.] The observed significance level .0013 (shaded in Figure 12.5) corresponds to the x_2 term; this implies that we would reject H_0: $\beta_2 = 0$ in favor of H_a: $\beta_2 \neq 0$ at any α level larger than .0013. Thus, there is very strong evidence of at least a linear relationship between sale price (y) and appraised improvements value (x_2) for these residential properties.

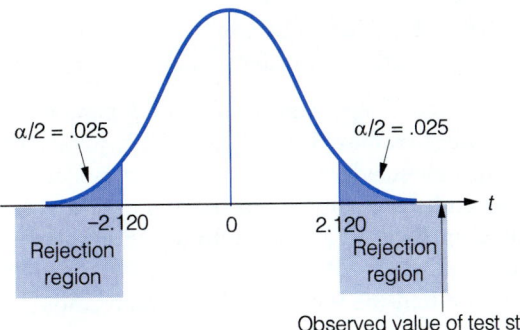

$\alpha/2 = .025$

$\alpha/2 = .025$

-2.120 0 2.120

Rejection region   Rejection region

t

Observed value of test statistic
$t = 3.885$

EXAMPLE 12.8   Refer to the SAS printout shown in Figure 12.5. Notice that the p-value for appraised land value (x_1) is .1314. Interpret this result.

Solution   Since the p-value is larger than $\alpha = .05$, we would fail to reject the null hypothesis H_0: $\beta_1 = 0$. Our first inclination might be to assume that $\beta_1 = 0$ and, therefore, that appraised land value is *not* useful for predicting sale price. However, such a decision is dangerous for two reasons. First, the fact that we are accepting H_0: $\beta_1 = 0$ leaves us vulnerable to making a Type II error with unknown probability of occurrence. Second, if $\beta_1 = 0$, the most we can say is that appraised land value (x_1) is not a useful *linear* predictor of sale price (y) once x_2 and x_3 are in the model. A strong curvilinear (second-order) relationship between y and x_1 may exist. We explain in Section 12.10 how to determine whether such a relationship exists.

WARNING

It is dangerous to conduct t tests on the individual β parameters in a *first-order linear model* for the purpose of determining which independent variables are useful for predicting y and which are not. If you fail to reject H_0: $\beta_i = 0$, several conclusions are possible:

1. There is no relationship between y and x_i.
2. A straight-line relationship between y and x_i exists (holding the other x's in the model fixed), but a Type II error occurred.
3. A relationship between y and x_i (holding the other x's in the model fixed) exists, but it is more complex than a straight-line relationship (e.g., a curvilinear relationship may be appropriate).

The most you can say about a β parameter test is that there is either sufficient (if you reject H_0: $\beta_i = 0$) or insufficient (if you do not reject H_0: $\beta_i = 0$) evidence of a *linear* (*straight-line*) relationship between y and x_i.

EXAMPLE 12.9   Refer to Example 12.2. Form a 95% confidence interval for the parameter β_2 in the first-order model. Interpret the result.

Solution   A confidence interval for any β parameter in a general linear model is given in the next box. From Figure 12.5, we see that $\hat{\beta}_2 = .820445$ (shaded). The estimated standard deviations of the model coefficients appear in the SAS printout under the column labeled **Standard Error**. The value $s_{\hat{\beta}_2} = .21118494$ is shaded in Figure 12.5. Substituting the values of $\hat{\beta}_2$, $s_{\hat{\beta}_2}$, and $t_{.025} = 2.120$ (based on $n - 4 = 16$ degrees of freedom) into the formula for a confidence interval, we find the 95% confidence interval for β_2 to be

$$\hat{\beta}_2 \pm t_{\alpha/2}s_{\hat{\beta}_2} = .820445 \pm (2.120)(.21118494)$$

or $(.372733, 1.268157)$. This interval can be used to estimate the change in mean sale price as appraised improvements value (x_2) is increased, holding the values of x_1 and x_3 fixed. It appears that for every \$1 increase in appraised improvements value, the mean sale price will increase by an amount between \$.37 and \$1.27. Note that all values in the interval are positive, reconfirming our test conclusion that β_2 is nonzero.

A $(1 - \alpha)$ 100% CONFIDENCE INTERVAL FOR AN INDIVIDUAL PARAMETER COEFFICIENT IN THE GENERAL LINEAR MODEL

$$\hat{\beta}_i \pm t_{\alpha/2}s_{\beta_i}$$

where $n =$ Number of observations

$s_{\hat{\beta}_i} =$ Estimated standard deviation of the repeated sampling distribution of $\hat{\beta}_i$

and the distribution of t has degrees of freedom equal to $(n -$ Number of β parameters in the model), and $t_{\alpha/2}$ is the t value such that $P(t > t_{\alpha/2}) = \alpha/2$.

THE COEFFICIENT OF DETERMINATION

12.6

Recall from Chapter 11 that the coefficient of determination, r^2, is a measure of how well a straight-line model fits a set of data. To measure how well a general linear model (for example, a first-order linear model) fits a set of data, we compute the multiple regression equivalent of r^2, called the **multiple coefficient of determination**, and denoted by the symbol R^2.

Just as for the simple linear model, R^2 represents the proportion of the sum of squares of deviations (SS_{yy}) of the y values about $\bar{y}$ that can be attributed to the regression model. Thus, $R^2 = 0$ implies a complete lack of fit of the model to the data and $R^2 = 1$ implies a perfect fit, with the model passing through every data point. In general, the larger the value of R^2, the better the model fits the data.

DEFINITION 12.1

The **multiple coefficient of determination**, R^2, is defined as

$$R^2 = 1 - \frac{\text{SSE}}{\text{SS}_{yy}} \qquad 0 \le R^2 \le 1$$

where $\text{SSE} = \Sigma(y - \hat{y})^2$, $\text{SS}_{yy} = \Sigma(y - \bar{y})^2$, and $\hat{y}$ is the predicted value of y for the multiple regression model.

EXAMPLE 12.10 Refer to Example 12.2. Locate the value of R^2 on the SAS printout and interpret its value. Does the first-order linear model appear to provide a good fit to the sales data for the 20 randomly selected properties?

Solution The SAS printout for the multiple regression is reproduced in Figure 12.7. The value of R^2 (shaded) is shown to the right of the entry labeled **R-Square**. This value, $R^2 = .8974$, implies that, by using the three independent variables (appraised land value, appraised improvements value, and home size) in a first-order model instead of $\bar{y}$ to predict y, we can reduce the sum of squared prediction errors by approximately 90%. Thus, this large value of R^2 indicates that the model provides a good fit to the $n = 20$ sample data points.

Analysis of Variance

Source	DF	Sum of Squares	Mean Square	F Value	Prob>F
Model	3	8779676740.6	2926558913.5	46.662	0.0001
Error	16	1003491259.4	62718203.714		
C Total	19	9783168000.0			

Root MSE	7919.48254	R-Square	0.8974
Dep Mean	56660.00000	Adj R-Sq	0.8782
C.V.	13.97720		

Parameter Estimates

Variable	DF	Parameter Estimate	Standard Error	T for H0: Parameter=0	Prob > |T|
INTERCEP	1	1470.275919	5746.3245832	0.256	0.8013
X1	1	0.814490	0.51221871	1.590	0.1314
X2	1	0.820445	0.21118494	3.885	0.0013
X3	1	13.528650	6.58568006	2.054	0.0567

FIGURE 12.7
SAS Printout for the Sale Price Model, Example 12.2

A large value of R^2 computed from the *sample* data does not necessarily mean that the model provides a good fit to all of the data points in the *population*. For example, a first-order linear model that contains three parameters will provide a perfect fit to a sample of three data points and R^2 will equal 1. Likewise, you will always obtain a perfect fit ($R^2 = 1$) to a set of n data points if the model contains exactly n parameters. Consequently, if you want to use the value of R^2 as a measure of how useful the model will be for predicting y, it should be based on a sample that contains substantially more data points than the number of parameters in the model.

> **WARNING**
>
> ---
>
> In a multiple regression analysis, use the value of R^2 as a measure of how useful a linear model will be for predicting y only if the sample contains substantially more data points than the number of β parameters in the model.

As an alternative to using R^2 as a measure of model adequacy, the **adjusted multiple coefficient of determination**, denoted R_a^2, is often reported. The formula for R_a^2 is shown in the box.

> **THE ADJUSTED MULTIPLE COEFFICIENT OF DETERMINATION**
>
> ---
>
> The **adjusted multiple coefficient of determination** is given by
> $$R_a^2 = 1 - \frac{n-1}{n-(k+1)}\left(\frac{\text{SSE}}{\text{SS}_{yy}}\right)$$
> $$= 1 - \frac{n-1}{n-(k+1)}(1 - R^2)$$

Unlike R^2, R_a^2 takes into account ("adjusts" for) both the sample size n and the number of β parameters in the model. R_a^2 will always be smaller than R^2, and more important, R_a^2 cannot be "forced" to 1 by simply adding more and more independent variables to the model. Consequently, analysts prefer the more conservative R_a^2 when choosing a measure of model adequacy.

EXAMPLE 12.11   Refer to Example 12.2. Locate the value of R_a^2 on the printout, Figure 12.7. Interpret its value.

Solution   The value of R_a^2 is shown on the SAS printout (Figure 12.7) directly underneath the value of R^2. Note that $R_a^2 = .8782$, a value only slightly smaller than R^2. Our interpretation is that after adjusting for sample size and the number of parameters in the model, approximately 85% of the sample variation in sale price can be "explained" by the first-order model.

Despite their utility, R^2 and R_a^2 are only sample statistics. Consequently, it is dangerous to judge the usefulness of the model based solely on these values. We discuss a more formal method of checking the predictive ability of a general linear model—a statistical test of hypothesis—in the following section.

12.7

The object of step 4 in a multiple regression analysis is to conduct a test of the utility of a general linear model—that is, a test to determine whether the model is really useful for predicting y. Conducting individual t tests on each β parameter in a model (Section 12.5) is generally *not* a good way to determine whether a model is contributing information for the prediction of y. For example, suppose you fit a first-order linear model in five quantitative independent variables and decide to conduct t tests on all five of the individual β's in the model, each at $\alpha = .05$. Even if all the β parameters (except β_0) in the model are in fact equal to 0, you will incorrectly reject the null hypothesis at least once and conclude that some β parameter is nonzero approximately 23% of the time (proof omitted). In other words, the overall probability of a Type I error is about .23, not .05!

A better way to test the overall utility of a linear model in step 4 is to conduct a test involving *all* the β parameters (except β_0) simultaneously. The null and alternative hypotheses for this test of model utility are given in the box.

> **HYPOTHESES FOR TESTING WHETHER A GENERAL LINEAR MODEL IS USEFUL FOR PREDICTING y**
>
> H_0: $\beta_1 = \beta_2 = \cdots = \beta_k = 0$
>
> H_a: At least one of the β parameters in H_0 is nonzero.

Practically speaking, this test for model utility is a comparison of the predictive ability of the estimated general linear model (which uses the predictor $\hat{y} = \hat{\beta}_0 + \hat{\beta}_1 x_1 + \hat{\beta}_2 x_2 + \cdots + \hat{\beta}_k x_k$) with a model that contains no x's (which uses the predictor $\hat{y} = \bar{y}$). If the test shows that at least one of the β's is nonzero, then the value of $\hat{y}$ obtained from the estimated linear model will generally predict a future of y more accurately than the sample mean $\bar{y}$. We illustrate a test of model utility in the following example.

EXAMPLE 12.12   Refer to Example 12.2. Test (using $\alpha = .05$) whether the first-order linear model in the three quantitative independent variables is useful for predicting sale price, y, by testing the null hypothesis

H_0: $\beta_1 = \beta_2 = \beta_3 = 0$

against the alternative hypothesis,

H_a: At least one of the model parameters, β_1, β_2, and β_3, differs from 0.

Solution   The test statistic used in the test for model utility is an F statistic. The formula for computing the F statistic is given in the next box. However, most statistical computer software packages with regression analysis routines give this F value. The F value for the sale price model (shaded in the SAS computer printout shown in Figure 12.7) is $F = 46.662$. To determine whether this F value is statistically significant, we read the value of the observed significance level given in the SAS printout. (For details on the test procedure, see Example 12.13.) The observed significance level for this test, .0001, is shaded in Figure 12.7 in the column headed **Prob > F**. This implies that we would reject the null hypothesis for any α level larger than .0001. Thus, we have strong evidence to reject H_0 and to conclude that at least one of the model coefficients, β_1, β_2, and β_3, is nonzero. Since the observed significance level is so small, there is ample evidence to indicate that the first-order model is useful for predicting the sale price of residential properties.

Some software packages do not compute the value of the F statistic for testing the model and some do not compute its observed significance level. In such cases, you can calculate the F statistic directly using the formula shown in the box and reject H_0: $\beta_1 = \beta_2 = \cdots = \beta_k$ for a given value of α if $F > F_\alpha$. The F test statistic is based on k numerator (where k is the number of β parameters in the model, excluding β_0) and $n - (k + 1)$ denominator degrees of freedom. The values of F_α for $\alpha = .10, .05, .025,$ and $.01$ are given in Tables 7, 8, 9, and 10 of Appendix G. The test procedure is summarized in the next box and illustrated in Example 12.13.

PROCEDURE FOR TESTING WHETHER THE OVERALL MODEL IS USEFUL FOR PREDICTING y

H_0: $\beta_1 = \beta_2 = \cdots = \beta_k = 0$

H_a: At least one of the parameters, $\beta_1, \beta_2, \ldots, \beta_k$, differs from 0.

Test statistic: $F = \dfrac{\text{Mean square for model}}{\text{Mean square for error}}$

$$= \left[\frac{\text{SS(Model)}}{\text{SSE}}\right]\left[\frac{n - (k + 1)}{k}\right]$$

$$= \left(\frac{R^2}{1 - R^2}\right)\left[\frac{n - (k + 1)}{k}\right]$$

Rejection region: $F > F_\alpha$

where n = Number of observations

k = Number of parameters in the model (excluding β_0)

R^2 = Multiple coefficient of determination

and the distribution of F depends on k numerator degrees of freedom and $n - (k + 1)$ denominator degrees of freedom. Values of F_α for $\alpha = .10, .05, .025,$ and $.01$ are given in Tables 7, 8, 9, and 10 of Appendix G.

EXAMPLE 12.13   Refer to Example 12.12. Use the procedure outlined in the box to perform a test (with $\alpha = .05$) of the null hypothesis

$$H_0: \quad \beta_1 = \beta_2 = \beta_3 = 0$$

against the alternative hypothesis

$$H_a: \quad \text{At least one of the parameters, } \beta_1, \beta_2, \text{ and } \beta_3, \text{ is nonzero.}$$

Solution   For this example, the number of data points is $n = 20$ and the number of parameters involved in H_0 is $k = 3$. Therefore, the rejection region for the test is

$$F > F_\alpha = 3.24$$

where $\alpha = .05$ and F is based on $k = 3$ numerator and $n - (k + 1) = 20 - (3 + 1) = 16$ denominator degrees of freedom, and $F_{.05}$ (given in Table 8 of Appendix G) is 3.24.

To compute the value of the F statistic, we need to locate SS(Model) and SSE on the SAS printout shown in Figure 12.7. The value of SS(Model), shown in the **Model** row, is SS(Model) $= 8,779,676,740.6$. The corresponding value of SSE is SSE $= 1,003,491,259.4$. Then the computed value of the test statistic is

$$
\begin{aligned}
F &= \frac{\text{SS(Model)}}{\text{SSE}} \left[\frac{n - (k + 1)}{k} \right] \\
&= \frac{8,779,676,740.6}{1,003,491,259.4} \left(\frac{16}{3} \right) \\
&= 46.662
\end{aligned}
$$

[This F value could also be calculated by finding MS(Model)/MS(Error), or by using the formula that involves R^2.]

Since the computed value of the test statistic, $F = 46.662$, exceeds the critical value, $F_{.05} = 3.24$, we reject H_0 and conclude that at least one of the parameters (β_1, β_2, and β_3) is nonzero. In other words, the model appears to be useful for predicting residential property sale prices.

After we have determined that the overall model is useful for predicting y using the F test, we may elect to conduct one or more t tests on the individual β parameters (see Section 12.5). However, the test (or tests) to be conducted should be decided a priori, i.e., prior to fitting the model. Also, we should limit the number of t tests conducted to avoid the potential problem of making too many Type I errors. Generally, the regression analyst will conduct t tests on only the "most important" β's. These are usually the β's associated with higher-order terms (x_1^2, x_1x_2, etc.). We provide insight in identifying the most important β's in a linear model in Sections 12.9–12.10.

USING THE MODEL FOR ESTIMATION AND PREDICTION

12.8 After checking the utility of the linear model and finding it to be useful for prediction and estimation, we may decide to use it for those purposes (step 6). Our methods for prediction and estimation using any general linear model are identical to those discussed in Section 11.9 for the simple straight-line model. We will use the model to form a confidence interval for the mean $E(y)$ for given values of x_1, x_2, x_3, etc., or a prediction interval for a future value of y for given values of x_1, x_2, x_3, etc.

EXAMPLE 12.14 Refer to Example 12.2. Using the first-order least squares model

$$\hat{y} = 1{,}470.28 + .8145x_1 + .8204x_2 + 13.53x_3$$

estimate the mean sale price, $E(y)$, for a property with an appraised land value of $x_1 = \$15{,}000$, an appraised improvements value of $x_2 = \$50{,}000$, and a home size of $x_3 = 1{,}800$ square feet. Use a 95% confidence interval.

Solution   Substituting $x_1 = 15{,}000$, $x_2 = 50{,}000$, and $x_3 = 1{,}800$ into the least squares prediction equation yields the following estimate of $E(y)$:

$$\hat{y} = 1{,}470.28 + .8145(15{,}000) + .8204(50{,}000) + 13.53(1{,}800)$$
$$= 79{,}061.4$$

To form a confidence interval for the mean, we need to know the standard deviation of the sampling distribution for the estimator $\hat{y}$. For general linear models, the form of this standard deviation is very complex. However, most regression packages (including SAS) allow us to obtain the confidence intervals for mean values of y at any given setting of the independent variables. The relevant portion of the SAS printout for the sale price example is shown in Figure 12.8. The 95% confidence interval for $E(y)$, the mean sale price for all properties with an appraised land value of $x_1 = \$15{,}000$, an appraised improvements value of $x_2 = \$50{,}000$, and a home size of $x_3 = 1{,}800$ square feet, is shown (shaded) to be \$73,380.70 to \$84,742.10.

FIGURE 12.8
SAS Printout for Estimated Mean and Corresponding Confidence Interval for $x_1 = 15{,}000$, $x_2 = 50{,}000$, and $x_3 = 1{,}800$

Obs	X1	X2	X3	Y	Predict Value	Residual	Lower95% Mean	Upper95% Mean
1	5960	44967	1873	68900.0	68556.7	343.3	59404.1	77709.4
2	9000	27860	928	48500.0	44212.9	4287.1	37904.0	50521.8
3	9500	31439	1126	55500.0	50235.2	5264.8	45337.9	55132.4
4	10000	39592	1265	62000.0	59212.0	2788.0	54284.8	64139.2
5	18000	72827	2214	116500	105834	10665.9	95659	116009
6	8500	27317	912	45000.0	43143.7	1856.3	36932.3	49355.0
7	8000	29856	899	38000.0	44643.6	-6643.6	38243.5	51043.8
8	23000	47752	1803	83000.0	83773.6	-773.6	71575.2	95972
9	8100	39117	1204	59000.0	56449.5	2550.5	50998.3	61900.6
10	9000	29349	1725	47500.0	56216.8	-8716.8	48549.9	63883.7
11	7300	40166	1080	40500.0	54981.0	-14481.0	47863.4	62098.5
12	8000	31679	1529	40000.0	54662.4	-14662.4	49554.8	59770.0
13	20000	58510	2455	97000	98977	-1977.1	88618.6	109336
14	8000	23454	1151	45500.0	42800.4	2699.6	37649.4	47951.4
15	8000	20897	1173	40900.0	41000.1	-100.1	35078.5	46921.8
16	10500	56248	1960	80000.0	82686.9	-2686.9	74642.8	90731.1
17	4000	20859	1344	56000.0	40024.4	15975.6	32814.2	47234.6
18	4500	22610	988	37000.0	37052.0	-52.0421	31826.5	42277.6
19	3400	35948	1076	50000.0	48289.7	1710.3	40503.6	56075.8
20	1500	5779	962	22400.0	20447.9	1952.1	11231.5	29664.3
21	15000	50000	1800	.	79061.4	.	73380.7	84742.1

EXAMPLE 12.15   Refer to Example 12.2. Construct a 95% prediction interval for y, the sale price of a particular property with an appraised land value of $x_1 = \$15{,}000$, an appraised improvements value of $x_2 = \$50{,}000$, and a home size of $x_3 = 1{,}800$ square feet.

Solution   When $x_1 = 15{,}000$, $x_2 = 50{,}000$, and $x_3 = 1{,}800$, the predicted value for y is again $\hat{y} = 79{,}061.4$. However, the prediction interval for a particular value of y will be wider than the confidence interval for the mean value. This is reflected in the SAS printout shown in Figure 12.9 (page 618). The prediction interval (shaded) extends from \$61,337.90 to \$96,785.00.

Obs	X1	X2	X3	Y	Predict Value	Residual	Lower95% Predict	Upper95% Predict
1	5960	44967	1873	68900.0	68556.7	343.3	49435.5	87678.0
2	9000	27860	928	48500.0	44212.9	4287.1	26278.1	62147.6
3	9500	31439	1126	55500.0	50235.2	5264.8	32747.0	67723.3
4	10000	39592	1265	62000.0	59212.0	2788.0	41715.4	76708.5
5	18000	72827	2214	116500	105834	10665.9	86203.0	125465
6	8500	27317	912	45000.0	43143.7	1856.3	25243.0	61044.3
7	8000	29856	899	38000.0	44643.6	-6643.6	26676.6	62610.7
8	23000	47752	1803	83000.0	83773.6	-773.6	63021.4	104526
9	8100	39117	1204	59000.0	56449.5	2550.5	38798.2	74100.8
10	9000	29349	1725	47500.0	56216.8	-8716.8	37760.6	74673.1
11	7300	40166	1080	40500.0	54981.0	-14481.0	36746.1	73215.9
12	8000	31679	1529	40000.0	54662.4	-14662.4	37114.1	72210.6
13	20000	58510	2455	97000	98977	-1977.1	79250.2	118704
14	8000	23454	1151	45500.0	42800.4	2699.6	25239.5	60361.3
15	8000	20897	1173	40900.0	41000.1	-100.1	23197.9	58802.3
16	10500	56248	1960	80000.0	82686.9	-2686.9	64070.8	101303
17	4000	20859	1344	56000.0	40024.4	15975.6	21753.1	58295.7
18	4500	22610	988	37000.0	37052.0	-52.0421	19469.1	54635.0
19	3400	35948	1076	50000.0	48289.7	1710.3	29783.6	66795.8
20	1500	5779	962	22400.0	20447.9	1952.1	1296.0	39599.8
21	15000	50000	1800	.	79061.4	.	61337.9	96785

FIGURE 12.9

SAS Printout for Predicted Value and Corresponding Prediction Interval for $x_1 = 15{,}000$, $x_2 = 50{,}000$, and $x_3 = 1{,}800$

Just as in simple linear regression, it is dangerous to use any general linear model for making predictions outside the region in which the sample data fall. Checking the sample data given in Table 12.1, we see that appraised land value (x_1) ranges from \$1,500 to \$23,000; appraised improvements (x_2) ranges from \$5,779 to \$72,827; and home size (x_3) ranges from 899 to 2,455 square feet. Consequently, in Examples 12.14 and 12.15, we would not use the estimated model to make estimates or predictions for properties with values of the independent variables outside their respective ranges.* In general, the fitted model might not provide a good model for the relationship between the mean y and the value of x when stretched over a wider range of x values.

WARNING

Do not use the least squares model to predict a value of y outside the region in which the sample data fall. In other words, do not predict y for values of the independent variables $x_1, x_2, \ldots, x_k$ that are not within the range of the sample data.

In the preceding sections, we have demonstrated the methods of multiple regression analysis by fitting a first-order linear model to a set of data. In the next three sections (Sections 12.9–12.11), we will introduce other, more complex models that are useful for relating a response variable y to a set of independent

*With two or more independent variables in the model, the values of x_1, x_2, etc., *jointly* define the experimental region. An observation with values of the x's that fall within their respective sample ranges may still fall outside the experimental region. For more information on this "hidden extrapolation" problem, consult the references given at the end of this chapter.

variables. Then, in Section 12.12, we show you how to compare models to determine which is "best" for predicting y.

EXERCISES
LEARNING THE MECHANICS

12.1   Write a first-order linear model relating the mean value of y, $E(y)$, to two quantitative independent variables.

12.2   Write a first-order linear model relating the mean value of y, $E(y)$, to four quantitative independent variables.

12.3   Consider the following first-order equation in two quantitative independent variables:

$$E(y) = 1 + 2x_1 + x_2$$

a. Graph the relationship between y and x_1 for $x_2 = 0$, 1, and 2.
b. How do the graphed lines in part **a** relate to each other? What is the slope of each line?
c. If a linear model is first-order in two independent variables, what type of geometric relationship will you obtain when $E(y)$ is graphed as a function of one of the independent variables for various values of the other independent variable?

12.4   Consider the first-order equation in three quantitative independent variables

$$E(y) = 1 + 2x_1 + x_2 - 3x_3$$

a. Graph the relationship between y and x_1 for $x_2 = 1$ and $x_3 = 3$.
b. Repeat part **a** for $x_2 = -1$ and $x_3 = 1$.
c. How do the graphed lines in parts **a** and **b** relate to each other? What is the slope of each line?
d. If a linear model is first-order in three independent variables, what type of geometric relationship will you obtain when $E(y)$ is graphed as a function of one of the independent variables for various combinations of values of the other independent variables?

12.5   Suppose $E(y)$ is related to four quantitative independent variables, x_1, x_2, x_3, and x_4, by the model

$$E(y) = \beta_0 + \beta_1 x_1 + \beta_2 x_2 + \beta_3 x_3 + \beta_4 x_4$$

Suppose you fit this model to a set of $n = 15$ data points and found $R^2 = .74$, SS(Total) $= 1.690$, and SSE $= .439$.

a. Calculate s^2, the estimate of the variance of the random error.
b. Calculate the F statistic for testing H_0: $\beta_1 = \beta_2 = \beta_3 = \beta_4 = 0$.
c. Do the data provide sufficient evidence to indicate that the model contributes information for predicting y? Test using $\alpha = .05$.

12.6   Suppose you fit the first-order multiple regression model

$$y = \beta_0 + \beta_1 x_1 + \beta_2 x_2 + \varepsilon$$

to $n = 25$ data points and obtain the prediction equation

$$\hat{y} = 6.4 + 3.1x_1 + .92x_2$$

The estimated standard deviations of the sampling distributions of $\hat{\beta}_1$ and $\hat{\beta}_2$ are 2.3 and .27, respectively.

a. Test H_0: $\beta_1 = 0$ against H_a: $\beta_1 > 0$. Use $\alpha = .05$.
b. Test H_0: $\beta_2 = 0$ against H_a: $\beta_2 \neq 0$. Use $\alpha = .05$.
c. Find a 90% confidence interval for β_1. Interpret the interval.
d. Find a 99% confidence interval for β_2. Interpret the interval.

12.7   Suppose you fit the first-order multiple regression model

$$y = \beta_0 + \beta_1 x_1 + \beta_2 x_2 + \beta_3 x_3 + \varepsilon$$

to $n = 20$ data points and obtain $R^2 = .2623$. Test the null hypothesis $H_0: \beta_1 = \beta_2 = \beta_3 = 0$ against the alternative hypothesis that at least one of the β parameters is nonzero. Use $\alpha = .05$.

APPLYING THE CONCEPTS

12.8   Perfectionists are persons who set themselves standards and goals that cannot be reasonably met or accomplished. One theory suggests that those individuals who are depressed have a tendency toward perfectionism. To study this phenomenon, 76 members of an introductory psychology class completed four questionnaires: (1) the ASO scale, designed to measure self-acceptance, (2) the Burns scale, designed to measure perfectionism, (3) the Zung scale, designed to measure depression, and (4) the Rotter scale, designed to measure perceptions between actions and reinforcement (*The Journal of Adlerian Theory, Research, and Practice*, Mar. 1986).

    a. Write a first-order model relating depression (Zung scale) to self-acceptance (ASO scale), perfectionism (Burns scale), and reinforcement (Rotter scale).

    b. The model, part **a**, was fit to the $n = 76$ points and resulted in a coefficient of determination of $R^2 = .70$. Interpret this value.

    c. Is there sufficient evidence to indicate that the model is useful for predicting depression (Zung scale) score? Test using $\alpha = .05$.

    d. A t test for the perfectionism (Burns scale) variable resulted in a (two-tailed) p-value of .87. Interpret this value.

12.9   Residential property appraisers make extensive use of multiple regression in their evaluation of property. Typically, the sale price (y) of a property is modeled as a function of several home-related conditions (e.g., gross living area, location, number of bedrooms). However, appraisers are not interested in the predicted price, $\hat{y}$. Rather, they use the regression model as a tool for making value adjustments to the property. These adjustments are derived from the parameter estimates of the model. The *Real Estate Appraiser* (Apr. 1992) reported the results of a multiple regression on the price (y) of $n = 157$ residential properties recently sold in a northern Virginia subdivision. A table showing the results of the SAS anlaysis is reproduced on the next page. Note that there are 27 independent variables in the model.

    a. Interpret the values of **F Value**, **Root MSE**, **R-Square**, and **Adj. R-Sq** shown on the printout.

    b. One of the independent variables in the model is gross living area (GLA), measured in square feet. A 95% confidence interval for the β coefficient associated with GLA is shown on the printout. Interpret this interval.

    c. Note that the independent variables with β coefficients significantly different from 0 (at $\alpha = .05$) are highlighted in bold on the printout. The nonsignificant variables are not highlighted. Would you advise the property appraiser to ignore any value adjustments based on nonsignificant independent variables? Explain.

12.10   In a production facility, an accurate estimate of man-hours needed to complete a task is crucial to management in making such decisions as the proper number of workers to hire, and accurate deadline to quote a client, or cost-analysis decisions regarding budgets. A manufacturer of boiler drums wants to use regression to predict the number of man-hours needed to erect the drums in future projects. To accomplish this, data for 35 boilers were collected. In addition to man-hours (y), the variables measured were boiler capacity ($x_1 =$ pounds per hour), boiler design pressure ($x_2 =$ pounds per square inch), boiler type ($x_3 = 1$ if industry field erected, 0 if utility field erected), and drum type ($x_4 = 1$ if steam, 0 if mud). The data are provided in the accompanying table (page 622). A Minitab printout for the model $E(y) = \beta_0 + \beta_1 x_1 + \beta_2 x_2 + \beta_3 x_3 + \beta_4 x_4$ appears on page 623.

DEPENDENT VARIABLE: SALE PRICE

ANALYSIS OF VARIANCE

Source	DF	Sum of Squares	Mean Square	F Value	Prob>F
Model	27	24,184,211,898	895,711,551.79	20.914	.0001
Error	129	5,524,834,283	42,828,172.73		
C Total	156	29,709,046,181			

Root MSE	6544.324	R-Square	.8140	
Dep Mean	173157.5	Adj R-Sq	.7751	
C.V.	3.779404			

PARAMETER ESTIMATES

Variable	Parameter Estimate	Std Error	95% Confidence Interval (@129df=1.98)	T for H_0: Parameter=0	Prob>|T|
Intercept	96,603	12,530	(71,794 to 121,412)	7.710	.0001
Time	150	123	(−94 to 394)	1.220	.2248
Lot Size	.60	.30	(0.01 to 1.19)	2.022	.0452*
Age	381	502	(−613 to 1,375)	.758	.4501
G.L.A.	22.40	3.67	(15.13 to 29.67)	6.099	.0001*
Bedrooms	2,263	1,609	(−923 to 5,499)	1.407	.1619
Half Baths	5,962	2,934	(153 to 11,771)	2.032	.0442*
Corner Lot	−1,481	1,692	(−4,831 to 1,869)	−.876	.3829
Cul-de-Sac	−56	2,557	(−5,119 to 5,007)	−.022	.9825
Back to Woods	4,086	2,044	(39 to 8,133)	1.999	.0477*
Deck	2,408	2,167	(−1,883 to 6,699)	1.111	.2686
Fence	2,896	1,271	(379 to 5,413)	2.279	.0243*
Shed	70	1,343	(−2,589 to 2,729)	.052	.9588
Patio	2,377	1,671	(−932 to 5,686)	1.423	.1572
Portico	−906	2,963	(−6,773 to 4,961)	−.306	.7603
Screen Porch	5,021	2,038	(986 to 9,056)	2.463	.0151*
In-grnd Pool	7,570	3,028	(1,575 to 13,565)	2.500	.0137*
Garage	2,989	1,446	(126 to 5,852)	2.068	.0407*
Driveway	−1,844	3,222	(−8,224 to 4,536)	−.572	.5681
Fireplace	1,290	1,277	(−1,238 to 3,818)	1.010	.3144
Brick Facade	−2,140	2,369	(−6,381 to 2,551)	−.903	.3680
Updated Kit.	4,171	1,470	(1,260 to 7,082)	2.837	.0053*
Remodel Kit.	6,091	2,367	(1,404 to 10,778)	2.574	.0112*
Intercom	1,933	2,146	(−2,316 to 6,182)	.901	.3693
Cen. Vacuum	−4,636	2,166	(−8,925 to −347)	−2.140	.0342*
Skylights	7,744	2,622	(2,552 to 12,936)	−2.954	.0037*
Air Filter	874	2,506	(−4,088 to 5,836)	−.349	.7280
Bay Window	−3,174	2,086	(−7,304 to 956)	−1.522	.1305

*Indicates significance at the 5% significance level.

Source: Gilson, S. J. "A Case Study—Comparing the results: Multiple regression analysis vs. matched pairs in residential subdivision." *The Real Estate Appraiser*, Apr. 1992, p. 37 (Table 4).

12.12 As a result of the U.S. surgeon general's warnings about the health hazards of smoking, Congress banned television and radio advertising of cigarettes in January 1971. The banning of prosmoking messages, however, also led to the virtual elimination of antismoking messages. In theory, if these antismoking commercials are more effective than prosmoking commercials, the net effect of the Congressional ban will be to increase the consumption of cigarettes and, therefore, benefit the tobacco industry. To test this hypothesis, researchers at the University of Houston built a cigarette demand model based on data collected from 46 states over the 18-year period 1963–1980 (*The Review of Economics & Statistics*, Feb. 1986). For each state–year, the following independent variables were recorded:

x_1 = Natural logarithm of price of a carton of cigarettes

x_2 = Natural log of minimum price of a carton of cigarettes in any neighboring state   (This variable was included to measure the effect of "bootlegging" cigarettes in nearby states with lower tax rates.)

x_3 = Natural log of real disposable income per capita

x_4 = Per capita index of expenditures for cigarette advertising on television and radio   (This value is 0 for the years 1971–1980, when the ban was in effect.)

The dependent variable of interest is y, the natural log of per capita consumption of cigarettes by persons of smoking age (14 years and older). The multiple regression model

$$E(y) = \beta_0 + \beta_1 x_1 + \beta_2 x_2 + \beta_3 x_3 + \beta_4 x_4$$

was fit to the $n = 828$ observations (46 states $\times$ 18 years) with the following results.

$$R^2 = .95 \qquad s = .047$$

a. Test the hypothesis that the model is useful for predicting y. (Use $\alpha = .05$.)
b. Interpret the value of s.
c. Give the null and alternative hypotheses appropriate for testing whether a decrease in per capita cigarette advertising expenditures is accompanied by an increase in per capita consumption of cigarettes over the period 1963–1980.
d. The value of $\hat{\beta}_4$ was determined to be .033. Interpret this value.
e. Does the value $\hat{\beta}_4 = .033$ support the alternative hypothesis of part c? Explain.

12.13 Refer to the *IEEE Transactions on Software Engineering* (Apr. 1985) study on identifying error-prone software, described in Exercise 11.58. A multiple regression analysis was conducted to identify the computer-module-related variables (called *metrics*) useful for predicting the number y of discovered module defects. For a certain product written in PL/S language, the following model was fit to data collected for $n = 253$ modules:

$$E(y) = \beta_0 + \beta_1 x_1 + \beta_2 x_2$$

where x_1 = Number of unique operands in the module

x_2 = Number of conditional statements, loops, and Boolean operators in the module

The multiple coefficient of determination of the model was $R^2 = .78$. Is there sufficient evidence to indicate that the model is useful for predicting the number y of defects in modules of the software product? Test using $\alpha = .05$.

12.14 Personal computer (PC) technology is changing at a phenomenal rate. As such, the retail price of a PC may vary dramatically depending on when it is purchased and what features it includes. Retail price data were recently collected for IBM and IBM-compatible PCs. The data for $n = 60$ PCs, shown in the accompanying table, were used to fit the multiple regression model

$$E(y) = \beta_0 + \beta_1 x_1 + \beta_2 x_2$$

The printout from an SPSS analysis follows the table.

RETAIL PRICE y	SPEED, MHz	CHIP	RETAIL PRICE y	SPEED, MHz	CHIP
$5099	33	386	$3249	25	386
3995	25	386	2995	20	386
2230	20	386	3419	20	386
4395	33	386	1590	20	386
6299	25	386	3899	20	386
2549	16	386	2249	12	286
3499	16	386	5796	25	386
2995	16	386	4330	16	286
1649	10	286	2699	16	386
5499	20	386	5579	20	386
1695	12	286	2095	16	386
2595	20	386	2695	25	386
3695	33	386	2295	20	386
3499	33	386	3445	25	386
2845	20	386	2445	16	386
4195	33	386	3795	25	386
2895	20	386	2395	16	386
2195	12	286	1595	12	286
5625	25	386	2095	16	386
2495	20	386	2995	25	386
3795	33	386	2895	20	386
3295	25	386	3995	33	386
1995	16	386	2595	20	386
2795	25	386	4995	25	386
5795	33	386	2695	25	386
3995	33	386	3990	33	386
1850	12	286	2795	20	386
1895	16	386	1995	20	286
1795	16	286	1595	16	286
2645	16	386	2875	20	386

Source: *Computer Monthly*, *Computer Shopper*, and IBM Corporation flyers. Data compiled by Jerasimos M. Mantas, University of South Florida business student.

```
* * * *   M U L T I P L E   R E G R E S S I O N   * * * *

Equation Number 1    Dependent Variable..   Y

Multiple R           .63263
R Square             .40022
Adjusted R Square    .37918
Standard Error    953.66516

Analysis of Variance
                    DF      Sum of Squares        Mean Square
Regression           2      34592103.00773     17296051.50386
Residual            57      51840202.92561       909477.24431

F =      19.01757      Signif F =  .0000

------------------ Variables in the Equation ------------------

Variable            B        SE B        Beta         T   Sig T

X2           357.184971  389.422935    .110908       .917  .3629
X1           104.838940   22.362982    .566873      4.688  .0000
(Constant)   648.022624  431.494302               1.502  .1387
```

where

y = Retail price ($)

x_1 = Microprocessor speed (megahertz)

$$x_2 = \begin{cases} 1 & \text{if 386 CPU chip} \\ 0 & \text{if 286 CPU chip} \end{cases}$$

a. Write the least squares prediction equation.
b. Is the model adequate for predicting y? Test using $\alpha = .10$.
c. Construct a 90% confidence interval for β_1. Interpret the interval.
d. Is CPU chip (x_2) a useful predictor of price (y) in this model? Test using $\alpha = .10$.

12.15   J. Vuorinen carried out a series of experiments to gather information on the coefficient of permeability of concrete (*Magazine of Concrete Research*, Sept. 1985). In one experiment, the outflow of water from the pores of a concrete specimen after it had been under saturating water pressure for a period of time was recorded for different combinations of concrete permeability and porosity. The resulting water quantities after different lapses of time for one permeability–porosity combination were used to estimate the water outflow–time slope coefficient. Vuorinen used these results to develop a model*

$$E(y) = \beta_0 + \beta_1 x_1 + \beta_2 x_2$$

where

x_1 = Porosity of the cement

x_2 = Estimated slope coefficient of the corresponding water outflow–time regression line

The data are reproduced here, and the SAS printout for the analysis is shown on the next page.

COEFFICIENT OF PERMEABILITY y, (meters per second) $\times 10^{-11}$	POROSITY x_1	ESTIMATED WATER OUTFLOW– TIME SLOPE COEFFICIENT x_2
1.00	.050	.903
1.00	.035	.722
1.00	.025	.590
.10	.050	.345
.10	.035	.282
.10	.025	.233
.01	.050	.103
.01	.035	.091
.01	.025	.078

Source: Vuorinen, J. "Applications of diffusion theory to permeability tests on concrete, Part II: Pressure-saturation test on concrete and coefficient of permeability." *Magazine of Concrete Research*, Vol. 37, No. 132, Sept. 1985, p. 156, Table II.1.

a. Give the least squares prediction equation.
b. Conduct a test of overall model utility. Interpret the p-value of the test.
c. Is there evidence that concrete porosity x_1 is a useful predictor of coefficient of permeability y? Test using $\alpha = .05$.
d. Is there evidence that the estimated water outflow–time slope is a useful predictor of coefficient of permeability y? Test using $\alpha = .05$.

---

*In actuality, Vuorinen fit the logarithmic model

$\log(y) = \beta_0 + \beta_1 \log(x_1) + \beta_2 \log(x_2) + \varepsilon$

Analysis of Variance

Source	DF	Sum of Squares	Mean Square	F Value	Prob>F
Model	2	1.65932	0.82966	35.843	0.0005
Error	6	0.13888	0.02315		
C Total	8	1.79820			

Root MSE	0.15214	R-square	0.9228	
Dep Mean	0.37000	Adj R-sq	0.8970	
C.V.	41.11920			

Parameter Estimates

Variable	DF	Parameter Estimate	Standard Error	T for H0: Parameter=0	Prob > |T|
INTERCEP	1	0.132021	0.19005130	0.695	0.5133
X1	1	-9.307122	5.05702529	-1.840	0.1153
X2	1	1.557563	0.18396157	8.467	0.0001

e. Locate R^2 on the printout and interpret its value.
f. Locate the estimate of σ on the printout and interpret its value.

12.16 Refer to Exercise 11.82. To improve the ability of the model to predict urban travel times, Cook and Russell added a second independent variable—weighted average speed limit between the two urban locations. The proposed model takes the form

$$y = \beta_0 + \beta_1 x_1 + \beta_2 x_2 + \varepsilon$$

where

y = Urban travel time (minutes)

x_1 = Distance between locations (miles)

x_2 = Weighted speed limit between locations (miles per hour)

This model was fit to the car and truck data sets, with the results shown in the accompanying table.

PASSENGER CARS	TRUCKS
$\hat{y} = 5.46 + 2.15x_1 - .09x_2$	$\hat{y} = 4.84 + 3.92x_1 - .09x_2$
$R^2 = .687$; $n = 567$	$R^2 = .771$; $n = 918$

Source: Cook, T. M. and Russell, R. A. "Estimating urban travel times: A comparative study." *Transportation Research*, Vol. 14A, June 1980, pp. 173–175. Reprinted with permission. Copyright 1980, Pergamon Press, Ltd.

a. Is the model useful for predicting the urban travel times of passenger cars? Use $\alpha = .05$.
b. Interpret the β estimates of the model for passenger cars.
c. Is the model useful for predicting urban travel times of trucks? Use $\alpha = .05$.
d. Interpret the β estimates of the model for trucks.

12.17 Because the coefficient of determination R^2 always increases when a new independent variable is added to the model, it may be tempting to include many variables in a model to force R^2 to be near 1. However, doing so reduces the degrees of freedom available for estimating σ^2, which adversely affects our ability to make reliable inferences. As an example, suppose you want to predict the CPU time of a computer job using 18 independent variables (such as size of job, time of submission, and estimated lines of print). You fit the model

$$y = \beta_0 + \beta_1 x_1 + \beta_2 x_2 + \cdots + \beta_{17} x_{17} + \beta_{18} x_{18} + \varepsilon$$

where y = CPU time and $x_1, x_2, \ldots, x_{18}$ are the predictor variables. Using the relevant information on $n = 20$ jobs to fit the model, you obtain $R^2 = .95$. Test to determine whether this value of R^2 is large enough for you to infer that this model is useful—i.e., that at least one term in the model is important for predicting CPU time. Use $\alpha = .05$.

12.18   Marketers are keenly interested in the factors that motivate coupon usage by consumers. Three dominant motivational factors are thought to be (1) price reduction, (2) time and effort required to collect coupons, and (3) self-satisfaction. Using questionnaire data collected for a sample of $n = 290$ shoppers, a trio of marketing researchers examined the relationship between coupon usage and these factors (*The Journal of Consumer Marketing*, Spring 1988). The multiple regression model took the form

$$E(y) = \beta_0 + \beta_1 x_1 + \beta_2 x_2 + \beta_3 x_3$$

where

$\quad y$ = Coupon redemption rate

$\quad x_1$ = Price consciousness score

$\quad x_2$ = Time value score

$\quad x_3$ = Satisfaction/pride score

The results are summarized as follows (t values for testing β's in parentheses):

$\hat{\beta}_1 = \quad .09784 \quad (1.444) \qquad R^2 = .11671$

$\hat{\beta}_2 = -.13134 \quad (-1.695) \qquad F = 9.6893$

$\hat{\beta}_3 = \quad .20019 \quad (2.571)$

a. Conduct an overall test of model adequacy. Use $\alpha = .10$.
b. In theory, coupon users are more price conscious than nonusers. Test the theory using $\alpha = .10$.
c. Interpret the negative β estimate for time value score (x_2).

MODEL BUILDING: INTERACTION MODELS

12.9

In Section 12.3 we demonstrated the relationship between $E(y)$ and the independent variables in a first-order linear model,

$$E(y) = \beta_0 + \beta_1 x_1 + \beta_2 x_2$$

When $E(y)$ is graphed against any one variable (say, x_1) for fixed values of the other variable (x_2), the result is a set of *parallel* straight lines (see Figure 12.10).

FIGURE 12.10

Graphs of $E(y)$ Versus x_1 for Fixed Values of x_2: First-Order Model

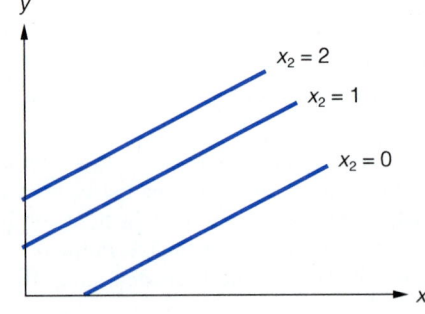

When this situation occurs (as it always does for a first-order model), we say that the relationship between $E(y)$ and any one independent variable *does not depend* on the value of the other independent variable(s) in the model—that is, we say that the independent variables **do not interact**.

However, if the relationship between $E(y)$ and x_1 does, in fact, depend on the value of x_2 held fixed, then the first-order model is not appropriate for predicting y. In this case, we need another model that will take into account this dependence. Such a model is illustrated in the next example.

EXAMPLE 12.16 Refer to Example 12.3, where we graphed the first-order model, $E(y) = 1 + 2x_1 + x_2$. Now suppose that the mean value $E(y)$ of a response y is related to two quantitative independent variables, x_1 and x_2, by the model

$$E(y) = \beta_0 + \beta_1 x_1 + \beta_2 x_2 + \beta_3 x_1 x_2$$

where $\beta_0 = 1$, $\beta_1 = 2$, $\beta_2 = -1$, and $\beta_3 = 1$, i.e.,

$$E(y) = 1 + 2x_1 - x_2 + x_1 x_2$$

Graph the relationship between $E(y)$ and x_1 for $x_2 = 0$, 1, and 2. Interpret the graph. [*Note:*   Since this model contains the terms $x_1 x_2$, in addition to all of the terms of the model in Example 12.3, we will be able to see the effect of adding the second-order cross-product term $(x_1 x_2)$ to a first-order model.]

Solution   We obtain three response curves relating $E(y)$ to x_1—one for each of the values $x_2 = 0$, 1, and 2. For $x_2 = 0$:

$$E(y) = 1 + 2x_1 - (0) + x_1(0) = 1 + 2x_1 \quad \text{(slope = 2)}$$

For $x_2 = 1$:

$$E(y) = 1 + 2x_1 - (1) + x_1(1) = 3x_1 \quad \text{(slope = 3)}$$

For $x_2 = 2$:

$$E(y) = 1 + 2x_1 - (2) + x_1(2) = -1 + 4x_1 \quad \text{(slope = 4)}$$

A careful examination reveals that the slope of each line is represented by $\beta_1 + \beta_3 x_2 = 2 + x_2$. Graphs of these three straight lines are shown in Figure 12.11 (page 630). The effect of adding a term involving the cross-product $x_1 x_2$ can be seen in Figure 12.11. In contrast to Figure 12.10, the lines relating $E(y)$ to x_1 are no longer parallel. The effect on $E(y)$ of a change in x_1 (i.e., the slope) now *depends* on the value of x_2. When this situation occurs, we say that x_1 and x_2 **interact**. The cross-product term, $\beta_3 x_1 x_2$, is called an **interaction term**, and the model $E(y) = \beta_0 + \beta_1 x_1 + \beta_2 x_2 + \beta_3 x_1 x_2$ is called an **interaction model** with two quantitative variables (see the box on the next page).

FIGURE 12.11
Graphs of $E(y) = 1 + 2x_1 - x_2 + x_1x_2$ for $x_2 = 0, 1, 2$

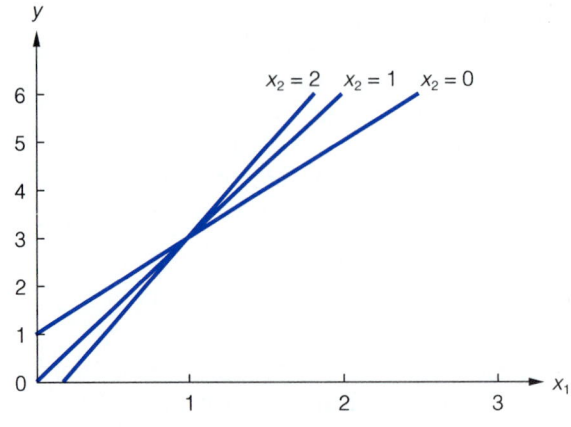

AN INTERACTION MODEL RELATING $E(y)$ TO TWO QUANTITATIVE INDEPENDENT VARIABLES

$$E(y) = \beta_0 + \beta_1 x_1 + \beta_2 x_2 + \beta_3 x_1 x_2$$

where $(\beta_1 + \beta_3 x_2)$ represents the change in $E(y)$ for every 1-unit increase in x_1, holding x_2 fixed;

$(\beta_2 + \beta_3 x_1)$ represents the change in $E(y)$ for every 1-unit increase in x_2, holding x_1 fixed.

A three-dimensional graph (generated by computer) of an interaction model in two quantitative x's is shown in Figure 12.12. Unlike the planar surface displayed in Figure 12.3, the interaction model traces a ruled surface (twisted plane) in three-dimensional space. If we slice the twisted plane at a fixed value of x_2, we obtain a straight line relating $E(y)$ to x_1; however, the slope of the line will change as we change the value of x_2.

EXAMPLE 12.17   Although a regional express delivery service bases the charge for shipping a package on the package weight and distance shipped, its profit per package depends on the package size (volume of space that it occupies) and the size and nature of the load on the delivery truck. The company recently conducted a study to investigate the relationship between the cost y of shipment (in dollars) and the variables that control the shipping charge—package weight, x_1 (in pounds), and distance shipped, x_2 (in miles). Twenty packages were randomly selected from among the large number received for shipment. A detailed analysis of the cost of shipment was made for each package, with the results shown in Table 12.2.

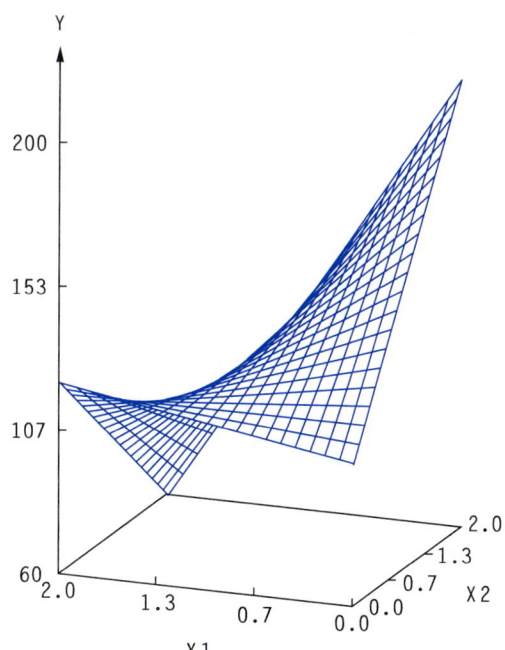

TABLE 12.2
Cost of Shipment Data for Example 12.17

PACKAGE	x_1	x_2	y	PACKAGE	x_1	x_2	y
1	5.9	47	2.60	11	5.1	240	11.00
2	3.2	145	3.90	12	2.4	209	5.00
3	4.4	202	8.00	13	.3	160	2.00
4	6.6	160	9.20	14	6.2	115	6.00
5	.75	280	4.40	15	2.7	45	1.10
6	.7	80	1.50	16	3.5	250	8.00
7	6.5	240	14.50	17	4.1	95	3.30
8	4.5	53	1.90	18	8.1	160	12.10
9	.60	100	1.00	19	7.0	260	15.50
10	7.5	190	14.00	20	1.1	90	1.70

a. Suppose the company believes that straight-line relationships exist between
mean shipment cost, $E(y)$, and package weight (x_1), and between $E(y)$ and
distance shipped (x_2), but that the rate of change of $E(y)$ with x_1 depends on
x_2, and vice versa. Hypothesize an appropriate linear model for $E(y)$.

b. Fit the model to the data and graph the prediction equation.

c. Give the estimated slope of the line relating $E(y)$ to x_2 for a fixed package
weight of $x_1 = 2$ pounds. Interpret this value.

d. Find the value of s and interpret it.

e. Find the value of R^2 and interpret it.

c. Conduct a test to determine whether the relationship between subordinate performance (y) and manager's use of a group decision method (x_1) depends on a manager's legitimization of conflict (x_2). Use $\alpha = .10$.

d. Based on the result of part c, would you recommend that the researchers conduct t tests on β_1 and β_2? Explain.

12.24   To what degree do the attitudes of your peers influence your behavior? There is general agreement among sociologists and psychologists that your behavior is dependent on the attitudes of and social support from your friends, neighbors, etc. However, it is unclear whether the effects of attitude and social support are additive or interactive. An attempt to resolve this attitude–behavior issue was presented in *Social Psychology Quarterly* (Vol. 50, 1987). The study included a sample of $n = 143$ adult drinkers in an urban setting characterized by high physical availability of alcoholic beverages. The goal of the study was to build a model relating frequency of drinking alcoholic beverages, y, to attitude toward drinking (x_1) and social support (x_2). Consider the interaction model

$$E(y) = \beta_0 + \beta_1 x_1 + \beta_2 x_2 + \beta_3 x_1 x_2$$

a. Interpret the phrase "x_1 and x_2 interact" in terms of the problem.

b. Write the null and alternative hypotheses for determining whether attitude (x_1) and social support (x_2) interact.

c. The reported p-value for the test, part b, was $p < .001$. Interpret this result.

12.25   Stock market analysts are continually searching for reliable predictors of stock price. Consider the problem of modeling the price per share, y, of electric utility stocks. Two variables thought to influence stock price are return on average equity, x_1, and annual rate of dividend, x_2. The stock prices, returns on equity, and dividend rates for a sample of 12 nuclear and 16 nonnuclear electric utility stocks are shown in the table. The interaction model

$$E(y) = \beta_0 + \beta_1 x_1 + \beta_2 x_2 + \beta_3 x_1 x_2$$

was fit to the data on each type of stock (nuclear and nonnuclear). The resulting SAS printouts are provided on the next page.

NUCLEAR STOCKS			NONNUCLEAR STOCKS		
y	x_1	x_2	y	x_1	x_2
21	15.1	2.36	25	15.2	2.60
31	15.0	3.00	20	13.9	2.14
26	11.2	3.00	15	15.8	1.52
11	12.1	1.96	34	12.8	3.12
24	16.3	3.00	20	6.9	2.48
8	11.9	1.40	33	14.6	3.08
18	14.9	1.80	28	15.4	2.92
23	11.8	2.56	30	17.3	2.76
13	13.4	2.06	23	13.7	2.36
14	16.2	1.94	24	12.7	2.36
35	17.1	2.96	25	15.3	2.56
13	13.3	2.20	26	15.2	2.80
			26	12.0	2.72
			20	15.3	1.92
			20	13.7	1.92
			13	13.3	1.60

Source: *United Business Investment Report*, Apr. 23, 1984.

Nuclear Stocks

Analysis of Variance

Source	DF	Sum of Squares	Mean Square	F Value	Prob>F
Model	3	640.93485	213.64495	13.217	0.0018
Error	8	129.31515	16.16439		
C Total	11	770.25000			

Root MSE	4.02050	R-square	0.8321	
Dep Mean	19.75000	Adj R-sq	0.7692	
C.V.	20.35695			

Parameter Estimates

Variable	DF	Parameter Estimate	Standard Error	T for H0: Parameter=0	Prob > |T|
INTERCEP	1	-17.556340	40.05327713	-0.438	0.6727
X1	1	0.518988	2.93598713	0.177	0.8641
X2	1	10.889423	15.57141358	0.699	0.5042
X1X2	1	0.132215	1.12491646	0.118	0.9093

Nonnuclear Stocks

Analysis of Variance

Source	DF	Sum of Squares	Mean Square	F Value	Prob>F
Model	3	478.30855	159.43618	60.851	0.0001
Error	12	31.44145	2.62012		
C Total	15	509.75000			

Root MSE	1.61868	R-square	0.9383	
Dep Mean	23.87500	Adj R-sq	0.9229	
C.V.	6.77981			

Parameter Estimates

Variable	DF	Parameter Estimate	Standard Error	T for H0: Parameter=0	Prob > |T|
INTERCEP	1	-44.681773	25.23972659	-1.770	0.1021
X1	1	2.879579	1.74113100	1.654	0.1241
X2	1	25.062181	10.02876655	2.499	0.0280
X1X2	1	-0.959006	0.69103996	-1.388	0.1904

a. Write the least squares prediction equations for the two types of electric utility stock.
b. Is the model useful for predicting the price of nuclear stocks? Nonnuclear stocks? Test each hypothesis using $\alpha = .05$.
c. Is there evidence of interaction between return on equity and dividend rate in the nuclear stock model? The nonnuclear stock model? Perform each test using $\alpha = .05$.
d. The SAS printout with 95% prediction intervals for price per share (y) of nuclear stocks is reproduced on page 638. Locate the lower and upper limits for a 95% prediction interval for y when $x_1 = 13.3$ and $x_2 = 2.20$ (observation #12).
e. Interpret the interval obtained in part d.
f. Would you recommend using the model to predict price per share of a nuclear stock with a dividend rate of 1.10? Explain.

FIGURE 12.16

Scattergram of the Data of
Table 12.3

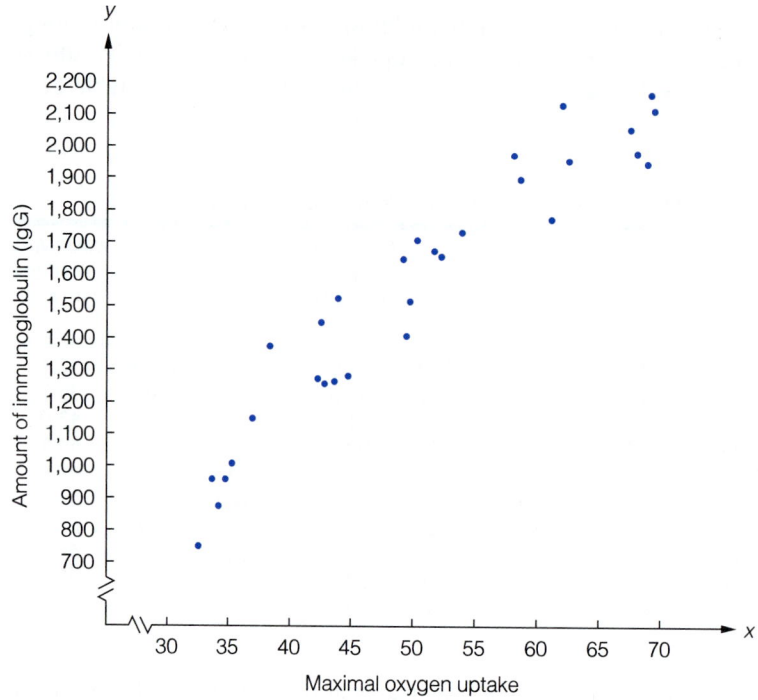

Solution

a. The quadratic model was fit to the data using SPSS. Part of the output of the SPSS multiple regression routine is reproduced in Figure 12.17. The least

FIGURE 12.17

SPSS Printout for the Quadratic
Model, Example 12.19

```
Equation Number 1    Dependent Variable..   Y


Variable(s) Entered on Step Number
    1..    XX
    2..    X



Multiple R            .96834
R Square              .93768
Adjusted R Square     .93306
Standard Error      106.43791

Analysis of Variance
                    DF      Sum of Squares       Mean Square
Regression           2        4602145.21691     2301072.60846
Residual            27         305883.74976       11329.02777

F =     203.11298      Signif F =   .0000
----------------------------------------------------------------

------------------ Variables in the Equation ------------------

Variable             B          SE B        Beta         T    Sig T

XX              -.535378      .158228   -1.625509     -3.384  .0022
X              88.214044    16.477361    2.571954      5.354  .0000
(Constant)  -1462.141458   411.482093               -3.553  .0014
```

squares estimates of the β parameters appear (shaded) in the SPSS column labeled B. You can see that $\hat{\beta}_0 = -1,462.141458$, $\hat{\beta}_1 = 88.214044$, and $\hat{\beta}_2 = -.535378$. Therefore, the equation that minimizes SSE for this data set is

$$\hat{y} = -1,462.14 + 88.21x - .535378x^2$$

b. From Figure 12.18 (page 644), we see that the graph of the quadratic regression model provides a good fit to the data of Table 12.3.

c. According to the box, the β estimates (rounded) have the following interpretations:

$\hat{\beta}_0 = -1,462$ is the estimated y-intercept of the least squares curve shown in Figure 12.18. We also know that $\hat{\beta}_0$ represents the estimate of $E(y)$ when $x = 0$, which is meaningless in this example because a minimal oxygen uptake value of 0 is not practical.

$\hat{\beta}_1 = 88.21$ is an estimate of the amount the curve in Figure 12.18 is shifted along the x-axis. This value rarely has a practical interpretation. Note that $\hat{\beta}_1$ is not a slope and should not be interpreted as such.

$\hat{\beta}_2 = -.535378$ is an estimate of the amount of curvature in the least squares curve in Figure 12.18. The sign of $\hat{\beta}_2$ is negative, which implies downward curvature in the IgG–maximal oxygen uptake relationship. We note here that the small value of $\hat{\beta}_2$ does not imply that the curvature is not significant, since the numerical scale of $\hat{\beta}_2$ depends on the scale of measurement. We will test the contribution of the quadratic coefficient in part e.

d. To determine whether the overall model is useful for predicting y, we test

H_0: $\beta_1 = \beta_2 = 0$
H_a: At least one $\beta_i \neq 0$

The test statistic, shaded on the printout, is $F = 203.11$; its associated p-value (also shaded) is approximately 0. Therefore, there is strong evidence of the utility of the model.

e. As noted earlier, β_2 measures the amount of curvature in the response curve. Thus, to determine whether the curvature exists in the population we test the null hypothesis H_0: $\beta_2 = 0$. Since we want to detect downward curvature, the alternative hypothesis is H_a: $\beta_2 < 0$. The t statistic is given on the SPSS printout (Figure 12.17), under the column labeled **T** in the **XX** row. This value (shaded) is $t = -3.38$. The two-tailed observed significance level (shaded), given under the **Sig T** column, is .0022. Recall that the p-value for a one-tailed test is half this value: $p = \frac{.0022}{2} = .0011$. This implies that we will reject H_0 for any α value larger than .0011; thus, there is strong evidence of downward curvature in the population.

FIGURE 12.20
SAS Graph of $E(y) = 1 + 2x_1 - x_2 + x_1x_2 + x_1^2 + 3x_2^2$ for $x_2 = 0, 1, 2$

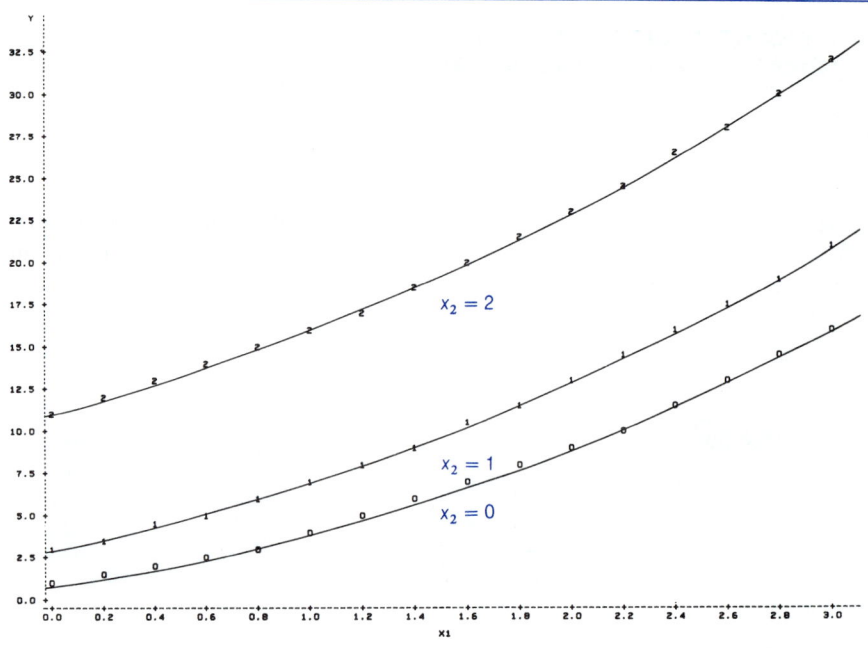

How can you choose an appropriate linear model with one or more quantitative independent variables to fit to a set of data? Since most relationships in the real world are curvilinear (at least to some extent), a good first choice would be a second-order linear model. If you are fairly certain that the relationships between $E(y)$ and the individual independent variables are approximately first-order and that the independent variables do not interact, you could select a first-order model for the data. If you have prior information that suggests there is moderate or very little curvature over the region in which the independent variables are measured, you could use the interaction model. However, keep in mind that for all multiple regression models, the number of data points must exceed the number of parameters in the model. Thus, you may be forced to use a first-order model rather than a second-order model simply because you do not have sufficient data to estimate all of the parameters in the second-order model.

Linear models can also be written to include qualitative independent variables. This topic is the subject of the next section.

EXERCISES

LEARNING THE MECHANICS

12.28   Graph the following quadratic models side by side on the same sheet of graph paper.

    a. $y = 5 + x^2$
    b. $y = -5 + x^2$
    c. What effect does the change in the constant (β_0) have on the graph of a quadratic model?

12.29   Graph the following quadratic (second-order) models side by side on the same sheet of graph paper.

   a. $y = 1 - 2x + x^2$
   b. $y = 1 + 2x + x^2$
   c. $y = x^2$
   d. What effect does the inclusion of the first-order term $(2x)$ have on the graph of the response curve?
   e. What effect does the sign of the first-order term have on the graph of the response curve?

12.30   Graph the following quadratic models side by side on the same sheet of graph paper.

   a. $y = x^2$
   b. $y = 3x^2$
   c. $y = -x^2$
   d. What effect does the coefficient of x^2 have on the graph of a quadratic model?
   e. What effect does the sign of the coefficient of x^2 have on the graph of a quadratic model?

12.31   Suppose you fit the quadratic model

$$E(y) = \beta_0 + \beta_1 x + \beta_2 x^2$$

to a set of $n = 20$ data points and found $R^2 = .91$, SS(Total) $= 29.24$, and SSE $= 2.63$.

   a. Is there sufficient evidence to indicate that the model contributes information for predicting y? Test using $\alpha = .05$.
   b. What null and alternative hypotheses would you test to determine whether upward curvature exists?
   c. What null and alternative hypotheses would you test to determine whether downward curvature exists?

12.32   Write a second-order linear model relating the mean value of y, $E(y)$, to:

   a. Two quantitative independent variables
   b. Three quantitative independent variables [*Hint:* Include all possible two-way cross-product terms and squared terms.]

12.33   Consider the second-order model

$$E(y) = 1 + x_1 - x_2 + x_1 x_2 + 2x_1^2 + x_2^2$$

   a. Graph the relationship between $E(y)$ and x_1 for $x_2 = 0$, 1, and 2.
   b. Are the graphed curves in part **a** first-order or second-order?
   c. How do the graphed curves in part **a** relate to each other?

12.34   Suppose you fit the second-order model

$$y = \beta_0 + \beta_1 x_1 + \beta_2 x_2 + \beta_3 x_1 x_2 + \beta_4 x_1^2 + \beta_5 x_2^2 + \varepsilon$$

to $n = 25$ data points and obtain the following values:

$\hat{\beta}_0 = 1.26$ $s_{\hat{\beta}_1} = 1.21$ SSE $= .41$

$\hat{\beta}_1 = -2.43$ $s_{\hat{\beta}_2} = .16$ $R^2 = .83$

$\hat{\beta}_2 = .05$ $s_{\hat{\beta}_3} = .26$

$\hat{\beta}_3 = .62$ $s_{\hat{\beta}_4} = 1.49$

$\hat{\beta}_4 = -1.81$ $s_{\hat{\beta}_5} = 3.65$

$\hat{\beta}_5 = -2.94$

   a. Is there sufficient evidence to indicate that at least one of the parameters, β_1, β_2, β_3, β_4, and β_5, is nonzero? Test using $\alpha = .05$.
   b. Test H_0: $\beta_4 = 0$ against H_a: $\beta_4 \neq 0$. Use $\alpha = .05$.
   c. Test H_0: $\beta_5 = 0$ against H_a: $\beta_5 \neq 0$. Use $\alpha = .05$.
   d. Use graphs to explain the consequences of the tests in parts b and c.

12.35 In the pharmaceutical industry, a new chemical entity (NCE) is defined as a new chemical or biological compound tested in humans for therapeutic purposes for the first time. A study published in *Managerial & Decision Economics* (Sept. 3, 1988) reported that expenditures on research and development (R&D) of NCEs in the United Kingdom has increased dramatically over the 20 years 1964–1984. A plot of R&D expenditures (y) versus year (x) is shown here.

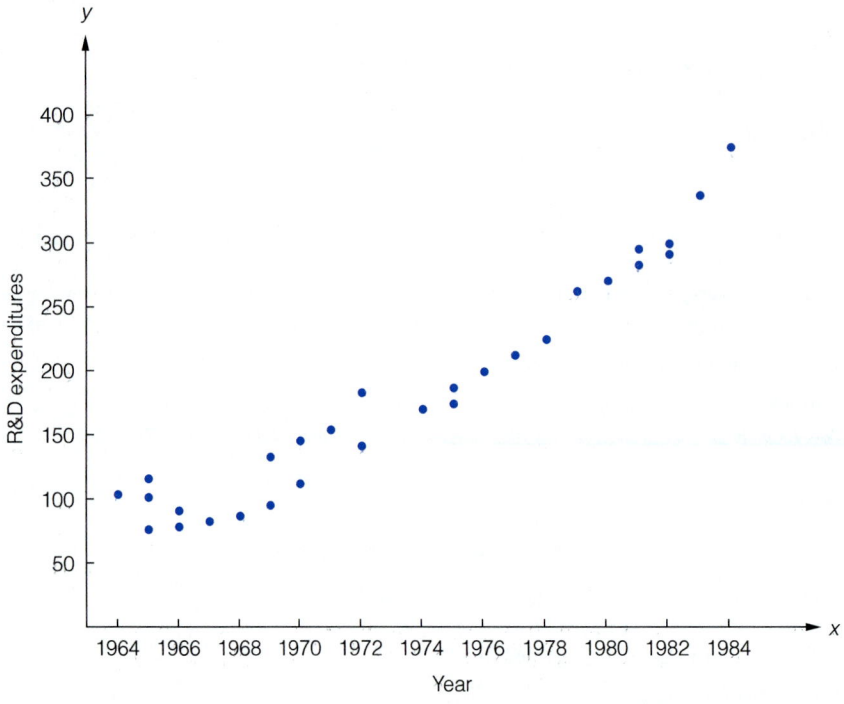

Source: Prentis, R. A., et al. "Pharmaceutical innovation and R&D investment in the UK." *Managerial & Decision Economics*, Sept. 3, 1988, p. 198 (Figure 1).

a. Propose a model for $E(y)$ that would seem to fit the data well.
b. What are the expected signs of the β's in the model, part **a**?

12.36 Newspaper cartoons, although designed to be funny, often invoke hostility, pain, and/or aggression in readers, especially those cartoons that are violent. A study was undertaken to determine how violence in cartoons is related to aggression or pain (*Motivation and Emotion*, Vol. 10, 1986). A group of volunteers (psychology students) rated each of 32 violent newspaper cartoons (16 "Herman" and 16 "Far Side" cartoons) on three dimensions:

y = Funniness   (0 = not funny, . . . , 9 = very funny)

x_1 = Pain   (0 = none, . . . , 9 = a very great deal)

x_2 = Aggression/hostility   (0 = none, . . . , 9 = a very great deal)

The ratings of the students on each dimension were averaged and the resulting n = 32 observations were subjected to a multiple regression analysis. Based on the underlying theory (called the *inverted-U theory*) that the funniness of a joke will increase at low levels of aggression or pain, level off, and then decrease at high levels of aggressiveness or pain, the following quadratic models were proposed:

Model 1: $E(y) = \beta_0 + \beta_1 x_1 + \beta_2 x_1^2$, $R^2 = .099$, $F = 1.60$

Model 2: $E(y) = \beta_0 + \beta_1 x_2 + \beta_2 x_2^2$, $R^2 = .100$, $F = 1.61$

a. According to the theory, what is the expected sign of β_2 in either model?
b. Is there sufficient evidence to indicate that the quadratic model relating pain to funniness rating is useful? Test at $\alpha = .05$.
c. Is there sufficient evidence to indicate that the quadratic model relating aggression/hostility to funniness rating is useful? Test at $\alpha = .05$.

12.37   Research on the relationship between job performance and job turnover has yielded conflicting results. Some early studies found a negative relationship (i.e., the lower the performance, the greater the likelihood of turnover) among all types of workers, whereas others detected a positive relationship (i.e., the higher the performance, the greater the likelihood of turnover) among those employed in white-collar positions. These early studies, however, focused on the linear (first-order) relationship between these variables. The possibility of a curvilinear (second-order) relationship between job performance and turnover was recently investigated both for white-collar workers (accountants) and for blue-collar workers (truck drivers). For each sample of workers the quadratic model $E(y) = \beta_0 + \beta_1 x + \beta_2 x^2$ was fit, where

$x = $ Performance rating ($1 = $ poor, . . . , $4 = $ outstanding)

$y = $ Probability of turnover (i.e., likelihood of worker leaving his or her job within 1 year)

The results are shown in the accompanying table.

ACCOUNTANTS ($n = 169$)	TRUCK DRIVERS ($n = 107$)
$\hat{\beta}_1 = -1.40$ ($t = -3.88$)	$\hat{\beta}_1 = 1.50$ ($t = -3.83$)
$\hat{\beta}_2 = 1.13$ ($t = 3.23$)	$\hat{\beta}_2 = 1.22$ ($t = 4.70$)
$R^2 = .114$	$R^2 = .298$

Source: Jackofsky, E. F., Ferris, K. R., and Breckenridge, B. G. "Evidence for a curvilinear relationship between job performance and turnover." *Journal of Management*, Vol. 12, No. 1, 1986, pp. 105–111.

a. Conduct a test of model adequacy for each of the two groups of workers. Use $\alpha = .05$.
b. Interpret the β estimates for each of the two groups of workers. Which of the $\hat{\beta}$'s have practical interpretations?
c. Is there evidence of upward curvature in the relationship between turnover and performance for accountants? Use $\alpha = .05$. What is the practical implication of this result?
d. Repeat part c for truck drivers.

12.38   A study reported in *Human Factors* (Apr. 1990), investigated the effects of recognizer accuracy and vocabulary size on the performance of a computerized speech recognition device. Accuracy (x_1) of the device, measured as the percentage of correctly recognized spoken utterances, was set at three levels: 90%, 95%, and 99%. Vocabulary size (x_2), measured as the percentage of words needed for the task, was also set at three levels: 75%, 87.5%, and 100%. The dependent variable of primary interest was task completion time (y, in minutes), measured from when a user of the recognition device spoke the first input until the recognizer displayed the last spoken word of the task. Data collected for $n = 162$ trials were used to fit a complete second-order model for task completion time (y), as a function of the quantitative independent variables accuracy (x_1) and vocabulary (x_2). The coefficient of determination for the model was $R^2 = .75$.

a. Write the complete second-order model for $E(y)$.

b. Interpret the value of R^2.

c. Conduct a test of overall model adequacy. Use $\alpha = .05$.

12.39 "Zoning" is defined as the distribution of vacant land to residential and nonresidential uses via policy set by local governments. Although the negative effects of zoning have been studied (e.g., distorting urban property markets, creating barriers to residential mobility, and impeding economic and social integration), little empirical evidence exists identifying the factors that encourage restrictive zoning practices. A recent study, reported in the *Journal of Urban Economics* (Vol. 21, 1987) developed a series of multiple regression models that hypothesize several determinants of zoning. One of the models studied took the following form:

$$E(y) = \beta_0 + \beta_1 x_1 + \beta_2 x_1^2 + \beta_3 x_2$$

where

y = Percentage of vacant land zoned for residential use

x_1 = Proportion of existing land in nonresidential use

x_2 = Proportion of total tax base derived from nonresidential property

The model was fit to data collected for $n = 185$ municipal communities in northeastern New Jersey, with the following results:

INDEPENDENT VARIABLE	PARAMETER ESTIMATE	STANDARD ERROR OF ESTIMATE	t VALUE	p VALUE
Intercept	92.26	3.07	30.05	$p < .01$
x_1	−96.35	46.59	−2.07	$p < .05$
x_1^2	166.80	120.88	1.38	$p > .10$
x_2	−75.51	13.35	−5.66	$p < .01$

Adjusted $R^2 = .25$ $F = 21.86$ $(p < .01)$

Source: Rolleston, B. S. "Determinants of restrictive suburban zoning: An empirical analysis." *Journal of Urban Economics*, Vol. 21, 1987, p. 15 (Table 4).

a. Construct a 95% confidence interval for β_3. Interpret the result.

b. Test the hypothesis that a curvilinear relationship exists between percentage (y) of land zoned for residential use and proportion (x_1) of existing land in nonresidential use.

c. Is the overall model statistically useful for predicting y?

d. Interpret the adjusted R^2 value.

12.40 The *Canadian Geotechnical Journal* (Aug. 1985) reported on a study to investigate the reliability of the use of fragmented Queenston Shale, a compaction shale, as a rockfill construction material. In particular, the researchers

DEVIATORIC STRESS y, kPa	AXIAL STRAIN x, %	DEVIATORIC STRESS y, kPa	AXIAL STRAIN x, %
500	1.0	6,000	13.5
2,000	2.8	6,625	16.7
2,750	4.3	7,000	19.8
3,500	6.0	7,125	23.0
4,375	7.5	7,000	26.0
4,875	9.0	7,125	27.5
5,250	10.5		

Source: Caswell, R. H. and Trak, B. "Some geotechnical characteristics of fragmented Queenston Shale." *Canadian Geotechnical Journal*, Vol. 22, No. 3, Aug. 1985, pp. 403–408.

wanted to estimate the stress–strain relationship of the fragmented material. Based on a graph shown in the paper, the accompanying data were reproduced on deviatoric stress and axial strain for wet shale specimens.

a. Plot the data on a scattergram. What type of relationship appears to exist?
b. The quadratic model $E(y) = \beta_0 + \beta_1 x + \beta_2 x^2$ was fit to the data, with the results shown in the accompanying SAS printout. Test the hypothesis that deviatoric stress y increases with axial strain x at a decreasing rate. Use $\alpha = .05$.
c. Give the observed significance level for the test of part b and interpret its value.
d. Locate the estimate of σ on the printout and interpret its value.

Dependent Variable: Y

Analysis of Variance

Source	DF	Sum of Squares	Mean Square	F Value	Prob>F
Model	2	57287428.971	28643714.485	802.791	0.0001
Error	10	356801.79868	35680.17987		
C Total	12	57644230.769			

Root MSE	188.89198	R-square	0.9938
Dep Mean	4932.69231	Adj R-sq	0.9926
C.V.	3.82939		

Parameter Estimates

Variable	DF	Parameter Estimate	Standard Error	T for H0: Parameter=0	Prob > |T|
INTERCEP	1	248.635810	147.80770576	1.682	0.1235
X	1	619.763109	25.80667772	24.016	0.0001
XX	1	-13.752316	0.87153259	-15.779	0.0001

12.41 In the mid 1800s, the U.S. census inquired about the real property and personal wealth of individual households. Using census information from 1860 and 1870, J. R. Kearl and C. L. Pope examined the mobility of Utah households as measured by their wealth holdings (*The Review of Economics and Statistics*, May 1984). Holding occupation, time of entry into the economy, nativity, sex, place of residence, and internal migration constant, Kearl and Pope fit the quadratic model $E(y) = \beta_0 + \beta_1 x + \beta_2 x^2$, where y is the personal wealth (in dollars) of a Utah houshold and x is the age (in years) of the head of household. The results of the regression are summarized as follows:

$$\hat{y} = 52.39 + 74.21x - .71x^2$$

$$n > 20,000 \qquad t(\text{for } \beta_1) = 13.79 \qquad t(\text{for } \beta_2) = -.15$$

a. Graph the least squares prediction equation.
b. Is there evidence of a quadratic relationship in the wealth–age relationship for Utah households during 1860–1870? Test using $\alpha = .10$.

12.11

Linear models can also be written to include **qualitative** (or **categorical**) **independent variables**. Qualitative variables, unlike quantitative variables, cannot be measured on a numerical scale. Therefore, we need to code the values of the qualitative variable (called **levels**) as numbers before we can fit the model. These coded qualitative variables are called **dummy variables** since the numbers assigned to the various levels are arbitrarily selected.

MODEL BUILDING: QUALITATIVE (DUMMY) VARIABLES

EXAMPLE 12.21 A female executive at a certain company claims that male executives earn higher salaries, on average, than female executives with the same education, experience, and responsibilities. To support her claim, she wants to model the salary y of an executive using a qualitative independent variable representing the sex of an executive (male or female).

a. Write a model for mean executive salary, $E(y)$, using a dummy variable for the sex of an executive.
b. Interpret the β parameters in the model.

Solution

a. A convenient method of coding the values of a qualitative variable at two levels involves assigning a value of 1 to one of the levels and a value of 0 to the other. For example, the dummy variable used to describe gender could be coded as follows:

$$x = \begin{cases} 1 & \text{if male} \\ 0 & \text{if female} \end{cases}$$

The choice of which level is assigned to 1 and which is assigned to 0 is arbitrary. The model then takes the following form:

$$E(y) = \beta_0 + \beta_1 x$$

b. The advantage of using a 0–1 coding scheme is that the β coefficients are easily interpreted. The model in part **a** allows us to compare the mean executive salary $E(y)$ for males with the corresponding mean for females:

Males $(x = 1)$: $E(y) = \beta_0 + \beta_1(1) = \beta_0 + \beta_1$
Females $(x = 0)$: $E(y) = \beta_0 + \beta_1(0) = \beta_0$

First note that β_0 represents the mean salary for females (say, μ_F). When a 0–1 coding convention is used, β_0 will always represent the mean response associated with the level of the qualitative variable assigned the value 0 (called the **base level**). The difference between the mean salary for males and the mean salary for females, $\mu_M - \mu_F$, is represented by β_1—that is,

$$\mu_M - \mu_F = (\beta_0 + \beta_1) - (\beta_0) = \beta_1$$

Therefore, with the 0–1 coding convention, β_1 will always represent the difference between the mean response for the level assigned the value 1 and the mean for the base level. Thus, for the executive salary model we have

$$\beta_0 = \mu_F$$
$$\beta_1 = \mu_M - \mu_F$$

The model relating a mean response $E(y)$ to a qualitative independent variable at two levels is shown in the box.

$$E(y) = \beta_0 + \beta_1 x$$

where $\quad x = \begin{cases} 1 & \text{if level A} \\ 0 & \text{if level B} \end{cases}$

Interpretation of β's: $\quad \beta_0 = \mu_B \quad$ (Mean for base level)

$$\beta_1 = \mu_A - \mu_B$$

For models that involve qualitative independent variables at more than two levels, additional dummy variables must be created. In general, the number of dummy variables used to describe a qualitative variable will be one less than the number of levels of the qualitative variable. The accompanying box presents a model that includes a qualitative independent variable at three levels.

$$E(y) = \beta_0 + \beta_1 x_1 + \beta_2 x_2$$

where

$x_1 = \begin{cases} 1 & \text{if level A} \\ 0 & \text{if not} \end{cases} \qquad x_2 = \begin{cases} 1 & \text{if level B} \\ 0 & \text{if not} \end{cases} \qquad$ Base level = Level C

Interpretation of β's: $\quad \beta_0 = \mu_C \quad$ (Mean for base level)

$$\beta_1 = \mu_A - \mu_C$$
$$\beta_2 = \mu_B - \mu_C$$

EXAMPLE 12.22   Refer to the problem of modeling the shipment cost, y, of a regional express delivery service, described in Example 12.17. Suppose we want to model $E(y)$ as a function of cargo type, where cargo type has three levels— fragile, semifragile, and durable.

a. Write a linear model relating $E(y)$ to cargo type.
b. Interpret the β coefficients in the model.
c. Explain the practical significance of the F test for overall model utility.

Solution

a. Since the qualitative variable of interest, cargo type, has three levels, we need to create $(3 - 1) = 2$ dummy variables. First, select (arbitrarily) one of the levels to be the base level—say, durable cargo. Then each of the remaining levels is assigned the value 1 in one of the two dummy variables as follows:

$$x_1 = \begin{cases} 1 & \text{if fragile} \\ 0 & \text{if not} \end{cases} \qquad x_2 = \begin{cases} 1 & \text{if semifragile} \\ 0 & \text{if not} \end{cases}$$

[Note that for the base level, durable cargo, $x_1 = x_2 = 0$.] Then the appropriate model is

$$E(y) = \beta_0 + \beta_1 x_1 + \beta_2 x_2$$

b. To interpret the β's, first write the mean shipment cost $E(y)$ for each of the three cargo types as a function of the β's:

Fragile $(x_1 = 1, x_2 = 0)$:
$$E(y) = \beta_0 + \beta_1(1) + \beta_2(0) = \beta_0 + \beta_1 = \mu_F$$

Semifragile $(x_1 = 0, x_2 = 1)$:
$$E(y) = \beta_0 + \beta_1(0) + \beta_2(1) = \beta_0 + \beta_2 = \mu_S$$

Durable $(x_1 = 0, x_2 = 0)$:
$$E(y) = \beta_0 + \beta_1(0) + \beta_2(0) = \beta_0 = \mu_D$$

Then we have

$$\beta_0 = \mu_D \quad \text{(Mean of the base level)}$$
$$\beta_1 = \mu_F - \mu_D$$
$$\beta_2 = \mu_S - \mu_D$$

Note that the β's associated with the nonbase levels of cargo type (fragile and semifragile) represent differences between a pair of means. As always, β_0 represents a single mean—the mean response for the base level (durable).

c. The F test for overall model utility tests the null hypothesis

$$H_0: \quad \beta_1 = \beta_2 = 0$$

Note that $\beta_1 = 0$ implies that $\mu_F = \mu_D$ and $\beta_2 = 0$ implies that $\mu_S = \mu_D$. Therefore, $\beta_1 = \beta_2 = 0$ implies that $\mu_F = \mu_S = \mu_D$. Thus, a test for model utility is equivalent to a test for equality of means, i.e.,

$$H_0: \quad \mu_F = \mu_S = \mu_D$$

If there is evidence of a difference between any two of the three mean shipment costs, then cargo type is a useful predictor of shipment cost y.

The linear models described in Sections 12.9–12.11 form the basis for building models with quantitative independent variables and models with qualitative independent variables. More complex models, such as those with interactions between qualitative variables and those with both quantitative and qualitative variables (including interactions), may be required in practice, however.

LEARNING THE MECHANICS

12.42 Write a model relating $E(y)$ to a qualitative independent variable with two levels, A and B. Interpret the β parameters.

12.43 Write a model relating $E(y)$ to a qualitative independent variable with four levels, A, B, C, and D. Interpret the β parameters.

12.44 Consider the model relating $E(y)$ to a qualitative variable with three levels:

$$E(y) = \beta_0 + \beta_1 x_1 + \beta_2 x_2$$

where

$$x_1 = \begin{cases} 1 & \text{if level 1} \\ 0 & \text{if not} \end{cases} \qquad x_2 = \begin{cases} 1 & \text{if level 2} \\ 0 & \text{if not} \end{cases} \qquad \text{Base level = Level 3}$$

The model was fit to n to 100 data points with the following result:

$$\hat{y} = 42.7 + 18.3x_1 - 7.7x_2$$

a. Estimate $E(y)$ when the qualitative variable is set at level 1.
b. Estimate $E(y)$ when the qualitative variable is set at level 2.
c. Estimate $E(y)$ when the qualitative variable is set at level 3.
d. Interpret $\hat{\beta}_0$, $\hat{\beta}_1$, and $\hat{\beta}_2$.
e. How would you test the hypothesis that $E(y)$ is the same for all three levels of the qualitative variable?

12.45 Consider the model relating $E(y)$ to a qualitative variable with five levels:

$$E(y) = \beta_0 + \beta_1 x_1 + \beta_2 x_2 + \beta_3 x_3 + \beta_4 x_4$$

where

$$x_1 = \begin{cases} 1 & \text{if level 1} \\ 0 & \text{if not} \end{cases} \qquad x_2 = \begin{cases} 1 & \text{if level 2} \\ 0 & \text{if not} \end{cases} \qquad x_3 = \begin{cases} 1 & \text{if level 3} \\ 0 & \text{if not} \end{cases}$$

$$x_4 = \begin{cases} 1 & \text{if level 4} \\ 0 & \text{if not} \end{cases} \qquad \text{Base level = Level 5}$$

The model was fit to $n = 20$ data points with the following results:

$$\hat{y} = 20 - 5.6x_1 + 11.2x_2 - 1.7x_3 - 9.0x_4$$

$$\text{SSE} = 662 \quad \text{SS(Total)} = 1{,}043$$

a. Interpret the estimates of the β parameters.
b. Interpret the following hypotheses in terms of the means of the five levels:

$$H_0: \quad \beta_1 = \beta_2 = \beta_3 = \beta_4 = 0$$
$$H_a: \quad \text{At least one } \beta \neq 0$$

c. Conduct the test specified in part **b**. Use $\alpha = .05$.

12.46 Refer to the *Academy of Management* (Mar. 1989) study of the relationship between wives' employment and husbands' well-being, Exercise 10.31. The researchers also used regression to analyze the data. The model $E(y) = \beta_0 + \beta_1 x$ was fit to data collected for $n = 413$ professional accountants, where y = husband's satisfaction (measured on a 5-point scale) and x is a dummy variable for employment status of wife (1 = employed, 0 = unemployed).

a. The estimate of β_1 was negative and statistically significant at $\alpha = .01$. Interpret these results.
b. The value of the coefficient of determination was $R^2 = .02$. Interpret this result.

12.47 The *Sociology of Sport Journal* (Vol. 3, 1986) investigated the problem of racial discrimination in professional baseball. The objective was to determine if the salary of a major league baseball player is influenced by race and, if so, whether black players are paid, on average, less than white players, taking into account levels of performance. Using data collected on the salaries of 212 players (nonpitchers) who started on Opening Day 1977, the following multiple regression model was fit:

$$E(y) = \beta_0 + \beta_1 x_1 + \beta_2 x_2 + \beta_3 x_3 + \beta_4 x_4 + \beta_5 x_5 + \beta_6 x_6 + \beta_7 x_7 + \beta_8 x_8$$

where

y = Natural log of 1977 salary

x_1 = Years of experience as a major league ball player

x_2 = Number of home runs hit during previous season

x_3 = Batting average (i.e., ratio of hits to times-at-bat) during previous season

$x_4 = \begin{cases} 1 & \text{if player's team made the playoffs during the previous year} \\ 0 & \text{if not} \end{cases}$

$x_5 = \begin{cases} 1 & \text{if infielder or catcher} \\ 0 & \text{if outfielder or designated hitter} \end{cases}$

$x_6 = \begin{cases} 1 & \text{if white} \\ 0 & \text{if black} \end{cases}$

x_7 = Amount ($ millions) paid by local radio and TV to broadcast rights to carry games of player's team

$x_8 = \begin{cases} 1 & \text{if bats left-handed or switch hitter} \\ 0 & \text{if not} \end{cases}$

The results are shown in the printout on the next page.

a. Is the model adequate for predicting y? Interpret the p-value of the test.
b. Calculate s, the estimated standard deviation of the random error. Interpret the result.
c. Calculate R^2 and interpret the result.
d. Calculate a 99% confidence interval for β_6. Interpret the result.
e. Is there evidence of salary discrimination against black baseball players? Test using $\alpha = .01$.

12.48 The liquefaction of coal is a major contributor of synthetic fuels. An experiment was conducted to evaluate the performances of a diesel engine run on synthetic (coal-derived) and petroleum-derived fuel oil (*Journal of Energy Resources Technology*, Mar. 1990). The petroleum-derived fuel used was a number 2 diesel fuel (DF-2) obtained from Phillips Chemical Company. Two synthetic fuels were used: a blended fuel (50% coal-derived and 50% DF-2) and a blended fuel with advanced timing. The brake power (kW) and fuel type were varied in test runs, and engine performance was measured. The table on the next page gives the experimental results for the performance measure, mass burning rate per degree of crank angle. Initially, the researchers fit the first-order,

SOURCE	DF	SS	MS	F	p-VALUE
Model	8	14.710	1.839	53.2	.001
Error	203	7.018	.035		
Total	211	21.728			

VARIABLE	PARAMETER ESTIMATE	STANDARD ERROR OF ESTIMATE	t-VALUE	p-VALUE (TWO-TAILED)
x_1	.037	.004	9.25	.001
x_2	.014	.002	7.00	.001
x_3	2.009	.303	6.63	.001
x_4	.108	.037	2.92	.010
x_5	.078	.029	2.69	.010
x_6	-.046	.028	-1.64	.100
x_7	.042	.034	1.24	.250
x_8	-.027	.027	-1.00	.350

Source: Christiano, K. J. "Salary discrimination in Major League baseball: The effect of race." *Sociology of Sport Journal*, Vol. 3, 1986, p. 148.

main effects model

$$E(y) = \beta_0 + \beta_1 x_1 + \beta_2 x_2 + \beta_3 x_3$$

where y = Mass burning rate

x_1 = Brake power

$$x_2 = \begin{cases} 1 & \text{if DF-2 fuel} \\ 0 & \text{if not} \end{cases}$$

$$x_3 = \begin{cases} 1 & \text{if blended fuel} \\ 0 & \text{if not} \end{cases}$$

Interpret the results shown in the Minitab printout on the next page.

BRAKE POWER x_1	FUEL TYPE	MASS BURNING RATE y
4	DF-2	13.2
4	Blended	17.5
4	Advanced Timing	17.5
6	DF-2	26.1
6	Blended	32.7
6	Advanced Timing	43.5
8	DF-2	25.9
8	Blended	46.3
8	Advanced Timing	45.6
10	DF-2	30.7
10	Blended	50.8
10	Advanced Timing	68.9
12	DF-2	32.3
12	Blended	57.1

Source: Litzinger, T. A. and Buzza, T. G. "Performance and emissions of a diesel engine using a coal-derived fuel." *Journal of Energy Resources Technology*, Vol. 112, Mar. 1990, p. 32 (Table 3).

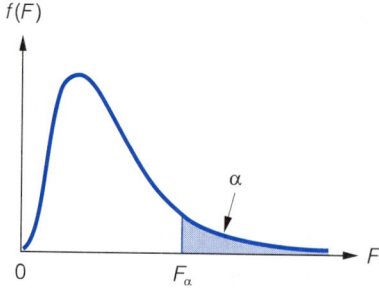

the SAS printout for the second-order (quadratic) model fit to the same $n = 20$ data points. Referring to the printouts, we find the following:

Straight-line interaction model (reduced model):

$$SSE_R = 6.63331 \quad \text{(see Figure 12.13)}$$

Second-order model (complete model):

$$SSE_C = 2.74474 \quad \text{(see Figure 12.22)}$$

FIGURE 12.22

SAS Printout for the Second-Order Model of Example 12.23

Analysis of Variance

Source	DF	Sum of Squares	Mean Square	F Value	Prob>F
Model	5	449.34076	89.86815	453.388	0.0001
Error	14	2.74474	0.19605		
C Total	19	452.08550			

Root MSE	0.44278	R-square	0.9939	
Dep Mean	6.33500	Adj R-sq	0.9918	
C.V.	6.98940			

Parameter Estimates

Variable	DF	Parameter Estimate	Standard Error	T for H0: Parameter=0	Prob > |T|
INTERCEP	1	0.827016	0.70228935	1.178	0.2586
X1	1	-0.609137	0.17990408	-3.386	0.0044
X2	1	0.004021	0.00799842	0.503	0.6230
X1X2	1	0.007327	0.00063743	11.495	0.0001
X1SQ	1	0.089751	0.02020542	4.442	0.0006
X2SQ	1	0.000015070	0.00002243	0.672	0.5127

Test the hypothesis that the quadratic terms do not contribute information for the prediction of y.

Solution The test statistic is

$$F = \frac{(SSE_R - SSE_C)/2}{SSE_C/(20 - 6)} = \frac{(6.63331 - 2.74474)/2}{2.74474/14} = \frac{1.94428}{.19605} = 9.92$$

The critical value of F for $\alpha = .05$, $\nu_1 = 2$, and $\nu_2 = 14$ is found in Table 8 (Appendix G) to be

$$F_{.05} = 3.74$$

Since the calculated $F = 9.92$ exceeds 3.74, we are confident in concluding that the quadratic terms contribute to the prediction of y, shipment cost per package. The curvature terms should be retained in the model.

The F test can be used to determine whether *any* set of terms should be included in a model by testing the null hypothesis that a particular set of β parameters simultaneously equal 0. For example, we may want to test to determine whether a set of interaction terms for quantitative variables or a set of main effect terms for a qualitative variable should be included in a model. The F test appropriate for testing the null hypothesis that all of a set of β parameters are equal to 0 is summarized in the box.

F TEST FOR TESTING THE NULL HYPOTHESIS:
SUBSET OF β PARAMETERS EQUAL ZERO

Reduced model:
$$E(y) = \beta_0 + \beta_1 x_1 + \cdots + \beta_g x_g$$

Complete model:
$$E(y) = \beta_0 + \beta_1 x_1 + \cdots + \beta_g x_g + \beta_{g+1} x_{g+1} + \cdots + \beta_k x_k$$

H_0: $\beta_{g+1} = \beta_{g+2} = \cdots = \beta_k = 0$

H_a: At least one of the β parameters under test is nonzero.

Test statistic:
$$F = \frac{(SSE_R - SSE_C)/(k - g)}{SSE_C/[n - (k + 1)]}$$
$$= \frac{(SSE_R - SSE_C)/(\text{\# } \beta\text{'s tested in } H_0)}{MSE_C}$$

where

SSE_R = Sum of squared errors for the reduced model

SSE_C = Sum of squared errors for the complete model

MSE_C = Mean square error (s^2) for the complete model

$k - g$ = Number of β parameters specified in H_0
(i.e., number of β parameters tested)

$k + 1$ = Number of β parameters in the complete model (including β_0)

n = Total sample size

Rejection region: $F > F_\alpha$

where F is based on $\nu_1 = k - g$ numerator degrees of freedom and
$\nu_2 = n - (k + 1)$ denominator degrees of freedom.

Suppose the F test in Example 12.23 yielded a test statistic that did not fall in the rejection region. Although we must be cautious about accepting H_0, most practitioners of regression analysis adopt the principle of **parsimony**. That is, in

x_3 = Manager–subordinate work relationship rating (1 = unsatisfactory, ..., 5 = excellent)

x_4 = Effort level (average number of hours per week invested in job)

$$x_5 = \begin{cases} 1 & \text{if middle/upper-level manager} \\ 0 & \text{if lower-level manager} \end{cases}$$

x_6 = Subordinate-related managerial behavior score (low scores indicate little or no effort spent on counseling, evaluating, and training subordinates)

The data collected on the 100 managers were used to fit several regression models of managerial performance.

a. Initially, the model

$$E(y) = \beta_0 + \beta_1 x_1 + \beta_2 x_2 + \beta_3 x_3 + \beta_4 x_4$$

was considered to account for the influence of sex, job tenure, manager–subordinate work relationship, and effort level on performance rating. For this model, SSE = 352 and R^2 = .11. Calculate the F statistic for testing model adequacy. Is the model useful for predicting performance rating y? (Use α = .05.)

b. Terms for managerial level and subordinate-related behavior (i.e., $\beta_5 x_5 + \beta_6 x_6$) were added to the model of part a, resulting in SSE = 341 and R^2 = .14. Do these terms contribute additional information for the prediction of performance rating y? (Test using α = .05.)

c. A third model was also considered:

$$E(y) = \beta_0 + \beta_1 x_1 + \beta_2 x_2 + \beta_3 x_3 + \beta_4 x_4 + \beta_5 x_5 + \beta_6 x_6 + \beta_7 x_5 x_6$$

The model resulted in SSE = 321 and R^2 = .19. Test the hypothesis that the interaction between managerial level (x_5) and subordinate-related behavior (x_6) is not important, i.e., test H_0: β_7 = 0. Use α = .05.

d. Interpret the result of part c in terms of the problem.

12.57 Refer to the *Journal of Personal Selling & Sales Management* (Summer 1990) study of gender differences in the industrial sales force, Exercise 11.46. Recall that a sample of 244 male sales managers and a sample of 153 female sales managers participated in the survey. One objective of the research was to assess how supervisory behavior affects intrinsic job satisfaction. Initially, the researchers fit the following reduced model to the data on each gender group:

$$E(y) = \beta_0 + \beta_1 x_1 + \beta_2 x_2 + \beta_3 x_3 + \beta_4 x_4$$

where

y = Intrinsic job satisfaction (measured on a scale of 0 to 40)

x_1 = Age (years)

x_2 = Education level (years)

x_3 = Firm experience (months)

x_4 = Sales experience (months)

To determine the effects of supervisory behavior, four variables (all measured on a scale of 0 to 50) were added to the model: x_5 = contingent reward behavior, x_6 = noncontingent reward behavior, x_7 = contingent punishment behavior, and x_8 = noncontingent punishment behavior. Thus, the complete model is

$$E(y) = \beta_0 + \beta_1 x_1 + \beta_2 x_2 + \beta_3 x_3 + \beta_4 x_4 + \beta_5 x_5 + \beta_6 x_6 + \beta_7 x_7 + \beta_8 x_8$$

a. For each gender, specify the null hypothesis and rejection region (α = .05) for testing whether any of the four supervisory behavior variables affect intrinsic job satisfaction.

b. The R^2 values for the four models (reduced and complete model for both samples) are given in the accompanying table. Interpret the results. For each gender, does it appear that the supervisory behavior variables have an impact on intrinsic job satisfaction? Explain.

MODEL	R^2	
	Males	Females
Reduced	.218	.268
Complete	.408	.496

Source: Schul, P. L., et al. "Assessing gender differences in relationships between supervisory behaviors and job-related outcomes in industrial sales force." *Journal of Personal Selling & Sales Management*, Vol. X, Summer 1990, p. 9 (Table 4).

c. The F statistics for comparing the two models are: $F_{\text{Males}} = 13.00$ and $F_{\text{Females}} = 9.05$. Conduct the tests, part a, and interpret the results.

12.58 Refer to the *Journal of Energy and Resources Technology* study of diesel engines, Exercise 12.48. Recall that the researchers fit the model

$$E(y) = \beta_0 + \beta_1 x_1 + \beta_2 x_2 + \beta_3 x_3$$

where

y = Mass burning rate

x_1 = Brake power

$x_2 = \begin{cases} 1 & \text{if DF-2 fuel} \\ 0 & \text{if not} \end{cases}$

$x_3 = \begin{cases} 1 & \text{if blended fuel} \\ 0 & \text{if not} \end{cases}$

The interaction model

$$E(y) = \beta_0 + \beta_1 x_1 + \beta_2 x_2 + \beta_3 x_3 + \beta_4 x_1 x_2 + \beta_5 x_1 x_3$$

was also fit using Minitab, with the results shown in the following printout. Conduct a test to determine whether brake power and fuel type interact. Test using $\alpha = .01$.

```
The regression equation is
Y = - 10.8 + 7.82 X1 + 19.4 X2 + 12.8 X3 - 5.68 X1X2 - 2.95 X1X3

Predictor       Coef      Stdev     t-ratio       p
Constant     -10.830      8.277      -1.31     0.227
X1             7.815      1.126       6.94     0.000
X2            19.35      10.69        1.81     0.108
X3            12.79      10.69        1.20     0.266
X1X2          -5.675      1.380      -4.11     0.003
X1X3          -2.950      1.380      -2.14     0.065

s = 5.037      R-sq = 94.1%      R-sq(adj) = 90.5%

Analysis of Variance

SOURCE       DF          SS         MS       F         p
Regression    5     3253.98     650.80   25.65   0.000
Error         8      203.01      25.38
Total        13     3456.99

SOURCE       DF      SEQ SS
X1            1     1603.93
X2            1     1086.22
X3            1      117.76
X1X2          1      330.04
X1X3          1      116.03
```

The residual plot, shown in Figure 12.30, indicates that the logarithmic transformation has stabilized the error variances. Note that the cone shape is gone; there is no apparent tendency of the residual variance to increase as mean salary increases. We therefore are confident that inferences using the log model are more reliable than those using the untransformed model.

b. Because we are using the logarithm of salary as the dependent variable, the β estimates have slightly different interpretations than previously discussed. In general, a parameter β in a log-transformed model represents the percentage increase (or decrease) in the dependent variable for a 1-unit increase in the corresponding independent variable. The percentage change is calculated by taking the antilogarithm of the β estimate and subtracting 1, i.e., $e^{\hat{\beta}} - 1$ (proof omitted). For example, the percentage change in auditor's salary associated with a 1-unit (i.e., 1-year) increase in years of experience x is $(e^{\hat{\beta}_1} - 1) = (e^{.05} - 1) = .051$. Thus, when all other independent variables are held constant, we estimate an auditor's salary to increase 5.1% for each additional year of experience.

FIGURE 12.30

Minitab Residual Plot for the Data in Example 12.26

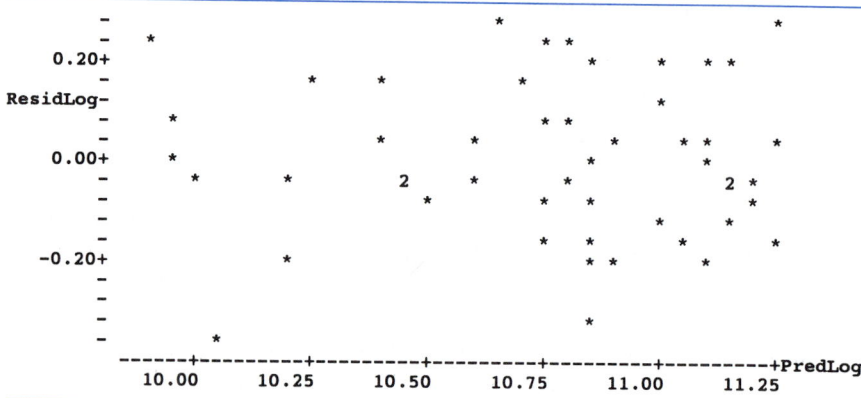

DETECTING NONNORMALITY

Of the four assumptions listed in Section 12.2, the assumption that the random error is normally distributed is the least restrictive when we apply regression analysis in practice. That is, moderate departures from the assumption of normality have very little effect on the validity of the statistical tests, confidence intervals, and prediction intervals. In this case we say that regression is **robust** with respect to nonnormality. However, great departures from normality cast doubt on any inferences derived from the regression analysis.

The simplest way to determine whether the data grossly violate the assumption of normality is to construct either a relative frequency histogram (Section 2.5) or a stem-and-leaf display (Section 2.3) of the residuals, as illustrated in Example 12.27.

EXAMPLE 12.27 Refer to the complete second-order model for shipping cost (y), fit in Example 12.23. The residuals for the model (obtained by computer) are provided in Table 12.8. Construct a stem-and-leaf display for the residuals. Interpret the plot.

Solution A Minitab stem-and-leaf display for the residuals is shown in Figure 12.31. Recall from Section 2.3 that if you turn the stem-and-leaf display on its side, it will look very much like a frequency histogram. You can see from Figure 12.31 that the distribution of the residuals is mound-shaped and reasonably symmetric about 0. Consequently, it is unlikely that the normality assumption would be violated using these data.

TABLE 12.8

Residuals for Complete Second-Order Model Fit to $n = 20$ Data Points, Example 12.23

RESIDUALS			
−.011	.230	.144	−.153
−.196	.577	−.056	.090
.176	−.006	−.033	−.207
−.283	−.486	−.386	−.860
.133	.943	.162	.225

```
Stem-and-leaf of RESID    N  = 20
Leaf Unit = 0.10

  1     -0 8
  1     -0
  2     -0 4
  5     -0 322
 (6)    -0 110000
  9      0 01111
  4      0 22
  2      0 5
  1      0
  1      0 9
```

FIGURE 12.31

Minitab Stem-and-Leaf Display of the Residuals, Table 12.8

When nonnormality of the random error term is detected, it can often be rectified by applying one of the transformations listed in Table 12.6. For example, if the relative frequency distribution (or stem-and-leaf display) of the residuals is highly skewed to the right (as it is for Poisson data), the square-root transformation on y will stabilize (approximately) the variance and, at the same time, will reduce skewness in the distribution of residuals. Nonnormality may also be due to outliers, discussed next.

Residual plots can also be used to detect **outliers**, values of y that appear to be in disagreement with the model. Since almost all values of y should lie within 3σ of $E(y)$, the mean value of y, we would expect most of them to lie within $3s$ of $\hat{y}$. If a residual is larger than $3s$ (in absolute value), we consider it an outlier and seek background information that might explain the reason for its large value.

DETECTING OUTLIERS

> **DEFINITION 12.4**
>
> A residual that is larger than $3s$ (in absolute value) is considered to be an **outlier**.

12.59 Identify the problem(s) in each of the following residual plots:

a. $(y - \hat{y})$

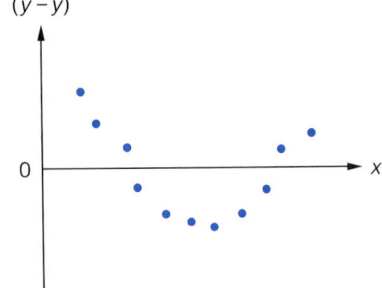

b. $(y - \hat{y})$

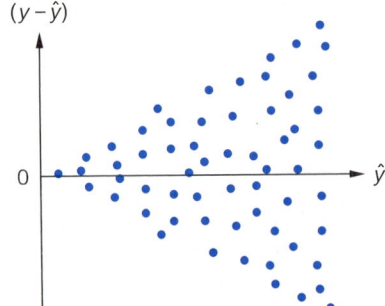

c. $(y - \hat{y})$

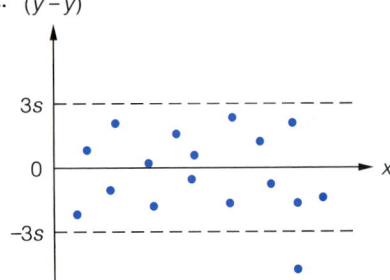

d. $(y - \hat{y})$

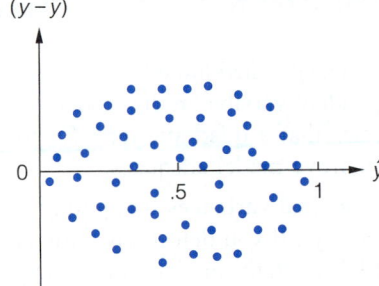

e.

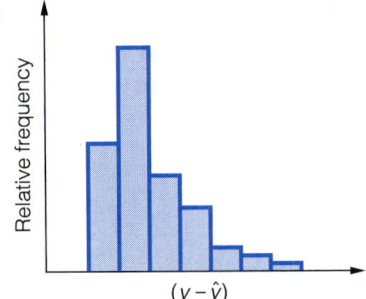

12.60 A first-order model is fit to the data shown in the table, with the following results:

$$\hat{y} = 2.588 + .541x \qquad s = .356$$

x	−2	−2	−1	−1	0	0	1	1	2	2	3	3
y	1.1	1.3	2.0	2.1	2.7	2.8	3.4	3.6	4.0	3.9	3.8	3.6

a. Calculate the residuals for the model.
b. Plot the residuals versus x. Do you detect any trends? If so, what does the pattern suggest about the model?
c. Plot the residuals versus $\hat{y}$. Identify any outliers on the plot.
d. Refer to the residual plot constructed in part c. Do you detect any trends? If so, what does the pattern suggest about the model?

12.61 A first-order model is fit to the data shown in the table, with the following results:

$$\hat{y} = -3.179 + 2.491x \qquad s = 4.154$$

x	2	4	7	10	12	15	18	20	21	25
y	5	10	12	22	25	27	39	50	47	65

a. Calculate the residuals for the model.
b. Plot the residuals versus x. Do you detect any trends? If so, what does the pattern suggest about the model?
c. Plot the residuals versus $\hat{y}$. Identify any outliers on the plot.
d. Refer to the residual plot constructed in part c. Do you detect any trends? If so, what does the pattern suggest about the model?

APPLYING THE CONCEPTS

12.62 Refer to *The New England Journal of Medicine* study of passive exposure to environmental tobacco smoke in children with cystic fibrosis, Exercise 11.45. Recall that the researchers investigated the correlation between a child's weight percentile (y) and the number of cigarettes smoked per day in the child's home (x). The accompanying table lists the data for the 25 boys. A SAS regression printout for the straight-line model relating y to x is shown on the next page.

WEIGHT PERCENTILE y	NUMBER OF CIGARETTES SMOKED PER DAY x	WEIGHT PERCENTILE y	NUMBER OF CIGARETTES SMOKED PER DAY x
6	0	43	0
6	15	49	0
2	40	50	0
8	23	49	22
11	20	46	30
17	7	54	0
24	3	58	0
25	0	62	0
17	25	66	0
25	20	66	23
25	15	83	0
31	23	87	44
35	10		

Source: Rubin, B. K. "Exposure of children with cystic fibrosis to environmental tobacco smoke." *The New England Journal of Medicine*, Sept. 20, 1990, Vol. 323, No. 12, p. 785 (data extracted from Figure 3).

12.65 A certain type of rare gem serves as a status symbol for many of its owners. In theory, as the price of the gem increases, the demand will decrease at low prices, level off at moderate prices, and increase at high prices as a result of the status the owners believe they gain by obtaining the gem. Although a quadratic model would seem to match the theory, the model proposed to explain the demand for the gem by its price is the first-order model

$$y = \beta_0 + \beta_1 x + \varepsilon$$

where y is the demand (in thousands) and x is the retail price per carat (dollars). This model was fit to the 12 data points given in the table. The SPSS printout of the analysis is shown here.

x	100	700	450	150	500	800	70	50	300	350	750	700
y	130	150	60	120	50	200	150	160	50	40	180	130

```
Multiple R            .23064
R Square              .05319
Adjusted R Square    -.04149
Standard Error       56.20658

Analysis of Variance
                    DF      Sum of Squares      Mean Square
Regression           1         1774.86785       1774.86785
Residual            10        31591.79882       3159.17988

F =        .56181      Signif F =   .4708


------------------ Variables in the Equation ------------------

Variable            B          SE B        Beta        T     Sig T

X                .04516      .06025      .23064      .750    .4708
(Constant)     99.81690    29.55565                 3.377   .0070
```

a. Use the least squares prediction equation to calculate the regression residuals.
b. Plot the residuals against retail price per carat, x.
c. Can you detect any trends in the residual plot? What does this imply?

12.66 In Hawaii, condemnation proceedings are under way to enable private citizens to own the property on which their homes are built. Prior to 1980, only estates were permitted to own land, and home owners leased the land from the estate (a law that dates back to the feudal period in Hawaii). To comply with the new law, a large Hawaiian estate wants to use regression analysis to estimate the fair market value of its land. A first proposal is the quadratic model

$$E(y) = \beta_0 + \beta_1 x + \beta_2 x^2$$

where

y = Leased fee value (i.e., sale price of property)

x = Size of property in thousands of square feet

Data collected for 20 property sales in a particular neighborhood, given in the accompanying table, were used to fit the model. The least squares prediction equation is

$$\hat{y} = -44.0947 + 11.5339x - .06378x^2$$

PROPERTY	LEASED FEE VALUE y, thousands of dollars	SIZE x, thousands of sq. ft.	PROPERTY	LEASED FEE VALUE y, thousands of dollars	SIZE x, thousands of sq. ft.
1	70.7	13.5	11	148.0	14.5
2	52.7	9.6	12	85.0	10.2
3	87.6	17.6	13	171.2	18.7
4	43.2	7.9	14	97.5	13.2
5	103.8	11.5	15	158.1	16.3
6	45.1	8.2	16	74.2	12.3
7	86.8	15.2	17	47.0	7.7
8	73.3	12.0	18	54.7	9.9
9	144.3	13.8	19	68.0	11.2
10	61.3	10.0	20	75.2	12.4

a. Calculate the predicted values and corresponding residuals for the model.
b. Plot the residuals versus $\hat{y}$. Do you detect any trends? If so, what does the pattern suggest about the model?
c. Based on your results, how should the estate proceed?

12.67   A recent study investigated the geopolitical and socioeconomic processes that shape the urban size distributions of the world's nations (*Economic Development and Cultural Change*, Oct. 1984). One of the goals of the study was to determine the factors that influence population size in each nation's largest city. Based on data collected for a sample of 126 countries, the following model was fit:

$$E(y) = \beta_0 + \beta_1 x_1 + \beta_2 x_2 + \beta_3 x_3 + \beta_4 x_4 + \beta_5 x_5 + \beta_6 x_6 + \beta_7 x_7 + \beta_8 x_8 + \beta_9 x_9 + \beta_{10} x_{10}$$

where

y = Logarithm of population (in thousands) of largest city in country
x_1 = Log of area (in thousands of square kilometers) of country
x_2 = Log of radius (in hundred kilometers) of city limits
x_3 = Log of national population (in thousands)
x_4 = Percentage annual change in national population (1960–1970)
x_5 = Log of energy consumption per capita (in kilograms of coal equivalent)
x_6 = Percentage of nation's population in urban areas
x_7 = Log of population (in thousands) of second largest city in country
$x_8 = \begin{cases} 1 & \text{if seaport city} \\ 0 & \text{if not} \end{cases}$

$x_9 = \begin{cases} 1 & \text{if capital city} \\ 0 & \text{if not} \end{cases}$

$x_{10} = \begin{cases} 1 & \text{if city data are for metropolitan area} \\ 0 & \text{if not} \end{cases}$

[*Note:*   All logarithms are log(base 10).]

The regression resulted in $R^2 = .879$ and MSE = .036.

a. Conduct a test of model adequacy. (Use $\alpha = .05$.)
b. A SAS stem-and-leaf plot of all the residuals is shown on the next page. Does it appear that the assumption of normal errors is satisfied?

Dependent Variable: PCB85

Analysis of Variance

Source	DF	Sum of Squares	Mean Square	F Value	Prob>F
Model	1	462349.85512	462349.85512	21.772	0.0001
Error	35	743254.18292	21235.83380		
C Total	36	1205604.0380			

Root MSE	145.72520	R-square	0.3835
Dep Mean	107.13351	Adj R-sq	0.3659
C.V.	136.02205		

Parameter Estimates

Variable	DF	Parameter Estimate	Standard Error	T for H0: Parameter=0	Prob > |T|
INTERCEP	1	85.013798	24.42159452	3.481	0.0014
PCB84	1	0.040454	0.00866978	4.666	0.0001

Obs	BAY	PCB84	Dep Var PCB85	Predict Value	Residual
1	Casco	95.28	77.5500	88.8682	-11.3182
2	Merrmack	52.97	29.2300	87.1566	-57.9266
3	Salem	533.58	403.1	106.6	296.5
4	Boston	17104.86	736.0	777.0	-40.9695
5	Buzzards	308.46	192.2	97.5	94.7
6	Narragan	159.96	220.6	91.4848	129.1
7	ELongIsl	10	8.6200	85.4183	-76.7983
8	WLongIsl	234.43	174.3	94.5	79.8126
9	Raritan	443.89	529.3	103.0	426.3
10	Delaware	2.5	130.7	85.1149	45.5551
11	LChesapk	51	39.7400	87.0769	-47.3369
12	Pamilico	0	0	85.0138	-85.0138
13	Charlest	9.1	8.4300	85.3819	-76.9519
14	Sapelo	0	0	85.0138	-85.0138
15	StJohns	140	120.0	90.6773	29.3627
16	Tampa	0	0	85.0138	-85.0138
17	Apalach	12	11.9300	85.4992	-73.5692
18	Mobile	0	0	85.0138	-85.0138
19	RoundIsl	0	0	85.0138	-85.0138
20	MissRiv	34	30.1400	86.3892	-56.2492
21	Baratara	0	0	85.0138	-85.0138
22	SanAnton	0	0	85.0138	-85.0138
23	CorpusCh	0	0	85.0138	-85.0138
24	SDiegoHa	422.1	531.7	102.1	429.6
25	SDiegoBa	6.74	9.3000	85.2865	-75.9865
26	DanaPt	7.06	5.7400	85.2994	-79.5594
27	SealBch	46.71	46.4700	86.9034	-40.4334
28	SanPedro	159.56	176.9	91.4686	85.4314
29	SantaMon	14	13.6900	85.5802	-71.8902
30	Bodega	4.18	4.8900	85.1829	-80.2929
31	Coos	3.19	6.6000	85.1428	-78.5428
32	Columbia	8.77	6.7300	85.3686	-78.6386
33	Nisquall	4.23	4.2800	85.1849	-80.9049
34	Commence	20.6	20.5000	85.8471	-65.3471
35	Elliot	329.97	414.5	98.4	316.1
36	Lutak	5.5	5.8000	85.2363	-79.4363
37	Nahku	6.6	5.0800	85.2808	-80.2008

d. An alternative approach is to use the log transformations $y^* =$ natural $\log(y + 1)$ and $x^* =$ natural $\log(x + 1)$, and fit the model $E(y^*) = \beta_0 + \beta_1 x^*$. A SAS printout for this model follows. Conduct a test for model adequacy and perform a residual analysis. Interpret the results. In particular, comment on the residual value for Boston Harbor.

Dependent Variable: LNPCB85

Analysis of Variance

Source	DF	Sum of Squares	Mean Square	F Value	Prob>F
Model	1	145.58169	145.58169	251.172	0.0001
Error	35	20.28631	0.57961		
C Total	36	165.86800			

Root MSE	0.76132	R-square	0.8777	
Dep Mean	2.94451	Adj R-sq	0.8742	
C.V.	25.85556			

Parameter Estimates

Variable	DF	Parameter Estimate	Standard Error	T for H0: Parameter=0	Prob > |T|
INTERCEP	1	0.425110	0.20232699	2.101	0.0429
LNPCB84	1	0.850826	0.05368523	15.848	0.0001

Obs	BAY	LNPCB84	Dep Var LNPCB85	Predict Value	Residual
1	Casco	4.567261	4.3637	4.3111	0.0527
2	Merrmack	3.988428	3.4088	3.8186	-0.4097
3	Salem	6.281481	6.0017	5.7696	0.2321
4	Boston	9.747176	6.6026	8.7183	-2.1157
5	Buzzards	5.734829	5.2635	5.3045	-0.0410
6	Narragan	5.081156	5.4009	4.7483	0.6526
7	ELongIsl	2.397895	2.2638	2.4653	-0.2015
8	WLongIsl	5.461414	5.1666	5.0718	0.0947
9	Raritan	6.097827	6.2734	5.6133	0.6601
10	Delaware	1.252763	4.8803	1.4910	3.3893
11	LChesapk	3.951244	3.7072	3.7869	-0.0797
12	Pamilico	0	0	0.4251	-0.4251
13	Charlest	2.312535	2.2439	2.3927	-0.1488
14	Sapelo	0	0	0.4251	-0.4251
15	StJohns	4.94876	4.7961	4.6356	0.1605
16	Tampa	0	0	0.4251	-0.4251
17	Apalach	2.564949	2.5596	2.6074	-0.0479
18	Mobile	0	0	0.4251	-0.4251
19	RoundIsl	0	0	0.4251	-0.4251
20	MissRiv	3.555348	3.4385	3.4501	-0.0116
21	Baratara	0	0	0.4251	-0.4251
22	SanAnton	0	0	0.4251	-0.4251
23	CorpusCh	0	0	0.4251	-0.4251
24	SDiegoHa	6.047609	6.2779	5.5706	0.7073
25	SDiegoBa	2.046402	2.3321	2.1662	0.1659
26	DanaPt	2.086914	1.9081	2.2007	-0.2926
27	SealBch	3.865141	3.8601	3.7137	0.1464
28	SanPedro	5.078668	5.1812	4.7462	0.4351
29	SantaMon	2.70805	2.6872	2.7292	-0.0420
30	Bodega	1.644805	1.7733	1.8246	-0.0513
31	Coos	1.432701	2.0281	1.6441	0.3841
32	Columbia	2.279316	2.0451	2.3644	-0.3193
33	Nisquall	1.654411	1.6639	1.8327	-0.1688
34	Commence	3.072693	3.0681	3.0394	0.0286
35	Elliot	5.802028	6.0295	5.3616	0.6679
36	Lutak	1.871802	1.9169	2.0177	-0.1008
37	Nahku	2.028148	1.8050	2.1507	-0.3457

12.69 The manager of a retail appliance store wants to model the proportion of appliance owners who decide to purchase a service contract for a specific major appliance. Since the manager believes that the proportion y decreases as the age x of the appliance (in years) increases, he will fit the first-order model

$$E(y) = \beta_0 + \beta_1 x$$

A sample of 50 purchasers of new appliances are contacted about the possibility of purchasing a service contract. Fifty owners of 1-year-old machines and 50 owners each of 2-, 3-, and 4-year-old machines are also contacted. One year later, another survey is conducted in a similar manner. The proportions y of owners deciding to purchase the service policy are shown in the table.

x	0	0	1	1	2	2	3	3	4	4
y	.94	.96	.70	.76	.60	.40	.24	.30	.12	.10

a. Fit the first-order model to the data.
b. Calculate the residuals and construct a residual plot versus $\hat{y}$.
c. What does the plot constructed in part b suggest about the variance of y?
d. Explain how you could stabilize the variances.

12.70 Refer to the *Communications of the Association for Computing Machinery* study, Exercise 11.16, in which the straight-line model was used to predict instructor grade (y) from AUTOMARK grade (x). Use a graphical technique to check the normality assumption.

CHAPTER CASE: IT NEVER RAINS IN CALIFORNIA

12.14 In the previous sections we learned how multiple regression analysis can be used to model the relationship among two or more variables. Since the problem of describing the interrelationships among variables is one that arises in almost all fields of research, it is not surprising that multiple regression analysis is one of the most frequently used statistical tools. In this section, we focus on the application of multiple regression in the science of geography.

Taylor (1980) sought to describe the method of multiple regression to the research geographer "in a completely nontechnical manner." For the purposes of illustration, he chose to investigate the variation in average annual precipitation in California—"a typical research problem which would be tackled using multiple regression analysis."*

Data on the average annual precipitation (y), altitude (x_1), latitude (x_2), and distance from the Pacific coast (x_3) were obtained for 30 meteorological stations scattered throughout the state. The data set is reproduced in Table 12.9. As a first attempt at explaining the average annual precipitation in California, Taylor proposed the first-order model

$$E(y) = \beta_0 + \beta_1 x_1 + \beta_2 x_2 + \beta_3 x_3 + \varepsilon$$

Taylor's interpretations of the β parameters in the model (for the benefit of the geographer with little or no background in regression analysis) are as follows:

[β_0] is the base constant and is an estimate of the value of the dependent variable when all the independent variables are zero. In our example it is level of precipitation associated with zero altitude, latitude, and distance from the coast. [β_1, β_2, and β_3] are regression coefficients which relate each independent variable to the dependent variable . . . . They tell how much change in the dependent variable is associated with a change of one unit of an independent variable. In our example the regression coefficient (β_1) that relates altitude to precipitation is an estimate of how much precipitation increased in inches for an increase of one foot of altitude.

*Taylor, P. J. "A pedagogic application of multiple regression analysis." *Geography*, July 1980, Vol. 65, pp. 203–212.

The least squares model relating to y to x_1, x_2, and x_3 was found to be

$$\hat{y} = -102.5314 + .004x_1 + 3.4536x_2 - .1426x_3$$

with a multiple coefficient of determination of $R^2 = .5942$. Recall from Section 12.6 that R^2 is a measure of the adequacy of the model. The sample value tells us that approximately 59% of the sample variation in annual precipitation (measured as the sum of squares of deviations of y about $\bar{y}$) can be explained by a station's altitude (x_1), latitude (x_2), and distance (x_3) from the Pacific coast. Does this imply that the model is a useful predictor of annual precipitation (y)? Possibly. But we cannot be sure (at a reasonable level of confidence) that the model is useful until we conduct a statistical test of hypothesis.

For this model, the appropriate test to conduct is

H_0: $\beta_1 = \beta_2 = \beta_3 = 0$ (Model is not useful)

TABLE 12.9
Data for 30 Meteorological Stations in California

STATION	AVERAGE ANNUAL PRECIPITATION y, inches	ALTITUDE x_1, feet	LATITUDE x_2, degrees	DISTANCE FROM COAST x_3, miles
1. Eureka	39.57	43	40.8	1
2. Red Bluff	23.27	341	40.2	97
3. Thermal	18.20	4,152	33.8	70
4. Fort Bragg	37.48	74	39.4	1
5. Soda Springs	49.26	6,752	39.3	150
6. San Francisco	21.82	52	37.8	5
7. Sacramento	18.07	25	38.5	80
8. San Jose	14.17	95	37.4	28
9. Giant Forest	42.63	6,360	36.6	145
10. Salinas	13.85	74	36.7	12
11. Fresno	9.44	331	36.7	114
12. Pt. Piedras	19.33	57	35.7	1
13. Pasa Robles	15.67	740	35.7	31
14. Bakersfield	6.00	489	35.4	75
15. Bishop	5.73	4,108	37.3	198
16. Mineral	47.82	4,850	40.4	142
17. Santa Barbara	17.95	120	34.4	1
18. Susanville	18.20	4,152	40.3	198
19. Tule Lake	10.03	4,036	41.9	140
20. Needles	4.63	913	34.8	192
21. Burbank	14.74	699	34.2	47
22. Los Angeles	15.02	312	34.1	16
23. Long Beach	12.36	50	33.8	12
24. Los Banos	8.26	125	37.8	74
25. Blythe	4.05	268	33.6	155
26. San Diego	9.94	19	32.7	5
27. Daggett	4.25	2,105	34.09	85
28. Death Valley	1.66	-178	36.5	194
29. Crescent City	74.87	35	41.7	1
30. Colusa	15.95	60	39.2	91

against

H_a:　At least one β is nonzero　(Model is useful)

Since we do not have access to a computer printout, we must compute the test statistic using the formula given in Section 12.7:

$$\text{Test statistic:}\quad F = \left(\frac{R^2}{1 - R^2}\right)\left[\frac{n - (k + 1)}{k}\right]$$

Substituting $R^2 = .5942$, $n = 30$ (meterological stations), and $k = 3$ (number of β parameters in the model excluding β_0), we have

$$F = \frac{.5942}{(1 - .5942)}\left[\frac{30 - (3 + 1)}{3}\right]$$

$$= \frac{.5942}{.4058}\left(\frac{26}{3}\right) = 12.69$$

For $\alpha = .05$, we will reject H_0 if $F > F_{.05}$, where $F_{.05}$ is based on $k = 3$ numerator df and $n - (k + 1) = 26$ denominator df. From Table 8 of Appendix G, $F_{.05} = 2.98$. Then the rejection region is

Rejection region:　$F > 2.98$

Since the computed value of F, 12.69, falls in the rejection region, we reject H_0. There is sufficient evidence (at $\alpha = .05$) to indicate that the first-order model is useful for predicting average annual precipitation.

Although the first-order model appears to be adequate, how can we be certain that additional variables or higher-order terms will not improve the prediction equation? One method, as explained in Section 12.13, is to examine the **residuals** for the regression model. Taylor used a graphical examination of the regression residuals for all 30 stations to investigate the possibility of including an additional independent variable. He found that the residuals exhibited a fairly consistent pattern. A plot of the residuals against the predicted values for the model is shown in Figure 12.34. Notice that stations located on the westward-facing slopes of the California mountains (identified by the symbol "W") invariably had positive residuals (i.e., the least squares model underpredicted the level of precipitation), whereas stations on the leeward side of the mountains (identified by the symbol "L") had negative residuals (i.e., the least squares model overpredicted the level of precipitation). In Taylor's words, "This suggests a very clear shadow effect of the mountains, for which California is known. We can add this to the model by incorporating a further variable (x_4), which we will term shadow effect. This will be what statisticians refer to as a 'dummy variable' taking only the values 0 and 1. All stations in the lee of mountains will score 1, other stations score 0." Stations 1, 4, 5, 6, 9, 12, 16, 17, 21, 22, 23, 26, and 29 were assigned a value of $x_4 = 0$; the remaining stations were assigned a value of $x_4 = 1$. The model with shadow effect takes the form

$$E(y) = \beta_0 + \beta_1 x_1 + \beta_2 x_2 + \beta_3 x_3 + \beta_4 x_4$$

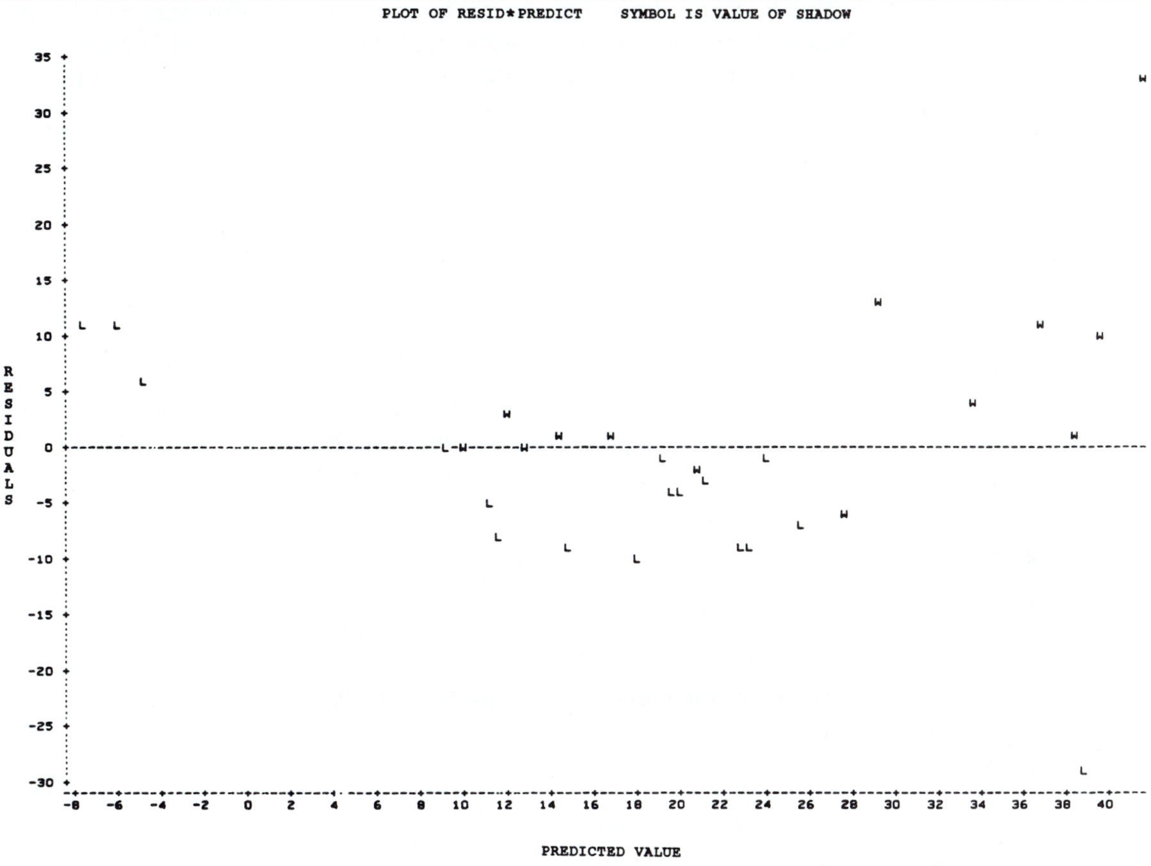

PLOT OF RESID*PREDICT   SYMBOL IS VALUE OF SHADOW

Taylor obtained the least squares equation

$$\hat{y} = -99.1909 + .0021x_1 + 3.4893x_2 - .6518x_3 - 16.1660x_4$$

The multiple coefficient of determination for the shadow-effect model was found to be $R^2 = .7374$. Note that the R^2 value for the shadow-effect model is larger than the R^2 value for the no-shadow-effect model by an amount equal to $.7374 - .5942 = .1432$. This implies that the shadow-effect model explains approximately 14% more of the variation in average annual precipitation than the no-shadow-effect model. However, it is dangerous to conclude that the shadow-effect model is a better predictor based on the value of R^2 alone. When independent variables are added to a model, R^2 will automatically increase. In fact, if enough independent variables are added to the model, you can force R^2 to equal 1. The important question is not whether R^2 has increased, but whether R^2 has increased significantly to warrant the addition of the new independent variables.

FIGURE 12.34
Residual Plot for the First-Order Model

In this case study, we can conduct a statistical test to determine whether the shadow-effect model is a more useful predictor of average annual precipitation. If there is no shadow effect, then β_4 will be 0. Therefore, we should test

$$H_0: \quad \beta_4 = 0$$

against

$$H_a: \quad \beta_4 \neq 0$$

The test statistic is

$$t = \frac{\hat{\beta}_4}{s_{\hat{\beta}_4}}$$

where t is based on $n - (k + 1) = 30 - (4 + 1) = 25$ degrees of freedom. The value of the standard error of the estimate of β_4, although not provided in Taylor's article, is $s_{\hat{\beta}_4} = 4.378$. [This value is obtained from a computer printout for the model.] Thus, the test statistic is calculated as

$$\textit{Test statistic:} \quad t = \frac{\hat{\beta}_4}{s_{\hat{\beta}_4}} = \frac{-16.1660}{4.378}$$

$$= -3.69$$

For $\alpha = .05$, the rejection region for the two-tailed test is

$$|t| > t_{.025}$$

From Table 5 of Appendix G, the critical t value with df $= 25$ is $t_{.025} = 2.060$. Thus, we have the following rejection region:

$$\textit{Rejection region:} \quad |t| > 2.060$$

Since the calculated value of t, -3.69, falls in the rejection region, we have sufficient evidence (at $\alpha = .05$) to conclude that $\beta_4 \neq 0$, i.e., the shadow-effect term contributes to the prediction of average annual precipitation. The model with the shadow-effect term is a more useful predictor of average annual precipitation than the no-shadow-effect model.

MULTICOLLINEARITY

12.15 Often, two or more of the independent variables used in the model for $E(y)$ will contribute redundant information. That is, the independent variables will be correlated with each other. For example, suppose we want to construct a model to predict the gasoline mileage rating, y, of a truck as a function of its load, x_1, and the horsepower, x_2, of its engine. In general, you would expect heavier loads to require greater horsepower and to result in lower mileage ratings. Thus, although both x_1 and x_2 contribute information for the prediction of mileage rating, some of the information is overlapping, because x_1 and x_2 are correlated. When the independent variables are correlated, we say that **multicollinearity** exists. In practice, it is not uncommon to observe correlations among the independent variables. However, a few problems arise when serious multicollinearity is present in the regression analysis.

DEFINITION 12.5

Multicollinearity exists when two or more of the independent variables used in regression are correlated.

First, high correlations among the independent variables increase the likelihood of rounding errors in the calculations of the β estimates, standard errors, and so forth. Second, and more important, the regression results may be confusing and misleading.

To illustrate, if the gasoline mileage rating model

$$E(y) = \beta_0 + \beta_1 x_1 + \beta_2 x_2$$

were fit to a set of data, we might find that the t values for both β_1 and β_2 are nonsignificant. However, the F test for $H_0: \beta_1 = \beta_2 = 0$ would probably be highly significant. The tests may seem to be contradictory, but really they are not. The t tests indicate that the contribution of one variable, say, $x_1 = $ load, is not significant after the effect of $x_2 = $ horsepower has been accounted for (because x_2 is also in the model). The significant F test, on the other hand, tells us that at least one of the two variables is making a contribution to the prediction of y (i.e., β_1, β_2, or both differ from 0). In fact, both are probably contributing, but the contribution of one overlaps with that of the other.

Multicollinearity can also have an effect on the signs of the parameter estimates. More specifically, a value of $\hat{\beta}_i$ may have the opposite sign from what is expected. For example, we expect the signs of both of the parameter estimates for the gasoline mileage rating model to be negative, yet the regression analysis for the model might yield the estimates $\hat{\beta}_1 = .2$ and $\hat{\beta}_2 = -.7$. The positive value of $\hat{\beta}_1$ seems to contradict our expectation that heavy loads will result in lower mileage ratings. This is the danger of interpreting a β coefficient when the independent variables are correlated. Because the variables contribute redundant information, the effect of load x_1 on mileage rating is measured only partially by β_1.

How can you avoid the problems of multicollinearity in regression analysis? One way is to conduct a designed experiment so that the levels of the x variables are uncorrelated.* Unfortunately, time and cost constraints may prevent you from collecting data in this manner. For these and other reasons, most data collected in business studies are observational. That is, the sample is selected, and the values of the independent variables are observed with no attempt to control their values. Since observational data frequently consist of correlated independent variables, you will need to recognize when multicollinearity is present and, if necessary, make modifications in the analysis.

Several methods are available for detecting multicollinearity in regression. A simple technique is to calculate the coefficient of correlation, r, between each pair of independent variables in the model and use the procedure outlined in

---

*Designed experiments are the topic of Chapter 13.

Section 11.7 to test for evidence of positive or negative correlation. If one or more of the r values is statistically different from 0, the variables in question are correlated and a severe multicollinearity problem may exist.* Other indications of the presence of multicollinearity include those mentioned in the beginning of this section—namely, nonsignificant t tests for the individual β parameters when the F test for overall model adequacy is significant, and estimates with opposite signs from what is expected.[†]

DETECTING MULTICOLLINEARITY IN THE REGRESSION MODEL

$$E(y) = \beta_0 + \beta_1 x_1 + \beta_2 x_2 + \cdots + \beta_k x_k$$

The following are indicators of multicollinearity:

1. Significant correlations between pairs of independent variables in the model
2. Nonsignificant t tests for all (or nearly all) of the individual β parameters when the F test for overall model adequacy $H_0 : \beta_1 = \beta_2 = \cdots = \beta_k = 0$ is significant
3. Opposite signs (from what is expected) in the estimated parameters

---

EXAMPLE 12.28   The Federal Trade Commission (FTC) annually ranks varieties of domestic cigarettes according to their tar, nicotine, and carbon monoxide contents. The U.S. surgeon general considers each of these three substances hazardous to a smoker's health. Past studies have shown that increases in the tar and nicotine contents of a cigarette are accompanied by an increase in the carbon monoxide emitted from the cigarette smoke. Table 12.10 presents data on tar, nicotine, and carbon monoxide contents (in milligrams) and weight (in grams) for a sample of 25 (filter) brands tested in a recent year. Suppose we want to model carbon monoxide content, y, as a function of tar content, x_1, nicotine content, x_2, and weight, x_3, using the model

$$E(y) = \beta_0 + \beta_1 x_1 + \beta_2 x_2 + \beta_3 x_3$$

The model is fit to the 25 data points in Table 12.10; a portion of the SAS printout is shown in Figure 12.35. Examine the printout. Do you detect any signs of multicollinearity?

---

*Remember that r measures only the pairwise correlation between x values. Three variables, x_1, x_2, and x_3, may be highly correlated as a group, but may not exhibit large pairwise correlations. Thus, multicollinearity may be present even when all pairwise correlations are not significantly different from 0.

[†]More formal methods for detecting multicollinearity, such as variance-inflation factors, are beyond the scope of this text. Consult the references for a discussion of these methods.

TABLE 12.10
FTC Cigarette Data for Example 12.28

BRAND	TAR x_1, milligrams	NICOTINE x_2, milligrams	WEIGHT x_3, grams	CARBON MONOXIDE y, milligrams
Alpine	14.1	.86	.9853	13.6
Benson & Hedges	16.0	1.06	1.0938	16.6
Bull Durham	29.8	2.03	1.1650	23.5
Camel Lights	8.0	.67	.9280	10.2
Carlton	4.1	.40	.9462	5.4
Chesterfield	15.0	1.04	.8885	15.0
Golden Lights	8.8	.76	1.0267	9.0
Kent	12.4	.95	.9225	12.3
Kool	16.6	1.12	.9372	16.3
L&M	14.9	1.02	.8858	15.4
Lark Lights	13.7	1.01	.9643	13.0
Marlboro	15.1	.90	.9316	14.4
Merit	7.8	.57	.9705	10.0
Multifilter	11.4	.78	1.1240	10.2
Newport Lights	9.0	.74	.8517	9.5
Now	1.0	.13	.7851	1.5
Old Gold	17.0	1.26	.9186	18.5
Pall Mall Light	12.8	1.08	1.0395	12.6
Raleigh	15.8	.96	.9573	17.5
Salem Ultra	4.5	.42	.9106	4.9
Tareyton	14.5	1.01	1.0070	15.9
True	7.3	.61	.9806	8.5
Viceroy Rich Lights	8.6	.69	.9693	10.6
Virginia Slims	15.2	1.02	.9496	13.9
Winston Lights	12.0	.82	1.1184	14.9

Source: Federal Trade Commission.

Dependent Variable: CO

Analysis of Variance

Source	DF	Sum of Squares	Mean Square	F Value	Prob>F
Model	3	495.25781	165.08594	78.984	0.0001
Error	21	43.89259	2.09012		
C Total	24	539.15040			

Root MSE	1.44573	R-square	0.9186	
Dep Mean	12.52800	Adj R-sq	0.9070	
C.V.	11.53996			

Parameter Estimates

Variable	DF	Parameter Estimate	Standard Error	T for H0: Parameter=0	Prob > |T|
INTERCEP	1	3.202190	3.46175473	0.925	0.3655
TAR	1	0.962574	0.24224436	3.974	0.0007
NICOTINE	1	-2.631661	3.90055745	-0.675	0.5072
WEIGHT	1	-0.130482	3.88534182	-0.034	0.9735

FIGURE 12.35
Portion of the SAS Printout

Solution   First, notice that a test of

$$H_0: \quad \beta_1 = \beta_2 = \beta_3 = 0$$

is highly significant. The F value (shaded on the printout) is very large ($F = 78.984$) and the observed significant level of the test (also shaded) is small ($p = .0001$). Therefore, we can reject H_0 for any α greater than .0001 and conclude that at least one of the parameters β_1, β_2, and β_3 is nonzero. The t tests for two of the three individual β's, however, are nonsignificant. (The p-values for these tests are shaded on the printout.) Unless tar is the only one of the three variables useful for predicting carbon monoxide content, these results are the first indication of a potential multicollinearity problem.

The negative values for $\hat{\beta}_2$ and $\hat{\beta}_3$ (shaded on the printout) are a second clue to the presence of multicollinearity.

$$\hat{\beta}_2 = -2.63 \qquad \hat{\beta}_3 = -.130$$

From past studies, the FTC expects carbon monoxide content, y, to increase when either nicotine content, x_2, or weight, x_3, increases—that is, the FTC expects *positive* relationships between y and x_2 and between y and x_3, not negative ones.

All signs indicate that a serious multicollinearity problem exists. To confirm our suspicions, we calculated the coefficient of correlation, r, for each of the three pairs of independent variables in the model. These values are given in Table 12.11. You can see that tar content, x_1, and nicotine content, x_2, appear to be highly correlated ($r = .977$), whereas weight, x_3, appears to be moderately correlated with both tar content ($r = .491$) and nicotine content ($r = .500$). In fact, all three sample correlations exceed the critical t value for a two-tailed test of $H_0 : \rho = 0$ conducted at $\alpha = .05$ with $n - 2 = 23$ df. (See Section 11.7.)

TABLE 12.11

Correlation Coefficients for the Three Pairs of Independent Variables in Example 12.28

PAIR	r
x_1, x_2	.977
x_1, x_3	.491
x_2, x_3	.500

Once you have detected that a multicollinearity problem exists, several alternative measures are available for solving the problem. The appropriate measure to take depends on the severity of the multicollinearity and the ultimate goal of the regression analysis.

Some researchers, when confronted with highly correlated independent variables, choose to include only one of the correlated variables in the final model. If you are interested only in using the model for estimation and prediction (step 6), you may decide not to drop any of the independent variables from the model. In the presence of multicollinearity, we have seen that it is dangerous to interpret the individual β's. However, confidence intervals for $E(y)$ and prediction intervals for y generally remain unaffected *as long as the values of the independent variables used to predict y follow the same pattern of multicollinearity exhibited in the sample data.* That is, you must take strict care to ensure that the values of the x variables fall within the range of the sample data.

When fitting higher-order regression models [e.g., the second-order model $E(y) = \beta_0 + \beta_1 x + \beta_2 x^2$], the independent variables $x_1 = x$ and $x_2 = x^2$ will often be correlated. If the correlation is high, the computer solution may result in extreme rounding errors. For this model, the solution is not to drop one of the independent variables but to transform the x variable in such a way that the correlation between the coded x and x^2 values is substantially reduced. One transformation that works reasonably well is the z transform. That is, replace the variable x with its (approximate) z score

$$z_x \approx \frac{x - \bar{x}}{s}$$

More sophisticated procedures (e.g., ridge regression) for dealing with the multicollinearity problem are available, but they are beyond the scope of this text. Consult the references for details on how to use these procedures.

SOLUTIONS TO SOME PROBLEMS CREATED BY MULTICOLLINEARITY

1. Drop one or more of the correlated independent variables from the final model.

2. If you decide to keep all the independent variables in the model:

   a. Avoid making inferences about the individual β parameters based on the t statistics.

   b. Restrict inferences about $E(y)$ and future y values to values of the independent variables that fall within the range of the sample data.

3. To reduce rounding errors in higher-order regression models, code the independent variables so that first-, second-, and higher-order terms for a particular x variable are not highly correlated.

EXERCISES
APPLYING THE CONCEPTS

12.71 A bioengineer wants to model the amount (y) of carbohydrate solubilized during steam processing of peat as a function of temperature (x_1), exposure time (x_2), and pH value (x_3). Data collected for each of 15 peat samples were used to fit the model

$$E(y) = \beta_0 + \beta_1 x_1 + \beta_2 x_2 + \beta_3 x_3$$

A summary of the regression results follows:

$$\hat{y} = -3,000 + 3.2x_1 - .4x_2 - 1.1x_3 \qquad R^2 = .93$$

$$s_{\hat{\beta}_1} = 2.4 \qquad s_{\hat{\beta}_2} = .6 \qquad s_{\hat{\beta}_3} = .8$$

$$r_{12} = .92 \qquad r_{13} = .87 \qquad r_{23} = .81$$

Based on these results, the bioengineer concludes that none of the three independent variables, x_1, x_2, and x_3, is a useful predictor of carbohydrate amount, y. Do you agree with this statement? Explain.

12.72 Refer to the FTC cigarette data of Example 12.28. The data are reproduced here for convenience.

BRAND	TAR x_1, milligrams	NICOTINE x_2, milligrams	WEIGHT x_3, grams	CARBON MONOXIDE y, milligrams
Alpine	14.1	.86	.9853	13.6
Benson & Hedges	16.0	1.06	1.0938	16.6
Bull Durham	29.8	2.03	1.1650	23.5
Camel Lights	8.0	.67	.9280	10.2
Carlton	4.1	.40	.9462	5.4
Chesterfield	15.0	1.04	.8885	15.0
Golden Lights	8.8	.76	1.0267	9.0
Kent	12.4	.95	.9225	12.3
Kool	16.6	1.12	.9372	16.3
L&M	14.9	1.02	.8858	15.4
Lark Lights	13.7	1.01	.9643	13.0
Marlboro	15.1	.90	.9316	14.4
Merit	7.8	.57	.9705	10.0
Multifilter	11.4	.78	1.1240	10.2
Newport Lights	9.0	.74	.8517	9.5
Now	1.0	.13	.7851	1.5
Old Gold	17.0	1.26	.9186	18.5
Pall Mall Light	12.8	1.08	1.0395	12.6
Raleigh	15.8	.96	.9573	17.5
Salem Ultra	4.5	.42	.9106	4.9
Tareyton	14.5	1.01	1.0070	15.9
True	7.3	.61	.9806	8.5
Viceroy Rich Lights	8.6	.69	.9693	10.6
Virginia Slims	15.2	1.02	.9496	13.9
Winston Lights	12.0	.82	1.1184	14.9

Source: Federal Trade Commission.

a. Fit the model $E(y) = \beta_0 + \beta_1 x_1$ to the data. Is there evidence that tar content (x_1) is useful for predicting carbon monoxide content (y)?

b. Fit the model $E(y) = \beta_0 + \beta_2 x_2$ to the data. Is there evidence that nicotine content (x_2) is useful for predicting carbon monoxide content (y)?

c. Fit the model $E(y) = \beta_0 + \beta_3 x_3$ to the data. Is there evidence that weight (x_3) is useful for predicting carbon monoxide content (y)?

d. Compare the signs of $\hat{\beta}_1$, $\hat{\beta}_2$, and $\hat{\beta}_3$ in the models of parts a, b, and c, respectively, to the signs of the $\hat{\beta}$'s in the multiple regression model fit in Example 12.28. The fact that the $\hat{\beta}$'s change dramatically when the independent variables are removed from the model is another indication of a serious multicollinearity problem.

12.73 Hamilton (1987) illustrated the multicollinearity problem with an example using the data shown in the accompanying table. The values of x_1, x_2, and y in the table represent appraised land value, appraised improvements value, and sale price, respectively, of a randomly selected residential property. (All measurements are in thousands of dollars.)

a. Calculate the coefficient of correlation between y and x_1. Is there evidence of a linear relationship between sale price and appraised land value?

b. Calculate the coefficient of correlation between y and x_2. Is there evidence of a linear relationship between the sale price and appraised improvements?

c. Based on the results in parts **a** and **b**, do you think the model $E(y) = \beta_0 + \beta_1 x_1 + \beta_2 x_2$ will be useful for predicting sale price?

x_1	x_2	y	x_1	x_2	y
22.3	96.6	123.7	30.4	77.1	128.6
25.7	89.4	126.6	32.6	51.1	108.4
38.7	44.0	120.0	33.9	50.5	112.0
31.0	66.4	119.3	23.5	85.1	115.6
33.9	49.1	110.6	27.6	65.9	108.3
28.3	85.2	130.3	39.0	49.0	126.3
30.2	80.4	131.3	31.6	69.6	124.6
21.4	90.5	114.4			

Source: Hamilton, D. "Sometimes $R^2 > r_{yx_1}^2 + r_{yx_2}^2$: Correlated variables are not always redundant." *The American Statistician*, Vol. 41, No. 2, May 1987, pp. 129–132.

d. Use a statistical computer software package to fit the model in part c, and conduct a test of model adequacy. In particular, note the value of R^2. Does the result agree with your answer to part c?

e. Calculate the coefficient of correlation between x_1 and x_2. What does the result imply?

f. Many researchers avoid the problems of multicollinearity by always omitting all but one of the "redundant" variables from the model. Would you recommend this strategy for this example? Explain. (Hamilton notes that in this case, such a strategy "can amount to throwing out the baby with the bathwater.")

SUMMARY

In this chapter we have discussed some of the methodology of *multiple regression analysis*. Throughout, we presented several different types of *general linear models* of the form

$$y = \beta_0 + \beta_1 x_1 + \beta_2 x_2 + \cdots + \beta_k x_k + \varepsilon$$

The steps employed in a multiple regression analysis are much the same as those employed in a simple linear regression analysis (Chapter 11):

1. The form of the probabilistic model is hypothesized.
2. The appropriate model assumptions are made.
3. The model coefficients are estimated using the method of least squares.
4. The utility of the model is checked using the overall F test, t tests on individual β parameters, the F test for a partial set of β's, R^2, and adjusted R^2.
5. A residual analysis is conducted to determine whether the data comply with the assumptions of step 2. Modifications to the model are made, if necessary.
6. If the model is deemed useful and the assumptions are satisfied, it may be used to make estimates and to predict values of y to be observed in the future.

We stress that this is not intended to be a complete coverage of multiple regression analysis. Whole texts have been devoted to this topic. However, we have presented the core necessary for a basic understanding of multiple regression and general linear models. If you are interested in a more extensive coverage, consult the references at the end of this chapter.

Adjusted R^2
Dummy variable
F test for determining whether the overall model is useful for predicting y
F test for testing portions of a model
First-order model
General linear model
Interaction
Interaction model
Multicollinearity
Multiple coefficient of determination: R^2

Multiple regression analysis
Outlier
Parsimonious model
Quadratic model
Qualitative independent variable
Quantitative independent variable
Residual
Residual analysis
Second-order model
t test for testing an individual β parameter of a model
Variance-stabilizing transformation

KEY FORMULAS

MULTIPLE COEFFICIENT OF DETERMINATION

$$R^2 = 1 - \frac{\text{SSE}}{\text{SS}_{yy}} = \frac{\text{SS(Model)}}{\text{SS(Total)}} \qquad R_a^2 = 1 - \left[\frac{(n-1)}{n-(k+1)} \right](1 - R^2)$$

ESTIMATED VARIANCE OF RANDOM ERROR TERM

$$s^2 = \frac{\text{SSE}}{n-(k+1)} = \text{MSE}$$

TEST OF OVERALL MODEL UTILITY

H_0: $\beta_1 = \beta_2 = \cdots = \beta_k = 0$

Test statistic: $F = \dfrac{\text{MS(Model)}}{\text{MSE}} = \dfrac{\text{SS(Model)}}{\text{SSE}}\left[\dfrac{n-(k+1)}{k}\right)\right] = \left[\dfrac{R^2}{1-R^2}\right]\left[\dfrac{n-(k+1)}{k}\right)\right]$

where F depends on $v_1 = k$ and $v_2 = n - (k+1)$df

TESTING PORTIONS OF A MODEL

H_0: $\beta_{g+1} = \beta_{g+2} = \cdots = \beta_k = 0$

Test statistic: $F = \dfrac{(\text{SSE}_R - \text{SSE}_C)/(\# \beta\text{'s tested in } H_0)}{\text{MSE}_C}$

where SSE_R = SSE for reduced model (with g parameters)

SSE_C = SSE for complete model (with k parameters)

MSE_C = MSE for complete model

F depends on $v_1 = (k - g)$ and $v_2 = [n - (k + 1)]$df

REGRESSION RESIDUAL

$(y - \hat{y})$

12.74 Refer to the *Chemosphere* (1990) study of 103 production workers exposed to chemicals contaminated with the dioxin 2,3,7,8-TCDD, Exercise 11.35. Multiple regression analysis was used to relate the TCDD level (y) of a worker (measured in parts per trillion) to four independent variables in the model:

$$E(y) = \beta_0 + \beta_1 x_1 + \beta_2 x_2 + \beta_3 x_3 + \beta_4 x_4$$

where

  x_1 = Logarithm of years of exposure to TCDD

  x_2 = Number of years since last exposure

  x_3 = Age (in years)

  x_4 = Body mass index

The results of the analysis are summarized in the accompanying table. Interpret these results.

INDEPENDENT VARIABLE	PARAMETER ESTIMATE	STANDARD ERROR OF ESTIMATE	t VALUE	p VALUE
Intercept	1.721	.770	2.24	—
x_1	.566	.054	10.48	< .001
x_2	−.085	.018	−4.72	< .001
x_3	.044	.010	4.40	< .001
x_4	.075	.021	3.57	< .001
$R^2 = .742$				

Source: Sweeney, M. H., et al. "Comparison of serum levels of 2,3,7,8-TCDD in TCP production workers and in an unexposed comparison group." *Chemosphere*, Vol. 20, Nos. 7–9, 1990, p. 997 (Table 2).

12.75 An extensive study was undertaken at Utah State University to find the strongest predictor of a student-athlete's academic performance (*Sociology of Sport Journal*, Vol. 3, 1986). Based on data collected for a sample of $n = 519$ student-athletes, the following regression model was fit:

$$y = \beta_0 + \beta_1 x_1 + \beta_2 x_2 + \beta_3 x_3 + \beta_4 x_4 + \beta_5 x_5$$

where

  y = University grade point average

  x_1 = Ratio of number of years the athlete lettered to number of years participated in sport

  x_2 = Standardized score on college entrance exam (SAT or ACT)

  x_3 = High school grade point average (GPA)

  $x_4 = \begin{cases} 1 & \text{if female} \\ 0 & \text{if male} \end{cases}$

  $x_5 = \begin{cases} 1 & \text{if minority} \\ 0 & \text{if white} \end{cases}$

The β estimates, standard errors, and corresponding t ratios are given in the table.

VARIABLE		$\hat{\beta}$	$s_{\hat{\beta}}$	t
Letter ratio	x_1	.152	.05	3.04
College entrance score	x_2	1.653	.22	7.51
High school GPA	x_3	.331	.06	5.52
Gender	x_4	.120	.05	2.40
Race	x_5	−.249	.06	−4.15

a. Interpret the β estimates in the model.
b. Construct a 95% confidence interval for β_3. Interpret the interval.
c. Test the hypothesis that letter ratio (x_1) is a useful predictor of university GPA (y). Use $\alpha = .05$.
d. The coefficient of determination for the model is $R^2 = .279$. Interpret this value.
e. Test the hypothesis that the overall model is useful for predicting university GPA (y). Use $\alpha = .05$.

12.76 Researchers at the Upjohn Company used multiple regression analysis in the development of a sustained release tablet.* One of the objectives of the research was to develop a model relating the dissolution y of a tablet (i.e., the percentage of the tablet dissolved over a specified period of time) to the following independent variables:

x_1 = Excipient level (i.e., amount of nondrug ingredient in the tablet)

x_2 = Process variable (e.g., machine setting under which tablet is processed)

a. Write the complete second-order model for $E(y)$.
b. Write a model that hypothesizes straight-line relationships between $E(y)$, x_1, and x_2. Assume that x_1 and x_2 do not interact.
c. Repeat part b, but add interaction to the model.
d. For the model in part c, what is the slope of the $E(y) - x_1$ line for fixed x_2?
e. For the model in part c, what is the slope of the $E(y) - x_2$ line for fixed x_1?

12.77 R. H. Brockhaus (St. Louis University) conducted a study to determine whether entrepreneurs, newly hired (transferred) managers, and newly promoted managers differ in their risk-taking propensities (*Academy of Management Journal*, Sept. 1980). For the purposes of this study, entrepreneurs were defined as individuals who, within 3 months before the study, had ceased working for their employers to own and manage business ventures. Thirty-one individuals from each of the three groups were randomly selected to participate in the study. Each was administered a questionnaire that required the respondent to choose between a safe alternative and a more attractive but risky one. Test scores were designed to measure risk-taking propensity. (Lower scores are associated with greater conservatism in risk-taking situations.) The test scores for the three groups are summarized in the table.

GROUP	SAMPLE SIZE	SAMPLE MEAN
Entrepreneurs	31	71.00
Transferred managers	31	72.52
Promoted managers	31	66.97
TOTAL	93	

*Source: Klassen, R. A. "The application of response surface methods to a tablet formulation problem." Paper presented at Joint Statistical Meetings, American Statistical Association, and Biometric Society, Aug. 1986, Chicago, Ill.

Suppose you were to fit the following model to the $n = 93$ data points:

$$E(y) = \beta_0 + \beta_1 x_1 + \beta_2 x_2$$

where

y = Test score

$x_1 = \begin{cases} 1 & \text{if entrepreneur} \\ 0 & \text{if not} \end{cases}$ $x^2 = \begin{cases} 1 & \text{if transferred manager} \\ 0 & \text{if not} \end{cases}$

Base level = Promoted manager

a. Use the information in the table to find the least squares prediction equation.
b. How would you test the hypothesis that there are no differences among the mean risk-taking propensities of the three groups of managers?

12.78   One of the provisions of the Rehabilitation Act of 1973 is that programs offered by colleges and universities that receive federal funds must be readily available to severely disabled students. Researchers conducted a study to determine the variables related to disabled college students' achievement in terms of grade point average (GPA) [*Journal of Rehabilitation*, Apr./May/June 1985]. Using data collected on 60 disabled college students (30 severely and 30 nonseverely disabled) from two southern universities, the researchers fit the multiple regression model

$$y = \beta_0 + \beta_1 x_1 + \beta_2 x_2 + \beta_3 x_3 + \beta_4 x_4 + \beta_5 x_2 x_3 + \beta_6 x_3 x_4 + \varepsilon$$

where

y = College GPA

x_1 = High school GPA

$x_2 = \begin{cases} 1 & \text{if severely disabled} \\ 0 & \text{if nonseverely disabled} \end{cases}$

x_3 = Internal–External Scale score (a measure of the degree to which the disabled student believes that life events are controlled internally or externally)

x_4 = Counseling Relationship Inventory Scale score (a measure of the overall quality of the counseling relationship as perceived by the disabled student)

a. The coefficient of determination for the multiple regression analysis is $R^2 = .579$. Interpret this value.
b. Conduct a test of model adequacy. Use $\alpha = .05$.
c. The t statistic for the high school GPA parameter, β_1, is $t = 3.19$. Is this sufficient evidence to indicate that college GPA is related to high school GPA for disabled students, holding all other variables constant? Test using $\alpha = .05$.

12.79   Refer to Exercise 11.76 and the consumer attitude (telephone) survey conducted by the University of Florida's Bureau of Economic and Business Research (BEBR). Suppose we want to model the refusal rate, y (i.e., the percentage of dialed households in a county that refuse to take part in the survey), as a function of the county's personal income per capita, x_1, and percentage of residents with a college education, x_2.

a. Write a first-order linear model for refusal rate y.
b. Write an interaction model for refusal rate y.
c. Write a second-order model for refusal rate y.
d. The second-order model of part **c** was fit to data on 12 Florida counties (data supplied by BEBR). The resulting SAS printout is reproduced on page 702. Find R^2 on the printout and interpret its value.
e. Is there evidence that the model is useful for predicting refusal rate y? Test using $\alpha = .05$.
f. The SAS printout also gives a 95% prediction interval for the refusal rate for each of the 12 counties. County 8 had a per capita income of $x_1 = \$10,000$ and $x_2 = 29.74\%$ of residents with a college education. Interpret the prediction interval for this county. How do you explain the large width of the interval?

ANALYSIS OF VARIANCE

SOURCE	DF	SUM OF SQUARES	MEAN SQUARE	F VALUE	PROB>F
MODEL	5	0.06174787	0.01234957	1.803	0.2465
ERROR	6	0.04109780	0.006849633		
C TOTAL	11	0.10284567			

ROOT MSE	0.08276251	R-SQUARE	0.6004	
DEP MEAN	0.3961667	ADJ R-SQ	0.2674	
C.V.	20.89083			

PARAMETER ESTIMATES

VARIABLE	DF	PARAMETER ESTIMATE	STANDARD ERROR	T FOR H0: PARAMETER=0	PROB > |T|
INTERCEP	1	1.98524969	0.83441322	2.379	0.0548
X1	1	-0.000431242	0.000255467	-1.688	0.1424
X2	1	0.03724751	0.08995885	0.414	0.6932
X1X2	1	-0.000002983	0.0000069142	-0.431	0.6812
X1SQ	1	2.72363E-08	2.20347E-08	1.236	0.2626
X2SQ	1	-0.000294703	0.000580117	-0.508	0.6296

OBS	ID	ACTUAL	PREDICT VALUE	STD ERR PREDICT	LOWER95% PREDICT	UPPER95% PREDICT	RESIDUAL
1	7737	0.2960	0.2630	0.0814	-0.0211	0.5470	0.0330
2	12330	0.4980	0.5042	0.0748	0.2313	0.7771	-.006186
3	12058	0.3860	0.4166	0.0712	0.1495	0.6838	-0.0306
4	9927	0.3270	0.3410	0.0400	0.1161	0.5659	-0.0140
5	6904	0.5000	0.5234	0.0690	0.2598	0.7871	-0.0234
6	9463	0.3330	0.4112	0.0757	0.1366	0.6857	-0.0782
7	11466	0.4290	0.4034	0.0368	0.1818	0.6249	0.0256
8	10000	0.4220	0.3564	0.0361	0.1354	0.5774	0.0656
9	10052	0.4410	0.3700	0.0378	0.1474	0.5927	0.0710
10	8636	0.1910	0.3234	0.0498	0.0871	0.5597	-0.1324
11	7445	0.5260	0.4634	0.0598	0.2135	0.7133	0.0626
12	9059	0.4050	0.3780	0.0397	0.1534	0.6026	0.0270

SUM OF RESIDUALS   -1.94289E-16
SUM OF SQUARED RESIDUALS   0.0410978
PREDICTED RESID SS (PRESS) 1.363713

12.80 Refer to the *Sociology of Sport Journal* investigation of racial discrimination in major league baseball, described in Exercise 12.47. Recall that the race variable (x_6) was not statistically significant (at $\alpha = .01$) in the model for predicting natural logarithm of a player's salary. However, when the researcher fit the model to separate subsamples of white and black players by position, the β estimates for the home run variable (x_2) were dramatically different, as shown here (standard errors of β estimates in parentheses):

Black infielder/catcher: $\hat{y} = 4.048 + .036x_1 + .007x_2 + 2.120x_3 + .143x_4$ $(n = 35)$
                                        (.009)   (.005)   (.601)   (.083)

White infielder/catcher: $\hat{y} = 3.860 + .043x_1 + .016x_2 + 2.436x_3 + .166x_4$ $(n = 91)$
                                        (.007)   (.003)   (.611)   (.061)

a. Interpret the value of $\hat{\beta}_2$ for black infielders/catchers.
b. Interpret the value of $\hat{\beta}_2$ for white infielders/catchers.
c. Test the null hypothesis $H_0: \beta_2 = 0$ against $H_a: \beta_2 > 0$ for each of the two groups of players. Use $\alpha = .05$.
d. The researcher concluded that "white infielders are apparently paid more for each home run they hit than their black counterparts." Do you agree?
e. Can you think of a reason why the results of this analysis differ from the results of the regression model in Exercise 12.47, with respect to the race variable? [*Hint:* Consider the fact that the model in Exercise 12.47 did not include interactions between race and the other independent variables in the model.]

12.81 A naval base is considering modifying or adding to its fleet of 48 standard aircraft. The final decision regarding the type and number of aircraft to be added depends on a comparison of cost versus effectiveness of the modified fleet. Consequently, the naval base would like to model the projected percentage increase y in fleet effectiveness by the end of the decade as a function of the cost x of modifying the fleet. A first proposal is the quadratic model

$$E(y) = \beta_0 + \beta_1 x + \beta_2 x^2$$

The data provided in the table were collected on 10 naval bases of similar size that recently expanded their fleets. The data were used to fit the model; the SAS printout of the multiple regression analysis is also reproduced here.

PERCENTAGE IMPROVEMENT AT END OF DECADE y	COST OF MODIFYING FLEET x, millions of dollars
18	125
32	160
9	80
37	162
6	110
3	90
30	140
10	85
25	150
2	50

Dependent Variable: Y

Analysis of Variance

Source	DF	Sum of Squares	Mean Square	F Value	Prob>F
Model	2	1368.77501	684.38750	33.079	0.0003
Error	7	144.82499	20.68928		
C Total	9	1513.60000			

Root MSE	4.54855	R-square	0.9043	
Dep Mean	17.20000	Adj R-sq	0.8770	
C.V.	26.44504			

Parameter Estimates

Variable	DF	Parameter Estimate	Standard Error	T for H0: Parameter=0	Prob > |T|
INTERCEP	1	10.659036	14.55009061	0.733	0.4876
X	1	-0.281606	0.28087588	-1.003	0.3494
XX	1	0.002672	0.00125383	2.131	0.0706

a. Interpret the value of R^2 on the printout.
b. Find the value of s and interpret it.
c. Perform a test of overall model adequacy. Use $\alpha = .05$.
d. Is there sufficient evidence to conclude that the percentage improvement y increases more quickly for more costly fleet modifications than for less costly fleet modifications? Test with $\alpha = .05$.
e. Calculate the regression residuals and construct a plot of the residuals versus x.

f. Examine the residual plot constructed in part **e**. Do you detect any outliers? Are there any trends?

g. Now consider the model

$$E(y) = \beta_0 + \beta_1 x_1 + \beta_2 x_1^2 + \beta_3 x_2 + \beta_4 x_1 x_2$$

where

$x_1 = $ Cost of modifying the fleet

$$x_2 = \begin{cases} 1 & \text{if American base} \\ 0 & \text{if foreign base} \end{cases}$$

The model is to fit the $n = 10$ data points and resulted in SSE $= 97.645$. Is there sufficient evidence to indicate that type of base (American or foreign) is a useful predictor of percentage improvement y? Test using $\alpha = .05$.

12.82 The 1986 EPA gas mileage guide gives the engine size and estimated city miles per gallon ratings, shown in the table, for 11 gasoline-fueled subcompact and compact cars. (The engine sizes are in total cubic inches of cylinder volume.) To predict gas mileage from the engine size of subcompact and compact cars, the first-order model

$$y = \beta_0 + \beta_1 x + \varepsilon$$

is fit to the data. The resulting least squares model is $\hat{y} = 37.677 - .0724x$.

CAR	CYLINDER VOLUME x	MILES PER GALLON y
VW Golf	97	37
Chevy Cavalier	173	19
Plymouth Horizon	97	31
Pontiac Firebird	151	23
Corvette	350	17
Honda Accord	119	27
Dodge Omni	97	31
Renault Alliance	85	35
Olds Firenza	173	19
Nissan Sentra	97	31
Ford Escort	114	32

Sources: 1986 Mileage Guide, EPA Fuel Economy Estimates, U.S. Dept. of Energy. Wards Automotive Yearbook, 1986.

a. Calculate the regression residuals for this model.

b. Verify that the sum of the residuals is 0.

c. Plot these residuals against cylinder volume, x.

d. Do you detect any distinctive patterns or trends in this plot?

e. What does your answer to part **c** suggest about model adequacy or the usual assumptions made about the error term?

12.83 Each year *Business Week* reports the total cash compensations (salary plus bonus) for the top corporate executives (see Case Study 3.1). The data in the table (in thousands of dollars) were extracted from *Business Week*'s 1990 Executive Compensation Scoreboard. To compare the mean 1990 cash compensation, $E(y)$, of executives in the four groups, the following model was fit to the data:

$$E(y) = \beta_0 + \beta_1 x_1 + \beta_2 x_2 + \beta_3 x_3$$

The Minitab printout is also provided.

where

$$x_1 = \begin{cases} 1 & \text{if consumer products} \\ 0 & \text{if not} \end{cases}$$

$$x_2 = \begin{cases} 1 & \text{if utilities} \\ 0 & \text{if not} \end{cases}$$

$$x_3 = \begin{cases} 1 & \text{if industrial–high tech} \\ 0 & \text{if not} \end{cases}$$

Base level = Financial services

CONSUMER PRODUCTS	UTILITIES	INDUSTRIAL–HIGH TECH	FINANCIAL SERVICES
1,567	1,862	2,925	3,125
3,313	1,390	3,409	4,143
2,058	1,115	1,767	4,013
25,216	1,105	4,097	6,583
4,634	1,272	3,196	3,169
5,214	2,849	4,042	5,217
20,795	1,732	2,601	3,447
9,162	1,474	8,286	4,469

Source: "Executive compensation scoreboard." *Business Week*, May 7, 1990, pp. 65–108.

```
The regression equation is
Y = 4271 + 4724 X1 - 2671 X2 - 480 X3

Predictor       Coef      Stdev     t-ratio        p
Constant        4271       1651        2.59    0.015
X1              4724       2334        2.02    0.053
X2             -2671       2334       -1.14    0.262
X3              -480       2334       -0.21    0.838

s = 4669      R-sq = 27.6%     R-sq(adj) = 19.8%

Analysis of Variance

SOURCE        DF          SS         MS        F       p
Regression     3   232505648   77501880     3.56   0.027
Error         28   610272512   21795446
Total         31   842778176

SOURCE        DF      SEQ SS
X1             1   200071984
X2             1    31510622
X3             1      923041

Unusual Observations
Obs.     X1        Y     Fit Stdev.Fit  Residual   St.Resid
  4    1.00    25216    8995      1651     16221       3.71R
  7    1.00    20795    8995      1651     11800       2.70R

R denotes an obs. with a large st. resid.
```

a. Is there sufficient evidence to indicate that the model is useful for predicting cash compensation? Test using $\alpha = .01$.

b. What does the result from part **a** imply about the mean cash compensation for the four groups of executives?

c. Find a 99% confidence interval for the difference between the mean 1990 cash compensations of executives in the consumer products and financial services industries.

12.84   Poly (perfluoropropyleneoxide), i.e., PPFPO, is a viscous liquid used extensively in the electronics industry as a lubricant. In a study reported in *Applied Spectroscopy* (Jan. 1986), the infrared reflectance spectra properties of PPFPO were examined. The optical density (y) for the prominent infrared absorption of PPFPO was recorded for different experimental settings of band frequency (x_1) and film thickness (x_2) in a Perkin-Elmer Model 621 infrared spectrometer. The results are given in the accompanying table.

OPTICAL DENSITY y	BAND FREQUENCY x_1, cm^{-1}	FILM THICKNESS x_2, milligrams
.231	740	1.1
.107	740	.62
.053	740	.31
.129	805	1.1
.069	805	.62
.030	805	.31
1.005	980	1.1
.559	980	.62
.321	980	.31
2.948	1,235	1.1
1.633	1,235	.62
.934	1,235	.31

Source: Pacansky, J., England, C. D., and Waltman, R. "Infrared spectroscopic studies of poly (perfluoropropyleneoxide) on gold substrates: A classical dispersion analysis for the refractive index." *Applied Spectroscopy*, Vol. 40, No. 1, Jan. 1986, p. 9, Table I.

a. If you have access to a multiple regression program package, fit the model

$$E(y) = \beta_0 + \beta_1 x_1 + \beta_2 x_2$$

b. Is the model useful for predicting optical density y?

c. Use the program package to find a 95% prediction interval for the optical density of PPFPO with band frequency $x_1 = 1,000$ cm^{-1} and film thickness $x_2 = .50$ mg.

12.85   To operate effectively, power companies have to be able to predict the peak power load at their various stations. The peak power load is the maximum amount of power that must be generated each day to meet demand. Suppose a power company located in the southern part of the United States decides to model daily peak power load, y, as a function of the daily high temperature, x, and the model is to be constructed for the summer months when demand is greatest. Although we would expect the peak power load to increase as the high temperature increases, the *rate* of increase in $E(y)$ might also increase as x increases. That is, a 1-unit increase in high temperature from 100°F to 101°F might result in a larger increase in power demand than would a 1-unit increase from 80°F to 81°F. Therefore, we postulate the second-order model

$$E(y) = \beta_0 + \beta_1 x + \beta_2 x^2$$

and we expect β_2 to be positive. A random sample of 25 summer days is selected. The data are shown in the accompanying table. The SAS printout for the second-order model is also given.

a. Give the least squares prediction equation.

b. Find R^2 and s on the printout and interpret their values.

c. Is there evidence that the model is useful for predicting peak power load y? Test using $\alpha = .05$.

d. Test the hypothesis that the power load increases at an increasing rate with temperature. Use $\alpha = .05$.

e. Use the least squares prediction equation to calculate the residual for each of the peak loads given in the table.

f. Calculate the mean and the variance of the residuals. The mean should equal 0 and the variance should be close to the value of MSE given in the SAS printout.

TEMPERATURE x, °F	PEAK LOAD y, megawatts	TEMPERATURE x, °F	PEAK LOAD y, megawatts	TEMPERATURE x, °F	PEAK LOAD y, megawatts
94	136.0	106	178.2	76	100.9
96	131.7	67	101.6	68	96.3
95	140.7	71	92.5	92	135.1
108	189.3	100	151.9	100	143.6
67	96.5	79	106.2	85	111.4
88	116.4	97	153.2	89	116.5
89	118.5	98	150.1	74	103.9
84	113.4	87	114.7	86	105.1
90	132.0				

Analysis of Variance

Source	DF	Sum of Squares	Mean Square	F Value	Prob>F
Model	2	15011.7720	7505.89	259.69	0.0001
Error	22	635.8784	28.90		
C Total	24	15647.6504			

Root MSE	5.3762	R-square	.9594	
Dep Mean	125.43	Adj R-sq	.9524	
C.V.	4.2863			

Parameter Estimates

Variable	Parameter Estimate	Standard Error	T for H0: Parameter=0	Prob > |T|
INTERCEP	385.0481	55.1724	6.98	0.0001
X	-8.2925	1.2990	-6.38	0.0001
XX	0.0598	0.0075	7.93	0.0001

g. Determine the proportion of the residuals that fall outside 3 estimated standard deviations ($3s$) of 0.

h. Plot the residuals against daily high temperature, x, and examine the graph for trends. What can you conclude about the assumptions concerning the random error term?

i. Construct a stem-and-leaf display for the residuals. Interpret the graph.

12.86   A large manufacturing firm wants to determine whether a relationship exists between the number of work-hours an employee misses per year, y, and the employee's annual wage, x (in thousands of dollars). A sample of 15 employees produced the data in the accompanying table. A first-order model was fit to the data with the following results:

$$\hat{y} = 222.64 - 9.60x \qquad r^2 = .073$$

EMPLOYEE	y	x	EMPLOYEE	y	x
1	49	12.8	9	191	7.8
2	36	14.5	10	6	15.8
3	127	8.3	11	63	10.8
4	91	10.2	12	79	9.7
5	72	10.0	13	543	12.1
6	34	11.5	14	57	21.2
7	155	8.8	15	82	10.9
8	11	17.2			

a. Interpret the value of r^2.

b. Calculate and plot the regression residuals. What do you notice?

c. After searching through its employees' files, the firm has found that employee 13 had been fired but that his name had not been removed from the active employee payroll. This explains the large accumulation of work-hours missed (543) by that employee. In view of this fact, what is your recommendation concerning this outlier?

d. Refit the model to the data, excluding the outlier, and find the least squares line. Calculate r^2 and comment on model adequacy.

CASE STUDY 12.1
THE SALARY RACE: MALES VERSUS FEMALES

Upon graduation from college, you will embark upon a career in your chosen field of study. Once you accept a job, you'll join the ranks of workers who are preoccupied with the size of their paychecks. Are you being fairly compensated? Why does your friend in another city receive a larger salary for a less demanding job? What can you do to get a raise?

Certainly, we expect our compensation to be tied to our qualifications. Graduates with engineering degrees expect to be paid more than nursing graduates. PhD.'s expect a higher starting salary than graduates with bachelor degrees. But, will your starting salary depend on your gender?

We can obtain a partial answer to this question by examining the data set in Appendix B. Recall that Appendix B contains the starting salaries of approximately 900 recent bachelor's degree graduates of the University of Florida in five different colleges: Business Administration, Engineering, Journalism, Liberal Arts and Sciences, and Nursing. In addition to starting salary and college, the gender (male or female) of the graduate was recorded.

Consider a multiple regression model relating starting salary y to the two qualitative independent variables, college (at 5 levels) and gender (at 2 levels). From Section 12.11, we require four dummy variables for college and one for gender. These are defined as follows:

$$\text{College} \quad x_1 = \begin{cases} 1 & \text{if Business Administration} \\ 0 & \text{if not} \end{cases}$$

$$x_2 = \begin{cases} 1 & \text{if Engineering} \\ 0 & \text{if not} \end{cases}$$

$$x_3 = \begin{cases} 1 & \text{if Liberal Arts and Sciences} \\ 0 & \text{if not} \end{cases}$$

$$x_4 = \begin{cases} 1 & \text{if Journalism} \\ 0 & \text{if not} \end{cases}$$

$$\text{Gender} \quad x_5 = \begin{cases} 1 & \text{if female} \\ 0 & \text{if male} \end{cases}$$

Note that we have arbitrarily selected nursing and male as the base levels for college and gender, respectively. A model relating mean starting salary, $E(y)$, to these two independent variables takes the form:

$$E(y) = \beta_0 + \underbrace{\beta_1 x_1 + \beta_2 x_2 + \beta_3 x_3 + \beta_4 x_4}_{\text{College terms}} + \underbrace{\beta_5 x_5}_{\text{Gender term}}$$

a. Write the equation relating mean starting salary, $E(y)$, to college, for male graduates only.
b. Interpret β_1 in the model, part a.
c. Interpret β_2 in the model, part a.
d. Interpret β_3 in the model, part a.
e. Interpret β_4 in the model, part a.
f. Write the equation relating mean starting salary, $E(y)$, to college, for female graduates only.
g. Interpret β_1 in the model, part f. Compare to your answer, part b.
h. Interpret β_2 in the model, part f. Compare to your answer, part c.
i. Interpret β_3 in the model, part f. Compare to your answer, part d.
j. Interpret β_4 in the model, part f. Compare to your answer, part e.
k. For a given college, interpret the value of β_5 in the model.
l. A SAS printout of the multiple regression analysis is displayed in Figure 12.36. Interpret the results. Part of your answer should include a statement about whether gender has an effect on average starting salary.

FIGURE 12.36
SAS Printout for the Model Relating Salary to College and Gender

Dependent Variable: SALARY

Analysis of Variance

Source	DF	Sum of Squares	Mean Square	F Value	Prob>F
Model	5	13609836905	2721967380.9	90.022	0.0001
Error	896	27092165483	30236791.833		
C Total	901	40702002387			

Root MSE	5498.79913	R-square	0.3344
Dep Mean	26020.20510	Adj R-sq	0.3307
C.V.	21.13280		

Parameter Estimates

Variable	DF	Parameter Estimate	Standard Error	T for H0: Parameter=0	Prob > |T|
INTERCEP	1	29215	739.33031785	39.515	0.0001
X1	1	-3928.938467	731.13739384	-5.374	0.0001
X2	1	1845.020611	778.46344679	2.370	0.0180
X3	1	-8375.343226	902.00089218	-9.285	0.0001
X4	1	-7349.696165	795.38316872	-9.240	0.0001
X5	1	-1142.171471	419.57634068	-2.722	0.0066

THE STOCK MARKET EFFECTS OF AIRLINE DEREGULATION

During the 1970s, several industries, including the airline, trucking, natural gas, and cable television industries, were deregulated as a result of pressure from special interest groups and legislators. In the case of the airline industry, federal regulations on fares, schedules, and routes were thought to protect the airlines from price competition and generally prohibit new entry into the market. Thus, in theory, deregulation was expected to benefit consumers while having a negative impact on the airlines and their shareholders.

To test this theory, Davidson, Chandy, and Walker (1984) analyzed the impact of the Airline Deregulation Act of 1978 on the security returns in the airline industry. Specifically, they examined the daily rates of returns of a sample of airline common stocks both prior to and following deregulation.

TABLE 12.12
Regression Results for 26 of 32 Airline Stocks

STOCK	$\hat{\beta}_0$	$\hat{\beta}_1$	$\hat{\beta}_2$	$\hat{\beta}_3$
1. AMR Corporation	.0005	2.8852*	−.0031	−.4106
2. Airborne Freight	.0024	.9748*	−.0012	−.6367*
3. Braniff International	.0007	1.4487*	−.0022	.5717
4. Canadian Pacific	.0011	.7687*	.0003	.3906
5. Continental Airlines	−.0021	2.1702*	.0004	−.7175*
6. Delta Airlines	−.0002	1.8557*	−.0005	−.6667*
7. Eastern Airlines	.0037	1.6353*	−.0069*	.9261*
8. Frontier	.0006	2.2188*	−.0033	−1.0355*
9. Greyhound	.0000	.3830*	.0020	.2163
10. Royal Dutch Airlines	.0004	1.3281	−.0024	−.5674
11. Lockheed Corporation	.0042	1.9337*	−.0053*	.6399
12. Northwest Airlines	−.0009	2.6056*	−.0005	−.1698
13. Ozark Airlines	.0032	1.1591*	−.0017	−.2921
14. PSA Airlines	.0033	.2555*	−.0036	1.4888*
15. Pan American	.0012	1.2136*	−.0025	−.0146
16. Piedmont	.0012	1.2138*	−.0025	−.0146
17. Purolator, Inc.	.0009	.6829*	.0002	−.3787
18. Republic Airlines	.0023	.9524*	−.0031	1.3359*
19. Southwest Airlines	.0030	.8600*	.0016	−.2357
20. Tiger International	.0003	2.4979*	−.0009	−.7462*
21. Trans World Corp.	.0007	2.6399*	.0000	.1699
22. United Airlines	.0023	1.8137*	−.0039	.1444
23. US Air, Inc.	.0041	.8628	−.0029	1.4039*
24. W.A.F., Inc.	.0024	1.9271*	.0035	1.7067*
25. Western Airlines	.0012	1.5750*	−.0014	−.3987
26. World Airways	−.0014	3.2765*	.0015	−1.4218*

*Asterisk identifies β coefficients significant at $\alpha = .10$.
Source: Davidson, W. N., Chandy, P. R., and Walker, M. "The stock market effects of airline deregulation." *Quarterly Journal of Business and Economics*, Autumn 1984, Vol. 23, No. 4, pp. 31–45.

Thirty-two airlines engaged in air transportation for at least 1 year and listed on either the New York or the American Stock Exchange were selected for analysis. For each airline, daily stock returns were recorded for each of 120 days prior to deregulation and 120 days after deregulation. Data for the total of $n = 240$ observations (days) were then used to fit the model

$$E(y) = \beta_0 + \beta_1 x_1 + \beta_2 x_2 + \beta_3 x_1 x_2$$

where

y = Daily rate of return on the airline stock

x_1 = Average daily rate of return on the market

$$x_2 = \begin{cases} 1 & \text{if after deregulation} \\ 0 & \text{if prior to deregulation} \end{cases}$$

Thus, 32 regression analyses were conducted, one for each airline in the sample.

a. Write the equation of the line relating daily stock return y to average daily market return x_1 prior to deregulation (i.e., when $x_2 = 0$). Identify the y-intercept and slope of the line. [The slope is used to measure the stock's systematic risk and is often called the β *risk index* or β *value* for the stock. When the β value is greater than 1, the stock is classified as an *aggressive* or *risky* security since its daily rate of return is expected to move (upward or downward) faster than the market rate of return. In contrast, when the β value is less than 1, the stock is classified as a *defensive* or *stable* security since its daily rate of return moves slower than the market. A stock with a β value near 1 is called a *neutral* security for its daily rate of return mirrors the market.]

b. Write the equation of the line relating daily stock return y to average daily market return x_1 after deregulation (i.e., when $x_2 = 1$). Identify the y-intercept and slope of the line.

c. What hypothesis would you test to determine whether the model is adequate for predicting y?

d. What hypothesis would you test to determine whether deregulation had an effect on the measure of systematic risk (i.e., β value) associated with the airline stock? [*Hint:* Use your answers to parts **a** and **b**.]

e. Of the 32 airline stocks analyzed, 26 had significant regressions (i.e., we could reject H_0: $\beta_1 = \beta_2 = \beta_3 = 0$) at $\alpha = .10$. The least squares prediction equations for these stocks are given in Table 12.12. Identify those stocks that have significant (at $\alpha = .10$) interaction between average market rate of return x_1 and deregulation x_2. For each of these stocks, how has deregulation affected the stock's β value?

FACTORS AFFECTING THE SALE PRICE OF CONDOMINIUM UNITS

This case study deals with an actual investigation of the factors that affect the sale price of oceanside condominium units. The sales data were obtained for a new oceanside condominium complex consisting of two adjacent and connected eight-floor buildings. The complex contains 200 equal-size (approximately 500 square feet) units. The locations of the buildings relative to the ocean, the swimming pool, the parking lot, etc., are shown in Figure 12.37.

The only elevator in the complex is located at the east end of building 1, as are the office and the game room. People moving to or from the higher floor units in building 2 would likely use the elevator and move through the passages to their units. Thus, units on the higher floors and at a greater distance from the elevator would be less convenient; they would require greater effort in moving baggage, groceries, etc., and would be farther away from the game room, the office, and the swimming pool. These units also have an advantage: They are the most private because they have the least amount of traffic through the hallways.

Lower-floor oceanside units are most suited to active people. These units, which open onto the beach, ocean, and pool, are within easy reach of the game room and the parking area. Checking Figure 12.37, you will see that some of the units in the center of the complex—units with numbers ending in digits 11 or 14—have part of their view blocked. One would expect this to be a disadvantage.

The condominium complex was completed at the time of a recession; sales were slow and the developer was forced to sell most of the units at auction approximately 18 months after opening. Many unsold units were furnished by the developer and rented before the auction. Because of the uniqueness of the condominium complex, the data provide a good opportunity to investigate the relationship between sale price, height of the unit (floor number), distance of the unit from the elevator, and presence or absence of an ocean view. The presence or absence of furniture in each of the units also permits an investigation of the effect of the availability of furniture on sale price. Finally, the auction data are completely buyer-specified and hence consumer-oriented in contrast to most other real estate sales data, which are, to a high degree, seller- and broker-specified.

In addition to the sale price (recorded in hundreds of dollars), the following data were recorded for each of the $n = 105$ units sold at the auction:

1. *Floor height*   The floor location of the unit; this variable, x_1, could take values 1, 2, . . . , 8.
2. *Distance from elevator*   This distance, measured along the length of the complex, was expressed in number of condominium units. An additional two units of distance were added to the units in building 2 to account for the walking distance in the connecting area between the two buildings. Thus, the distance of unit 105 from the elevator would be 3, and the distance between unit 113 and the elevator would be 9. This variable, x_2, could take values 1, 2, . . . , 15.

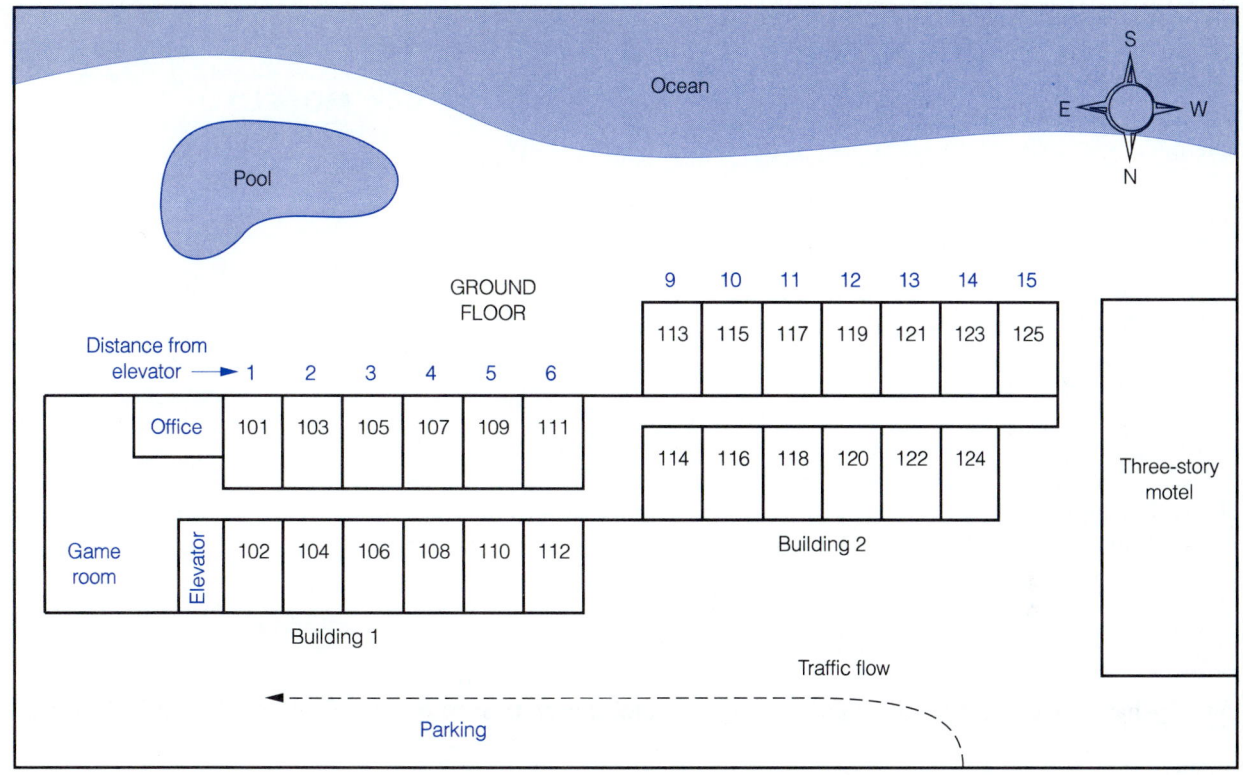

FIGURE 12.37
Layout of Condominium Complex

3. *View of ocean*   The presence or absence of an ocean view was recorded for each unit and entered into the model with a dummy variable, x_3, where $x_3 = 1$ if the unit had an ocean view and $x_3 = 0$ if not. Note that units not possessing an ocean view would face the parking lot.

4. *End unit*   The view of end units on the ocean side (numbers ending in 11) is partially blocked by building 2. To account for this potential disadvantage, this qualitative variable was entered into the model with a dummy variable, x_4, where $x_4 = 1$ if the unit has a unit number ending in 11 and $x_4 = 0$ if not.

5. *Furniture*   The presence or absence of furniture was recorded for each unit. This qualitative variable was entered into the model using a single dummy variable, x_5, where $x_5 = 1$ if the unit was furnished and $x_5 = 0$ if not.

For this case study, we postulate the following three models:

$$E(y) = \beta_0 + \beta_1 x_1 + \beta_2 x_2 + \beta_3 x_3 + \beta_4 x_4 + \beta_5 x_5$$

where

x_1 = Floor height (x_1 = 1, 2, ... , 8)

x_2 = Distance from elevator (x_2 = 1, 2, ... , 15)

$x_3 = \begin{cases} 1 & \text{if ocean view} \\ 0 & \text{if not} \end{cases}$

$x_4 = \begin{cases} 1 & \text{if an end unit} \\ 0 & \text{if not} \end{cases}$

$x_5 = \begin{cases} 1 & \text{if furnished} \\ 0 & \text{if not} \end{cases}$

Model 1 hypothesizes that the five factors affect the price in an independent manner and that the effect of the two quantitative factors on sale price is linear.

$$E(y) = \beta_0 + \underbrace{\beta_1 x_1 + \beta_2 x_2 + \beta_3 x_1 x_2 + \beta_4 x_1^2 + \beta_5 x_2^2}_{\text{Second-order model in } x_1 \text{ and } x_2}$$

$$+ \underbrace{\beta_6 x_3}_{\substack{\text{View of} \\ \text{ocean}}} + \underbrace{\beta_7 x_4}_{\substack{\text{End} \\ \text{unit}}} + \underbrace{\beta_8 x_5}_{\text{Furniture}}$$

Model 2 hypothesizes a second-order relationship between sale price and the quantitative factors and assumes that the shape of the curvilinear relationship is the same, regardless of the view, whether the unit is an end unit and whether the unit is furnished. Thus, model 2 assumes that there is no interaction between any of the qualitative factors (view of ocean, end unit, and furniture) and the quantitative factors (floor height and distance from elevator).

$$E(y) = \beta_0 + \underbrace{\beta_1 x_1 + \beta_2 x_2 + \beta_3 x_1 x_2 + \beta_4 x_1^2 + \beta_5 x_2^2}_{\text{Second-order model in } x_1 \text{ and } x_2} + \underbrace{\beta_6 x_3}_{\substack{\text{View of} \\ \text{ocean}}} +$$

$$\underbrace{\beta_7 x_4}_{\substack{\text{End} \\ \text{unit}}} + \underbrace{\beta_8 x_5}_{\text{Furniture}} + \underbrace{\beta_9 x_1 x_3 + \beta_{10} x_2 x_3 + \beta_{11} x_1 x_2 x_3 + \beta_{12} x_1^2 x_3 + \beta_{13} x_2^2 x_3}_{\substack{\text{Interaction of the second-order} \\ \text{model with view of ocean}}}$$

Model 3 provides for two completely different second-order relationships—one for ocean-view units and one for bay-view units. Further, the model assumes that the effects of the two qualitative factors, end unit and furniture, are additive; i.e., their presence or absence will simply shift the mean sale price response surface up or down by a fixed amount.

The SAS regression analysis printouts for fitting models 1, 2, and 3 to the data are shown in Figures 12.38, 12.39, and 12.40, respectively. A summary containing SSE values for these models, their respective degrees of freedom, R^2 values, and Root MSE values is provided in Table 12.13 (page 716).

page 716

FIGURE 12.38
SAS Printout for Model 1

```
Model: MODEL1
Dependent Variable: PRICE

                      Analysis of Variance

                            Sum of        Mean
    Source        DF       Squares       Square     F Value    Prob>F

    Model          5    21059.51941   4211.90388    60.633     0.0001
    Error         99     6877.10916     69.46575
    C Total      104    27936.62857

         Root MSE        8.33461     R-square     0.7538
         Dep Mean      191.11429     Adj R-sq     0.7414
         C.V.            4.36106

                      Parameter Estimates

                   Parameter     Standard     T for H0:
    Variable  DF    Estimate       Error     Parameter=0    Prob > |T|

    INTERCEP   1   178.528277    3.50998073    50.863        0.0001
    X1         1    -0.802615    0.44683728    -1.796        0.0755
    X2         1    -0.775970    0.20666247    -3.755        0.0003
    X3         1    29.866076    1.88998132    15.802        0.0001
    X4         1   -16.797365    3.35438026    -5.008        0.0001
    X5         1     8.496748    1.74141868     4.879        0.0001
```

FIGURE 12.39
SAS Printout for Model 2

```
Model: MODEL2
Dependent Variable: PRICE

                      Analysis of Variance

                            Sum of        Mean
    Source        DF       Squares       Square     F Value    Prob>F

    Model          8    22230.35558   2778.79445    46.749     0.0001
    Error         96     5706.27299     59.44034
    C Total      104    27936.62857

         Root MSE        7.70976     R-square     0.7957
         Dep Mean      191.11429     Adj R-sq     0.7787
         C.V.            4.03411

                      Parameter Estimates

                   Parameter     Standard     T for H0:
    Variable  DF    Estimate       Error     Parameter=0    Prob > |T|

    INTERCEP   1   198.743497    6.14526957    32.341        0.0001
    X1         1    -8.237009    1.97016582    -4.181        0.0001
    X2         1    -2.879996    0.98153582    -2.934        0.0042
    X1X2       1     0.022253    0.10840032     0.205        0.8378
    X1SQ       1     0.726163    0.19097653     3.802        0.0003
    X2SQ       1     0.149139    0.06180764     2.413        0.0177
    X3         1    28.721405    1.76927801    16.233        0.0001
    X4         1   -15.401569    3.28423568    -4.690        0.0001
    X5         1    10.014268    1.64947840     6.071        0.0001

Test: SECORDER Numerator:   390.2787  DF:   3   F value:   6.5659
                Denominator:  59.44034  DF:  96   Prob>F:    0.0004
```

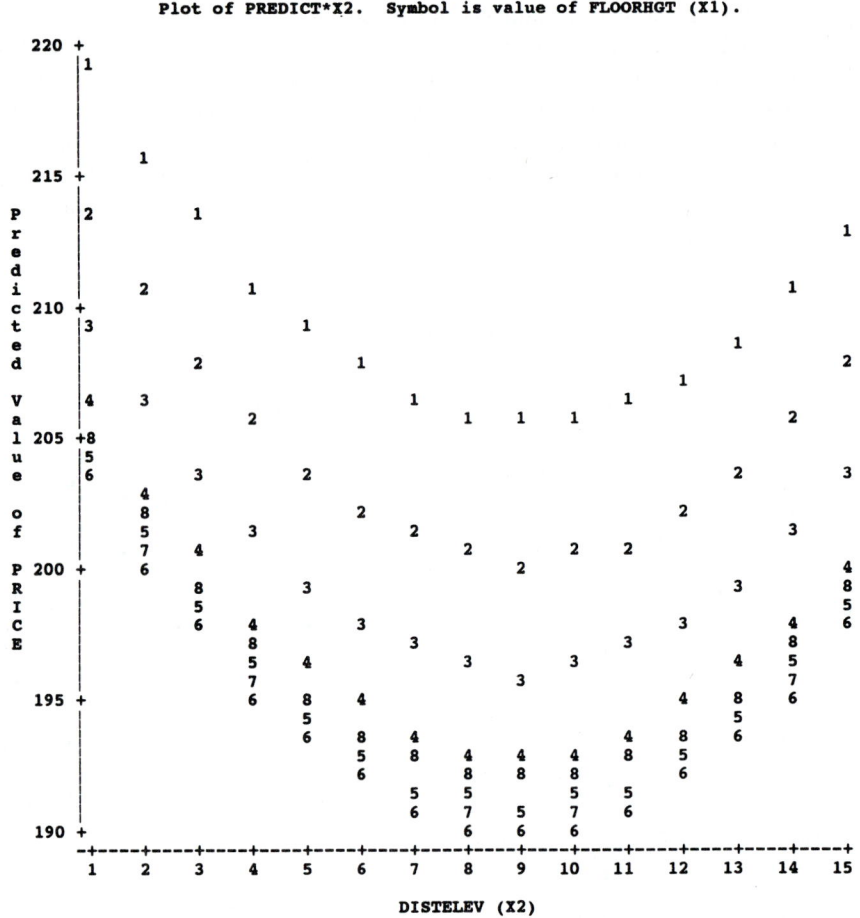

Plots of Model 3

------------------------------------ VIEW (X3)=1 ------------------------------------

Plot of PREDICT*X2. Symbol is value of FLOORHGT (X1).

FIGURE 12.42
Graph of Predicted Price for Model 3: Ocean-View Units

COMPUTER LAB
MULTIPLE REGRESSION AND RESIDUAL ANALYSIS

All three of the statistical software packages discussed throughout this text have routines for performing a multiple regression analysis and a graphical residual analysis. The sample programs in this section give the commands or instructions for analyzing the model

$$E(y) = \beta_0 + \beta_1 x_1 + \beta_2 x_2 + \beta_3 x_1 x_2 + \beta_4 x_1^2$$

where

$$y = \text{Shipping cost}$$

$$x_1 = \text{Package weight}$$

$$x_2 = \begin{cases} 1 & \text{if fragile product} \\ 0 & \text{if not} \end{cases}$$

Note that in each package, the higher-order terms specified in the model (e.g., x_1^2 and $x_1 x_2$) must be created through data transformation statements.

SAS

```
Command
Line
  1    DATA SHIP;                    ⎫
  2    INPUT Y X1 X2 @@;             ⎬  Data entry instructions
  3    X1X2 = X1*X2;                 ⎪
  4    X1SQ = X1*X1;                 ⎪
  5    CARDS;                        ⎭

     [Input data values]

  6    PROC REG;
  7    MODEL Y = X1 X2 X1X2 X1SQ/P CLI;   ⎫  Regression analysis
  8    ID X1 X2;                          ⎬  95% prediction limits
  9    OUTPUT OUT=RESIDS P=YHAT R=RESID;  ⎭
 10    PROC PLOT;                    ⎫  Residual plots
 11    PLOT RESID*(YHAT X1);         ⎭
 12    PROC CHART;                   ⎫  Histogram of residuals
 13    VBAR RESID;                   ⎭
```

COMMAND 6 The REG procedure fits general linear models.

COMMAND 7 In the MODEL statement, the dependent variable is listed to the left of the equals sign and the independent variables to the right. The option P (following the slash) prints predicted values and residuals, and the option CLI prints corresponding lower and upper 95% prediction limits. Specify CLM to obtain 95% confidence intervals for $E(y)$.

COMMANDS 10–11 Two plots are produced: residuals versus predicted ($\hat{y}$) and residuals versus quantitative x (x_1).

COMMANDS 12–13 The CHART procedure is used to produce a histogram for the regression residuals.

ANALYSIS OF VARIANCE

CHAPTER 13

Consider the problem of comparing the mean starting salaries of University of Florida graduates in five colleges: Business Administration, Engineering, Journalism, Liberal Arts/Sciences, and Nursing. Recall that Appendix B contains starting salary data, obtained from a recent CRC survey, for the graduates of the five colleges. How would you decide whether the data indicate differences among the mean starting salaries for the five colleges? In this chapter, we will consider the general problem of comparing more than two population means. We will examine the data in Appendix B in greater detail in Section 13.8.

INTRODUCTION

13.1 As we have seen in the preceding chapters, the solutions to many statistical problems are based on inferences about population means. For example, a sociologist interested in comparing the mean grade point averages of two groups of high school students from different socioeconomic classes might choose to apply the independent-samples procedure of Section 8.5. Or, the sociologist may decide to use matched pairs (Section 10.6) to test the mean difference between high school and college grade point averages. This chapter extends the methods of Chapters 8–10 to the comparison of more than two means.

When the data have been obtained according to certain specified sampling procedures, they are easy to analyze and also may contain more information pertinent to the population means than could be obtained using simple random sampling. The procedure for selecting sample data is called the **design of the experiment**, and the statistical procedure for comparing the population means is called an **analysis of variance**. The objective of this chapter is to introduce some aspects of experimental design and the analysis of data from such experiments using an analysis of variance. In optional Section 13.9, we illustrate the relationship between this method of analysis and multiple regression.

EXPERIMENTAL DESIGN: TERMINOLOGY

13.2 The study of experimental design originated in England and, in its early years, was associated solely with agricultural experimentation. The need for experimental design in agriculture was very clear: It takes a full year to obtain a single observation on the yield of a new variety of wheat. Consequently, the need to save time and money led to a study of ways to obtain more information using smaller samples. Similar motivation led to its subsequent acceptance and wide use in all fields of scientific experimentation. Despite this fact, the terminology associated with experimental design clearly indicates its early assocation with the biological sciences.

We will call the process of collecting sample data an **experiment** and the variable to be measured the **response**. (In this chapter we consider only experiments in which the response is a quantitative variable. Experiments with qualitative responses are beyond the scope of this text.) The planning of the sampling procedure is called the **design** of the experiment. The object upon which the response measurement is taken is called an **experimental** (or **sampling**) **unit**.

DEFINITION 13.1

The process of collecting sample data is called an **experiment**.

DEFINITION 13.2

The plan for collecting the sample is called the **design** of the experiment.

> **DEFINITION 13.3**
>
> The variable measured in the experiment is called the **response variable**. (In this chapter, all response variables will be quantitative variables.)

> **DEFINITION 13.4**
>
> The object upon which the response variable is measured is called an **experimental** (or **sampling**) **unit**.

Variables that may be related to a response variable are called **factors**. The value—that is, the intensity setting—assumed by a factor in an experiment is called a **level**. The combinations of levels of the factors for which the response will be observed are called **treatments**.

EXAMPLE 13.1 A marketing study is conducted to investigate the effects of brand and shelf location on weekly coffee sales. Coffee sales are recorded for each of the two brands (brand A and brand B) and three shelf locations (bottom, middle, and top) each week for a period of 20 weeks. For this experiment, identify

a. the experimental unit   b. the response
c. the factors   d. the factor levels
e. the treatments

Solution

a. Since our data will be collected each week for a period of 20 weeks, the experimental unit is one week.
b. The variable of interest, i.e., the response, is weekly coffee sales. Note that weekly coffee sales is a quantitative variable.
c. Since we are interested in investigating the effect of brand and shelf location on sales, *brand* and *shelf location* are the factors. Note that both factors are qualitative.
d. For this experiment, brand is measured at two levels (A and B) and shelf location at three levels (bottom, middle, and top).
e. Since coffee sales are recorded for each of the six brand–shelf location combinations (brand A, bottom), (brand A, middle), (brand A, top), (brand B, bottom), (brand B, middle), and (brand B, top), the experiment involves six treatments. The term *treatments* is used to describe the factor level combinations to be included in an experiment because many experiments involve "treating" or doing something to alter the nature of the experimental unit. Thus, we might view the six brand–shelf location combinations as treatments on the experimental units in the marketing study involving coffee sales.

> **DEFINITION 13.5**
>
> The variables, quantitative or qualitative, that are related to a response variable are called **factors**.

> **DEFINITION 13.6**
>
> The intensity setting of a factor (i.e., the value assumed by a factor in an experiment) is called a **factor level**.

> **DEFINITION 13.7**
>
> A **treatment** is a particular combination of levels of the factors involved in an experiment.

Now that you understand some of the terminology, it is helpful to think of the design of an experiment in four steps:

1. Select the factors to be included in the experiment and identify the parameters that are the object of the study. Usually, the target parameters are the population means associated with the factor level combinations (i.e., treatments).
2. Choose the treatments (the factor level combinations) to be included in the experiment.
3. Determine the number of observations (sample size) to be made for each treatment. [This will usually depend on the standard error(s) that you desire. See Section 8.8.]
4. Plan how the treatments will be assigned to the experimental units. That is, decide on which design to employ.

Entire texts are devoted to properly executing these steps for various experimental designs. (See the references given at the end of this chapter.) The main objective of this chapter, however, is to show how to analyze the data that are collected in a designed experiment.

In Sections 13.4–13.6, we consider three popular experimental designs and demonstrate how to analyze the data for each. First, in Section 13.3, we present a short discussion of the logic behind the analysis of data collected from such experiments.

THE LOGIC BEHIND AN ANALYSIS OF VARIANCE

13.3 Once the data for a designed experiment have been collected, we will want to use the sample information to make inferences about the population means associated with the various treatments. The method used to compare the treatment means is known as **analysis of variance**, or ANOVA.

The concept behind an analysis of variance can be explained using the following simple example.

EXAMPLE 13.2 Suppose we want to compare the means (μ_1 and μ_2) of two populations using independent random samples of size $n_1 = n_2 = 5$ from each of the populations. The sample observations and the sample means are listed in Table 13.1 and shown on a line plot in Figure 13.1.

TABLE 13.1
Data for Example 13.2a

SAMPLE FROM POPULATION 1	SAMPLE FROM POPULATION 2
6	8
−1	1
0	3
4	7
1	6
$\bar{y}_1 = 2$	$\bar{y}_2 = 5$

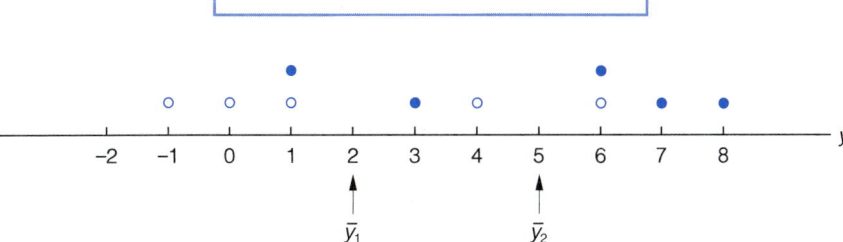

FIGURE 13.1
Line Plot of Data in Table 13.1

a. Do you think these data provide sufficient evidence to indicate a difference between the population means μ_1 and μ_2?
b. Now look at two more samples of $n_1 = n_2 = 5$ measurements from the populations, as listed in Table 13.2 and plotted in Figure 13.2 (page 728). Do these data appear to provide evidence of a difference between μ_1 and μ_2?

TABLE 13.2
Data for Example 13.2b

SAMPLE FROM POPULATION 1	SAMPLE FROM POPULATION 2
2	5
3	5
2	5
2	4
1	6
$\bar{y}_1 = 2$	$\bar{y}_2 = 5$

critical value, F_α, we reject H_0 and conclude that at least two of the treatment means differ.

TEST TO COMPARE k POPULATION MEANS FOR A COMPLETELY RANDOMIZED DESIGN

H_0: $\mu_1 = \mu_2 = \cdots = \mu_k$ [i.e., there is no difference in the treatment (population) means]

H_a: At least two treatment means differ

Test statistic: $F = \dfrac{MST}{MSE}$

Rejection region: $F > F_\alpha$

where the distribution of F is based on $(k - 1)$ numerator df and $(n - k)$ denominator df, and F_α is the F value found in Tables 7, 8, 9, and 10 of Appendix G such that $P(F > F_\alpha) = \alpha$.

Assumptions: 1. All k population probability distributions are normal.

2. The k population variances are equal.

3. The samples from each population are random and independent.

The results of an analysis of variance are usually summarized and presented in an **analysis of variance (ANOVA) table**. Such a table shows the sources of variation, their respective degrees of freedom, sums of squares, mean squares, and computed F statistic. The results of the analysis of variance for Example 13.3 (calculations not shown) are given in Table 13.4, and the general form of the ANOVA table for a completely randomized design is shown in the box.

TABLE 13.4

Analysis of Variance Table for Example 13.3

SOURCE	df	SS	MS	F
Among treatment means	2	2.33	1.165	4.21
Within samples (Error)	17	4.71	.277	
Total	19	7.04		

EXAMPLE 13.4 Refer to Example 13.3 and the ANOVA presented in Table 13.4. Conduct the test of the hypotheses

H_0: $\mu_1 = \mu_2 = \mu_3$

H_a: At least two population means are different

where μ_1 is the true mean GPA for lower-class college freshmen, μ_2 is the true mean GPA for middle-class college freshmen, and μ_3 is the true mean GPA for upper-class college freshmen. Use $\alpha = .05$.

ANALYSIS OF VARIANCE TABLE FOR A COMPLETELY RANDOMIZED DESIGN

SOURCE	df	SS	MS	F
Treatments	$k - 1$	SST	$MST = \dfrac{SST}{k - 1}$	$F = \dfrac{MST}{MSE}$
Error	$n - k$	SSE	$MSE = \dfrac{SSE}{n - k}$	
Total	$n - 1$	SS(Total)		

Solution The test statistic appropriate for carrying out the test is given in Table 13.4: $F = MST/MSE = 4.21$. To find the rejection region for the test, we first need to find the appropriate values of v_1 and v_2, the numerator and denominator degrees of freedom, respectively, of the F distribution.

Since we are comparing $k = 3$ means in Example 13.3, the numerator degrees of freedom is $(k - 1) = (3 - 1) = 2$. There are $n = 20$ measurements in the combined samples, so the denominator degrees of freedom is $(n - k) = (20 - 3) = 17$. Thus, $v_1 = 2$ and $v_2 = 17$. Note that these values are given in Table 13.4 as df(Treatments) and df(Error), respectively. Using $\alpha = .05$, we will reject the null hypothesis that the three means are equal if

$$F > F_{.05}$$

where from Table 8 of Appendix G, the F value associated with 2 numerator df and 17 denominator df is $F_{.05} = 3.59$. This rejection region is shown in Figure 13.4. Since the computed value of the test statistic, $F = 4.21$, exceeds the tabulated value, $F_{.05} = 3.59$, it lies in the rejection region. Consequently, we have sufficient evidence (at significance level $\alpha = .05$) to conclude that the true mean freshmen GPAs differ for at least two of the three socioeconomic classes. The chance that this procedure will result in a Type I error (conclude that there are differences when no differences among the means exist), is, at most, $\alpha = .05$.

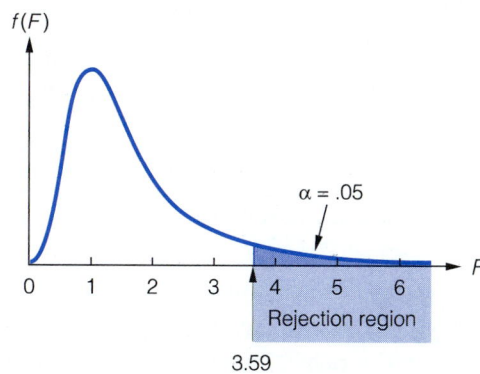

FIGURE 13.4
Rejection Region for Example 13.4; Numerator df = 2, Denominator df = 17

Virtually all statistical computer software packages available today have routines that perform the analysis of data collected from a designed experiment. In this chapter we emphasize the interpretation of the results of the ANOVA rather than the tedious calculations that generated the results, as the previous example illustrates. For those who are interested, the computational formulas for each design are provided in optional Section 13.11.

EXAMPLE 13.5 Consider the problem of comparing the mean 1990 sale prices of residential properties contained in seven different neighborhoods (A, B, C, D, E, F, and G) of Tampa, Florida. Many of these properties were not sold, so we will never know their sale prices, but we can still imagine that a sale price exists for each property in a neighborhood and that the totality of these prices constitutes a population that characterizes property values in that neighborhood. Thus, we want to compare the mean sale prices corresponding to the seven neighborhood sale price populations, i.e., we wish to test

H_0: $\mu_A = \mu_B = \mu_C = \mu_D = \mu_E = \mu_F = \mu_G$

H_a: At least two means are different

To perform the comparison, independent random samples of sale prices were selected from the actual sale prices for the seven neighborhoods obtained from the data files of the city's property appraiser. The data (recorded in thousands of dollars and listed in Table 13.5) were subjected to an analysis of variance using SAS. A portion of the SAS printout is reproduced in Figure 13.5. Interpret the results.

TABLE 13.5

Sale Prices (in Thousands of Dollars) of Properties in the Seven Neighborhoods

			NEIGHBORHOOD			
A	B	C	D	E	F	G
191.5	100.5	35.3	86.0	69.9	46.0	14.5
208.5	147.5	50.0	140.0	159.5	42.0	50.0
255.0	115.0	160.0	120.0	84.5	48.0	39.0
375.0	210.0	59.9	72.9	57.7	57.0	20.0
205.0	155.0	38.5	72.0	45.4	34.5	32.2
191.5	127.5	105.9	68.7	43.9	74.5	54.0
189.0	115.0	312.0	90.0	205.0	66.9	21.5
139.0	110.0	84.5	100.0	59.9	54.5	55.0
138.0	183.0	68.5	97.0	37.0	46.8	43.0
162.0	107.5	160.0	84.0	68.0	6.9	13.3

Source: Hillsborough County (Fla.) property appraiser's office, 1990.

Solution You can see that the SAS printout (Figure 13.5) presents the results in the form of an ANOVA table. The source of variation attributable to treatments, i.e., to the variability among the sample means for the seven neighborhoods, is

```
                Analysis of Variance Procedure

Dependent Variable: SALEPRIC
                                 Sum of           Mean
Source                 DF        Squares         Square     F Value    Pr > F

Model                   6     199341.2469     33223.5411     13.34     0.0001
Error                  63     156930.9230      2490.9670
Corrected Total        69     356272.1699

           R-Square          C.V.        Root MSE        SALEPRIC Mean

           0.559520        49.33797      49.90959          101.158571


Source                 DF       Anova SS     Mean Square    F Value    Pr > F

NBRHOOD                 6     199341.2469     33223.5411     13.34     0.0001
```

FIGURE 13.5
SAS Printout of the ANOVA for Example 13.5

labeled **Model**, and the source of variation attributable to error, i.e., to the **within-sample variability**, is labeled **Error**. Their corresponding sum of squares and mean squares (rounded) are:

$$SST = 199{,}341.2469$$
$$SSE = 156{,}930.9230$$
$$MST = 33{,}223.5411$$
$$MSE = 2{,}490.9670$$

The computed value of the test statistic, given under the column heading **F Value**, is

$$F = 13.34$$

To determine whether to reject the null hypothesis

$$H_0: \quad \mu_A = \mu_B = \cdots = \mu_G$$

in favor of the alternative

H_a: At least two population means are different

we may consult Appendix G for tabulated values of the F distribution corresponding to an appropriately chosen significance level α. However, since the SAS printout gives the observed significance level (p-value) of the test, we will use this quantity to assist us in reaching a conclusion.

The observed significance level of the test is under the column headed **Pr > F**. This value, .0001, implies that H_0 will be rejected at any chosen level of α larger than .0001. Thus, there is very strong evidence of a difference among the mean 1990 sale prices of residential properties in the seven neighborhoods. The probability that this procedure will lead to a Type I error (conclude that there is a difference among the means when in fact they are all equal) is .0001.

An analysis of variance for a completely randomized design may include the construction of confidence intervals for a single mean or for the difference between two means. Because the independent sampling design involves the selection of independent random samples, we can find a confidence interval for a single mean using the method of Section 8.3 and for the difference between two population means using the method of Section 8.5.*

However, the researcher is usually interested in making all possible pairwise comparisons of the means. Consequently, several intervals must be constructed, inflating the overall probability of a Type I error. We discuss a better, alternative approach to constructing confidence intervals for means in ANOVA in Section 13.7.

Before ending our discussion of completely randomized designs, we make the following comment. The proper application of the ANOVA procedure requires that certain assumptions be satisfied, i.e., all k populations are approximately normal with equal variances. In Section 13.10, we provide details on how to use residuals to determine if these assumptions are satisfied to a reasonable degree. If the residual analysis, for example, reveals that one or more of the populations are nonnormal (e.g., highly skewed), then any inferences derived from the ANOVA of the data are suspect. In this case, we can apply a nonparametric technique (Chapter 15).

WARNING

When the assumptions for analyzing data collected from a completely randomized design are violated, any inferences derived from the ANOVA are suspect. An alternative technique to use in this situation is the nonparametric Kruskal–Wallis test (see Section 15.5).

EXERCISES
LEARNING THE MECHANICS

13.1   Independent random samples were selected from two populations, with the results shown in the table.

SAMPLE 1	SAMPLE 2
10	12
7	8
8	13
11	10
10	10
9	11
9	

---

*The only modification we will make in these two procedures is that we will use an estimate of σ^2 based on the information contained in all k samples—namely, the pooled measure of variability within the k samples: MSE $= s^2 = \dfrac{\text{SSE}}{(n-k)}$.

a. Construct a line plot of the data similar to Figures 13.1 and 13.2. Do you think the data provide evidence of a difference between the population means?
b. Calculate MST for the data using the formula on page 729. What type of variability is measured by this quantity?
c. Calculate MSE for the data using the formula on page 728. What type of variability is measured by this quantity?
d. How many degrees of freedom are associated with MST?
e. How many degrees of freedom are associated with MSE?
f. Compute the test statistic appropriate for testing H_0: $\mu_1 = \mu_2$ against the alternative hypothesis that the two means differ.
g. Summarize the results of parts b–f in an ANOVA table.
h. Specify the rejection region, using a significance level of $\alpha = .05$.
i. Make the proper conclusion. How does this compare to your answer to part a?
j. Construct 95% confidence intervals for each of the two population means.
k. Construct a 95% confidence interval for $(\mu_2 - \mu_1)$.

13.2   Independent random samples were selected from three populations. The data are shown in the accompanying table, followed by a SAS printout of the analysis of variance.

SAMPLE 1	SAMPLE 2	SAMPLE 3
2.1	4.4	1.1
3.3	2.6	.2
.2	3.0	2.0
	1.9	

Analysis of Variance Procedure

Dependent Variable: Y

Source	DF	Sum of Squares	Mean Square	F Value	Pr > F
Model	2	6.22183333	3.11091667	2.21	0.1798
Error	7	9.83416667	1.40488095		
Corrected Total	9	16.05600000			

R-Square	C.V.	Root MSE	Y Mean
0.387508	56.98446	1.185277	2.08000000

Source	DF	Anova SS	Mean Square	F Value	Pr > F
SAMPLE	2	6.22183333	3.11091667	2.21	0.1798

a. Locate the value of MST. What type of variability is measured by this quantity?
b. Locate the value of MSE. What type of variability is measured by this quantity?
c. How many degrees of freedom are associated with MST?
d. How many degrees of freedom are associated with MSE?
e. Locate the value of the test statistic for testing H_0: $\mu_1 = \mu_2 = \mu_3$ against the alternative hypothesis that at least one population mean is different from the other two.
f. Summarize the results of parts a–e in an ANOVA table.
g. Specify the rejection region, using a sigificance level of $\alpha = .05$.
h. State the proper conclusion.
i. Locate and interpret the p-value for the test of part e. Does this agree with your answer to part h?
j. Calculate 90% confidence intervals for each of the three population means.
k. Find a 90% confidence interval for $(\mu_1 - \mu_2)$.

a. Identify the response, the treatments, and the experimental units in this design.

b. Is there evidence of a difference among the means of the 1990 total cash compensations for the three groups of corporate executives? Test using $\alpha = .01$.

c. Find a 90% confidence interval for the mean 1990 total cash compensation executives in the consumer products industry.

13.8 Because of the growing concern over problems related to alcohol consumption, the Federal Trade Commission has considered banning alcohol advertising. It is unclear, however, whether alcohol advertising actually increases alcohol consumption. A study was recently undertaken to examine the effect of price advertising on sales of beer in Lower Michigan. The state of Michigan was selected because it has prohibited retailers from advertising the price of beer products since 1975, except for a brief period (March 1982–May 1983) when the ban was temporarily lifted. The data in the table are the bimonthly total sales of brewed beverages (in thousands of 31-gallon barrels) over the period May 1981–April 1984. The data allow us to compare total sales of beer in three periods, before (period 1), during (period 2), and after (period 3) the lifting of the price advertising restrictions.

PERIOD 1: PRICE ADVERTISING RESTRICTED (MAY/JUNE 1981–JAN./FEB. 1982)	PERIOD 2: NO RESTRICTIONS (MAR./APR. 1982–MAY/JUNE 1983)	PERIOD 3: PRICE ADVERTISING RESTRICTED (JULY/AUG. 1983–MAR./APR. 1984)
462	522	433
417	508	470
516	427	609
605	477	442
654	603	446
	692	
	584	
	496	

Source: Wilcox, G. B. "The effect of price advertising on alcoholic beverage sales." *Journal of Advertising Research*, Vol. 25, No. 5, Oct./Nov. 1985, pp. 33–37.

a. Treating this as a completely randomized design, identify the treatments for this experiment.

b. The data were subjected to an ANOVA using Minitab, which generated the accompanying printout. Identify the key elements on the printout.

```
ANALYSIS OF VARIANCE ON SALES
SOURCE     DF        SS        MS       F        p
PERIOD      2     11358      5679     0.78    0.476
ERROR      15    109153      7277
TOTAL      17    120510
                                    INDIVIDUAL 95 PCT CI'S FOR MEAN
                                    BASED ON POOLED STDEV
LEVEL       N      MEAN     STDEV    ----+---------+---------+---------+--
  1         5    530.80     98.22              (-------------*-------------)
  2         8    538.63     83.68                 (----------*---------)
  3         5    480.00     73.40    (-------------*-------------)
                                     ----+---------+---------+---------+--
POOLED STDEV =     85.30              420       480       540       600
```

c. Is there sufficient evidence to indicate differences in the average total sales of beer in the three periods? Test using $\alpha = .10$.

d. Find a 95% confidence interval for the mean sales of period 2 on the printout, and interpret the result.

13.9 When marketing its products in a foreign country, should a company use its own salespeople or salespeople from the target market country? To research this question, a study was designed to investigate the effect of salesperson nationality on buyer attitudes (*Journal of Business Research*, Vol. 22, 1991). A sample of American MBA students were divided into two groups and shown a videotape of an advertisement for forklift trucks made in India. For group 1, an Indian sales representative made the presentation; for group 2, an American sales representative made the presentation. After viewing the tape, the subjects were asked whether the salesperson was trustworthy (measured on a 50-point scale). The mean scores were compared using an ANOVA.

a. The ANOVA resulted in an F value of 2.32, with an observed significance level of .13. Is there evidence of a difference between the mean trustworthiness scores of the two groups of MBA students? Use $\alpha = .10$.

b. The sample mean scores for the two groups are $\bar{y}_1 = 3.12$ and $\bar{y}_2 = 3.49$. Suppose you were to test $H_0: \mu_1 = \mu_2$ against $H_a: \mu_1 < \mu_2$ at $\alpha = .10$. Use the result, part a, to make the proper conclusion. [*Hint:* Use Exercise 13.4 and the fact that the p-value for a two-tailed t test is double the p-value for a one-tailed test.]

13.10 In the past decade, many county-run local jails have acquired major responsibility for treating mentally ill patients. Historically, state mental hospitals have experienced conflicts between the correctional and mental health staffs. A study reported in *Criminology* was conducted to determine whether the conflict between correctional and mental health staffs found in state mental hospitals also exists in local jails. Staff members at 43 jails with mental health programs were mailed questionnaires addressing the conflict in the day-to-day delivery of mental health services in the jails. A total of $n = 167$ questionnaires were returned. In one portion of the study, respondents were classified into one of four groups according to staff affiliation: (1) jail correctional, (2) jail mental health, (3) county mental health, and (4) other mental health. The variable in question was "perceived level of conflict in providing treatment to inmates/patients," measured on a scale of 1 (little or no conflict) to 7 (extreme conflict). The data were subjected to a one-way analysis of variance on staff affiliation with the results summarized in the accompanying table.

SOURCE	df	SS	MS	F
Staff affiliation	—	21.31	7.10	—
Error	163	—	—	
Total	—	120.62		

Source: Steadman, H. J., Morrissey, J. P., and Robbins, P. C. "Reevaluating the custody–therapy conflict paradigm in correctional mental health settings." *Criminology*, Vol. 23, No. 1, 1985, pp. 165–179.

a. Complete the ANOVA table.

b. Is there sufficient evidence of a difference among the mean levels of conflict perceived by the four groups of jail staff members? Test using $\alpha = .05$.

13.11 The display consoles of modern computer-based systems use many abbreviated words to accommodate the large volume of information to be displayed. Therefore, operators must learn to decode each abbreviation quickly and accurately. An experiment was conducted to determine the optimal method for abbreviating any specific set of words on the sonar consoles used at the Naval Submarine Medical Research Laboratory in Groton, Connecticut (*Human Factors*, Feb. 1984). Of the 20 Navy and civilian personnel who took part in the study, five were highly familiar with the sonar system. The 15 subjects unfamiliar with the system were randomly divided into three groups of five. Thus, the study consisted of a total of four groups (one experienced and three inexperienced groups), with five subjects per group. The experienced group and one inexperienced group (denoted TE and TI, respectively) were assigned to learn the simple method of abbreviation. One of the remaining inexperienced groups was assigned the conventional single abbreviation method (denoted CS), whereas the other was assigned the conventional multiple abbreviation method (denoted CM). Each subject was then given a list of 75 abbreviations to learn, one at a time, through the display console of a minicomputer. The number of trials until the subject accurately decoded at least 90% of the words on the list was recorded. Do the data provide sufficient evidence to indicate differences among the mean numbers of trials required for the four groups? Test using $\alpha = .05$. (Use the accompanying SAS printout on page 742 to solve this problem.)

CM	CS	TE	TI
4	6	5	8
7	9	5	4
5	5	7	8
6	7	8	10
8	6	7	3

Source: Data are simulated values based on the group means reported in *Human Factors*, Feb. 1984. Copyright 1984 by the Human Factors Society, Inc. and reproduced by permission.

```
                        Analysis of Variance Procedure

Dependent Variable: TRIALS
                                  Sum of              Mean
Source                    DF      Squares            Square      F Value      Pr > F

Model                      3    1.20000000        0.40000000        0.10      0.9566

Error                     16   61.60000000        3.85000000

Corrected Total           19   62.80000000

                 R-Square              C.V.        Root MSE              TRIALS Mean

                 0.019108         30.658464       1.9621417             6.40000000

Source                    DF      Anova SS     Mean Square      F Value      Pr > F

GROUP                      3    1.20000000      0.40000000         0.10      0.9566
```

RANDOMIZED BLOCK DESIGNS

13.5 In Chapters 8 and 10, we learned that for the same number of observations, paired samples often provide more information on the difference between a pair of population means than do independent samples. In Section 10.6, we compared the length of wear of two types of jogging shoes. Instead of randomly assigning 10 joggers to each of the shoe types A and B (an independent sampling design), we used only 10 joggers in the experiment. Each jogger tested both types of shoes and then the difference in the lengths of wear was computed. By making the comparison of length of wear for shoe types A and B *within* a single jogger (a matched pair of observations), we were able to eliminate the error created by the large variation in wear among the different joggers.

Suppose we wanted to compare three types of jogging shoes—A, B, and C. We could use the same type of matching procedure as used for a matched-pairs design, except that we would match the experimental units in groups of three. For the jogging shoe example, we would assign to each of the 10 joggers the task of testing three pairs of jogging shoes—one each of types A, B, and C (see Figure 13.6). In the language of statistics, this is called a **randomized block design**, one consisting of $b = 10$ blocks, with three experimental units in each, and $k = 3$ treatments. A *treatment* is what makes the experimental units in one population differ from those in another population. The blocks always contain matched groups of k experimental units, one unit for each treatment.

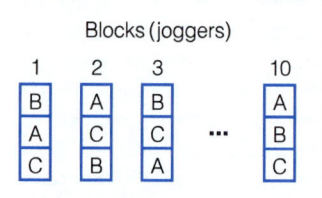

FIGURE 13.6

Diagram for a Randomized Block Design Containing $b = 10$ Blocks (Joggers) and $k = 3$ Treatments (Shoe Types)

EXAMPLE 13.6   A supermarket advertisement states: "You'll save up to 21% with Albertson's lower prices" (*Gainesville Sun*, Mar. 18, 1984). To substantiate the claim, Albertson's supermarket compared the prices of 49 grocery items at three competing supermarkets with the prices at its store on a given day. The survey results for seven items randomly selected from the 49 are shown in Table 13.6. How could we determine whether the mean prices of grocery items differ among the four supermarkets?

Solution   To determine whether the mean prices charged at the supermarkets differ, we will need to conduct an analysis of variance for a randomized block design. The columns of Table 13.6 correspond to the $k = 4$ treatments (i.e., the supermarkets) and the rows correspond to $b = 7$ blocks of grocery items, each consisting of $k = 4$ observations. The observations within a block are matched because all prices within a block are for the same item on the same day. A randomized block design is necessary to assure that the same items are compared at the four supermarkets.

TABLE 13.6
Supermarket Survey Results

GROCERY ITEM	ALBERTSON'S	KASH 'N KARRY	PUBLIX	FOOD 4 LESS
Cheerios cereal	$1.10	$1.18	$1.39	$1.18
JELL-O gelatin	.24	.24	.31	.26
Dial soap	.52	.60	.63	.55
Crisco oil	1.26	1.70	2.27	1.29
Kleenex	.67	.70	.79	.70
Star-Kist tuna	.63	.66	.79	.63
Del Monte peas	.43	.47	.65	.47

The analysis of variance for a randomized block design partitions SS(Total) into three portions: the sum of squares for treatments (SST), the sum of squares for blocks (SSB), and the sum of squares for error (SSE). The analysis of variance is shown diagrammatically in Figure 13.7 (page 744). The sum of squares for treatments measures the variation between treatment means. As in the case of the analysis of variance for a completely randomized design, we compare the mean square for treatments,

$$MST = \frac{SST}{k-1}$$

As with completely randomized designs, you should always check to determine whether the assumptions required to properly conduct an ANOVA for a randomized block design are satisfied (see Section 13.10). In situations where one or more of the assumptions are clearly violated, you will want to employ one of the nonparametric methods discussed in Chapter 15.

WARNING

When the assumptions for analyzing data collected from a randomized block design are violated, any inferences derived from the ANOVA are suspect. An alternative technique to use in this situation is the nonparametric Friedman test (see Section 15.6).

EXERCISES
LEARNING THE MECHANICS

13.12   A randomized block design was conducted to compare the mean responses for three treatments, A, B, and C, in four blocks. The data are shown in the accompanying table, followed by a partial summary ANOVA table.

	BLOCK			
TREATMENT	1	2	3	4
A	3	6	1	2
B	5	7	4	6
C	2	3	2	2

SOURCE	df	SS	MS	F
Treatments	—	23.167	—	—
Blocks	—	14.250	4.750	—
Error	—	—	.917	
Total	—	42.917		

a. Complete the ANOVA table.
b. Do the data provide sufficient evidence to indicate a difference among treatment means? Test using $\alpha = .05$.
c. Do the data provide sufficient evidence to indicate that blocking was effective in reducing the experimental error? Test using $\alpha = .05$.
d. What assumptions must the data satisfy to make the F tests in parts b and c valid?
e. Find a 95% confidence interval for the difference between the means of treatments A and C.

13.13 The analysis of variance for a randomized block design produced the ANOVA table entries shown in the table.

SOURCE	df	SS	MS	F
Treatments	3	27.1	—	—
Blocks	5	—	14.90	—
Error	—	33.4		
Total	—			

The sample means for the four treatments are as follows:

$$\bar{x}_A = 9.7 \qquad \bar{x}_B = 12.1 \qquad \bar{x}_C = 6.2 \qquad \bar{x}_D = 9.3$$

b. Do the data provide sufficient evidence to indicate a difference among the treatment means? Test using $\alpha = .01$.

c. Do the data provide sufficient evidence to indicate that blocking was a useful design strategy to employ for this experiment? Explain.

d. Find a 95% confidence interval for $(\mu_A - \mu_B)$.

e. Find a 95% confidence interval for $(\mu_B - \mu_D)$.

APPLYING THE CONCEPTS

13.14 *Physical Therapy* (Aug. 1986) reported on a study to "determine whether the medial rotation that accompanies flexion of the shoulder took place during the performance of the flexion-abduction-lateral-rotation proprioceptive neuromuscular facilitation pattern (D_2F)." Ten college students, who exhibited no evidence of disease or limitation of movement in their shoulders, served as the subjects for the study. For each subject, the medial rotation was measured (in degrees) at each of three positions in the D_2F pattern: (1) beginning position, (2) point at which rotation changed directions, and (3) ending position. The goal of the analysis is to compare the mean medial rotation measurements of the three positions.

a. Identify the treatments in this experiment.

b. Identify the blocks in this experiment.

c. Identify the response variable.

d. Explain why a randomized block design is appropriate for this experiment.

13.15 Refer to the comparison of prices of 60 food items at Winn-Dixie and Publix supermarkets, Exercises 8.60 and 10.45. In addition to the prices at Winn-Dixie and Publix, the newspaper advertisement also reported the prices for the same 60 grocery items at a third supermarket, Kash 'N Karry. The data for all three supermarkets are provided in the accompanying table on the next page.

a. Suppose we want to use the data to compare the mean prices of grocery items at the three supermarkets. Identify the treatments and blocks for this randomized block design.

b. The data were subjected to an ANOVA using SAS. Use the information in the SAS printout on page 750 to construct an ANOVA table for the data.

c. Test to determine whether the mean prices of grocery items differ among the three supermarkets. Use $\alpha = .01$.

d. Construct a 95% confidence interval for the difference between the mean prices per item at Winn-Dixie and Kash 'N Karry.

e. Does the interval obtained in part **d** provide evidence of a significant difference in mean prices per item between Winn-Dixie and Kash 'N Karry?

ITEM	WINN-DIXIE	PUBLIX	KASH 'N KARRY	ITEM	WINN-DIXIE	PUBLIX	KASH 'N KARRY
Big Thirst Towel	1.21	1.49	1.59	Keb Graham Crust	.79	1.29	1.28
Camp Crm/Broccoli	.55	.67	.67	Spiffits Glass	1.98	2.19	2.59
Royal Oak Charcoal	2.99	3.59	3.39	Prog Lentil Soup	.79	1.13	1.12
Combo Chdr/Chz Snk	1.29	1.29	1.39	Lipton Tea Bags	2.07	2.17	2.17
Sure Sak Trash Bag	1.29	1.79	1.89	Carnation Hot Coco	1.59	1.89	1.99
Dow Handi Wrap	1.59	2.39	2.29	Crystal Hot Sauce	.70	.87	.89
White Rain Shampoo	.96	.97	1.39	C/F/N/ Coffee Bag	1.17	1.15	1.55
Post Golden Crisp	2.78	2.99	3.35	Soup Start Bf Veg	1.39	2.03	1.94
Surf Detergent	2.29	1.89	1.89	Camp Pork & Beans	.44	.49	.58
Sacramento T/Juice	.79	.89	.99	Sunsweet Pit Prune	.98	1.33	1.10
SS Prune Juice	1.36	1.61	1.48	DM Vgcls Grdn Duet	1.07	1.13	1.29
V-8 Cocktail	1.18	1.29	1.28	Argo Corn Starch	.69	.89	.79
Rodd Kosher Dill	1.39	1.79	1.79	Sno Drop Bowl Clnr	.53	1.15	.99
Bisquick	2.09	2.19	2.09	Cadbury Milk Choc	.79	1.29	1.28
Kraft Italian Drs	.99	1.19	1.00	Andes Crm/De Ment	1.09	1.30	1.09
BC Hamburger Helper	1.46	1.75	1.75	Combat Ant & Roach	2.33	2.39	2.79
Comstock Chrry Pie	1.29	1.69	1.69	Joan/Arc Kid Bean	.45	.56	.38
Dawn Liquid King	2.59	2.29	2.58	La Vic Salsa Pican	1.22	1.75	1.49
DelMonte Ketchup	1.05	1.25	.59	Moist N Beef/Chz	2.39	3.19	2.99
Silver Floss Kraut	.77	.81	.69	Ortega Taco Shells	1.08	1.33	1.09
Trop Twist Beverag	1.74	2.15	2.25	Fresh Step Cat Lit	3.58	3.79	3.81
Purina Kitten Chow	1.09	1.05	1.29	Field Trial Dg/Fd	3.49	3.79	3.49
Niag Spray Starch	.89	.99	1.39	Tylenol Tablets	5.98	5.29	5.98
Soft Soap Country	.97	1.19	1.19	Rolaids Tablets	1.88	2.20	2.49
Northwood Syrup	1.13	1.37	1.37	Plax Rinse	2.88	3.14	2.53
Bumble Bee Tuna	.58	.65	.65	Correctol Laxative	3.44	3.98	3.59
Mueller Elbow/Mac	2.09	2.69	2.69	Tch Scnt Potpourri	1.50	1.89	1.89
Kell Nut Honey Crn	2.95	3.25	3.23	Chld Enema 2.250	.98	1.15	1.19
Cutter Spray	3.09	3.95	3.69	Gillette Atra Plus	5.00	5.24	5.59
Lawry Season Salt	2.28	2.97	2.85	Colgate Shave	.94	1.10	1.19

```
                    Analysis of Variance Procedure

Dependent Variable: PRICE
                                    Sum of           Mean
Source                    DF        Squares         Square    F Value     Pr > F

Model                     61     218.2361989      3.5776426    106.27     0.0001
Error                    118       3.9725322      0.0336655
Corrected Total          179     222.2087311

                   R-Square          C.V.        Root MSE        PRICE Mean

                   0.982123       9.989324       0.183482        1.83677778

Source                    DF      Anova SS     Mean Square    F Value     Pr > F

SUPERMKT                   2     2.6412678      1.3206339      39.23      0.0001
ITEM                      59   215.5949311      3.6541514     108.54      0.0001

                   Level of       ------------PRICE------------
                   SUPERMKT    N      Mean             SD

                   KNKarry    60   1.92533333      1.14118503
                   Publix     60   1.91950000      1.10339142
                   WinnDix    60   1.66550000      1.09622376
```

13.16 The Perth (Australia) Metropolitan Water Authority recently completed construction of land pipeline for transporting domestic wastewaters from a primary treatment plant. During construction, the cement mortar lining of the pipeline was tested for cracking to determine whether autogenous healing will seal the cracks. Otherwise, expensive epoxy filling repairs would be necessary. After the cracks were observed in the pipeline, it was kept full of water for a period of 14 weeks. At each of the 12 crack locations, crack widths were measured (in millimeters) after the 2nd, 6th, and 14th week of the wet period, as shown in the accompanying table. The data were subjected to an analysis of variance using SPSS. The SPSS printout is shown here.

CRACK LOCATION	CRACK WIDTH AFTER WETTING			
	0 Weeks	2 Weeks	6 Weeks	14 Weeks
1	.50	.20	.10	.10
2	.40	.20	.10	.10
3	.60	.30	.15	.10
4	.80	.40	.10	.10
5	.80	.30	.05	.05
6	1.00	.40	.05	.05
7	.90	.25	.05	.05
8	1.00	.30	.05	.10
9	.70	.25	.10	.10
10	.60	.25	.10	.05
11	.30	.15	.10	.05
12	.30	.14	.05	.05

Source: Cox, B. G. and Kelsall, K. J. "Construction of Cape Peron ocean outlet Perth, Western Australia." *Proceedings of the Institution of Civil Engineers*, Part I, Vol. 80, Apr. 1986, p. 479 (Table 1).

* * * A N A L Y S I S   O F   V A R I A N C E   * * *

WIDTH
BY PERIOD
    LOCATION

Source of Variation	Sum of Squares	DF	Mean Square	F	Signif of F
Main Effects	2.962	14	.212	13.708	.000
PERIOD	2.685	3	.895	57.988	.000
LOCATION	.277	11	.025	1.632	.135
Explained	2.962	14	.212	13.708	.000
Residual	.509	33	.015		
Total	3.471	47	.074		

a. Conduct a test to determine whether the mean crack widths differ for the four time periods. Test using $\alpha = .05$.
b. Construct a 95% confidence interval for the difference between the initial mean crack width (0 weeks) and the mean crack width after wetting for 14 weeks. Interpret the interval.

13.17 Plant therapists believe that plants can reduce the stress levels of humans. A Kansas State University study was conducted to investigate this phenomenon. Two weeks prior to final exams, 10 undergraduate students took part in an experiment to determine what effect the presence of a live plant, a photo of a plant, or the absence of a plant has on the student's ability to relax while isolated in a dimly lit room. Each student participated in

FIGURE 13.10

Hypothetical Plot of the Means
for the Six Machine–Material
Combinations

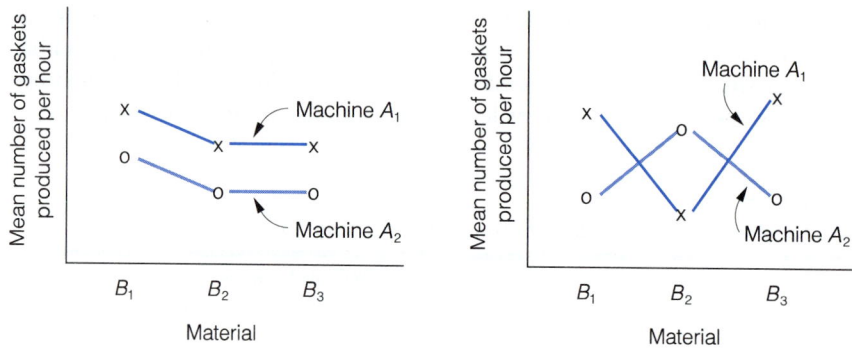

a. No interaction   **b.** Interaction

Solution Figure 13.10a suggests that machine A_1 produces a larger number of gaskets per hour, regardless of the gasket material, and is therefore superior to machine A_2. On the average, machine A_1 stamps more cork (B_1) gaskets per hour than rubber or plastic, but the *difference* in the mean numbers of gaskets produced by the two machines remains approximately the same, regardless of the gasket material. Thus, the difference between the mean numbers of gaskets produced by the two machines is *independent* of the gasket material used in the stamping process.

In contrast to Figure 13.10a, Figure 13.10b shows the productivity of machine A_1 to be larger than the productivity of machine A_2 when the gasket material is cork (B_1) or plastic (B_3). But the means are reversed for rubber (B_2) gasket material. For this material, machine A_2 produces, on the average, more gaskets per hour than machine A_1. Thus, Figure 13.10b illustrates a situation where the mean value of the response variable *depends* on the combination of the factor levels. When this situation occurs, we say that the factors **interact**. Thus, one of the most important objectives of a factorial experiment is to detect factor interaction if it exists.

DEFINITION 13.12

In a factorial experiment, when the difference between the mean levels of factor A depends on the different levels of factor B, we say that the factors A and B **interact**. If the difference is independent of the levels of B, then there is **no interaction** between factors A and B.

The analysis of variance for a two-factor factorial experiment is very similar to the analysis of variance for a randomized block design. The sums of squares for the qualitative variables, blocks and treatments in the randomized block design, are now replaced by the sums of squares for the two factors, SS(A) and SS(B), called **main effects sums of squares**. The failure of the difference in the mean levels of factor A to be the same for all levels of factor B is reflected in the interaction sum of squares, SS(AB). Finally, because we have more than one observation per cell for the two-way table, we calculate a sum of squares designated as SSE. The partitioning of SS(Total) into its parts is shown in Figure 13.11.

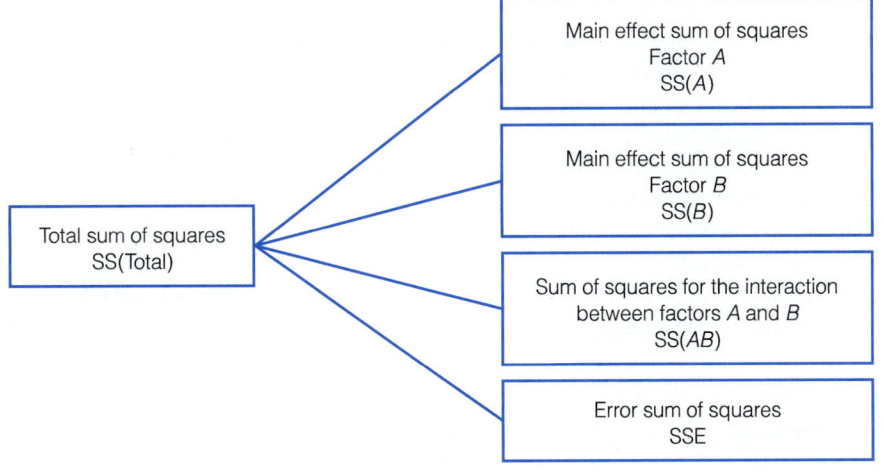

FIGURE 13.11
Partitioning of the Total Sum of Squares for a Two-Factor Factorial Experiment

The formulas for the sums of squares and the corresponding mean squares for a factorial experiment are given in optional Section 13.11. As in the previous two sections, we will resort to one of the many statistical software packages available to perform the analysis of the data, and we will summarize the ANOVA results in a table. In general, the ANOVA table for a two-factor factorial experiment, with factor A at a levels, factor B at b levels, and with r replications, appears as shown in the box.

ANALYSIS OF VARIANCE TABLE FOR A TWO-FACTOR FACTORIAL EXPERIMENT

SOURCE	df	SS	MS	F
Main effect, A	$a - 1$	SS(A)	$MS(A) = \dfrac{SS(A)}{a - 1}$	$F = \dfrac{MS(A)}{MSE}$
Main effect, B	$b - 1$	SS(B)	$MS(B) = \dfrac{SS(B)}{b - 1}$	$F = \dfrac{MS(B)}{MSE}$
AB interaction	$(a - 1)(b - 1)$	SS(AB)	$MS(AB) = \dfrac{SS(AB)}{(a - 1)(b - 1)}$	$F = \dfrac{MS(AB)}{MSE}$
Error	$ab(r - 1)$	SSE	$MSE = \dfrac{SSE}{ab(r - 1)}$	
Total	$abr - 1$	SS(Total)		

You can see that the number of degrees of freedom for a main effect is always equal to 1 less than the number of levels for the factor. The number of degrees of freedom for the AB interaction is always equal to the product of the main effects degrees of freedom—namely, $(a - 1)(b - 1)$. Finally, note that the degrees of freedom for the sources sum to the $(abr - 1) = (n - 1)$ degrees of freedom for SS(Total).

The mean squares are computed in the same manner as for all analyses of variance, by dividing the respective sums of squares by their degrees of freedom. Also, the F statistics for testing the three sources of variation (i.e., the source corresponding to main effect factor A, the source corresponding to main effect factor B, and the source corresponding to AB interaction) are formed by dividing a source mean square by MSE. The numerator degrees of freedom for an F statistic is the number associated with the mean square used in the numerator of the F statistic. The denominator degrees of freedom is the number, $ab(r - 1)$, of degrees of freedom for MSE.

A summary of the ANOVA F tests for a factorial experiment is provided in the box on the next page.

WARNING

When the assumptions for analyzing data collected from a factorial experiment (see box) are violated, any inferences derived from the ANOVA are suspect. Nonparametric methods are available for analyzing factorial experiments, but they are beyond the scope of this text. Consult the references given at the end of this chapter if you want to learn about such techniques.

EXAMPLE 13.12 Refer to Examples 13.10 and 13.11. A SAS printout for the 2×3 factorial ANOVA of the data in Table 13.8 is shown in Figure 13.12. Identify the pertinent elements of the printout.

FIGURE 13.12

Portion of the SAS Printout for Factorial ANOVA of Example 13.12

Analysis of Variance Procedure

Dependent Variable: NUMBER

Source	DF	Sum of Squares	Mean Square	F Value	Pr > F
Model	5	1.68122778	0.33624556	76.52	0.0001
Error	12	0.05273333	0.00439444		
Corrected Total	17	1.73396111			

R-Square	C.V.	Root MSE	NUMBER Mean
0.969588	1.734095	0.066291	3.82277778

Source	DF	Anova SS	Mean Square	F Value	Pr > F
MACHINE	1	0.10125000	0.10125000	23.04	0.0004
MATERIAL	2	0.81194444	0.40597222	92.38	0.0001
MACHINE*MATERIAL	2	0.76803333	0.38401667	87.39	0.0001

ANOVA *F* TESTS FOR TWO-FACTOR FACTORIAL EXPERIMENT

TEST FOR FACTOR INTERACTION

H_0:   No interaction between factors A and B

H_a: Factors A and B interact

Test statistic: $F = \dfrac{\text{MS}(AB)}{\text{MSE}} = \dfrac{\text{MS}(AB)}{s^2}$

Rejection region: $F > F_\alpha$, where F is based on $v_1 = (a - 1)(b - 1)$ and $v_2 = ab(r - 1)$df

TEST FOR MAIN EFFECTS FOR FACTOR *A*

H_0:   There are no differences among the means for main effect A

H_a: At least two of the main effect A means differ

Test statistic: $F = \dfrac{\text{MS}(A)}{\text{MSE}} = \dfrac{\text{MS}(A)}{s^2}$

Rejection region: $F > F_\alpha$, where F is based on $v_1 = (a - 1)$ and $v_2 = ab(r - 1)$df

TEST FOR MAIN EFFECTS FOR FACTOR *B*

H_0:   There are no differences among the means for main effect B

H_a: At least two of the main effect B means differ

Test statistic: $F = \dfrac{\text{MS}(B)}{\text{MSE}} = \dfrac{\text{MS}(B)}{s^2}$

Rejection region: $F > F_\alpha$, where F is based on $v_1 = (b - 1)$ and $v_2 = ab(r - 1)$df

Assumptions: 1. The population probability distribution of the observations for any factor level combination is approximately normal.

2. The variance of the probability distribution is constant and the same for all factor level combinations.

3. The treatments (factor level combinations) are randomly assigned to the experimental units.

4. The observations for each factor level combination represent independent random samples.

Solution   The SAS printout for the factorial ANOVA is separated into two shaded sections. The upper shaded area in Figure 13.12 shows the partitioning of SS(Total) into two sources of variation, **Model** and **Error**. The source designated as **Model** represents the sum of the factor main effects and interaction. You can see that SSE = .05273333 and MSE = .00439444 in the **Error** row.

The bottom shaded section of Figure 13.12 gives the breakdown of the **Model** source sum of squares into the sum of squares due to the factor main effects (**MACHINE** and **MATERIAL**) and the sum of squares due to factor interaction (**MACHINE*MATERIAL**). These sums of squares are $SS(A) = .10125000$, $SS(B) = .81194444$, and $SS(AB) = .76803333$. Note that the sum $SS(A) + SS(B) + SS(AB)$ is equal to the sum of squares for **Model** shown in the upper shaded section. The F values 23.04 and 92.38 represent the ANOVA F statistics for testing the corresponding factor main effects. The F value 87.39 represents the ANOVA F statistic for testing interaction between the factors. Note that the degrees of freedom associated with the three sources are given in the respective rows in the **DF** column; $df(A) = 1$, $df(B) = 2$, and $df(AB) = 2$.

EXAMPLE 13.13 Refer to Example 13.12. Use the information in the SAS printout in Figure 13.12 to construct an ANOVA summary table for the factorial experiment.

Solution The only ANOVA quantities that are not already calculated and shown in Figure 13.12 are the mean squares for the three sources A, B, and AB interaction. From the formulas in the box, we have

$$MS(A) = \frac{SS(A)}{a - 1} = \frac{.101}{1} = .101$$

$$MS(B) = \frac{SS(B)}{b - 1} = \frac{.812}{2} = .406$$

$$MS(AB) = \frac{SS(AB)}{(a - 1)(b - 1)} = \frac{.768}{(1)(2)} = .384$$

The complete ANOVA table for the 2×3 factorial experiment is shown in Table 13.9.

TABLE 13.9
ANOVA Table for the Two-Factor Factorial
Experiment of Examples 13.12–13.13

SOURCE	df	SS	MS	F
Machines, A	1	.101	.101	22.95
Materials, B	2	.812	.406	92.27
AB Interaction	2	.768	.384	87.27
Error	12	.053	.0044	
Total	17	1.734		

EXAMPLE 13.14 A plot of the six means corresponding to the six machine–material combinations is shown in Figure 13.13. Do you think that the factors A and B interact? Perform a test of hypothesis to determine whether the data provide sufficient evidence to indicate that the more productive stamping machine depends on the gasket material. Test using $\alpha = .05$.

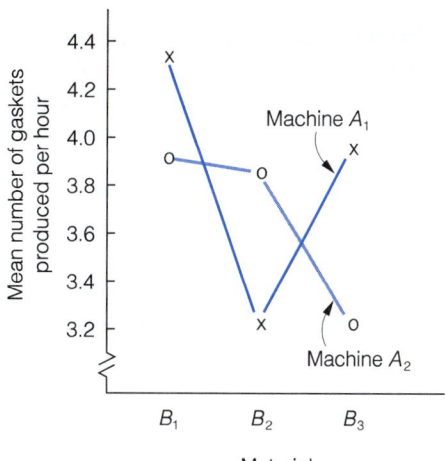

FIGURE 13.13
Plot of the Means for the Six
Machine–Material Combinations

Solution   The test statistic for testing the null hypothesis of no interaction between machines and materials is

$$F = \frac{MS(AB)}{MSE} = 87.27$$

where the F statistic has 2 numerator df and 12 denominator df. When $MS(AB)$ is substantially larger than MSE, i.e., when F is large, we will reject H_0 and conclude that there is an interaction between machines and materials. Referring to Table 8 of Appendix G, we find the critical value to be $F_{.05} = 3.89$. Therefore, the rejection region for the test is

$$F > F_{.05} = 3.89$$

Since the computed value $F = 87.27$ exceeds $F_{.05} = 3.89$, we conclude that there is an interaction between machines and materials. Therefore, there is sufficient evidence to indicate that neither machine is the more productive for all three materials. (This result could also be obtained by noting that the p-value for the test, given on Figure 13.12 in the **MACHINE*MATERIAL** row, is .0001.) If the differences in mean productivity are large enough, the manufacturer should use both machines, selecting the machine that gives the greater productivity for a specific material.

Tests for differences in the mean levels of the main effects in a factorial experiment are relevant *only when the factors do not interact*. When there is no factor interaction, the differences in the mean levels of factor A are the same for all levels of factor B. The test for main effect A tests the significance of these differences. In the presence of interaction, however, the main effect test is irrelevant since the differences in the mean levels of factor A are not the same at each level of factor B. The following example demonstrates the mechanics of the tests for main effects and discusses the practical implications of the tests.

13.22 The data for a 3 × 4 factorial experiment with two observations per treatment are shown in the accompanying table.

		FACTOR B 1	2	3	4
	1	5 4	7 9	6 5	5 7
FACTOR A	2	6 4	10 9	5 8	9 7
	3	8 10	7 6	5 8	6 5

An analysis of variance was conducted on the data. The results are reported in the accompanying ANOVA summary table.

SOURCE	df	SS	MS	F
A	2	6.58333	3.2917	1.84
B	3	13.79167	4.5972	2.57
AB	6	35.08333	5.8472	3.26
Error	12	21.50000	1.7917	
Total	23	76.95833		

a. Do the data provide sufficient evidence of interaction between factor A and factor B? Test using $\alpha = .05$.
b. Given the results of the test in part **a**, would you recommend that tests for main effects be conducted? Explain.
c. Find a 90% confidence interval for the mean of the factor level combination A_2B_3.
d. Find a 95% confidence interval for the difference between the means of factor level combinations A_2B_3 and A_3B_3.

APPLYING THE CONCEPTS

13.23 Video games have revolutionized children's leisure time activities. However, many parents, including the U.S. surgeon general, believe that video games are a bad influence on their children. A study was conducted to examine the effect of playing video games on fifth-graders' free play (*Journal of Applied Social Psychology*, Vol. 16, 1986). Eighty-four fifth-graders were paired randomly, and each pair was randomly assigned to one of three types of games, an aggressive video game (Missile Command), a nonaggressive video game (Pac-Man), or a pen-and-paper maze-solving game (control) in equal numbers. One member of each pair was then randomly chosen to play the designated game (player) for 8 minutes, while the other member watched (observer). Thus, 14 fifth-graders were assigned to each of the 3 × 2 = 6 experimental conditions. After video play was concluded, the children were sent to a toy room for free play. The goal of the experiment was to investigate the effect of type of game (Missile Command, Pac-Man, and control) and position (player or observer) on degree of aggressive play in the toy room.

a. Identify the factors in this factorial experiment.
b. Identify the levels of the factors.
c. What are the treatments?
d. Give the sources of variation and their respective degrees of freedom in an ANOVA table for this experiment.
e. A significant interaction was found between type of game and position (*p*-value < .01). Interpret this result in the words of the problem.

13.24 The *Accounting Review* (Jan. 1991) reported on a study of the effect of two factors, confirmation of accounts receivable and verification of sales transactions, on account misstatement risk by auditors. Both factors were held at the same two levels: completed or not completed. Thus, the experimental design is a 2 × 2 factorial design.

    a. Identify the factors, factor levels, and treatments for this experiment.

    b. Explain what factor interaction means for this experiment.

    c. A graph of the hypothetical mean misstatement risks for each of the 2 × 2 = 4 treatments is displayed here. In this hypothetical case, does it appear that interaction exists?

Source: Brown, C. E. and Solomon, I. "Configural information processing in auditing: The role of domain-specific knowledge." *The Accounting Review*, Vol. 66, No. 1, Jan. 1991, p. 105 (Figure 1).

13.25 Computer-based management information systems (MIS) is one of the fastest growing industries in the United States, yet little is known about the ethical decision-making processes of persons involved in creating and maintaining these systems. An empirical investigation was conducted to determine whether MIS majors, on average, exhibit ethical decision-making processes that differ from other business students (*Journal of Business Ethics*, Vol. 10, 1991). A large sample of business students were divided into two groups based on their major (MIS or non-MIS). Within each group, half of the students were administered the regular form of the Defining Issues Test (DIT) and the other half, a form of the DIT modified to incorporate MIS. Thus, a 2 × 2 factorial experimental design was used, with the two factors as major and form. The dependent (response) variable measured was the "principled morality," or *P* score, expressed as a percentage. (High scores are indicative of morally conscious decisions.) Interpret the results of the ANOVA summarized in the table.

SOURCE	df	SS	MS	F	p-VALUE
Form	1	.13	.13	0	.98
Major	1	1,290.10	1,290.10	6.25	.01
Form × Major	1	56.38	56.38	.27	.60
Error	233	48,120.15	206.52		
Total	236	49,466.76			

Source: Paradice, D. B. and Dejoie, R. M. "The ethical decision-making processes of information systems workers." *Journal of Business Ethics*, Vol. 10, 1991, p. 9 (Table V).

13.26 The chemical element antimony is sometimes added to tin–lead solder to replace the more expensive tin and to reduce the cost of soldering. A factorial experiment was conducted to determine how antimony affects the strength of the tin–lead solder joint (*Journal of Materials Science*, May 1986). Tin–lead solder specimens were prepared using one of four possible cooling methods (water-quenched, WQ; oil-quenched, OQ; air-blown, AB; and furnace-cooled, FC) and with one of four possible amounts of antimony (0%, 3%, 5%, and 10%) added to the composition. Three solder joints were randomly assigned to each of the $4 \times 4 = 16$ treatments and the shear strength of each measured. The experimental results, shown in the accompanying table, were subjected to an ANOVA using SAS. The SAS printout is also reproduced here.

AMOUNT OF ANTIMONY % weight	COOLING METHOD	SHEAR STRENGTH MPa
0	WQ	17.6, 19.5, 18.3
0	OQ	20.0, 24.3, 21.9
0	AB	18.3, 19.8, 22.9
0	FC	19.4, 19.8, 20.3
3	WQ	18.6, 19.5, 19.0
3	OQ	20.0, 20.9, 20.4
3	AB	21.7, 22.9, 22.1
3	FC	19.0, 20.9, 19.9
5	WQ	22.3, 19.5, 20.5
5	OQ	20.9, 22.9, 20.6
5	AB	22.9, 19.7, 21.6
5	FC	19.6, 16.4, 20.5
10	WQ	15.2, 17.1, 16.6
10	OQ	16.4, 19.0, 18.1
10	AB	15.8, 17.3, 17.1
10	FC	16.4, 17.6, 17.6

Source: Tomlinson, W. J. and Cooper, G. A. "Fracture mechanism of brass/Sn-Pb-Sb solder joints and the effect of production variables on the joint strength." *Journal of Materials Science*, Vol. 21, No. 5, May 1986, p. 1731, Table II. Copyright 1986 Chapman and Hall.

```
                    Analysis of Variance Procedure

Dependent Variable: STRENGTH
                                 Sum of          Mean
Source                DF         Squares         Square      F Value    Pr > F

Model                 15    157.95250000     10.53016667       6.10     0.0001

Error                 32     55.24666667      1.72645833

Corrected Total       47    213.19916667

             R-Square          C.V.       Root MSE               STRENGTH Mean

             0.740868     6.7195275      1.3139476                  19.55416667


Source                DF      Anova SS   Mean Square    F Value    Pr > F

AMOUNT                 3    104.194167     34.731389      20.12     0.0001
METHOD                 3     28.627500      9.542500       5.53     0.0036
AMOUNT*METHOD          9     25.130833      2.792315       1.62     0.1523
```

a. Construct an ANOVA summary table for the experiment.
b. Conduct a test to determine whether the two factors, amount of antimony and cooling method, interact. Use $\alpha = .01$.
c. Interpret the result obtained in part b.
d. If appropriate, conduct the tests for main effects. Use $\alpha = .01$.
e. Find a 99% confidence interval for the mean shear strength of tin–lead solder joints composed of 5% antimony and cooled with the air-blown method.
f. Find a 99% confidence interval for the difference between the mean shear strengths of 5% antimony–tin–lead solder joints cooled under two methods, water-quenched and oil-quenched.

13.27 How do women compare with men in their ability to perform laborious tasks that require strength? Some information on this quesiton is provided in a study of the firefighting ability of men and women ("Shipboard Fire-Fighting Performance of Females and Males," *Human Factors*, Vol. 24, 1982). The researchers conducted a 2 × 2 factorial experiment to investigate the effect of the factor sex (male or female) and the factor weight (light or heavy) on the length of time required for a person to perform a particular firefighting task. Eight persons were selected for each of the 2 × 2 = 4 sex–weight categories of the 2 × 2 factorial experiment, and the length of time needed to complete the task was recorded for each of the 32 persons. The means and standard deviations of the four samples are shown in the table.

| | LIGHT WEIGHT | | HEAVY WEIGHT | |
	Mean	Standard Deviation	Mean	Standard Deviation
Female	18.30	6.81	14.50	2.93
Male	13.00	5.04	12.25	5.70

Source: *Human Factors*, 1982, Vol. 24. Copyright 1982 by the Human Factors Society, Inc. and reproduced by permission.

```
ANALYSIS OF VARIANCE   TIME

SOURCE          DF        SS         MS
SEX              1    114.005    114.005
WEIGHT           1     41.005     41.005
INTERACTION      1     18.605     18.605
ERROR           28    789.968     28.213
TOTAL           31    963.983
```

a. The data were subjected to an analysis of variance using Minitab. Use the accompanying Minitab printout to construct an ANOVA for the data.
b. Plot the eight means from the table on a graph similar to Figure 13.9. Does it appear that the two factors, sex and weight, interact? Explain.
c. Do the data provide sufficient evidence of interaction between sex and weight? Test using $\alpha = .05$.
d. Do the data provide sufficient evidence that mean time to complete the task differs for male and female firefighters?
e. When is the test of part d appropriate?
f. Construct a 95% confidence interval for the difference in mean time to complete the task between light men and light women. Interpret the interval.
g. Construct a 95% confidence interval for the difference in mean time to complete the task between heavy men and heavy women. Interpret the interval.

13.28 Many television and radio commercials utilize high-intensity stimulation (e.g., rapid changes in visual imagery, quick movements, bright lights, and loud sounds) to increase the general arousal of the viewer or listener. In theory, the more aroused the viewer, the more likely he or she is to remember the advertised product. *Psychology & Marketing* (Summer 1986) reported on a study designed to examine the effect of high-intensity advertising on the audience. Two groups of psychology students (10 introverts and 10 extroverts) at a small midwestern liberal arts college took part in the experiment. (Scores on the Eysenck Personality Questionnaire were used to classify the students according to the two groups.) Five subjects from each group were then randomly assigned to one of two experimental conditions: high-volume commercial or normal-volume commercial. Thus, the experiment consists of two factors, personality type and commercial volume, each at two levels. All students listened to a 5-minute tape-recorded "radio program" that included a commercial for a fictitious brand of chewing gum played at the assigned volume; each subject's attitude toward the product was measured on a 14-point scale (1 = strongly dislike, 14 = strongly like). Since five responses were recorded for each of the 2 × 2 = 4 factor level combinations, the factorial experiment includes five replicates of each treatment.

TABLE 13.11

Sample Means for the $k = 6$ Treatments of the Factorial Experiment, Example 13.18

		MATERIAL		
		B_1	B_2	B_3
MACHINE	A_1	4.33	3.42	3.95
	A_2	3.91	3.85	3.48

Solution From Examples 13.12–13.13, we have $s = \sqrt{MSE} = .066$, $\nu = 12$ df for error, and $n_i = n_j = 3$ observations for all treatment pairs (i, j). For $k = 6$ means, the number of pairwise comparisons to be made is

$$g = \frac{k(k-1)}{2} = \frac{6(5)}{2} = 15$$

Also, $\alpha^* = \alpha/15 = \frac{.03}{15} = .002$. Thus, according to the box, we need to find the critical value, $t_{\alpha^*/2} = t_{.002/2} = t_{.001}$, for the t distribution based on $\nu = 12$ df. This value, shown in Table 5 of Appendix G, is 3.930. Substituting $t_{.001} = 3.930$ into the equation for Bonferroni's critical difference B_{ij}, we have

$$B_{ij} = (t_{.001})s\sqrt{\frac{1}{n_i} + \frac{1}{n_j}} = (3.930)(.066)\sqrt{\frac{1}{3} + \frac{1}{3}} = .212$$

for any treatment pair (i, j). Therefore, population means corresponding to pairs of sample means that differ by more than .212 will be judged to be significantly different. The six treatment means are ranked as follows:

Sample means:	3.42	3.48	3.85	3.91	3.95	4.33
Treatments (A_iB_j):	(A_1B_2)	(A_2B_3)	(A_2B_2)	(A_2B_1)	(A_1B_3)	(A_1B_1)

Using $B_{ij} = .212$ as a yardstick to determine differences between pairs of treatments, we have placed connecting horizontal bars over those means that *do not* significantly differ.

You can see that the treatment corresponding to machine A_1 and material B_1 has a significantly higher mean number of gaskets produced per hour than the other five treatments. The bar over the three means for treatments (A_2B_2), (A_2B_1), and (A_1B_3) indicates that we are unable to detect differences between any pair of these treatments. However, these treatments have significantly larger means than those for treatments (A_1B_2) and (A_2B_3). In summary, the treatment means appear to fall into three groups, as shown here:

	TREATMENTS
Group 1: (highest mean number produced per hour)	A_1B_1
Group 2:	A_2B_2, A_2B_1, A_1B_3
Group 3: (lowest mean number produced per hour)	A_1B_2, A_2B_3

Bonferroni's method guarantees that all inferences derived from this analysis can be made at an overall confidence level of at least $(1 - \alpha) = (1 - .03) = .97$.

The technique presented in the previous two examples will not always lead to distinct groupings of the treatment means. For example, suppose the rankings of six treatment means (denoted A, B, C, ..., F) appear as follows:

Treatments: A B C D E F
                  Low   Middle High

Because of the overlapping bars above treatment E, it is unclear into which group we should place this treatment—the group with the highest mean (treatment F) or the group with the middle means (treatments C and D). The most we can say is that, although the sample mean associated with treatment F is significantly larger than the corresponding means for treatments C and D, the sample mean for treatment E does not differ significantly from C, D, or F when compared in a pairwise fashion.

In this section we have presented a multiple comparisons procedure that could be employed in a follow-up analysis to ANOVA. Keep in mind, however, that many other methods of making multiple comparisons are available and one or more of these techniques may be more appropriate to use in your particular application. Consult the references given at the end of this chapter for details on other techniques.

In closing, we remind you that multiple comparisons of treatment means should be performed only as a follow-up analysis to the ANOVA, i.e., only after we have conducted the appropriate analysis of variance F test(s) and determined that sufficient evidence exists of differences among the treatment means. Be wary of conducting multiple comparisons when the ANOVA F test indicates no evidence of a difference among treatment means—this may lead to confusing and contradictory results.*

> **WARNING**
>
> In practice, it is advisable to avoid conducting multiple comparisons of a small number of treatment means when the corresponding ANOVA F test is nonsignificant; otherwise, confusing and contradictory results may occur.

EXERCISES
LEARNING THE MECHANICS

13.30   Consider a completely randomized design with k treatments and total sample size n. For each of the following, find the critical value of t used in Bonferroni's multiple comparisons procedure if all pairwise comparisons of treatments are to be made.

*When a large number of treatments are to be compared, a borderline, nonsignificant F value (e.g., $.05 < p\text{-value} < .10$) may mask differences between some of the means. In this situation, it is better to ignore the F test and proceed directly to a multiple comparisons procedure.

a. $k = 3$, $n = 15$, $\alpha = .06$ b. $k = 5$, $n = 20$, $\alpha = .10$
c. $k = 4$, $n = 40$, $\alpha = .06$ d. $k = 5$, $n = 25$, $\alpha = .05$

13.31 Consider a randomized block design with k treatments and b blocks. For each of the following, find the critical value of t used in Bonferroni's multiple comparisons procedure if all pairwise comparisons of treatments are to be made.

a. $k = 5$, $b = 4$, $\alpha = .02$ b. $k = 3$, $b = 5$, $\alpha = .03$
c. $k = 6$, $b = 10$, $\alpha = .10$ d. $k = 5$, $b = 6$, $\alpha = .10$

13.32 Consider a 2×2 factorial experiment with r replications. For each of the following, find the critical value of t used in Bonferroni's multiple comparisons procedure if all pairwise comparisons of treatments are to be made.

a. $r = 2$, $\alpha = .06$ b. $r = 4$, $\alpha = .03$ c. $r = 10$, $\alpha = .05$

13.33 Calculate Bonferroni's critical difference for each pairwise comparison of the three treatment means of Exercise 13.2. Use $\alpha = .09$.

13.34 Calculate Bonferroni's critical difference for each pairwise comparison of the three treatment means of Exercise 13.12. Use $\alpha = .06$.

13.35 Calculate Bonferroni's critical difference for each pairwise comparison of the $3 \times 4 = 12$ treatment means of Exercise 13.22. Use $\alpha = .15$.

APPLYING THE CONCEPTS

13.36 Videocassette recorders (VCRs) are currently being used in over half of American households, and the percentage is expected to reach 85% by 1995. Are all VCR users alike or can they be segmented into subgroups with different motives and behaviors? This question was the topic of research reported in the *Journal of Advertising Research* (Apr./May 1988). A sample of 371 members of a large videotape rental club in a southeastern city were surveyed about their VCR use. Based on their responses, each member was categorized into one of five groups:

	GROUP	NUMBER IN SAMPLE
(1) *Videophile:*	Record TV programs often, and rent/buy videotapes often	61
(2) *Time shifter:*	Record TV programs often, rarely rent/buy videotapes	74
(3) *Source shifter:*	Rarely record TV programs, rent/buy videotapes often	50
(4) *Low user:*	Rarely record TV programs or rent/buy videotapes	58
(5) *Regular user:*	Periodically record TV programs and/or rent/buy videotapes	128
Total		371

One of the dependent variables measured was degree to which the user "zipped" (i.e., fast-forwarded) through commercials while replaying a taped TV program. This "ad avoidance" variable was measured on a 7-point scale, where $1 =$ almost always and $7 =$ never.

a. The F value for testing the hypothesis H_0: $\mu_1 = \mu_2 = \mu_3 = \mu_4 = \mu_5$ is $F = 5.4$. Interpret this value (use $\alpha = .05$).
b. The mean "ad avoidance" levels of the five groups are listed here, as well as the results of a multiple comparisons analysis (at $\alpha = .05$). Interpret the results.

Mean "ad avoidance":	1.8	2.6	2.8	2.9	3.4
VCR user segment:	Time Shifter	Videophile	Regular User	Low User	Source Shifter

13.37 Refer to the *Journal of Abnormal Psychology* study of self-esteem, eating disorders, and depression in women described in Exercise 13.5. A follow-up analysis, similar to the Bonferroni multiple comparisons procedure, was conducted on the three group means. The ranked means are listed here, with overbars connecting the means that are not significantly different at $\alpha = .01$. Interpret the results.

Group:	Depression	Eating Disorders	Normal
Mean self-esteem rating:	3.6	3.9	4.9

13.38 Refer to Exercise 13.6. The mean levels of polysaccharides found in the five groups of Dutchmen are provided in the accompanying table. Use Bonferroni's method to rank the group means. Use $\alpha = .01$.

FISH CONSUMPTION Grams/Day	SAMPLE SIZE	MEAN LEVEL OF POLYSACCHARIDES Percentage of Energy
0	159	27.0
1–14	283	27.0
15–29	215	26.6
30–44	116	25.7
45 or more	79	24.4

13.39 Refer to the *Criminology* study described in Exercise 13.10. The mean levels of conflict perceived by the four groups of jail staff members are provided in the table. Compare the means (at $\alpha = .06$) using Bonferroni's procedure. Interpret the result.

STAFF AFFILIATION	SAMPLE SIZE	MEAN LEVEL OF CONFLICT
Jail correctional	58	1.81
Jail mental health	29	2.32
County mental health	52	2.32
Other mental health	28	2.84

13.40 Refer to the video display terminals study of Exercise 13.19. Use Bonferroni's method to rank the mean preference scores for the seven video display color combinations. Use $\alpha = .10$.

13.41 Refer to the *Journal of Material Science* study, Exercise 13.26. Use a multiple comparisons procedure to compare the mean shear strengths for the four antimony amounts. Identify the means that appear to differ. Use $\alpha = .01$.

13.42 Refer to the *Human Factors* (Apr. 1990) study of recognizer accuracy at three levels (90%, 95%, and 99%) and vocabulary size at three levels (75%, 87.5%, and 100%) on the performance of a computerized speech recognizer, Exercise 12.38. The data on task completion times (minutes) were subjected to an analysis of variance for a 3×3 factorial design. The F test for accuracy–vocabulary interaction resulted in a p-value less than .0003.

a. Interpret the result of the test for interaction.
b. As a follow-up to the test for interaction, the mean task completion times for the three levels of accuracy were compared under each level of vocabulary. Comment on this method of analysis.
c. Refer to part b. Bonferroni's multiple comparison method was used to compare the three accuracy means within each level of vocabulary at an overall error rate of $\alpha = .05$. The results are summarized on page 776. Interpret these results.

MEAN TASK COMPLETION TIME			
	ACCURACY LEVEL		
VOCABULARY SIZE	99%	95%	90%
75%	15.49	19.29	22.19
87.5%	12.77	14.31	16.48
100%	8.67	9.68	11.88

Source: Casali, S. P., Williges, B. H., and Dryden, R. D. "Effects of recognition accuracy and vocabulary size of a speech recognition system on task performance and user acceptance." *Human Factors*, Vol. 32, No. 2, Apr. 1990, p. 190 (Figure 2).

CHAPTER CASE: COMPARING THE MEAN STARTING SALARIES OF COLLEGE GRADUATES

13.8

We have given practical examples of analysis of variance for the completely randomized design, the randomized block design, and factorial experiment in the previous sections. We will now apply the techniques of this chapter to the starting salary data provided in Appendix B.

Consider the problem of comparing the mean starting salaries of Fall 1989–Spring 1991 University of Florida graduates in the five colleges (Business Administration, Journalism, Engineering, Liberal Arts and Sciences, and Nursing) discussed in Chapter 2. Many of these graduates did not return the CRC questionnaire, so we will never know their starting salaries, but we can still imagine that a starting salary exists for each graduate of a college and that the totality of these salaries constitutes a population that characterizes the first-year financial compensations of graduates of that college.

We want to compare the mean starting salaries corresponding to the starting salary populations of the five colleges, i.e., we wish to test

H_0: $\mu_1 = \mu_2 = \mu_3 = \mu_4 = \mu_5$
H_a: At least two means are different

where

μ_1 = Mean starting salary of graduates in the College of Business Administration

μ_2 = Mean starting salary of graduates in the College of Journalism

μ_3 = Mean starting salary of graduates in the College of Engineering

μ_4 = Mean starting salary of graduates in the College of Liberal Arts and Sciences

μ_5 = Mean starting salary of graduates in the College of Nursing

To perform the comparison, we assume that the data in Appendix B represent independent random samples of starting salaries for the five colleges. The data were then subjected to an analysis of variance for a completely randomized design using SAS. A portion of the SAS printout is reproduced in Figure 13.14.

```
                Analysis of Variance Procedure

Dependent Variable: SALARY
                                    Sum of           Mean
Source                   DF        Squares         Square      F Value       Pr > F

Model                     4     13385770707     3346442677      109.89       0.0001
Error                   897     27316231680       30452878
Corrected Total         901     40702002387

              R-Square             C.V.        Root MSE            SALARY Mean

              0.328873         21.20818        5518.413            26020.2051


Source                   DF        Anova SS    Mean Square      F Value       Pr > F

COLLEGE                   4     13385770707     3346442677      109.89       0.0001
```

The computed value of the test statistic for testing H_0: $\mu_1 = \mu_2 = \mu_3 = \mu_4 = \mu_5$, shown under the column heading **F Value** in Figure 13.14, is

$$F = 109.89$$

and the observed significance level (p-value) of the test, given under the column headed **Pr > F**, is .0001. This implies that H_0 will be rejected at any chosen level of α larger than .0001. Thus, there is very strong evidence of a difference among the mean starting salaries of graduates of the five colleges. The probability that this procedure will lead to a Type I error (concluding that there is a difference among the means if in fact they are all equal) is only .0001.

Once we have determined that the mean starting salaries differ, we may want to estimate the difference between a pair of means with a confidence interval.

If we are interested in all pairwise comparisons of the five means, the Bonferroni technique of Section 13.7 is recommended. With this multiple comparison of means procedure, we are able to control the probability of at least one Type I error in the entire experiment.[†]

The portion of the SAS printout showing the Bonferroni pairwise comparisons is given in Figure 13.15 (page 778). SAS calculates a confidence interval for the difference between each pair of college means using the adjusted critical t value given in the box in Section 13.7. For "simultaneous" 95% confidence intervals, we specify $\alpha = .05$. For five means, the total number of pairwise comparisons to be made is $g = 5(5 - 1)/2 = 10$. Consequently, the "adjusted" α is $\alpha^* = \alpha/10 = .05/10 = .005$ and the critical t value to be used in forming the 10 intervals is $t_{\alpha^*/2} = t_{.005/2} = t_{.0025}$, where the distribution of t is based on $\nu = 897$ degrees of freedom (i.e., the number of degrees of freedom associated with MSE). This value, $t_{.0025} = 2.814$, is shaded on the SAS printout, Figure 13.15.

The confidence intervals that do not contain 0 are identified on the SAS printout by asterisks (***). Note that only one pair of means does not differ significantly from 0—the Colleges of Liberal Arts/Sciences and Journalism. All

[†]The probability of making at least one Type I error in the experiment is known as the **experimentwise error rate** (in contrast to a **comparisonwise error rate** for a single comparison).

other pairs of colleges have significantly different mean starting salaries. For example, the confidence interval for the difference between the mean starting salaries of engineering and nursing graduates is (732.6, 4,727.6). Thus, we estimate that the mean starting salary for engineering graduates exceeds the corresponding mean for nursing graduates by at least $732.60 and by at most $4,727.60. Similar statements can be made about the other pairs of means. The overall level of confidence associated with all inferences drawn from the Bonferroni follow-up analysis is at least 95%.

FIGURE 13.15

SAS Printout: Bonferroni Confidence Intervals for Comparing the Five Means

```
                      Analysis of Variance Procedure

               Bonferroni (Dunn) T tests for variable: SALARY

       NOTE: This test controls the type I experimentwise error rate but
             generally has a higher type II error rate than Tukey's for all
             pairwise comparisons.

            Alpha= 0.05   Confidence= 0.95  df= 897  MSE= 30452878
                        Critical Value of T= 2.81400

       Comparisons significant at the 0.05 level are indicated by '***'.


                                      Simultaneous              Simultaneous
                                         Lower     Difference      Upper
                         COLLEGE      Confidence    Between     Confidence
                        Comparison      Limit        Means         Limit

       ENGINEERING    - NURSING           732.6      2730.1        4727.6     ***
       ENGINEERING    - BUSINESS ADM     4795.1      6062.8        7330.5     ***
       ENGINEERING    - LIB ARTS/SCIENCE 7982.3      9556.1       11129.8     ***
       ENGINEERING    - JOURNALISM/COMM  8842.4     10882.3       12922.3     ***

       NURSING        - BUSINESS ADM     1362.8      3332.7        5302.6     ***
       NURSING        - LIB ARTS/SCIENCE 4646.5      6825.9        9005.4     ***
       NURSING        - JOURNALISM/COMM  5615.5      8152.2       10689.0     ***

       BUSINESS ADM   - LIB ARTS/SCIENCE 1954.6      3493.2        5031.9     ***
       BUSINESS ADM   - JOURNALISM/COMM  2806.5      4819.5        6832.5     ***

       LIB ARTS/SCIENCE - JOURNALISM/COMM -892.2     1326.3        3544.8
```

As illustrated in Section 13.7, the results of the Bonferroni multiple comparisons can be summarized by the means according to pairs that are not significantly different. Figure 13.16 is a SAS printout of the Bonferroni groupings. According to the printout, "means with the same letter are not significantly different." You can see that the College of Engineering (group A) has the highest mean starting salary, one that is significantly larger (at overall $\alpha = .05$) than the mean starting salaries of the other four colleges. The College of Nursing (group B) mean is significantly larger than the means for the Colleges of Business Administration, Liberal Arts/Sciences, and Journalism, whereas the College of Business Administration (group C) is significantly larger than the means for the Colleges of

Liberal Arts/Sciences and Journalism. Note that the Colleges of Liberal Arts/Sciences and Journalism are grouped with the same letter (D); this implies that their means are not significantly different.

FIGURE 13.16
SAS Printout: Bonferroni Groupings for Comparing the Five Means

```
                 Analysis of Variance Procedure

           Bonferroni (Dunn) T tests for variable: SALARY

     NOTE: This test controls the type I experimentwise error rate, but
           generally has a higher type II error rate than REGWQ.

                 Alpha= 0.05  df= 897  MSE= 30452878
                      Critical Value of T= 2.81
                 Minimum Significant Difference= 1965.8
                    WARNING: Cell sizes are not equal.
                 Harmonic Mean of cell sizes= 124.8085

     Means with the same letter are not significantly different.

          Bon Grouping          Mean      N   COLLEGE

                   A          30876.9    281   ENGINEERING

                   B          28146.8     77   NURSING

                   C          24814.1    322   BUSINESS ADM

                   D          21320.8    149   LIB ARTS/SCIENCE
                   D
                   D          19994.5     73   JOURNALISM/COMM
```

THE RELATIONSHIP BETWEEN ANALYSIS OF VARIANCE AND REGRESSION (OPTIONAL)

13.9 The preceding sections illustrated the analysis of variance approach to analyzing data collected from designed experiments. Although we utilized the computer to conduct the analysis, the ANOVA sum of squares can be computed with the aid of a pocket or desk calculator (see optional Section 13.11). By forming a ratio of mean squares, we are able to test the hypothesis that a set of population means (treatment means, block means, or factor main effect means) are equal.

The same analysis can also be conducted using a multiple regression analysis. Each experimental design is associated with a general linear model for the response y, called the **complete model**. The analysis of variance F test for testing a set of means is equivalent to either an overall test of model utility or a partial F test in which the complete model is fit and compared to a **reduced model**. The proper test will depend on the specific design used. Consequently, before you can apply regression analysis in an analysis of variance, you need to learn the appropriate complete and reduced models to fit for each type of experimental design.

The regression approach to analyzing data for all three designs covered in this chapter is summarized in the boxes.

H_0: $\beta_{a+b-1} = \beta_{a+b} = \cdots = \beta_{ab-1} = 0$
(i.e., H_0: No interaction between factors A and B)

H_a: At least one of the β parameters listed in H_0 differs from 0
(i.e., H_a: Factors A and B interact)

Complete model:
$$E(y) = \beta_0 + \overbrace{\beta_1 x_1 + \cdots + \beta_{a-1} x_{a-1}}^{\text{Main effect } A \text{ terms}} + \overbrace{\beta_a x_a + \cdots + \beta_{a+b-2} x_{a+b-2}}^{\text{Main effect } B \text{ terms}}$$

$$+ \overbrace{\beta_{a+b-1} x_1 x_a + \beta_{a+b} x_1 x_{a+1} + \cdots + \beta_{ab-1} x_{a-1} x_{a+b-2}}^{AB \text{ interaction terms}}$$

where*

$$x_1 = \begin{cases} 1 & \text{if level 2 of factor } A \\ 0 & \text{if not} \end{cases} \quad \cdots \quad x_{a-1} = \begin{cases} 1 & \text{if level } a \text{ of factor } A \\ 0 & \text{if not} \end{cases}$$

$$x_a = \begin{cases} 1 & \text{if level 2 of factor } B \\ 0 & \text{if not} \end{cases} \quad \cdots \quad x_{a+b-2} = \begin{cases} 1 & \text{if level } b \text{ of factor } B \\ 0 & \text{if not} \end{cases}$$

Reduced model:
$$E(y) = \beta_0 + \overbrace{\beta_1 x_1 + \cdots + \beta_{a-1} x_{a-1}}^{\text{Main effect } A \text{ terms}} + \overbrace{\beta_a x_a + \cdots + \beta_{a+b-2} x_{a+b-2}}^{\text{Main effect } B \text{ terms}}$$

Test statistic:
$$F = \frac{(SSE_R - SSE_C)/[(a-1)(b-1)]}{SSE_C[ab(r-1)]}$$

where $SSE_R = $ SSE for reduced model

$SSE_C = $ SSE for complete model

$r = $ Number of replications

Rejection region: $F > F_\alpha$, where F is based on $\nu_1 = (a-1)(b-1)$ and $\nu_2 = ab(r-1)$df.

EXAMPLE 13.19   Refer to Example 13.3, where we compared the mean GPAs of freshmen in three socioeconomic classes. The experiment is a completely randomized design with $k = 3$ treatments: lower, middle, and upper classes. Analyze the data shown in Table 13.3 using a regression analysis.

Note: The independent variables, $x_1, x_2, \ldots, x_{a+b-2}$, are defined for an experiment in which both factors represent *qualitative* variables. When a factor is *quantitative*, you may choose to represent the main effects with quantitative terms such as x, x^2, x^3, and so forth.

Solution According to the box, the appropriate complete model for $k = 3$ treatments is

$$E(y) = \beta_0 + \beta_1 x_1 + \beta_2 x_2$$

where $x_1 = \begin{cases} 1 & \text{if lower class} \\ 0 & \text{if not} \end{cases}$

$x_2 = \begin{cases} 1 & \text{if middle class} \\ 0 & \text{if not} \end{cases}$

The Minitab regression analysis for the complete model is shown in Figure 13.17. Note that the SSE shown in the printout, 4.7111, agrees (except for rounding) with the value of SSE calculated previously.

FIGURE 13.17

Minitab Regression Printout for the Completely Randomized Design, Example 13.19

```
The regression equation is
GPA = 2.54 - 0.022 X1 + 0.705 X2

Predictor       Coef       Stdev     t-ratio        p
Constant       2.5433      0.2149      11.83     0.000
X1            -0.0219      0.2929      -0.07     0.941
X2             0.7052      0.2929       2.41     0.028

s = 0.5264      R-sq = 33.2%      R-sq(adj) = 25.3%

Analysis of Variance

SOURCE         DF          SS          MS         F        p
Regression      2      2.3409      1.1704      4.22    0.032
Error          17      4.7111      0.2771
Total          19      7.0520
```

The null hypothesis that the three population means are equal, i.e.,

$$H_0: \quad \mu_1 = \mu_2 = \mu_3$$

is equivalent to the null hypothesis

$$H_0: \quad \beta_1 = \beta_2 = 0$$

[This is because (according to the 0–1 system of coding) $\beta_1 = \mu_2 - \mu_1$ and $\beta_2 = \mu_3 - \mu_1$. Thus, when both β_1 and β_2 are 0, all treatment mean differences equal 0.] According to the box, the test statistic for a completely randomized design is the global F value for the complete model. This value, shaded on the Minitab printout in Figure 13.17, is $F = 4.22$. This value is identical to the analysis of variance F statistic computed in Example 13.3. The observed significance level of the test, also shaded, is $p = .032$. Since the p-value is less than $\alpha = .05$, there is sufficient evidence to indicate that the means for the three socioeconomic classes differ.

EXAMPLE 13.20 Refer to the machine–material 2×3 factorial experiment of Examples 13.12–13.13.

a. Write the complete model for the experiment.

b. What hypothesis would you test to determine whether machine and material interact?

a. Since both factors are qualitative, we need to set up dummy variables as follows:

$$x_1 = \begin{cases} 1 & \text{if machine } A_2 \\ 0 & \text{if machine } A_1 \end{cases} \quad x_2 = \begin{cases} 1 & \text{if material } B_2 \\ 0 & \text{if not} \end{cases} \quad x_3 = \begin{cases} 1 & \text{if material } B_3 \\ 0 & \text{if not} \end{cases}$$

Then, according to the box, the complete factorial model is

$$E(y) = \beta_0 + \beta_1 x_1 + \beta_2 x_2 + \beta_3 x_3 + \beta_4 x_1 x_2 + \beta_5 x_1 x_3$$

where y = number of gaskets produced. Note that the interaction terms for the model are constructed by taking the products of the various main effect dummy variables, one from each factor.

b. To test the null hypothesis that machine and material do not interact, we must test the null hypothesis that the interaction terms are not needed in the complete model of part **a**, i.e.,

$$H_0: \quad \beta_4 = \beta_5 = 0$$

This requires that we fit the reduced model

$$E(y) = \beta_0 + \beta_1 x_1 + \beta_2 x_2 + \beta_3 x_3$$

and perform the partial F test outlined in Section 12.12. The test statistic is

$$F = \frac{(SSE_R - SSE_C)/2}{MSE_C}$$

where SSE_R = SSE for reduced model

   SSE_C = SSE for complete model

   MSE_C = MSE for complete model

EXERCISES

APPLYING THE CONCEPTS

13.43   Refer to Exercise 13.7.

  a. Give the complete and reduced models for conducting the ANOVA.
  b. If you have access to a statistical software package, fit the models specified in part **a** and compute the value of the F statistic. The value you obtain should agree with the value calculated in Exercise 13.7.

13.44   Refer to Exercise 13.8.

  a. Give the complete and reduced models for conducting the ANOVA.
  b. If you have access to a statistical software package, fit the models specified in part **a** and compute the value of the F statistic. The value you obtain should agree with the value calculated in Exercise 13.8.

13.45   Refer to Exercise 13.15.

    a. Give the complete and reduced models for conducting the ANOVA.

    b. If you have access to a statistical software package, fit the models specified in part **a** and compute the value of the F statistic. The value you obtain should agree with the value calculated in Exercise 13.15.

13.46   Refer to Exercise 13.17.

    a. Give the complete and reduced models for conducting the ANOVA.

    b. If you have access to a statistical software package, fit the models specified in part **a** and compute the value of the F statistic. The value you obtain should agree with the value calculated in Exercise 13.17.

13.47   Refer to Exercise 13.26.

    a. Give the complete and reduced models for conducting the ANOVA.

    b. If you have access to a statistical software package, fit the models specified in part **a** and compute the value of the F statistic. The value you obtain should agree with the value calculated in Exercise 13.26.

CHECKING ANOVA ASSUMPTIONS

13.10

For each of the experiments and designs discussed in this chapter, we listed in the relevant boxes the assumptions underlying the analysis in the terminology of ANOVA. For example, the assumptions for a completely randomized design are as follows: (1) the k probability distributions of the response y corresponding to the k treatments are normal; and (2) the population variances of the k treatments are equal. Similarly, for randomized block designs and factorial designs, the data for the treatments must come from normal probability distributions with equal variances.

These assumptions are equivalent to those required for a regression analysis (see Section 12.2). The reason, of course, is that the probabilistic model for the response y that underlies each design is the familiar general linear regression model of Chapters 11 and 12. Thus, checks on the ANOVA assumptions can be performed by examining the regression residuals, as described in Section 12.13. A brief overview of these techniques is given in the box on the next page.

EXAMPLE 13.21   Refer to the completely randomized ANOVA, Examples 13.3 and 13.4. Check the ANOVA assumptions of normal populations with equal variances.

TABLE 13.12

Residuals for Completely Randomized ANOVA, Examples 13.3–13.4

SOCIOECONOMIC CLASS		
Lower	Middle	Upper
−10	6.7	−26.2
−77	1.7	−10.2
−45	−23.3	53.8
35	12.7	−54.2
20	−6.3	−32.2
12	41.7	68.8
65	−33.3	

Solution   To check the ANOVA assumptions, we need to calculate the regression residuals for the data in Table 13.3. These residuals (obtained from a computer printout of the complete model) are listed in Table 13.12.

With such a small number (6 or 7) of observations per treatment, a stem-and-leaf display to check for normality will not be very revealing. Not knowing whether the residuals are normally distributed is not a major concern, since ANOVA, like regression, is robust with respect to nonnormal data. That is, the procedure yields valid inferences even when the residuals deviate from normality.

To check for equal variances, we plotted the residuals for each treatment in Figure 13.18. The residual frequency plot shows the spread of the residuals for each treatment. You can see that the variability of the residuals is about the same; thus, the assumption of equal variances appears to be satisfied.

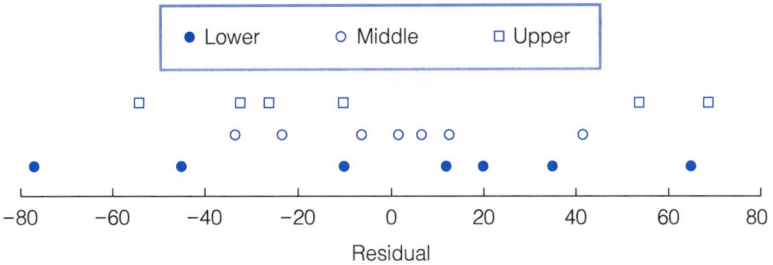

FIGURE 13.18
Residual Frequency Plot for
Example 13.21

Residual

CHECKING ANOVA ASSUMPTIONS

DETECTING NONNORMAL POPULATIONS

1. For each treatment, construct a histogram or stem-and-leaf display of the residuals. Look for highly skewed distributions. (Remember, ANOVA, like regression, is robust with respect to the normality assumption. That is, slight departures from normality will have little impact on the validity of the inferences derived from the analysis.) [*Note:* If the sample size for each treatment is small, then these graphs will probably be of limited use.]

2. Formal statistical tests of normality are also available. The null hypothesis is that the probability distribution of the response is normal. These tests, however, are sensitive to slight departures from normality. Since in most scientific applications the normality assumption will not be satisfied exactly, these tests will likely result in a rejection of the null hypothesis and, consequently, are of limited use in practice. Consult the references for more information on these formal tests.

3. If the distribution of the residuals departs greatly from normality, a *normalizing transformation* may be necessary. For example, for highly skewed distributions, transformations on the response y such as $\log(y)$ or $\sqrt{y}$ tend to "normalize" the data since these functions "pull" the observations in the tail of the distribution back toward the mean.

DETECTING UNEQUAL VARIANCES

1. For each treatment, construct a **residual frequency plot** and look for differences in the spread (variability) of the residuals shown in the plots. (See Figure 13.18.)

2. When the sample sizes are small for each treatment, only a few points are plotted on the residual frequency plots, making it difficult to detect differences in variation. In this situation, you may want to use one of several formal statistical tests of homogeneity of variances that are available. Consult the references at the end of the chapter for information on these tests.

3. When unequal variances are detected, use one of the **variance stabilizing** transformations of the response y discussed in Section 12.13.

where AB_{ij} is the sum of all observations in the cell corresponding to the ith level of factor A and the jth level of factor B.

[*Note:* To find SS(AB), square each cell total, sum the squares for all cell totals, divide by r, and then subtract SS(A), SS(B), and CM.]

6. $\text{SSE} = \text{SS(Total)} - \text{SS}(A) - \text{SS}(B) - \text{SS}(AB)$

7. $\text{MS}(A) = \dfrac{\text{SS}(A)}{a - 1}$

$\text{MS}(B) = \dfrac{\text{SS}(B)}{b - 1}$

$\text{MS}(AB) = \dfrac{\text{SS}(AB)}{(a - 1)(b - 1)}$

$\text{MSE} = s^2 = \dfrac{\text{SSE}}{ab(r - 1)}$

8. F statistic for testing AB interaction: $F = \dfrac{\text{MS}(AB)}{\text{MSE}}$

9. F statistic for testing A main effect: $F = \dfrac{\text{MS}(A)}{\text{MSE}}$

10. F statistic for testing B main effect: $F = \dfrac{\text{MS}(B)}{\text{MSE}}$

EXERCISES

LEARNING THE MECHANICS

13.51 Use the ANOVA calculation formulas for a completely randomized design to construct the ANOVA summary table for Exercise 13.2.

13.52 Use the ANOVA calculation formulas for a completely randomized design to construct the ANOVA summary table for Exercise 13.7.

13.53 Use the ANOVA calculation formulas for a randomized block design to construct the ANOVA summary table for Exercise 13.12.

13.54 Use the ANOVA calculation formulas for a randomized block design to construct the ANOVA summary table for Exercise 13.16.

13.55 Use the ANOVA calculation formulas for a factorial design to construct the ANOVA summary table for Exercise 13.22.

13.56 Use the ANOVA calculation formulas for a factorial design to construct the ANOVA summary table for Exercise 13.26.

SUMMARY

This chapter presented an extension of the methods for comparing two population means to allow for the comparison of more than two means. The *completely randomized design* uses independent random samples selected from each of k populations. The comparison of the population means is made by comparing the variance among the sample means, as measured by the *mean square for treatments* (MST), to the variation attributable to differences within the samples, as measured by the *mean square for error* (MSE). If the ratio of MST to MSE is large, we conclude that a difference exists between the means of at least two of the k populations.

We also presented an analysis of variance for a comparison of two or more population means using matched groups of experimental units in a *randomized block design*, an extension of the matched-pairs design. The design not only allows us to test for differences among the treatment means, but also enables us to test for differences among block means. By testing for differences among block means, we can determine whether blocking is effective in reducing the variation present when comparing the treatment means.

Finally, we discussed an analysis of variance for a *factorial experiment*. Factorial experiments allow us to investigate the effect of two or more *factors* on the mean value of a response variable. The combinations of the factor levels represent the *treatments*, and an analysis of variance enables us to compare the means of the treatments. One of the most important objectives of the factorial experiment is to test for the presence of *factor interaction*.

We remind you that the proper application of these ANOVA techniques requires that certain assumptions must be satisfied. An examination of the ANOVA residuals can reveal violations of these assumptions. In most practical applications, the assumptions will not be satisfied exactly. However, these analysis of variance procedures are fairly robust in the sense that slight departures from the assumptions will not significantly affect the analysis or the validity of the resulting inferences.

There are various methods of collecting data and designing experiments for the purpose of comparing more than two population means; the completely randomized design, the randomized block design, and the two-factor factorial experiment are the simplest of these preconceived plans. If you would like to study other types of experimental designs and the ANOVA techniques associated with each, consult the references listed at the end of the chapter.

KEY WORDS

Analysis of variance	Factorial experiment
Analysis of variance (ANOVA) table	Mean square
Bonferroni multiple comparisons	Randomized block design
Completely randomized design	Replication
F distribution	Sum of squares
Factor	Variability among sample means
Factor interaction	Within-sample variability
Factor level	

Analysis of variance: ANOVA
Total sum of squares of deviations: SS(Total)
Sum of squares for treatments: SST
Sum of squares for blocks: SSB
Sum of squared errors: SSE
Mean square for treatments: MST
Mean square for blocks: MSB
Mean square for error: MSE
Ratio of mean squares: F
Sum of squares for main effect A: SS(A)
Sum of squares for interaction between factors A and B: SS(AB)
Mean square for main effect A: MS(A)
Mean square for AB interaction: MS(AB)

SUPPLEMENTARY EXERCISES

[*Note:*   List the assumptions necessary to ensure the validity of the procedure you use to solve each problem. Starred (*) exercises are from the optional sections of this chapter.]

13.57   In social psychology, researchers have found that when individuals have positive attitudes about an issue, a less credible source has a higher impact on positive attitude change than a highly credible source. Does this phenomenon exist in a personal selling situation? The *Journal of Personal Selling & Sales Management* (Fall 1990) reported on a study to examine the effects of salesperson credibility on buyer persuasion. The experiment involved two factors, each at two levels: brand quality (high versus low) and salesperson credibility (high versus low). Each of 64 undergraduate students were randomly assigned to one of the $2 \times 2 = 4$ experimental treatments (8 students per treatment). After viewing a presentation on laptop computers, the students' intentions to buy were measured with a questionnaire.

*a.   Treating intention to buy as the dependent variable, write the appropriate model for this experiment.

b.   A portion of the ANOVA table for this experiment is shown here. Interpret the results.

SOURCE OF VARIATION	df	SS	MS	F	p-VALUE
Brand quality (A)	1	59.30	59.30	39.74	.001
Salesperson credibility (B)	1	3.51	3.51	2.35	.130
A × B	1	15.13	15.13	10.14	.002

Source: Sharma, A. "The persuasive effect of salesperson credibility: Conceptual and empirical examination." *Journal of Personal Selling & Sales Management*, Fall 1990, Vol. 10, pp. 71–80 (Table 2).

13.58   Refer to the *Journal of Personal Selling & Sales Management* study, Exercise 13.57. A multiple comparisons procedure (with $\alpha = .01$) was used to compare the mean buyer intentions for the $2 \times 2 = 4$ experimental conditions. Since the two factors (brand quality and salesperson credibility) interact, the means for high and low salesperson credibility were compared for each of the two levels of brand quality.

a. The results for high brand quality are shown here. Interpret the results.

Mean buyer intention:	4.88	5.27
Sales credibility:	High	Low

b. The results for low brand quality are shown here. Interpret the results.

Mean buyer intention:	2.35	3.92
Sales credibility:	Low	High

13.59 Patients recovering from myocardial infarctions (i.e., heart attacks) and other cardiovascular diseases frequently undergo cardiac rehabilitation during their hospital stay. The rehabilitation usually involves supervised low-intensity exercise. A clinical trial was conducted to study the physiological effects on patients of five active exercises used in the earliest stages of cardiac rehabilitation (*Physical Therapy*, Aug. 1986). Twelve healthy female physical therapy students participated in the study, performing each of the five exercises (exercises A, B, C, D, E) in a randomized order. Thus, the experiment was set up as a randomized block design, with the subjects representing the blocks and the exercises the treatments. After each exercise set, the subject's heart rate (in beats per minute) was recorded. The results of the ANOVA on heart rate are summarized in the table. [*Note:* In addition to the five exercises, the heart rate was also measured when the students were at rest. Thus, the experiment involves a total of six treatments: rest and the five exercises.] Interpret the results.

SOURCE	df	SS	MS	F
Exercises	5	873	174.60	30.21
Subjects	11	3,963	360.27	62.33
Error	55	318	5.78	
Total	71	5,154		

Source: Dehne, P. R. and Protas, E. J. "Oxygen consumption and heart rate responses during five exercises." *Physical Therapy*, Vol. 66, No. 8, Aug. 1986, pp. 1215–1219 (Table 2). Reprinted with the permission of the American Physical Therapy Association.

13.60 Refer to the *Physical Therapy* study of heart rates after exercise in Exercise 13.59. A follow-up analysis, similar to the Bonferroni multiple comparisons procedure, was conducted on the six treatment means. The ranked treatment means are listed below, with overbars connecting the means that are not significantly different at $\alpha = .05$. Interpret the results.

Treatment:	Rest	A	B	C	D	E
Mean heart rate (beats/minute):	61	62	66	68	69	69

13.61 Refer to the *American Journal of Psychology* 4×2 factorial experiment, Exercise 13.29. Recall that group was the only significant factor. The mean proportions of ideas recalled correctly for the four groups were compared with a multiple comparisons procedure (at $\alpha = .05$); the results are shown here. Interpret the results.

Mean:	.252	.361	.379	.589
Group:	LD-Low SES	ND-Low SES	LD-High SES	ND-High SES

13.62 Studies conducted at the University of Melbourne (Australia) indicate that there may be a difference in the pain thresholds of blonds and brunettes. Men and women of various ages were divided into four categories according to hair color: light blond, dark blond, light brunette, and dark brunette. The purpose of the experiment was to determine whether hair color is related to the amount of pain produced by common types of mishaps and assorted types of trauma. Each person in the experiment was given a pain threshold score based on his or her performance in a pain sensitivity test (the higher the score, the lower the person's pain tolerance). Consider the results shown in the accompanying table and a Minitab printout of the analysis of variance.

LIGHT BLOND	DARK BLOND	LIGHT BRUNETTE	DARK BRUNETTE
62	63	42	32
60	57	50	39
71	52	41	51
55		37	30
48			35

```
ANALYSIS OF VARIANCE ON PAIN
SOURCE     DF        SS        MS         F         p
COLOR       3     1566.6     522.2      9.46     0.001
ERROR      13      717.7      55.2
TOTAL      16     2284.2
                                    INDIVIDUAL 95 PCT CI'S FOR MEAN
                                    BASED ON POOLED STDEV
LEVEL       N      MEAN      STDEV   ----------+---------+---------+------
    1       5     59.200     8.526                        (------*-----)
    2       3     57.333     5.508                     (--------*---------)
    3       4     42.500     5.447        (--------*-------)
    4       5     37.400     8.325   (------*-------)
                                    ----------+---------+---------+------
POOLED STDEV =     7.430                    40        50        60
```

a. Is there evidence of a difference among the mean pain thresholds for people with the four hair color types? Use $\alpha = .05$.

b. Conduct a multiple comparisons analysis of the mean pain thresholds for the four types. Use $\alpha = .06$.

13.63 Refer to study on the effect of price advertising on beer sales in Lower Michigan, described in Exercise 13.8. In that exercise you conducted the analysis as a completely randomized design. The fact that the experimental unit is a month (actually, a pair of successive months since we are measuring total bimonthly beer sales) could introduce an unwanted source of variation into the analysis—namely, the month-to-month variation in beer sales.

MONTHS	PERIOD 1	PERIOD 2	PERIOD 3
Jan./Feb.	654	692	442
Mar./Apr.	—	522, 584	446
May/June	462	508, 496	—
July/Aug.	417	427	433
Sept./Oct.	516	477	470
Nov./Dec.	605	603	609

Source: Wilcox, G. B., "The effect of price advertising on alcoholic beverage sales." *Journal of Advertising Research*, Vol. 25, No. 5, Oct./Nov. 1985, pp. 33–37.

```
ANALYSIS OF VARIANCE  SALES

SOURCE     DF        SS        MS
PERIOD      2       9726      4863
MONTH       3      68258     22753
ERROR       6      27947      4658
TOTAL      11     105932
```

a. Explain how to set up a randomized block design to reduce this unwanted source of variation.

b. The data of Exercise 13.8 are reorganized in the accompanying table. Notice that the months in which the beer sales are recorded are now identified for each period. A randomized block ANOVA was conducted on only the data for those months that appear across all three periods (why is this necessary?). Use the information in the Minitab printout to construct an ANOVA summary table.

c. Is there sufficient evidence to indicate that the mean total bimonthly beer sales differ among the three periods? Test using $\alpha = .10$.

d. Calculate the value of s for this experiment and compare it to the value for the completely randomized design of Exercise 13.8. Does it appear that blocking on months was effective in reducing the month-to-month variation in beer sales?

e. Conduct a test to determine whether blocking was effective in reducing month-to-month variation in beer sales. Use $\alpha = .10$. Does the result support your answer to part d?

13.64 A trade-off study regarding the inspection and test of transformer parts was conducted by the quality department of a major defense contractor. The investigation was structured to examine the effects of varying inspection levels and incoming test times to detect early part failure or fatigue. The levels of inspection selected were full military inspection (A), reduced military specification level (B), and commercial grade (C). Operational burn-in test times chosen for this study were at 1-hour increments from 1 hour to 9 hours. The response was failures per 1,000 pieces obtained from samples taken from lot sizes inspected to a specified level and burned-in over a prescribed time length. Three replications were randomly sequenced under each condition, making this a complete 3×9 factorial experiment (a total of 81 observations). The data for the study, shown in the accompanying table, were subjected to an ANOVA using SAS. The SAS printout is reproduced here. Analyze and interpret the results.

| | | INSPECTION LEVELS | | | | | | | | |
		Full Military Inspection A			Reduced Military Specification B			Commercial C		
BURN-IN HOURS	1	7.60	7.50	7.67	7.70	7.10	7.20	6.16	6.13	6.21
	2	6.54	7.46	6.84	5.85	6.15	6.15	6.21	5.50	5.64
	3	6.53	5.85	6.38	5.30	5.60	5.80	5.41	5.45	5.35
	4	5.66	5.98	5.37	5.38	5.27	5.29	5.68	5.47	5.84
	5	5.00	5.27	5.39	4.85	4.99	4.98	5.65	6.00	6.15
	6	4.20	3.60	4.20	4.50	4.56	4.50	6.70	6.72	6.54
	7	3.66	3.92	4.22	3.97	3.90	3.84	7.90	7.47	7.70
	8	3.76	3.68	3.80	4.37	3.86	4.46	8.40	8.60	7.90
	9	3.46	3.55	3.45	5.25	5.63	5.25	8.82	9.76	9.52

Source: Danny La Nuez, College of Business Administration, graduate student, University of South Florida, 1989–1990.

```
                     Analysis of Variance Procedure

Dependent Variable: FAILURES
                                    Sum of           Mean
Source                    DF        Squares         Square    F Value    Pr > F

Model                     26     168.6120667      6.4850795    101.31     0.0001
Error                     54       3.4565333      0.0640099
Corrected Total           80     172.0686000


                     R-Square            C.V.       Root MSE        FAILURE Mean
                     0.979912         4.405990      0.253002          5.74222222


Source                    DF        Anova SS    Mean Square    F Value    Pr > F

BURNIN                     8     27.97440000     3.49680000     54.63     0.0001
INSLEVEL                   2     43.08411852    21.54205926    336.54     0.0001
BURNIN*INSLEVEL           16     97.55354815     6.09709676     95.25     0.0001
```

13.65 In business, the prevailing theory is that companies can be categorized into one of four types based on their strategic profile: reactors (marginal competitors, unstable, victims of industry forces); defenders (specialize in established products, lower costs while maintaining quality); prospectors (develop new/improved products); and analyzers (operate in two product areas—one stable, one dynamic). The *American Business Review* (Jan. 1990) reported on a study that proposes a fifth organization type, balancers, who operate in three product spheres—one stable and two dynamic. Each firm in a sample of 78 glassware firms was categorized into one of these five types; the level of performance (process research and development ratio) of each was measured.

    a. A completely randomized design ANOVA of the data resulted in a significant (at $\alpha = .05$) F value for treatments (organization types). Interpret this result.

    b. Multiple comparisons of the five mean performance levels (using a procedure similar to Bonferroni, at $\alpha = .05$) are summarized in the following table. Interpret the results.

Mean:	.138	.235	.820	.826	.911
Type:	Reactor	Prospector	Defender	Analyzer	Balancer

Source: Wright, P., et al. "Business performance and conduct of organization types: A study of select special-purpose and laboratory glassware firms." *American Business Review*, Jan. 1990, p. 95 (Table 4).

*13.66 Vanadium (V) is a recently recognized essential trace element. An experiment was conducted to compare the concentrations of V in biological materials using isotope dilution mass spectrometry. The accompanying table gives the quantities of V (measured in nanograms per gram) in dried samples of oyster tissue, citrus leaves, bovine liver, and human serum.

OYSTER TISSUE	CITRUS LEAVES	BOVINE LIVER	HUMAN SERUM
2.35	2.32	.39	.10
1.30	3.07	.54	.17
.34	4.09	.30	.14
			.16
			.16

Source: Fassett, J. D. and Kingston, H. M. "Determination of nanogram quantities of vanadium in biological material by isotope dilution thermal ionization mass spectrometry with ion counting detection." *Analytical Chemistry*, Vol. 57, No. 13, Nov. 1985, p. 2475 (Table II). Copyright 1985 American Chemical Society. Reprinted with permission.

    a. Construct an ANOVA table for the data.

    b. Is there sufficient evidence (at $\alpha = .05$) to indicate that the mean V concentrations differ among the four biological materials?

    c. Estimate the mean V concentration in human serum with a 95% confidence interval.

    d. Estimate the difference between the mean V concentrations in oyster tissue and citrus leaves with a 95% confidence interval.

13.67 How does a worker's sex or rank in a company affect other people's evaluations of the worker's performance or qualifications? W. R. Morrow, G. Lowenberg, S. Larson, M. Redfearn, and J. Schoone conducted a study to determine the effects of gender and organizational position on evaluations of business memos (*Personnel Psychology*, Spring 1983). In one portion of the study, each in a sample of approximately 100 subjects (all members of the Wisconsin Personnel and Industrial Relations Association) was asked to rate a poorly written memo on a 7-point scale (1 = very poor, . . . , 7 = very good). The attributed sex (male or female) of the supposed author of the memo and the attributed memo author's organizational position (executive or assistant) were varied from memo to memo. Thus, the experiment consists of two factors, author sex and author position, with each at two levels. Twenty-five subjects were sampled for each of the four memo types, and the rating scores for the subjects were determined. A partial ANOVA summary table for the experiment is shown here.

SOURCE	df	SS	MS	F
Sex (S)	—	1.21	—	.58
Position (P)	—	—	.01	—
SP	—	6.25	—	—
Error	96	—	2.10	
Total	—	209.07		

a. Complete the ANOVA table.
b. What do we mean when we say that the two factors, sex and organizational position, interact? Illustrate with a graph.
c. The sample mean rating scores for the four memo types are given in the accompanying table. Plot the means. How does this graph compare with the graph of part b?

		AUTHOR SEX	
		Male	Female
AUTHOR POSITION	Executive	3.04	3.76
	Assistant	3.56	3.30

d. Is there sufficient evidence of interaction between author sex and organizational position? Test using $\alpha = .01$.
e. Based on the result in part d, would you recommend that additional tests be conducted? Explain.
f. Conduct the tests for main effects.

*13.68 A simulation study was conducted to investigate the machine performance of several new algorithms for functions in the FORTRAN computer program library (*IBM Journal of Research and Development*, Mar. 1986). The accompanying table gives the time per call (in microseconds) for several randomly selected scalar functions (averaged over 10,000 random arguments) on each of three different IBM System/370 machines.

FUNCTION	IBM 4331	IBM 4361	IBM 4341
EDUM	9.90	3.07	4.88
ACOS CIRC(O,PI)	179.62	33.28	33.23
SIN LINEAR(−PI,PI)	105.72	24.13	27.08
EXP LINEAR(−16,16)	254.82	39.14	37.46
D2DUM	13.47	4.63	5.72

Source: Agarwal, R. C., et al. "New scalar and vector elementary functions for the IBM System/370." *IBM Journal of Research and Development*, Vol. 30, No. 2, Mar. 1986, p. 139 (Table 4). Copyright 1986 by International Business Machines Corporation; reprinted with permission.

a. Treating the functions as blocks, construct an ANOVA summary table for this randomized block experiment.
b. Is there sufficient evidence to indicate that the mean function call times differ for the three IBM System/370 machines? Test using $\alpha = .10$.
c. Conduct a test to determine if blocking on functions was effective in removing an extraneous source of variation. Use $\alpha = .10$.

13.69 In increasingly severe oil well environments, oil producers are interested in high strength nickel alloys that are corrosion-resistant. Since nickel alloys are especially susceptible to hydrogen embrittlement, an experiment was conducted to compare the yield strengths of nickel alloy tensile specimens cathodically charged in a 4% sulfuric acid solution saturated with carbon disulfide, a hydrogen recombination poison. The alloys were tested under two material conditions (cold rolled and cold drawn), each at three different charging times (0, 25, and 50 days). Thus, a 2 × 3 factorial experiment was conducted, with material condition at two levels and charging time at three levels. Two hydrogen-charged tensile specimens were prepared for each of the six factor level combinations.

Their yield strengths (kilograms per square inch) are recorded in the accompanying table. The SAS ANOVA printout for the 2×3 factorial is also reproduced here.

		MATERIAL TYPE			
		Cold rolled		Cold drawn	
	0 days	53.4	52.6	47.1	49.3
CHARGING TIME	25 days	55.2	55.7	50.8	51.4
	50 days	51.0	50.5	45.2	44.0

```
                    Analysis of Variance Procedure

Dependent Variable: YIELD
                                Sum of            Mean
Source                DF        Squares          Square    F Value    Pr > F

Model                  5      142.5466667     28.5093333     43.97    0.0001
Error                  6        3.8900000      0.6483333
Corrected Total       11      146.4366667

               R-Square            C.V.        Root MSE           YIELD Mean

               0.973436         1.593913        0.805191          50.5166667

Source                DF       Anova SS     Mean Square    F Value    Pr > F

TIME                   2      62.76166667    31.38083333     48.40    0.0002
MATERIAL               1      78.03000000    78.03000000    120.35    0.0001
TIME*MATERIAL          2       1.75500000     0.87750000      1.35    0.3272
```

*a. Use the ANOVA calculation formulas to construct an ANOVA table. Compare your results with those shown on the SAS printout.
b. Is there evidence of interaction between material condition and charging time?
c. Based on the result of part b, what tests would you conduct next?
d. Conduct the tests for main effects and interpret the results.
e. Compare the mean yield strengths for the three charging times using Bonferroni's method. Use $\alpha = .06$.
*f. Give the complete and reduced models appropriate for conducting the analysis with regression.

13.70 In an *Ecology* study of the lifespan of a certain species of predatory, ciliate protozoan, 166 individuals of this species were captured from the Sacramento River (California) and randomly divided into five groups. All individuals were fed on a predetermined schedule; however, the food level varied among groups. The first group was fed 3 paramecia per day; the second group, 1 paramecium per day; the third group, $\frac{1}{2}$ paramecium per day; the fourth, $\frac{1}{4}$ paramecium per day; and the fifth group was starved. The lifespan, in days, was recorded for each individual. An analysis of variance was performed with the results shown in the table.

SOURCE	df	SS	MS	F
Food levels	4	—	196.14	—
Error	111	—	59.12	
Total	115	—		

Source: Kent, E. B. "Life history responses to resource variation in a sessile predator, the ciliate protozoan *Tokophyra Lemnarum* Stein." *Ecology*, Apr. 1981, Vol. 62, pp. 296–302. Copyright © 1981, the Ecological Society of America. Reprinted by permission.

a. Complete the ANOVA summary table.
b. Is there evidence of a difference (at significance level $\alpha = .025$) among the mean lifespans for the five food levels?

MORAL DEVELOPMENT OF TEENAGERS

In recent years there has been an increase in the number of psychological studies of the moral development of adolescents. However, almost all research has been carried out with children or teenagers attending regular schools; very few experiments have used juvenile delinquents as subjects. Sagi and Eisikovits (1981) addressed the issue in a study whose purpose was to "compare delinquent and nondelinquent populations, using concepts and measures taken from the area of moral development."

Sagi and Eisikovits randomly sampled 249 adolescents, representing both males and females between the ages of 13 and 17, drawn from three "sociological affiliations"—middle-class nondelinquents, lower-class nondelinquents, and lower-class delinquents. The study employed a version of MOTEC, a morality test for children that has been widely acclaimed for its high degree of reliability and validity. The test consisted of seven situations presented in the form of a booklet. Sagi and Eisikovits* explain:

Each situation represented a young protagonist facing a moral dilemma, and the subject was asked to indicate how the protagonist would solve the dilemma. The dilemmas represented a range of seven immoral acts, as follows: stealing, cheating, expropriation, minor violence, major violence, damaging of property, and lying.

The following situation (damaging of property) is an example: "The picture you see represents two young persons lighting a fire for a barbecue party. As they are running short of wood, one of them is looking for more in the area and discovers a wooden shack. No one else is around." Upon exposure to the dilemma, the subject was asked to indicate whether the protagonist would or would not resist the temptation. The answer was either "Yes" or "No." A response of "Yes" (the subject did not resist temptation) resulted in assignment of 0 points to the subject. A "No" response (the subject did resist temptation) granted one point.

The subject was then instructed to turn to the appropriate page in the booklet, depending on the answer. If the answer was "Yes," the subject was asked to turn to a page where four possible reasons for not resisting temptation were provided. If the answer was "No," the subject was asked to turn to a page where there were four possible reasons for resisting temptation. The subject was then asked to indicate which of the four reasons best explained the protagonist's behavior.

The four reasons (with points in parentheses) for each answer are as follows:

Yes: Because he likes to eat barbecued meat very much (1)
Because no one sees him (2)
Because he promised to bring wood to his friends (3)
Because there were clear signs the shack was abandoned (4)

---

*Reprinted from Sagi, A. and Eisikovits, Z. "Juvenile delinquency and moral development." *Criminal Justice and Behavior*, Mar. 1981, Vol. 8, pp. 79–93. Copyright © 1981 by American Association of Correctional Psychologists. Reprinted with permission of Sage Publications, Inc.

No: Because he may get hurt (1)
Because he is afraid of cops hanging out in the area (2)
His friends may get punished because of him (3)
Because one shouldn't damage property belonging to others (4)

Test scores were tabulated for five moral dimensions. These dimensions (with range of scores in parentheses) are as follows: resistance to temptation (0–7); moral reasoning (7–28); feelings after offense (7–21); severity of punishment (7–28); confession (0–7). The mean moral dimension scores for the three groups of teenagers are reported in Table 13.13.

TABLE 13.13
Mean Moral Dimension Scores

MORAL DIMENSION	SOCIOLOGICAL AFFILIATION		
	Lower-class Delinquents	Lower-class Nondelinquents	Middle-class Nondelinquents
Resistance to temptation	3.39	4.82	4.48
Moral development	18.74	22.85	21.71
Feelings after offense	13.60	15.95	14.85
Severity of punishment	16.34	17.10	15.56
Confession	2.51	3.47	3.70

a. Consider the moral dimension, resistance to temptation. Do you think it is possible, based on the sample means, to determine whether differences exist among the true mean MOTEC test scores for the three sociological affiliations? Why is a statistical test needed?

b. Set up the appropriate null and alternative hypotheses to determine whether the mean MOTEC test scores for "resistance to temptation" differ for at least two of the three sociological affiliations.

c. Sagi and Eisikovits conducted an analysis of variance on MOTEC test scores for each of the five moral dimensions. The corresponding test statistics and attained significance levels (*p*-values) are given in Table 13.14. Interpret these results.

TABLE 13.14
Test Statistics and *p*-Values

MORAL DIMENSION	F	p-VALUE
Resistance to temptation	15.80	.001
Moral development	22.47	.001
Feelings after offense	13.91	.001
Severity of punishment	2.79	.070
Confession	10.55	.001

IDENTIFYING MANAGEMENT POTENTIAL: THE IN-TRAY EXERCISE

R. W. T. Gill considered the role of the "in-tray" exercise in assessing the management potential of future administrators and executives.* The in-tray or in-basket exercise was developed 25 years ago as a training tool for officers in the U.S. Air Force. In describing how the technique works, Gill writes:

> The in-tray is a simulation representing the typical contents of an executive's in-tray with a variety of everyday problems in a written form—letters, memoranda, notes, reports, and telephone messages—both expected and unexpected requiring decisions and action. [Trainees] are provided with instructions, information on the company, its organization and the role to be played, and the in-tray contents. There is a fixed time allowed, usually $1\frac{1}{2}$ hours, during which they write letters, memoranda, and notes on their decisions and actions as if they were really doing the job.

After completing the tasks, the trainees' performances are assessed by one or more expert raters. However, "the utility of the in-tray exercise," says Gill, "rests on the logical assumption that it must be realistic. The 'content validity' of the in-tray exercise has to be established for a given situation, both in terms of perceived relevance, realism and representativeness and in terms of actual representativeness of items on which [trainees'] relative performance can be discriminated."

Ratings of overall in-tray performance are usually given on a scale of 1 (high performance) to 6 (low performance). This overall rating is based on performance in the following areas: planning/organizing, communication/control, judgment/analysis, decision making, work attitude, and output. Because of the difficulty of scoring performance in the in-tray exercise, the need for experienced, trained assessors or raters is evident. "Insufficient assessor training," writes Gill, "can lead to unreliable evaluations by assessors and thus lower exercise validity. A possible tendency to the mean must be checked, especially where assessors are inexperienced, overcautious, or overanxious." That is, the reliability of assessors' ratings should be determined before using the in-tray results as a measure of managerial effectiveness.

To investigate the phenomenon of rater reliability, Gill obtained data for seven subjects who were given the in-tray test. The subjects, all candidates for a general management position in a manufacturing company in the British motor industry, were from a variety of backgrounds—engineering, finance, production, and marketing. Overall in-tray performance of each candidate was assessed by three different raters. The results are given in Table 13.15.

---

*Gill, R. W. T. "The in-tray (in-basket) exercise as a measure of management potential." *Journal of Occupational Psychology*, 1979, Vol. 52, pp. 185–195.

TABLE 13.15
Ratings of Candidates by Individual Assessors on Overall In-Tray Performance

CANDIDATE	RATER 1	RATER 2	RATER 3
A	4.5	4.5	5.0
B	2.5	4.5	4.5
C	5.0	3.0	4.0
D	4.0	4.5	4.5
E	1.5	2.0	4.5
F	3.5	4.5	4.5
G	4.0	4.0	4.0

a. One way to determine whether the overall in-tray ratings are reliable is to compare the mean performance scores assessed by the three raters. A difference among means most likely indicates that the raters are in disagreement on the candidates' overall performances. Visually inspect the data of Table 13.15. Do you think that the mean score assessed to the candidates differs among the three raters? What null hypothesis should you test?

b. Note that each of the three raters judges the overall performance of all seven candidates. Thus, the candidates represent blocks and the sampling design used is a randomized block design. In contrast, a completely randomized design would consist of, say, 21 different candidates, with seven judged by each rater. Why is the randomized block design more appropriate than the completely randomized design for comparing the raters' abilities to score candidates?

c. The SPSS ANOVA printout for the data of Table 13.15 is provided in Figure 13.19. Is there evidence of a difference among the mean performance scores assessed by the three raters? Use a significance level of $\alpha = .05$.

```
* * *  A N A L Y S I S   O F   V A R I A N C E  * * *

              SCORE
         BY   RATER
              CANDIDAT

                            Sum of            Mean            Signif
Source of Variation         Squares   DF      Square     F    of F

Main Effects                 9.786    8       1.223    1.797   .174
  RATER                      2.667    2       1.333    1.959   .184
  CANDIDAT                   7.119    6       1.187    1.743   .194

Explained                    9.786    8       1.223    1.797   .174

Residual                     8.167   12        .681

Total                       17.952   20        .898


  21 Cases were processed.
   0 Cases (   .0 PCT) were missing.
```

FIGURE 13.19
SPSS ANOVA Printout for the Randomized Block Design of Case Study 13.2

d. Is there evidence of a difference among the mean performance scores of the candidates? Test at $\alpha = .05$.

e. Conduct a multiple comparisons analysis of the three rater means using Bonferroni's method. Use $\alpha = .06$.

f. Find a 95% confidence interval for the difference between the mean performance scores assessed by raters 2 and 3. Interpret the interval.

CASE STUDY 13.3

RELUCTANCE TO TRANSMIT BAD NEWS: THE MUM EFFECT

In a 1970 experiment, psychologists S. Rosen and A. Tesser found that people were reluctant to transmit bad news to peers. Rosen and Tesser termed this phenomenon the "MUM effect."* Since that time, numerous studies have been conducted to determine why people have a tendency to keep mum when given the opportunity to transmit bad news to others. Two theories have emerged from this research. The first maintains that the MUM effect is an aversion to private discomfort. To avoid discomforts such as empathy with the victim's distress or guilt feelings for their own good fortune, would-be communicators of bad news keep mum. The second theory is that the MUM effect is a public display. People experience little or no discomfort when transmitting bad news, but keep mum to avoid an unfavorable impression or to pay homage to a social norm.

The subject of this case study is an article by C. F. Bond and E. L. Anderson (*Journal of Experimental Social Psychology*, Vol. 23, 1987). Bond and Anderson conducted a controlled experiment to determine which of the two explanations for the MUM effect is more plausible. "If the MUM effect is an aversion to private discomfort," they state, "subjects should show the effect whether or not they are visible [to the victim]. If the effect is a public display, it should be stronger if the subject is visible than if the subject cannot be seen."

Forty undergraduates at Duke University participated in the experiment. Each subject was asked to administer an IQ test to another student and then provide the test taker with his or her percentile score. Unknown to the subject, the test taker was a confederate student who was working with the researchers. The experiment manipulated two factors, *subject visibility* and *confederate success*, each at two levels. Subject visibility was either visible (i.e., the subject was visible to the test taker) or not visible. Confederate success was either success (i.e., the test taker supplied a set of answers that placed him in the top 20% of all Duke undergraduates) or failure (i.e., the bottom 20%). Ten subjects were randomly assigned to each of the $2 \times 2 = 4$ experimental conditions; thus, a 2×2 factorial design with 10 replications was employed.

---

*Rosen, S. and Tesser, A. "On reluctance to communicate undesirable information: The MUM effect," *Journal of Communication*, Vol. 22, 1970, pp. 124–141.

One of several behavioral variables that were measured during the experiment was *latency to feedback*, defined as the time (in seconds) between the end of the test and the delivery of feedback (i.e., the percentile score) from the subject to the test taker. Table 13.16 summarizes the results of the ANOVA on latency to feedback for the 40 subjects.

TABLE 13.16

Partial ANOVA Table for the 2 × 2 Factorial Experiment

SOURCE	df	SS	MS	F
Subject visibility	—	—	1,380.24	4.26
Confederate success	—	—	1,325.16	4.09
Visibility × Success	—	—	3,385.80	10.45
Error	—	11,664	324.00	
Total	39	—		

Source: Bond, C. F. and Anderson, E. L. "The reluctance to transmit bad news: Private discomfort or public display?" *Journal of Experimental Social Psychology*, Vol. 23, 1987, pp. 176–187.

a. Calculate the missing degrees of freedom in Table 13.16.
b. Interpret the F values given in Table 13.16. What conclusions can you draw from this analysis? (Use $\alpha = .05$.)
c. The sample means (in seconds) for each of the four experimental conditions are provided next. Conduct an appropriate follow-up analysis (i.e., multiple comparison of means) using $\alpha = .06$. Interpret the results.

		CONFEDERATE SUCCESS	
		Success	Failure
SUBJECT	Visible	73.1	147.2
VISIBILITY	Not Visible	89.6	72.5

d. Bond and Anderson conclude that "subjects appear reluctant to transmit bad news—but only when they are visible to the news recipient." Do you agree? [*Hint:* Use your answer to part c.]

All three statistical software packages described in this text contain ANOVA routines for designed experiments, ranging from simple completely randomized designs to the more sophisticated factorial experiments. The key to using these packages is identifying the source(s) of variation for the experiment, i.e., treatments for completely randomized designs, treatments and blocks for randomized designs, and main effects and factor interaction for factorial designs.

SAS   a. Completely randomized design

```
Command
  Line
    1    DATA CR;                        ⎫
    2    INPUT TRTMENT $ RESPONSE;       ⎬  Data entry instructions
    3    CARDS;                          ⎭

         [Input data values (1 observation per line)]

    4    PROC ANOVA;                     ⎫
    5    CLASSES TRTMENT;                ⎬  ANOVA instructions
    6    MODEL RESPONSE=TRTMENT;         ⎭
    7    MEANS TRTMENT/BON;              ⎬  Bonferroni multiple comparisons
```

COMMAND 5   The CLASSES statement identifies the sources of variation (in addition to ERROR) for the experiment.

COMMAND 6   The sources of variation are specified to the right of the equals sign ($=$) in the MODEL statement, the dependent (response) variable to the left.

COMMAND 7   The MEANS command produces a multiple comparisons analysis of the means of the specified source. The BON option selects the Bonferroni multiple comparisons procedure.

b. Randomized block design

```
Command
  Line
    1    DATA RB;                             ⎫
    2    INPUT TRTMENT $ BLOCK $ RESPONSE;    ⎬  Data entry instructions
    3    CARDS;                               ⎭

         [Input data values (1 observation per line)]

    4    PROC ANOVA;                          ⎫
    5    CLASSES TRTMENT BLOCK;               ⎬  ANOVA instructions
    6    MODEL RESPONSE=TRTMENT BLOCK;        ⎭
    7    MEANS TRTMENT/BON;                   ⎬  Bonferroni multiple
                                                 comparisons
```

c. Factorial experiment

Command Line		
1	`DATA FACT;`	
2	`INPUT A $ B $ RESPONSE;`	} Data entry instructions
3	`CARDS;`	

[Input data values (1 observation per line)]

4	`PROC ANOVA;`	
5	`CLASSES A B;`	} ANOVA instructions
6	`MODEL RESPONSE=A B A*B;`	
7	`MEANS A*B/BON;`	} Bonferroni multiple comparisons

COMMAND 6 Interactions are specified by placing an asterisk (*) between the factors (e.g., A*B).

COMMAND 7 When A*B is specified in the MEANS statement, multiple comparisons are made on the means of all possible combinations of A and B.

a. Completely randomized design

SPSS

Command Line		
1	`DATA LIST FREE/RESPONSE TRTMENT.`	} Data entry instructions
2	`BEGIN DATA.`	

[Input data values (1 observation per line)]

3	`END DATA.`	
4	`ANOVA RESPONSE BY TRTMENT(1,3).`	} ANOVA instruction

INPUT DATA VALUES All values of TRTMENT must be coded as numerical whole numbers (e.g., 1, 2, and 3 for treatment 1, treatment 2, and treatment 3, respectively).

COMMAND 4 The dependent (response) variable is listed to the left of BY in the ANOVA command; the sources of variation are listed to the right. The range of the coded values of the sources must be specified in parentheses after each source. [*Note:* In mainframe SPSS (i.e., SPSS^X), the sources of variation must also be specified in a DESIGN subcommand. For this completely randomized design, use the subcommand DESIGN=TRTMENT/.]

GENERAL Multiple comparisons are not available in the SPSS ANOVA procedure. Remember to omit the periods at the end of each command when using SPSS in the mainframe environment.

b. Randomized block design

```
Command
Line
  1    DATA LIST FREE/RESPONSE TRTMENT BLOCK.    ⎫  Data entry
  2    BEGIN DATA.                               ⎬  instructions

       [Input data values (1 observation per line)]

  3    END DATA.
  4    ANOVA RESPONSE BY TRTMENT(1,3),BLOCK(1,5)/  ⎫  ANOVA
  5          OPTIONS=3.                            ⎬  instructions
```

COMMAND 5   The subcommand OPTIONS=3 *excludes* the interaction between treatments and blocks in the ANOVA. [*Note:* In the mainframe environment, DESIGN=TRTMENT, BLOCK/.]

c. Factorial experiment

```
Command
Line
  1    DATA LIST FREE/RESPONSE A B.    ⎫
  2    BEGIN DATA.                     ⎬  Data entry instructions

       [Input data values (1 observation per line)]

  3    END DATA.
  4    ANOVA RESPONSE BY A(0,2) B(1,4).    ⎬  ANOVA instruction
```

COMMAND 4   By default, interactions between the factors are automatically included in the ANOVA. [*Note:* In the mainframe environment, you must also specify the subcommand DESIGN=A,B,A BY B/.]

MINITAB   a. Completely randomized design

```
Command
Line
  1    READ RESPONSE IN C1, TRTMENT IN C2  }  Data entry instruction

       [Input data values (1 observation per line)]

  2    ONEWAY ANOVA ON C1, SOURCE IN C2    }  ANOVA instruction
```

GENERAL   Multiple comparisons of treatment means are not available in Minitab.

b. Randomized block design

Command Line		
1	`READ RESPONSE IN C1, TRTMENT IN C2, BLOCK IN C3` }	Data entry instruction
	[Input data values (1 observation per line)]	
2	`TWOWAY ANOVA ON C1, SOURCES IN C2 C3;` }	ANOVA instructions
3	`ADDITIVE.`	

COMMAND 3   When the ADDITIVE subcommand is specified, interaction between the sources in C2 and C3 (i.e., treatments and blocks) is *excluded*.

c. Factorial experiment

Command Line		
1	`READ RESPONSE IN C1, A IN C2, B IN C3` }	Data entry instruction
	[Input data values (1 observation per line)]	
2	`TWOWAY ANOVA ON C1, SOURCES IN C2 C3` }	ANOVA instruction

GENERAL   Since the ADDITIVE subcommand is omitted, the ANOVA will include interaction between the factors (i.e., A*B interaction).

COMPUTER ACTIVITIES

Refer to the data on DDT measurements in fish, given in Appendix D. Suppose we want to compare the mean DDT concentrations of the three species of fish (channel catfish, largemouth bass, and smallmouth buffalo) collected from the Tennessee River and its creek tributaries. Use an available statistical software package to perform an analysis of variance on DDT concentrations of the three species. If appropriate, conduct a Bonferroni comparison of the mean DDT concentrations of the three species. Interpret the results.

Cochran, W. G. and Cox, G. M. *Experimental Designs*, 2nd ed. New York: Wiley, 1957.

Davies, O. L. *Statistical Methods in Research and Production*, 3rd ed. London: Oliver and Boyd, 1958.

Davies, O. L. *The Design and Analysis of Industrial Experiments*, 2nd ed. New York: Hafner, 1956.

Dunn, O. J. and Clark, V. *Applied Statistics: Analysis of Variance and Regression*. New York: Wiley, 1974.

Johnson, N. and Leone, F. *Statistics and Experimental Design in Engineering and the Physical Sciences*, Vol. II, 2nd ed. New York: Wiley, 1977.

Mason, R. L., Gunst, R. F., and Hess, J. L. *Statistical Design and Analysis of Experiments*. New York: Wiley, 1989.

Mendenhall, W. *Introduction to Linear Models and the Design and Analysis of Experiments*. Belmont, Calif.: Wadsworth, 1968.

Mendenhall, W., Scheaffer, R., and Wackerly, D. *Mathematical Statistics with Applications*. 3rd ed. Boston: Duxbury, 1989.

Neter, J., Wasserman, W., and Kutner, M. H. *Applied Linear Statistical Models*, 3rd ed. Homewood, Ill: Richard D. Irwin, 1989.

Ott, L. *An Introduction to Statistical Methods and Data Analysis*. Boston: Duxbury, 1978.

Scheffé, H. *The Analysis of Variance*. New York: Wiley, 1959.

Snedecor, G. W. and Cochran, W. G. *Statistical Methods*, 7th ed. Ames, Iowa: Iowa State University Press, 1980.

REFERENCES

CATEGORICAL DATA ANALYSIS

CHAPTER 14

I s your lifestyle based on pleasing, achieving, suppressing, avoiding, or outdoing? Based on your personality priority, psychologists can determine your lifestyle type. But more important to Adlerian psychologists is whether your choice of a marriage partner depends on your lifestyle type. To answer this question, we need to be able to test for a dependence between two qualitative variables—wife's lifestyle and husband's lifestyle. In this chapter we will show you how to conduct this test, and we will deal more thoroughly with the lifestyle—marriage problem in the chapter case in Section 14.4.

CONTENTS

CATEGORICAL DATA AND THE MULTINOMIAL EXPERIMENT

14.1

Recall from Section 2.2 that observations on a qualitative variable can only be categorized. For example, consider the highest level of education attained by each in a group of conficted felons. Level of education is a qualitative variable; each felon would fall into one and only one of the following five categories: some high school; high school diploma; some college; college degree; and graduate degree. The result of the categorization would be a count of the numbers of felons falling in the respective categories. When the qualitative variable results in one of the two responses (yes or no, success or failure, favor or do not favor, etc.) the data (i.e., the counts) can be analyzed using the binomial probability distribution discussed in Chapter 5. However, qualitative variables, such as level of education, that allow for more than two categories for a response are much more common, and these must be analyzed using a different method.

Qualitative data that fall into more than two categories often result from a **multinomial experiment**. The characteristics for a multinomial experiment with k outcomes are described in the box. You can see that the binomial experiment of Chapter 5 is a multinomial experiment with $k = 2$.

PROPERTIES OF THE MULTINOMIAL EXPERIMENT

1. The experiment consists of n identical trials.
2. There are k possible outcomes to each trial.
3. The probabilities of the k outcomes, denoted by $\pi_1, \pi_2, \ldots, \pi_k$, remain the same from trial to trial, where $\pi_1 + \pi_2 + \cdots + \pi_k = 1$.
4. The trials are independent.

EXAMPLE 14.1 Consider the problem of determining the highest level of education attained by each of $n = 100$ convicted felons at a large state prison. Suppose we want to categorize level of education into one of five categories: some high school, high school diploma, some college, college degree, and graduate degree. Is this a multinomial experiment to a reasonable degree of approximation?

Solution Checking the four properties of a multinomial experiment shown in the box, we have:

1. The experiment consists of $n = 100$ identical trials, where each trial is to determine the highest level of education of a felon.
2. There are $k = 5$ possible outcomes to each trial corresponding to the five education level categories.
3. The probabilities of the $k = 5$ outcomes, $\pi_1, \pi_2, \pi_3, \pi_4$, and π_5, remain (to a reasonable degree of approximation) the same from trial to trial, where π_i represents the true probability that a convicted felon attains level of education i.
4. The trials are independent, i.e., the education level attained by one felon does not affect the level attained by any other felon.

Thus, the properties of a multinomial experiment are satisfied.

This chapter is concerned with the analysis of categorical data—specifically, the data that represent the counts for each category of a multinomial experiment. In Section 14.2 we will learn how to make inferences about category probabilities for data classified according to a single qualitative (or categorical) variable; in Section 14.3 we consider inferences about category probabilities for data classified according to two qualitative variables; and, in optional Section 14.5, we discuss a method of modeling the categorical variable responses. The statistic used for these inferences is one that possesses, approximately, the familiar chi-square distribution.

14.2 TESTING CATEGORICAL PROBABILITIES: ONE-WAY TABLE

In this section we consider a multinomial experiment with k outcomes that correspond to categories of a *single* qualitative variable. The results of such an experiment are summarized in a **one-way table**. Typically, we want to make inferences about the true percentages that occur in the k categories based on the sample information in the one-way table.

EXAMPLE 14.2 Jobs submitted to a university computer center may run under three different priority classes: urgent, normal priority, and low priority. The computer center estimates that 20% of the jobs are submitted as urgent, 50% as normal priority, and 30% as low priority. To verify these percentages, a random sample of $n = 100$ recently submitted jobs were selected and the priority class recorded. The number of jobs falling into each of the three priority class categories is shown in the one-way table in Table 14.1.

Do the data given in Table 14.1 disagree with the percentages of 20%, 50%, and 30% estimated by the computer center? The first step in answering this question is to find the number of jobs in the sample of 100 that would be expected to fall in each of the three categories of Table 14.1, assuming that the computer center's percentages are accurate.

TABLE 14.1
One-Way Table of Category Counts Corresponding to Priority Class

PRIORITY CLASS			
Urgent	Normal	Low	TOTAL
15	61	24	100

Solution Each computer job in the sample was assigned to one and only one of the three categories listed in Table 14.1. If the computer center's percentages are correct, then the probabilities that a job will fall in the three priority categories are as shown in Table 14.2.

TABLE 14.2
Category Probabilities Based on Computer Center Percentages

	PRIORITY CLASS			
	Urgent	Normal	Low	TOTAL
Cell Number	1	2	3	
Cell Probability	$\pi_1 = .2$	$\pi_2 = .5$	$\pi_3 = .3$	1.0

Consider first the "urgent" cell of Table 14.2. If we assume that the priority class of any one job is independent of the priority class of any other, then the observed number O_1 of responses falling in cell 1 is a binomial random variable and its expected value is

$$e_1 = n\pi_1 = (100)(.2) = 20$$

Similarly, the expected observed numbers of responses in cells 2 and 3 (categories 2 and 3) are

$$e_2 = n\pi_2 = (100)(.5) = 50$$

and

$$e_3 = n\pi_3 = (100)(.3) = 30$$

The observed numbers of responses and the corresponding expected numbers (in parentheses) are shown in Table 14.3.

TABLE 14.3
Observed (and Expected) Number of Responses Falling in the Cell Categories for Example 14.2

PRIORITY CLASS			
Urgent	Normal	Low	TOTALS
15	61	24	100
(20)	(50)	(30)	100

FORMULA FOR CALCULATING EXPECTED CELL COUNTS: ONE-WAY TABLE

$$e_i = n\pi_i$$

where e_i = Expected count for cell i
n = Sample size
π_i = Hypothesized probability that an observation will fall in cell i

---

Do the observed responses for the sample of 100 computer jobs in Example 14.2 disagree with the category probabilities based on the computer center's estimates? If they do, we say that the theorized computer center probabilities do not fit the data or, alternatively, that a **lack of fit** exists. The relevant null and alternative hypotheses are:

H_0:   The category (cell) probabilities are $\pi_1 = .2$, $\pi_2 = .5$, $\pi_3 = .3$.

H_a: At least two of the probabilities, π_1, π_2, π_3, differ from the values specified in the null hypothesis.

To find the value of the test statistic, we first calculate

$$\frac{(\text{Observed cell count} - \text{Expected cell count})^2}{\text{Expected cell count}} = \frac{(O_i - e_i)^2}{e_i}$$

for each of the cells, $i = 1, 2, 3$. The sum of these quantities is the test statistic used for the goodness-of-fit test:

$$\chi^2 = \frac{(O_1 - e_1)^2}{e_1} + \frac{(O_2 - e_2)^2}{e_2} + \frac{(O_3 - e_3)^2}{e_3}$$

$$= \sum_{i=1}^{3} \frac{(O_i - e_i)^2}{e_i}$$

The procedure for calculating the χ^2 (chi-square) statistic and testing the goodness of fit of the survey data to the computer center percentages will be illustrated in the following examples.

EXAMPLE 14.3 Refer to Example 14.2. Calculate the value of χ^2 needed to test

$$H_0: \quad \pi_1 = .2, \ \pi_2 = .5, \ \pi_3 = .3$$

Solution   Substituting the values of the observed and expected cell counts (from Table 14.3) into the formula for calculating χ^2, we obtain

$$\chi^2 = \frac{(O_1 - e_1)^2}{e_1} + \frac{(O_2 - e_2)^2}{e_2} + \frac{(O_3 - e_3)^2}{e_3}$$

$$= \frac{(15 - 20)^2}{20} + \frac{(61 - 50)^2}{50} + \frac{(24 - 30)^2}{30}$$

$$= 1.25 + 2.42 + 1.20 = 4.87$$

EXAMPLE 14.4 Specify the rejection region for the test described in Examples 14.2 and 14.3. Use $\alpha = .05$. Test to determine whether the sample data disagree with the computer center's estimated percentages.

Solution   The greater the difference between the observed and expected cell counts, the greater is the evidence to indicate a lack of fit between the theorized cell probabilities and those that pertain to your computer center. Since the value of chi-square increases as the differences between the observed and expected cell counts increase, we will reject

$$H_0: \quad \pi_1 = .2, \ \pi_2 = .5, \ \pi_3 = .3$$

for values of chi-square larger than some critical value, say, χ_α^2, i.e.,

Rejection region: $\chi^2 > \chi_\alpha^2$

To find the value of χ_α^2, we need to find the sampling distribution of χ^2. Based on the properties of a multinomial experiment described in Section 14.1, the sampling distribution of the χ^2 statistic has approximately a **chi-square (χ^2) probability distribution**. Recall that tabulated values of the chi-square distribution are given in Table 6 of Appendix G and that it depends on a quantity called the "degrees of freedom" (see optional Section 8.10).

The degrees of freedom for the chi-square statistic used to test the goodness of fit of a set of cell probabilities will always be 1 less than the number of cells. For example, if k cells were used in the categorization of the sample data, then

Degrees of freedom: df $= k - 1$

For our example, df $= (k - 1) = (3 - 1) = 2$ and $\alpha = .05$. From Table 6 of Appendix G, the tabulated value of $\chi_{.05}^2$ corresponding to df $= 2$ is 5.99147.

The rejection region for the test, $\chi^2 > \chi_{.05}^2$, is illustrated in Figure 14.1 (page 814). We will reject H_0 if $\chi^2 > 5.99147$. Since the calculated value of the test statistic, $\chi^2 = 4.87$, is less than $\chi_{.05}^2$, we cannot reject H_0. There is insufficient information to indicate a lack of fit of the sample data to the percentages estimated by the computer center.

POLICY ENFORCEMENT METHOD	NUMBER OF COMPANIES
1. Do not take any action	10
2. Internal audits	49
3. Honor system	28
4. Manager audits/random checks	12
5. Others	22
TOTAL	121

Source: Athey, S. A. "Software copying policies of the Fortune 500." *Journal of Systems Management*, July 1989, p. 33 (Table 6).

14.4 A *New England Journal of Medicine* study (Nov. 13, 1986) found that a substantial portion of acute hospital care is reported to be unnecessary. The physicians who conducted the study reviewed the medical records of 1,132 patients hospitalized at six different locations across the country. Overall, 60% of the admissions in the sample were judged to be appropriate and 23% were deemed inappropriate, whereas the remaining 17% could have been avoided by the use of ambulatory surgery. Let π_1, π_2, and π_3 represent the true percentages of hospital admissions in the three categories: appropriate, inappropriate, and avoidable by ambulatory surgery, respectively. Test the null hypothesis H_0: $\pi_1 = .8$, $\pi_2 = .1$, $\pi_3 = .1$. Use $\alpha = .10$.

14.5 Refer to the *Human Factors* (Dec. 1988) study of color brightness as a body orientation clue, Exercise 8.34. Ninety college students, reclining on their backs in the dark, were disoriented when positioned on a rotating platform under a slowly rotating disk that blocked their field of vision. The subjects were asked to say "stop" when they felt as if they were right-side up. The position of the brightness pattern on the disk in relation to each student's body orientation was then recorded. Subjects selected only three disk brightness patterns as subjective vertical clues: (1) brighter side up, (2) darker side up, and (3) brighter and darker side aligned on either side of the subject's head. The frequency counts for the experiment are given in the accompanying table. Conduct a test to compare the proportions of subjects that fall in the three disk orientation categories. Assume you want to determine whether the three proportions differ. Use $\alpha = .05$.

DISK ORIENTATION		
Brighter Side Up	Darker Side Up	Bright and Dark Side Aligned
58	15	17

14.6 J. A. Breaugh investigated employees' reactions to compressed work weeks (*Personnel Psychology*, Summer 1983). *Compressed work weeks* are defined as "alternative work schedules in which a trade is made between the number of hours worked per day, and the number of days worked per week, in order to work the standard number of weekly hours in less than 5 days." A field study was conducted at a large midwestern continuous-processing (7 days/24 hours) chemical plant that had experimented with four different work schedules, two of which were compressed:

Three 8-hour fixed shifts (day, evening, midnight)

Three 8-hour rotating shifts

Two 12-hour fixed shifts (12 A.M.–12 P.M., 12 P.M.–12 A.M.)

Two 12-hour rotating shifts

Six hundred seventy-one hourly employees were asked to rank the four work schedules in order of preference. The accompanying table gives the number of first-place rankings for each schedule. Is there sufficient evidence to indicate that the hourly employees have a preference for one of the work schedules? Test using $\alpha = .01$. [*Hint:* If the employees have no preference for any one of the schedules, then the true percentages in each of the four categories will be equal to $\frac{1}{4}$, or .25.]

8-HOUR FIXED	8-HOUR ROTATING	12-HOUR FIXED	12-HOUR ROTATING
389	54	208	20

14.7 In March 1981, a waterborne nonbacterial gastroenteritis outbreak occurred in Colorado as the result of a long-standing filter deficiency and malfunction of a sewage treatment plant. A study was conducted to determine whether the incidence of gastrointestinal disease during the epidemic was related to water consumption (*American Water Works Journal*, Jan. 1986). A telephone survey of households yielded the accompanying information on daily consumption of 8-ounce glasses of water for a sample of 40 residents who exhibited gastroenteritis symptoms during the epidemic. Conduct a test to determine whether the incidence of gastrointestinal disease during the epidemic is related to water consumption. Use $\alpha = .01$.

	DAILY CONSUMPTION OF 8-OUNCE GLASSES OF WATER				TOTAL
	0	1–2	3–4	5 or more	
Number of respondents with symptoms	6	11	13	10	40

Source: Hopkins, R. S., et al. "Gastroenteritis: Case study of a Colorado outbreak." *American Water Works Journal*, Vol. 78, No. 1, Jan. 1986, p. 42, Table 1, Copyright © 1986, American Water Works Association. Reprinted by permission.

14.8 A recent survey of deans at AACSB-accredited business schools found that the microcomputer software application with the highest level of exposure for both undergraduate and graduate marketing majors is statistical analysis (*Journal of Marketing Education*, Fall 1988). The same survey reported on the degree of usage of microcomputer statistical software packages for marketing undergraduates. The responses for the 51 deans in the survey are summarized in the table. Test to determine whether the percentage in the five response categories are equal. Use $\alpha = .05$.

DEGREE OF USAGE OF MICROCOMPUTER STATISTICAL SOFTWARE				
High	Moderate	Low	Trial	Not Using
21	10	10	5	5

14.3

Data are often categorized according to two qualitative variables. As a practical example of a two-variable classification of data, we will consider a study reported in *The American Naturalist* on the competition for survival between parent and offspring herring sea gulls.

The objects observed in the study were dead young herring gulls. The dead gulls were collected at each of three locations; therefore, each dead gull was categorized according to location.* In addition, each gull was categorized according to the time period (month) in which the dead bird was found. Therefore, the study involved the classification of the dead gulls according to each of two qualitative variables, location (three categories) and time period (four categories). The numbers of dead gulls falling into the $4 \times 3 = 12$ month–location categories are shown in the **two-way table**, Table 14.4 (page 818). For example, the number of dead gulls found during the July–August time period in New Jersey was 11.

TESTING CATEGORY PROBABILITIES: TWO-WAY (CONTINGENCY) TABLE

---

*The researcher gives data collected at five locations. To simplify our analysis, we present the data from only three locations.

FIGURE 14.2

SAS Printout for Example 14.6

TABLE OF TIME BY LOCATION

TIME   LOCATION

```
Frequency|
Expected |
Percent  |
Row Pct  |
Col Pct  |ENG      |NETH     |NJ       |  Total
---------+---------+---------+---------+
JULAUG   |      70 |      32 |      11 |    113
         |  39.525 |  66.586 |  6.8882 |
         |   10.16 |    4.64 |    1.60 |  16.40
         |   61.95 |   28.32 |    9.73 |
         |   29.05 |    7.88 |   26.19 |
---------+---------+---------+---------+
NOVDEC   |      43 |      94 |       3 |    140
         |   48.97 |  82.496 |  8.5341 |
         |    6.24 |   13.64 |    0.44 |  20.32
         |   30.71 |   67.14 |    2.14 |
         |   17.84 |   23.15 |    7.14 |
---------+---------+---------+---------+
OCT      |      39 |     150 |       9 |    198
         |  69.257 |  116.67 |   12.07 |
         |    5.66 |   21.77 |    1.31 |  28.74
         |   19.70 |   75.76 |    4.55 |
         |   16.18 |   36.95 |   21.43 |
---------+---------+---------+---------+
SEPT     |      89 |     130 |      19 |    238
         |  83.248 |  140.24 |  14.508 |
         |   12.92 |   18.87 |    2.76 |  34.54
         |   37.39 |   54.62 |    7.98 |
         |   36.93 |   32.02 |   45.24 |
---------+---------+---------+---------+
Total          241       406        42      689
             34.98     58.93      6.10   100.00
```

STATISTICS FOR TABLE OF TIME BY LOCATION

Statistic	DF	Value	Prob
Chi-Square	6	75.891	0.000
Likelihood Ratio Chi-Square	6	78.231	0.000
Mantel-Haenszel Chi-Square	1	9.854	0.002
Phi Coefficient		0.332	
Contingency Coefficient		0.315	
Cramer's V		0.235	

Sample Size = 689

EXAMPLE 14.7 Refer to Examples 14.5 and 14.6. Calculate the value of the chi-square statistic, and test the null hypothesis that the two directions of classification, time period in which the young gulls die and location, are independent against the alternative hypothesis that the two directions of classification are dependent.

Solution The chi-square test statistic is computed in the same manner as shown in Section 14.2, i.e.,

$$\chi^2 = \frac{(O_{11} - e_{11})^2}{e_{11}} + \frac{(O_{12} - e_{12})^2}{e_{12}} + \cdots + \frac{(O_{43} - e_{43})^2}{e_{43}}$$

$$= \frac{(11 - 6.888)^2}{6.888} + \frac{(32 - 66.586)^2}{66.586} + \cdots + \frac{(43 - 48.970)^2}{48.970}$$

$$= 75.89$$

This value is also shown (shaded) at the bottom of the SAS printout, Figure 14.2. Is the value $\chi^2 = 75.89$ large enough for us to conclude that the proportions of dead herring gulls falling in the time period categories depend on the location? To determine how large χ^2 must be before it is too large to be attributed to chance, we make use of the fact that, under certain conditions (see the box on page 822), the sampling distribution of χ^2 is approximately a chi-square probability distribution if the null hypothesis is true. For a significance level of $\alpha = .05$, we need to find the tabulated value of $\chi^2_{.05}$ in Table 6 of Appendix G. If the computed test statistic is larger than this critical value, i.e., if $\chi^2 > \chi^2_{.05}$, then we will reject the null hypothesis.

The appropriate degrees of freedom for a contingency table analysis will always be $(r - 1)(c - 1)$, where r is the number of rows and c is the number of columns in the table. For the herring gull data, we have $r = 4$ rows and $c = 3$ columns; hence, the appropriate number of degrees of freedom for χ^2 is

$$df = (r - 1)(c - 1) = (3)(2) = 6$$

The tabulated value of $\chi^2_{.05}$ corresponding to 6 df is 12.5916; therefore, the rejection region (shaded in Figure 14.3) is

Rejection region: $\chi^2 > 12.5916$

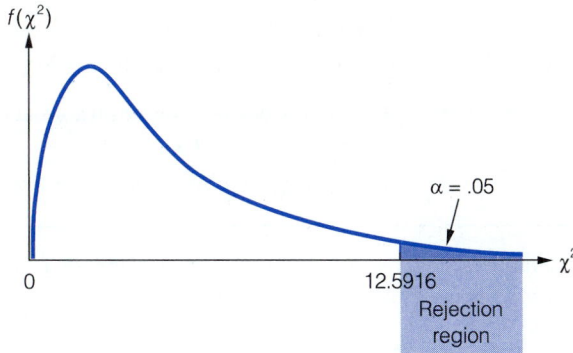

FIGURE 14.3
Rejection Region for the Dead Herring Gull Young Contingency Table, Example 14.7

Since the computed value $\chi^2 = 75.89$ exceeds the critical value 12.5916, we reject the null hypothesis in favor of the alternative. At significance level $\alpha = .05$, the data of Table 14.4 indicate that time period and location are dependent, i.e., the distributions of the percentages of responses in the categories corresponding to time period are different among the three locations.

We can arrive at the same conclusion by observing that the p-value for the test, shaded in Figure 14.2, is less than $\alpha = .05$.

The elements of χ^2 test for category probabilities in a contingency table are summarized in the box on the next page.

The data of Table 14.4 could also be used to obtain an estimate of the percentage of responses in a specific category of the population or to test a hypothesis about the value of a particular percentage. The techniques are identical to those

> ## A TEST OF HYPOTHESIS ABOUT CATEGORY PROBABILITIES: TWO-WAY TABLE
>
> H_0: The two directions of classification in the contingency table are independent.
>
> H_a: The two directions of classification in the contingency table are dependent.
>
> Test statistic: $\chi^2 = \sum_{i=1}^{r} \sum_{j=1}^{c} \dfrac{(O_{ij} - e_{ij})^2}{e_{ij}}$
>
> where r = Number of rows in the table
>
> c = Number of columns in the table
>
> O_{ij} = Observed number of responses in the cell in row i and column j
>
> e_{ij} = Estimated expected number of responses in cell (ij)
>
> $= \dfrac{(R_i)(C_j)}{n}$
>
> Rejection region: $\chi^2 > \chi^2_\alpha$
>
> where χ^2_α is the tabulated value of the chi-square distribution based on $(r - 1)(c - 1)$ degrees of freedom such that $P(\chi^2 > \chi^2_\alpha) = \alpha$.
>
> Assumptions: 1. The properties of a multinomial experiment are satisfied (see Section 14.1).
>
>   2. The expected count for each of the $r \times c$ cells is at least 5.

of Chapters 8, 9, and 10, where we considered large-sample inferences about binomial proportions. We illustrate with two examples.

EXAMPLE 14.8   Use the data of Table 14.4 to estimate the proportion of deaths of herring gull young in The Netherlands that might be expected to occur during September. Use a 95% confidence interval.

Solution   Note that the data in Table 14.4 represent three independent random samples of dead herring gull young—one from New Jersey, one from The Netherlands, and one from England. We will restrict our attention to the sample of 406 dead gulls found in The Netherlands.

Let the true proportion of young gulls found in The Netherlands that died during September be denoted by π. Since we are now interested in only one population proportion, we may treat the data of Table 14.4 as binomial data. We can think of a dead gull found in The Netherlands as being classified into one of only two categories: (1) died in September, or (2) did not die in September. The proportion π then represents the probability of success in a binomial experiment consisting of $n = 406$ trials (i.e., 406 observations of dead herring gulls in The Netherlands), where a "success" is defined as observing a dead herring gull in September.

The 95% confidence interval for π is

$$p \pm 1.96 \sqrt{\frac{pq}{n}}$$

where p is the sample proportion of successes in n trials and $q = 1 - p$. From Table 14.4, the number of successes (i.e., the number of dead gulls in September in The Netherlands) is 130. Thus, our estimate is

$$p = \frac{\text{Number of successes in the sample}}{n} = \frac{130}{406} = .32$$

Substituting $p = .32$, $q = 1 - .32 = .68$, and $n = 406$ into the confidence interval for π, we obtain

$$p \pm 1.96 \sqrt{\frac{pq}{n}} = .32 \pm 1.96 \sqrt{\frac{(.32)(.68)}{406}}$$

$$= .32 \pm (1.96)(.02315)$$

$$= .32 \pm .045 = (.275, .365)$$

We estimate, with 95% confidence, that the percentage of dead herring gulls found in The Netherlands that died in September falls between 27.5% and 36.5%.

EXAMPLE 14.9   Refer to Table 14.4. Test to determine whether the proportion of dead herring gull young found in New Jersey that died in October is larger than the corresponding proportion in England. Use $\alpha = .01$.

Solution   Let π_N and π_E be the true proportions of dead herring gull young found in New Jersey and England, respectively, that died during the month of October. To determine whether π_N is larger than π_E, we test the following hypotheses:

H_0: $(\pi_N - \pi_E) = 0$

H_a: $(\pi_N - \pi_E) > 0$

Again, we recognize that the portion of the data of Table 14.4 in which we are interested may be treated as data from two independent binomial experiments; consequently, the procedure outlined in Chapters 9 and 10 for testing the difference between two binomial proportions may be applied. The appropriate test statistic is given by

$$z = \frac{(p_N - p_E) - 0}{\sigma_{p_N - p_E}} = \frac{(p_N - p_E) - 0}{\sqrt{pq\left(\frac{1}{n_N} + \frac{1}{n_E}\right)}}$$

We need to obtain the estimates p_N and p_E, and the pooled estimate p from Table 14.4. Consider first the sample of 42 dead herring gull young found in New Jersey. We can think of a dead gull as being classified into one of two categories: (1) died in October, or (2) did not die in October. The population proportion π_N then represents the probability of success in a binomial experiment consisting of $n_N = 42$ trials, where "success" is defined as observing a dead herring gull in October. The sample proportion of successes used to estimate π_N is obtained from Table 14.4 as follows:

$$p_N = \frac{\left(\begin{array}{c}\text{Number of the sample of 42 dead gulls}\\ \text{found in New Jersey that died in October}\end{array}\right)}{42}$$

$$= \frac{9}{42} = .214$$

Similarly, π_E is the probability of success in a binomial experiment consisting of $n_E = 241$ trials and is estimated by

$$p_E = \frac{\left(\begin{array}{c}\text{Number in the sample of 241 dead gulls}\\ \text{found in England that died in October}\end{array}\right)}{241}$$

$$= \frac{39}{241} = .162$$

Then the pooled estimate of π required for $\sigma_{(p_N - p_E)}$ is

$$p = \frac{\left(\begin{array}{c}\text{Number of dead gulls in the combined New}\\ \text{Jersey and England samples that died in October}\end{array}\right)}{\left(\begin{array}{c}\text{Total number of dead gulls found in}\\ \text{New Jersey and England}\end{array}\right)}$$

$$= \frac{9 + 39}{42 + 241} = \frac{48}{283} = .170$$

Substituting these values into the test statistic, we have

$$z = \frac{(p_N - p_E) - 0}{\sqrt{pq\left(\dfrac{1}{n_N} + \dfrac{1}{n_E}\right)}} = \frac{(.214 - .162) - 0}{\sqrt{(.170)(.830)\left(\dfrac{1}{42} + \dfrac{1}{241}\right)}}$$

$$= \frac{.052}{.0628} = .828$$

The rejection region for this test is given by

$$z > z_{.01} = 2.33 \quad \text{(from Table 4 of Appendix G)}$$

Since the computed value $z = .828$ does not exceed the critical value 2.33, we fail to reject the null hypothesis at $\alpha = .01$. There is insufficient evidence to claim that the proportion of dead herring gulls that died in October is greater for those dead gulls found in New Jersey than for those found in England. However, it is important to note that the data were collected at the two locations in different years. Consequently, the percentages π_N and π_E can be compared only if the environmental conditions as well as food supplies at the two locations were similar at the times the samples were collected.

LEARNING THE MECHANICS

14.9 Refer to the accompanying 2 × 3 (two rows and three columns) contingency table.

		COLUMNS			TOTALS
		1	2	3	
ROWS	1	14	37	23	74
	2	21	32	38	91
TOTALS		35	69	61	165

a. Calculate the estimated expected cell counts for the contingency table.
b. Calculate the chi-square statistic for the table.

14.10 Refer to the 2 × 2 contingency table shown here.

		COLUMNS		TOTALS
		1	2	
ROWS	1	133	219	352
	2	201	247	448
TOTALS		334	466	800

a. Calculate the estimated expected cell counts for the contingency table.
b. Calculate the chi-square statistic for the table.

14.11 Give the degrees of freedom for a test of independence of the two directions of classification in a contingency table with:

a. $r = 2$ rows and $c = 2$ columns **b.** $r = 4$ rows and $c = 2$ columns
c. $r = 3$ rows and $c = 3$ columns **d.** $r = 3$ rows and $c = 4$ columns

14.12 Test the null hypothesis of independence of the two directions of classification for the 2 × 2 contingency table shown here. Use $\alpha = .05$.

		COLUMNS	
		1	2
ROWS	1	6	24
	2	16	36

14.13 Refer to the 3 × 4 contingency table given here.

		COLUMNS				TOTALS
		1	2	3	4	
	1	18	12	21	37	88
ROWS	2	7	10	15	31	63
	3	9	6	14	30	59
TOTALS		34	28	50	98	210

a. Calculate the estimated expected cell counts for the contingency table.
b. Calculate the chi-square statistic for the table.
c. Test the hypothesis of independence of the two directions of classification for the 3 × 4 table. Use $\alpha = .01$.

APPLYING THE CONCEPTS

14.14　According to research reported in the *Journal of the National Cancer Institute* (Apr. 1991), eating foods high in fiber may help protect against breast cancer. The researchers randomly divided 120 laboratory rats into four groups of 30 each. All rats were injected with a drug that causes breast cancer, then each rat was fed a diet of fat and fiber for 15 weeks. However, the levels of fat and fiber varied from group to group. At the end of the feeding period, the number of rats with cancer tumors was determined for each group. The data are summarized in the accompanying contingency table.

		DIET				
		High fat/ no fiber	High fat/ fiber	Low fat/ no fiber	Low fat/ fiber	TOTALS
CANCER TUMORS	Yes	27	20	19	14	80
	No	3	10	11	16	40
TOTALS		30	30	30	30	120

Source: Tampa Tribune, Apr. 3, 1991.

a. Does the sampling appear to satisfy the assumptions for a multinomial experiment (see Section 14.1)? Explain.
b. Calculate the expected cell counts for the contingency table.
c. Calculate the χ^2 statistic.
d. Is there evidence to indicate that diet and presence/absence of cancer are independent? Test using $\alpha = .05$.
e. Compare the percentage of rats on a high fat/no fiber diet with cancer to the percentage of rats on a high fat/fiber diet with cancer using a 95% confidence interval. Interpret the result.

14.15　Refer to the *American Journal on Mental Retardation* (Jan. 1992) study of the social interactions of two groups of children, Exercise 6.27. Independent random samples of 15 children with and 15 children without developmental delays (i.e., mild mental retardation) were the subjects of the experiment. After observing the children during "freeplay," the researchers recorded the number of children for each group who exhibited disruptive behavior (e.g., ignoring or rejecting other children, taking toys from another child). The data are summarized in the two-way table. Analyze the data and interpret the results.

	DISRUPTIVE BEHAVIOR	NONDISRUPTIVE BEHAVIOR	TOTALS
With developmental delays	12	3	15
Without developmental delays	5	10	15
TOTALS	17	13	30

Source: Kopp, C. B., Baker, B. and Brown, K. W. "Social skills and their correlates: Preschoolers with developmental delays." American Journal on Mental Retardation, Vol. 96, No. 4, Jan. 1992.

14.16　An Ernst and Young survey of 126 warehouses operated by retail stores was conducted for *Stores* magazine (Feb. 1991). The data in the table are the number of responses in the categories for two variables: frequency of deliveries to stores by warehouses per week, and size of warehouse. A Minitab printout of a contingency table analysis of the data is also shown. Is there evidence that frequency of deliveries and size of warehouse are dependent? Test using $\alpha = .01$.

		WAREHOUSE SIZE (THOUSANDS SQ. FT.)			
		<100	100–249.9	250–400	>400
FREQUENCY OF DELIVERIES	≤ 1 Time per week	5	13	9	5
	2–3 Times per week	12	11	13	6
	4–5 Times per week	9	14	13	16

```
Expected counts are printed below observed counts

          <100   100-249  250-400    >400    Total
   1          5       13        9        5       32
           6.60     9.65     8.89     6.86

   2         12       11       13        6       42
           8.67    12.67    11.67     9.00

   3          9       14       13       16       52
          10.73    15.68    14.44    11.14

Total        26       38       35       27      126

ChiSq =   0.389 +  1.162 +  0.001 +  0.503 +
          1.282 +  0.219 +  0.152 +  1.000 +
          0.279 +  0.181 +  0.144 +  2.117 = 7.431

df = 6
```

14.17 According to University of Florida sociologist Michael Radelet, "if you kill a white person (in Florida), the chances of getting the death penalty are three times greater than if you kill a black person" (*Gainesville Sun*, Oct. 20, 1986). Radelet formed his opinion by studying over 1,000 murder cases in Florida—a study that was eventually reviewed by the U.S. Supreme Court and used to overturn a Florida murder conviction. Concentrating only on crimes against strangers, Radelet classified the data of 326 murders, according to race of the victim and death sentence, as shown in the accompanying table.

		DEATH SENTENCE		TOTALS
		Yes	No	
RACE OF VICTIM	White	30	184	214
	Black	6	106	112
TOTALS		36	290	326

a. Conduct a test to determine whether the two directions of classification, victim's race and death sentence, are independent. Use $\alpha = .05$.

b. Give a practical interpretation of the result, part a.

c. Construct a 95% confidence interval for $(\pi_1 - \pi_2)$, where π_1 is the percentage of murderers of whites who received the death penalty, and π_2 is the percentage of murderers of blacks who received the death penalty. Interpret the interval.

14.18 One of the biggest problems with conducting mail surveys is poor response rates. To reduce nonresponse, several different techniques for formatting questionnaires have been proposed. An experiment was conducted to study the effect of questionnaire layout and page size on response rates in a mail survey. Approximately 850 students enrolled at the University of Leyden (The Netherlands) were questioned about their attitudes toward suicide. Four different questionnaire formats were used: (1) typewriting on small (15 × 21 cm) page; (2) typewriting on large (18.5 × 25.5 cm) page; (3) typeset on small page; and (4) typeset on large page. The numbers of students mailed each type of questionnaire and the numbers responding are given in the accompanying table.

		NUMBER OF RESPONSES	NUMBER OF NONRESPONSES	TOTAL NUMBER MAILED
QUESTIONNAIRE FORMAT	Typewritten, small page	86	57	143
	Typewritten, large page	191	97	288
	Typeset, small page	72	69	141
	Typeset, large page	192	92	284
TOTALS		541	315	856

Source: Reprinted with permission of author and publisher from: Jansen, J. H. "Effect of questionnaire layout and size and issue-involvement on response rates in mail surveys." *Perceptual and Motor Skills*, Vol. 61, 1985, pp. 139–142.

a. Scan the data. Do the response rates appear to differ among the four questionnaire formats?
b. Why is a statistical test useful in answering part **a**?
c. Calculate the number of students you would expect to fall in each of the eight cells of the contingency table if, in fact, the response rates are identical for the four questionnaire types.
d. Find the difference between the observed and the (estimated) expected numbers for each of the eight cells.
e. Calculate and interpret the value of the chi-square statistic for the contingency table.

14.19 Marketing strategists recognize that the success or failure of new and established products often varies regionally. Typically, marketers segment the United States into four geographic regions: East, Midwest, South, and West. An alternative conceptualization of regions within the United States, based on cultural rather than political boundaries, has been proposed. The "nine nations of North America" are New England (capital city, Boston); Quebec (Quebec City); The Foundry (Detroit); Dixie (Atlanta); The Islands (Miami); Empty Quarter (Denver); Breadbasket (Kansas City); MexAmerica (Los Angeles); and Ecotopia (San Francisco). A study was conducted to compare attitudes of consumers in the eight U.S. "nations" (*Journal of Marketing*, Apr. 1986). Each in a random sample of 2,235 Americans was classified according to "nation" and his or her response to the following question: "What is your most important value?" The results (numbers in each of the response categories) are given in the accompanying table.

MOST IMPORTANT VALUE	NEW ENGLAND	THE FOUNDRY	DIXIE	THE ISLANDS	BREADBASKET	MEXAMERICA	EMPTY QUARTER	ECOTOPIA
Self-respect	27	154	147	8	55	34	12	34
Security	26	147	152	5	62	26	6	37
Warm relationships	17	125	90	3	63	27	2	35
Sense of accomplishment	17	88	65	3	38	17	3	23
Self-fulfillment	11	74	55	1	23	24	2	24
Being well-respected	10	65	72	5	31	4	1	8
Sense of belonging	6	63	49	4	24	10	6	15
Fun-enjoyment	6	34	23	3	11	8	2	13

Source: Kahle, L. R. "The nine nations of North America and the value basis of geographic segmentation." *Journal of Marketing*, Vol. 50, No. 4, Apr. 1986, pp. 37–47. Reprinted from the *Journal of Marketing*, published by the American Marketing Association.

a. What hypothesis would you test to determine whether the distributions of percentages in the "most important value" categories differ for the eight U.S. "nations"?
b. A computer printout of the analysis is shown on the next page. Interpret the results.
c. Compare the proportions of Americans in New England and Dixie whose most important value is self-respect. Use a 95% confidence interval.

14.20 *Dear enemy recognition* is the term used by naturalists and ecologists for the aggressive behavior of birds, mammals, and ants when their territorial boundaries are violated by one of their own species. Dear enemy recognition is often followed by escalated attacks on the invading animal. *The American Naturalist* (June 1981) explored the

TABLE OF VALUE BY NATION

VALUE   NATION

Frequency Expected	1NewEngl	2Foundry	3Dixie	4Islands	5Breadbk	6MexAmer	7EmptyQt	8Ecotopi	Total
1SelfRespect	27	154	147	8	55	34	12	34	471
	25.323	157.63	137.8	6.7527	64.784	31.653	7.1747	39.883	
2Security	26	147	152	5	62	26	6	37	461
	24.785	154.29	134.87	6.6093	63.408	30.981	7.0224	39.036	
3WarmRelation	17	125	90	3	63	27	2	35	362
	19.462	121.15	105.91	5.19	49.791	24.328	5.5143	30.653	
4Accomplish	17	88	65	3	38	17	3	23	254
	13.656	85.008	74.311	3.6416	34.936	17.07	3.8692	21.508	
5SelfFullfil	11	74	55	1	23	24	2	24	214
	11.505	71.621	62.608	3.0681	29.435	14.382	3.2599	18.121	
6WellRespect	10	62	72	5	31	4	1	8	193
	10.376	64.593	56.465	2.767	26.546	12.97	2.94	16.343	
7Belonging	6	63	49	4	24	10	6	15	177
	9.5161	59.238	51.784	2.5376	24.345	11.895	2.6962	14.988	
8Fun	6	34	23	3	11	8	2	13	100
	5.3763	33.468	29.256	1.4337	13.754	6.7204	1.5233	8.4677	
Total	120	747	653	32	307	150	34	189	2232

Statistic	DF	Value	Prob
Chi-Square	49	68.758	0.033
Likelihood Ratio Chi-Square	49	69.935	0.026
Mantel-Haenszel Chi-Square	1	0.370	0.543
Phi Coefficient		0.176	
Contingency Coefficient		0.173	
Cramer's V		0.066	

Sample Size = 2232

possibility that the red-backed salamander employs dear enemy recognition by using chemical signals to distinguish familiar from unfamiliar salamanders. In escalated contests, a salamander will attempt to bite an opponent's snout—an injury that could reduce a salamander's ability to locate prey, mates, and territorial competitors. One part of the study focused on a comparison of the proportions of males and females exhibiting wounds in the snout. One hundred forty-four salamanders were collected from a forest, killed, and inspected for scar tissue in the snout. The results are shown in the table. Use a chi-square test to determine if there is a difference between the proportions of males and females with scar tissue in the snout. Use $\alpha = .01$.

	MALE	FEMALE	TOTALS
Scar tissue in snout	5	12	17
No scar tissue in snout	76	51	127
TOTALS	81	63	144

Source: Jaeger, R. G. "Dear enemy recognition and the costs of aggression between salamanders." *The American Naturalist*, June 1981, Vol. 117, pp. 962–973. Reprinted by permission of the University of Chicago Press. © 1981 The University of Chicago.

14.21   In the travel industry, destination-specific travel literature (DSTL) refers to booklets, brochures, and pamphlets that describe a destination in detail (e.g., information on activities, facilities, and prices). DSTL is made available to travelers free of charge upon request. A study reported in the *Journal of Travel Research* investigated the differences between information seekers (i.e., those who request DSTL) and nonseekers on a variety of consumer travel dimensions, including education. The accompanying contingency table gives the breakdown of the number of seekers and nonseekers at each level of education. Conduct a test to determine if the two directions of classification, tourist type and education level, are independent. Test using $\alpha = .01$. What are the practical implications of this result to a marketer of tourist information?

		TOURIST TYPE	
		Information Seeker	Nonseeker
	Some high school	13	27
EDUCATION	High school degree	64	118
LEVEL	Some college	100	123
	College degree	59	69
	Graduate degree	67	46
TOTALS		303	383

Source: Etzel, M. J. and Wahlers, R. G. "The use of requested promotional material by pleasure travelers." *Journal of Travel Research*, Vol. 23, No. 4, 1985, pp. 2–6.

14.22   The nuclear mishap on Three Mile Island near Harrisburg, Pennsylvania, on March 28, 1979, forced many local residents to evacuate their homes. To assess the impact of the accident on the area population, a questionnaire was designed and mailed to a sample of 150 households within 2 weeks after the accident occurred. One question concerned residents' attitudes toward a full evacuation: "Should there have been a full evacuation of the immediate area?" Respondents were grouped according to the distance (in miles) of the community in which they reside from Three Mile Island and their responses were recorded. A table showing a summary of the results, adapted from the survey report, follows the SAS printout of the contingency table analysis. Conduct a test to determine if local residents' attitudes toward a full evacuation is independent of distance of residence from Three Mile Island. Use $\alpha = .10$.

```
                    TABLE OF EVACUATE BY DISTANCE

EVACUATE       DISTANCE

Frequency|
Expected |1-3     |10-12   |13-15   |15+     |4-6     |7-9     | Total
---------+--------+--------+--------+--------+--------+--------+
no       |      9 |      6 |      6 |     39 |     11 |     13 |    84
         |   8.96 |   6.16 |    5.6 |  38.08 |  12.32 |  12.88 |
---------+--------+--------+--------+--------+--------+--------+
yes      |      7 |      5 |      4 |     29 |     11 |     10 |    66
         |   7.04 |   4.84 |    4.4 |  29.92 |   9.68 |  10.12 |
---------+--------+--------+--------+--------+--------+--------+
Total          16       11       10       68       22       23     150


              STATISTICS FOR TABLE OF EVACUATE BY DISTANCE

       Statistic                       DF     Value       Prob
       --------------------------------------------------------
       Chi-Square                       5     0.449      0.994
       Likelihood Ratio Chi-Square      5     0.448      0.994
       Mantel-Haenszel Chi-Square       1     0.021      0.884
       Phi Coefficient                        0.055
       Contingency Coefficient                0.055
       Cramer's V                             0.055

       Sample Size = 150
```

| | | DISTANCE FROM THREE MILE ISLAND (MILES) | | | | | | TOTALS |
		1–3	4–6	7–9	10–12	13–15	15+	
FULL EVACUATION	Yes	7	11	10	5	4	29	66
	No	9	11	13	6	6	39	84
TOTALS		16	22	23	11	10	68	150

Source: Brown, S., et al. "Final report on a survey of Three Mile Island area residents." Department of Geography, Michigan State University, Aug. 1979.

14.23 The accompanying table, extracted from the *Journal of Speech and Hearing Disorders* (Feb. 1981), presents information on 469 stutterers 14 years or younger from speech-language pathologists selected from each of the 50 states. Each of the stutterers had one other accompanying problem and was classified according to therapeutic intervention, as shown in the table.

| ACCOMPANYING PROBLEM | THERAPEUTIC INTERVENTION | | | TOTALS |
	Stuttering therapy only	Both stuttering therapy and other therapy	Other therapy only	
Articulation disorder	10	155	5	170
Language disorder	8	89	7	104
Language disability	6	60	9	75
Others	4	62	54	120
TOTALS	28	366	75	469

Source: Blood, G. W. and Seider, R. "The concomitant problems of young stutterers." *Journal of Speech and Hearing Disorders*, Feb. 1981, Vol. 46, pp. 31–33.

a. Use a chi-square test to determine if the distributions of percentages of therapeutic intervention categories differ among the categories of accompanying problems of young stutterers. Use $\alpha = .05$.

b. Construct a 95% confidence interval for the true proportion of young stutterers with an accompanying language disorder who are receiving stuttering therapy only.

14.4 CHAPTER CASE: LIFESTYLES OF THE MARRIED AND NOT FAMOUS

In this section, we examine a psychological study reported in *The Journal of Adlerian Theory, Research and Practice*. According to Adlerian psychology theory, the choice of a spouse is not accidental but purposeful, and one that reflects a person's lifestyle. The main purpose of the study was to investigate the role lifestyles play in forming a marriage.

The data were obtained by distributing questionnaires to 202 married couples living in family housing at the University of Georgia. Each spouse in the sample completed the Langenfield Inventory of Personality Profiles (LIPP), a 75-item paper-and-pencil test designed to assess "personality priorities." In Adlerian theory, a personality priority is a *dominant* behavior pattern consisting of a set of convictions that a person uses to gain a sense of belonging. Adlerian psychologists use personality priorities to group people into one of the following five lifestyle types:

1. *Pleasing priority*—a strong desire to make others happy and win their approval
2. *Achieving priority*—industrious, responsible, orderly, with an active approach to life (particularly in the workplace)

3. *Outdoing priority*—a strong desire to be on top, in a position of superiority over others; likes to be considered the best at whatever he/she does

4. *Suppressing (control) priority*—maintains control over his/her emotions; discloses very little of himself/herself

5. *Avoiding priority*—is easily hurt; avoids any potentially painful (physical or emotional) situation

First, the personality priority of each respondent was determined based on their answers to the LIPP. The researchers then classified the 202 couples according to both wife's personality and husband's personality priority. The results are shown in Table 14.5.

TABLE 14.5

Contingency Table for Study of Married Couples

		WIFE'S PERSONALITY PRIORITY					
		Pleasing	Outdoing	Avoiding	Control	Achieving	TOTALS
	Pleasing	9	6	6	3	9	33
HUSBAND'S	Outdoing	7	11	12	11	5	46
PERSONALITY	Avoiding	8	8	6	11	5	38
PRIORITY	Control	8	10	7	15	11	51
	Achieving	4	6	7	10	7	34
TOTALS		36	41	38	50	37	202

Source: Evans, T. D. and Bozarth, J. "Pairing of personality profiles in marriage." *The Journal of Adlerian Theory, Research and Practice*, Vol. 42, No. 1, March 1986, pp. 59–64.

The key question of the study is: "Do the personality profiles contribute to marriage pairings?" In other words, is there a relationship between the personality priorities of the husband and wife pairs? If so, then the distribution of the percentages of responses corresponding to the husband's personality profile categories will depend on the wife's personality priority.

The null hypothesis that we want to test is that the cell frequencies of husband's personality priorities are independent of wife's personality priorities. The alternative hypothesis is that these two classifications of married couples are dependent. Thus, we have

H_0: The two directions of classification, husband's personality and wife's personality priority, are independent.

H_a: The two directions of classification, husband's personality and wife's personality priority, are dependent.

The SAS printout of the contingency table analysis for the data of Table 14.5 is shown in Figure 14.4. The test statistic, $\chi^2 = 13.715$, is shaded as well as the associated degrees of freedom, $(r - 1)(c - 1) = 4 \times 4 = 16$. The p-value for the test (also shaded) is .62. Since the p-value greatly exceeds any reasonable choice of α (say, $\alpha = .05$ or $\alpha = .10$), we cannot reject H_0 in favor of H_a. Thus, there is insufficient evidence to indicate that the two directions of classification in the contingency table are dependent.

FIGURE 14.4

```
                    TABLE OF HUSBAND BY WIFE

     HUSBAND       WIFE

     Frequency|
     Expected |Achievng|Avoiding|Control |Outdoing|Pleasing|  Total
     ---------+--------+--------+--------+--------+--------+
     Achievng |    7   |    7   |   10   |    6   |    4   |    34
              | 6.2277 | 6.396  | 8.4158 | 6.901  | 6.0594 |
     ---------+--------+--------+--------+--------+--------+
     Avoiding |    5   |    6   |   11   |    8   |    8   |    38
              | 6.9604 | 7.1485 | 9.4059 | 7.7129 | 6.7723 |
     ---------+--------+--------+--------+--------+--------+
     Control  |   11   |    7   |   15   |   10   |    8   |    51
              | 9.3416 | 9.5941 | 12.624 | 10.351 | 9.0891 |
     ---------+--------+--------+--------+--------+--------+
     Outdoing |    5   |   12   |   11   |   11   |    7   |    46
              | 8.4257 | 8.6535 | 11.386 | 9.3366 | 8.198  |
     ---------+--------+--------+--------+--------+--------+
     Pleasing |    9   |    6   |    3   |    6   |    9   |    33
              | 6.0446 | 6.2079 | 8.1683 | 6.698  | 5.8812 |
     ---------+--------+--------+--------+--------+--------+
     Total         37       38       50       41       36      202


            STATISTICS FOR TABLE OF HUSBAND BY WIFE

     Statistic                      DF     Value      Prob
     ---------------------------------------------------------
     Chi-Square                     16    13.715     0.620
     Likelihood Ratio Chi-Square    16    14.604     0.554
     Mantel-Haenszel Chi-Square      1     0.124     0.725
     Phi Coefficient                       0.261
     Contingency Coefficient               0.252
     Cramer's V                            0.130

     Sample Size = 202
```

FIGURE 14.4

SAS Contingency Table Analysis for Personality Priority Marriage Study

For this sample, the Adlerian theory that personality priorities (lifestyles) influence whom we marry is not supported. However, the researchers warn that we should not rule out the possibility that personality priorities are related to marriage. "The lack of significant results may have occurred because the true effect in the population was not identified. This could be due to the possibility that personality priorities are not identifiable with paper and pencil instruments. In addition, the results could have been affected by some artifact in the (sampling) design." It is important to recognize, as the researchers do, the dangers of blindly accepting H_0 when the χ^2 test (or any other statistical test of hypothesis) leads to a nonsignificant result.

14.5 MODELING CATEGORY PROBABILITIES: LOGISTIC REGRESSION (OPTIONAL)

In Chapters 11 and 12 we presented an in-depth discussion of regression. For all models considered, the dependent variable y was restricted to being a *quantitative* variable. In this optional section, we consider a model for a *qualitative* dependent variable y at two levels.

For example, an entrepreneur may want to relate the success or failure of a new business to the characteristics (such as age, years of experience, and years of education) of the owner. The value of the response of interest to the entrepreneur is either *yes*, the new business is a success, or *no*, the new business is a failure. (A success implies the business did not fail.) Similarly, a state attorney

TABLE 14.6
Data for a Sample of 31 Road Construction Bids

CONTRACT	BID STATUS y	NUMBER OF BIDDERS x_1	DIFFERENCE BETWEEN WINNING BID AND ENGINEER'S ESTIMATE x_2, %	CONTRACT	BID STATUS y	NUMBER OF BIDDERS x_1	DIFFERENCE BETWEEN WINNING BID AND ENGINEER'S ESTIMATE x_2, %
1	1	4	19.2	17	0	10	6.6
2	1	2	24.1	18	1	5	−2.5
3	0	4	−7.1	19	0	13	24.2
4	1	3	3.9	20	0	7	2.3
5	0	9	4.5	21	1	3	36.9
6	0	6	10.6	22	0	4	−11.7
7	0	2	−3.0	23	1	2	22.1
8	0	11	16.2	24	1	3	10.4
9	1	6	72.8	25	0	2	9.1
10	0	7	28.7	26	0	5	2.0
11	1	3	11.5	27	0	6	12.6
12	1	2	56.3	28	1	5	18.0
13	0	5	−.5	29	0	3	1.5
14	0	3	−1.3	30	1	4	27.3
15	0	3	12.9	31	0	10	−8.4
16	0	8	34.1				

properties.* Many of the available statistical software packages employ ML estimation to fit logistic regression models. The SAS printout for the data of Table 14.6 is shown in Figure 14.6.

The maximum likelihood estimates of β_0, β_1, and β_2 are given in Figure 14.6 under the column heading **Parameter Estimate**. These estimates (shaded in the printout) are $\hat{\beta}_0 = 1.4212$, $\hat{\beta}_1 = -.7553$, and $\hat{\beta}_2 = .1122$. Therefore, the prediction equation for the probability of a fixed bid (i.e., the probability that $y = 1$) is

$$\hat{y} = \frac{\exp(1.4212 - .7553x_1 + .1122x_2)}{1 + \exp(1.4212 - .7553x_1 + .1122x_2)}$$

The standard errors of the β estimates are given under the column **Standard Error** and the (squared) ratios of the β estimates to their respective standard errors are given under the column **Wald Chi-Square**. As in a regression with a linear model, this ratio[†] provides a test statistic for testing the contribution of each variable to the model (i.e., testing $H_0: \beta_i = 0$). The observed significance levels of the tests (i.e., the p-values) are given under the column **Pr > Chi-Square**. Note that both independent variables, **NUMBIDS** (x_1) and **DOTEST** (x_2), have p-values less than .03 (implying that we would reject $H_0: \beta_1 = 0$ and $H_0: \beta_2 = 0$ for $\alpha = .03$).

---

*For details on how to obtain maximum likelihood estimators and their distributional properties, consult the references given at the end of the chapter.
[†]In the logistic regression model, the ratio $(\hat{\beta}_i/s_{\hat{\beta}_i})^2$ has an approximate χ^2 distribution with 1 degree of freedom.

FIGURE 14.6
A Portion of the SAS Printout for
Example 14.10

Criteria for Assessing Model Fit

Criterion	Intercept Only	Intercept and Covariates	Chi-Square for Covariates
AIC	43.381	28.843	.
SC	44.815	33.145	.
-2 LOG L	41.381	22.843	18.538 with 2 DF (p=0.0001)
Score	.	.	13.466 with 2 DF (p=0.0012)

Analysis of Maximum Likelihood Estimates

Variable	Parameter Estimate	Standard Error	Wald Chi-Square	Pr > Chi-Square	Standardized Estimate
INTERCPT	1.4212	1.2867	1.2199	0.2694	.
NUMBIDS	-0.7553	0.3388	4.9708	0.0258	-1.231128
DOTEST	0.1122	0.0514	4.7670	0.0290	1.143067

Predicted Probabilities and 95% Confidence Limits

OBS	NUMBIDS	DOTEST	STATUS	PRED	CLLOWER	CLUPPER
1	4	19.2	1	0.63510	0.32984	0.86023
2	2	24.1	1	0.93180	0.53648	0.99384
3	4	-7.1	0	0.08342	0.01043	0.44010
4	3	3.9	1	0.39958	0.15868	0.70132
5	9	4.5	0	0.00760	0.00016	0.26825
6	6	10.6	0	0.12770	0.02582	0.44708
7	2	-3.0	0	0.39506	0.10273	0.78836
8	11	16.2	0	0.00624	0.00007	0.36810
9	6	72.8	1	0.99368	0.35201	0.99998
10	7	28.7	0	0.34391	0.06138	0.80776
11	3	11.5	1	0.60958	0.31579	0.84081
12	2	56.3	1	0.99803	0.69696	0.99999
13	5	-0.5	0	0.08229	0.01253	0.38782
14	3	-1.3	0	0.27078	0.07452	0.63131
15	3	12.9	0	0.64626	0.34076	0.86589
16	8	34.1	0	0.31103	0.03168	0.86168
17	10	6.6	0	0.00453	0.00006	0.26602
18	5	-2.5	1	0.06686	0.00852	0.37403
19	13	24.2	0	0.00339	0.00001	0.45712
20	7	2.3	0	0.02639	0.00166	0.30640
21	3	36.9	1	0.96428	0.54751	0.99834
22	4	-11.7	0	0.05152	0.00412	0.41602
23	2	22.1	1	0.91608	0.51886	0.99103
24	3	10.4	1	0.57984	0.29466	0.82011
25	2	9.1	0	0.71740	0.33904	0.92627
26	5	2.0	0	0.10611	0.02005	0.40784
27	6	12.6	0	0.15485	0.03485	0.48180
28	5	18.0	1	0.41683	0.17873	0.70127
29	3	1.5	0	0.33704	0.11480	0.66587
30	4	27.3	1	0.81200	0.40060	0.96541
31	10	-8.4	0	0.00085	0.00000	0.15843

The test statistic for testing the overall adequacy of the logistic model, i.e., for testing H_0: $\beta_1 = \beta_2 = 0$, is given in the upper portion of the printout (shaded) as $\chi^2 = 18.538$, with observed significance level (shaded) $p = 0.0001$.* Based on the p-value of the test, we can reject H_0 and conclude that at least one of the β coefficients is nonzero. Thus, the model is adequate for predicting bid status, y.

Finally, the bottom portion of the printout gives predicted values and lower and upper 95% prediction limits for each observation used in the analysis in

---

*The test statistic has an approximate χ^2 distribution with $k = 2$ degrees of freedom, where k is the number of β parameters in the model (excluding β_0).

the columns titled **PRED, CLLOWER,** and **CLUPPER,** respectively. The 95% prediction interval for y for a contract with $x_1 = 3$ bidders and winning bid amount $x_2 = 11.5\%$ above the engineer's estimate is shaded on the printout. We estimate the probability of this particular contract being fixed to fall between .3158 and .8408. Note that all the predicted values and limits lie between 0 and 1, a property of the logistic model.

This optional section should be viewed only as an overview of logistic regression. Many of the details of logistic regression models, including interpretation of the β parameters and models with responses with three or more categories, have been omitted. Before conducting a logistic regression analysis, we strongly recommend that you consult the references given at the end of this chapter.

EXERCISES
APPLYING THE CONCEPTS

14.24   A retailer of home personal computers (PCs) conducted a study to relate PC ownership with annual income of heads of households. Data collected for a random sample of 20 households were used to fit the logit model

$$E(y) = \frac{\exp(\beta_0 + \beta_1 x)}{1 + \exp(\beta_0 + \beta_1 x)}$$

The data is shown in the table. An SPSS printout of the logistic regression is shown below. Interpret the results.

```
Dependent Variable..   Y

                      Chi-Square    df Significance
 -2 Log Likelihood       22.969     18        .1918
 Model Chi-Square          2.929      1        .0870
 Improvement               2.929      1        .0870
 Goodness of Fit          19.343     18        .3710

Classification Table for Y
                      Predicted
                   .00      1.00      Percent Correct
                    0   |    1
Observed
   .00     0      12   |    1          92.31%
                      --|--
  1.00     1       5   |    2          28.57%
                      --|--
                     Overall   70.00%

----------------------- Variables in the Equation -----------------------

Variable            B        S.E.      Wald      df      Sig       R      Exp(B)

X              5.47E-05  3.491E-05    2.4533      1     .1173    .1323    1.0001
Constant        -2.0188   1.0314      3.8314      1     .0503
-------------------------------------------------------------------------
```

HOUSEHOLD	y	x	HOUSEHOLD	y	x
1	0	$16,300	11	1	$22,400
2	0	11,200	12	0	10,600
3	0	36,500	13	0	21,400
4	1	21,700	14	0	8,300
5	1	40,200	15	1	27,500
6	0	12,400	16	0	15,700
7	0	15,000	17	0	12,100
8	0	9,200	18	1	59,600
9	1	36,700	19	1	20,200
10	0	62,000	20	0	33,100

14.25 Suppose you are investigating allegations of sex discrimination in the hiring practice of a particular firm. An equal-rights group claims that females are less likely to be hired than males with the same background, experience, and other qualifications. Data collected on 28 former applicants (shown in the table) will be used to fit the logit model:

$$E(y) = \frac{\exp(\beta_0 + \beta_1 x_1 + \beta_2 x_2 + \beta_3 x_3)}{1 + \exp(\beta_0 + \beta_1 x_1 + \beta_2 x_2 + \beta_3 x_3)}$$

where

$$y = \begin{cases} 1 & \text{if hired} \\ 0 & \text{if not} \end{cases}$$

x_1 = Years of higher education (4, 6, or 8)

x_2 = Years of experience

$$x_3 = \begin{cases} 1 & \text{if male applicant} \\ 0 & \text{if female applicant} \end{cases}$$

A SAS printout of the logistic regression follows on the next page.

HIRING STATUS y	EDUCATION x_1, years	EXPERIENCE x_2, years	SEX x_3	HIRING STATUS y	EDUCATION x_1, years	EXPERIENCE x_2, years	SEX x_3
0	6	2	0	1	4	5	1
0	4	0	1	0	6	4	0
1	6	6	1	0	8	0	1
1	6	3	1	1	6	1	1
0	4	1	0	0	4	7	0
1	8	3	0	0	4	1	1
0	4	2	1	0	4	5	0
0	4	4	0	0	6	0	1
0	6	1	0	1	8	5	1
1	8	10	0	0	4	9	0
0	4	2	1	0	8	1	0
0	8	5	0	0	6	1	1
0	4	2	0	1	4	10	1
0	6	7	0	1	6	12	0

Criteria for Assessing Model Fit

Criterion	Intercept Only	Intercept and Covariates	Chi-Square for Covariates
AIC	37.165	22.735	.
SC	38.497	28.064	.
-2 LOG L	35.165	14.735	20.430 with 3 DF (p=0.0001)
Score	.	.	15.032 with 3 DF (p=0.0018)

Analysis of Maximum Likelihood Estimates

Variable	Parameter Estimate	Standard Error	Wald Chi-Square	Pr > Chi-Square	Standardized Estimate
INTERCPT	-14.2483	6.0805	5.4909	0.0191	.
EDUC	1.1549	0.6023	3.6767	0.0552	1.001936
EXP	0.9098	0.4293	4.4919	0.0341	1.690596
SEX	5.6037	2.6028	4.6352	0.0313	1.569063

Predicted Probabilities and 95% Confidence Limits

OBS	EDUC	EXP	SEX	HIRED	PRED	CLL	CLU
1	6	6	1	1	0.97688	0.42319	0.99959
2	6	3	1	1	0.73385	0.26804	0.95405
3	8	3	0	1	0.09282	0.00485	0.68232
4	8	10	0	1	0.98352	0.27405	0.99989
5	4	5	1	1	0.62813	0.11439	0.95669
6	6	1	1	1	0.30886	0.07490	0.71155
7	8	5	1	1	0.99420	0.50305	0.99997
8	4	10	1	1	0.99378	0.29208	0.99998
9	6	12	0	1	0.97338	0.19876	0.99981
10	6	2	0	0	0.00407	0.00005	0.27086
11	4	0	1	0	0.01755	0.00048	0.40027
12	4	1	0	0	0.00016	0.00000	0.15324
13	4	2	1	0	0.09927	0.00894	0.57370
14	4	4	0	0	0.00250	0.00002	0.24465
15	6	1	0	0	0.00164	0.00001	0.24049
16	4	2	1	0	0.09927	0.00894	0.57370
17	8	5	0	0	0.38699	0.04394	0.89661
18	4	2	0	0	0.00041	0.00000	0.17559
19	6	7	0	0	0.27888	0.04027	0.78091
20	6	4	0	0	0.02461	0.00108	0.37067
21	8	0	1	0	0.64439	0.11457	0.96209
22	4	7	0	0	0.03698	0.00141	0.50996
23	4	1	1	0	0.04248	0.00221	0.47030
24	4	5	0	0	0.00618	0.00009	0.30151
25	6	0	1	0	0.15248	0.02129	0.59808
26	4	9	0	0	0.19153	0.01081	0.83708
27	8	1	0	0	0.01631	0.00026	0.51595
28	6	1	1	0	0.30886	0.07490	0.71155

a. Conduct a test of model adequacy. Use $\alpha = .05$.
b. Is there sufficient evidence to indicate that sex is an important predictor of hiring status? Test using $\alpha = .05$.
c. Calculate a 95% confidence interval for the mean response $E(y)$ when $x_1 = 4$, $x_2 = 0$, and $x_3 = 1$. Interpret the interval.

SUMMARY

Surveys that allow for more than two categories for a single response (i.e., a one-way table) can be analyzed using the *chi-square goodness-of-fit test* outlined in this chapter. The appropriate test statistic, called the χ^2 *statistic*, has a sampling distribution that is (approximately) a *chi-square probability distribution* and measures the amount of disagreement between the observed number of responses and the expected number of responses in each category.

A *contingency table analysis* is an application of the χ^2 test for a *two-way* (or *two-variable*) *classification of data*. The test allows us to determine whether the two directions of classification are independent. Caution should be exercised to avoid misuse of the χ^2 procedure. The underlying distribution of the response data should have the properties of a multinomial experiment outlined in the box in Section 14.1. Also, the estimated number of responses in any cell should not be too small.

When the dependent variable in regression is a categorical variable, a *logistic regression model* is more appropriate than a linear regression model. The distribution of each of the test statistics for logistic regression follows an approximate χ^2 distribution.

KEY WORDS

Chi-square probability distribution
Chi-square statistic: χ^2
Expected cell counts
Goodness of fit
Independence of two directions of
   classification

Logistic regression model*
Logit model*
Multinomial experiment
Observed cell counts
One-way table
Two-way (contingency) table

KEY FORMULAS

GOODNESS-OF-FIT TEST (ONE-WAY TABLE)

Test statistic: $\quad \chi^2 = \sum_{i=1}^{k} \dfrac{(O_i - e_i)^2}{e_i}$

where $\quad k = $ Number of categories (cells)

$\quad\quad O_i = $ Observed count for cell i

$\quad\quad e_i = $ Expected cell count for cell $i = n\pi_i$

$\quad\quad \pi_i = $ Hypothesized probability for cell i

$\quad\quad n = $ Sample size

---

*From the optional section in this chapter.

Pascal. The reported results include a distribution of the relative frequency of occurrence of the different types of statements used in typical Algol and Pascal programs of approximately the same size. The reported percentages were used to tabulate the information given in the table.

		ALGOL	PASCAL
	IF	125	2,045
TYPE OF STATEMENT	FOR	968	350
	IO	135	1,847
	Assignment	8,923	4,763
	Other	261	465
TOTALS		10,412	9,470

Source: Adapted from De Prycker, M. "On the development of a measurement system for high-level language program statistics." *IEEE Transactions on Computers*, Vol. C-31, No. 9, Sept. 1982, pp. 888–890. © 1982 IEEE.

a. Conduct a test to determine whether the percentages of the different types of programming statements differ for the two languages. Test using $\alpha = .05$.

b. Construct a 95% confidence interval for the difference in the percentages of assignment statements used in the two languages.

14.36 *Toehold acquisition* is a term used by financial analysts to describe the purchase of a relatively small proportion (often less than 10%) of the shares of a corporation by a corporate raider whose ultimate objective is to take over control of the firm. For each toehold acquisition, the potential raider of the firm must report the transaction to the Securities and Exchange Commission by filing form 13D. Experts hypothesize that the filing of the form 13D provides a clear signal to investors of the corporate raiders' intention. To test this theory, a sample of 461 transactions was drawn from a list of firms in which corporate raiders acquired at least 5% of the outstanding shares (*Akron Business and Economic Review*, Spring 1990). A contingency table relating transaction type and transaction timing for the 461 transactions is shown here.

		TIME RELATIVE TO 13D FILING	
		Before	After
TRANSACTION TYPE	Purchase	184	83
	Sale	105	89

a. If, in fact, the experts' theory is true, describe the relationship between transaction type and transaction timing.

b. The value of the χ^2 statistic for testing whether transaction type and transaction timing are independent was reported in the article to be $\chi^2 = 13.28$. Also, the p-value of the test was reported to be less than .01. Interpret the results.

14.37 In recent years, many companies have converted to the metric system of measurements. To investigate this phenomenon, researchers analyzed data collected on 757 small manufacturers for a U.S. Metric Board study (*Technological Forecasting and Social Change*, Apr. 1984). The firms were cross-classified according to metric conversion (converters or nonconverters) and level of technology (high- or non–high-technology). The contingency table for the data is shown here.

	HIGH-TECH	NON–HIGH-TECH
Metric converters	81	296
Nonconverters	80	300

a. Calculate the estimated expected number of firms in each of the four cells of the table.

b. Calculate the χ^2 statistic.

c. Is there sufficient evidence to indicate that the distributions of percentages of metric converters and non-converters differ for high-tech and non–high-tech firms? Test using $\alpha = .01$.

14.38 The advances of medical technology have led to an increased survival rate of infants born with genetic diseases. Since no true cure for genetic disease is currently available, some physicians provide genetic counseling for their patients to prevent the birth of genetically defective infants. However, genetic counseling has faced a certain amount of resistance from both physicians and patients. In a survey of general and family practitioners, pediatricians, and obstetrician-gynecologists in the cities of Phoenix and Tucson, Arizona, each physician was classified according to religion and opinion on genetic counseling. A summary of the responses for Jewish, Protestant, and Catholic physicians is shown in the accompanying table.

| | RELIGION | | | TOTALS |
	Jewish	Protestant	Catholic	
Strongly support genetic counseling	21	36	10	67
Do not strongly support genetic counseling	26	142	52	220
TOTALS	47	178	62	287

Source: Weitz, R. "Barriers to acceptance of genetic counseling among primary care physicians." *Social Biology*, Fall 1979, Vol. 26, p. 192.

a. Scan the data. Do you believe there is evidence of a difference in the proportions of physicians who strongly support genetic counseling for the three religions?

b. Why is a statistical test useful in answering part **a**?

c. Calculate the number of physicians you would expect to fall in each of the six cells of the contingency table.

d. Find the difference between the observed and the (estimated) expected numbers for each of the six cells.

e. Calculate the value of the chi-square statistic for the contingency table.

14.39 Research was conducted "to assess whether college students—tomorrow's executives—believe others use certain questionable behaviors in pursuit of advancement" (*Journal of Business Ethics*, Vol. 3, 1984). A sample of over 900 students at a large university, with majors in liberal arts, business administration, education, engineering, pharmacy, and law, were administered questionnaires during selected classes. The questionnaire was designed to measure student perceptions of a number of traits and practices that might describe a person seeking advancement in a large industrial enterprise. Students' response to the statement, "To progress, one has to develop the philosophy that *winning is everything*," were classified into one of three categories: frequently or more often (FM), occasionally (OCC), infrequently or less often (IL). To check for cultural differences, the researchers also classified the responses according to citizenship of student (native-born American, naturalized American, or other). The response percentages and corresponding sample sizes for the cross-tabulation are shown in the accompanying table.

	NATIVE-BORN	NATURALIZED	OTHER
FM	75.9	65.2	79.5
OCC	16.0	26.1	12.8
IL	8.1	8.7	7.7
n	855	23	39

Source: Pressley, M. M. and Blevin, D. E. "Students perceptions of 'job politics' as practiced by those climbing the corporate ladder." *Journal of Business Ethics*, Vol. 3, 1984, pp. 127–138.

a. Use the percentages in the table to compute expected frequencies for each cell.
b. Is there sufficient evidence to indicate that the percentages of responses in the three categories, FM, OCC, and IL, are identical for the three citizenship categories? Test using $\alpha = .01$.

14.40 With the recent growth in the number of U.S. households wired for cable television, major advertisers are now beginning to show an increased interest in cable advertising. Since its inception, cable television has offered advertisers relatively inexpensive rates and selective target audiences. Is advertising on cable really different from that on the three major networks? For purposes of comparison, A. J. Bush and J. H. Leigh conducted a content analysis of television commercials on the three major networks, ABC, CBS, and NBC, and three of the more popular, well-established cable networks, CNN, ESPN, and TBS (*Journal of Advertising Research*, Apr./May 1984). One of the variables measured was the number of advertisements shown during prime-time (7:00–10:00 P.M.) and late-night (10:00 P.M.–12:00 A.M.) viewing segments on two weekday evenings during November 1981. The results are summarized in the table.

		NUMBER OF ADVERTISEMENTS		TOTALS
		Prime-time	Late-night	
MAJOR NETWORKS	ABC	94	69	163
	CBS	90	84	174
	NBC	86	94	180
CABLE NETWORKS	TBS	66	47	113
	CNN	95	60	155
	ESPN	87	45	132
TOTALS		518	399	917

a. Is there evidence (at $\alpha = .05$) that the distribution of proportions of television advertisements during prime-time and late-night viewing segments differ among the six networks? Use the accompanying SPSS printout to make the inference.

```
VIEWTIME   by   NETWORK

                    NETWORK
          Count
          Exp Val
                                                            Row
                    ABC    CBS    CNN    ESPN   NBC    TBS   Total
VIEWTIME
          LATE        69     84     60     45     94     47     399
                    70.9   75.7   67.4   57.4   78.3   49.2   43.5%

          PRIME       94     90     95     87     86     66     518
                    92.1   98.3   87.6   74.6  101.7   63.8   56.5%

          Column     163    174    155    132    180    113     917
          Total     17.8%  19.0%  16.9%  14.4%  19.6%  12.3% 100.0%


       Chi-Square                 Value          DF         Significance
   -----------------------      -----------      ----       ------------

   Pearson                      13.64541          5            .01803
   Likelihood Ratio             13.70626          5            .01759

   Minimum Expected Frequency -    49.168

   Number of Missing Observations:   0
```

b. One of the objectives of the analysis is to compare the three major networks, as a group, to the three cable networks, as a group. To do this, you need to form a 2 × 2 contingency table, where the two directions of classification are network (major or cable) and viewing segment (prime-time or late-night). Construct the contingency table by summing the numbers of the three individual networks in each network–viewing segment category.

c. Use the contingency table formed in part b to determine whether the two directions of classification, network and viewing segment, are dependent. Test using $\alpha = .05$. Interpret your results.

CASE STUDY 14.1

THE RELATIONSHIP BETWEEN NEIGHBORHOOD DESIGN AND CRIME

One of the major problems confronting urban planners is crime prevention. How should the buildings, sites, and neighborhoods in an urban (or urban renewal) area be physically designed to minimize crime levels? Over the past two decades, two planning theories have emerged—the *defensible space* approach and the *opportunity* approach. The defensible space theory suggests that land use characteristics and the design of streets, buildings, and building sites affect crime through informal social (i.e., nonpolice) control. Urban areas are designed to maximize the use of public areas, both day and night. The more people who use the streets, the more "eyes" there are for informal surveillance, which, in turn, discourages criminal activity. The opportunity theory, on the other hand, suggests that proximity to criminal opportunities is the most important determinant of crime. The physical characteristics of the urban area are designed so that the degree of access, ease of entrance and exit by potential offenders, and the supply of potential targets are minimized.

To assess the validity of the two different planning perspectives, S. W. Greenberg and W. M. Rohe (1984) examined differences in physical characteristics and various dimensions of informal social control in six neighborhoods in Atlanta, Georgia. The neighborhoods were grouped into three pairs—(a) white middle income pair, (b) black lower middle income pair, and (c) black lower income pair—where one member of each pair had a relatively high crime rate, and the other had a relatively low crime rate. The results of their study showed that "physical characteristics distinguished between high- and low-crime neighborhoods to a much greater extent than did differences in informal social control." Consequently, "the results fail to support the [defensible space] theory."*

Greenberg and Rohe used contingency tables to summarize and analyze much of the data. The number of census blocks (or residential properties) in each neighborhood pair were classified according to crime rate (high or low) and several physical characteristics. The classifications for three of these characteristics—street front, nonresidential use, and housing—are shown, respectively, in Tables 14.7–14.9.

---

*Greenberg, S. W. and Rohe, W. M. "Neighborhood design and crime: A test of two perspectives." *Journal of the American Planning Association,* Vol. 50, No. 1, 1984, pp. 48–60. Reprinted by permission.

TABLE 14.7
Distribution of Street Types Fronted by Blocks (Number of Blocks)

STREET CHARACTERISTICS	(a) WHITE MIDDLE INCOME PAIR		(b) BLACK LOWER MIDDLE INCOME PAIR		(c) BLACK LOWER INCOME PAIR	
	High	Low	High	Low	High	Low
Major thoroughfare	20	9	25	1	22	30
Small neighborhood street	7	9	25	27	8	42
Other	21	15	36	14	3	23
TOTALS	48	33	86	42	33	95

TABLE 14.8
Distribution of Nonresidential Land Use (Number of Blocks)

LEVEL OF NONRESIDENTIAL USE	(a) WHITE MIDDLE INCOME PAIR		(b) BLACK LOWER MIDDLE INCOME PAIR		(c) BLACK LOWER INCOME PAIR	
	High	Low	High	Low	High	Low
High	12	6	34	10	29	43
Moderate	23	11	37	15	3	46
Low	13	16	15	17	1	6
TOTALS	48	33	86	42	33	95

TABLE 14.9
Distribution of Housing Units per Structure (Number of Residential Properties)

NUMBER OF HOUSING UNITS	(a) WHITE MIDDLE INCOME PAIR		(b) BLACK LOWER MIDDLE INCOME PAIR		(c) BLACK LOWER INCOME PAIR	
	High	Low	High	Low	High	Low
1	650	764	1,581	947	178	931
2–3	272	225	228	46	114	359
4–9	81	21	26	16	23	19
10 or more	92	4	21	15	12	9
TOTALS	1,095	1,014	1,856	1,024	327	1,318

[*Note:* Greenberg and Rohe reported the results in terms of percentages. For the reader's convenience, we have converted all percentages to cell counts.]

a. Refer to Table 14.7. For each neighborhood pair, conduct a chi-square test for independence of classifications. (Use $\alpha = .05$.) Interpret your results.

b. Refer to Table 14.8. For each neighborhood pair, conduct a chi-square test for independence of classifications. (Use $\alpha = .05$.) Interpret your results.

c. Refer to Table 14.9. For each neighborhood pair, conduct a chi-square test for independence of classifications. (Use $\alpha = .05$.) Interpret your results.

d. Refer to part **a**. For each neighborhood pair, construct a 95% confidence interval for the difference between the percentages of high-crime and low-crime neighborhood blocks that front a major thoroughfare.

e. Refer to part **b**. For each neighborhood pair, construct a 95% confidence interval for the difference between the percentages of high-crime and low-crime neighborhood blocks with a high level of nonresidential use.

f. Refer to part **c**. For each neighborhood pair, construct a 95% confidence interval for the difference between the percentages of high-crime and low-crime neighborhood residential properties with 2–3 housing units.

CASE STUDY 14.2

BIRTH ORDER AND THE CAR SALESMAN: IS LAST BEST?

The following is an excerpt from *Personal Selling & Sales Management* (Aug. 1988):

Listed below are characteristics of three individuals. Check the one that best seems to describe yourself.

[   ]   *Person A*:   perfectionist, reliable, conscientious, list maker, well organized, critical, serious, scholarly

[   ]   *Person B*:   mediator, fewest pictures in the family photo album, avoids conflict, independent, extreme loyalty to peer group, many friends, a maverick

[   ]   *Person C*:   manipulative, charming, blames others, shows off, people person, good salesperson, engaging

Readers whose personal characteristics are closest to those described for person A are likely to be the firstborn in their households. Characteristics of person B are most often shared by those who are middle children, while the youngest sibling is most likely to manifest the personality traits listed for person C.

The personality traits of person C above are thought, by many, to be those characteristics required of a good salesperson. Consequently, lastborn have traditionally been viewed as most likely to achieve success as sales professionals. Is this stereotype based on fact or fantasy?

The *Personal Selling & Sales Management* article reported on a study designed to answer this question. A systematic sample of 138 automobile dealerships was selected from the telephone directory listings of three cities in Alabama, Florida, and Washington. A personal interview was conducted with one salesperson from each of the dealerships to determine the salesperson's birth order. Thus, the data consisted of the birth orders of the 138 salespeople.

Table 14.10 (page 852) gives the number of salespeople in the sample who fell into each of the three birth order categories. Note that lastborn did not constitute the predominant birth order among respondents. In fact, both firstborn and middle children outnumbered the last born.

g. Plot the predicted probabilities in Table 14.11 on a two-dimensional graph with $\hat{y}$ on the vertical axis and x_1 on the horizontal axis. Connect the points associated with $x_2 = 1$ with a dotted line; connect the points associated with $x_2 = 0$ in a similar fashion. Use the pattern of points on the graph to interpret interaction in the logistic regression model.

COMPUTER LAB
CONTINGENCY TABLE ANALYSIS AND LOGISTIC REGRESSION

In this lab section, we provide instructions on using SAS, SPSS, and Minitab to analyze one-way tables, two-way (contingency) tables, and logistic regression models.

a. **Analysis of One-Way Table** (not available in SAS)   **SAS**

b. **Analysis of Contingency Tables** (data in Table 14.4)

```
Command
Line
  1    DATA GULL;                                    ⎫
  2    INPUT TIME $ LOCATION $ NUMBER;   ⎬ Data entry instructions
  3    CARDS;                                        ⎭
  4    JULAUG   NJ    11     ⎫
  5    JULAUG   NETH  32     │
  6    JULAUG   ENG   70     │
  7    SEPT     NJ    19     │
  .      •       •    •      ⎬ Input data values (1 observation per line)
  .      •       •    •      │
  .      •       •    •      │
 15    NOVDEC   ENG   43     ⎭
 16    PROC FREQ;                                    ⎫
 17    TABLES TIME*LOCATION/EXPECTED CHISQ;   ⎬ Contingency
 18    WEIGHT NUMBER;                         ⎭ table analysis
```

COMMAND 16   The FREQ procedure generates a frequency (or contingency) table for the data.

COMMAND 17   The TABLES statement defines the two classification variables for the contingency table. Variable names are separated by an asterisk (*). The options EXPECTED and CHISQ (following the slash) request that expected cell frequencies and the χ^2 statistic for the contingency table be printed.

COMMAND 18   The WEIGHT statement defines the weighting variable of the contingency table (e.g., NUMBER). This statement is necessary when the cell counts have already been tabulated, as in Table 14.4. If the cell counts have *not* been pretabulated (i.e., if the data are "raw"), omit the WEIGHT statement.

c. **Logistic regression** (data in Table 14.6)

```
Command
  Line
    1    DATA BID;                                          ⎫
    2    INPUT STATUS NUMBIDS DOTEST @@;                    ⎬ Data entry instructions
    3    CARDS;                                             ⎭
    4    1   4 19.2   1 2 24.1   0 4 -7.1                   ⎫
    5    1   3  3.9   0 9  4.5   0 6 10.6                   ⎪
    .    .   .  .     . .  .     . . .                      ⎬ Input data values
    .    .   .  .     . .  .     . . .                      ⎪ (3 observations per line)
    .    .   .  .     . .  .     . . .                      ⎭
   14    0 10  8.4
   15    PROC LOGISTIC;                                     ⎫ Logistic
   16    MODEL STATUS=NUMBIDS DOTEST;                       ⎬ regression
   17    OUTPUT OUT=NEW P=PRED L=CLLOWER U=CLUPPER;         ⎭ analysis
   18    PROC PRINT DATA=NEW;              } 95% prediction intervals
```

COMMAND 15   The LOGISTIC procedure fits a logit model.

COMMAND 16   The MODEL statement specifies the dependent variable to the left of the equals sign and the independent variables to the right.

COMMAND 17   The OUTPUT statement specifies an output data set that will contain the predicted values and the lower and upper 95% confidence limits.

COMMAND 18   The PRINT procedure prints the predicted values and confidence limits saved in the OUTPUT data set.

SPSS   a. **Analysis of one-way table** (data in Table 14.1)

```
Command
  Line
    1    DATA LIST FREE/NUMBER PRIORITY (A6).     ⎫
    2    WEIGHT BY NUMBER.                         ⎬ Data entry instructions
    3    BEGIN DATA.                               ⎭
    4    15  URGENT   61 NORMAL   24 LOW           } Input data values
    5    END DATA.
    6    NPAR TESTS CHISQUARE=PRIORITY/            ⎫ One-way χ² analysis
    7              EXPECTED=2,5,3.                  ⎭
```

COMMAND 1   Remember that when using mainframe SPSS (i.e., SPSS^x), you must (1) separate numeric character variables in the DATA LIST statement by an asterisk (*) and (2) omit the period at the end of all commands.

COMMAND 2   The WEIGHT statement defines the weighting variable of the one-way table (e.g., NUMBER). Use this statement when the cell counts have already been tabulated (as in Table 14.1). If the cell counts have *not* been tabulated (if the data are "raw"), omit the WEIGHT statement.

COMMAND 6 The NPAR tests procedure with subcommand CHISQUARE produces a one-way, chi-square analysis of the qualitative variable specified (e.g., PRIORITY).

COMMAND 7 The EXPECTED subcommand gives the hypothesized proportions for each cell of the table. The proportions are obtained by dividing each value specified by the sum of the values. (For equal proportions, specify all "1's" in the EXPECTED subcommand.)

b. **Analysis of contingency tables** (data in Table 14.4)

```
Command
  Line
    1    DATA LIST FREE/NUMBER TIME (A6) LOCATION (A4).    ⎫  Data entry
    2    WEIGHT BY NUMBER.                                  ⎬  instructions
    3    BEGIN DATA.                                        ⎭
    4    11   JULAUG   NJ      ⎫
    5    32   JULAUG   NETH    ⎪
    6    70   JULAUG   ENG     ⎪
    7    19   SEPT     NJ      ⎬  Input data values
    .     .    .       .       ⎪  (1 observation per line)
    .     .    .       .       ⎪
    .     .    .       .       ⎪
   15    43   NOVDEC   ENG     ⎭
   16    END DATA.
   17    CROSSTABS TABLES=TIME BY LOCATION/    ⎫
   18              CELLS=COUNT EXPECTED/        ⎬  Contingency table analysis
   19              STATISTICS=CHISQ.            ⎭
```

COMMAND 2 The WEIGHT statement defines the weighting variable of the contingency table (e.g., NUMBER). This statement is necessary when the cell counts have already been tabulated, as in Table 14.4. If the cell counts have *not* been pretabulated (i.e., if the data are "raw"), omit the WEIGHT statement.

COMMAND 17 The CROSSTABS procedure generates a contingency (or cross-tabulation) table for the data. The classification variables are specified following TABLES= and separated by the keyword BY.

COMMAND 18 CELLS=COUNT EXPECTED generates expected cell frequencies as well as cell counts for the contingency table.

COMMAND 19 STATISTICS=CHISQ generates the χ^2 statistic (and p-value) for the contingency table.

c. **Logistic regression** (data in Table 14.6)

```
Command
 Line
  1    DATA LIST FREE/STATUS BIDDERS PCTDOT,     ⎫  Data entry insructions
  2    BEGIN DATA,                               ⎭
  3    1   4   19,2   1   2   24,1   0   4   -7,1   ⎫
  4    1   3    3,9   0   9    4,5   0   6   10,6   ⎪
  .    .   .    .     .   .    .     .   .    .      ⎬  Input data values
  .    .   .    .     .   .    .     .   .    .      ⎪ (3 observations per line)
  .    .   .    .     .   .    .     .   .    .      ⎭
 13    0  10    8,4
 14    END DATA,
 15    LOGISTIC REGRESSION STATUS WITH BIDDERS, PCTDOT,   ⎫ Logit
                                                          ⎭ regression
```

COMMAND 15   The LOGISTIC REGRESSION command fits a logit model. The 0–1 dependent variable (STATUS) is specified before the WITH subcommand, whereas the independent variables are specified after WITH.

MINITAB   a. **Analysis of one-way table** (not available in Minitab)

b. **Analysis of contingency table** (data in Table 14.4)

```
Command
 Line
  1    READ TABLE IN C1 C2 C3   ⎫ Data entry instruction
  2    11    32    70           ⎫
  3    19   130    89           ⎪ Input data values (each row represents
  4     9   150    39           ⎬ one row of the contingency table)
  5     3    94    43           ⎭
  6    CHISQUARE ANALYSIS ON TABLE C1-C3   ⎫ Contingency table analysis
```

COMMAND 2–5   The pretabulated cell counts of the contingency table are read into the columns of the Minitab worksheet. Each row represents one row and each column represents one column of the contingency table.

COMMAND 6   The CHISQUARE command generates expected cell frequencies and the χ^2 statistic for the contingency table stored in the columns specified.

c. **Logistic regression** (not available in Minitab)

COMPUTER ACTIVITIES

Analyze the data in Tables 14.7–14.9, Case Study 14.1, using an available statistical software package. Interpret the results.

REFERENCES

Agresti, A. *Categorical Data Analysis*. New York: Wiley, 1990.

Cochran, W. G. "Some Methods for Strengthening Common χ^2 Tests." *Biometrics*, Vol. 10.

Cochran, W. G. "The χ^2 test of goodness of fit." *Annals of Mathematical Statistics*, 1952, Vol. 23, pp. 315–345.

Hosmer, D. W. and Lemeshow, S. *Applied Logistic Regression*. New York: Wiley, 1989.

Tsiatis, A. A. "A note on the goodness-of-fit test for the logistic regression model." *Biometrika*, 1980, Vol. 67, pp. 250–251.

Walker, S. H. and Duncan, D. B. "Estimation of the probability of an event as a function of several independent variables." *Biometrika*, 1967, Vol. 54, pp. 167–179.

NONPARAMETRIC STATISTICS

CHAPTER 15

R ecently, data was collected on the dioxin levels of Vietnam veterans possibly exposed to the defoliant Agent Orange. An analysis reveals that this data is not normally distributed. Consequently, one of the assumptions required to conduct the parametric tests described in earlier chapters is violated. In this chapter we learn how to use **nonparametric (distribution-free) statistical tests** to analyze data of this type. We will apply these techniques to the data on dioxin levels in Vietnam veterans in the chapter case of Section 15.5.

15.1 Sometimes the sample data do not satisfy the various assumptions required for the statistical methods of Chapters 8–13. In other sampling situations, such as taste-testing of foods or other types of consumer product evaluations, we can determine that we like product A better than product B, and B better than C, but we cannot obtain exact quantitative values for the respective observations. We can only rank them.

Statistical test procedures that use the ranks of observations to perform tests of hypotheses are called **nonparametric statistical tests**. They are used when the sample data are in the form of ranks, or when we are concerned that the data do not satisfy the assumptions of the various tests described in the preceding chapters. As you will subsequently see, nonparametric tests are *distribution-free*; that is, they rely on very few underlying assumptions about the probability distribution of the sampled population.

TESTING FOR LOCATION OF A SINGLE POPULATION

15.2 Recall that small-sample procedures for estimating a population mean (Section 8.3) or for testing a hypothesis about a population mean (Section 10.2) require that the population have an approximately normal distribution. For situations in which we collect a small sample ($n < 30$) from a nonnormal population, the t test is not valid, and we must resort to a **nonparametric procedure**. The simplest nonparametric technique to apply in this situation is the **sign test**. The sign test is specifically designed for testing hypotheses about the **median** of any continuous population. Like the mean, the median is a measure of the center, or location, of the distribution; consequently, the sign test is sometimes referred to as a **test for location**.

EXAMPLE 15.1 Consider a population with unknown median η, and suppose we want to test the null hypothesis H_0: $\eta = 100$ against the one-sided alternative H_a: $\eta > 100$. From Definition 3.2 we know that the median is a number such that half the area under the probability distribution lies to the left of η and half lies to the right (see Figure 15.1). Therefore, the probability that an x value selected from the population is larger than η is .5, i.e., $P(x_i > \eta) = .5$. If, in fact, the null hypothesis is true, then we would expect to observe approximately half the sample x values greater than $\eta = 100$.

FIGURE 15.1

Location of the Population Median, η

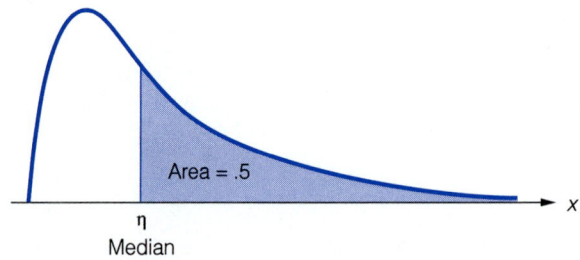

The sign test utilizes the test statistic S, where

$$S = \text{Number of sample observations } (x\text{'s}) \text{ that exceed } 100$$

Find the rejection region for the test.

Solution Notice that S depends only on the *sign* (positive or negative) of the difference between each sample value x_i and 100. That is, we are simply counting the number of positive $(+)$ signs among the sample differences $(x_i - 100)$. If S is "too large" (i.e., if we observe an unusual number of x_i's exceeding 100), then we will reject H_0 in favor of the alternative H_a: $\eta > 100$.

The rejection region for the sign test is derived as follows. Let each sample difference $(x_i - 100)$ denote the outcome of a single trial in an experiment consisting of n identical trials. If we call a positive difference a "success" and a negative difference a "failure," then S is the number of successes in n trials. Under H_0, the probability of observing a success on any one trial is

$$\begin{aligned} \pi = P(\text{Success}) &= P(x_i - 100 > 0) \\ &= P(x_i > 100) \\ &= .5 \end{aligned}$$

Since the trials are independent, the properties of a binomial experiment, listed in Chapter 5, are satisfied. Therefore, S has a binomial distribution with parameters n and $\pi = .5$. We can use this fact to calculate the observed significance level (p-value) of the sign test.

EXAMPLE 15.2 The Environmental Protection Agency (EPA) sets certain pollution guidelines for major industries. For a particular company that discharges waste into a nearby river, the EPA criterion is that the median amount of pollution in water samples collected from the river may not exceed 5 parts per million (ppm). Responding to numerous complaints, the EPA takes 10 water samples from the river at the discharge point and measures the pollution level in each sample. The results (in ppm) are given here:

| 5.1 | 4.3 | 5.3 | 6.2 | 5.6 | 4.7 | 8.4 | 5.9 | 6.8 | 3.0 |

Do the data provide sufficient evidence to indicate that the median pollution level in water discharged at the plant exceeds 5 ppm? Test using $\alpha = .05$.

Solution Letting η represent the population median pollution level, we want to test

H_0: $\eta = 5$
H_a: $\eta > 5$

using the sign test. The test statistic is

$$\begin{aligned} S &= \text{Number of sample observations that exceed } 5 \\ &= 7 \end{aligned}$$

where S has a binomial distribution with parameters $n = 10$ and $\pi = .5$.

From Section 9.6, the observed significance level (p-value) of the test is the probability that we observe a value of the test statistic S that is at least as contradictory to the null hypothesis as the computed value. For this one-sided case, the p-value is the probability that we observe a value of S greater than or equal to 7. We find this probability using the binomial table for $n = 10$ and $\pi = .5$ in Table 3 of Appendix G:

$$p\text{-value} = P(S \geq 7) = 1 - P(S \leq 6)$$
$$= 1 - .8281 = .1719$$

The p-value is also shown (shaded) on the Minitab printout of the analysis shown in Figure 15.2. Since the p-value, .1719, is larger than $\alpha = .05$, we cannot reject the null hypothesis. That is, there is insufficient evidence to indicate that the median pollution level of water discharged from the plant exceeds 5.

FIGURE 15.2

Minitab Printout for Example 15.2

```
SIGN TEST OF MEDIAN = 5.000 VERSUS G.T. 5.000

            N   BELOW  EQUAL  ABOVE   P-VALUE    MEDIAN
pollevel   10     3      0      7     0.1719     5.450
```

A summary of the sign test for one-sided and two-sided alternatives is provided in the upper box on the facing page.

For a two-tailed test, you may calculate the test statistic as either

$$S_1 = \text{Number of sample observations greater than } \eta_0$$
$$= \text{Number of successes in } n \text{ trials}$$

or

$$S_2 = \text{Number of sample observations less than } \eta_0$$
$$= \text{Number of failures in } n \text{ trials}$$

Note that $S_1 + S_2 = n$; therefore, $S_2 = n - S_1$. In either case, the p-value of the test is double the corresponding one-sided p-value.

Recall from Section 6.2 that a normal distribution with mean $\mu = n\pi$ and $\sigma = \sqrt{n\pi(1 - \pi)}$ can be used to approximate the binomial distribution for large n. When $\pi = .5$, the normal approximation performs reasonably well even for n as small as 10. Thus, for $n \geq 10$, we can conduct the sign test using the familiar standard normal z statistic of Section 10.2. Since S has a binomial distribution with mean $\mu = n\pi = .5n$ and standard deviation $\sigma = \sqrt{n\pi(1 - \pi)} = \sqrt{n(.5)^2} = .5\sqrt{n}$, the large-sample test statistic is

$$z = \frac{S - \mu}{\sigma} = \frac{S - .5n}{.5\sqrt{n}}$$

This large-sample sign test is summarized in the box (opposite).

SIGN TEST FOR A POPULATION MEDIAN, η, OF A SINGLE POPULATION

ONE-TAILED TEST

H_0: $\eta = \eta_0$

H_a: $\eta > \eta_0$

[or H_a: $\eta < \eta_0$]

Test statistic:

S = Number of sample observations greater than η_0

[or S = Number of sample observations less than η_0]

Observed significance level:

p-value = $P(x \geq S)$

TWO-TAILED TEST

H_0: $\eta = \eta_0$

H_a: $\eta \neq \eta_0$

Test statistic:

S = Larger of S_1, S_2,

where

S_1 = Number of sample observations greater than η_0

and

S_2 = Number of sample observations less than η_0

Observed significance level:

p-value = $2P(x \geq S)$

where x has a binomial distribution with parameters n and $\pi = .5$.

Rejection region: Reject H_0 if p-value $< \alpha$.

Assumption: The sample is randomly selected from a continuous probability distribution. [*Note:* No assumptions have to be made about the shape of the probability distribution.]

SIGN TEST FOR A LARGE SAMPLE ($n \geq 10$)

ONE-TAILED TEST

H_0: $\eta = \eta_0$

H_a: $\eta > \eta_0$

[or H_a: $\eta < \eta_0$]

TWO-TAILED TEST

H_0: $\eta = \eta_0$

H_a: $\eta \neq \eta_0$

Test statistic: $z = \dfrac{S - .5n}{.5\sqrt{n}}$

[*Note:* The value of S is calculated as shown in the previous box.]

Rejection region: $z > z_\alpha$

Rejection region: $z > z_{\alpha/2}$

where tabulated values of z_α and $z_{\alpha/2}$ are given in Table 4 of Appendix G.

EXERCISES
LEARNING THE MECHANICS

15.1 Suppose you want to use the sign test to test the null hypothesis that the population median equals 75, i.e., H_0: $\eta = 75$. Use Table 3 of Appendix G to find the observed significance level (p-value) of the test for each of the following situations:

a. H_a: $\eta > 75$, $n = 5$, $S = 2$
b. H_a: $\eta \neq 75$, $n = 20$, $S = 16$
c. H_a: $\eta < 75$, $n = 10$, $S = 8$

15.2 A random sample of six observations from a continuous population resulted in the following:

18.2   21.3   20.5   19.4   19.6   17.7

Is there sufficient evidence to indicate that the population median differs from 20? Test using $\alpha = .05$. [*Hint:* Use the binomial formula for $p(x)$ given in Section 5.4 to calculate the p-value of the test.]

APPLYING THE CONCEPTS

15.3 In zoology, the phenomenon of fish moving excessively from one confined area to another is known as excessive transitory migration (ETM). To investigate the ETM of guppy populations, 40 adult female guppies were placed into the left compartment of an experimental aquarium tank that was divided in half by a glass plate. After the plate was removed, the numbers of fish passing through the slit from the left compartment to the right one, and vice versa, was monitored every minute for 30 minutes (*Zoological Science*, Vol. 6, 1989). If equilibrium is reached, the researchers would expect the median number of fish remaining in the left compartment to be 20. The data for the 30 observations (i.e., numbers of fish in the left compartment at the end of the minute interval) are shown here. Use the large-sample sign test to determine whether the median is less than 20. Test using $\alpha = .05$.

16	11	12	15	14	16	18	15	13	15
14	14	16	13	17	17	14	22	18	19
17	17	20	23	18	19	21	17	21	17

Source: Terami, H. and Watanabe, M. "Excessive transitory migration of guppy populations. III. Analysis of perception of swimming space and a mirror effect." *Zoological Science*, Vol. 6, 1989, p. 977 (Figure 2).

15.4 Refer to the *Risk Management* study on victims who attempted to evacuate fires at compartmented fire-resistive buildings, described in Exercise 3.13. The numbers of victims who died attempting to evacuate for a sample of 14 recent fires are shown in the accompanying table.

a. In part **a** of Exercise 3.13 you constructed a stem-and-leaf display for the data. Does it appear that the sample data are from a normally distributed population?
b. Based on your answer to part **a**, why is a nonparametric test for location preferred over a parametric test? Explain.
c. Conduct a test (at $\alpha = .01$) to determine whether the median number of victims who attempt to evacuate fires at compartmented fire-resistive buildings differs from 6. Use the accompanying Minitab printout to make the proper inference.

FIRE	NUMBER OF VICTIMS
Las Vegas Hilton (Las Vegas)	5
Inn on the Park (Toronto)	5
Westchase Hilton (Houston)	8
Holiday Inn (Cambridge, OH)	10
Conrad Hilton (Chicago)	4
Providence College (Providence)	8
Baptist Towers (Atlanta)	7
Howard Johnson (New Orleans)	5
Cornell University (Ithaca, New York)	9
Westport Central Apartments (Kansas City, MO)	4
Orrington Hotel (Evanston, IL)	0
Hartford Hospital (Hartford, CT)	16
Milford Plaza (New York, NY)	0
MGM Grand (Las Vegas)	36

Source: Macdonald, J. N. "Is evacuation a fatal flaw in fire-fighting philosophy?" *Risk Management,* Vol. 33, No. 2, Feb. 1986, p. 37.

```
SIGN TEST OF MEDIAN = 6.000 VERSUS  N.E.  6.000

            N  BELOW  EQUAL  ABOVE  P-VALUE   MEDIAN
victims    14    7      0      7    1.0000    6.000
```

15.5   The data on student loan default rates for a sample of 66 colleges, Exercise 2.28, is reproduced on the next page. Suppose you want to determine whether the median default rate of all U.S. colleges is less than 15%. See the accompanying SPSS printout of the analysis.

```
- - - - - Sign Test

     DEFRATE
with MEDIAN

        Cases

        21  - Diffs (MEDIAN Lt DEFRATE)        Z =      2.8311
        45  + Diffs (MEDIAN Gt DEFRATE)
         0    Ties                        2-tailed P =      .0046
        --
        66    Total
```

a. Interpret the results.
b. Why might the results, part a, be biased? [*Hint:* The 66 colleges in the sample are all located in Florida.]

15.6   An important property of certain manufactured products that are in powder or granular form is their particle-size distribution. For example, refractory cements are adversely affected by too high a proportion of coarse granules, which can lead to weakness because of poor packing. The following data, extracted from the *Journal of Quality Technology* (July 1985), represent the percentages of coarse granules for a random sample of eight refractory cement specimens from a large lot:

   1.7   .9   3.4   2.5   3.1   .6   1.0   2.1

Is there sufficient evidence to indicate that fewer than half of the cement specimens in the lot have more than 2% coarse granules? Test using $\alpha = .05$.

Data for Exercise 15.5

COLLEGE/UNIVERSITY	DEFAULT RATE	COLLEGE/UNIVERSITY	DEFAULT RATE
Florida College of Business	76.2	Brevard Community College	9.4
Fort Lauderdale College	48.5	College of Boca Raton	9.1
Florida Career College	48.3	Florida International University	8.7
United College	46.8	Santa Fe Community College	8.6
Florida Memorial College	46.2	Edison Community College	8.5
Bethune Cookman College	43.0	Palm Beach Junior College	8.0
Edward Waters College	38.3	Eckerd College	7.9
Florida College of Medical and		University of Tampa	7.6
Dental Careers	32.6	Lakeland College of Business	7.2
International Fine Arts College	26.5	Pensacola Junior College	6.8
Tampa College	23.9	University of Miami	6.7
Miami Technical College	23.3	Florida Institute of Technology	6.7
Tallahassee Community College	20.6	University of West Florida	6.3
Charron Williams College	20.2	Palm Beach Atlantic College	6.0
Florida Community College	19.1	University of Central Florida	5.7
Miami-Dade Community College	19.0	Seminole Community College	5.6
Broward Community College	18.4	Polk Community College	5.6
Daytona Beach Community College	16.9	Phillips Junior College	5.6
Lake Sumter Community College	16.7	Nova University	5.5
Forida Technical College	16.6	Rollins College	5.5
Florida A & M University	15.8	St. Leo College	5.5
Prospect Hall College	15.1	Gulf Coast Community College	5.4
Hillsborough Community College	14.4	Southern College	5.3
Pasco-Hernando Community College	13.5	Flagler College	4.7
Orlando College	13.5	Florida Atlantic University	4.4
Jones College	13.1	University of South Florida	4.2
Webber College	11.8	Manatee Junior College	4.1
Warner Southern College	11.8	Florida State University	4.0
Central Florida Community College	11.8	University of North Florida	3.9
Indian River Community College	11.8	Barry University	3.1
St. Petersburg Community College	11.3	University of Florida	3.1
Valencia Community College	10.8	Stetson University	2.9
Florida Southern College	10.3	Jacksonville University	1.5
Lake City Community College	9.8		

15.7   Refer to the data on the total 1990 compensations (in thousands of dollars) for the top 20 corporate executives (as determined by *Business Week*) in Exercise 3.17. The data are reproduced in the table on the facing page.

a. Construct a stem-and-leaf display for the data. Based on the graph, would you be willing to assume that the sample data come from a normal population?

b. Based on the result from part a, would you advise using the small-sample t statistic to analyze the data?

c. Test the hypothesis that the median total 1990 compensation of the highest-paid corporate executives in the United States exceeds $5,000,000. Use $\alpha = .10$.

15.8   The data on the results from hematology tests administered to a sample of 50 black workers, Exercise 3.55, is reproduced on the facing page. Test the hypothesis that the median lymphocyte (LYMPHO) count of all black workers exceeds 20. Use $\alpha = .05$.

CORPORATE EXECUTIVE (COMPANY)	TOTAL 1989 CASH COMPENSATION (in thousands)
Craig O. McCaw (McCaw Cellular)	$53,944
Steven J. Ross (Time Warner)	34,200
Donald A. Pels (Lin Broadcasting)	22,791
Jim P. Manzi (Lotus Development)	16,363
Paul Fireman (Reebok International)	14,606
Ronald K. Richey (Torchmark)	12,666
Martin S. Davis (Paramount)	11,635
Roberto C. Goizueta (Coca-Cola)	10,715
Michael D. Eisner (Walt Disney)	9,589
August A. Busch III (Anheuser-Busch)	8,861
William G. McGowan (MCI)	8,666
James R. Moffett (Freeport McMoRan)	7,300
Donald E. Petersen (Ford Motor)	7,147
P. Roy Vagelos (Merck)	6,764
W. Michael Blumenthal (Unisys)	6,511
S. Parker Gilbert (Morgan Stanley)	5,510
Harry A. Merlo (Louisiana-Pacific)	5,314
Reuben Mark (Colgate-Palmolive)	5,004
Robert J. Pfeiffer (Alexander & Baldwin)	4,943
William P. Stiritz (Ralston Purina)	4,854

Source: "Pay stubs of the rich and corporate." *Business Week*, May 7, 1990, p. 57.

Data for Exercise 15.8

CASE NUMBER	WBC	LYMPHO	CASE NUMBER	WBC	LYMPHO
1	4,100	14	26	4,300	9
2	5,000	15	27	5,200	16
3	4,500	19	28	3,900	18
4	4,600	23	29	6,000	17
5	5,100	17	30	4,700	23
6	4,900	20	31	7,900	43
7	4,300	21	32	3,400	17
8	4,400	16	33	6,000	23
9	4,100	27	34	7,700	31
10	8,400	34	35	3,700	11
11	5,600	26	36	5,200	25
12	5,100	28	37	6,000	30
13	4,700	24	38	8,100	32
14	5,600	26	39	4,900	17
15	4,000	23	40	6,000	22
16	3,400	9	41	4,600	20
17	5,400	18	42	5,500	20
18	6,900	28	43	6,200	20
19	4,600	17	44	4,900	26
20	4,200	14	45	7,200	40
21	5,200	8	46	5,800	22
22	4,700	25	47	8,400	61
23	8,600	37	48	3,100	12
24	5,500	20	49	4,000	20
25	4,200	15	50	6,900	35

Source: Royston, J. P. "Some techniques for assessing multivariate normality based on the Shapiro–Wilk W." *Applied Statistics*, Vol. 32, No. 2, 1983, pp. 121–133.

15.3 In Section 10.4 we presented *parametric* tests (tests about population parameters), based on the z and the t statistics, to test for a difference between two population means, μ_1 and μ_2. Recall that the mean of a population measures the *location* of the population distribution. Thus, if the data provide sufficient evidence to indicate that μ_1 is larger than μ_2, we envision the distribution for population 1 shifted to the right of population 2.

The equivalent nonparametric test is not a test about the difference between population means. Rather, it is a test to detect whether distribution 1 is shifted to the right of distribution 2 or vice versa. The test, based on independent random samples of n_1 and n_2 observations from the respective populations, is known as the **Wilcoxon rank sum test**.

To use the Wilcoxon rank sum test, we first rank all $(n_1 + n_2)$ observations, assigning a rank of 1 to the smallest, 2 to the second smallest, and so on. The sum of the ranks, called a **rank sum**, is then calculated for each sample. If the two distributions are identical, we would expect the sample rank sums, designated as T_1 and T_2, to be nearly equal. In contrast, if one rank sum—say, T_1—is much larger than the other, T_2, then the data suggest that the distribution for population 1 is shifted to the right of the distribution for population 2. The procedure for conducting a Wilcoxon rank sum test is summarized in the box (opposite) and illustrated in Example 15.3.

EXAMPLE 15.3   A mail-order house wanted to compare the number of customer complaints received daily before and after instituting a new quality control program. After sufficient time had elapsed to make certain that customers were receiving shipments of goods produced after implementation of the program, the number of complaints was recorded each day for a period of 10 days. Similarly, the number of complaints per day was recorded for each of a random sample of 10 days selected prior to implementation of the new quality control procedure. The data are shown in Table 15.1. Is there sufficient evidence to indicate that the number of customer complaints has been reduced since the new quality control procedure was instituted? Test using $\alpha = .05$.

TABLE 15.1
Number of Complaints per Day

BEFORE IMPLEMENTATION	AFTER IMPLEMENTATION
17	10
14	15
12	7
16	6
23	13
18	11
10	12
8	9
19	17
22	14

Solution   The ranks of the 20 observations, from lowest to highest, are shown in Table 15.2 (page 870). Tied ranks receive the average of the ranks the scores would have received if they had not been tied.

Since we want to determine whether the daily number of complaints has decreased since implementation of the quality control program, we will test

H_0: The two distributions of numbers of daily complaints are identical.

H_a: The distribution of population 1 (the number of daily complaints *before*) is shifted to the right of the distribution for population 2 (the number of daily complaints *after*).

The test statistic is (according to the box) $T_1 = 132$.

Table 11 of Appendix G gives lower- and upper-tailed critical values of the rank sum distribution, denoted T_L and T_U, respectively, for values $n_1 \leq 10$ and $n_2 \leq 10$. The portion of Table 11 for a one-tailed test with $\alpha = .05$ and for a two-tailed test with $\alpha = .10$ is reproduced in Table 15.3 (page 870). Values of n_1 are given across the top of the table; values of n_2 are given at the left.

TABLE 15.2

Calculation of Rank Sums for Example 15.3

BEFORE IMPLEMENTATION		AFTER IMPLEMENTATION	
Raw Data	Rank	Raw Data	Rank
17	15.5	10	5.5
14	11.5	15	13
12	8.5	7	2
16	14	6	1
23	20	13	10
18	17	11	7
10	5.5	12	8.5
8	3	9	4
19	18	17	15.5
22	19	14	11.5
	$T_1 = 132$		$T_2 = 78$

TABLE 15.3

A Portion of the Wilcoxon Rank Sum Table, Table 11 of Appendix G

		n_1															
		3		4		5		6		7		8		9		10	
		T_L	T_U	T_L	T_U	T_L	T_U	T_L	T_U	T_L	T_U	T_L	T_U	T_L	T_U	T_L	T_U
	3	6	15	7	17	7	20	8	22	9	24	9	27	10	29	11	31
	4	7	17	12	24	13	27	14	30	15	33	16	36	17	39	18	42
	5	7	20	13	27	19	36	20	40	22	43	24	46	25	50	26	54
n_2	6	8	22	14	30	20	40	28	50	30	54	32	58	33	63	35	67
	7	9	24	15	33	22	43	30	54	39	66	41	71	43	76	46	80
	8	9	27	16	36	24	46	32	58	41	71	52	84	54	90	57	95
	9	10	29	17	39	25	50	33	63	43	76	54	90	66	105	69	111
	10	11	31	18	42	26	54	35	67	46	80	57	95	69	111	83	127

Examining Table 15.3, you will find that the critical values (shaded) corresponding to $n_1 = n_2 = 10$ are $T_L = 83$ and $T_U = 127$. Therefore, for a one-tailed (upper-tailed) test at $\alpha = .05$, we will reject H_0 if $T_1 \geq T_U$, i.e., reject H_0 if $T_1 \geq 127$ (see Figure 15.3). Since the observed value of the test statistic, $T_1 = 132$, is greater than 127, we reject H_0 and conclude (at $\alpha = .05$) that the distribution of the numbers of complaints before implementation of the quality control procedure is shifted to the right of the distribution of the numbers of complaints after implementation of the procedure.

FIGURE 15.3

Rejection Region for One-Tailed Test, $\alpha = .05$, Example 15.3

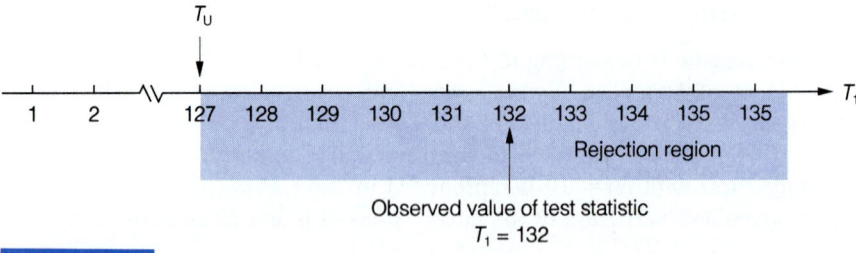

Many nonparametric test statistics have sampling distributions that are approximately normal when n_1 and n_2 are large. For these situations we can test hypotheses using the large-sample z test of Chapter 10.

THE WILCOXON RANK SUM TEST FOR LARGE SAMPLES ($n_1 \geq 10$ AND $n_2 \geq 10$)

Let D_1 and D_2 represent the relative frequency distributions for populations 1 and 2, respectively.

ONE-TAILED TEST

H_0: D_1 and D_2 are identical.

H_a: D_1 is shifted to the right of D_2.
[or H_a: D_1 is shifted to the left of D_2].

TWO-TAILED TEST

H_0: D_1 and D_2 are identical.

H_a: D_1 is shifted either to the right or to the left of D_2.

Test statistic: $z = \dfrac{T_1 - \left[\dfrac{n_1 n_2 + n_1(n_1 + 1)}{2} \right]}{\sqrt{\dfrac{n_1 n_2 (n_1 + n_2 + 1)}{12}}}$

Rejection region:

$z > z_\alpha$ [or $z < -z_\alpha$]

Rejection region:

$|z| > z_{\alpha/2}$

[*Note:*   The sample sizes n_1 and n_2 must both be at least 10.]

EXAMPLE 15.4   Refer to Example 15.3.

a. Show that, for this situation, the large-sample Wilcoxon rank sum z test gives the same result as the exact test performed in Example 15.3.

b. Find the p-value of the large-sample test.

Solution

a. The value of the large-sample Wilcoxon rank sum z test statistic is

$$z = \dfrac{T_1 - \left[\dfrac{n_1 n_2 + n_1(n_1 + 1)}{2} \right]}{\sqrt{\dfrac{n_1 n_2 (n_1 + n_2 + 1)}{12}}} = \dfrac{132 - \left[\dfrac{(10)(10) + 10(11)}{2} \right]}{\sqrt{\dfrac{(10)(10)(10 + 10 + 1)}{12}}}$$

$$= \dfrac{132 - 105}{\sqrt{2{,}100/12}}$$

$$= 2.04$$

Since we want to detect a possible shift in the distribution of population 1 to the right of population 2, we will reject H_0 for values of z in the upper tail of the z distribution. For $\alpha = .05$, the rejection region is $z > z_{.05}$, or $z > 1.645$ (see Figure 15.4 on the next page).

FIGURE 15.4
Rejection Region for
Example 15.4

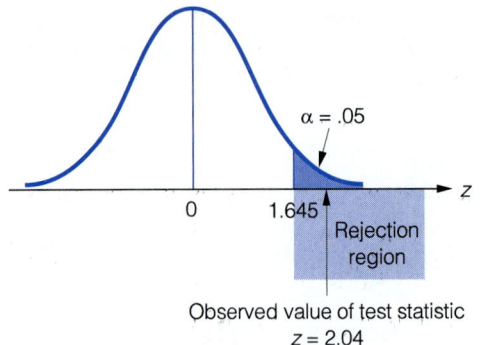

Since the observed value of the test statistic, $z = 2.04$, falls in the rejection region, our conclusion is the same as that for Example 15.3—reject H_0 and conclude that the distribution of the numbers of complaints before implementation of the procedure (population 1) is shifted to the right of the distribution of the numbers of complaints after implementation of the procedure (population 2).

b. From Table 4 of Appendix G, the observed significance level (p-value) of the upper-tailed large-sample test is

$$p\text{-value} = P(z > 2.04) = .5 - .4793 = .0207$$

Consequently, we will reject H_0 for any value of α larger than .0207.

We can also obtain this p-value from a computer printout. The SAS printout of the analysis is shown in Figure 15.5. Note that both rank sums and the large-sample z values are shaded on the printout, as well as the two-tailed p-value of the Wilcoxon rank sum test. The one-tailed p-value is half this value, $p\text{-value} = \frac{.0448}{2} = .0224$. [*Note:* The z value and p-value on the printout differ slightly from the corresponding calculated values because SAS uses the continuity correction factor in its calculations.]

FIGURE 15.5

SAS Printout for Example 15.4

```
            N P A R 1 W A Y   P R O C E D U R E

        Wilcoxon Scores (Rank Sums) for Variable NUMBER
                 Classified by Variable IMPLEMNT

                       Sum of      Expected      Std Dev        Mean
   IMPLEMNT     N      Scores      Under H0      Under H0       Score

   before      10      132.0        105.0      13.2088487    13.2000000
   after       10       78.0        105.0      13.2088487     7.8000000
                 Average Scores were used for Ties
        Wilcoxon 2-Sample Test (Normal Approximation)
        (with Continuity Correction of .5)

   S=  132.000      Z=  2.00623      Prob > |Z|  =   0.0448

   T-Test approx. Significance =      0.0593

   Kruskal-Wallis Test (Chi-Square Approximation)
   CHISQ=  4.1783      DF=  1     Prob > CHISQ=      0.0409
```

LEARNING THE MECHANICS

15.9 Specify the rejection region for the Wilcoxon rank sum test for independent samples in each of the following situations. Assume the test statistic is T_1.

    a. H_0: Two probability distributions, 1 and 2, are identical.
        H_a: Probability distribution 1 is shifted either to the right or to the left of probability distribution 2.

        $n_1 = 5, \quad n_2 = 10, \quad \alpha = .05$

    b. H_0: Two probability distributions, 1 and 2, are identical.
        H_a: Probability distribution 1 is shifted to the right of probability distribution 2.

        $n_1 = 8, \quad n_2 = 8, \quad \alpha = .05$

    c. H_0: Two probability distributions, 1 and 2, are identical.
        H_a: Probability distribution 1 is shifted to the left of probability distribution 2.

        $n_1 = 5, \quad n_2 = 7, \quad \alpha = .05$

15.10 Repeat Exercise 15.9, but assume the test statistic is T_2.

15.11 For each of the following situations, find the large-sample Wilcoxon rank sum test statistic and corresponding rejection region:

    a. Lower-tailed test, $T_1 = 71, \quad n_1 = 10, \quad n_2 = 14, \quad \alpha = .05$
    b. Upper-tailed test, $T_1 = 750, \quad n_1 = 25, \quad n_2 = 25, \quad \alpha = .10$
    c. Two-tailed test, $\quad T_1 = 430, \quad n_1 = 20, \quad n_2 = 15, \quad \alpha = .05$

15.12 Independent random samples were selected from two populations. The data are shown in the table. Suppose you want to determine whether population 1 is shifted to the right of population 2.

SAMPLE FROM POPULATION 1		SAMPLE FROM POPULATION 2	
15	12	5	7
17	16	9	4
12		13	5
14		10	10

    a. Compute the rank sums, T_1 and T_2.
    b. Show that $T_1 + T_2 = n(n + 1)/2$.
    c. Give the rejection region for the test using $\alpha = .05$. (Use T_1 as a test statistic.)
    d. State the appropriate conclusion.

APPLYING THE CONCEPTS

15.13 The data in the table, extracted from *Technometrics* (Feb. 1986), represent daily accumulated stream flow and precipitation (in inches) for two U.S. Geological Survey stations in Colorado. Conduct a test to determine whether the distribution of daily accumulated stream flow and precipitation for the two stations differ in location. Use $\alpha = .10$.

STATION 1			STATION 2		
127.96	108.91	100.85	114.79	85.54	280.55
210.07	178.21	85.89	109.11	117.64	145.11
203.24	285.37		330.33	302.74	95.36

Source: Gastwirth, J. L and Mahmoud, H. "An efficient robust nonparametric test for scale change for data from a gamma distribution." *Technometrics*, Vol. 28, No. 1, Feb. 1986, p. 83 (Table 2).

15.14 In Exercise 2.23, you graphically compared the effect sizes for two independent samples of studies, brand name studies and store name studies. (Recall that effect size measures the magnitude of the effect of brand name or store name on product quality.) The stem-and-leaf displays for the two samples are reproduced here. Conduct a large-sample nonparametric test to compare the distributions of effect sizes for the two types of studies. Test using $\alpha = .05$.

Brand Name (15 studies)

STEM	LEAF
.6	0
.5	7
.4	
.3	4
.2	5 5
.1	0 1 1 2 4
.0	3 3 5 5 7

Store Name (17 studies)

STEM	LEAF
.6	
.5	
.4	3 4
.3	
.2	
.1	2
.0	0 0 0 1 1 2 2 3 3 4 6 7 8 8

Source: Rao, A. R. and Monroe, K. B. "The effect of price, brand name, and store name on buyers' perceptions of product quality: An integrative review." *Journal of Marketing Research*, Vol. 26, Aug. 1989, p. 354 (Table 2).

15.15 An experiment was conducted to study the effect of reinforced flanges on the torsional capacity of reinforced concrete T-beams (*Journal of the American Concrete Institute*, Jan.–Feb. 1983). Several different types of T-beams were used in the experiment, each type having a different flange width. The beams were tested under combined torsion and bending until failure (i.e., cracking). One variable of interest is the cracking torsion moment at the top of the flange of the T-beam. Cracking torsion moments for eight beams with 70-cm slab widths and eight beams with 100-cm slab widths are recorded here:

70-cm slab width: 6.00, 7.20, 10.20, 13.20, 11.40, 13.60, 9.20, 11.20

100-cm slab width: 6.80, 9.20, 8.80, 13.20, 11.20, 14.90, 10.20, 11.80

Is there evidence of a difference in the locations of the cracking torsion moment distributions for the two types of T-beams? Test using $\alpha = .10$.

15.16 The Clearing House Interbank Payment System (CHIPS) is a federal operation that clears about 90% of dollar-denominated foreign exchange transactions. Prior to October 1, 1981, settlements through CHIPS used "next-day" funds. That is, funds credited to an account one day did not become available until the following business day. On October 1, 1981, CHIPS switched to "same-day" settlement. An empirical study was conducted to determine the impact of the CHIPS same-day settlement policy on the variability of the federal funds rate (*Quarterly Journal of Business and Economics*, Winter 1985). The daily federal funds rate was recorded for a sample of 101 Fridays prior to October 1, 1981 (i.e., before CHIPS converted to same-day settlement), and for a sample of 50 Fridays after October 1, 1981. The large-sample Wilcoxon rank sum test was used to compare the distributions of the daily federal funds rate for the two periods because "the effective rates are grossly nonnormal." The large-sample test statistic was calculated as $z = -1.03$.

a. Is there sufficient evidence to indicate a shift in the distribution of the Friday federal funds rate following the switch to same-day settlement by CHIPS? Test using $\alpha = .10$.

b. Calculate the p-value for the test of part **a.** Interpret the result.

15.17 A preliminary study was conducted to obtain information on the background levels of the toxic substance polychlorinated biphenyl (PCB) in soil samples in the United Kingdom (*Chemosphere*, Feb. 1986). Such information could then be used as a benchmark against which PCB levels at waste disposal facilities in the United Kingdom can be compared. The accompanying table contains the measured PCB levels of soil samples taken at 14 rural and 15 urban locations in the United Kingdom (PCB concentration is measured in .0001 gram per kilogram of soil). From these preliminary results, the researchers reported "a significant difference between (the PCB levels) for rural areas . . . and for urban areas." Do the data support the researchers' conclusions? Test using $\alpha = .05$.

RURAL			URBAN		
3.5	23.0	12.0	24.0	94.0	22.0
8.1	1.5	8.2	29.0	141.0	13.0
1.8	5.3	9.7	16.0	11.0	18.0
9.0	9.8	1.0	21.0	11.0	12.0
1.6	15.0		107.0	49.0	18.0

Source: Badsha, K. and Eduljee, G. "PCB in the U.K. environment—a preliminary survey." *Chemosphere*, Vol. 15, No. 2, Feb. 1986, p. 213, Table 1. Copyright 1986, Pergamon Press, Ltd. Reprinted with permission.

15.18 The Customer Satisfaction Index, computed by J. D. Powers & Associates, is designed to measure customer satisfaction with new automobiles and automakers. The results of a recent customer satisfaction survey are shown in the table. Suppose you want to determine whether the distribution of customer satisfaction indexes for foreign automakers is shifted to the right of the distribution for domestic automakers. A Minitab printout of the analysis follows. Interpret the results of the test.

AUTO MANUFACTURER	FOREIGN (F) OR DOMESTIC (D)	CUSTOMER SATISFACTION INDEX
Toyota	F	137
Subaru	F	135
Honda	F	124
Mazda	F	118
Lincoln-Mercury	D	114
Ford	D	107
Mitsubishi	F	103
Nissan	F	101
Dodge	D	93
Oldsmobile	D	92
Chrysler-Plymouth	D	90
Chevrolet	D	83
Isuzu	F	81
Cadillac	D	81
Buick	D	80
Pontiac	D	77
American Motors	D	76

Source: J. D. Powers & Associates.

```
Mann-Whitney Confidence Interval and Test

F          N =   7     Median =      118.00
D          N =  10     Median =       86.50
Point estimate for ETA1-ETA2 is         25.50
95.5 pct c.i. for ETA1-ETA2 is (5.01,45.00)
W = 87.5
Test of ETA1 = ETA2  vs.  ETA1 n.e. ETA2 is significant at 0.0192
The test is significant at 0.0191 (adjusted for ties)
```

Note: The Mann-Whitney test is equivalent to the Wilcoxon rank sum test.

15.19   An educational psychologist claims that the order in which test questions are asked affects a student's ability to answer correctly. To investigate this assertion, a professor randomly divides a class of 13 students into two groups—seven in one group and six in the other. The professor prepares one set of test questions but arranges the questions in two different orders. On test A, the questions are arranged in order of increasing difficulty (that is, from easiest to most difficult), whereas on test B the order is reversed. One group of students is given test A, the other test B, and the test score is recorded for each student. The results are as follows:

Test A: 90, 71, 83, 82, 75, 91, 65

Test B: 66, 78, 50, 68, 80, 60

Do the data provide sufficient evidence to indicate a difference between the two tests in a student's ability to answer the questions? Test using $\alpha = .05$.

15.20   In the early 1960s, the air-conditioning systems of a fleet of Boeing 720 jet airplanes came under investigation. The accompanying table presents the lifelengths (in hours) of the air-conditioning systems in two different Boeing 720 planes. Assuming the data represent random samples from the respective populations, is there evidence of a shift in the location of the lifelength distributions of air-conditioning systems for the two Boeing 720 planes? Test using $\alpha = .05$.

PLANE 1			PLANE 2		
23	156	76	59	66	67
118	49	62	32	230	34
90	10		14	54	
29	310		102	152	

Source: Hollander, M., Park, D. H., and Proschan, F. "Testing whether F is 'more NBU' than is G." Microelectronics and Reliability, Vol. 26, No. 1, 1986, p. 43, Table 1. Copyright 1986, Pergamon Press, Ltd. Reprinted with permission.

15.21   Studies have shown that in a nonbusiness (e.g., academic) setting, those who have job mobility are generally better performers. To examine the performance turnover relationship in a business setting, G. F. Dreher examined the personnel records of a large national oil company (Academy of Management Journal, Mar. 1982). The company's employees were divided into two groups: "stayers"—those employees who stayed with the company from 1964 to 1979, and "leavers"—those former employees who left the company at varying points during the 15-year period. The company's annual performance appraisals corresponding to the initial years of service were used to form an initial performance rating for each of a sample of nine stayers and 10 leavers. The performance ratings are shown in the table. [Note: Dreher's sample consisted of 174 stayers and 355 leavers; we consider smaller samples for ease of computation. Actual data for individual employees were not provided in the article. The data shown here are simulated, based on the means and standard deviations given by Dreher. Smaller values of performance rating correspond to better performance.] Is there evidence that leavers are better performers than stayers at the oil company? Test using $\alpha = .05$.

STAYERS		LEAVERS	
3	4	4	5
5	3	3	5
2	2	3	3
3	5	2	4
4		3	3

15.4 COMPARING TWO POPULATIONS: MATCHED-PAIRS DESIGN

Recall from Sections 8.6 and 10.6 that the analysis of matched-pairs data is based on the differences within the matched pairs of observations. The **Wilcoxon signed ranks test** is a nonparametric test to detect shifts in locations for population relative frequency distributions. To perform the test, we assign ranks to the absolute values of the differences and then base the comparison on the rank sums of the negative (T^-) and the positive (T^+) differences. Differences equal to 0 are eliminated, and the number n of differences is reduced accordingly. Tied absolute differences receive ranks equal to the average of the ranks they would have received had they not been tied. The test is summarized in the box on page 878 and is illustrated in Example 15.5.

EXAMPLE 15.5 In Example 10.12 (Section 10.6), we used the parametric t test to determine whether experimental data supported the contention that type A jogging shoes are more durable than type B shoes, where durability is measured in number of weeks of use. Test the hypothesis of no difference in the relative frequency distributions of length of use for types A and B jogging shoes. Use the Wilcoxon signed ranks test with $\alpha = .05$.

Solution   The numbers of weeks of use for the two types of jogging shoes and the differences within pairs for $n = 10$ joggers (given in Table 10.5) are repeated in Table 15.4.

TABLE 15.4
Numbers of Weeks of Use and Differences for Example 15.5

JOGGER	SHOE TYPE A	B	DIFFERENCE	ABSOLUTE VALUE OF DIFFERENCE	RANK OF ABSOLUTE VALUE
1	27	23	4	4	7
2	35	28	7	7	9
3	19	16	3	3	4.5
4	39	31	8	8	10
5	34	38	−4	4	7
6	32	30	2	2	2.5
7	15	17	−2	2	2.5
8	26	22	4	4	7
9	18	15	3	3	4.5
10	17	16	1	1	1

THE WILCOXON SIGNED RANKS TEST: MATCHED PAIRS

Let D_1 and D_2 represent the relative frequency distributions for populations 1 and 2, respectively.

ONE-TAILED TEST

H_0: D_1 and D_2 are identical.

H_a: D_1 is shifted to the right of D_2.
[or H_a: D_1 is shifted to the left of D_2.]

TWO-TAILED TEST

H_0: D_1 and D_2 are identical.

H_a: D_1 is shifted either to the left or to the right of D_2.

Calculate the difference within each of the n matched pairs of observations. Then rank the absolute values of the n differences from the smallest (rank 1) to the highest (rank n) and calculate the rank sum T^- of the negative differences and the rank sum T^+ of the positive differences.

Test statistic:

   T^-, the rank sum of the negative differences

   [or T^+, the rank sum of the positive differences]

Test statistic:

   T, the smaller of T^- or T^+

Rejection region:

   $T^- \leq T_0$ [or $T^+ \leq T_0$]

Rejection region:

   $T \leq T_0$

where T_0 is given in Table 12 of Appendix G.

[*Note:* Differences equal to 0 are eliminated and the number n of differences is reduced accordingly. Tied absolute differences receive ranks equal to the average of the ranks they would have received had they not been tied.]

The rank sums of the positive and negative (shaded) differences are $T^+ = 45.5$ and $T^- = 9.5$, respectively. Since we wish to determine whether the distribution of length of use for shoe type A is shifted to the right of the distribution for shoe type B, we will use T^- as the test statistic and reject H_0 if $T^- \leq T_0$.

The critical values of the Wilcoxon signed rank statistic are provided in Table 12 of Appendix G, which gives the value of T_0 for one-tailed tests for values of α equal to .05, .025, .01, and .005, and for two-tailed tests for values of α equal to .10, .05, .02, and .01. A portion of Table 12 of Appendix G is reproduced in Table 15.5.

To illustrate the use of Table 12 of Appendix G, we will look at the portion of the table corresponding to $n = 10$ differences. Since we want to conduct a one-tailed test, we move down the one-tailed test column to the desired value of $\alpha = .05$. We find the value $T_0 = 11$ (shaded in Table 15.5). Thus, we would reject H_0 if $T^- \leq 11$. This rejection region is shown in Figure 15.6.

TABLE 15.5

A Portion of the Wilcoxon Signed Ranks Table, Table 12 of Appendix G

ONE-TAILED	TWO-TAILED	$n = 5$	$n = 6$	$n = 7$	$n = 8$	$n = 9$	$n = 10$
$\alpha = .05$	$\alpha = .10$	1	2	4	6	8	11
$\alpha = .025$	$\alpha = .05$		1	2	4	6	8
$\alpha = .01$	$\alpha = .02$			0	2	3	5
$\alpha = .005$	$\alpha = .01$				0	2	3
		$n = 11$	$n = 12$	$n = 13$	$n = 14$	$n = 15$	$n = 16$
$\alpha = .05$	$\alpha = .10$	14	17	21	26	30	36
$\alpha = .025$	$\alpha = .05$	11	14	17	21	25	30
$\alpha = .01$	$\alpha = .02$	7	10	13	16	20	24
$\alpha = .005$	$\alpha = .01$	5	7	10	13	16	19

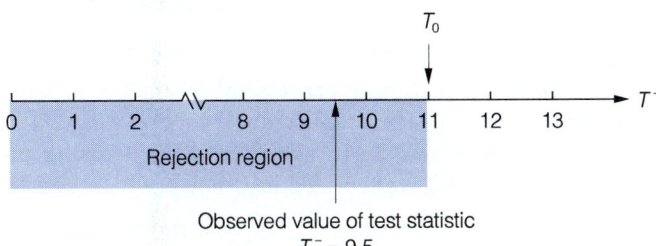

FIGURE 15.6

Rejection Region for the Wilcoxon Signed Ranks Test of Example 15.5

For our example, the computed value of $T^- = 9.5$ is less than $T_0 = 11$. Therefore, we reject H_0 and conclude that there is sufficient evidence to indicate that the distribution of the number of weeks of usable life for shoe type A is shifted to the right of the distribution for shoe type B. This is exactly the same conclusion we obtained by using the Student's t test in Example 10.12.

A Minitab printout of the identical analysis is shown in Figure 15.7. Note that Minitab uses $T^+ = 45.5$ as the test statistic. The two-tailed p-value of the test (shaded) is .074. Thus, the one-tailed p-value is $\frac{.074}{2} = .037$; since this value is less than $\alpha = .05$, we reject H_0.

```
TEST OF MEDIAN = 0.000000 VERSUS MEDIAN N.E. 0.000000

                    N FOR    WILCOXON              ESTIMATED
            N       TEST     STATISTIC  P-VALUE    MEDIAN
AminusB     10      10       45.5       0.074      3.000
```

FIGURE 15.7

Minitab Printout for Example 15.5

The Wilcoxon signed ranks statistic has a sampling distribution that is approximately normal when the number n of pairs is large—say, $n \geq 25$. This large-sample nonparametric matched-pairs test is summarized in the following box.

EXERCISES
LEARNING THE MECHANICS

15.22 Specify the test statistic and the rejection region for the Wilcoxon signed ranks test for the matched-pairs design in each of the following situations:

a. H_0: Two probability distributions, A and B, are identical.
   H_a: Probability distribution for population A is shifted to the right or left of the probability distribution for population B.

   $n = 19$, $\alpha = .05$

b. H_0: Two probability distributions, A and B, are identical.
   H_a: Probability distribution for population A is shifted to the right of the probability distribution for population B.

   $n = 36$, $\alpha = .01$

c. H_0: Two probability distributions, A and B, are identical.
   H_a: Probability distribution for population A is shifted to the left of the probability distribution for population B.

   $n = 50$, $\alpha = .005$

15.23 Suppose you want to test the hypothesis that two treatments, A and B, are equivalent against the alternative that the responses for A tend to be larger than those for B.

a. if $n = 8$ and $\alpha = .01$, give the rejection region for a Wilcoxon signed ranks test.

b. Suppose you want to detect a difference in the locations of the distributions of the responses for A and B if such a difference exists. If $n = 7$ and $\alpha = .10$, give the rejection region for the Wilcoxon signed ranks test.

15.24 A random sample of nine pairs of measurements is shown in the table.

PAIR	SAMPLE DATA FROM POPULATION 1	SAMPLE DATA FROM POPULATION 2
1	8	7
2	10	1
3	6	4
4	10	10
5	7	4
6	8	3
7	4	6
8	9	2
9	8	4

a. Use the Wilcoxon signed ranks test to determine whether the data provide sufficient evidence to indicate that the probability distribution for population 1 is shifted to the right of the probability distribution for population 2. Test using $\alpha = .05$.

b. Use the Wilcoxon signed ranks test to determine whether the data provide sufficient evidence to indicate that the probability distribution for population 1 is shifted either to the right or to the left of the probability distribution for population 2. Test using $\alpha = .05$.

APPLYING THE CONCEPTS

15.25 Dental researchers have developed a new material for preventing cavities—a plastic sealant that is applied to the chewing surfaces of teeth. To determine whether the sealant is effective, it was applied to half of the teeth of each of 12 school-age children. After 5 years, the numbers of cavities in the sealant-coated teeth and untreated teeth were counted. The results are given in the accompanying table. Is there sufficient evidence to indicate that sealant-coated teeth are less prone to cavities than are untreated teeth? Test using $\alpha = .05$.

CHILD	SEALANT-COATED	UNTREATED	CHILD	SEALANT-COATED	UNTREATED
1	3	3	7	1	5
2	1	3	8	2	0
3	0	2	9	1	6
4	4	5	10	0	0
5	1	0	11	0	3
6	0	1	12	4	3

15.26 Medical researchers believe that exposure to dust from cotton bract induces respiratory disease in susceptible field workers. An experiment was conducted to determine the effect of air-dried green cotton bract extract (GBE) on the cells of non–dust-exposed individuals (*Environmental Research*, Feb. 1986). Blood samples taken on six workers were incubated with varying concentrations of GBE. After a short period of time, the cyclic AMP level (measure of cell activity expressed in picomoles per million cells) of each blood sample was measured. The data for two GBE concentrations, 0 mg/ml (salt solution) and .2 mg/ml, are reproduced in the table. [Note that one blood sample was taken from each worker, with one aliquot exposed to the salt buffer solution and the other to the GBE.] Conduct a test to detect a shift in the locations of the cyclic AMP level distributions for the two GBE concentrations. Test using $\alpha = .10$.

WORKER	GBE CONCENTRATION (mg/ml)	
	0	.2
A	8.8	4.4
B	13.0	5.7
C	9.2	4.4
D	6.5	4.1
F	9.1	4.4
H	17.0	7.9

Source: Butcher, B. T, Reed, M. A., and O'Neil, C. E. "Biochemical and immunologic characterization of cotton bract extract and its effect on *in vitro* cyclic AMP production." *Environmental Research*, Vol. 39, No. 1, Feb. 1986, p. 119.

15.27 Traditionally, jobs in the United States have required employees to perform their work during a fixed 8-hour work day. A recent job-scheduling innovation that is helping managers to overcome the motivation and absenteeism problems associated with the fixed work day is a concept called *flextime*. Flextime is a flexible working-hours program that permits employees to design their own 40-hour work week. The management of a large manufacturing firm is considering adopting a flextime program for its hourly employees and has decided to base the decision on the success or failure of a pilot flextime program. Ten employees were randomly selected and given a questionnaire designed to measure their attitudes toward their jobs. These people were then permitted to design and follow a flextime workday. After 6 months, attitudes towards their jobs were again measured. The resulting attitude scores are displayed in the table. The higher the score, the more favorable is the employee's attitude toward his or her work. Use a nonparametric test procedure to evaluate the success of the pilot flextime program. Test using $\alpha = .05$.

EMPLOYEE	BEFORE FLEXTIME	AFTER FLEXTIME	EMPLOYEE	BEFORE FLEXTIME	AFTER FLEXTIME
1	54	68	6	82	88
2	25	42	7	94	90
3	80	80	8	72	81
4	76	91	9	33	39
5	63	70	10	90	93

15.28 Refer to the paired comparison of 60 grocery items at Winn-Dixie and Publix supermarkets, Exercises 8.60 and 10.45. The data were subjected to a nonparametric analysis in SPSS; the SPSS printout is displayed here. Interpret the results. Is the conclusion consistent with the parametric test conducted in Exercise 10.45?

```
- - - - - Wilcoxon Matched-pairs Signed-ranks Test

      WINNDIX
with PUBLIX


   Mean Rank    Cases

      27.90        5  - Ranks  (PUBLIX Lt WINNDIX)
      30.19       54  + Ranks  (PUBLIX Gt WINNDIX)
                   1    Ties   (PUBLIX Eq WINNDIX)
                  --
                  60    Total

     Z =   -5.6270          2-tailed P =   .0000
```

15.29 A study was conducted to test the significance of the measurement errors in the U.S. Department of Treasury's bond yield series (*Quarterly Journal of Business and Economics*, Autumn 1986). The sample data consisted of 164 new government debt offerings that were issued at exact maturities (e.g., 2-year, 3-year, etc.). The actual yields for these bond issues were compared to the estimated yields from the Treasury's published yield series using the Wilcoxon signed ranks test for matched pairs. The results of the test (2-year and 30-year bonds) are shown in the table. [*Note:* For each issue, the difference was calculated as Difference = (Actual yield − Estimated yield).]

	BOND MATURITY	
	2-year	30-year
Number of bonds	92	15
Number of positive differences	56	8
Rank sum, T^+	2,549.68	73.52
Number of negative differences	36	7
Rank sum, T^-	1,728.36	73.01

a. Test the hypothesis that the distribution of exact 2-year bond yields is shifted to the right or left of the distribution of estimated 2-year bond yields. (Use the large-sample version of the test.)
b. Repeat part **a** for the 30-year bonds.

15.30 Tetrachlorodibenzo-p-dioxin (TCDD) is a highly toxic substance found in industrial wastes. A study was conducted to determine the amount of TCDD present in the tissues of bullfrogs inhabiting the Rocky Branch Creek in central Arkansas, an area known to be contaminated by TCDD. The level of TCDD (in parts per trillion) was measured in several specific tissues of four female bullfrogs and the ratio of TCDD in the tissue to TCDD in the leg muscle of the frog was recorded for each. The relative ratios of contaminant for two tissues, the liver and the ovaries, are given for each of the four frogs in the accompanying table. According to the researchers, "the data set suggests that the relative level of TCDD in the ovaries of female frogs is higher than the level in the liver of the frogs." Test this claim using $\alpha = .05$. [*Hint:* Find the approximate rejection region by using the value of T_0 given in Table 12 of Appendix G for $n = 5$.]

	FROG			
	A	B	C	D
Liver	11.0	14.6	14.3	12.2
Ovaries	34.2	41.2	32.5	26.2

Source: Korfmacher, W. A., Hansen, E. B., Jr., and Rowland, K. L. "Tissue distribution of 2,3,7,8-TCDD in bullfrogs obtained from a 2,3,7,8-TCDD-contaminated area." *Chemosphere*, Vol. 15, No. 2, Feb. 1986, p. 125. Copyright 1986, Pergamon Press, Ltd. Reprinted with permission.

15.5 During the Vietnam War, American soldiers were exposed to Agent Orange, a defoliant used by the U.S. Armed Forces to destroy the dense plant and tree cover of the Asian jungle. As a result of this exposure, many Vietnam veterans have dangerously high levels of the dioxin 2,3,7,8-TCDD in blood and fat tissue.

A study published in *Chemosphere* (Vol. 20, 1990) reported on the TCDD levels of 20 Massachusetts Vietnam vets who were possibly exposed to Agent Orange. The amounts of TCDD (measured in parts per trillion) in fat tissue drawn from each veteran are shown in Table 15.6 (page 883). The data provide us with an opportunity to apply the nonparametric tests of the previous sections.

CHAPTER CASE: DEADLY EXPOSURE: AGENT ORANGE AND VIETNAM VETS

TABLE 15.6

TCDD Levels in Fat Tissue of Vietnam Veterans

4.9	6.9	10.0	4.4	4.6	1.1	2.3
5.9	7.0	5.5	7.0	1.4	11.0	2.5
4.4	4.2	41.0	2.9	7.7	2.5	

Source: Schecter, A., et al. "Partitioning of 2,3,7,8-chlorinated dibenzo-p-dioxins and dibenzofurans between adipose tissue and plasma lipid of 20 Massachusetts Vietnam veterans." *Chemosphere*, Vol. 20, Nos. 7–9, 1990, pp. 954–955 (Table II).

Suppose we want to use this small sample ($n = 20$) to test whether the median TCDD level in fat tissue of Vietnam vets exceeds 3 parts per trillion, i.e., we want to test

$$H_0: \quad \eta = 3$$
$$H_a: \quad \eta > 3$$

If the population of TCDD levels is normally distributed, then the median η and the mean μ are identical and we can apply the parametric small-sample t test of Section 10.2. A Minitab stem-and-leaf plot of the sample TCDD levels is displayed in Figure 15.8. Note that the distribution is highly skewed. This casts doubt on the validity of the normality assumption required to conduct the t test. Consequently, we will analyze the data using the nonparametric sign test.

FIGURE 15.8

Minitab Stem-and-Leaf Display for the Data of Table 15.6

```
Stem-and-leaf of TCDD      N  = 20
Leaf Unit = 1.0

 (11)   0 11222244444
  9     0 556777
  3     1 01
  1     1
  1     2
  1     2
  1     3
  1     3
  1     4 1
```

To test the hypothesis, we count the number of sample TCDD levels that exceed $\eta = 3$. Examining Table 15.6, we see that 14 of the 20 TCDD levels exceed 3; thus, the test statistic for the sign test is $S = 14$.

The p-value for the test is

$$p\text{-value} = P(x \geq S) = P(x \geq 14)$$

where x has a binomial distribution with parameters $n = 20$ and $\pi = .5$. From Table 3, Appendix G, we have

$$p\text{-value} = P(x \geq 14) = 1 - P(x \leq 13) = 1 - .9423 = .0577$$

This value implies that we will reject H_0 in favor of H_a for any selected α level that exceeds .0577.

Clearly, the choice of α is a critical one. If we select $\alpha = .10$, for example, we have sufficient evidence to reject H_0 and will conclude that the median TCDD level in fat tissue of Vietnam vets exceeds 3 parts per trillion. However, if we select $\alpha = .05$ or $\alpha = .01$, we have insufficient evidence to reject H_0. Thus, the sign test yields a result that is "borderline" significant.

Although the sign test is easy to apply, it is not the most powerful test available. In Section 9.7 (optional), we defined the power of a test as the probability that the test rejects the null hypothesis when it is false. In this case, the power is the probability that the test rejects H_0: $\eta = 3$ when, in fact, η exceeds 3. Thus, if η exceeds 3, the probability that the sign test will detect it is not as high as when using several other nonparametric tests.

A nonparametric test with higher power than the sign test is the familiar Wilcoxon signed ranks test of Section 15.4. Although we presented the signed ranks procedure as a test to compare two populations in a matched-pairs design, it can also be used to test the location (median) of a single population. To conduct the test on TCDD level, we calculate the differences, $(y_i - 3)$ for the sample, where y_i is the TCDD level for veteran i and 3 is the hypothesized value of the median η. The absolute values of the differences are ranked; the rank sum of the negative differences represents the test statistic.

THE WILCOXON SIGNED RANKS TEST FOR THE MEDIAN, η, OF A SINGLE POPULATION

ONE-TAILED TEST

H_0: $\eta = \eta_0$

H_a: $\eta > \eta_0$

   [or, H_a: $\eta < \eta_0$]

Test statistic:

   T^-, the negative rank sum

   [or, T^+, the positive rank sum]

TWO-TAILED TEST

H_0: $\eta = \eta_0$

H_a: $\eta \neq \eta_0$

Test statistic:

   T, the smaller of the positive and negative rank sums, T^+ and T^-

[*Note:* The sample differences are computed as $(y_i - \eta_0)$.]

Rejection region:

   $T^- \leq T_0$

   [or, $T^+ \leq T_0$]

Rejection region:

   $T \leq T_0$

where T_0 is found in Table 12 of Appendix G.

Assumptions: 1. A random sample of observations has been selected from the population.

      2. The absolute differences $y_i - \eta_0$ can be ranked. [No assumptions must be made about the form of the population probability distribution.]

      3. Differences equal to 0 are eliminated and n is reduced accordingly. Tied differences are assigned ranks equal to the average of the ranks of the tied observations.

The 20 differences are computed and ranked in Table 15.7, with the positive differences shown on the left side of the table and the negative differences shown on the right. Note that we use the usual ranking convention for ties.

The rank sum of the negative differences, $T^- = 29$, represents the test statistic. According to the box on page 885, we will reject H_0 if T^- falls below the signed ranks critical value T_0 obtained from Table 12, Appendix G.

TABLE 15.7
Signed Ranks for the Data of Table 15.6

TCDD LEVEL	POSITIVE DIFFERENCE	RANK	TCDD LEVEL	NEGATIVE DIFFERENCE	RANK
4.9	1.9	10.5	2.9	−.1	1
5.9	2.9	13	1.4	−1.6	8.5
4.4	1.4	6.5	1.1	−1.9	10.5
6.9	3.9	14	2.5	−.5	2.5
7.0	4.0	15.5	2.3	−.7	4
4.2	1.2	5	2.5	−.5	2.5
10.0	7.0	18			
5.5	2.5	12	Rank Sum:		$T^- = 29$
41.0	38.0	20			
4.4	1.4	6.5			
7.0	4.0	15.5			
4.6	1.6	8.5			
7.7	4.7	17			
11.0	8.0	19			
Rank Sum:		$T^+ = 181$			

The smallest α level shown in the table is $\alpha = .005$. If we select $\alpha = .005$, the one-tailed critical value is $T_0 = 37$. Since T^- falls below this value, there is sufficient evidence to reject H_0 and conclude that the median TCDD level exceeds 3 ppt. Further, the p-value of the test, obtained from a computer printout, is $p = .005$ (see Figure 15.9). Thus, for an α level as small as $\alpha = .005$, the signed ranks test will yield a "significant" result.

To summarize, this case illustrates the application of the sign test and the signed ranks procedure to a practical problem involving Agent Orange and TCDD levels in Vietnam veterans. Although it is very easy to apply, the sign test is not as powerful as the signed ranks test. As illustrated in this case, the sign test failed to detect a median TCDD level exceeding 3 at $\alpha = .05$. The signed ranks test, on the other hand, detected $\eta > 3$ at an α level as small as $\alpha = .005$.

FIGURE 15.9
Minitab Printout of Signed Ranks Analysis

```
TEST OF MEDIAN = 0.000000 VERSUS MEDIAN N.E. 0.000000

                    N FOR    WILCOXON              ESTIMATED
            N       TEST    STATISTIC   P-VALUE      MEDIAN
TCDD_3      20       20        181.0     0.005        2.100
```

15.6 In Section 15.3, we showed how the Wilcoxon rank sum test can be used in place of the Student's t test for comparing the locations of two population relative frequency distributions. The advantage of the Wilcoxon rank sum test over the t test is that we do not need to make restrictive assumptions (e.g., normality, equal variances) about the sampled populations.

The **Kruskal–Wallis H test** provides a nonparametric alternative to the analysis of variance F test (Section 13.4) for comparing the locations of more than two populations based on independent random samples, i.e., the completely randomized design. As with the Wilcoxon rank sum test, no assumptions regarding the normality or variances of the sampled populations are required.

The sample observations are ranked from the smallest to the largest and the rank sums are calculated for each sample. For example, if you had three samples with $n_1 = 8$, $n_2 = 6$, and $n_3 = 7$, you would rank the $n_1 + n_2 + n_3 = 21$ observations from the smallest (rank 1) to the largest (rank 21) and then calculate the rank sums, T_1, T_2, and T_3, for the three samples. The Kruskal–Wallis H test uses these rank sums to calculate an H test statistic that possesses an approximate chi-square sampling distribution of the type encountered in Chapter 14. The elements of the test are summarized in the box and illustrated in Example 15.6.

THE KRUSKAL–WALLIS H TEST FOR COMPARING k POPULATION RELATIVE FREQUENCY DISTRIBUTIONS: COMPLETELY RANDOMIZED DESIGN

H_0: The k population relative frequency distributions are identical.

H_a: At least two of the population relative frequency distributions differ in location (shifted either to the left or to the right of one another).

Test statistic: $\quad H = \dfrac{12}{n(n+1)} \displaystyle\sum_{i=1}^{k} \dfrac{T_i^2}{n_i} - 3(n+1)$

where $\quad n_i$ = Number of observations in sample i

T_i = Rank sum of sample i

n = Total sample size = $n_1 + n_2 + \cdots + n_k$

Rejection region:

$H > \chi_\alpha^2$ where χ_α^2 is based on $(k-1)$ degrees of freedom

Assumptions:
1. The k samples have been independently and randomly selected from their respective populations.
2. For the chi-square approximation to be adequate, there should be five or more observations in each sample.
3. Tied observations are assigned ranks equal to the average of the ranks that would have been assigned to the observations had they not been tied.

EXAMPLE 15.6 In Examples 13.3 and 13.4, we used an analysis of variance to compare the mean grade point averages of college freshmen from three different socioeconomic backgrounds. Three independent random samples of $n_1 = 7$, $n_2 = 7$, and $n_3 = 6$ freshmen were selected from the three populations. The data are reproduced in Table 15.8.

TABLE 15.8
Grade Point Averages for Three
Socioeconomic Groups

LOWER CLASS	MIDDLE CLASS	UPPER CLASS
2.87	3.23	2.25
2.16	3.45	3.13
3.14	2.78	2.44
2.51	3.77	3.27
1.80	2.97	2.81
3.01	3.53	1.36
2.16	3.01	

Use the Kruskal–Wallis H test to determine whether the data provide sufficient evidence to indicate that the mean grade point average of freshmen depends on the students' socioeconomic backgrounds. Test using $\alpha = .05$. Compare the results of this test with those of the analysis of variance F test in Section 13.4.

Solution The first step in conducting the Kruskal–Wallis H test is to rank the $n_1 + n_2 + n_3 = 7 + 7 + 6 = 20$ observations from the smallest to the largest. Thus, we give the smallest observation (1.36) a rank of 1, the next smallest (1.80) a rank of 2, ... , and the largest observation (3.77) a rank of 20. The original data, their associated ranks, and the sample rank sums, T_1, T_2, and T_3, are shown in Table 15.9.

TABLE 15.9
Rank Sums for Grade Point Average Data

LOWER CLASS		MIDDLE CLASS		UPPER CLASS	
GPA	RANK	GPA	RANK	GPA	RANK
2.87	10	3.23	16	2.25	5
2.16	3.5	3.45	18	3.13	14
3.14	15	2.78	8	2.44	6
2.51	7	3.77	20	3.27	17
1.80	2	2.97	11	2.81	9
3.01	12.5	3.53	19	1.36	1
2.16	3.5	3.01	12.5	$T_3 = 52$	
	$T_1 = 53.5$		$T_2 = 104.5$		

The null hypothesis for the test is that the three population grade point average distributions are identical. The alternative hypothesis is that they differ in location, i.e., that one (or more) of the distributions is shifted to the right or left of the others.

The test statistic for testing the null hypothesis is

$$H = \frac{12}{n(n + 1)} \sum \frac{T_i^2}{n_i} - 3(n + 1)$$

where $n_1 = 7$, $n_2 = 7$, $n_3 = 6$, $n = 20$, $T_1 = 53.5$, $T_2 = 104.5$, and $T_3 = 52$. Substituting these values into the formula for H, we obtain

$$H = \frac{12}{20(20 + 1)} \left[\frac{(53.5)^2}{7} + \frac{(104.5)^2}{7} + \frac{(52)^2}{6} \right] - 3(20 + 1)$$

$$= \frac{12}{420} \left(\frac{2,862.25}{7} + \frac{10,920.25}{7} + \frac{2,704}{6} \right) - 63$$

$$= 6.13$$

The rejection region for the H test is $H > \chi_\alpha^2$, where χ_α^2 is based on $(k - 1)$ degrees of freedom. Since we have selected $\alpha = .05$ and we want to compare $k = 3$ population relative frequency distributions, we need the value of $\chi_{.05}^2$ for $(k - 1) = (3 - 1) = 2$ degrees of freedom. This value is given in Table 6 of Appendix G as $\chi_{.05}^2 = 5.99147$. Therefore, the rejection region for the test includes all values of H larger than 5.99147 (see Figure 15.10).

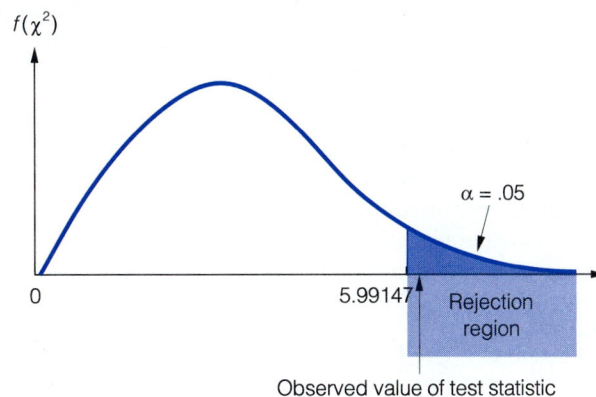

FIGURE 15.10
Rejection Region for the
Kruskal–Wallis H Test of
Example 15.6

The final step in conducting the test is to determine whether the value of the test statistic falls in the rejection region. Since the value computed for the grade point data, $H = 6.13$, is larger than $\chi_{.05}^2 = 5.99147$, we reject the null hypothesis, i.e., there is sufficient evidence to indicate differences in location for the freshmen grade point average distributions associated with the three socioeconomic populations. This conclusion is the same as that reached by the analysis of variance F test performed in Example 13.4.

A SAS printout of the analysis is shown in Figure 15.11 (page 890). The test statistic (rounded) is shaded, as is the p-value of the test. Note that the p-value (.0464) is less than $\alpha = .05$, resulting in our conclusion of "reject H_0."

FIGURE 15.11

SAS Printout for Example 15.6

N P A R 1 W A Y P R O C E D U R E

Wilcoxon Scores (Rank Sums) for Variable GPA
Classified by Variable CLASS

CLASS	N	Sum of Scores	Expected Under H0	Std Dev Under H0	Mean Score
LOWER	7	53.500000	73.5000000	12.6099376	7.6428571
MIDDLE	7	104.500000	73.5000000	12.6099376	14.9285714
UPPER	6	52.000000	63.0000000	12.1152362	8.6666667

Average Scores were used for Ties

Kruskal-Wallis Test (Chi-Square Approximation)
CHISQ= 6.1405   DF= 2   Prob > CHISQ= 0.0464

EXERCISES

LEARNING THE MECHANICS

15.31 Use Table 6 in Appendix G to find each of the following χ^2 values:

a. $\chi^2_{.05}$, df = 10   b. $\chi^2_{.025}$, df = 23   c. $\chi^2_{.01}$, df = 5

d. $\chi^2_{.10}$, df = 10   e. $\chi^2_{.05}$, df = 3   f. $\chi^2_{.005}$, df = 3

15.32 Suppose you want to use the Kruskal–Wallis H test to compare the probability distributions of three populations. The following are independent random samples selected from the three populations:

I: 45 33   55 88 58

II: 22 31   16 25 30 33

III: 91 96 102 75 88

a. What type of experimental design was used?
b. Specify the null and alternative hypotheses you would test.
c. Specify the rejection region that would be used for your hypothesis test at $\alpha = .05$.
d. Calculate the rank sums for the three samples.
e. Calculate the test statistic.
f. State the appropriate conclusion of the test.

APPLYING THE CONCEPTS

15.33 B. L. Davis and M. K. Mount conducted a study to evaluate the effectiveness of performance appraisal training in an organizational setting (*Personnel Psychology*, Aug. 1984). A sample of middle-level managers were randomly assigned to one of three training conditions: no training, computer-assisted training, or computer-assisted training plus a behavior modeling workshop. After the formal training, the managers were administered a 25-question multiple-choice test of managerial knowledge and the number of correct answers was recorded for each. The data in the table are adapted from summary information provided in the article. Is there sufficient evidence to indicate that the relative frequency distributions of scores differ in location for the three types of performance appraisal training? Test using $\alpha = .01$.

NO TRAINING	COMPUTER-ASSISTED TRAINING	COMPUTER TRAINING PLUS WORKSHOP
16	19	12
18	22	19
11	13	18
14	15	22
23	20	16
	18	25
	21	

15.34 Phosphoric acid is chemically produced by reacting phosphate rock with sulfuric acid. An important consideration in the chemical process is the length of time required for the chemical reaction to reach a specified temperature. The shorter the length of time, the higher the reactivity of the phosphate rock. An experiment was conducted to compare the reactivity of phosphate rock mined in north, central, and south Florida. Rock samples were collected from each location and placed in vacuum bottles with a 56%-strength sulfuric acid solution. The time (in seconds) for the chemical reaction to reach 200°F was recorded for each sample. Do the data provide sufficient evidence to indicate a difference in the reactivity of phosphoric rock mined at the three locations? Test using $\alpha = .05$.

SOUTH	CENTRAL	NORTH
40.6	41.1	25.6
42.0	38.3	36.4
37.5	40.2	28.2
38.1	33.5	31.3
41.9	35.7	29.5
		22.8
		27.5

15.35 Refer to Exercise 13.7 and the data on total cash compensation (in thousands of dollars) for corporate executives in each of four industries. The data, extracted from *Business Week*'s 1989 Executive Compensation Scoreboard, is reproduced in the table.

CONSUMER PRODUCTS	UTILITIES	INDUSTRIAL–HIGH TECH	FINANCIAL SERVICES
1,567	1,862	2,925	3,125
3,313	1,390	3,409	4,143
2,058	1,115	1,767	4,013
25,216	1,105	4,097	6,583
4,634	1,272	3,196	3,169
5,214	2,849	4,042	5,217
20,795	1,732	2,601	3,447
9,162	1,474	8,286	4,469

Source: "Executive compensation scoreboard." *Business Week*, May 7, 1990.

a. In Exercise 13.7 you conducted an analysis of variance F test to compare the mean compensation of executives in the three industries. Why might this parametric test be invalid?

b. Conduct a nonparametric test to determine whether the distributions of the total 1989 compensations for executives in the three industry groups differ in location. Test using $\alpha = .01$.

15.36 A *modulator/demodulator*, or *modem*, is a device that converts electrical impulses sent from a computer into audio tones that travel over telephone lines to a remote terminal. The performance of a modem varies, depending on the speed with which it can send and receive signals (called the *baud rate*) and whether it transmits and receives data at the same time (*full duplex*) or must take turns with the computer-transmitting data (*half duplex*). A new type of modem, called a *smart modem*, has been developed. Smart modems have built-in microprocessors with advanced features that improve overall modem performance and efficiency. Four new modems with self-contained microprocessors are currently on the market: Bizcomp 1012, Cermetek 212A, Hayes Smartmodem 1200, and Vadic 3451 Auto-Dial. Suppose five users of each type of modem are randomly selected and asked to rate modem performance (measured on a scale from 1 to 100). Based on the data shown in the table, is there sufficient evidence to indicate a difference among the performance ratings of the four smart modems? Test using $\alpha = .10$.

BIZCOMP 1012	CERMETEK 212A	SMARTMODEM 1200	VADIC 3451
87	81	69	98
80	66	72	78
91	52	70	94
63	90	83	90
72	75	80	86

15.37   Vanadium (V) is a recently recognized essential trace element. An experiment was conducted to compare the concentrations of V in biological materials using isotope dilution mass spectrometry (*Analytical Chemistry*, Nov. 1985). The accompanying table gives the quantities of V (measured in nanograms per gram) in dried samples of oyster tissue, citrus leaves, bovine liver, and human serum. Conduct a nonparametric test to determine whether the distributions of V concentrations for the four biological materials differ in location. Test using $\alpha = .05$.

OYSTER TISSUE	CITRUS LEAVES	BOVINE LIVER	HUMAN SERUM
2.35	2.32	.39	.10
1.30	3.07	.54	.17
.34	4.09	.30	.14
			.16
			.16

Source: Fassett, J. D. and Kingston, H. M. "Determination of nanogram quantities of vanadium in biological material by isotope dilution thermal ionization mass spectrometry with ion counting detection." *Analytical Chemistry*, Vol. 57, No. 13, Nov. 1985, p. 2475, Table II. Copyright 1985, American Chemical Society. Reprinted with permission.

15.38   The ratio of stockholders' equity to total assets is widely used as a measure of a firm's financial leverage. G. N. Naidu examined the degree of financial leverage of industries in four Asian countries: Hong Kong, Malaysia, Philippines, and India (*Management International Review*, 1st Quarter, 1984). The samples for the study consisted of five industries in Hong Kong, eight industries in Malaysia, nine in the Philippines, and nine in India. The 1980 financial leverage of each industry (computed as the mean ratio of stockholders' equity to total assets of firms in the industry) is recorded in the table. Do the distributions of financial leverages for industries in the four Asian countries differ? Use the accompanying Minitab printout to make the proper inference.

HONG KONG	MALAYSIA		PHILIPPINES		INDIA	
.5415	.6736	.4311	.3964	.4854	.2898	.3700
.6367	.6202	.6529	.3382	.3872	.4776	.3446
.4896	.7815	.5103	.3569	.3741	.3162	.3924
.5231	.5375		.1237	.2963	.2948	.3861
.5918	.5202		.1577		.3860	

```
LEVEL     NOBS     MEDIAN   AVE. RANK   Z VALUE
  1         5      0.5415      24.4       2.26
  2         8      0.5789      25.1       3.30
  3         9      0.3569       9.3      -2.61
  4         9      0.3700       9.9      -2.39
OVERALL    31                  16.0

H = 21.23   d.f. = 3   p = 0.000
```

15.7 COMPARING THREE OR MORE POPULATIONS: RANDOMIZED BLOCK DESIGN

15.7 In Section 13.5, we explained how it is often possible to obtain more information about the differences in a set of population means by matching observations, with one observation from each population in each matched set. This method for comparing the means of k populations (i.e., k *treatments*) employs blocks of k matched experimental units. Then the k treatments to be compared are randomly assigned, one to each of the k experimental units in a block. This is repeated for each of the blocks (we will suppose that there are b blocks) included in the experiment.

The **Friedman F_r test** for detecting shifts in location of the k population relative frequency distributions is accomplished by ranking the k observations *in each block*, from the smallest (rank 1) to the largest (rank k). The rank sums, $T_1, T_2, \ldots, T_k$, are then calculated for each of the k treatments and used to compute the Friedman F_r statistic. This statistic, like the Kruskal–Wallis H statistic, has a sampling distribution that can be approximated by a chi-square distribution with $(k - 1)$ degrees of freedom. The test procedure is specified in the box on the next page and illustrated in Example 15.7.

EXAMPLE 15.7 A food products company conducted a *single-blind experiment* to compare the tastes of three similarly priced brands of coffee. Each of six experienced taste testers was presented three cups of coffee—one of each brand—in random order, and asked to rate the cups from the poorest taste (rank 1) to the best taste (rank 3). The experiment is called *single-blind* because the brands contained in the cups were unknown to the taste testers, thereby eliminating the possibility of a bias, i.e., that the testers might favor one particular brand if the brands were known. The results of the taste test are shown in Table 15.10.

TABLE 15.10
Taste Tester Rankings for Three Brands of Coffee

TASTE TESTER	BRAND		
	A	B	C
1	1	3	2
2	2	3	1
3	1	2	3
4	1	3	2
5	2	3	1
6	1	3	2
Rank sums	$T_A = 8$	$T_B = 17$	$T_C = 11$

Do the data provide sufficient evidence to indicate differences in taste appeal among the three brands of coffee? Test using the Friedman F_r test with $\alpha = .05$.

THE FRIEDMAN F_r TEST FOR A RANDOMIZED BLOCK DESIGN

H_0: The relative frequency distributions for the k populations are identical.

H_a: At least two of the k populations differ in location (shifted either to the left or to the right of one another).

Test statistic:

Rank each of the k observations within each block from the smallest (rank 1) to the largest (rank k). Calculate the treatment rank sums, T_1, $T_2, \ldots, T_k$. Then the test statistic is

$$F_r = \frac{12}{bk(k+1)} \sum_{i=1}^{k} T_i^2 - 3b(k+1)$$

where b = Number of blocks employed in the experiment

k = Number of treatments

T_i = Sum of the ranks for the ith treatment

Rejection region:

$F_r > \chi_\alpha^2$ where χ_α^2 is based on $(k-1)$ degrees of freedom.

Assumptions: 1. The k treatments were randomly assigned to the k experimental units within each block.
2. For the chi-square approximation to be adequate, either the number b of blocks or the number k of treatments should exceed five.
3. Tied observations are assigned ranks equal to the average of the ranks that would have been assigned to the observations had they not been tied.

Solution This experiment was conducted according to a randomized block design. Comparisons of the taste of the three brands of coffee were made by each taste tester (a block) with the cups assigned for tasting in random order.

The null hypothesis is that there are no differences in taste among the three brands of coffee. The alternative hypothesis is that at least one brand is preferred over the others. The rank sums for the three brands, shown in Table 15.10, are $T_A = 8$, $T_B = 17$, and $T_C = 11$. Substituting these values along with $b = 6$ and $k = 3$ into the formula for the test statistic, we obtain

$$F_r = \frac{12}{bk(k+1)} \sum T_i^2 - 3b(k+1)$$

$$= \frac{12}{(6)(3)(3+1)} [(8)^2 + (17)^2 + (11)^2] - 3(6)(3+1)$$

$$= \frac{12}{72} (64 + 289 + 121) - 72$$

$$= 7.0$$

The rejection region for the test contains all values of F_r larger than χ_α^2 with $(k-1)$ degrees of freedom. The value of $\chi_{.05}^2$ (given in Table 6 of Appendix G) that corresponds to $\alpha = .05$ and $(k-1) = (3-1) = 2$ degrees of freedom is 5.99147. Therefore, the rejection region for the test contains all values of F_r larger than $\chi_{.05}^2 = 5.99147$ (see Figure 15.12). Since the value of F_r computed for this test, $F_r = 7.0$, falls in the rejection region, we reject the null hypothesis and conclude that at least one of the brands of coffee is preferred over the others.

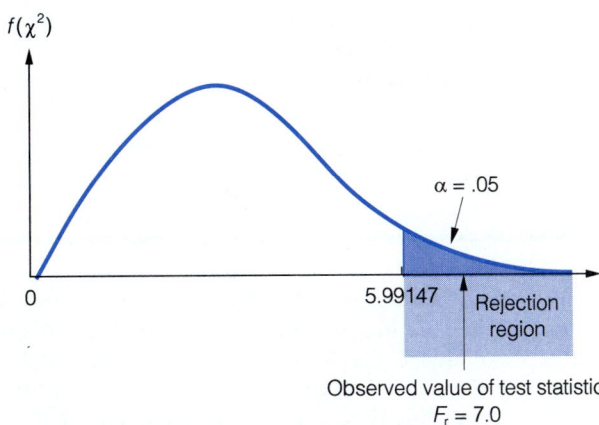

FIGURE 15.12
Rejection Region for the Friedman F_r Test of Example 15.7

The same conclusion can be reached by interpreting the results shown on an SPSS printout of the analysis, Figure 15.13. The test statistic (shaded) agrees with the computed value, and the p-value of the test (also shaded) is less than $\alpha = .05$; thus, there is evidence to reject H_0.

```
- - - - - Friedman Two-way ANOVA

   Mean Rank    Variable

       1.33     A
       2.83     B
       1.83     C

       Cases        Chi-Square       D.F.   Significance
         6           7.0000            2        .0302
```

FIGURE 15.13
SPSS Printout for Example 15.7

EXERCISES

LEARNING THE MECHANICS

15.39   Suppose you have used a randomized block design to help you compare the effectiveness of three different treatments, A, B, and C. You obtained the following data and plan to conduct a Friedman F_r test.

BLOCK	A	B	C
1	10	13	18
2	12	12	13
3	9	10	12
4	13	15	16
5	11	9	10
6	15	11	16

a. Specify the null and alternative hypotheses you would test.
b. Specify the rejection region for the test, using $\alpha = .05$.
c. Calculate the rank sums for the three treatments.
d. Calculate the test statistic.
e. State the appropriate conclusion.

APPLYING THE CONCEPTS

15.40   Refer to the Kansas State study designed to investigate the effects of plants on stress levels of humans, Exercise 13.17. The data, reproduced in the table, are finger temperatures for each of 10 students in a dimly lit room under three experimental conditions: presence of a live plant, presence of only a plant photo, and absence of a plant (either live or photo). Analyze the data using a nonparametric procedure. Do the students' finger temperatures depend on the experimental condition?

STUDENT	LIVE PLANT	PLANT PHOTO	NO PLANT (CONTROL)
1	91.4	93.5	96.6
2	94.9	96.6	90.5
3	97.0	95.8	95.4
4	93.7	96.2	96.7
5	96.0	96.6	93.5
6	96.7	95.5	94.8
7	95.2	94.6	95.7
8	96.0	97.2	96.2
9	95.6	94.8	96.0
10	95.6	92.6	96.6

Source: Elizabeth Schreiber, Department of Statistics, Kansas State University, Manhattan, Kansas.

15.41   Refer to the comparison of prices of 60 items at Winn-Dixie, Publix, and Kash N' Karry supermarkets, Exercise 13.15. An SPSS printout of the nonparametric analysis of the data is shown here. Interpret the results.

```
- - - - - Friedman Two-way ANOVA


    Mean Rank     Variable

        1.23      WINNDIX
        2.43      PUBLIX
        2.34      KASHNKAR

        Cases          Chi-Square        D.F.    Significance
          60             54.3083            2         .0000
```

15.42 The data for the *Computers & Electronics* study, Exercise 13.19, is reproduced in the accompanying table. Recall that 10 test subjects were asked to specify which color combination for a video display terminal they preferred by ranking each of seven color combinations on a scale from 0 (no preference) to 10.

SUBJECT	GREEN/ BLACK	WHITE/ BLACK	YELLOW/ WHITE	ORANGE/ WHITE	YELLOW	YELLOW/ AMBER	YELLOW/ ORANGE
1	7	6	7	2	8	9	3
2	8	6	9	4	9	8	1
3	5	5	7	1	6	8	2
4	3	4	2	0	2	6	0
5	9	8	8	3	9	9	2
6	7	5	6	2	7	7	1
7	6	7	8	4	6	9	5
8	6	5	8	1	8	9	1
9	9	9	8	2	9	8	0
10	9	8	8	3	9	10	1

Source: Adapted from Solomon, L. and Burawa, A. "Maximize your computing comfort & efficiency." *Computers & Electronics*, Apr. 1983, pp. 35–40.

a. Do the data provide sufficient evidence to indicate a difference among the preference scores for the seven video display color combinations? Test using $\alpha = .05$.

b. Do the data provide sufficient evidence to indicate that video display terminal users prefer a yellow/amber color combination over a green/black color combination? Test using $\alpha = .05$.

c. Do the data provide sufficient evidence to indicate that video display terminal users prefer a yellow/amber color combination over yellow? Test using $\alpha = .05$.

15.43 A study was conducted to explore the sources of occupational stress for engineers (*IEEE Transactions on Engineering Management*, Feb. 1986). One of the objectives was to determine "if there are consistent and significant differences among engineers at different levels of the organizational hierarchy in the degree to which they consider the different factors as sources of stress." A sample of male engineers from different types of organizations in Ontario, Canada, was administered the Stress Diagnostic Survey (SDS). The SDS provides stress ratings for each of 15 categories of work stressors. The researchers ranked 15 stress categories from 1 (highest stress) to

STRESS CATEGORY	NONSUPERVISORS	1ST LEVEL	2ND LEVEL	3RD LEVEL
Politics	5	6	4	7
Underutilization	3	5	7	5
Human resources development	4	3	2	3
Supervisory style	6	7	8	8
Rewards	1	1	1	4
Organizational structure	9	9	10	12
Participation	2	4	5	6
Role ambiguity	10	13	12	14
Overload/quantitative	15	15	15	15
Overload/qualitative	12	8	6	2
Time pressure	8	2	3	1
Role conflict	13	10	13	10
Career progression	7	10	9	13
Job scope	11	12	14	11
Responsibility for people	14	14	11	9

Source: Saleh, S. D. and Desai, K. "Occupational stress for engineers." *IEEE Transactions on Engineering Management*, Vol. EM-33, No. 1, Feb. 1986, p. 8, Table II. © 1986 IEEE.

15 (lowest stress) for each of four groups of engineers—nonsupervisors, first-level supervisors, second-level supervisors, and third-level supervisors—as shown in the table. Conduct a test to determine whether the rank orderings of the stress categories differ among the four groups of engineers. Test using $\alpha = .01$.

15.44 *Acid rain* is considered by some environmentalists to be the nation's most serious environmental problem. It is formed by the combination of water vapor in clouds with nitrogen oxide and sulfuric dioxide emissions from the burning of coal, oil, and natural gas. The acidity of rain in central and northern Florida consistently ranges from 4.5 to 5 on the pH scale, a decidedly acid condition. To determine the effects of acid rain on the acidity of soils in a natural ecosystem, engineers at the University of Florida's Institute of Food and Agricultural Sciences irrigated experimental plots near Gainesville, Florida, with acid rain at five pH levels: 3.7, 4.0, 4.5, 4.7, and 5.0. The acidity of the soil was then measured at three different depths: 0–15, 15–30, and 30–46 centimeters. Tests were conducted during three different time periods in 1981. The resulting soil pH values are shown in the table. Suppose the main objective of the experiment is to compare the acidity of soil irrigated with acid rain at the five different pH levels, and that the different soil depths are to be treated as blocks. Use a nonparametric test to compare the soil pH values of the five treatments.

SOIL DEPTH	ACID RAIN pH				
	3.7	4.0	4.5	4.7	5.0
0–15 cm	5.20	5.19	5.13	5.10	5.01
15–30 cm	5.33	5.32	5.20	5.21	5.15
30–46 cm	5.33	5.30	5.17	5.20	5.10

Source: Adapted from "Acid rain linked to growth of coal-fired power." *Florida Agricultural Research 83*, Vol. 2, No. 1, Winter 1983.

15.45 Over the past decade there has been an explosion of computer software available to the human resource community. This rapid growth of human resource software has dramatically changed the ways in which organizations conduct their personnel operations. A nationwide study of nearly 700 organizations was recently conducted to determine the current use and future interest for computer software in human resource management (*Personnel Administrator*, Aug. 1984). The accompanying table gives the rank order of the 10 areas most preferred for computerization by each of three groups of organizations: small (fewer than 500 employees), medium (between 500 and 2,000 employees), and large (2,000 or more employees).

SOFTWARE APPLICATION	RANKS BY COMPANY SIZE		
	Small	Medium	Large
Management information systems	2	1	1
Personnel inventories	1	2	2
EEO/AA records	3	3	3
Recruitment/tracking	4	4	4
Performance appraisal	5	5	5
Job analysis	6	6	6
Career pathing	10	8	7
Attitude questionnaires	9	7	10
Person-job matching	7	10	8
Training needs	8	9	9

Source: Zurakowski, D. S. and Harris, W. G. "Software applications in human resource management." Reprinted from *HRMagazine*, formerly *Personnel Administrator*, Aug. 1984, p. 81.

a. Conduct a nonparametric test to compare the software applications rankings of the three groups. Test using $\alpha = .05$.
b. Why is a nonparametric test more appropriate for analyzing this data than a parametric test?

15.8

We learned in Section 12.13 how to modify the regression analysis when the assumptions about the random error term ε are violated. For example, if the variance σ^2 of ε is not constant, we transform the dependent variable, y, using one of the variance-stabilizing transformations discussed in Section 12.13. An alternative procedure is to conduct a nonparametric regression analysis of the data.

In nonparametric regression, tests of model adequacy do not require any assumptions about the probability distribution of ε; thus, they are distribution-free. Although the tests are intuitively appealing, they can become quite difficult to apply in practice, especially when the number of observations is large. For this reason, and the fact that residual diagnostics are readily available via the computer, most analysts prefer to use the techniques of Section 12.13 when the standard regression assumptions are violated.

For those who are interested, we provide brief descriptions of the nonparametric alternatives to the parametric simple linear regression tests of Chapter 11. Specifically, we discuss a nonparametric test for (1) linear correlation and (2) the slope parameter of the straight-line model.

As an alternative to the Pearson product moment correlation coefficient, r (Section 11.7), we can compute a correlation coefficient based on ranks. **Spearman's rank correlation coefficient**, denoted r_s, can then be used to test for linear correlation between two variables, y and x.

The formulas for computing r_s and the nonparametric test of hypothesis for rank correlation are shown in the box. Example 15.8 illustrates the procedures.

SPEARMAN'S NONPARAMETRIC TEST FOR RANK CORRELATION

ONE-TAILED TEST

H_0: $\rho = 0$

H_a: $\rho > 0$

[or H_a: $\rho < 0$]

TWO-TAILED TEST

H_0: $\rho = 0$

H_a: $\rho \neq 0$

$$\text{Test statistic:} \quad r_s = 1 - \frac{6 \sum_{i=1}^{n} d_i^2}{n(n^2 - 1)}$$

where d_i is the difference between the y rank and x rank for the ith observation.

[*Note:* In the case of ties, calculate r_s by substituting the ranks of the y's and the ranks of the x's for the actual y values and x values in the formula for r given in Section 11.7.]

Rejection region:

$r_s > r_\alpha$ (or $r_s < -r_\alpha$)

Rejection region:

$|r_s| > r_{\alpha/2}$

where the values of r_α and $r_{\alpha/2}$ are given in Table 13 of Appendix G.

Assumptions: None

EXAMPLE 15.8 A large manufacturing firm wants to determine whether a relationship exists between the number of work-hours an employee misses per year and the employee's annual wages (in thousands of dollars). A sample of 15 employees produced the data shown in Table 15.11.

TABLE 15.11
Work-Hours Missed and Annual
Wages for Example 15.8

EMPLOYEE	WORK-HOURS MISSED y	ANNUAL WAGES x
1	49	15.8
2	36	17.5
3	127	11.3
4	91	13.2
5	72	13.0
6	34	14.5
7	155	11.8
8	11	20.2
9	191	10.8
10	6	18.8
11	63	13.8
12	79	12.7
13	43	15.1
14	57	24.2
15	82	13.9

a. Calculate Spearman's rank correlation coefficient as a measure of the strength of the relationship between work-hours missed and annual wages.
b. Is there sufficient evidence to indicate that work-hours missed decreases as annual wages increases, i.e., that work-hours and annual wages are negatively correlated? Test using $\alpha = .01$.

Solution

a. Spearman's rank correlation coefficient is found by first ranking the values of each variable separately. (Ties are treated by averaging the tied ranks.) Then r_s is computed in exactly the same way as the Pearson correlation coefficient r; the only difference is that the values of x and y that appear in the formula for r are replaced by their ranks. That is, the *ranks* of the raw data are used to compute r_s rather than the raw data themselves. When there are no (or few) ties in the ranks, this formula reduces to the simple expression

$$r_s = 1 - \frac{6 \sum d_i^2}{n(n^2 - 1)}$$

where d_i is the difference between the rank of y and x for the ith observation.

The ranks of y and x, the differences between the ranks, and the squared differences for each of the 15 employees are also shown in Table 15.12. Note that the sum of the squared differences is $\sum d_i^2 = 1{,}038$. Substituting this value into the formula for r_s, we obtain

$$r_s = 1 - \frac{6 \sum d_i^2}{n(n^2 - 1)} = 1 - \frac{6(1{,}038)}{15(224)} = -.854$$

This large negative value of r_s implies that a fairly strong negative correlation exists between work-hours missed y and annual wages x in the sample.

TABLE 15.12
Calculation Table for Example 15.8

EMPLOYEE	WORK-HOURS MISSED	RANK	ANNUAL WAGES	RANK	d_i	d_i^2
1	49	6	15.8	11	−5	25
2	36	4	17.5	12	−8	64
3	127	13	11.3	2	11	121
4	91	12	13.2	6	6	36
5	72	9	13.0	5	4	16
6	34	3	14.5	9	−6	36
7	155	14	11.8	3	11	121
8	11	2	20.2	14	−12	144
9	191	15	10.8	1	14	196
10	6	1	18.8	13	−12	144
11	63	8	13.8	7	1	1
12	79	10	12.7	4	6	36
13	43	5	15.1	10	−5	25
14	57	7	24.2	15	−8	64
15	82	11	13.9	8	3	9
					$\sum d_i^2 =$	1,038

b. To determine whether a negative correlation exists in the population, we would test H_0: $\rho = 0$ against H_a: $\rho < 0$ using r_s as a test statistic. As you would expect, we reject H_0 for small values of r_s. Upper-tailed critical values of Spearman's r_s are provided in Table 13 of Appendix G. This table is partially reproduced in Table 15.13. Since the distribution of r_s is symmetric around 0, the lower-tailed critical value is the negative of the corresponding upper-tailed critical value. For $\alpha = .01$ and $n = 15$, the critical value (shaded in Table 15.13) is $r_{.01} = .623$. Thus, the rejection region for the test is

Reject H_0 if $r_s < -.623$.

Since the test statistic, $r_s = -.854$, falls in the rejection region, there is sufficient evidence (at $\alpha = .01$) of negative correlation between work-hours missed y and annual wages x in the population.

TABLE 15.13

A portion of the Spearman's r_s Table, Table 13 of Appendix G

	$\alpha = .10$	$\alpha = .05$	$\alpha = .02$	$\alpha = .01$	← TWO-TAILED
n	$\alpha = .05$	$\alpha = .025$	$\alpha = .01$	$\alpha = .005$	← ONE-TAILED
15	.441	.525	.623	.689	
16	.425	.507	.601	.666	
17	.412	.490	.582	.645	
18	.399	.476	.564	.625	
19	.388	.462	.549	.608	
20	.377	.450	.534	.591	
21	.368	.438	.521	.576	
22	.359	.428	.508	.562	
23	.351	.418	.496	.549	
24	.343	.409	.485	.537	
25	.336	.400	.475	.526	

Alternatively, we could test for linear correlation in the population by testing the slope parameter β_1 in the simple linear regression model

$$y = \beta_0 + \beta_1 x + \varepsilon$$

That is, we could test $H_0\colon \beta_1 = 0$ against $H_a\colon \beta_1 \neq 0$. A distribution-free test for the slope is the **Theil C test**.

This nonparametric test is described in the box.

THEIL'S TEST FOR ZERO SLOPE IN THE STRAIGHT-LINE MODEL $y = \beta_0 + \beta_1 x + \varepsilon$

ONE-TAILED TEST

$H_0\colon \ \beta_1 = 0$

$H_a\colon \ \beta_1 > 0$
  (or $H_a\colon \ \beta_1 < 0$)

TWO-TAILED TEST

$H_0\colon \ \beta_1 = 0$

$H_a\colon \ \beta_1 \neq 0$

Test statistic: $C = (-1)$(Number of negative $y_j - y_i$ differences)
  $+ (1)$(Number of positive $y_j - y_i$ differences)

where i and j are the ith and jth observations ranked in increasing order of the x values, $i < j$.

Observed significance level:

$$p\text{-value} = \begin{cases} P(x \geq C) & \text{for } H_a\colon \beta_1 > 0 \\ P(x \leq C) & \text{for } H_a\colon \beta_1 < 0 \end{cases}$$

Observed significance level:

$$p\text{-value} = 2 \cdot P(x \geq |C|)$$

where the values of $P(x \geq C) = P(x \leq -C)$ are given in Table 14 of Appendix G.

Assumption: The random errors ε are independent.

EXAMPLE 15.9 Refer to the work-hours missed–annual wages data, Example 15.8. Conduct the Theil C test for zero slope on the data in Table 15.12.

Solution   To conduct this nonparametric test, we first rank the x values in increasing order and list the ordered x–y pairs, as shown in Table 15.14. Next, we calculate all possible differences $y_j - y_i$, $i < j$ (where i and j represent the ith and jth ranked observations), and note the sign (positive or negative) of each difference.

TABLE 15.14
Data of Table 15.12

RANKED EMPLOYEE	HOURS MISSED y	ANNUAL WAGES x	DIFFERENCES, $y_j - y_i$ $(i < j)$ # negatives	# positives
1	191	10.8	—	—
2	127	11.3	1	0
3	155	11.8	1	1
4	79	12.7	3	0
5	72	13.0	4	0
6	91	13.2	3	2
7	63	13.8	6	0
8	82	13.9	4	3
9	34	14.5	8	0
10	43	15.1	8	1
11	49	15.8	8	2
12	36	17.5	10	1
13	6	18.8	12	0
14	11	20.2	12	1
15	57	24.2	8	6
			Totals: 98	17

For example, the y-value for employee ranked #2, $y_2 = 127$, is compared to the y value for each employee with a lower rank. In this case, the only employee ranked lower is employee #1, with $y_1 = 191$ (see Table 15.14). The difference

$$y_2 - y_1 = 127 - 191 = -64$$

is negative, and is noted as such in Table 15.14. Similarly, we compare the y value of employee ranked #3, $y_3 = 155$, to the y values of employees of lower rank, $y_2 = 127$ and $y_1 = 191$, by the differences

$$y_3 - y_2 = 155 - 127 = 28$$

and

$$y_3 - y_1 = 155 - 191 = -36$$

This results in one positive and one negative difference. Continuing in this manner, we obtain a total of 17 positive differences and 98 negative differences, as shown in Table 15.14.

The test statistic C is obtained by scoring each positive difference as a $+1$ and each negative difference as a -1 (0 differences are assigned a score of 0) and summing the scores. Therefore, for the data of Table 15.14, we obtain the test statistic

$$C = (+1)(17) + (-1)(98) = -81$$

The observed significance level (p-value) of the test is obtained from Table 14 of Appendix G. For this lower-tailed test, i.e., a test for a negative slope, the p-value is $P(C < -81)$. Searching the $n = 15$ column and the $x = 81$ row of Table 14 of Appendix G, we obtain the p-value ≈ 0. Thus, there is strong evidence to reject H_0 and conclude that work-hours missed, y, is negatively linearly related to annual wages, x, at this firm.

Nonparametric tests are also available for multiple regression models. These tests are very sophisticated, however, and require the use of specialized statistical computer software not yet available on a commercial basis. Consult the references if you want to learn more about these nonparametric techniques.

EXERCISES

LEARNING THE MECHANICS

15.46 Specify the rejection region for Spearman's nonparametric test for rank correlation in each of the following situations:

a. H_a: $\rho > 0$, $n = 15$, $\alpha = .05$
b. H_a: $\rho < 0$, $n = 20$, $\alpha = .01$
c. H_a: $\rho \neq 0$, $n = 10$, $\alpha = .05$

15.47 A random sample of seven pairs of observations is recorded on two variables, x and y. The data are shown in the table.

PAIR	VALUE OF x	VALUE OF y
1	65	59
2	57	61
3	55	58
4	38	23
5	29	34
6	43	38
7	49	37

a. Rank the values of each variable, x and y. Note that there are no tied ranks.
b. Compute the test statistic, r_s, using the shortcut formula.
c. Do the data provide sufficient evidence to conclude that the ranked pairs are correlated? Test using $\alpha = .05$.

15.48 Find the observed significance level of the Theil test for zero slope for each of the following situations:

a. $n = 5$, $C = 8$, H_a: $\beta_1 > 0$
b. $n = 10$, $C = 25$, H_a: $\beta_1 \neq 0$
c. $n = 20$, $C = -40$, H_a: $\beta_1 < 0$

15.49 Conduct Theil's test for zero slope on the data of Exercise 15.47. Test using $\alpha = .05$.

APPLYING THE CONCEPTS

15.50 Housing planners, insurance companies, and credit institutions have shown great interest in measuring factors related to the risk of default in home mortgages. Typically, the default rate (i.e., the probability of defaulting on a home loan) is used as the standard measure of risk. Another, possibly superior, measure of risk is the expected loss for a home loan (i.e., the dollar loss expected on a default). A study was conducted to compare the two measures of mortgage default risk. The default rate and expected loss (measured as a percentage of loan value) for 16 different FHA categories of home mortgages are recorded in the accompanying table.

	LOAN CATEGORY			DEFAULT RATE	EXPECTED LOSS
Loan Size	Loan Value	Race	Location	%	%
GT $20,750	GT 95%	White	Other city	3.44	.948
GT $20,750	GT 95%	White	Suburb	5.20	1.181
GT $20,750	GT 95%	Black	Other city	12.50	3.178
GT $20,750	GT 95%	Black	Suburb	16.95	2.696
LT $20,750	GT 95%	White	Other city	3.89	1.256
LT $20,750	GT 95%	White	Suburb	4.85	1.578
LT $20,750	GT 95%	Black	Other city	13.40	6.453
LT $20,750	GT 95%	Black	Suburb	14.67	4.025
GT $20,750	90–95%	White	Other city	2.15	.558
GT $20,750	90–95%	White	Suburb	2.66	.630
GT $20,750	90–95%	Black	Other city	5.60	1.448
GT $20,750	90–95%	Black	Suburb	7.95	2.304
LT $20,750	90–95%	White	Other city	2.41	.674
LT $20,750	90–95%	White	Suburb	2.60	.897
LT $20,750	90–95%	Black	Other city	10.27	4.836
LT $20,750	90–95%	Black	Suburb	5.66	1.579

Source: Evans, R. D., Maris, B. A., and Weinstein, R. I. "Expected loss and mortgage default risk." *Quarterly Journal of Business and Economics*, Vol. 24, No. 1, Winter 1985, pp. 75–92.

a. Calculate Spearman's rank correlation coefficient between the two measures of mortgage default risk.
b. Is there sufficient evidence to indicate that default rate and expected loss are positively correlated? Test using $\alpha = .01$.
c. Analyze the data in the table using Theil's test for zero slope. Do the results agree with part b?

15.51 Health maintenance organizations (HMOs) provide and monitor a variety of short-term outpatient services, including crisis intervention mental health services. To investigate the standards used by HMOs in interpreting crisis intervention, Cheifetz and Salloway (1985) conducted a comprehensive questionnaire survey of 145 national HMOs. Each HMO was asked to "write a brief, descriptive definition of situations or states you would include as qualifying for 'crisis intervention.'" The researchers sorted these situations into 10 categories and then asked three experienced clinicians to rate each category for two criteria: validity of crisis intervention (i.e., is the situation defined really a "crisis") and clarity of guidelines for offering service. A 4-point rating scale was provided for both criteria. The mean ratings for the 10 categories on both the crisis intervention and clarity scales are given in the table. Is there evidence of a positive relationship between the mean crisis intervention and mean clarity ratings? Test using $\alpha = .05$.

CATEGORY (SITUATION)	CRISIS INTERVENTION RATING (1 = definitely a crisis, 4 = definitely not a crisis)	CLARITY RATING (1 = very clear guideline, 4 = very unclear guideline)
Psychosis	1.31	1.33
Drug/alcohol abuse	1.33	1.29
Depression/anxiety	1.48	1.59
Emphasis on acuteness	1.76	2.50
Insistence on "short-term" response	2.48	3.22
Suicide	1.13	1.32
Family problems	2.59	2.30
Violence/harm	1.06	1.86
Miscellaneous	2.60	2.33
Nondefinition	3.57	3.57

Source: Cheifetz, D. I. and Salloway, J. C. "Crisis intervention: Interpretation and practice by HMO." *Medical Care*, Vol. 23, No. 1, Jan. 1985, pp. 89–93.

15.52 Each year Storyboard Tests conducts a survey of over 20,000 television viewers to determine the most popular TV advertisement. Pepsi had the top-rated ad campaign of 1990, which included music legend Ray Charles plugging Diet Pepsi with the tune "You got the right one baby—uh huh." The top 10 TV ad campaigns of 1990 are listed in the table along with their 1989 ranks. Use a nonparametric procedure to test for a positive correlation between the 1989 and 1990 ranks of the TV ads. Use $\alpha = .01$.

1990 RANK	1989 RANK	PRODUCT
1	2	Pepsi/Diet Pepsi
2	10	Nike
3	4	Energizer
4	7	Coca-Cola
5	1	McDonald's
6	17	Little Caesar
7	8	Miller Lite
8	3	California Raisins
9	15	Budweiser
10	9	Infiniti

Source: Video Storyboard Tests' *Commercial Break*, Mar. 1991, Vol. 3, No. 1, p. 1.

15.53 What are the top graduate business schools in the United States? *Business Week* polled a random sample of approximately 3,000 graduates of the schools that most often make top-20 lists. The graduates assessed the quality of the teaching, curriculum, environment, and job placement efforts on a scale of 1 to 10, and the schools were awarded an average score. Based on those average scores, *Business Week* compiled the following list of the top 20 business schools in the United States. In addition, the table reports the average starting pay of the schools' graduates. Is there evidence of a positive correlation between the business school's rank and average starting salary of its graduates? Test using $\alpha = .05$.

BUSINESS SCHOOL	BUSINESS WEEK RANK	AVERAGE STARTING PAY
Northwestern	1	$53,031
Harvard	2	64,112
Dartmouth	3	62,681
Wharton	4	55,183
Cornell	5	52,339
Michigan	6	43,976
Virginia	7	50,554
North Carolina	8	44,941
Stanford	9	65,176
Duke	10	48,740
Chicago	11	54,772
Indiana	12	38,407
Carnegie-Mellon	13	49,109
Columbia	14	49,397
MIT	15	60,860
UCLA	16	45,378
California (Berkeley)	17	45,083
NYU	18	47,037
Yale	19	46,455
Rochester	20	39,990

Source: *Business Week*, November 28, 1988, pp. 78–79.

15.54 The Federal Communications Commission (FCC) specifies that radiated electromagnetic emissions from digital devices are to be measured in an open-field test site. To verify test-site acceptability, the site attenuation (i.e., the transmission loss from the input of one half-wave dipole to the output of another when both dipoles are positioned over the ground plane) must be evaluated. A study conducted at a test site in Fort Collins, Colorado, yielded the accompanying data on site attenuation (in decibels) and transmission frequency (in megahertz) for dipoles at a distance of 3 meters.

TRANSMISSION FREQUENCY x, MHz	SITE ATTENUATION y, dB
50	11.5
100	15.8
200	18.2
300	22.6
400	26.2
500	27.1
600	29.5
700	30.7
800	31.3
900	32.6
1,000	34.9

Source: Bennett, W. S. "An error analysis of the FCC site-attenuation approximation." *IEEE Transactions on Electromagnetic Compatibility*, Vol. EMC-27, No. 3, Aug. 1985, p. 113 (Table IV). © 1985, IEEE.

a. Find the Spearman coefficient of correlation between transmission frequency (x) and site attenuation level (y). Interpret the result.

b. Conduct a test to determine if site attenuation level (y) is positively correlated with transmission frequency (x). Test using $\alpha = .10$.

15.55 Refer to Exercise 11.14 and *Fortune* magazine's ranking of the top 10 American cities with respect to their ability to provide high-quality, low-cost labor. The data on labor market stress index (y) and unemployment rate (x) is reproduced in the table. Consider the straight-line model $E(y) = \beta_0 + \beta_1 x$.

RANK	CITY	LABOR MARKET STRESS INDEX y	UNEMPLOYMENT RATE $x\%$
1	Salt Lake City	107	4.5
2	Minneapolis-St. Paul	107	3.8
3	Atlanta	100	5.1
4	Sacramento	100	4.9
5	Austin (Texas)	80	5.4
6	Columbus (Ohio)	100	4.8
7	Dallas/Fort Worth	100	5.5
8	Phoenix	93	4.3
9	Jacksonville (Florida)	87	5.7
10	Oklahoma City	80	4.6

Source: *Fortune*, Oct. 22, 1990, pp. 58–63.

a. Calculate Theil's test statistic for a test of zero slope.
b. Is there evidence of a linear relationship between labor market stress index and unemployment rate? Test using $\alpha = .01$.

15.56 Civil engineers often use the straight-line equation $E(y) = \beta_0 + \beta_1 x$ to model the relationship between the mean shear strength $E(y)$ of masonry joints and precompression stress, x. To test this model, a series of stress tests were performed on solid bricks arranged in triplets and joined with mortar (*Proceedings of the Institute of Civil Engineers*, Mar. 1990). The precompression stress was varied for each triplet, and the ultimate shear load just before failure (called the shear strength) was recorded. The stress results for seven triplets (measured in N/mm^2) are shown in the accompanying table. Conduct a nonparametric test of $H_0: \beta_1 = 0$ against the alternative $H_a: \beta_1 > 0$. Test using $\alpha = .05$.

TRIPLET TEST	1	2	3	4	5	6	7
Shear strength, y	1.00	2.18	2.24	2.41	2.59	2.82	3.06
Precompression stress, x	0	.60	1.20	1.33	1.43	1.75	1.75

Source: Riddington, J. R. and Ghazali, M. Z. "Hypothesis for shear failure in masonry joints." *Proceedings of the Institute of Civil Engineers*, Part 2, Mar. 1990, Vol. 89, p. 96 (Figure 7).

SUMMARY

We have presented several useful *nonparametric techniques* as alternatives to the parametric tests discussed in Chapters 10, 11, and 13. Nonparametric techniques are useful when the underlying assumptions for their parametric counterparts are not justified or when it is impossible to assign specific values to the observations. Nonparametric methods provide more general comparisons of populations than parametric methods, because they compare the probability distributions of the populations rather than specific parameters. Consequently, they are *distribution-free*, i.e., they require no assumptions about the distributions of the sampled data.

The *sign test* can be used to test for a shift in the median of a population. Since the median is a measure of location, the sign test is a test for location of a single population.

Rank sums are the primary tools of nonparametric statistics. The *Wilcoxon rank sum test* can be used to compare two populations based on an independent sampling experiment, and the *Wilcoxon signed ranks test* can be used for a *matched-pairs experiment*. The *Kruskal–Wallis H test* is applied when comparing *k* populations using a *completely randomized design*. The *Friedman F_r test* is used to compare *k* populations when a randomized block design is conducted. Other nonparametric tests are *Spearman's test* for correlation and *Theil's slope test*.

The strength of nonparametric statistics lies in their general applicability. Few restrictive assumptions are required, and they may be used for observations that can be ranked but not exactly measured. Therefore, nonparametric methods provide useful alternatives to parametric tests.

KEY WORDS

Distribution-free tests

Friedman F_r test

Kruskal–Wallis H test

Median

Nonparametric
  methods

Rank sum

Ranks

Spearman's rank correlation

Test for location

Theil's zero slope test

Wilcoxon rank sum test

Wilcoxon signed ranks test

KEY FORMULAS

1. Sign test

$$p\text{-value} = \begin{cases} P(x \geq S) & \text{if one-tailed test} \\ 2P(x \geq S) & \text{if two-tailed test} \end{cases}$$

where $S = \begin{cases} \text{Number of sample observations greater than hypothesized} \\ \text{median } \eta_0 \text{ if upper-tailed or two-tailed test} \\ \text{Number of sample observations less than } \eta_0 \text{ if} \\ \text{lower-tailed test} \end{cases}$

Large-sample z statistic: $z = \dfrac{S - .5n}{.5\sqrt{n}}$

2. Wilcoxon rank sum test

Large-sample z statistic: $z = \dfrac{T_1 - \left[\dfrac{n_1 n_2 + n_1(n_1 + 1)}{2}\right]}{\sqrt{\dfrac{n_1 n_2(n_1 + n_2 + 1)}{12}}}$

3. Wilcoxon signed ranks test

Large-sample z statistic: $\quad z = \dfrac{T^+ - [n(n + 1)/4]}{\sqrt{[n(n + 1)(2n + 1)]/24}}$

4. Kruskal–Wallis H test

$$H = \frac{12}{n(n + 1)} \sum \left(\frac{T_i^2}{n_i} \right) - 3(n + 1)$$

where T_i = Rank sum for sample i
  n_i = Number of observations in sample i
  $n = n_1 + n_2 + \cdots + n_k$

5. Friedman F_r test

$$F_r = \frac{12}{bk(k + 1)} \sum (T_i^2) - 3b(k + 1)$$

where b = Number of blocks
  k = Number of treatments
  T_i = Rank sum for treatment i

6. Spearman's rank correlation

$$r_s = 1 - \frac{6 \sum d_i^2}{n(n^2 - 1)} \quad \text{if no ties in the data}$$

7. Theil's zero slope test

$$C = (-1)(\# \text{ negative } y_j - y_i \text{ differences}) + (1)(\# \text{ positive } y_j - y_i \text{ differences})$$

SUPPLEMENTARY EXERCISES

15.57 *Scram* is the term used by nuclear engineers to describe a rapid emergency shutdown of a nuclear reactor. The nuclear industry has made a concerted effort to reduce significantly the number of unplanned scrams each year. The number of unplanned scrams at each of a random sample of 20 nuclear reactor units in 1984 are given here (*Transactions of the American Nuclear Society*, Vol. 50, 1985):

> 1 8 0 3 3   9 1 2 4 5
> 4 3 1 2 7 10 2 6 3 0

Test the hypothesis that the median number of unplanned scrams at nuclear reactor plants in 1984 is less than 5. Use $\alpha = .10$.

15.58 Refer to the *Chemosphere* (Vol. 20, 1990) study on the TCDD levels of 20 Massachusetts Vietnam veterans who were possibly exposed to Agent Orange, Section 15.5. The amounts of TCDD (measured in parts per trillion) in blood plasma and fat tissue drawn from each veteran are shown in the table. Is there sufficient evidence of a difference between the distributions of TCDD levels in plasma and fat tissue for Vietnam veterans exposed to Agent Orange? Test using $\alpha = .05$.

VETERAN	TCDD LEVELS IN PLASMA	TCDD LEVELS IN FAT TISSUE	VETERAN	TCDD LEVELS IN PLASMA	TCDD LEVELS IN FAT TISSUE
1	2.5	4.9	11	6.9	7.0
2	3.1	5.9	12	3.3	2.9
3	2.1	4.4	13	4.6	4.6
4	3.5	6.9	14	1.6	1.4
5	3.1	7.0	15	7.2	7.7
6	1.8	4.2	16	1.8	1.1
7	6.0	10.0	17	20.0	11.0
8	3.0	5.5	18	2.0	2.5
9	36.0	41.0	19	2.5	2.3
10	4.7	4.4	20	4.1	2.5

Source: Schecter, A., et al. "Partitioning of 2,3,7,8-chlorinated dibenzo-p-dioxins and dibenzofurans between adipose tissue and plasma lipid of 20 Massachusetts Vietnam veterans." *Chemosphere*, Vol. 20, Nos. 7–9, 1990, pp. 954–955 (Tables I and II).

15.59 Many successful corporations issue bonds to investors to raise capital for expansion. The sale of the bonds is usually handled by a bond underwriter at an investment bank. Since the price a corporation receives for a bond often depends on the skill of the underwriter, D. E. Logue and R. J. Rogalski considered the question, "Does it pay to shop for your bond underwriter?" (*Harvard Business Review*, July–Aug. 1979). In one portion of the study, they used the Kruskal–Wallis H test to compare the changes in the underwriting selling prices of bonds handled by different investment bankers. The accompanying table gives the changes in bond prices (in dollars) over a 12-month period for bonds underwritten by four firms—Morgan Stanley, First Boston, Goldman Sachs, and Merrill Lynch. Assume that six bonds are randomly selected from each firm. [*Note:* The actual data for the study were not given. We provide simulated data in the table, based on the means and variances provided in the *Harvard Business Review* article.]

MORGAN STANLEY	FIRST BOSTON	GOLDMAN SACHS	MERRILL LYNCH
.037	−.128	.025	−.047
−.016	−.054	−.080	.010
−.132	.007	−.031	−.003
−.148	−.011	.049	−.104
.022	.031	−.019	−.082
−.049	−.042	−.027	−.039

a. Give the null and alternative hypotheses appropriate for comparing the distributions of changes in bond prices for the four underwriters.
b. Calculate the nonparametric test statistic.
c. Is there sufficient evidence to indicate that differences in distributions of bond price changes exist among the four investment firms? Test using $\alpha = .01$.

15.60 The presence of lead in drinking water is cause for alarm in some older cities, many of which use lead pipes for water service lines. In one study, researchers attempted to document the levels of lead, copper, and iron in the Boston water supply for areas supplied by lead service lines. The data shown in the table are the mean concentrations of lead (milligrams per liter) for samples collected at nine locations over the period from 1976 to 1978. In May 1977, the city began treating the water with sodium hydroxide to reduce traces of metal in the water. Compare the mean lead concentration levels before and after May 1977 using a Wilcoxon signed ranks test. Do the data provide sufficient evidence to indicate a reduction in lead content in the city's water supply after the initiation of the sodium hydroxide water treatment? Test using $\alpha = .05$.

LOCATION	DATE	MEAN CONCENTRATION OF LEAD	DATE	MEAN CONCENTRATION OF LEAD
1	Feb. 1976	.074	June 1977	.035
2	Mar. 1976	.064	July 1977	.060
3	Apr. 1976	.069	Aug. 1977	.055
4	May 1976	.063	Oct. 1977	.035
5	July 1976	.077	Nov. 1977	.031
6	Sept. 1976	.095	Dec. 1977	.039
7	Oct. 1976	.092	Jan. 1978	.038
8	Nov. 1976	.091	Mar. 1978	.049
9	Dec. 1976	.067	Apr. 1978	.073

Source: Karalekas, P. C., Jr., Ryan, C. R., and Taylor, F. B. "Control of lead, copper, and iron pipe corrosion in Boston." *American Water Works Journal*, Vol. 75, No. 2, Feb. 1983, pp. 92–95. Reprinted by permission. Copyright © 1983, The American Water Works Association.

15.61 Bulk specimens of Chilean lumpy iron ore (95% particle size, 150 millimeters) were randomly sampled from a 35,323-long-ton shipload of ore; the percentage of iron in each ore specimen was determined (*Reports of Statistical Application Research*, Union of Japanese Scientists and Engineers, Vol. 18, No. 1, 1971). The data for 10 bulk specimens are given here:

> 63.01 61.75 63.22 62.38 62.80
> 63.92 62.94 63.71 62.10 64.34

Is there sufficient evidence to indicate that the median percentage of iron in bulk specimens from the shipload of ore differs from 63? Test using $\alpha = .05$.

15.62 As oil drilling costs rise at unprecedented rates, the task of measuring drilling performance becomes essential to a successful oil company. One method of lowering drilling costs is to increase drilling speed. Researchers at Cities Service Co. have developed a drill bit, called the PD-1, that they believe penetrates rock at a faster rate than any other bit on the market. It is decided to compare the speed of the PD-1 with the two fastest drill bits known, the IADC 1-2-6 and the IADC 5-1-7, at 15 drilling locations in Texas. Five drilling sites were randomly assigned to each bit, and the rate of penetration (RoP) in feet per hour (fph) was recorded after drilling 3,000 feet at each site. The data are given in the table. Based on this information, can Cities Service Co. conclude that the distributions of RoP for the three drill bits differ (in location)? Test at the $\alpha = .05$ level of significance.

PD-1	IADC 1-2-6	IADC 5-1-7
35.2	25.8	14.7
30.1	29.7	28.9
37.6	26.6	23.3
34.3	30.1	16.2
33.2	27.4	20.6

15.63 The thermogravimetric balance (TG) is a new technique developed to evaluate the thermal behavior of chemical compounds. Abou El Naga and Salem (1986) compared the TG technique to the standard method of evaluating the thermooxidation stability of base oils and their additive blends (e.g., transformer oils, turbine oils, transmission oils, and so forth). For each of a sample of 10 base oils, the amount of oxidative compounds formed at the oxidation point was determined using the TG technique, and the total percentage of oxidation products was determined by the standard method. The results of the experiment are shown in the accompanying table. Do the data provide sufficient evidence to indicate that the oxidation measurements of the two methods are positively correlated? Test using $\alpha = .01$.

BASE OIL	TG TECHNIQUE Amount of Oxidative Compounds, % weight	STANDARD METHOD Total Oxidation Products, %
1	25.4	2.3
2	27.11	2.5
3	28.0	2.65
4	17.9	1.3
5	18.9	1.45
6	22.9	1.9
7	30.8	3.3
8	18.6	1.4
9	24.4	2.1
10	29.8	2.9

Source: Abou El Naga, H. H. and Salem, A. E. M. "Base oils thermooxidation." *Lubrication Engineering*, Vol. 24, No. 4, Apr. 1986, p. 213. Reprinted by permission of the American Society of Lubrication Engineers. All rights reserved.

15.64 A state highway department has decided to investigate the increased severity of automobile accidents occurring at a particular urban intersection since the adoption of the right-turn-on-red law. From police records, they chose a random sample of 10 accidents that occurred at the intersection before the law was enacted, and a random sample of 10 accidents that occurred after the law was enacted. They used the total damage estimate for each accident as á measure of the accident's severity. The damage estimates are recorded in the table. Use the Wilcoxon rank sum test to determine whether the damages tended to increase after the enactment of the law. Test using $\alpha = .05$. Draw appropriate conclusions.

BEFORE RIGHT-TURN LAW	AFTER RIGHT-TURN LAW
$150	$ 145
500	390
250	680
301	560
242	899
435	1,250
100	290
402	963
716	180
200	550

15.65 Governmental agencies periodically monitor nuclear-powered electrical generating plants for the purpose of establishing baseline radiation guidelines. These guidelines then permit the detection of any changes resulting from operation that may endanger the surrounding environment. In 1978–1979, the Department of Health and Rehabilitative Services monitored three nuclear power plants in Florida for radiation in air particulates. The data shown in the table (page 914) represent mean gross beta values (pCi/m^3)—a measure of radioactive air particulates—recorded at each plant and an Orlando control site for the first 10 weeks in 1979.

a. Do the data provide sufficient evidence to indicate a difference in the radioactivity of air particulates at the four Florida sites? Test using $\alpha = .05$.

b. Do the data provide sufficient evidence to indicate a difference in the radioactivity of air particulates at the Crystal River plant site and at the Orlando control site? Test using $\alpha = .05$.

WEEK	ORLANDO	TURKEY POINT	ST. LUCIE	CRYSTAL RIVER
1	.048	.023	.023	.041
2	.019	.025	.020	.032
3	.022	.026	.022	.025
4	.015	.026	.028	.027
5	.027	.034	.031	.030
6	.122	.035	.025	.033
7	.013	.033	.020	.080
8	.007	.022	.022	.026
9	.025	.021	.013	.015
10	.025	.027	.026	.042

Source: "Monitoring of nuclear power plant environs in Florida: 1978–79," Dept. of Health and Rehabilitative Services, Health & Technical Support Services, Radiological Health Services.

15.66 A political scientist wants to determine if a voter's image of a particular liberal political candidate is correlated with the age of the voter. Each of 12 randomly selected voters rated the candidate on a scale of 1 to 20 (the higher the rating, the more favorable the candidate). The data are presented in the table.

VOTER	RATING	AGE OF VOTER
1	9	56
2	18	30
3	3	47
4	8	50
5	15	22
6	4	61
7	12	49
8	7	56
9	5	43
10	19	35
11	17	28
12	12	41

a. Calculate Spearman's rank correlation coefficient, r_s. Interpret its value in the context of this problem.

b. Does it appear that a voter's image of the liberal political candidate is correlated with the voter's age? Test using $\alpha = .01$.

15.67 Airplane pilots are trained to act quickly and decisively in the face of an emergency. Aeronautical specialists believe that by finding the most efficient arrangement of instruments on the control panel of a plane they can save the pilots precious seconds in an emergency. Two different arrangements were compared by simulating an emergency condition and then measuring the reaction time required to correct the problem. Twenty pilots were selected and randomly assigned to the two different arrangements. The following reaction times (in seconds) were measured:

Arrangement 1: 9, 4, 9, 5, 11, 4, 12, 5, 7, 6

Arrangement 2: 6, 13, 8, 11, 16, 9, 8, 10, 9, 12

Use a nonparametric test to check the specialists' claim that arrangement 1 leads to faster reaction times. Use $\alpha = .05$.

15.68 Financial analysts often use sophisticated statistical models to forecast 1-year-ahead cost earnings. *The Accounting Review* (Oct. 1985) conducted a study to compare the forecast accuracies of four models that incorporate analysts' forecasts of certain components of historical cost income (e.g., sales, depreciation, general expenses). The models were used to forecast 1-year-ahead cost earnings of each in a sample of 129 firms over the years 1977–1981.

The accompanying table summarizes the forecast accuracies of each model for the aggregate sample of firms. (Forecast accuracy is measured by mean absolute error as a percentage of sales.) Conduct a test to determine whether the distribution of 1-year-ahead forecast accuracies differ among the four models. Test using $\alpha = .10$.

YEAR	MODEL 1	MODEL 2	MODEL 3	MODEL 4
1977	.0190	.0190	.0246	.0190
1978	.0173	.0155	.0208	.0175
1979	.0190	.0196	.0218	.0196
1980	.0172	.0114	.0149	.0169.
1981	.0159	.0141	.0558	.0159

Source: Swanson, E. P., Shearon, W. T., and Thomas, L. R. "Predicting current cost operating profit using component models incorporating analysts' forecasts." *The Accounting Review*, Vol. 60, No. 4, Oct. 1985, pp. 681–691.

CASE STUDY 15.1

COST-EFFECTIVENESS OF CASE MANAGER PHYSICIANS

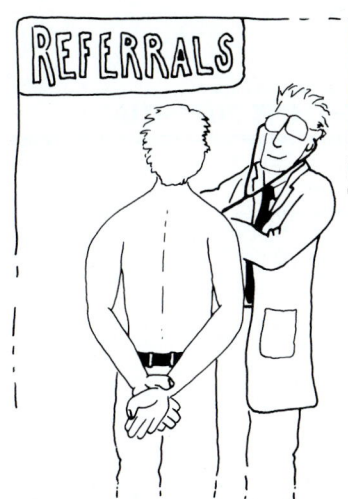

In Examples 2.3 and 2.4 we described a group of HMO physicians, The Tampa Bay Area Doctors (TBAD), who employ case management practices. A case manager, you will recall, authorizes any out-of-ordinary testing, referral to physicians other than the primary care physician, and referral to a hospital for admission. The question of interest to TBAD is whether some case managers are more cost-effective than others.

The data that will enable us to answer this question are in Appendix C, which contains the following variables measured on 186 TBAD physicians:

Primary specialty (general practice, internal medicine, pediatrics, family practice, or other)

Secondary specialty (yes or no)

Certification level (0, 1, or 2)

Gender (male or female)

Country of medical school (USA or foreign)

Country of medical residency (USA or foreign)

Years of experience

Total per-member per-month cost

Total patient-months

Several theories have been proposed concerning which physicians are the best case managers (i.e., the physicians that accrue the least total costs), including the following:

THEORY 1 Family practice physicians are more cost-effective than general internists.

THEORY 2 Primary care physicians are more cost-effective than physicians with a secondary specialty.

THEORY 3 Older (more experienced) physicians are more cost-effective than younger (less experienced) physicians.

THEORY 4 Foreign medical school graduates are less cost-effective than domestic medical school graduates.

The purpose of this case study is to determine whether the data in Appendix C support any of these four theories. [*Note:* If you have access to a statistical computer software package, then use all of the relevant data in Appendix C to answer the case study questions. (The data is available on floppy disk from the publisher.) Otherwise, use a manageable subset of the data in Appendix C (e.g., 10 measurements per sample) and conduct the analyses by hand.]

a. Construct a stem-and-leaf display (or histogram) for the data on all total physician costs in Appendix C. Discuss why it is more appropriate to analyze the data using nonparametric procedures.

b. Use a nonparametric method to test Theory 1.

c. Use a nonparametric method to test Theory 2.

d. Use a nonparametric method to test Theory 3.

e. Use a nonparametric method to test Theory 4.

CASE STUDY 15.2

REANALYZING THE IN-TRAY EXERCISE DATA FOR MEASURING MANAGEMENT POTENTIAL

In Case Study 13.2 we examined R. W. T. Gill's study* in which he used the "in-tray" or "in-basket" exercise to assess the management potential of future executives. Recall that seven candidates for a managerial position in a British motor industry were administered the in-tray test. Each candidate was given a fixed time $\left(1\frac{1}{2} \text{ hours}\right)$ to perform a variety of tasks that a manager would be expected to perform in his or her daily routine. Gill was concerned primarily with the reliability of the overall test ratings; therefore, the overall performances of the candidates were scored by each of three different raters. Ratings ranged from a score of 1 (high performance) to 6 (low performance). The results of the in-tray exercise are reproduced in Table 15.15.

In Case Study 13.2 you analyzed the data using the analysis of variance F test for a randomized block design, where the seven candidates represent the blocks and the three raters represent the treatments. The goal was to determine whether the raters are in disagreement on the candidates' overall performances.

a. Examine the sample data for the in-tray exercise. Note that the observations are ratings (or ranks) measured on a scale from 1 to 6. What does this suggest about the validity of the ANOVA F test? What nonparametric procedure is more appropriate?

b. Use a nonparametric test to determine whether the three raters are in disagreement on the candidates' overall in-tray performances. Test using $\alpha = .01$.

---

*Source: Gill, R. W. T. "The in-tray (in-basket) exercise as a measure of management potential." *Journal of Occupational Psychology*, 1979, Vol. 52, pp. 185–195.

TABLE 15.15
Ratings of Candidates by Individual Assessors
on Overall In-Tray Performance

CANDIDATE	RATER 1	RATER 2	RATER 3
A	4.5	4.5	5.0
B	2.5	4.5	4.5
C	5.0	3.0	4.0
D	4.0	4.5	4.5
E	1.5	2.0	4.5
F	3.5	4.5	4.5
G	4.0	4.0	4.0

CASE STUDY 15.3

THE PUBLIC'S PERCEPTION OF RISK: REAL OR IMAGINED?

Nuclear power continues to be one of mankind's biggest fears. The nuclear accidents at Pennsylvania's Three Mile Island and at Russia's Chernobyl, as well as the critically acclaimed television drama *The Day After*, brought into focus the dangers associated with nuclear power plants.

But is it fair to say that nuclear power is more dangerous than, say, cigarette smoking? Studies have found that antismoking campaigns can save more lives each year than nuclear power plant safety programs. To quantify the public's perception of risk, Decision Research asked three groups of people to rank 30 products or activities from most risky to least risky. The three groups who took part in the survey were the League of Women Voters, college students, and business and professional club members. Table 15.16 (page 918) gives the rankings for the three groups and, in parentheses, the number of esitmated deaths per year attributable to each activity.

Is the public's perception of risk consistent with risk based on estimated number of deaths per year? To answer this question, we can apply the nonparametric tests of this chapter.

a. Is there evidence of a positive correlation between the rankings of the League of Women Voters and the rankings of the number of deaths per year? Test using $\alpha = .05$.

b. Is there evidence of a positive correlation between the rankings of college students and the rankings of the number of deaths per year? Test using $\alpha = .05$.

c. Is there evidence of a positive correlation between the rankings of business and professional club members and the rankings of the number of deaths per year? Test using $\alpha = .05$.

d. Conduct a single nonparametric test to compare the rankings of the three groups: League of Women Voters, college students, and business and professional club members. Test using $\alpha = .05$.

TABLE 15.16
Results of Decision Research Survey

ACTIVITY AND DEATHS PER YEAR (EST.)	LEAGUE OF WOMEN VOTERS	COLLEGE STUDENTS	BUSINESS AND PROFESSIONAL CLUB MEMBERS
Smoking (150,000)	4	3	4
Alcoholic beverages (100,000)	6	7	5
Motor vehicles (50,000)	2	5	3
Handguns (17,000)	3	2	1
Electric power (14,000)	18	19	19
Motorcycles (3,000)	5	6	2
Swimming (3,000)	19	30	17
Surgery (2,800)	10	11	9
X-rays (2,300)	22	17	24
Railroads (1,950)	24	23	20
General (private) aviation (1,300)	7	15	11
Large construction (1,000)	12	14	13
Bicycles (1,000)	16	24	14
Hunting (800)	13	18	10
Home appliances (200)	29	27	27
Fire fighting (195)	11	10	6
Police work (160)	8	8	7
Contraceptives (150)	20	9	22
Commercial aviation (130)	17	16	18
Nuclear power (100)	1	1	8
Mountain climbing (30)	15	22	12
Power mowers (24)	27	28	25
High school & college football (23)	23	26	21
Skiing (18)	21	25	16
Vaccinations (10)	30	29	29
Food coloring*	26	20	30
Food preservatives*	25	12	28
Pesticides*	9	4	15
Prescription antibiotics*	28	21	26
Spray cans*	14	13	23

*Not available

Source: "What price safety? The 'zero-risk' debate." *Dun's Review*, Sept. 1979.

COMPUTER LAB
NONPARAMETRIC TESTS

In this section, we give the commands for accessing the nonparametric tests available in the three statistical software packages, SAS, SPSS, and Minitab. For illustration, we analyze the data sets used in the examples presented in this chapter.

a. **Sign test** (not available in SAS)

b. **Wilcoxon rank sum test** (data from Table 15.1)

Command
Line

```
1    DATA QC;                                        ⎫ Data entry instructions
2    INPUT IMPLEMNT $ NUMBER @@;                     ⎭
3    CARDS;
4    B  17 B  14 B  12 B  16 B  23 B  18 B  10 B  8 B  19 B  22    ⎫ Input data values
5    A  10 A  15 A   7 A   6 A  13 A  11 A  12 A  9 A  17 A  14    ⎬ (multiple observations
                                                                  ⎭  per line)
6    PROC NPAR1WAY WILCOXON;                         ⎫
7    VAR NUMBER;                                      ⎬ Wilcoxon rank sum test
8    CLASS IMPLEMNT;                                 ⎭
```

COMMAND 6 The NPAR1WAY procedure with the WILCOXON option performs a Wilcoxon rank sum test for comparing two populations.

COMMAND 7 The quantitative variable to be analyzed is specified in the VAR statement.

COMMAND 8 The quantitative variable containing the levels corresponding to the two different populations (treatments) is specified in the CLASS statement.

c. **Wilcoxon signed ranks test** (not available in SAS)

d. **Kruskal–Wallis test** (data from Table 15.8)

Command
Line

```
1    DATA CRD;                                       ⎫ Data entry instructions
2    INPUT SOCIO $ GPA @@;                           ⎭
3    CARDS;
4    L 2.87 L 2.16 L 3.14 L 2.51 L 1.80 L 3.01 L 216    ⎫ Input data values
5    M 3.23 M 3.45 M 2.78 M 3.77 M 2.97 M 3.53 M 3.01    ⎬ (multiple observations
6    U 2.25 U 3.13 U 2.44 U 3.27 U 2.81 U 1.36           ⎭  per line)
7    PROC NPAR1WAY WILCOXON;                         ⎫
8    VAR GPA;                                         ⎬ Kruskal–Wallis test
9    CLASS SOCIO;                                    ⎭
```

COMMAND 7 The WILCOXON option of NPAR1WAY is also used to conduct the Kruskal–Wallis test for a completely randomized design.

e. **Friedman test** (not available in SAS)

f. Spearman's rank correlation test (data from Table 15.12)

```
Command
 Line
   1    DATA FIRM;                        ⎫
   2    INPUT HOURS WAGES @@;            ⎬  Data entry instructions
   3    CARDS;                           ⎭
   4    49 15.8 36 17.5 127 11.3 91 13.2 72 13.0  ⎫
   .     .   .   .   .    .   .    .   .    .   .   ⎪  Input data values
   .     .   .   .   .    .   .    .   .    .   .   ⎬  (multiple observations
   .     .   .   .   .    .   .    .   .    .   .   ⎪  per line)
   8    63 13.8 79 12.7  43 15.1 57 24.2 82 13.9   ⎭
   9    PROC RANK OUT=RANKS;             
  10    VAR HOURS WAGES;                 ⎫
  11    PROC CORR DATA=RANKS;            ⎬  Spearman's rank correlation test
  12    VAR HOURS WAGES;                 ⎭
```

COMMANDS 9–10   The RANK procedure is used to rank the values of the variables specified on the VAR statement. These ranks are stored in the temporary SAS data set (RANKS) specified after OUT= and have the same names as the original variables.

COMMANDS 11–12   The CORR procedure calculates the correlation between the variables specified on the VAR statement. Since the data set RANKS is used, the correlation coefficient will equal Spearman's rank correlation coefficient.

a. Sign test (data from Example 15.2)

SPSS

```
Command
 Line
   1    DATE LIST FREE / POLLEVEL.    ⎫
   2    BEGIN DATA.                    ⎬  Data entry instructions
   3    5.1 4.3 5.3 6.2 5.6            ⎫  Input data values
   4    4.7 8.4 5.9 6.8 3.0            ⎬  (multiple observations per line)
   5    END DATA.                      
   6    COMPUTE MU=5.                  
   7    NPAR TESTS SIGN=POLLEVEL MU.   }  Sign test
```

COMMAND 6   The value of the hypothesized median is specified.

COMMAND 7   The NPAR TESTS procedure with subcommand SIGN conducts a sign test for determining whether the median of POLLEVEL differs from the constant MU.

GENERAL   Remember to omit the period at the end of each command when using SPSS on a mainframe computer.

b. Wilcoxon rank sum test (data from Table 15.1)

```
Command
  Line
    1    DATE LIST FREE / IMPLEMNT NUMBER,  ⎫  Data entry instructions
    2    BEGIN DATA,                        ⎭
    3    1 17 1 14 1 12 ...  ⎫  Input data values
    4    2 10 2 15 2 17 ...  ⎭  (multiple observations per line)
    5    END DATA,
    6    NPAR TESTS M-W=NUMBER BY IMPLEMNT(1,2),  }  Wilcoxon rank sum test
```

COMMAND 1   NUMBER is the variable of interest. IMPLEMNT is a grouping variable name that takes on the value 1 if the observation is from the first sample (before implementation), and the value 2 if the observation is from the second sample (after implementation).

COMMAND 6   The NPAR TESTS procedure with subcommand M-W conducts a Wilcoxon rank sum test for determining whether the distributions of NUMBER differ for the two groups identified by IMPLEMNT.

c. Wilcoxon signed ranks test (data from Table 15.4)

```
Command
  Line
    1    DATA LIST FREE / A B,   ⎫  Data entry instructions
    2    BEGIN DATA,             ⎭
    3    27 23 35 28           ⎫
    .     .   .   .   .         ⎪
    .     .   .   .   .         ⎬  Input data values
    .     .   .   .   .         ⎪  (multiple observations per line)
    7    18 15 17 16           ⎭
    8    END DATA,
    9    NPAR TESTS WILCOXON=A B,  }  Wilcoxon signed ranks test
```

COMMAND 1   The variables A and B contain the measurements (number of weeks of use) for each member of the matched pair.

COMMAND 9   The NPAR TESTS procedure with subcommand WILCOXON conducts a Wilcoxon signed ranks test for determining whether the distributions of A and B differ.

d. Kruskal–Wallis *H* test (data from Table 15.8)

```
Command
  Line
    1    DATA LIST FREE / SOCIO GPA,    ⎫  Data entry instructions
    2    BEGIN DATA,                    ⎭
    3    1 2,87 1 2,16 1 3,14 ...   ⎫  Input data values
    4    2 3,23 2 3,45 2 2,78 ...   ⎬  (multiple observations
    5    3 2,25 3 3,13 3 2,44 ...   ⎭  per line)
    6    END DATA,
    7    NPAR TESTS K-W=GPA BY SOCIO (1,3),  }  Kruskal–Wallis H test
```

COMMANDS 3–5   All values of the grouping variable SOCIO must be coded as numerical whole numbers (e.g., 1, 2, and 3 for lower, middle, and upper income groups, respectively).

COMMAND 7   The NPAR TESTS procedure with subcommand K–W conducts a Kruskal–Wallis H test for determining whether the distributions of GPA differ for the three classes identified by SOCIO. The numbers in parentheses represent the low and high values of the class variable (SOCIO) being analyzed.

e. **Friedman F_r test** (data from Table 15.10)

Command Line		
1	DATA LIST FREE / A B C.	Data entry instructions
2	BEGIN DATA.	
3	1 3 2	
.	. . .	Input data values
.	. . .	(one block per line)
.	. . .	
8	1 3 2	
9	END DATA	
10	NPAR TESTS FRIEDMAN=A B C.	Friedman F_r test

COMMANDS 3–8   Each row of the input data values represents one block in the experimental design.

COMMAND 10   The NPAR TESTS procedure with subcommand FRIEDMAN conducts a Friedman F_r test for determining whether the distributions of A, B, and C differ.

f. **Spearman's rank correlation test** (not available in SPSS)

a. **Sign test** (data from Example 15.2)

MINITAB

Command Line		
1	SET C1	Data entry instruction
2	5.1 4.3 5.3 6.2 5.6	Input data values (multiple
3	4.7 8.4 5.9 6.8 3.0	observations per line)
4	C1='POLLEVEL'	
5	STEST MEDIAN=5 DATA IN C1;	Sign test
6	ALTERNATIVE=+1.	

COMMAND 5   The STEST command performs a sign test on the data in the specified column. Optionally, the value of the hypothesized median (e.g., 5) is specified before the column. (If the median is not specified, the default is 0.)

COMMAND 6   The Minitab default is to conduct a two-tailed test. If a one-tailed test is desired, specify either $+1$ (upper-tailed test) or -1 (lower-tailed test) in the ALTERNATIVE subcommand.

b. Wilcoxon rank sum test (data from Table 15.1)

Command Line		
1	SET BEFORE IN C1	} Data entry instruction
2	17 14 12 ...	} Input data values (multiple observations per line)
3	SET AFTER IN C2	} Data entry instruction
4	10 15 7 ...	} Input data values (multiple observations per line)
5	NAME C1='BEFORE' C2='AFTER'	
6	MANN-WHITNEY C1 C2	} Wilcoxon rank sum test

COMMAND 6   The MANN–WHITNEY command performs a two-sample Wilcoxon rank sum test on the data specified in the two columns. [*Note:*   Use the ALTERNATIVE subcommand to perform a one-tailed test.]

c. Wilcoxon signed ranks test (data from Table 15.4)

Command Line		
1	READ C1 C2 C3	} Data entry instruction
2	1 27 23	
3	2 35 28	
.	. . .	Input data values (one observation per line)
.	. . .	
.	. . .	
11	10 17 16	
12	SUBTRACT C3 FROM C2, PUT IN C4	
13	NAME C1='JOGGER' C2='A' C3='B' C4='AminusB'	
14	WTEST C4	} Wilcoxon signed ranks test

COMMAND 12   To perform a Wilcoxon signed ranks test, first compute the differences between the two samples.

COMMAND 14   The WTEST command performs a Wilcoxon signed ranks test on the differences in the specified column. [*Note:*   Use the ALTERNATIVE subcommand to perform a one-tailed test.]

d. Kruskal–Wallis H test (data from Table 15.8)

Command Line		
1	READ C1 C2	} Data entry instruction
2	1 2.87	
3	1 2.16	
.	. .	Input data values (one observation per line)
.	. .	
.	. .	
21	3 1.36	
22	NAME C1='SOCIO' C2='GPA'	
23	KRUSKAL-WALLIS C2 C1	} Kruskal–Wallis H test

COMMAND 23   The KRUSKAL–WALLIS command performs a Kruskal–Wallis H test for a completely randomized design. The column containing the response variable (C2) is specified first, followed by the column containing the levels of the treatments (C1).

e. Friedman F_r test (data from Table 15.10)

```
Command
  Line
    1    READ C1 C2 C3  } Data entry instruction
    2     1 1 1         ⎫
    3     1 2 3         ⎪
    4     1 3 2         ⎬ Input data values
    .     . . .         ⎪ (one observation per line)
    .     . . .         ⎪
    .     . . .         ⎪
   19     6 3 2         ⎭
   20    NAME C1='TESTER' C2='BRAND' C3='RANK'
   21    FRIEDMAN C3 C1 C2                    } Friedman F_r test
```

COMMAND 21   The FRIEDMAN command performs a Friedman F_r test for a randomized block design. The column containing the response variable (C3) is specified first, followed by the treatment column (C1) and the block column (C2).

f. Spearman's rank correlation test (data from Table 15.12)

```
Command
  Line
    1    READ C1 C2  } Data entry instruction
    2    49 15.8      ⎫
    3    36 17.5      ⎪
    .     .  .        ⎬ Input data values
    .     .  .        ⎪ (one observation per line)
    .     .  .        ⎪
   16    82 13.9      ⎭
   17    NAME C1='HOURS' C2='WAGES'
   18    RANK DATA IN C1, PUT IN C3
   19    RANK DATA IN C2, PUT IN C4
   20    NAME C3='RANKHOUR' C4='RANKWAGE'      } Spearman's rank
   21    CORRELATION C3 C4                        correlation test
```

COMMANDS 18–19   The RANK command is used to rank the data in the specified columns. The second column listed will contain the ranks.

COMMAND 21   The CORRELATION command computes the correlation between the variables (columns) specified. When the ranks are specified, the result is Spearman's rank correlation coefficient.

Refer to the DDT in fish data listed in Appendix D. Recall that the U.S. Army Corps of Engineers collected DDT samples of three species of fish: channel catfish, smallmouth buffalo, and largemouth bass.

a. The Food and Drug Administration (FDA) has set the limit for DDT content of contaminated fish at 5 parts per million (ppm). For each species, use a nonparametric test in an available statistical software package to determine whether the median DDT level exceeds the FDA limit.

b. Use a nonparametric test in an available statistical software package to compare the DDT distributions for the three species.

REFERENCES

Friedman, M. "The use of ranks to avoid the assumption of normality implicit in the analysis of variance." *Journal of the American Statistical Association*, Vol. 32, 1937.

Gibbons, J. D. *Nonparametric Statistical Inference*. New York: McGraw-Hill, 1971.

Hollander, M. and Wolfe, D. A. *Nonparametric Statistical Methods*. New York: Wiley, 1973.

Kruskal, W. H. and Wallis, W. A. "Use of ranks in one-criterion variance analysis." *Journal of the American Statistical Association*, Vol. 47, 1952.

Lehmann, E. L. *Nonparametrics: Statistical Methods Based on Ranks*. San Francisco: Holden–Day, 1975.

Siegel, S. *Nonparametric Statistics for the Behavioral Sciences*. New York: McGraw-Hill, 1956.

Wilcoxon, F. and Wilcox, R. A. "Some rapid approximate statistical procedures." The American Cyanamid Co., 1964.

DATA SET: STARTING SALARIES OF UNIVERSITY OF FLORIDA GRADUATES, FALL 1989 TO SPRING 1991

Note: The following two pages present the data for the first 100 of the 1,795 total graduates in the study. The complete data set is available on $5\frac{1}{4}''$ floppy or $3\frac{1}{2}''$ micro diskette from the publisher.

OBS	DATE	GENDER	DEGREE	COLLEGE	MAJOR	JOBTYPE	SALARY
1	FALL89	F	PHD	EDUCATION	CURRICULUM	HIGHER ED	33000
2	FALL89	F	MASTERS	BUSINESS ADM	ACCOUNTING	PUBLIC AUDIT	30100
3	FALL89	M	BACHELOR	ENGINEERING	AEROSPACE		28900
4	FALL89	M	BACHELOR	ENGINEERING	CHEMICAL		28000
5	FALL89	F	MASTERS	LIB ARTS/SCIENCES	HISTORY	SECONDARY ED	25000
6	FALL89	M	MASTERS	ENGINEERING	NUCLEAR	NUCLEAR	36600
7	FALL89	M	BACHELOR	AGRICULTURE	ANIMAL SCI	ANIMAL SCI	27500
8	FALL89	M	BACHELOR	ENGINEERING	ELECTRICAL		30300
9	FALL89	M	BACHELOR	LIB ARTS/SCIENCES	SOCIOLOGY	REAL ESTATE	18000
10	FALL89	M	PHD	ENGINEERING	ENVIRONMENTAL	ENVIRONMENTAL	40100
11	FALL89	M	MASTERS	ENGINEERING	ELECTRICAL	ELECTRICAL	40000
12	FALL89	F	MASTERS	EDUCATION	EDUC ADMIN	ELEMENTARY ED	23000
13	FALL89	F	BACHELOR	ENGINEERING	COMPUTER SCI		31700
14	FALL89	M	BACHELOR	BUILDING CONSTR	BUILDING CONS	EXPEDITING	26000
15	FALL89	M	MASTERS	LIB ARTS/SCIENCES	POLITICAL SCI	PUBLIC AUDIT	46000
16	FALL89	M	BACHELOR	ENGINEERING	CIVIL	CIVIL	25000
17	FALL89	M	DOCTOR	LAW	LAW	CRIMINAL LAW	28000
18	FALL89	F	MASTERS	HEALTH	REHABILITATION	THERAPEUTIC	25000
19	FALL89	M	DOCTOR	LAW	LAW	CIVIL LAW	40000
20	FALL89	F	BACHELOR	JOURNALISM/COMM	ADVERTISING	RETAIL MANAGE	20000
21	FALL89	M	MASTERS	HEALTH	TOURISM		27000
22	FALL89	M	MASTERS	ENGINEERING	ELECTRICAL	TESTING	30400
23	FALL89	M	BACHELOR	LIB ARTS/SCIENCES	HISTORY		18000
24	FALL89	M	BACHELOR	BUSINESS ADM	FINANCE	COMMERCIAL	23000
25	FALL89	M	PHD	MEDICINE	IMM & MED MICRO	MOLECULAR BIO	18000
26	FALL89	M	BACHELOR	BUSINESS ADM	COMPUTER SCI		27600
27	FALL89	M	BACHELOR	ENGINEERING	MATERIAL SCI		32700
28	FALL89	F	MASTERS	EDUCATION	ELEMENTARY ED	ELEMENTARY ED	28300
29	FALL89	F	MASTERS	LIB ARTS/SCIENCES	SPANISH		23600
30	FALL89	F	MASTERS	BUSINESS ADM	ACCOUNTING	PUBLIC AUDIT	33600
31	FALL89	F	BACHELOR	JOURNALISM/COMM	PUBLIC RELATIONS	RETAIL MANAGE	21500
32	FALL89	M	BACHELOR	LIB ARTS/SCIENCES	CRIMINAL JUSTICE	LAW ENFORCE	25000
33	FALL89	F	BACHELOR	BUSINESS ADM	MARKETING		18000
34	FALL89	F	MASTERS	ENGINEERING	MATERIAL SCI	MATERIALS ENG	37000
35	FALL89	M	BACHELOR	ENGINEERING	MECHANICAL		38400
36	FALL89	M	PHD	AGRICULTURE	ANIMAL SCIENCE	BUSINESS/MGT	37500
37	FALL89	M	MASTERS	LIB ARTS/SCIENCES	GEOLOGY		25000
38	FALL89	F	BACHELOR	HEALTH	OCCUP THERAPY	OCCUP THERAPY	26500
39	FALL89	F	BACHELOR	LIB ARTS/SCIENCES	COMPUTER SCI	INFO SYSTEMS	26500
40	FALL89	F	BACHELOR	HEALTH	OCCUP THERAPY	OCCUP THERAPY	31000
41	FALL89	M	BACHELOR	ENGINEERING	ENVIRONMENTAL		29000
42	FALL89	M	BACHELOR	ENGINEERING	INDUS & SYSTEMS	TECH SALES	32000
43	FALL89	M	BACHELOR	ENGINEERING	ELECTRICAL	ELECTRICAL	33500
44	FALL89	M	BACHELOR	ENGINEERING	AEROSPACE		27000
45	FALL89	F	MASTERS	NURSING	NURSING	REGIST NURSE	32000
46	FALL89	F	BACHELOR	HEALTH	OCCUP THERAPY		29000
47	FALL89	M	MASTERS	ENGINEERING	CIVIL		35000
48	FALL89	M	PHD	ENGINEERING	MECHANICAL	MECHANICAL	44000
49	FALL89	M	BACHELOR	BUSINESS ADM	FINANCE		19000
50	FALL89	F	BACHELOR	FINE ARTS	MUSIC ED	ELEMENTARY ED	20900

OBS	DATE	GENDER	DEGREE	COLLEGE	MAJOR	JOBTYPE	SALARY
51	FALL89	M	MASTERS	ARCHITECTURE	BUILDING CONST	AIR FORCE	32000
52	FALL89	M	PHD	AGRICULTURE	ENTOMOLOGY	ENTOM BIOLOGY	24000
53	FALL89	F	BACHELOR	HEALTH	OCCUP THERAPY		29000
54	FALL89	F	MASTERS	LIB ARTS/SCIENCES	ENGLISH		25500
55	FALL89	M	MASTERS	ENGINEERING	CIVIL		43000
56	FALL89	F	PHD	FORESTRY	FOREST RESOURCES	WILDLIFE CONS	29900
57	FALL89	F	BACHELOR	LIB ARTS/SCIENCES	HISTORY	NON PROFIT	35300
58	FALL89	F	BACHELOR	JOURNALISM/COMM	JOURNALISM	EDITING	24200
59	FALL89	M	MASTERS	LIB ARTS/SCIENCES			12000
60	FALL89	M	MASTERS	LIB ARTS/SCIENCES	POLITICAL SCI	ARMY	50000
61	FALL89	M	BACHELOR	BUILDING CONSTR	BUILDING CONSTR	CONSTR MGT.	41000
62	FALL89	M	MASTERS	ENGINEERING	ENVIRONMENTAL	AQUATIC BIOLOGY	33000
63	FALL89	M	BACHELOR	ENGINEERING	ELECTRICAL		36300
64	FALL89	F	MASTERS	HEALTH	REHAB COUNSELING		32000
65	FALL89	F	BACHELOR	EDUCATION	SPECIAL ED		23000
66	FALL89	M	BACHELOR	BUSINESS ADM	MARKETING		25000
67	FALL89	M	MASTERS	ARCHITECTURE	ARCHITECTURE	COMMERCIAL	35000
68	FALL89	F	MASTERS	LIB ARTS/SCIENCES	CHEMISTRY	ORGANIC	22000
69	FALL89	M	BACHELOR	BUSINESS ADM	ACCOUNTING		18200
70	FALL89	M	PHD	AGRICULTURE	ANIMAL SCIENCES		8000
71	FALL89	M	BACHELOR	ENGINEERING	ELECTRICAL	ELECTRICAL	25400
72	FALL89	M	MASTERS	LIB ARTS/SCIENCES	POLITICAL SCI	ARMY	35000
73	FALL89	M	MASTERS	LIB ARTS/SCIENCES	POLITICAL SCI	ARMY	51000
74	FALL89	M	MASTERS	BUSINESS ADM	COMPUTER SCI	AIR FORCE	36000
75	FALL89	M	BACHELOR	JOURNALISM/COMM	PUBLIC RELATIONS	MARINES	24000
76	FALL89	F	BACHELOR	HEALTH	RECREATION	TOURISM	20000
77	FALL89	M	PHD	EDUCATION	HIGHER EDUCATION		42000
78	FALL89	F	BACHELOR	BUSINESS ADM	MANAGEMENT	WHOLESALE	22000
79	FALL89	F	MASTERS	ENGINEERING	ENVIRONMENTAL		19000
80	FALL89	F	BACHELOR	ENGINEERING	INDUSTRIA		32000
81	FALL89	M	BACHELOR	ENGINEERING	COMPUTER SCI	SOFTWARE DESIGN	28200
82	FALL89	M	MASTERS	NURSING	NURSING	REG NURSE	44200
83	FALL89	F	PHD	ENGINEERING	ENVIRONMENTAL		36400
84	FALL89	F	PHD	AGRICULTURE	BOTANY	MOLECULAR BIO	21100
85	FALL89	M	BACHELOR	LIB ARTS/SCIENCES	POLITICAL SCI	CRIMINAL LAW	15000
86	FALL89	F	BACHELOR	NURSING	NURSING	REG NURSE	27000
87	FALL89	M	MASTERS	ENGINEERING	ENG SCIENCE	MECHANICAL	39900
88	FALL89	M	BACHELOR	BUSINESS ADM	ACCOUNTING	PUBLIC AUDIT	30000
89	FALL89	F	BACHELOR	BUSINESS ADM	MARKETING		18800
90	FALL89	F	BACHELOR	BUSINESS ADM	MARKETING		21500
91	FALL89	M	BACHELOR	BUILDING CONSTR	BUILDING CONSTR		23000
92	FALL89	F	MASTERS	EDUCATION	SPECIAL ED		22000
93	FALL89	F	BACHELOR	ENGINEERING	ENVIRONMENTAL	ENVIRONMENTAL	25500
94	FALL89	M	BACHELOR	BUSINESS ADM	ACCOUNTING	PUBLIC TAX	25000
95	FALL89	F	SPECIAL	LAW	TAXATION	CIVIL LAW	60000
96	FALL89	F	MASTERS	EDUCATION	SCIENCE ED	SECONDARY ED	30000
97	FALL89	F	BACHELOR	LIB ARTS/SCIENCES	CRIMINAL JUSTICE		13500
98	FALL89	M	BACHELOR	BUSINESS ADM	FINANCE	BANK/FINANCE	23600
99	FALL89	F	BACHELOR	BUSINESS ADM	MARKETING	RETAIL MANAGE	19000
100	FALL89	M	BACHELOR	ENGINEERING	ELECTRICAL		30600

DATA SET: STARTING SALARIES OF UNIVERSITY OF FLORIDA BACHELOR'S DEGREE GRADUATES IN FIVE COLLEGES

Note: The following pages present information for the first 100 graduates in each college. The complete data set is available on $5\frac{1}{4}''$ floppy or $3\frac{1}{2}''$ micro diskette from the publisher.

OBS	GENDER	MAJOR	JOBTYPE	SALARY
1	M	FINANCE	BANKING/FINANCE COMMERCIAL	23000
2	M	COMPUTER & INFORMATION SCIENCE		27600
3	F	MARKETING		18000
4	M	FINANCE		19000
5	M	MARKETING		25000
6	M	ACCOUNTING		18200
7	F	MANAGEMENT	WHOLESALE SALES	22000
8	M	ACCOUNTING	PUBLIC AUDIT	30000
9	F	MARKETING		18800
10	F	MARKETING		21500
11	M	ACCOUNTING	PUBLIC TAX	25000
12	M	FINANCE	BANKING/FINANCE	23600
13	F	MARKETING	RETAIL MANAGEMENT	19000
14	F	FINANCE	CONSULTANT(MIS)	30000
15	M	FINANCE	BANKING/FINANCE COMMERCIAL	24000
16	M	ACCOUNTING		24500
17	F	FINANCE	BANKING/FINANCE COMMERCIAL	24500
18	F	FINANCE	BANKING/FINANCE CORPORATE	25800
19	F	MARKETING		19000
20	M	FINANCE	RETAIL SALES	26500
21	M	MARKETING		12500
22	F	MANAGEMENT	RETAIL SALES	19000
23	F	FINANCE	BANKING/FINANCE COMMERCIAL	22500
24	F	MARKETING	PURCHASING	13000
25	M	FINANCE		23000
26	M	FINANCE	INVESTMENT BROKERING	25000
27	M	ACCOUNTING	PUBLIC AUDIT	28000
28	M	MARKETING	INSURANCE CLAIMS	23600
29	F	FINANCE	MANAGEMENT TRAINEE (ALL)	19500
30	M	FINANCE	BANKING/FINANCE COMMERCIAL	25000
31	M	ACCOUNTING	PUBLIC AUDIT	27500
32	M	ACCOUNTING		23500
33	M	ACCOUNTING	PUBLIC TAX	28500
34	M	COMPUTER & INFORMATION SCIENCE	COMPUTER INFORMATION SYSTEMS	27000
35	M	ACCOUNTING	PUBLIC AUDIT	27000
36	M	MARKETING	RETAIL SALES	23500
37	M	FINANCE	BANKING/FINANCE	23000
38	F	FINANCE		23000
39	F	FINANCE	BANKING/FINANCE CONSUMER (RET)	23000
40	M	MARKETING		23000
41	M	MANAGEMENT		32000
42	F	MARKETING		20000
43	M	FINANCE	RETAIL SALES	19500
44	M	FINANCE	CONSULTING	26000
45	M	FINANCE		22500
46	F	FINANCE	RETAIL SALES	20000
47	M	MARKETING		21000
48	F	FINANCE	BANKING/FINANCE CONSUMER (RET)	23000
49	F	FINANCE	COST	28500
50	M	ACCOUNTING	AUDIT	25000
51	F	ACCOUNTING	INDUSTRIAL MANAGERIAL	25500
52	M	ACCOUNTING		20000
53	F	MARKETING	MANAGEMENT TRAINEE (ALL)	26000
54	M	ACCOUNTING		28500
55	M	ACCOUNTING	PUBLIC AUDIT	29000
56	F	COMPUTER & INFORMATION SCIENCE	COMPUTER SYSTEMS ENGINEERING	27600
57	M	ACCOUNTING	PUBLIC AUDIT	28500
58	F	ACCOUNTING		23000

OBS	GENDER	MAJOR	JOBTYPE	SALARY
59	M	ACCOUNTING	ACCOUNTING	43000
60	M	MARKETING		25025
61	M	FINANCE	BANKING/FINANCE CORPORATE	26000
62	M	MARKETING	NON-TECHNICAL SALES	18000
63	F	FINANCE	BANKING/FINANCE CORPORATE	26000
64	M	FINANCE	INSURANCE ACTUARIAL	25000
65	M	FINANCE	COMPUTER INFORMATION SYSTEMS	29000
66	M	ACCOUNTING		28000
67	M	MANAGEMENT	MANAGEMENT TRAINEE (ALL)	20800
68	F	FINANCE	BANKING/FINANCE	22500
69	F	FINANCE	RETAIL MANAGEMENT	20000
70	M	FINANCE		22000
71	M	FINANCE		22500
72	F	MARKETING		23000
73	M	FINANCE		25900
74	M	COMPUTER & INFORMATION SCIENCE	COMPUTER PROGRAMMING	25000
75	M	FINANCE	BANKING/FINANCE CORPORATE	22000
76	M	MANAGEMENT	CONSTRUCTION MGT.	22000
77	M	FINANCE	SALES	19000
78	M	FINANCE	BANKING/FINANCE CORPORATE	22000
79	M	FINANCE	BANKING/FINANCE MORTGAGE	35000
80	M	MANAGEMENT		15000
81	F	MARKETING		28800
82	M	ACCOUNTING		29000
83	F	ACCOUNTING	AUDIT	29500
84	M	MANAGEMENT	AIR FORCE	21000
85	F	ACCOUNTING		32500
86	F	ACCOUNTING		28000
87	F	FINANCE	RETAIL MANAGEMENT	20000
88	M	MANAGEMENT		21500
89	M	ACCOUNTING	PUBLIC AUDIT	22000
90	M	ACCOUNTING		29000
91	F	MARKETING	BUYING/MERCHANDISING (RETAIL)	40000
92	M	MARKETING	RETAIL MANAGEMENT	21000
93	M	REAL ESTATE	CONSULTING	40000
94	M	FINANCE		24000
95	F	ACCOUNTING	PUBLIC AUDIT	29000
96	M	MANAGEMENT		20000
97	M	FINANCE		22000
98	F	FINANCE		23000
99	F	ACCOUNTING		28000
100	M	MARKETING	INDUSTRIAL SALES	13000

OBS	GENDER	MAJOR	JOBTYPE	SALARY
323	M	AEROSPACE ENGINEERING		28900
324	M	CHEMICAL ENGINEERING		28000
325	M	ELECTRICAL ENGINEERING		30300
326	F	COMPUTER & INFORMATION SCIENCE		31700
327	M	CIVIL ENGINEERING	CIVIL ENGINEERING	25000
328	M	MATERIAL SCIENCE & ENGINEER		32700
329	M	MECHANICAL ENGINEERING		38400
330	M	ENVIRONMENTAL ENGINEERING		29000
331	M	INDUSTRIAL & SYSTEMS ENGINEER	TECHNICAL SALES	32000
332	M	ELECTRICAL ENGINEERING	ELECTRICAL ENGINEERING	33500

OBS	GENDER	MAJOR	JOBTYPE	SALARY
333	M	AEROSPACE ENGINEERING		27000
334	M	ELECTRICAL ENGINEERING		36300
335	M	ELECTRICAL ENGINEERING	ELECTRICAL ENGINEERING	25400
336	F	INDUSTRIAL & SYSTEMS ENGINEER		32000
337	M	COMPUTER & INFORMATION SCIENCE	COMPUTER SOFTWARE DESIGN	28200
338	F	ENVIRONMENTAL ENGINEERING	ENVIRONMENTAL ENGINEERING	25500
339	M	ELECTRICAL ENGINEERING		30600
340	M	CIVIL ENGINEERING	CONSULTING ENGINEERING	28000
341	M	ELECTRICAL ENGINEERING		27100
342	M	CIVIL ENGINEERING	CIVIL ENGINEERING	26400
343	M	CIVIL ENGINEERING	CIVIL ENGINEERING	24500
344	M	COMPUTER & INFORMATION SCIENCE		30100
345	M	ELECTRICAL ENGINEERING	ELECTRICAL ENGINEERING	32000
346	M	MECHANICAL ENGINEERING	MECHANICAL ENGINEERING	30600
347	M	MECHANICAL ENGINEERING		31800
348	M	CIVIL ENGINEERING		28400
349	M	ELECTRICAL ENGINEERING	COMPUTER ENGINEERING	31700
350	M	COMPUTER & INFORMATION SCIENCE		32400
351	M	MECHANICAL ENGINEERING		30000
352	M	MECHANICAL ENGINEERING	MANUFACTURING ENGINEERING	34300
353	F	INDUSTRIAL & SYSTEMS ENGINEER	INDUSTRIAL/SYSTEMS ENGIN.	26000
354	M	NUCLEAR ENGINEERING	NUCLEAR ENGINEERING	31000
355	F	COMPUTER & INFORMATION SCIENCE	COMPUTER SOFTWARE ENGINEERING	32300
356	M	CHEMICAL ENGINEERING	CHEMICAL ENGINEERING	36000
357	M	CIVIL ENGINEERING		22000
358	M	AGRICULTURAL ENGINEERING	AGRICULTURAL ENGINEERING	35000
359	M	MECHANICAL ENGINEERING	MECHANICAL ENGINEERING	32600
360	M	ELECTRICAL ENGINEERING	ELECTRICAL ENGINEERING	25300
361	M	MECHANICAL ENGINEERING	MECHANICAL ENGINEERING	23000
362	F	ELECTRICAL ENGINEERING		35400
363	M	ELECTRICAL ENGINEERING	COMPUTER SOFTWARE ENGINEERING	33600
364	M	MECHANICAL ENGINEERING		25500
365	M	COMPUTER & INFORMATION SCIENCE		30800
366	M	MECHANICAL ENGINEERING		36300
367	F	ELECTRICAL ENGINEERING		33600
368	M	MECHANICAL ENGINEERING	MECHANICAL ENGINEERING	32600
369	M	INDUSTRIAL & SYSTEMS ENGINEER		32000
370	F	ENVIRONMENTAL ENGINEERING		30600
371	M	MECHANICAL ENGINEERING		37200
372	M	MECHANICAL ENGINEERING		34000
373	M	INDUSTRIAL & SYSTEMS ENGINEER	SALES ENGINEERING	33000
374	M	INDUSTRIAL & SYSTEMS ENGINEER		20000
375	M	COMPUTER & INFORMATION SCIENCE	COMPUTER SOFTWARE ENGINEERING	31700
376	F	AEROSPACE ENGINEERING	AEROSPACE ENGINEERING	29500
377	F	COMPUTER & INFORMATION SCIENCE		32000
378	M	MECHANICAL ENGINEERING		35000
379	M	AGRICULTURAL ENGINEERING	CONSULTING ENGINEERING	34000
380	F	COMPUTER & INFORMATION SCIENCE		8500
381	M	ELECTRICAL ENGINEERING		33600
382	M	ELECTRICAL ENGINEERING		33500
383	M	ELECTRICAL ENGINEERING	ELECTRICAL ENGINEERING	32000
384	M	MECHANICAL ENGINEERING		31000
385	F	INDUSTRIAL & SYSTEMS ENGINEER		26800
386	F	ELECTRICAL ENGINEERING	ELECTRICAL ENGINEERING	25500
387	M	ELECTRICAL ENGINEERING		26500
388	F	INDUSTRIAL & SYSTEMS ENGINEER		32100
389	F	MECHANICAL ENGINEERING	MECHANICAL ENGINEERING	35500

```
-------------------------------- COLLEGE=ENGINEERING --------------------------------
```

OBS	GENDER	MAJOR	JOBTYPE	SALARY
390	M	ELECTRICAL ENGINEERING	ELECTRICAL ENGINEERING	44000
391	M	ELECTRICAL ENGINEERING		31200
392	M	ELECTRICAL ENGINEERING	ELECTRICAL ENGINEERING	32000
393	F	MATERIAL SCIENCE & ENGINEER		34000
394	M	MECHANICAL ENGINEERING	MECHANICAL ENGINEERING	28300
395	M	LAND SURVEYING	CIVIL ENGINEERING	21100
396	F	AEROSPACE ENGINEERING	AEROSPACE ENGINEERING	28500
397	M	LAND SURVEYING	SURVEYING AND CARTOGRAPHY	27400
398	M	CHEMICAL ENGINEERING	CHEMICAL ENGINEERING	36600
399	M	CIVIL ENGINEERING	CIVIL ENGINEERING	28000
400	M	ELECTRICAL ENGINEERING		38500
401	F	MECHANICAL ENGINEERING	NUCLEAR ENGINEERING	36200
402	M	AEROSPACE ENGINEERING		31800
403	M	MECHANICAL ENGINEERING	MECHANICAL ENGINEERING	32200
404	M	ELECTRICAL ENGINEERING		33000
405	M	MECHANICAL ENGINEERING		25600
406	M	NUCLEAR ENGINEERING		33000
407	M	COMPUTER & INFORMATION SCIENCE	COMPUTER SOFTWARE DESIGN	33000
408	M	MECHANICAL ENGINEERING	MECHANICAL ENGINEERING	30000
409	M	MECHANICAL ENGINEERING		33200
410	M	LAND SURVEYING		29100
411	M	MATERIAL SCIENCE & ENGINEER		30600
412	M	CIVIL ENGINEERING	CIVIL ENGINEERING	25000
413	M	CHEMICAL ENGINEERING		35000
414	M	AEROSPACE ENGINEERING	AEROSPACE ENGINEERING	32000
415	F	INDUSTRIAL & SYSTEMS ENGINEER	CONSULTING ENGINEERING	30000
416	M	ELECTRICAL ENGINEERING	SOCIAL WORK	12000
417	M	ELECTRICAL ENGINEERING		26300
418	M	MECHANICAL ENGINEERING		31000
419	M	CIVIL ENGINEERING	CIVIL ENGINEERING	26400
420	M	COMPUTER & INFORMATION SCIENCE		31500
421	M	MATERIAL SCIENCE & ENGINEER		32400
422	M	MECHANICAL ENGINEERING	MANUFACTURING/PRODUCTION MGT.	33600

```
-------------------------------- COLLEGE=JOURNALISM/COMM --------------------------------
```

OBS	GENDER	MAJOR	JOBTYPE	SALARY
604	F	ADVERTISING	RETAIL MANAGEMENT	20000
605	F	PUBLIC RELATIONS	RETAIL MANAGEMENT	21500
606	F	JOURNALISM	JOURNALISM EDITING	24200
607	M	PUBLIC RELATIONS	MARINES	24000
608	M	TELECOMMUNICATIONS	COMPUTER INFORMATION SYSTEMS	29000
609	F	JOURNALISM	REPORTING JOURNALISM	18700
610	F	ADVERTISING		16000
611	M	JOURNALISM		17400
612	M	PUBLIC RELATIONS		25000
613	M	TELECOMMUNICATIONS	PUBLIC RELATIONS	15000
614	M	JOURNALISM	REPORTING JOURNALISM	27000
615	M	JOURNALISM		14400
616	M	PUBLIC RELATIONS	MANAGEMENT TRAINEE (ALL)	22000
617	F	PUBLIC RELATIONS		22100
618	F	PUBLIC RELATIONS	TECHNICAL WRITING/EDITING	17000
619	F	JOURNALISM	REPORTING JOURNALISM	26000
620	F	ADVERTISING		25000
621	F	JOURNALISM	TELEVISION	23000

OBS	GENDER	MAJOR	JOBTYPE	SALARY
622	F	ADVERTISING	ADVERTISING	17000
623	M	ADVERTISING	ADVERTISING ACCOUNT EXECUTIVE	25200
624	M	TELECOMMUNICATIONS	BROADCASTING/TELECOMMS.	20000
625	F	TELECOMMUNICATIONS		18000
626	F	JOURNALISM		16000
627	F	TELECOMMUNICATIONS	MANAGEMENT TRAINEE (ALL)	20000
628	F	PUBLIC RELATIONS		17000
629	F	ADVERTISING	NON-TECHNICAL SALES	24200
630	F	ADVERTISING	RETAIL MANAGEMENT	20000
631	F	PUBLIC RELATIONS	PUBLIC RELATIONS	20200
632	F	TELECOMMUNICATIONS		16500
633	F	JOURNALISM		38000
634	F	PUBLIC RELATIONS	HUMAN RESOURCES DEVELOPMENT	17000
635	F	ADVERTISING		10800
636	F	PUBLIC RELATIONS		15000
637	F	ADVERTISING	MANAGEMENT TRAINEE (ALL)	20000
638	F	ADVERTISING	LAYOUT PUBLISHING	8400
639	F	ADVERTISING		24000
640	M	TELECOMMUNICATIONS	TELEVISION	12500
641	F	PUBLIC RELATIONS	TECHNICAL SALES	24200
642	F	PUBLIC RELATIONS	RETAIL MANAGEMENT	20000
643	M	TELECOMMUNICATIONS	BROADCASTING/TELECOMMS.	16000
644	F	ADVERTISING	BUYING/MERCHANDISING (RETAIL)	20000
645	F	JOURNALISM		26000
646	M	PUBLIC RELATIONS		20500
647	F	JOURNALISM	INSURANCE CLAIMS	24000
648	F	ADVERTISING	SALES	26500
649	F	TELECOMMUNICATIONS	BROADCASTING/TELECOMMS.	12000
650	F	ADVERTISING	MANAGERIAL	34000
651	F	ADVERTISING	ADVERTISING GRAPHIC DESIGN	10000
652	F	PUBLIC RELATIONS	PUBLIC RELATIONS	20000
653	F	PUBLIC RELATIONS	CONSULTING	15000
654	F	JOURNALISM	EDITING PUBLISHING	16000
655	F	JOURNALISM	RETAIL MANAGEMENT	19000
656	M	TELECOMMUNICATIONS	BROADCASTING/TELECOMMS.	18000
657	F	ADVERTISING		21000
658	F	TELECOMMUNICATIONS		16500
659	M	TELECOMMUNICATIONS	BROADCASTING/TELECOMMS.	22000
660	M	ADVERTISING		20000
661	F	ADVERTISING	RETAIL MANAGEMENT	21000
662	F	PUBLIC RELATIONS	PUBLIC RELATIONS	25000
663	F	PUBLIC RELATIONS	RETAIL SALES	20000
664	F	ADVERTISING	NON PROFIT ORGAN. ADMIN.	17500
665	M	ADVERTISING	CIVIL LAW	10400
666	M	ADVERTISING	ADVERTISING	17000
667	F	ADVERTISING	ADVERTISING	30000
668	F	PUBLIC RELATIONS	INDUSTRIAL SALES	26000
669	F	ADVERTISING	ADVERTISING SALES	13000
670	F	TELECOMMUNICATIONS	BROADCASTING/TELECOMMS.	9000
671	M	TELECOMMUNICATIONS	SECONDARY EDUCATION	27000
672	F	TELECOMMUNICATIONS	BROADCASTING/TELECOMMS.	15400
673	F	ADVERTISING	RETAIL MANAGEMENT	21000
674	F	JOURNALISM		19500
675	F	JOURNALISM		20000
676	F	ADVERTISING	ADVERTISING ACCOUNT EXECUTIVE	20000

OBS	GENDER	MAJOR	JOBTYPE	SALARY
677	M	SOCIOLOGY	REAL ESTATE PROPERTY MGT	18000
678	M	HISTORY		18000
679	M	CRIMINAL JUSTICE	LAW ENFORCEMENT AND JUSTICE	25000
680	F	COMPUTER & INFORMATION SCIENCE	COMPUTER INFORMATION SYSTEMS	26500
681	F	HISTORY	NON PROFIT ORGAN. ADMIN.	35300
682	M	POLITICAL SCIENCE	CRIMINAL LAW	15000
683	F	CRIMINAL JUSTICE		13500
684	M	SOCIOLOGY		27500
685	F	ENGLISH	PUBLIC RELATIONS	15000
686	M	COMPUTER & INFORMATION SCIENCE	COMPUTER INFORMATION SYSTEMS	27000
687	M	SOCIOLOGY	INVESTMENT BANKING	20000
688	M	POLITICAL SCIENCE	LAW ENFORCEMENT AND JUSTICE	17000
689	M	COMPUTER & INFORMATION SCIENCE	COMPUTER SOFTWARE DESIGN	32000
690	F	SOCIOLOGY	RETAIL MANAGEMENT	20000
691	F	MATHEMATICS		23800
692	F	CLASSICAL STUDIES	REAL ESTATE BROKERING (SALES)	22000
693	F	SPEECH THERAPY		19500
694	F	ZOOLOGY		11000
695	M	HISTORY	ARMY	22100
696	M	HISTORY	MARINES	21600
697	F	PSYCHOLOGY	OPERATIONS RESEARCH	14600
698	F	ENGLISH	TECHNICAL WRITING/EDITING	24000
699	M	MICROBIOLOGY		19000
700	F	POLITICAL SCIENCE		17000
701	M	POLITICAL SCIENCE	CIVIL LAW	21000
702	M	CRIMINAL JUSTICE	LAW ENFORCEMENT AND JUSTICE	24000
703	F	MATHEMATICS	COMPUTER SYSTEMS ENGINEERING	30200
704	M	ECONOMICS		21000
705	F	POLITICAL SCIENCE		20000
706	F	HISTORY		19000
707	F	HISTORY	SECONDARY EDUCATION	9000
708	M	ENGLISH	BANKING/FINANCE COMMERCIAL	24000
709	M	ECONOMICS		21000
710	F	SOCIOLOGY	BANKING/FINANCE	15600
711	F	SPEECH	SECONDARY EDUCATION	12000
712	M	ECONOMICS	INSURANCE CLAIMS	22500
713	F	SPEECH		22500
714	F	MATHEMATICS	AIR FORCE	22000
715	M	HISTORY		20000
716	M	POLITICAL SCIENCE		16300
717	M	HISTORY		19200
718	F	FRENCH	SECONDARY EDUCATION	20000
719	M	GEOGRAPHY	LAW ENFORCEMENT AND JUSTICE	24000
720	F	SPEECH		18000
721	F	ENGLISH		22000
722	M	SOCIOLOGY		25500
723	M	MATHEMATICS	ANALYTICAL CHEMISTRY	24000
724	F	ECONOMICS	RETAIL SALES	20000
725	M	ENGLISH	RETAIL MANAGEMENT	20000
726	M	POLITICAL SCIENCE	ARMY	23000
727	M	SPEECH	MANAGEMENT TRAINEE (ALL)	15000
728	F	BOTANY	RETAIL SALES	13000
729	M	POLITICAL SCIENCE	LAW ENFORCEMENT AND JUSTICE	20000
730	F	PSYCHOLOGY	RETAIL MANAGEMENT	20000
731	M	COMPUTER & INFORMATION SCIENCE		30000
732	M	INTERDISCIPLINARY STUDIES	MUSEUM/GALLERY SERVICES	15000
733	M	ECONOMICS	RETAIL MANAGEMENT	17000

```
------------------------- COLLEGE=LIB ARTS/SCIENCES -------------------------

OBS GENDER MAJOR                          JOBTYPE                         SALARY

734   F   PSYCHOLOGY                                                       20000
735   M   POLITICAL SCIENCE             AIR FORCE                          25000
736   F   PSYCHOLOGY                   RETAIL MANAGEMENT                   21000
737   F   ENGLISH                      PUBLIC RELATIONS                    17000
738   F   PSYCHOLOGY                                                       13800
739   M   CRIMINAL JUSTICE             LAW ENFORCEMENT AND JUSTICE         32000
740   M   CHEMISTRY                                                        17500
741   M   HISTORY                                                          25000
742   M   ECONOMICS                    REAL ESTATE APPRAISING              19000
743   F   PSYCHOLOGY                   RETAIL MANAGEMENT                   10000
744   F   INTERDISCIPLINARY STUDIES    FILM                               12000
745   M   CHEMISTRY                    ANALYTICAL CHEMISTRY                21000
746   M   ENGLISH                      CONSULTING                         20000
747   F   PSYCHOLOGY                                                       26500
748   F   PSYCHOLOGY                   COUNSELING EDUCATION                15000
749   F   PSYCHOLOGY                   COUNSELING EDUCATION                11200
750   M   HISTORY                                                          20000
751   M   HISTORY                                                          12000
752   F   POLITICAL SCIENCE                                                24000
753   M   MATHEMATICS                  INSURANCE ACTUARIAL                 28000
754   M   ECONOMICS                    MANAGEMENT TRAINEE (ALL)            18000
755   M   ECONOMICS                                                        24000
756   M   RELIGION                     RETAIL MANAGEMENT                   19000
757   M   RUSSIAN                      ARMY                               22300
758   F   PSYCHOLOGY                                                       13500
759   M   HISTORY                                                          15000
760   F   SPEECH                                                           39500
761   F   POLITICAL SCIENCE            INSURANCE CLAIMS                    27500
762   F   CHEMISTRY                                                        29500
763   F   SPANISH                                                          23700
764   F   ENGLISH                                                          21700
765   M   ANTHROPOLOGY                                                     18000
766   M   ECONOMICS                    INDUSTRIAL SALES                    20000
767   F   PSYCHOLOGY                   EARLY CHILDHOOD EDUCATION            9000
768   F   ENGLISH                      HUMAN RESOURCES DEVELOPMENT         11500
769   M   PSYCHOLOGY                   NAVY                               21500
770   F   SOCIOLOGY                                                        21000
771   F                                                                   30000
772   M   ENGLISH                      ARMY                               22300
773   M   CRIMINAL JUSTICE                                                 17400
774   M   SOCIOLOGY                    FILM                               20000
775   M   POLITICAL SCIENCE                                                31000
776   M   POLITICAL SCIENCE            NAVY                               22000


------------------------------ COLLEGE=NURSING ------------------------------

OBS GENDER  MAJOR                         JOBTYPE                         SALARY

826   F   NURSING                      REGISTERED NURSE                   27000
827   F   NURSING                      REGISTERED NURSE                   32000
828   F   NURSING                                                         30800
829   F   NURSING                      REGISTERED NURSE                   24000
830   F   NURSING                      REGISTERED NURSE                   24700
831   F   NURSING                                                         44000
832   F   NURSING                      REGISTERED NURSE                   28000
833   F   NURSING                                                         29900
834   F   NURSING                      REGISTERED NURSE                   27300
```

APPENDIX B   DATA SET: STARTING SALARIES OF GRADUATES IN FIVE COLLEGES

```
------------------------------- COLLEGE=NURSING -------------------------------
```

OBS	GENDER	MAJOR	JOBTYPE	SALARY
835	F	NURSING	REGISTERED NURSE	25000
836	F	NURSING	REGISTERED NURSE	27000
837	F	NURSING	REGISTERED NURSE	24600
838	F	NURSING		25000
839	F	NURSING		31200
840	M	NURSING	REGISTERED NURSE	28000
841	F	NURSING		27000
842	F	NURSING	REGISTERED NURSE	27900
843	F	NURSING	REGISTERED NURSE	30000
844	F	NURSING	REGISTERED NURSE	30000
845	F	NURSING	REGISTERED NURSE	33400
846	F	NURSING	REGISTERED NURSE	26000
847	F	NURSING		27000
848	F	NURSING	REGISTERED NURSE	26000
849	F	NURSING	REGISTERED NURSE	30600
850	F	NURSING	REGISTERED NURSE	22700
851	F	NURSING		24100
852	F	NURSING	REGISTERED NURSE	27900
853	F	NURSING		32000
854	M	NURSING	REGISTERED NURSE	32000
855	F	NURSING		25000
856	F	NURSING	REGISTERED NURSE	25000
857	F	NURSING	REGISTERED NURSE	30000
858	F	NURSING	REGISTERED NURSE	25000
859	M	NURSING	REGISTERED NURSE	25700
860	F	NURSING		20000
861	F	NURSING		22000
862	F	NURSING		31600
863	F	NURSING		24000
864	F	NURSING	REGISTERED NURSE	22000
865	F	NURSING	REGISTERED NURSE	24000
866	F	NURSING		37500
867	F	NURSING		40000
868	F	NURSING	REGISTERED NURSE	35300
869	F	NURSING	REGISTERED NURSE	33000
870	F	NURSING		24000
871	F	NURSING		25000
872	F	NURSING	REGISTERED NURSE	27500
873	M	NURSING	REGISTERED NURSE	28000
874	F	NURSING	REGISTERED NURSE	25000
875	F	NURSING	REGISTERED NURSE	31000
876	M	NURSING	REGISTERED NURSE	51500
877	F	NURSING	REGISTERED NURSE	29100
878	F	NURSING	REGISTERED NURSE	25000
879	F	NURSING		23000
880	F	NURSING	AIR FORCE	22000
881	F	NURSING	REGISTERED NURSE	25000
882	F	NURSING		28000
883	F	NURSING	REGISTERED NURSE	30000
884	F	NURSING	REGISTERED NURSE	24000
885	F	NURSING	REGISTERED NURSE	35000
886	F	NURSING	REGISTERED NURSE	36000
887	F	NURSING	REGISTERED NURSE	26000
888	F	NURSING		26000
889	F	NURSING		27000
890	F	NURSING	REGISTERED NURSE	38000
891	F	NURSING	REGISTERED NURSE	32000
892	F	NURSING	REGISTERED NURSE	23000

```
-------------------------------- COLLEGE=NURSING --------------------------------

         OBS    GENDER    MAJOR     JOBTYPE                              SALARY

         893      F      NURSING    REGISTERED NURSE                     26000
         894      F      NURSING    REGISTERED NURSE                     26000
         895      F      NURSING    REGISTERED NURSE                     26000
         896      F      NURSING                                         26000
         897      F      NURSING    REGISTERED NURSE                     30000
         898      F      NURSING    REGISTERED NURSE                     30000
         899      F      NURSING    REGISTERED NURSE                     24000
         900      F      NURSING    REGISTERED NURSE                     27000
         901      F      NURSING    REGISTERED NURSE                     25000
         902      F      NURSING    REGISTERED NURSE                     21000
```

DATA SET: CHARACTERISTICS OF HMO PHYSICIANS IN A MANAGED-CARE SYSTEM

APPENDIX C

Note: The complete data set is available on $5\frac{1}{4}''$ floppy or $3\frac{1}{2}''$ micro diskette from the publisher.

PRIMSPEC	OBS	SECSPEC	CERTIF	GENDER	MEDSCHL	MEDRESID	EXP	COST	MBMNTH
FAMILY	1	N	1	M	USA	FOR	22	20.5	88
	2	N	1	M	USA	USA	24	1.8	38
	3	N	1	M	USA	FOR	30	71.3	507
	4	N	1	F	FOR	FOR	0	69.7	123
	5	N	1	M	USA	USA	8	67.0	563
	6	N	2	M	FOR	USA	45	45.1	958
	7	N	1	M	FOR	FOR	28	48.9	3047
	8	N	1	M	USA	USA	16	66.8	679
	9	N	2	M	FOR	USA	26	9.8	199
	10	N	1	M	FOR	USA	0	74.1	515
	11	N	1	M	FOR	FOR	38	154.3	1255
	12	N	1	M	USA	FOR	33	71.8	1835
	13	N	1	M	USA	USA	19	242.0	18
	14	N	1	M	USA	USA	22	61.7	5459
	15	N	2	M	USA	FOR	31	45.5	1345
	16	N	2	M	FOR	FOR	27	122.5	654
	17	N	1	M	USA	USA	18	54.5	568
	18	N	1	M	USA	USA	9	39.0	898
	19	N	1	M	FOR	FOR	0	69.3	1237
	20	N	1	M	USA	USA	8	42.9	1771
	21	N	1	M	FOR	USA	0	80.8	431
	22	N	1	M	USA	USA	29	140.3	1049
	23	N	1	M	USA	USA	11	34.9	176
	24	N	1	M	FOR	FOR	12	431.2	36
	25	N	1	M	USA	USA	34	18.9	565
	26	N	1	M	USA	FOR	25	44.0	6716
	27	N	1	M	USA	FOR	35	24.1	471
	28	N	1	M	USA	USA	19	56.0	1141
	29	N	1	M	FOR	USA	12	11.1	152
	30	N	1	M	FOR	FOR	39	135.3	70
	31	N	1	M	FOR	USA	13	35.8	845
	32	N	0	M	USA	FOR	11	47.7	1412
	33	N	1	M	USA	USA	11	57.6	2648
	34	N	1	M	USA	USA	12	32.7	320
	35	N	1	M	USA	FOR	29	64.1	2646
	36	N	1	M	USA	USA	13	31.0	1413
	37	N	0	M	FOR	FOR	22	46.1	181
	38	N	1	M	USA	FOR	40	52.7	958
	39	N	1	M	USA	FOR	30	32.7	4
	40	N	1	M	USA	USA	12	30.4	87
	41	N	2	M	FOR	FOR	43	51.3	1969
	42	N	1	M	USA	USA	12	14.8	1
	43	N	1	M	USA	USA	42	18.4	220
	44	N	1	M	FOR	USA	0	52.2	2359
	45	N	2	M	USA	USA	21	52.5	1265
	46	N	1	M	USA	FOR	35	24.5	56Q
	47	N	1	M	FOR	USA	20	27.2	128
	48	N	0	F	USA	USA	6	78.6	1702
	49	N	1	M	USA	FOR	9	42.6	928
	50	Y	1	M	USA	USA	31	105.7	459
	51	N	1	M	USA	USA	16	36.2	1039
	52	N	1	M	FOR	FOR	23	1.7	41
GENERAL	53	N	0	M	FOR	FOR	32	20.0	69
	54	Y	0	M	FOR	USA	30	11.4	56
	55	N	0	M	FOR	USA	27	51.7	262
	56	N	0	M	FOR	FOR	0	136.5	76
	57	N	0	M	FOR	FOR	0	53.9	2822
	58	N	2	M	FOR	FOR	0	18.2	181
	59	N	1	M	FOR	FOR	11	61.0	763
	60	Y	2	M	FOR	USA	0	45.9	21
	61	Y	0	M	FOR	FOR	0	67.7	558
	62	Y	0	M	FOR	FOR	0	47.2	30
	63	Y	0	M	FOR	FOR	0	195.4	337
	64	N	2	M	USA	FOR	32	47.6	1558
	65	N	0	F	FOR	FOR	0	61.3	8
	66	Y	0	M	USA	USA	32	3.1	71
	67	Y	2	M	FOR	USA	0	69.3	74
	68	Y	1	M	FOR	FOR	0	33.4	103
	69	N	0	M	FOR	FOR	0	30.1	1568
	70	N	0	M	USA	FOR	0	43.2	187
	71	Y	1	M	FOR	FOR	0	164.1	190

PRIMSPEC	OBS	SECSPEC	CERTIF	GENDER	MEDSCHL	MEDRESID	EXP	COST	MBMNTH
INTERNAL	72	N	1	M	FOR	USA	8	18.2	12
	73	N	0	M	FOR	FOR	38	8.0	71
	74	N	1	M	FOR	USA	20	45.1	650
	75	N	1	M	USA	FOR	41	56.3	344
	76	N	1	M	FOR	USA	12	79.9	397
	77	N	1	M	USA	USA	16	103.7	1779
	78	N	1	M	USA	USA	14	24.2	37
	79	N	1	M	USA	USA	9	78.5	538
	80	N	1	M	USA	USA	13	83.5	1117
	81	N	1	M	FOR	FOR	28	61.2	1160
	82	N	2	F	FOR	USA	14	75.5	3305
	83	N	2	M	USA	USA	6	77.7	1260
	84	N	0	M	FOR	FOR	0	41.9	319
	85	N	2	M	FOR	USA	16	56.7	357
	86	Y	1	M	USA	USA	6	169.1	733
	87	N	2	M	FOR	FOR	11	47.3	226
	88	Y	1	M	USA	USA	22	66.1	795
	89	N	1	M	FOR	USA	0	27.1	111
	90	N	2	M	USA	USA	6	140.2	111
	91	N	1	F	FOR	USA	14	55.1	417
	92	N	1	M	FOR	USA	13	33.5	424
	93	N	1	M	USA	USA	22	117.5	354
	94	N	1	M	USA	USA	15	93.5	2250
	95	N	1	M	USA	USA	17	52.9	476
	96	N	2	M	FOR	USA	16	135.9	291
	97	N	1	M	USA	USA	38	50.0	982
	98	N	1	M	USA	USA	47	151.1	971
	99	Y	2	M	FOR	USA	0	17.1	48
	100	N	2	M	FOR	USA	8	16.8	21
	101	N	2	M	USA	USA	38	66.0	517
	102	N	2	M	FOR	USA	18	144.7	765
	103	N	1	F	USA	USA	15	225.9	451
	104	N	2	F	USA	USA	6	86.0	889
	105	N	0	M	USA	USA	6	40.6	51
	106	N	2	M	FOR	FOR	0	94.6	212
	107	N	1	M	USA	USA	13	50.5	435
OBSTETRICS	108	N	1	M	USA	USA	14	8.9	18
	109	N	1	M	USA	USA	9	413.3	7
	110	N	1	M	USA	USA	9	14.3	4
	111	N	1	F	USA	USA	18	1104.3	6
	112	N	1	M	FOR	USA	30	25.5	3
	113	N	1	M	FOR	USA	24	24.6	2
	114	N	1	M	USA	USA	20	65.9	3
	115	N	1	M	USA	USA	20	58.5	7
	116	N	2	M	FOR	FOR	18	133.9	2
	117	N	2	M	USA	USA	10	1121.0	5
OTHER	118	N	0	M	FOR	FOR	0	141.4	707
	119	Y	1	M	USA	USA	45	105.0	1036
	120	N	1	M	FOR	FOR	0	58.1	218
	121	Y	1	M	FOR	USA	40	81.5	2
	122	Y	1	M	USA	USA	18	29.7	396
	123	N	2	M	FOR	FOR	16	126.7	2
	124	Y	0	M	FOR	USA	28	32.8	85
	125	N	2	M	FOR	FOR	21	8.2	14
	126	Y	1	M	FOR	USA	0	113.7	22
	127	Y	1	M	FOR	FOR	14	37.8	42
	128	Y	0	M	USA	USA	24	817.3	32
	129	Y	1	M	USA	USA	16	4725.2	1
	130	Y	1	M	USA	FOR	32	28.6	740
	131	Y	2	M	FOR	USA	15	169.8	65

PRIMSPEC	OBS	SECSPEC	CERTIF	GENDER	MEDSCHL	MEDRESID	EXP	COST	MBMNTH
PEDIATRICS	132	N	1	M	USA	USA	37	47.6	2108
	133	Y	1	M	FOR	USA	35	25.9	18
	134	N	2	M	FOR	USA	0	53.7	430
	135	N	1	F	FOR	USA	21	74.1	255
	136	N	1	M	USA	FOR	23	21.6	1001
	137	N	1	F	FOR	USA	0	32.0	577
	138	N	1	M	FOR	USA	13	137.2	80
	139	N	1	F	FOR	FOR	26	12.1	427
	140	N	0	M	FOR	USA	0	7.7	82
	141	N	1	F	USA	USA	14	14.3	243
	142	N	1	M	USA	USA	12	39.3	511
	143	N	1	M	USA	USA	47	14.5	235
	144	N	1	M	USA	USA	29	50.6	103
	145	N	0	M	FOR	FOR	0	12.0	11
	146	N	0	M	FOR	FOR	0	37.3	344
	147	Y	0	M	FOR	FOR	0	79.1	561
	148	N	1	M	USA	USA	6	116.0	282
	149	N	2	M	FOR	USA	30	13.6	369
	150	N	1	M	USA	USA	39	18.2	74
	151	N	1	M	USA	USA	10	47.7	432
	152	N	0	M	FOR	FOR	0	79.1	103
	153	Y	1	M	USA	USA	41	117.3	534
	154	N	1	F	FOR	USA	27	35.9	322
	155	N	1	M	USA	FOR	9	35.7	129
	156	N	1	M	USA	USA	13	33.1	20
	157	N	1	M	FOR	FOR	13	59.5	759
	158	N	1	M	FOR	USA	29	2.1	36
	159	N	1	F	USA	USA	13	16.5	18
	160	N	2	F	FOR	USA	8	60.2	12
	161	N	2	M	FOR	USA	3	80.3	13
	162	N	1	M	USA	USA	14	53.4	211
	163	N	1	M	USA	USA	38	22.2	78
	164	N	1	M	FOR	USA	18	63.8	422
	165	N	2	F	FOR	USA	15	34.2	849
	166	N	1	M	USA	USA	34	52.4	387
	167	Y	1	M	USA	USA	25	159.6	276
	168	N	1	M	FOR	USA	16	32.7	975
	169	N	1	F	FOR	USA	0	24.3	262
	170	N	1	M	USA	USA	15	111.7	164
	171	N	0	M	FOR	FOR	0	23.4	81
	172	Y	1	M	FOR	USA	0	24.5	157
	173	N	2	M	USA	USA	37	42.5	520
	174	N	1	F	USA	USA	12	10.4	324
	175	N	1	F	FOR	USA	17	30.6	490
	176	N	2	M	USA	USA	46	10.4	363
	177	N	1	F	USA	USA	11	35.9	977
	178	N	1	M	USA	USA	27	80.8	261
	179	Y	1	M	FOR	USA	0	29.6	194
	180	N	2	F	FOR	USA	33	35.1	360
	181	N	1	M	USA	USA	23	31.6	1066
	182	N	1	M	FOR	USA	0	112.2	65
	183	N	1	F	USA	FOR	12	24.7	218
	184	N	2	F	FOR	USA	0	59.4	157
	185	N	2	M	FOR	USA	15	59.6	602
	186	N	1	M	FOR	USA	26	21.0	143

DATA SET: DDT ANALYSES ON FISH SAMPLES, TENNESSEE RIVER, ALABAMA

Note: The complete data set is available on $5\frac{1}{4}''$ floppy or $3\frac{1}{2}''$ micro diskette from the publisher.

OBS	LOCATION	SPECIES	LENGTH	WEIGHT	DDT
1	FCM5	CHANNELCATFISH	42.5	732	10.00
2	FCM5	CHANNELCATFISH	44.0	795	16.00
3	FCM5	CHANNELCATFISH	41.5	547	23.00
4	FCM5	CHANNELCATFISH	39.0	465	21.00
5	FCM5	CHANNELCATFISH	50.5	1252	50.00
6	FCM5	CHANNELCATFISH	52.0	1255	150.00
7	LCM3	CHANNELCATFISH	40.5	741	28.00
8	LCM3	CHANNELCATFISH	48.0	1151	7.70
9	LCM3	CHANNELCATFISH	48.0	1186	2.00
10	LCM3	CHANNELCATFISH	43.5	754	19.00
11	LCM3	CHANNELCATFISH	40.5	679	16.00
12	LCM3	CHANNELCATFISH	47.5	985	5.40
13	SCM1	CHANNELCATFISH	44.5	1133	2.60
14	SCM1	CHANNELCATFISH	46.0	1139	3.10
15	SCM1	CHANNELCATFISH	48.0	1186	3.50
16	SCM1	CHANNELCATFISH	45.0	984	9.10
17	SCM1	CHANNELCATFISH	43.0	965	7.80
18	SCM1	CHANNELCATFISH	45.0	1084	4.10
19	TRM275	CHANNELCATFISH	48.0	986	8.40
20	TRM275	CHANNELCATFISH	45.0	1023	15.00
21	TRM275	CHANNELCATFISH	49.0	1266	25.00
22	TRM275	CHANNELCATFISH	50.0	1086	5.60
23	TRM275	CHANNELCATFISH	46.0	1044	4.60
24	TRM275	CHANNELCATFISH	52.0	1770	8.20
25	TRM280	CHANNELCATFISH	48.0	1048	6.10
26	TRM280	CHANNELCATFISH	51.0	1641	13.00
27	TRM280	CHANNELCATFISH	48.5	1331	6.00
28	TRM280	CHANNELCATFISH	51.0	1728	6.60
29	TRM280	CHANNELCATFISH	44.0	917	5.50
30	TRM280	CHANNELCATFISH	51.0	1398	11.00
31	TRM280	SMALLMOUTHBUFF	49.0	1763	4.50
32	TRM280	SMALLMOUTHBUFF	46.0	1459	4.20
33	TRM280	SMALLMOUTHBUFF	52.0	2302	3.00
34	TRM280	SMALLMOUTHBUFF	46.0	1614	2.30
35	TRM280	SMALLMOUTHBUFF	46.0	1444	2.50
36	TRM280	SMALLMOUTHBUFF	48.0	2006	6.80
37	TRM285	CHANNELCATFISH	44.0	936	19.00
38	TRM285	CHANNELCATFISH	42.0	1058	7.20
39	TRM285	CHANNELCATFISH	42.5	800	6.00
40	TRM285	CHANNELCATFISH	45.5	1087	10.00
41	TRM285	CHANNELCATFISH	48.0	1329	12.00
42	TRM285	CHANNELCATFISH	44.0	897	2.80
43	TRM285	LARGEMOUTHBASS	28.5	778	0.48
44	TRM285	LARGEMOUTHBASS	26.0	532	0.18
45	TRM285	LARGEMOUTHBASS	25.5	441	0.34
46	TRM285	LARGEMOUTHBASS	25.0	544	0.11
47	TRM285	LARGEMOUTHBASS	23.0	393	0.22
48	TRM285	LARGEMOUTHBASS	28.0	733	0.80
49	TRM290	CHANNELCATFISH	41.0	961	8.70
50	TRM290	CHANNELCATFISH	44.0	886	22.00
51	TRM290	CHANNELCATFISH	41.0	678	13.00
52	TRM290	CHANNELCATFISH	42.0	1011	3.50
53	TRM290	CHANNELCATFISH	42.5	947	9.30
54	TRM290	CHANNELCATFISH	44.0	989	21.00
55	TRM290	SMALLMOUTHBUFF	43.5	1291	3.40
56	TRM290	SMALLMOUTHBUFF	46.5	1186	13.00
57	TRM290	SMALLMOUTHBUFF	43.0	1293	5.60
58	TRM290	SMALLMOUTHBUFF	47.0	1709	12.00
59	TRM290	SMALLMOUTHBUFF	46.0	1425	21.00

OBS	LOCATION	SPECIES	LENGTH	WEIGHT	DDT
60	TRM290	SMALLMOUTHBUFF	41.0	1176	8.00
61	TRM295	CHANNELCATFISH	36.0	980	12.00
62	TRM295	CHANNELCATFISH	47.5	1176	6.00
63	TRM295	CHANNELCATFISH	41.5	989	4.70
64	TRM295	CHANNELCATFISH	49.5	1084	31.00
65	TRM295	CHANNELCATFISH	46.0	1115	5.20
66	TRM295	CHANNELCATFISH	46.5	724	27.00
67	TRM300	CHANNELCATFISH	36.0	847	18.00
68	TRM300	CHANNELCATFISH	37.0	876	7.50
69	TRM300	CHANNELCATFISH	35.0	844	3.00
70	TRM300	CHANNELCATFISH	36.0	908	13.00
71	TRM300	CHANNELCATFISH	48.0	1358	7.30
72	TRM300	CHANNELCATFISH	49.0	1019	15.00
73	TRM300	SMALLMOUTHBUFF	35.5	1300	1.30
74	TRM300	SMALLMOUTHBUFF	46.0	1365	4.80
75	TRM300	SMALLMOUTHBUFF	45.0	1437	5.10
76	TRM300	SMALLMOUTHBUFF	44.5	1460	5.10
77	TRM300	SMALLMOUTHBUFF	49.0	1671	4.00
78	TRM300	SMALLMOUTHBUFF	47.5	1717	10.00
79	TRM305	CHANNELCATFISH	35.0	613	12.00
80	TRM305	CHANNELCATFISH	51.0	353	22.00
81	TRM305	CHANNELCATFISH	42.5	909	10.00
82	TRM305	CHANNELCATFISH	38.0	886	11.00
83	TRM305	CHANNELCATFISH	41.0	890	17.00
84	TRM305	CHANNELCATFISH	47.0	1031	9.70
85	TRM310	CHANNELCATFISH	45.0	1083	12.00
86	TRM310	CHANNELCATFISH	45.5	864	4.70
87	TRM310	CHANNELCATFISH	45.0	886	6.00
88	TRM310	CHANNELCATFISH	45.0	965	3.80
89	TRM310	CHANNELCATFISH	39.0	537	17.00
90	TRM310	CHANNELCATFISH	40.5	630	12.00
91	TRM310	SMALLMOUTHBUFF	46.0	1486	1.40
92	TRM310	SMALLMOUTHBUFF	47.0	1743	6.10
93	TRM310	SMALLMOUTHBUFF	48.5	2061	2.80
94	TRM310	SMALLMOUTHBUFF	48.0	1707	4.80
95	TRM310	SMALLMOUTHBUFF	38.0	862	5.70
96	TRM310	SMALLMOUTHBUFF	38.5	911	3.30
97	TRM315	CHANNELCATFISH	29.5	476	3.30
98	TRM315	CHANNELCATFISH	42.0	743	3.70
99	TRM315	CHANNELCATFISH	47.5	1128	9.90
100	TRM315	CHANNELCATFISH	43.5	848	6.80
101	TRM315	CHANNELCATFISH	47.5	1091	13.00
102	TRM315	CHANNELCATFISH	43.5	715	8.80
103	TRM320	CHANNELCATFISH	47.5	983	57.00
104	TRM320	CHANNELCATFISH	51.5	1251	96.00
105	TRM320	CHANNELCATFISH	49.5	1255	360.00
106	TRM320	CHANNELCATFISH	47.0	1152	130.00
107	TRM320	CHANNELCATFISH	47.5	1085	13.00
108	TRM320	CHANNELCATFISH	47.0	1118	61.00
109	TRM320	SMALLMOUTHBUFF	36.0	1285	12.00
110	TRM320	SMALLMOUTHBUFF	34.5	1178	33.00
111	TRM320	SMALLMOUTHBUFF	44.5	1492	48.00
112	TRM320	SMALLMOUTHBUFF	46.0	1524	10.00
113	TRM320	SMALLMOUTHBUFF	46.0	1473	44.00
114	TRM320	SMALLMOUTHBUFF	32.5	520	0.43
115	TRM325	CHANNELCATFISH	46.0	863	1100.00
116	TRM325	CHANNELCATFISH	40.0	549	9.40
117	TRM325	CHANNELCATFISH	43.5	810	4.10
118	TRM325	CHANNELCATFISH	46.5	908	2.80
119	TRM325	CHANNELCATFISH	43.0	804	0.74

OBS	LOCATION	SPECIES	LENGTH	WEIGHT	DDT
120	TRM325	CHANNELCATFISH	47.5	1179	14.00
121	TRM330	CHANNELCATFISH	32.0	556	22.00
122	TRM330	CHANNELCATFISH	40.5	659	9.10
123	TRM330	CHANNELCATFISH	51.5	1229	140.00
124	TRM330	CHANNELCATFISH	48.0	1050	4.20
125	TRM330	CHANNELCATFISH	47.0	952	12.00
126	TRM330	CHANNELCATFISH	41.0	826	2.00
127	TRM330	SMALLMOUTHBUFF	33.5	599	0.30
128	TRM330	SMALLMOUTHBUFF	47.0	1704	1.20
129	TRM340	CHANNELCATFISH	50.0	1207	7.10
130	TRM340	CHANNELCATFISH	45.0	911	180.00
131	TRM340	CHANNELCATFISH	49.0	1498	1.50
132	TRM340	CHANNELCATFISH	49.5	1496	2.40
133	TRM340	CHANNELCATFISH	50.0	1142	4.30
134	TRM340	CHANNELCATFISH	45.0	879	3.90
135	TRM340	SMALLMOUTHBUFF	32.5	525	0.99
136	TRM340	SMALLMOUTHBUFF	38.0	806	0.45
137	TRM340	SMALLMOUTHBUFF	38.5	694	2.50
138	TRM340	SMALLMOUTHBUFF	36.0	643	0.25
139	TRM345	LARGEMOUTHBASS	26.5	514	0.58
140	TRM345	LARGEMOUTHBASS	23.5	358	2.00
141	TRM345	LARGEMOUTHBASS	30.0	856	2.20
142	TRM345	LARGEMOUTHBASS	29.0	793	7.40
143	TRM345	LARGEMOUTHBASS	17.5	173	0.35
144	TRM345	LARGEMOUTHBASS	36.0	1433	1.90

DATA SET: SUPERMARKET CUSTOMER CHECKOUT TIMES

Note: The complete data set is available on $5\frac{1}{4}''$ floppy or $3\frac{1}{2}''$ micro diskette from the publisher.

Checkout Times (Seconds)

18	37	63	6	116	53	65	35	229	67
13	63	20	37	77	69	16	12	98	12
57	84	65	124	53	18	12	57	84	65
124	53	18	12	18	12	15	18	7	27
34	15	51	14	25	43	216	119	45	15
25	7	30	8	73	10	39	61	126	8
44	120	80	36	17	156	31	16	47	36
56	17	23	60	22	39	93	20	96	237
245	10	52	33	10	27	12	74	101	46
142	100	7	73	25	66	12	25	66	12
25	12	45	26	48	165	44	38	8	35
22	17	78	15	113	17	25	30	20	6
32	7	7	18	20	15	10	63	30	36
108	45	33	10	6	32	50	143	16	135
81	28	27	13	24	4	39	76	71	6
40	47	13	5	35	120	44	45	26	38
40	50	25	26	25	70	20	25	110	49
180	90	111	69	75	40	39	40	60	85
20	24	91	50	63	40	30	20	120	29
30	50	82	100	120	40	55	110	40	53
30	15	215	150	120	40	43	35	40	25
22	40	24	340	350	107	80	110	180	75
155	53	55	170	20	10	20	123	90	40
63	25	15	57	30	40	35	50	16	120
23	50	23	40	140	15	20	43	228	127
20	20	135	40	57	145	7	159	62	15
115	292	85	33	34	99	103	155	45	50
57	30	103	30	135	30	130	60	25	353
110	25	10	9	35	35	10	20	15	5
20	8	15	5	20	8	10	5	35	50
4	45	4	40	15	45	50	15	5	10
45	105	40	10	30	15	30	110	30	5
10	10	5	8	10	50	100	18	10	70
40	20	15	20	15	5	8	10	50	70
3	110	10	95	30	10	5	45	85	30
35	30	25	15	40	70	17	10	5	35
55	80	70	40	45	15	120	12	90	30
3	100	8	55	15	50	25	2	55	30
35	80	25	42	21	48	90	30	100	7
10	40	5	35	60	50	7	20	5	10
30	50	45	2	13	57	26	23	2	100
23	52	45	95	30	50	8	37	40	30
70	80	7	80	100	60	75	45	60	35
17	12	20	125	18	3	10	20	50	60
20	35	40	80	30	65	35	10	100	35
100	25	80	20	35	70	10	5	60	120
23	7	13	15	40	20	38	13	25	30
4	3	26	60	65	30	4	70	50	87
30	13	135	6	35	130	75	35	40	55
65	97	65	45	70	85	10	27	20	25

DATA SET: FEDERAL TRADE COMMISSION RANKINGS OF DOMESTIC CIGARETTE BRANDS

APPENDIX F

Note: The complete data set is available on $5\frac{1}{4}''$ floppy or $3\frac{1}{2}''$ micro diskette from the publisher.

FTC RANKINGS OF 372 CIGARETTE BRANDS

OBS	BRAND	LENGTH	MENTHOL	FILTER	LIGHT	PACK	TAR	NICOTINE	CO
1	Alpine	85	M	F	R	SP	17	1.1	16
2	Alpine	100	M	F	R	SP	15	1.1	15
3	Alpine	85	M	F	L	SP	9	0.7	11
4	Alpine	100	M	F	L	SP	9	0.8	11
5	Alpine	100	M	F	R	HP	15	0.9	15
6	Alpine	100	M	F	L	HP	10	0.7	10
7	American Filter	100	NM	F	R	SP	16	1.2	15
8	American Filter	85	NM	F	R	SP	16	1.2	14
9	American Lights	100	NM	F	L	SP	12	1.0	13
10	American Lights	85	NM	F	L	SP	12	0.9	13
11	American Lights	100	M	F	L	SP	12	1.0	12
12	Belair	85	M	F	R	SP	10	0.8	10
13	Belair	100	M	F	R	SP	8	0.6	9
14	Belair	85	M	F	L	SP	9	0.9	9
15	Belair	100	M	F	L	SP	10	0.9	10
16	Benson and Hedges	85	NM	F	R	HP	15	1.2	13
17	Benson and Hedges	100	NM	F	L	HP	11	0.9	12
18	Benson and Hedges	100	M	F	L	HP	10	0.8	11
19	Benson and Hedges	100	NM	F	R	SP	17	1.2	17
20	Benson and Hedges	100	NM	F	R	HP	16	1.2	15
21	Benson and Hedges	100	M	F	R	HP	16	1.2	16
22	Benson and Hedges	100	M	F	R	SP	16	1.2	16
23	Benson and Hedges	100	NM	F	L	SP	11	0.8	13
24	Benson and Hedges	100	M	F	L	SP	11	0.9	12
25	Benson and Hedges	100	NM	F	E	HP	5	0.5	7
26	Benson and Hedges	70	M	F	E	HP	5	0.5	7
27	Benson and Hedges	85	NM	F	R	SP	11	0.9	11
28	Bristol	85	NM	F	L	SP	11	0.9	12
29	Bristol	85	M	F	L	SP	11	0.9	12
30	Bristol	100	NM	F	L	SP	11	0.9	11
31	Bristol	100	M	F	L	SP	11	0.9	14
32	Bucks	85	NM	F	R	SP	14	1.0	13
33	Bucks	85	NM	F	L	SP	11	0.7	11
34	Cambridge	100	NM	F	E	SP	4	0.4	6
35	Cambridge	85	NM	F	L	SP	12	0.9	13
36	Cambridge	85	M	F	L	SP	12	0.9	13
37	Cambridge	100	NM	F	L	SP	12	0.9	14
38	Cambridge	100	M	F	L	SP	12	0.9	14
39	Cambridge	85	NM	F	R	SP	17	1.1	16
40	Cambridge	100	NM	F	R	SP	17	1.2	17
41	Camel	85	NM	F	R	SP	15	1.0	14
42	Camel	85	NM	F	R	HP	17	1.0	16
43	Camel	100	NM	F	L	SP	11	0.8	14
44	Camel	85	NM	F	L	SP	9	0.7	12
45	Camel	85	NM	F	L	HP	9	0.6	12
46	Camel	70	NM	NF	R	SP	22	1.4	14
47	Camel	100	NM	F	R	SP	17	1.0	19
48	Capri	100	NM	F	R	HP	10	0.8	7
49	Capri	100	M	F	R	HP	9	0.8	6
50	Capri	120	NM	F	R	HP	14	1.1	10
51	Capri	120	M	F	R	HP	12	1.0	8
52	Carlton	120	NM	F	R	SP	6	0.6	6
53	Carlton	85	NM	F	R	SP	1	0.1	2
54	Carlton	120	M	F	R	SP	6	0.6	5
55	Carlton	85	M	F	R	SP	0.5	0.1	1
56	Carlton	85	NM	F	R	HP	1	0.1	2
57	Carlton	100	NM	F	R	HP	0.5	0.1	1
58	Carlton	100	NM	F	R	SP	3	0.3	4
59	Carlton	100	M	F	R	HP	0.5	0.1	1
60	Carlton	100	M	F	R	SP	4	0.4	6
61	Carlton	85	NM	F	E	HP	0.5	0.05	0.5
62	Cartier Vendome	100	NM	F	R	HP	9	0.7	8
63	Cartier Vendome	100	M	F	R	HP	9	0.7	7
64	Century	85	NM	F	R	SP	15	1.0	17
65	Century	85	NM	F	L	SP	9	0.7	11
66	Century	100	NM	F	R	SP	17	1.1	20
67	Century	100	NM	F	L	SP	12	0.9	13
68	Century	100	M	F	L	SP	11	0.8	13
69	Chelsea	100	M	F	R	HP	11	0.9	12
70	Chelsea	100	NM	F	R	HP	11	0.9	12

OBS	BRAND	LENGTH	MENTHOL	FILTER	LIGHT	PACK	TAR	NICOTINE	CO
71	Chesterfield	70	NM	NF	R	SP	20	1.3	13
72	Chesterfield	85	NM	NF	R	SP	24	1.6	15
73	Chesterfield	85	NM	F	L	SP	10	0.8	12
74	Chesterfield	100	NM	F	L	SP	10	0.9	11
75	Class A	70	NM	NF	R	SP	19	1.3	13
76	Class A	85	NM	NF	R	SP	23	1.8	15
77	Class A	85	NM	F	R	SP	16	1.3	13
78	Class A	85	M	F	R	SP	16	1.3	14
79	Class A	100	NM	F	R	SP	17	1.4	14
80	Class A	100	M	F	R	SP	17	1.4	14
81	Class A	85	NM	F	L	SP	14	1.2	13
82	Class A	85	M	F	L	SP	14	1.2	13
83	Class A	100	NM	F	L	SP	14	1.2	14
84	Class A	100	M	F	L	SP	14	1.2	12
85	Class A	85	NM	F	E	SP	6	0.7	6
86	Class A	85	M	F	E	SP	6	0.7	7
87	Class A	100	NM	F	E	SP	6	0.7	5
88	Class A	100	M	F	E	SP	6	0.7	4
89	Class A Deluxe	85	NM	F	R	SP	16	1.3	13
90	Class A Deluxe	100	NM	F	R	SP	17	1.4	14
91	Class A Deluxe	85	NM	F	L	HP	14	1.2	13
92	Class A Deluxe	85	NM	F	L	SP	14	1.2	13
93	Class A Deluxe	85	M	F	L	SP	14	1.2	13
94	Class A Deluxe	100	NM	F	L	SP	14	1.2	14
95	Class A Deluxe	100	M	F	L	SP	14	1.2	12
96	Class A Deluxe	100	NM	F	E	SP	6	0.7	5
97	Doral	85	NM	F	L	SP	10	0.7	11
98	Doral	85	M	F	L	SP	10	0.7	11
99	Doral	100	NM	F	L	SP	10	0.8	10
100	Doral	100	M	F	L	SP	11	0.9	11
101	Doral	85	NM	F	R	SP	16	1.0	17
102	Doral	100	NM	F	R	SP	16	1.1	18
103	Doral	100	NM	F	E	SP	5	0.5	8
104	Doral	85	M	F	R	SP	16	1.1	16
105	English Oval	85	NM	NF	R	HP	23	1.8	15
106	Eve	120	NM	F	L	HP	13	1.1	11
107	Eve	120	M	F	L	HP	13	1.1	11
108	Eve	100	M	F	L	HP	13	1.0	11
109	Eve	100	NM	F	L	HP	13	1.0	11
110	Eve	120	NM	F	E	HP	5	0.6	4
111	Eve	120	M	F	E	HP	5	0.6	4
112	Falcon	85	NM	F	L	SP	11	0.7	11
113	Falcon	85	M	F	L	SP	10	0.8	12
114	Falcon	100	NM	F	L	SP	11	0.8	14
115	Falcon	100	M	F	L	SP	9	0.7	11
116	Golden Lights	85	NM	F	L	SP	8	0.7	8
117	Golden Lights	85	M	F	L	SP	8	0.7	9
118	Golden Lights	100	NM	F	L	SP	9	0.8	9
119	Golden Lights	100	M	F	L	SP	10	0.9	10
120	Golden Lights	85	NM	F	L	HP	8	0.7	9
121	Golden Lights	100	NM	F	L	HP	9	0.8	10
122	GPC Approved	70	NM	NF	R	SP	24	1.4	16
123	GPC Approved	85	NM	F	L	SP	11	0.7	12
124	GPC Approved	100	NM	F	L	SP	11	0.8	13
125	GPC Approved	85	M	F	L	SP	10	0.8	9
126	GPC Approved	100	M	F	L	SP	9	0.8	9
127	GPC Approved	85	NM	F	E	SP	6	0.5	6
128	GPC Approved	100	NM	F	E	SP	6	0.5	7
129	GPC Approved	85	NM	F	R	SP	17	1.1	17
130	GPC Approved	100	NM	F	R	SP	16	1.1	18
131	GPC Approved	85	M	F	R	SP	16	1.0	13
132	GPC Approved	100	M	F	R	SP	14	1.0	13
133	Harley Davidson	85	NM	F	R	SP	14	1.0	16
134	Harley Davidson	85	NM	F	L	SP	8	0.7	11
135	Herbert Tareyton	85	NM	NF	R	SP	25	1.6	18
136	Hi-Lite	100	NM	F	R	HP	13	1.1	13
137	Kent	85	NM	F	R	SP	12	0.9	14
138	Kent	100	NM	F	R	SP	14	1.0	15
139	Kent	85	NM	F	R	HP	11	0.8	13
140	Kent	100	M	F	R	SP	14	1.0	15

OBS	BRAND	LENGTH	MENTHOL	FILTER	LIGHT	PACK	TAR	NICOTINE	CO
281	Players	85	NM	F	L	SP	10	0.8	11
282	Players	85	M	F	L	SP	10	0.7	11
283	Players	100	NM	F	L	SP	12	0.9	14
284	Players	100	M	F	L	SP	12	0.9	14
285	Pyramid	85	NM	F	L	SP	14	1.1	13
286	Pyramid	100	NM	F	L	SP	13	1.1	14
287	Pyramid	100	M	F	L	SP	12	1.1	12
288	Pyramid	85	NM	NF	R	SP	23	1.5	15
289	Pyramid	100	NM	F	E	SP	6	0.6	5
290	Pyramid	85	NM	F	R	SP	16	1.3	13
291	Pyramid	85	M	F	R	SP	16	1.3	14
292	Pyramid	100	NM	F	R	SP	17	1.4	14
293	Pyramid	100	M	F	R	SP	17	1.4	14
294	Raleigh	85	NM	F	R	SP	15	1.0	14
295	Raleigh	100	NM	F	R	SP	16	1.0	16
296	Raleigh	85	NM	F	L	SP	11	0.9	13
297	Raleigh	100	NM	F	L	SP	12	0.9	15
298	Raleigh	70	NM	NF	R	SP	25	1.5	17
299	Richland	85	NM	F	R	SP	17	1.2	16
300	Richland	85	M	F	R	SP	16	1.0	16
301	Richland	85	NM	F	L	SP	12	0.9	13
302	Richland	100	NM	F	R	SP	18	1.3	17
303	Richland	100	NM	F	L	SP	12	0.9	14
304	Richland	100	M	F	R	SP	16	1.1	16
305	Ritz	100	M	F	R	HP	11	0.8	11
306	Ritz	100	NM	F	R	HP	10	0.8	11
307	Salem	85	M	F	R	SP	17	1.2	17
308	Salem	100	M	F	L	SP	8	0.7	10
309	Salem	85	M	F	L	SP	9	0.7	12
310	Salem	100	M	F	R	SP	16	1.2	16
311	Salem	100	M	F	L	HP	10	0.7	10
312	Salem	85	M	F	E	SP	5	0.4	8
313	Salem	100	M	F	E	SP	4	0.4	7
314	Salem Custom Case	100	M	F	L	HP	11	0.8	11
315	Saratoga	120	NM	F	R	HP	15	1.2	15
316	Saratoga	120	M	F	R	HP	15	1.1	14
317	Satin	100	NM	F	R	SP	11	0.9	13
318	Satin	100	M	F	R	SP	11	0.9	14
319	Savvy	100	NM	F	L	SP	10	0.9	11
320	Savvy	100	M	F	L	SP	10	0.9	11
321	Savvy	100	NM	F	E	SP	5	0.5	5
322	Silva Thins	100	NM	F	R	SP	12	1.0	10
323	Silva Thins	100	M	F	R	SP	12	1.1	10
324	Silva Thins	100	NM	F	R	HP	12	1.0	12
325	Silva Thins	100	M	F	R	HP	11	0.9	12
326	Spring	100	M	F	R	SP	18	1.4	18
327	Spring	85	NM	F	L	SP	9	0.9	12
328	Spring	85	M	F	L	SP	9	0.9	12
329	Spring	100	NM	F	L	SP	20	1.0	14
330	Spring	100	M	F	L	SP	10	1.0	14
331	Tall	120	NM	F	R	SP	19	1.6	19
332	Tall	120	M	F	R	SP	18	1.6	17
333	Tareyton	85	NM	F	R	SP	14	1.0	16
334	Tareyton	100	NM	F	R	SP	14	1.0	17
335	Tareyton	70	NM	F	L	SP	5	0.4	6
336	Tareyton	100	NM	F	L	SP	8	0.7	9
337	Triumph	85	NM	F	R	SP	3	0.3	4
338	Triumph	85	M	F	R	SP	3	0.4	4
339	Triumph	100	NM	F	R	SP	5	0.5	8
340	Triumph	100	M	F	R	SP	6	0.5	8
341	True	85	NM	F	R	SP	5	0.5	6
342	True	85	M	F	R	SP	5	0.5	6
343	True	100	NM	F	R	SP	6	0.6	7
344	True	100	M	F	R	SP	6	0.6	7
345	Vantage	100	NM	F	R	SP	9	0.6	11
346	Vantage	85	NM	F	R	SP	10	0.7	13
347	Vantage	100	NM	F	E	SP	5	0.4	8
348	Vantage	85	NM	F	E	SP	5	0.4	7
349	Vantage	85	M	F	R	SP	10	0.7	12
350	Vantage	100	M	F	R	SP	8	0.6	13

FTC RANKINGS OF 372 CIGARETTE BRANDS

OBS	BRAND	LENGTH	MENTHOL	FILTER	LIGHT	PACK	TAR	NICOTINE	CO
351	Vantage Excel	100	NM	F	R	SP	9	0.7	9
352	Viceroy	85	NM	F	R	SP	17	1.1	16
353	Viceroy	100	NM	F	R	SP	16	1.2	15
354	Viceroy	85	NM	F	L	SP	11	0.9	13
355	Viceroy	100	NM	F	L	SP	12	0.9	14
356	Virginia Slims	100	NM	F	R	SP	15	1.1	13
357	Virginia Slims	100	M	F	R	SP	15	1.2	13
358	Virginia Slims	100	NM	F	L	HP	9	0.7	10
359	Virginia Slims	100	M	F	L	HP	9	0.7	10
360	Virginia Slims	120	NM	F	L	HP	14	1.1	14
361	Virginia Slims	120	M	F	L	HP	14	1.1	13
362	Virginia Slims	100	NM	F	E	HP	5	0.5	5
363	Virginia Slims	100	M	F	E	HP	5	0.4	5
364	Winston	100	NM	F	R	SP	17	1.1	19
365	Winston	85	NM	F	R	SP	17	1.1	16
366	Winston	85	NM	F	R	HP	17	1.1	16
367	Winston	100	NM	F	L	SP	11	0.8	13
368	Winston	85	NM	F	L	SP	10	0.7	11
369	Winston	100	NM	F	L	HP	10	0.6	12
370	Winston	100	NM	F	E	SP	5	0.4	8
371	Winston	85	NM	F	E	SP	5	0.5	8
372	Winston	85	NM	F	L	HP	10	0.7	12

TABLE 1

Random Numbers

ROW	1	2	3	4	5	6	7	8	9	10	11	12	13	14
1	10480	15011	01536	02011	81647	91646	69179	14194	62590	36207	20969	99570	91291	90700
2	22368	46573	25595	85393	30995	89198	27982	53402	93965	34095	52666	19174	39615	99505
3	24130	48360	22527	97265	76393	64809	15179	24830	49340	32081	30680	19655	63348	58629
4	42167	93093	06243	61680	07856	16376	39440	53537	71341	57004	00849	74917	97758	16379
5	37570	39975	81837	16656	06121	91782	60468	81305	49684	60672	14110	06927	01263	54613
6	77921	06907	11008	42751	27756	53498	18602	70659	90655	15053	21916	81825	44394	42880
7	99562	72905	56420	69994	98872	31016	71194	18738	44013	48840	63213	21069	10634	12952
8	96301	91977	05463	07972	18876	20922	94595	56869	69014	60045	18425	84903	42508	32307
9	89579	14342	63661	10281	17453	18103	57740	84378	25331	12566	58678	44947	05585	56941
10	85475	36857	53342	53988	53060	59533	38867	62300	08158	17983	16439	11458	18593	64952
11	28918	69578	88231	33276	70997	79936	56865	05859	90106	31595	01547	85590	91610	78188
12	63553	40961	48235	03427	49626	69445	18663	72695	52180	20847	12234	90511	33703	90322
13	09429	93969	52636	92737	88974	33488	36320	17617	30015	08272	84115	27156	30613	74952
14	10365	61129	87529	85689	48237	52267	67689	93394	01511	26358	85104	20285	29975	89868
15	07119	97336	71048	08178	77233	13916	47564	81056	97735	85977	29372	74461	28551	90707
16	51085	12765	51821	51259	77452	16308	60756	92144	49442	53900	70960	63990	75601	40719
17	02368	21382	52404	60268	89368	19885	55322	44819	01188	65255	64835	44919	05944	55157
18	01011	54092	33362	94904	31273	04146	18594	29852	71585	85030	51132	01915	92747	64951
19	52162	53916	46369	58586	23216	14513	83149	98736	23495	64350	94738	17752	35156	35749
20	07056	97628	33787	09998	42698	06691	76988	13602	51851	46104	88916	19509	25625	58104
21	48663	91245	85828	14346	09172	30168	90229	04734	59193	22178	30421	61666	99904	32812
22	54164	58492	22421	74103	47070	25306	76468	26384	58151	06646	21524	15227	96909	44592
23	32639	32363	05597	24200	13363	38005	94342	28728	35806	06912	17012	64161	18296	22851
24	29334	27001	87637	87308	58731	00256	45834	15398	46557	41135	10367	07684	36188	18510
25	02488	33062	28834	07351	19731	92420	60952	61280	50001	67658	32586	86679	50720	94953
26	81525	72295	04839	96423	24878	82651	66566	14778	76797	14780	13300	87074	79666	95725
27	29676	20591	68086	26432	46901	20849	89768	81536	86645	12659	92259	57102	80428	25280

COLUMN

Continued

TABLE 2
Binomial Probabilities

Tabulated values are p(x).

a. n = 5

x							π							
	.01	.05	.1	.2	.3	.4	.5	.6	.7	.8	.9	.95	.99
0	.9510	.7738	.5905	.3277	.1681	.0778	.0313	.0102	.0024	.0003	.0000	.0000	.0000
1	.0480	.2036	.3280	.4096	.3601	.2592	.1563	.0768	.0283	.0064	.0005	.0000	.0000
2	.0010	.0214	.0729	.2048	.3087	.3456	.3125	.2304	.1323	.0512	.0081	.0011	.0000
3	.0000	.0011	.0081	.0512	.1323	.2304	.3125	.3456	.3087	.2048	.0729	.0214	.0010
4	.0000	.0000	.0004	.0064	.0283	.0768	.1563	.2592	.3601	.4096	.3280	.2036	.0480
5	.0000	.0000	.0000	.0003	.0024	.0102	.0313	.0778	.1681	.3277	.5905	.7738	.9510

b. n = 6

x							π							
	.01	.05	.1	.2	.3	.4	.5	.6	.7	.8	.9	.95	.99
0	.9415	.7351	.5314	.2621	.1176	.0467	.0156	.0041	.0007	.0001	.0000	.0000	.0000
1	.0571	.2321	.3543	.3932	.3025	.1866	.0938	.0369	.0102	.0015	.0001	.0000	.0000
2	.0014	.0305	.0984	.2458	.3241	.3110	.2344	.1382	.0595	.0154	.0012	.0001	.0000
3	.0000	.0021	.0146	.0819	.1852	.2765	.3125	.2765	.1852	.0819	.0146	.0021	.0000
4	.0000	.0001	.0012	.0154	.0595	.1382	.2344	.3110	.3241	.2458	.0984	.0305	.0014
5	.0000	.0000	.0001	.0015	.0102	.0369	.0938	.1866	.3025	.3932	.3543	.2321	.0571
6	.0000	.0000	.0000	.0001	.0007	.0041	.0156	.0467	.1176	.2621	.5314	.7351	.9415

c. n = 7

x							π							
	.01	.05	.1	.2	.3	.4	.5	.6	.7	.8	.9	.95	.99
0	.9321	.6983	.4783	.2097	.0824	.0280	.0078	.0016	.0002	.0000	.0000	.0000	.0000
1	.0659	.2573	.3720	.3670	.2471	.1306	.0547	.0172	.0036	.0004	.0000	.0000	.0000
2	.0020	.0406	.1240	.2753	.3177	.2613	.1641	.0774	.0250	.0043	.0002	.0000	.0000
3	.0000	.0036	.0230	.1147	.2269	.2903	.2734	.1935	.0972	.0287	.0026	.0002	.0000
4	.0000	.0002	.0026	.0287	.0972	.1935	.2734	.2903	.2269	.1147	.0230	.0036	.0000
5	.0000	.0000	.0002	.0043	.0250	.0774	.1641	.2613	.3177	.2753	.1240	.0406	.0020
6	.0000	.0000	.0000	.0004	.0036	.0172	.0547	.1306	.2471	.3670	.3720	.2573	.0659
7	.0000	.0000	.0000	.0000	.0002	.0016	.0078	.0280	.0824	.2097	.4783	.6983	.9321

TABLE 2 (Continued)

d. $n = 8$

x	π												
	.01	.05	.1	.2	.3	.4	.5	.6	.7	.8	.9	.95	.99
0	.9227	.6634	.4305	.1678	.0576	.0168	.0039	.0007	.0001	.0000	.0000	.0000	.0000
1	.0746	.2793	.3826	.3355	.1977	.0896	.0313	.0079	.0012	.0001	.0000	.0000	.0000
2	.0026	.0515	.1488	.2936	.2965	.2090	.1094	.0413	.0100	.0011	.0000	.0000	.0000
3	.0001	.0054	.0331	.1468	.2541	.2787	.2187	.1239	.0467	.0092	.0004	.0000	.0000
4	.0000	.0004	.0046	.0459	.1361	.2322	.2734	.2322	.1361	.0459	.0046	.0004	.0000
5	.0000	.0000	.0004	.0092	.0467	.1239	.2167	.2787	.2541	.1468	.0331	.0054	.0001
6	.0000	.0000	.0000	.0011	.0100	.0413	.1094	.2090	.2965	.2936	.1488	.0515	.0026
7	.0000	.0000	.0000	.0001	.0012	.0079	.0313	.0896	.1977	.3355	.3826	.2793	.0746
8	.0000	.0000	.0000	.0000	.0001	.0007	.0039	.0168	.0576	.1678	.4305	.6634	.9227

e. $n = 9$

x	π												
	.01	.05	.1	.2	.3	.4	.5	.6	.7	.8	.9	.95	.99
0	.9135	.6302	.3874	.1342	.0404	.0101	.0020	.0003	.0000	.0000	.0000	.0000	.0000
1	.0830	.2985	.3874	.3020	.1556	.0605	.0176	.0035	.0004	.0000	.0000	.0000	.0000
2	.0034	.0629	.1722	.3020	.2668	.1612	.0703	.0212	.0039	.0003	.0000	.0000	.0000
3	.0001	.0077	.0446	.1762	.2668	.2508	.1641	.0743	.0210	.0028	.0001	.0000	.0000
4	.0000	.0006	.0074	.0661	.1715	.2508	.2461	.1672	.0735	.0165	.0008	.0000	.0000
5	.0000	.0000	.0008	.0165	.0735	.1672	.2461	.2508	.1715	.0661	.0074	.0006	.0000
6	.0000	.0000	.0001	.0028	.0210	.0743	.1641	.2508	.2668	.1762	.0446	.0077	.0001
7	.0000	.0000	.0000	.0003	.0039	.0212	.0703	.1612	.2668	.3020	.1722	.0629	.0034
8	.0000	.0000	.0000	.0000	.0004	.0035	.0176	.0605	.1556	.3020	.3874	.2985	.0830
9	.0000	.0000	.0000	.0000	.0000	.0003	.0020	.0101	.0404	.1342	.3874	.6302	.9135

f. $n = 10$

x	π												
	.01	.05	.1	.2	.3	.4	.5	.6	.7	.8	.9	.95	.99
0	.9044	.5987	.3487	.1074	.0282	.0060	.0010	.0001	.0000	.0000	.0000	.0000	.0000
1	.0914	.3151	.3874	.2684	.1211	.0403	.0098	.0016	.0001	.0000	.0000	.0000	.0000
2	.0042	.0746	.1937	.3020	.2335	.1209	.0439	.0106	.0014	.0001	.0000	.0000	.0000
3	.0001	.0105	.0574	.2013	.2668	.2150	.1172	.0425	.0090	.0008	.0000	.0000	.0000
4	.0000	.0010	.0112	.0881	.2001	.2508	.2051	.1115	.0368	.0055	.0001	.0000	.0000
5	.0000	.0001	.0015	.0264	.1029	.2007	.2461	.2007	.1029	.0264	.0015	.0001	.0000
6	.0000	.0000	.0001	.0055	.0368	.1115	.2051	.2508	.2001	.0881	.0112	.0010	.0000
7	.0000	.0000	.0000	.0008	.0090	.0425	.1172	.2150	.2668	.2013	.0574	.0105	.0001
8	.0000	.0000	.0000	.0001	.0014	.0106	.0439	.1209	.2335	.3020	.1937	.0746	.0042
9	.0000	.0000	.0000	.0000	.0001	.0016	.0098	.0403	.1211	.2684	.3874	.3151	.0914
10	.0000	.0000	.0000	.0000	.0000	.0001	.0010	.0060	.0282	.1074	.3487	.5987	.9044

Continued

TABLE 2   (Continued)

g. $n = 15$

x	.01	.05	.1	.2	.3	.4	.5	.6	.7	.8	.9	.95	.99
							π						
0	.8601	.4633	.2059	.0352	.0047	.0005	.0000	.0000	.0000	.0000	.0000	.0000	.0000
1	.1303	.3658	.3432	.1319	.0305	.0047	.0005	.0000	.0000	.0000	.0000	.0000	.0000
2	.0092	.1348	.2669	.2309	.0916	.0219	.0032	.0003	.0000	.0000	.0000	.0000	.0000
3	.0004	.0307	.1285	.2501	.1700	.0634	.0139	.0016	.0001	.0000	.0000	.0000	.0000
4	.0000	.0049	.0423	.1876	.2186	.1268	.0417	.0074	.0006	.0000	.0000	.0000	.0000
5	.0000	.0006	.0105	.1032	.2061	.1859	.0916	.0245	.0030	.0001	.0000	.0000	.0000
6	.0000	.0000	.0019	.0430	.1472	.2066	.1527	.0612	.0116	.0007	.0000	.0000	.0000
7	.0000	.0000	.0003	.0138	.0811	.1771	.1964	.1181	.0348	.0035	.0000	.0000	.0000
8	.0000	.0000	.0000	.0035	.0348	.1181	.1964	.1771	.0811	.0138	.0003	.0000	.0000
9	.0000	.0000	.0000	.0007	.0116	.0612	.1527	.2066	.1472	.0430	.0019	.0000	.0000
10	.0000	.0000	.0000	.0001	.0030	.0245	.0916	.1859	.2061	.1032	.0105	.0006	.0000
11	.0000	.0000	.0000	.0000	.0006	.0074	.0417	.1268	.2186	.1876	.0428	.0049	.0000
12	.0000	.0000	.0000	.0000	.0001	.0016	.0139	.0634	.1700	.2501	.1285	.0307	.0004
13	.0000	.0000	.0000	.0000	.0000	.0003	.0032	.0219	.0916	.2309	.2669	.1348	.0092
14	.0000	.0000	.0000	.0000	.0000	.0000	.0005	.0047	.0305	.1319	.3432	.3658	.1303
15	.0000	.0000	.0000	.0000	.0000	.0000	.0000	.0005	.0047	.0352	.2059	.4633	.8601

h. $n = 20$

x	.01	.05	.1	.2	.3	.4	.5	.6	.7	.8	.9	.95	.99
							π						
0	.8179	.3585	.1216	.0115	.0008	.0000	.0000	.0000	.0000	.0000	.0000	.0000	.0000
1	.1652	.3774	.2702	.0576	.0068	.0005	.0000	.0000	.0000	.0000	.0000	.0000	.0000
2	.0159	.1887	.2852	.1369	.0278	.0031	.0002	.0000	.0000	.0000	.0000	.0000	.0000
3	.0010	.0596	.1901	.2054	.0716	.0123	.0011	.0000	.0000	.0000	.0000	.0000	.0000
4	.0000	.0133	.0898	.2182	.1304	.0350	.0046	.0003	.0000	.0000	.0000	.0000	.0000
5	.0000	.0022	.0319	.1746	.1789	.0746	.0148	.0013	.0000	.0000	.0000	.0000	.0000
6	.0000	.0003	.0089	.1091	.1916	.1244	.0370	.0049	.0002	.0000	.0000	.0000	.0000
7	.0000	.0000	.0020	.0545	.1643	.1659	.0739	.0146	.0010	.0000	.0000	.0000	.0000
8	.0000	.0000	.0004	.0222	.1144	.1797	.1201	.0355	.0039	.0001	.0000	.0000	.0000
9	.0000	.0000	.0001	.0074	.0654	.1597	.1602	.0710	.0120	.0005	.0000	.0000	.0000
10	.0000	.0000	.0000	.0020	.0308	.1171	.1762	.1171	.0308	.0020	.0000	.0000	.0000
11	.0000	.0000	.0000	.0005	.0120	.0710	.1602	.1597	.0654	.0074	.0001	.0000	.0000
12	.0000	.0000	.0000	.0001	.0039	.0355	.1201	.1797	.1144	.0222	.0004	.0000	.0000
13	.0000	.0000	.0000	.0000	.0010	.0146	.0739	.1659	.1643	.0545	.0020	.0000	.0000
14	.0000	.0000	.0000	.0000	.0002	.0049	.0370	.1244	.1916	.1091	.0089	.0003	.0000
15	.0000	.0000	.0000	.0000	.0000	.0013	.0148	.0746	.1789	.1746	.0319	.0022	.0000
16	.0000	.0000	.0000	.0000	.0000	.0003	.0046	.0350	.1304	.2182	.0898	.0133	.0000
17	.0000	.0000	.0000	.0000	.0000	.0000	.0011	.0123	.0716	.2054	.1901	.0596	.0010
18	.0000	.0000	.0000	.0000	.0000	.0000	.0002	.0031	.0278	.1369	.2852	.1887	.0159
19	.0000	.0000	.0000	.0000	.0000	.0000	.0000	.0005	.0068	.0576	.2702	.3774	.1652
20	.0000	.0000	.0000	.0000	.0000	.0000	.0000	.0000	.0008	.0115	.1216	.3585	.8179

TABLE 2 (Continued)

i. $n = 25$

x	.01	.05	.1	.2	.3	.4	.5	.6	.7	.8	.9	.95	.99
							π						
0	.7778	.2774	.0718	.0038	.0001	.0000	.0000	.0000	.0000	.0000	.0000	.0000	.0000
1	.1964	.3650	.1994	.0236	.0014	.0000	.0000	.0000	.0000	.0000	.0000	.0000	.0000
2	.0238	.2305	.2659	.0708	.0074	.0004	.0000	.0000	.0000	.0000	.0000	.0000	.0000
3	.0018	.0930	.2265	.1358	.0243	.0019	.0001	.0000	.0000	.0000	.0000	.0000	.0000
4	.0001	.0269	.1384	.1867	.0572	.0071	.0004	.0000	.0000	.0000	.0000	.0000	.0000
5	.0000	.0060	.0646	.1960	.1030	.0199	.0016	.0000	.0000	.0000	.0000	.0000	.0000
6	.0000	.0010	.0239	.1633	.1472	.0442	.0053	.0002	.0000	.0000	.0000	.0000	.0000
7	.0000	.0001	.0072	.1108	.1712	.0800	.0143	.0009	.0000	.0000	.0000	.0000	.0000
8	.0000	.0000	.0018	.0623	.1651	.1200	.0322	.0031	.0001	.0000	.0000	.0000	.0000
9	.0000	.0000	.0004	.0294	.1336	.1511	.0609	.0088	.0004	.0000	.0000	.0000	.0000
10	.0000	.0000	.0001	.0118	.0916	.1612	.0974	.0212	.0013	.0000	.0000	.0000	.0000
11	.0000	.0000	.0000	.0040	.0536	.1465	.1328	.0434	.0042	.0001	.0000	.0000	.0000
12	.0000	.0000	.0000	.0012	.0268	.1140	.1550	.0760	.0115	.0003	.0000	.0000	.0000
13	.0000	.0000	.0000	.0003	.0115	.0760	.1550	.1140	.0268	.0012	.0000	.0000	.0000
14	.0000	.0000	.0000	.0001	.0042	.0434	.1328	.1465	.0536	.0040	.0000	.0000	.0000
15	.0000	.0000	.0000	.0000	.0013	.0212	.0974	.1612	.0916	.0118	.0001	.0000	.0000
16	.0000	.0000	.0000	.0000	.0004	.0088	.0609	.1511	.1336	.0294	.0004	.0000	.0000
17	.0000	.0000	.0000	.0000	.0001	.0031	.0322	.1200	.1651	.0623	.0018	.0000	.0000
18	.0000	.0000	.0000	.0000	.0000	.0009	.0143	.0800	.1712	.1108	.0072	.0001	.0000
19	.0000	.0000	.0000	.0000	.0000	.0002	.0053	.0442	.1472	.1633	.0239	.0010	.0000
20	.0000	.0000	.0000	.0000	.0000	.0000	.0016	.0199	.1030	.1960	.0646	.0060	.0000
21	.0000	.0000	.0000	.0000	.0000	.0000	.0004	.0071	.0572	.1867	.1384	.0269	.0001
22	.0000	.0000	.0000	.0000	.0000	.0000	.0001	.0019	.0243	.1358	.2265	.0930	.0018
23	.0000	.0000	.0000	.0000	.0000	.0000	.0000	.0004	.0074	.0708	.2659	.2305	.0238
24	.0000	.0000	.0000	.0000	.0000	.0000	.0000	.0000	.0014	.0236	.1994	.3650	.1964
25	.0000	.0000	.0000	.0000	.0000	.0000	.0000	.0000	.0001	.0038	.0718	.2774	.7778

TABLE 3
Cumulative Binomial Probabilities

a. $n = 5$

x	.01	.05	.1	.2	.3	.4	.5	.6	.7	.8	.9	.95	.99
							π						
0	.9510	.7738	.5905	.3277	.1681	.0778	.0313	.0102	.0024	.0003	.0000	.0000	.0000
1	.9990	.9774	.9185	.7373	.5282	.3370	.1875	.0870	.0308	.0067	.0005	.0000	.0000
2	1.0000	.9988	.9914	.9421	.8369	.6826	.5000	.3174	.1631	.0579	.0086	.0012	.0000
3	1.0000	1.0000	.9995	.9933	.9692	.9130	.8125	.6630	.4718	.2627	.0815	.0226	.0010
4	1.0000	1.0000	1.0000	.9997	.9976	.9898	.9687	.9222	.8319	.6723	.4095	.2262	.0490

b. $n = 6$

x	.01	.05	.1	.2	.3	.4	.5	.6	.7	.8	.9	.95	.99
							π						
0	.9415	.7351	.5314	.2621	.1176	.0467	.0156	.0041	.0007	.0001	.0000	.0000	.0000
1	.9985	.9672	.8857	.6554	.4202	.2333	.1094	.0410	.0109	.0016	.0001	.0000	.0000
2	1.0000	.9978	.9841	.9011	.7443	.5443	.3437	.1792	.0705	.0170	.0013	.0001	.0000
3	1.0000	.9999	.9987	.9830	.9295	.8208	.6562	.4557	.2557	.0989	.0158	.0022	.0000
4	1.0000	1.0000	.9999	.9984	.9891	.9590	.8906	.7667	.5798	.3446	.1143	.0328	.0015
5	1.0000	1.0000	1.0000	.9999	.9993	.9959	.9844	.9533	.8824	.7379	.4686	.2649	.0585

c. $n = 7$

x	.01	.05	.1	.2	.3	.4	.5	.6	.7	.8	.9	.95	.99
							π						
0	.9321	.6983	.4783	.2097	.0824	.0280	.0078	.0016	.0002	.0000	.0000	.0000	.0000
1	.9980	.9556	.8503	.5767	.3294	.1586	.0625	.0188	.0038	.0004	.0000	.0000	.0000
2	1.0000	.9962	.9743	.8520	.6471	.4199	.2266	.0963	.0288	.0047	.0002	.0000	.0000
3	1.0000	.9998	.9973	.9667	.8740	.7102	.5000	.2898	.1260	.0333	.0027	.0002	.0000
4	1.0000	1.0000	.9998	.9953	.9712	.9037	.7734	.5801	.3529	.1480	.0257	.0038	.0000
5	1.0000	1.0000	1.0000	.9996	.9962	.9812	.9375	.8414	.6706	.4233	.1497	.0444	.0020
6	1.0000	1.0000	1.0000	1.0000	.9998	.9984	.9922	.9720	.9176	.7903	.5217	.3017	.0679

TABLE 3 (Continued)

d. $n = 8$

x	.01	.05	.1	.2	.3	.4	π .5	.6	.7	.8	.9	.95	.99
0	.9227	.6634	.4305	.1678	.0576	.0168	.0039	.0007	.0001	.0000	.0000	.0000	.0000
1	.9973	.9423	.8131	.5033	.2553	.1064	.0352	.0085	.0013	.0001	.0000	.0000	.0000
2	.9999	.9942	.9619	.7969	.5518	.3154	.1445	.0498	.0113	.0012	.0000	.0000	.0000
3	1.0000	.9996	.9950	.9437	.8059	.5941	.3633	.1737	.0580	.0104	.0004	.0000	.0000
4	1.0000	1.0000	.9996	.9896	.9420	.8263	.6367	.4059	.1941	.0563	.0050	.0004	.0000
5	1.0000	1.0000	1.0000	.9988	.9887	.9502	.8555	.6346	.4482	.2031	.0381	.0058	.0001
6	1.0000	1.0000	1.0000	.9999	.9987	.9915	.9648	.8936	.7447	.4967	.1869	.0572	.0027
7	1.0000	1.0000	1.0000	1.0000	.9999	.9993	.9961	.9832	.9424	.8322	.5695	.3366	.0773

e. $n = 9$

x	.01	.05	.1	.2	.3	.4	π .5	.6	.7	.8	.9	.95	.99
0	.9135	.6302	.3874	.1342	.0404	.0101	.0020	.0003	.0000	.0000	.0000	.0000	.0000
1	.9966	.9288	.7748	.4362	.1960	.0705	.0195	.0038	.0004	.0000	.0000	.0000	.0000
2	.9999	.9916	.9470	.7382	.4623	.2318	.0898	.0250	.0043	.0003	.0000	.0000	.0000
3	1.0000	.9994	.9917	.9144	.7297	.4826	.2539	.0994	.0253	.0031	.0001	.0000	.0000
4	1.0000	1.0000	.9991	.9804	.9012	.7334	.5000	.2666	.0988	.0196	.0009	.0000	.0000
5	1.0000	1.0000	.9999	.9969	.9747	.9006	.7461	.5174	.2703	.0856	.0083	.0006	.0000
6	1.0000	1.0000	1.0000	.9997	.9957	.9750	.9102	.7682	.5372	.2618	.0530	.0084	.0001
7	1.0000	1.0000	1.0000	1.0000	.9996	.9962	.9805	.9295	.8040	.5638	.2252	.0712	.0034
8	1.0000	1.0000	1.0000	1.0000	1.0000	.9997	.9980	.9899	.9596	.8658	.6126	.3698	.0865

f. $n = 10$

x	.01	.05	.1	.2	.3	.4	π .5	.6	.7	.8	.9	.95	.99
0	.9044	.5987	.3487	.1074	.0282	.0060	.0010	.0001	.0000	.0000	.0000	.0000	.0000
1	.9957	.9139	.7361	.3758	.1493	.0464	.0107	.0017	.0001	.0000	.0000	.0000	.0000
2	.9999	.9885	.9298	.6778	.3828	.1673	.0547	.0123	.0016	.0001	.0000	.0000	.0000
3	1.0000	.9990	.9872	.8791	.6496	.3823	.1719	.0548	.0106	.0009	.0000	.0000	.0000
4	1.0000	.9999	.9984	.9672	.8497	.6331	.3770	.1662	.0473	.0064	.0001	.0000	.0000
5	1.0000	1.0000	.9999	.9936	.9527	.8338	.6230	.3669	.1503	.0328	.0016	.0001	.0000
6	1.0000	1.0000	1.0000	.9991	.9894	.9452	.8281	.6177	.3504	.1209	.0128	.0010	.0000
7	1.0000	1.0000	1.0000	.9999	.9984	.9877	.9453	.8327	.6172	.3222	.0702	.0115	.0001
8	1.0000	1.0000	1.0000	1.0000	.9999	.9983	.9893	.9536	.8507	.6242	.2639	.0861	.0043
9	1.0000	1.0000	1.0000	1.0000	1.0000	.9999	.9990	.9940	.9718	.8926	.6513	.4013	.0956

Continued

TABLE 3 (Continued)

g. $n = 15$

							π						
x	.01	.05	.1	.2	.3	.4	.5	.6	.7	.8	.9	.95	.99
0	.8601	.4633	.2059	.0352	.0047	.0005	.0000	.0000	.0000	.0000	.0000	.0000	.0000
1	.9904	.8290	.5490	.1671	.0353	.0052	.0005	.0000	.0000	.0000	.0000	.0000	.0000
2	.9996	.9638	.8159	.3980	.1268	.0271	.0037	.0003	.0000	.0000	.0000	.0000	.0000
3	1.0000	.9945	.9444	.6482	.2969	.0905	.0176	.0019	.0001	.0000	.0000	.0000	.0000
4	1.0000	.9994	.9873	.8358	.5155	.2173	.0592	.0093	.0007	.0000	.0000	.0000	.0000
5	1.0000	.9999	.9978	.9389	.7216	.4032	.1509	.0338	.0037	.0001	.0000	.0000	.0000
6	1.0000	1.0000	.9997	.9819	.8689	.6098	.3036	.0950	.0152	.0008	.0000	.0000	.0000
7	1.0000	1.0000	1.0000	.9958	.9500	.7869	.5000	.2131	.0500	.0042	.0000	.0000	.0000
8	1.0000	1.0000	1.0000	.9992	.9848	.9050	.6964	.3902	.1311	.0181	.0003	.0000	.0000
9	1.0000	1.0000	1.0000	.9999	.9963	.9662	.8491	.5968	.2784	.0611	.0022	.0001	.0000
10	1.0000	1.0000	1.0000	1.0000	.9993	.9907	.9408	.7827	.4845	.1642	.0127	.0006	.0000
11	1.0000	1.0000	1.0000	1.0000	.9999	.9981	.9824	.9095	.7031	.3518	.0556	.0055	.0000
12	1.0000	1.0000	1.0000	1.0000	1.0000	.9997	.9963	.9729	.8732	.6020	.1841	.0362	.0004
13	1.0000	1.0000	1.0000	1.0000	1.0000	1.0000	.9995	.9948	.9647	.8329	.4510	.1710	.0096
14	1.0000	1.0000	1.0000	1.0000	1.0000	1.0000	1.0000	.9995	.9953	.9648	.7941	.5367	.1399

h. $n = 20$

							π						
x	.01	.05	.1	.2	.3	.4	.5	.6	.7	.8	.9	.95	.99
0	.8179	.3585	.1216	.0115	.0008	.0000	.0000	.0000	.0000	.0000	.0000	.0000	.0000
1	.9831	.7358	.3917	.0692	.0076	.0005	.0000	.0000	.0000	.0000	.0000	.0000	.0000
2	.9990	.9245	.6769	.2061	.0355	.0036	.0002	.0000	.0000	.0000	.0000	.0000	.0000
3	1.0000	.9841	.8670	.4114	.1071	.0160	.0013	.0000	.0000	.0000	.0000	.0000	.0000
4	1.0000	.9974	.9568	.6296	.2375	.0510	.0059	.0003	.0000	.0000	.0000	.0000	.0000
5	1.0000	.9997	.9887	.8042	.4164	.1256	.0207	.0016	.0000	.0000	.0000	.0000	.0000
6	1.0000	1.0000	.9976	.9133	.6080	.2500	.0577	.0065	.0003	.0000	.0000	.0000	.0000
7	1.0000	1.0000	.9996	.9679	.7723	.4159	.1316	.0210	.0013	.0000	.0000	.0000	.0000
8	1.0000	1.0000	.9999	.9900	.8867	.5956	.2517	.0565	.0051	.0001	.0000	.0000	.0000
9	1.0000	1.0000	1.0000	.9974	.9520	.7553	.4119	.1275	.0171	.0006	.0000	.0000	.0000
10	1.0000	1.0000	1.0000	.9994	.9829	.8725	.5881	.2447	.0480	.0026	.0000	.0000	.0000
11	1.0000	1.0000	1.0000	.9999	.9949	.9435	.7483	.4044	.1133	.0100	.0001	.0000	.0000
12	1.0000	1.0000	1.0000	1.0000	.9987	.9790	.8684	.5841	.2277	.0321	.0004	.0000	.0000
13	1.0000	1.0000	1.0000	1.0000	.9997	.9935	.9423	.7500	.3920	.0867	.0024	.0000	.0000
14	1.0000	1.0000	1.0000	1.0000	1.0000	.9984	.9793	.8744	.5836	.1958	.0113	.0003	.0000
15	1.0000	1.0000	1.0000	1.0000	1.0000	.9997	.9941	.9490	.7625	.3704	.0432	.0026	.0000
16	1.0000	1.0000	1.0000	1.0000	1.0000	1.0000	.9987	.9840	.8929	.5886	.1330	.0159	.0000
17	1.0000	1.0000	1.0000	1.0000	1.0000	1.0000	.9998	.9964	.9645	.7939	.3231	.0755	.0010
18	1.0000	1.0000	1.0000	1.0000	1.0000	1.0000	1.0000	.9995	.9924	.9308	.6083	.2642	.0169
19	1.0000	1.0000	1.0000	1.0000	1.0000	1.0000	1.0000	1.0000	.9992	.9885	.8784	.6415	.1821

TABLE 3 (Continued)

i. $n = 25$

x	.01	.05	.1	.2	.3	.4	.5	.6	.7	.8	.9	.95	.99
							π						
0	.7778	.2774	.0718	.0038	.0001	.0000	.0000	.0000	.0000	.0000	.0000	.0000	.0000
1	.9742	.6424	.2712	.0274	.0016	.0001	.0000	.0000	.0000	.0000	.0000	.0000	.0000
2	.9980	.8729	.5371	.0982	.0090	.0004	.0000	.0000	.0000	.0000	.0000	.0000	.0000
3	.9999	.9659	.7636	.2340	.0332	.0024	.0001	.0000	.0000	.0000	.0000	.0000	.0000
4	1.0000	.9928	.9020	.4207	.0905	.0095	.0005	.0000	.0000	.0000	.0000	.0000	.0000
5	1.0000	.9988	.9666	.6167	.1935	.0294	.0020	.0001	.0000	.0000	.0000	.0000	.0000
6	1.0000	.9998	.9905	.7800	.3407	.0736	.0073	.0003	.0000	.0000	.0000	.0000	.0000
7	1.0000	1.0000	.9977	.8909	.5118	.1536	.0216	.0012	.0000	.0000	.0000	.0000	.0000
8	1.0000	1.0000	.9995	.9532	.6769	.2735	.0539	.0043	.0001	.0000	.0000	.0000	.0000
9	1.0000	1.0000	.9999	.9827	.8106	.4246	.1148	.0132	.0005	.0000	.0000	.0000	.0000
10	1.0000	1.0000	1.0000	.9944	.9022	.5858	.2122	.0344	.0018	.0000	.0000	.0000	.0000
11	1.0000	1.0000	1.0000	.9985	.9558	.7323	.3450	.0778	.0060	.0001	.0000	.0000	.0000
12	1.0000	1.0000	1.0000	.9996	.9825	.8462	.5000	.1538	.0175	.0004	.0000	.0000	.0000
13	1.0000	1.0000	1.0000	.9999	.9940	.9222	.6550	.2677	.0442	.0015	.0000	.0000	.0000
14	1.0000	1.0000	1.0000	1.0000	.9982	.9656	.7878	.4142	.0978	.0056	.0000	.0000	.0000
15	1.0000	1.0000	1.0000	1.0000	.9995	.9868	.8852	.5754	.1894	.0173	.0001	.0000	.0000
16	1.0000	1.0000	1.0000	1.0000	.9999	.9957	.9461	.7265	.3231	.0468	.0005	.0000	.0000
17	1.0000	1.0000	1.0000	1.0000	1.0000	.9988	.9784	.8464	.4882	.1091	.0023	.0000	.0000
18	1.0000	1.0000	1.0000	1.0000	1.0000	.9997	.9927	.9264	.6593	.2200	.0095	.0002	.0000
19	1.0000	1.0000	1.0000	1.0000	1.0000	.9999	.9980	.9706	.8065	.3833	.0334	.0012	.0000
20	1.0000	1.0000	1.0000	1.0000	1.0000	1.0000	.9995	.9905	.9095	.5793	.0980	.0072	.0000
21	1.0000	1.0000	1.0000	1.0000	1.0000	1.0000	.9999	.9976	.9668	.7660	.2364	.0341	.0001
22	1.0000	1.0000	1.0000	1.0000	1.0000	1.0000	1.0000	.9996	.9910	.9018	.4629	.1271	.0020
23	1.0000	1.0000	1.0000	1.0000	1.0000	1.0000	1.0000	.9999	.9984	.9726	.7288	.3576	.0258
24	1.0000	1.0000	1.0000	1.0000	1.0000	1.0000	1.0000	1.0000	.9999	.9962	.9282	.7226	.2222

TABLE 4
Normal Curve Areas

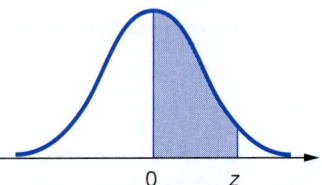

z	.00	.01	.02	.03	.04	.05	.06	.07	.08	.09
.0	.0000	.0040	.0080	.0120	.0160	.0199	.0239	.0279	.0319	.0359
.1	.0398	.0438	.0478	.0517	.0557	.0596	.0636	.0675	.0714	.0753
.2	.0793	.0832	.0871	.0910	.0948	.0987	.1026	.1064	.1103	.1141
.3	.1179	.1217	.1255	.1293	.1331	.1368	.1406	.1443	.1480	.1517
.4	.1554	.1591	.1628	.1664	.1700	.1736	.1772	.1808	.1844	.1879
.5	.1915	.1950	.1985	.2019	.2054	.2088	.2123	.2157	.2190	.2224
.6	.2257	.2291	.2324	.2357	.2389	.2422	.2454	.2486	.2517	.2549
.7	.2580	.2611	.2642	.2673	.2704	.2734	.2764	.2794	.2823	.2852
.8	.2881	.2910	.2939	.2967	.2995	.3023	.3051	.3078	.3106	.3133
.9	.3159	.3186	.3212	.3238	.3264	.3289	.3315	.3340	.3365	.3389
1.0	.3413	.3438	.3461	.3485	.3508	.3531	.3554	.3577	.3599	.3621
1.1	.3643	.3665	.3686	.3708	.3729	.3749	.3770	.3790	.3810	.3830
1.2	.3849	.3869	.3888	.3907	.3925	.3944	.3962	.3980	.3997	.4015
1.3	.4032	.4049	.4066	.4082	.4099	.4115	.4131	.4147	.4162	.4177
1.4	.4192	.4207	.4222	.4236	.4251	.4265	.4279	.4292	.4306	.4319
1.5	.4332	.4345	.4357	.4370	.4382	.4394	.4406	.4418	.4429	.4441
1.6	.4452	.4463	.4474	.4484	.4495	.4505	.4515	.4525	.4535	.4545
1.7	.4554	.4564	.4573	.4582	.4591	.4599	.4608	.4616	.4625	.4633
1.8	.4641	.4649	.4656	.4664	.4671	.4678	.4686	.4693	.4699	.4706
1.9	.4713	.4719	.4726	.4732	.4738	.4744	.4750	.4756	.4761	.4767
2.0	.4772	.4778	.4783	.4788	.4793	.4798	.4803	.4808	.4812	.4817
2.1	.4821	.4826	.4830	.4834	.4838	.4842	.4846	.4850	.4854	.4857
2.2	.4861	.4864	.4868	.4871	.4875	.4878	.4881	.4884	.4887	.4890
2.3	.4893	.4896	.4898	.4901	.4904	.4906	.4909	.4911	.4913	.4916
2.4	.4918	.4920	.4922	.4925	.4927	.4929	.4931	.4932	.4934	.4936
2.5	.4938	.4940	.4941	.4943	.4945	.4946	.4948	.4949	.4951	.4952
2.6	.4953	.4955	.4956	.4957	.4959	.4960	.4961	.4962	.4963	.4964
2.7	.4965	.4966	.4967	.4968	.4969	.4970	.4971	.4972	.4973	.4974
2.8	.4974	.4975	.4976	.4977	.4977	.4978	.4979	.4979	.4980	.4981
2.9	.4981	.4982	.4982	.4983	.4984	.4984	.4985	.4985	.4986	.4986
3.0	.4987	.4987	.4987	.4988	.4988	.4989	.4989	.4989	.4990	.4990

Source: Abridged from Table I of A. Hald, *Statistical Tables and Formulas*. New York: John Wiley & Sons, Inc., 1952. Reproduced by permission of A. Hald and the publisher.

TABLE 5

Critical Values for Student's t

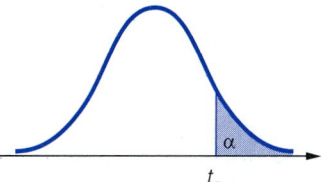

ν	$t_{.100}$	$t_{.050}$	$t_{.025}$	$t_{.010}$	$t_{.005}$	$t_{.001}$	$t_{.0005}$
1	3.078	6.314	12.706	31.821	63.657	318.31	636.62
2	1.886	2.920	4.303	6.965	9.925	22.326	31.598
3	1.638	2.353	3.182	4.541	5.841	10.213	12.924
4	1.533	2.132	2.776	3.747	4.604	7.173	8.610
5	1.476	2.015	2.571	3.365	4.032	5.893	6.869
6	1.440	1.943	2.447	3.143	3.707	5.208	5.959
7	1.415	1.895	2.365	2.998	3.499	4.785	5.408
8	1.397	1.860	2.306	2.896	3.355	4.501	5.041
9	1.383	1.833	2.262	2.821	3.250	4.297	4.781
10	1.372	1.812	2.228	2.764	3.169	4.144	4.587
11	1.363	1.796	2.201	2.718	3.106	4.025	4.437
12	1.356	1.782	2.179	2.681	3.055	3.930	4.318
13	1.350	1.771	2.160	2.650	3.012	3.852	4.221
14	1.345	1.761	2.145	2.624	2.977	3.787	4.140
15	1.341	1.753	2.131	2.602	2.947	3.733	4.073
16	1.337	1.746	2.120	2.583	2.921	3.686	4.015
17	1.333	1.740	2.110	2.567	2.898	3.646	3.965
18	1.330	1.734	2.101	2.552	2.878	3.610	3.922
19	1.328	1.729	2.093	2.539	2.861	3.579	3.883
20	1.325	1.725	2.086	2.528	2.845	3.552	3.850
21	1.323	1.721	2.080	2.518	2.831	3.527	3.819
22	1.321	1.717	2.074	2.508	2.819	3.505	3.792
23	1.319	1.714	2.069	2.500	2.807	3.485	3.767
24	1.318	1.711	2.064	2.492	2.797	3.467	3.745
25	1.316	1.708	2.060	2.485	2.787	3.450	3.725
26	1.315	1.706	2.056	2.479	2.779	3.435	3.707
27	1.314	1.703	2.052	2.473	2.771	3.421	3.690
28	1.313	1.701	2.048	2.467	2.763	3.408	3.674
29	1.311	1.699	2.045	2.462	2.756	3.396	3.659
30	1.310	1.697	2.042	2.457	2.750	3.385	3.646
40	1.303	1.684	2.021	2.423	2.704	3.307	3.551
60	1.296	1.671	2.000	2.390	2.660	3.232	3.460
120	1.289	1.658	1.980	2.358	2.617	3.160	3.373
∞	1.282	1.645	1.960	2.326	2.576	3.090	3.291

Source: From E. S. Pearson and H. O. Hartley (eds.), *The Biometrika Tables for Statisticians*, Vol. 1, 3rd ed., *Biometrika*, 1966. Reproduced by permission of the *Biometrika* Trustees.

TABLE 6

Critical Values for
the χ^2 Statistic

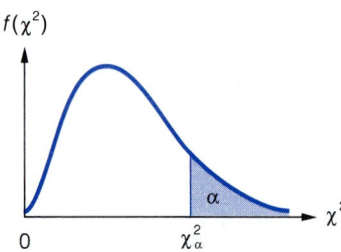

DEGREES OF FREEDOM	$\chi^2_{.995}$	$\chi^2_{.990}$	$\chi^2_{.975}$	$\chi^2_{.950}$	$\chi^2_{.900}$
1	.0000393	.0001571	.0009821	.0039321	.0157908
2	.0100251	.0201007	.0506356	.102587	.210720
3	.0717212	.114832	.215795	.351846	.584375
4	.206990	.297110	.484419	.710721	1.063623
5	.411740	.554300	.831211	1.145476	1.61031
6	.675727	.872085	1.237347	1.63539	2.20413
7	.989265	1.239043	1.68987	2.16735	2.83311
8	1.344419	1.646482	2.17973	2.73264	3.48954
9	1.734926	2.087912	2.70039	3.32511	4.16816
10	2.15585	2.55821	3.24697	3.94030	4.86518
11	2.60321	3.05347	3.81575	4.57481	5.57779
12	3.07382	3.57056	4.40379	5.22603	6.30380
13	3.56503	4.10691	5.00874	5.89186	7.04150
14	4.07468	4.66043	5.62872	6.57063	7.78953
15	4.60094	5.22935	6.26214	7.26094	8.54675
16	5.14224	5.81221	6.90766	7.96164	9.31223
17	5.69724	6.40776	7.56418	8.67176	10.0852
18	6.26481	7.01491	8.23075	9.39046	10.8649
19	6.84398	7.63273	8.90655	10.1170	11.6509
20	7.43386	8.26040	9.59083	10.8508	12.4426
21	8.03366	8.89720	10.28293	11.5913	13.2396
22	8.64272	9.54249	10.9823	12.3380	14.0415
23	9.26042	10.19567	11.6885	13.0905	14.8479
24	9.88623	10.8564	12.4011	13.8484	15.6587
25	10.5197	11.5240	13.1197	14.6114	16.4734
26	11.1603	12.1981	13.8439	15.3791	17.2919
27	11.8076	12.8786	14.5733	16.1513	18.1138
28	12.4613	13.5648	15.3079	16.9279	18.9392
29	13.1211	14.2565	16.0471	17.7083	19.7677
30	13.7867	14.9535	16.7908	18.4926	20.5992
40	20.7065	22.1643	24.4331	26.5093	29.0505
50	27.9907	29.7067	32.3574	34.7642	37.6886
60	35.5346	37.4848	40.4817	43.1879	46.4589
70	43.2752	45.4418	48.7576	51.7393	55.3290
80	51.1720	53.5400	57.1532	60.3915	64.2778
90	59.1963	61.7541	65.6466	69.1260	73.2912
100	67.3276	70.0648	74.2219	77.9295	82.3581
150	109.142	112.668	117.985	122.692	128.275
200	152.241	156.432	162.728	168.279	174.835
300	240.663	245.972	253.912	260.878	269.068
400	330.903	337.155	346.482	354.641	364.207
500	422.303	429.388	439.936	449.147	459.926

DEGREES OF FREEDOM	$\chi^2_{.100}$	$\chi^2_{.050}$	$\chi^2_{.025}$	$\chi^2_{.010}$	$\chi^2_{.005}$
1	2.70554	3.84146	5.02389	6.63490	7.87944
2	4.60517	5.99147	7.37776	9.21034	10.5966
3	6.25139	7.81473	9.34840	11.3449	12.8381
4	7.77944	9.48773	11.1433	13.2767	14.8602
5	9.23635	11.0705	12.8325	15.0863	16.7496
6	10.6446	12.5916	14.4494	16.8119	18.5476
7	12.0170	14.0671	16.0128	18.4753	20.2777
8	13.3616	15.5073	17.5346	20.0902	21.9550
9	14.6837	16.9190	19.0228	21.6660	23.5893
10	15.9871	18.3070	20.4831	23.2093	25.1882
11	17.2750	19.6751	21.9200	24.7250	26.7569
12	18.5494	21.0261	23.3367	26.2170	28.2995
13	19.8119	22.3621	24.7356	27.6883	29.8194
14	21.0642	23.6848	26.1190	29.1413	31.3193
15	22.3072	24.9958	27.4884	30.5779	32.8013
16	23.5418	26.2962	28.8454	31.9999	34.2672
17	24.7690	27.5871	30.1910	33.4087	35.7185
18	25.9894	28.8693	31.5264	34.8053	37.1564
19	27.2036	30.1435	32.8523	36.1908	38.5822
20	28.4120	31.4104	34.1696	37.5662	39.9968
21	29.6151	32.6705	35.4789	38.9321	41.4010
22	30.8133	33.9244	36.7807	40.2894	42.7956
23	32.0069	35.1725	38.0757	41.6384	44.1813
24	33.1963	36.4151	39.3641	42.9798	45.5585
25	34.3816	37.6525	40.6465	44.3141	46.9278
26	35.5631	38.8852	41.9232	45.6417	48.2899
27	36.7412	40.1133	43.1944	46.9630	49.6449
28	37.9159	41.3372	44.4607	48.2782	50.9933
29	39.0875	42.5569	45.7222	49.5879	52.3356
30	40.2560	43.7729	46.9792	50.8922	53.6720
40	51.8050	55.7585	59.3417	63.6907	66.7659
50	63.1671	67.5048	71.4202	76.1539	79.4900
60	74.3970	79.0819	83.2976	88.3794	91.9517
70	85.5271	90.5312	95.0231	100.425	104.215
80	96.5782	101.879	106.629	112.329	116.321
90	107.565	113.145	118.136	124.116	128.299
100	118.498	124.342	129.561	135.807	140.169
150	172.581	179.581	185.800	193.208	198.360
200	226.021	233.994	241.058	249.445	255.264
300	331.789	341.395	349.874	359.906	366.844
400	436.649	447.632	457.305	468.724	476.606
500	540.930	553.127	563.852	576.493	585.207

TABLE 7

Critical Values for the F Statistic: $F_{.10}$

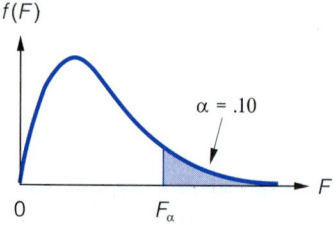

$f(F)$

$\alpha = .10$

F_α

		NUMERATOR DEGREES OF FREEDOM, ν_1							
	1	2	3	4	5	6	7	8	9
1	39.86	49.50	53.59	55.83	57.24	58.20	58.91	59.44	59.86
2	8.53	9.00	9.16	9.24	9.29	9.33	9.35	9.37	9.38
3	5.54	5.46	5.39	5.34	5.31	5.28	5.27	5.25	5.24
4	4.54	4.32	4.19	4.11	4.05	4.01	3.98	3.95	3.94
5	4.06	3.78	3.62	3.52	3.45	3.40	3.37	3.34	3.32
6	3.78	3.46	3.29	3.18	3.11	3.05	3.01	2.98	2.96
7	3.59	3.26	3.07	2.96	2.88	2.83	2.78	2.75	2.72
8	3.46	3.11	2.92	2.81	2.73	2.67	2.62	2.59	2.56
9	3.36	3.01	2.81	2.69	2.61	2.55	2.51	2.47	2.44
10	3.29	2.92	2.73	2.61	2.52	2.46	2.41	2.38	2.35
11	3.23	2.86	2.66	2.54	2.45	2.39	2.34	2.30	2.27
12	3.18	2.81	2.61	2.48	2.39	2.33	2.28	2.24	2.21
13	3.14	2.76	2.56	2.43	2.35	2.28	2.23	2.20	2.16
14	3.10	2.73	2.52	2.39	2.31	2.24	2.19	2.15	2.12
15	3.07	2.70	2.49	2.36	2.27	2.21	2.16	2.12	2.09
16	3.05	2.67	2.46	2.33	2.24	2.18	2.13	2.09	2.06
17	3.03	2.64	2.44	2.31	2.22	2.15	2.10	2.06	2.03
18	3.01	2.62	2.42	2.29	2.20	2.13	2.08	2.04	2.00
19	2.99	2.61	2.40	2.27	2.18	2.11	2.06	2.02	1.98
20	2.97	2.59	2.38	2.25	2.16	2.09	2.04	2.00	1.96
21	2.96	2.57	2.36	2.23	2.14	2.08	2.02	1.98	1.95
22	2.95	2.56	2.35	2.22	2.13	2.06	2.01	1.97	1.93
23	2.94	2.55	2.34	2.21	2.11	2.05	1.99	1.95	1.92
24	2.93	2.54	2.33	2.19	2.10	2.04	1.98	1.94	1.91
25	2.92	2.53	2.32	2.18	2.09	2.02	1.97	1.93	1.89
26	2.91	2.52	2.31	2.17	2.08	2.01	1.96	1.92	1.88
27	2.90	2.51	2.30	2.17	2.07	2.00	1.95	1.91	1.87
28	2.89	2.50	2.29	2.16	2.06	2.00	1.94	1.90	1.87
29	2.89	2.50	2.28	2.15	2.06	1.99	1.93	1.89	1.86
30	2.88	2.49	2.28	2.14	2.05	1.98	1.93	1.88	1.85
40	2.84	2.44	2.23	2.09	2.00	1.93	1.87	1.83	1.79
60	2.79	2.39	2.18	2.04	1.95	1.87	1.82	1.77	1.74
120	2.75	2.35	2.13	1.99	1.90	1.82	1.77	1.72	1.68
∞	2.71	2.30	2.08	1.94	1.85	1.77	1.72	1.67	1.63

DENOMINATOR DEGREES OF FREEDOM, ν_2

Source: From M. Merrington and C. M. Thompson, "Tables of percentage points of the inverted beta (F)-distribution," *Biometrika*, 1943, Vol. 33, pp. 73–88. Reproduced by permission of the *Biometrika* Trustees.

		NUMERATOR DEGREES OF FREEDOM, ν_1								
	10	**12**	**15**	**20**	**24**	**30**	**40**	**60**	**120**	**∞**
1	60.19	60.71	61.22	61.74	62.00	62.26	62.53	62.79	63.06	63.33
2	9.39	9.41	9.42	9.44	9.45	9.46	9.47	9.47	9.48	9.49
3	5.23	5.22	5.20	5.18	5.18	5.17	5.16	5.15	5.14	5.13
4	3.92	3.90	3.87	3.84	3.83	3.82	3.80	3.79	3.78	3.76
5	3.30	3.27	3.24	3.21	3.19	3.17	3.16	3.14	3.12	3.10
6	2.94	2.90	2.87	2.84	2.82	2.80	2.78	2.76	2.74	2.72
7	2.70	2.67	2.63	2.59	2.58	2.56	2.54	2.51	2.49	2.47
8	2.54	2.50	2.46	2.42	2.40	2.38	2.36	2.34	2.32	2.29
9	2.42	2.38	2.34	2.30	2.28	2.25	2.23	2.21	2.18	2.16
10	2.32	2.28	2.24	2.20	2.18	2.16	2.13	2.11	2.08	2.06
11	2.25	2.21	2.17	2.12	2.10	2.08	2.05	2.03	2.00	1.97
12	2.19	2.15	2.10	2.06	2.04	2.01	1.99	1.96	1.93	1.90
13	2.14	2.10	2.05	2.01	1.98	1.96	1.93	1.90	1.88	1.85
14	2.10	2.05	2.01	1.96	1.94	1.91	1.89	1.86	1.83	1.80
15	2.06	2.02	1.97	1.92	1.90	1.87	1.85	1.82	1.79	1.76
16	2.03	1.99	1.94	1.89	1.87	1.84	1.81	1.78	1.75	1.72
17	2.00	1.96	1.91	1.86	1.84	1.81	1.78	1.75	1.72	1.69
18	1.98	1.93	1.89	1.84	1.81	1.78	1.75	1.72	1.69	1.66
19	1.96	1.91	1.86	1.81	1.79	1.76	1.73	1.70	1.67	1.63
20	1.94	1.89	1.84	1.79	1.77	1.74	1.71	1.68	1.64	1.61
21	1.92	1.87	1.83	1.78	1.75	1.72	1.69	1.66	1.62	1.59
22	1.90	1.86	1.81	1.76	1.73	1.70	1.67	1.64	1.60	1.57
23	1.89	1.84	1.80	1.74	1.72	1.69	1.66	1.62	1.59	1.55
24	1.88	1.83	1.78	1.73	1.70	1.67	1.64	1.61	1.57	1.53
25	1.87	1.82	1.77	1.72	1.69	1.66	1.63	1.59	1.56	1.52
26	1.86	1.81	1.76	1.71	1.68	1.65	1.61	1.58	1.54	1.50
27	1.85	1.80	1.75	1.70	1.67	1.64	1.60	1.57	1.53	1.49
28	1.84	1.79	1.74	1.69	1.66	1.63	1.59	1.56	1.52	1.48
29	1.83	1.78	1.73	1.68	1.65	1.62	1.58	1.55	1.51	1.47
30	1.82	1.77	1.72	1.67	1.64	1.61	1.57	1.54	1.50	1.46
40	1.76	1.71	1.66	1.61	1.57	1.54	1.51	1.47	1.42	1.38
60	1.71	1.66	1.60	1.54	1.51	1.48	1.44	1.40	1.35	1.29
120	1.65	1.60	1.55	1.48	1.45	1.41	1.37	1.32	1.26	1.19
∞	1.60	1.55	1.49	1.42	1.38	1.34	1.30	1.24	1.17	1.00

DENOMINATOR DEGREES OF FREEDOM, ν_2

TABLE 8

Critical Values for the F Statistic: $F_{.05}$

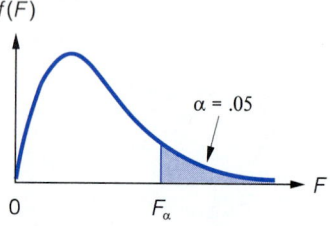

		1	2	3	4	5	6	7	8	9
		\multicolumn{9}{c}{NUMERATOR DEGREES OF FREEDOM, ν_1}
	1	161.4	199.5	215.7	224.6	230.2	234.0	236.8	238.9	240.5
	2	18.51	19.00	19.16	19.25	19.30	19.33	19.35	19.37	19.38
	3	10.13	9.55	9.28	9.12	9.01	8.94	8.89	8.85	8.81
	4	7.71	6.94	6.59	6.39	6.26	6.16	6.09	6.04	6.00
	5	6.61	5.79	5.41	5.19	5.05	4.95	4.88	4.82	4.77
	6	5.99	5.14	4.76	4.53	4.39	4.28	4.21	4.15	4.10
	7	5.59	4.74	4.35	4.12	3.97	3.87	3.79	3.73	3.68
	8	5.32	4.46	4.07	3.84	3.69	3.58	3.50	3.44	3.39
	9	5.12	4.26	3.86	3.63	3.48	3.37	3.29	3.23	3.18
	10	4.96	4.10	3.71	3.48	3.33	3.22	3.14	3.07	3.02
	11	4.84	3.98	3.59	3.36	3.20	3.09	3.01	2.95	2.90
	12	4.75	3.89	3.49	3.26	3.11	3.00	2.91	2.85	2.80
	13	4.67	3.81	3.41	3.18	3.03	2.92	2.83	2.77	2.71
	14	4.60	3.74	3.34	3.11	2.96	2.85	2.76	2.70	2.65
	15	4.54	3.68	3.29	3.06	2.90	2.79	2.71	2.64	2.59
	16	4.49	3.63	3.24	3.01	2.85	2.74	2.66	2.59	2.54
	17	4.45	3.59	3.20	2.96	2.81	2.70	2.61	2.55	2.49
	18	4.41	3.55	3.16	2.93	2.77	2.66	2.58	2.51	2.46
	19	4.38	3.52	3.13	2.90	2.74	2.63	2.54	2.48	2.42
	20	4.35	3.49	3.10	2.87	2.71	2.60	2.51	2.45	2.39
	21	4.32	3.47	3.07	2.84	2.68	2.57	2.49	2.42	2.37
	22	4.30	3.44	3.05	2.82	2.66	2.55	2.46	2.40	2.34
	23	4.28	3.42	3.03	2.80	2.64	2.53	2.44	2.37	2.32
	24	4.26	3.40	3.01	2.78	2.62	2.51	2.42	2.36	2.30
	25	4.24	3.39	2.99	2.76	2.60	2.49	2.40	2.34	2.28
	26	4.23	3.37	2.98	2.74	2.59	2.47	2.39	2.32	2.27
	27	4.21	3.35	2.96	2.73	2.57	2.46	2.37	2.31	2.25
	28	4.20	3.34	2.95	2.71	2.56	2.45	2.36	2.29	2.24
	29	4.18	3.33	2.93	2.70	2.55	2.43	2.35	2.28	2.22
	30	4.17	3.32	2.92	2.69	2.53	2.42	2.33	2.27	2.21
	40	4.08	3.23	2.84	2.61	2.45	2.34	2.25	2.18	2.12
	60	4.00	3.15	2.76	2.53	2.37	2.25	2.17	2.10	2.04
	120	3.92	3.07	2.68	2.45	2.29	2.17	2.09	2.02	1.96
	∞	3.84	3.00	2.60	2.37	2.21	2.10	2.01	1.94	1.88

Source: From M. Merrington and C. M. Thompson, "Tables of percentage points of the inverted beta (F)-distribution," *Biometrika*, 1943, Vol. 33, pp. 73–88. Reproduced by permission of the *Biometrika* Trustees.

		NUMERATOR DEGREES OF FREEDOM, ν_1								
		10	12	15	20	24	30	40	60	120	∞
DENOMINATOR DEGREES OF FREEDOM, ν_2	1	241.9	243.9	245.9	248.0	249.1	250.1	251.1	252.2	253.3	254.3
	2	19.40	19.41	19.43	19.45	19.45	19.46	19.47	19.48	19.49	19.50
	3	8.79	8.74	8.70	8.66	8.64	8.62	8.59	8.57	8.55	8.53
	4	5.96	5.91	5.86	5.80	5.77	5.75	5.72	5.69	5.66	5.63
	5	4.74	4.68	4.62	4.56	4.53	4.50	4.46	4.43	4.40	4.36
	6	4.06	4.00	3.94	3.87	3.84	3.81	3.77	3.74	3.70	3.67
	7	3.64	3.57	3.51	3.44	3.41	3.38	3.34	3.30	3.27	3.23
	8	3.35	3.28	3.22	3.15	3.12	3.08	3.04	3.01	2.97	2.93
	9	3.14	3.07	3.01	2.94	2.90	2.86	2.83	2.79	2.75	2.71
	10	2.98	2.91	2.85	2.77	2.74	2.70	2.66	2.62	2.58	2.54
	11	2.85	2.79	2.72	2.65	2.61	2.57	2.53	2.49	2.45	2.40
	12	2.75	2.69	2.62	2.54	2.51	2.47	2.43	2.38	2.34	2.30
	13	2.67	2.60	2.53	2.46	2.42	2.38	2.34	2.30	2.25	2.21
	14	2.60	2.53	2.46	2.39	2.35	2.31	2.27	2.22	2.18	2.13
	15	2.54	2.48	2.40	2.33	2.29	2.25	2.20	2.16	2.11	2.07
	16	2.49	2.42	2.35	2.28	2.24	2.19	2.15	2.11	2.06	2.01
	17	2.45	2.38	2.31	2.23	2.19	2.15	2.10	2.06	2.01	1.96
	18	2.41	2.34	2.27	2.19	2.15	2.11	2.06	2.02	1.97	1.92
	19	2.38	2.31	2.23	2.16	2.11	2.07	2.03	1.98	1.93	1.88
	20	2.35	2.28	2.20	2.12	2.08	2.04	1.99	1.95	1.90	1.84
	21	2.32	2.25	2.18	2.10	2.05	2.01	1.96	1.92	1.87	1.81
	22	2.30	2.23	2.15	2.07	2.03	1.98	1.94	1.89	1.84	1.78
	23	2.27	2.20	2.13	2.05	2.01	1.96	1.91	1.86	1.81	1.76
	24	2.25	2.18	2.11	2.03	1.98	1.94	1.89	1.84	1.79	1.73
	25	2.24	2.16	2.09	2.01	1.96	1.92	1.87	1.82	1.77	1.71
	26	2.22	2.15	2.07	1.99	1.95	1.90	1.85	1.80	1.75	1.69
	27	2.20	2.13	2.06	1.97	1.93	1.88	1.84	1.79	1.73	1.67
	28	2.19	2.12	2.04	1.96	1.91	1.87	1.82	1.77	1.71	1.65
	29	2.18	2.10	2.03	1.94	1.90	1.85	1.81	1.75	1.70	1.64
	30	2.16	2.09	2.01	1.93	1.89	1.84	1.79	1.74	1.68	1.62
	40	2.08	2.00	1.92	1.84	1.79	1.74	1.69	1.64	1.58	1.51
	60	1.99	1.92	1.84	1.75	1.70	1.65	1.59	1.53	1.47	1.39
	120	1.91	1.83	1.75	1.66	1.61	1.55	1.50	1.43	1.35	1.25
	∞	1.83	1.75	1.67	1.57	1.52	1.46	1.39	1.32	1.22	1.00

TABLE 9

Critical Values for the F Statistic: $F_{.025}$

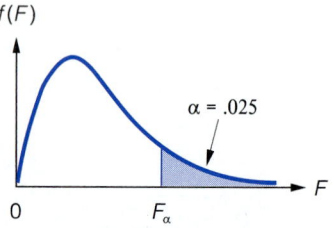

		NUMERATOR DEGREES OF FREEDOM, ν_1							
	1	2	3	4	5	6	7	8	9
1	647.8	799.5	864.2	899.6	921.8	937.1	948.2	956.7	963.3
2	38.51	39.00	39.17	39.25	39.30	39.33	39.36	39.37	39.39
3	17.44	16.04	15.44	15.10	14.88	14.73	14.62	14.54	14.47
4	12.22	10.65	9.98	9.60	9.36	9.20	9.07	8.98	8.90
5	10.01	8.43	7.76	7.39	7.15	6.98	6.85	6.76	6.68
6	8.81	7.26	6.60	6.23	5.99	5.82	5.70	5.60	5.52
7	8.07	6.54	5.89	5.52	5.29	5.12	4.99	4.90	4.82
8	7.57	6.06	5.42	5.05	4.82	4.65	4.53	4.43	4.36
9	7.21	5.71	5.08	4.72	4.48	4.32	4.20	4.10	4.03
10	6.94	5.46	4.83	4.47	4.24	4.07	3.95	3.85	3.78
11	6.72	5.26	4.63	4.28	4.04	3.88	3.76	3.66	3.59
12	6.55	5.10	4.47	4.12	3.89	3.73	3.61	3.51	3.44
13	6.41	4.97	4.35	4.00	3.77	3.60	3.48	3.39	3.31
14	6.30	4.86	4.24	3.89	3.66	3.50	3.38	3.29	3.21
15	6.20	4.77	4.15	3.80	3.58	3.41	3.29	3.20	3.12
16	6.12	4.69	4.08	3.73	3.50	3.34	3.22	3.12	3.05
17	6.04	4.62	4.01	3.66	3.44	3.28	3.16	3.06	2.98
18	5.98	4.56	3.95	3.61	3.38	3.22	3.10	3.01	2.93
19	5.92	4.51	3.90	3.56	3.33	3.17	3.05	2.96	2.88
20	5.87	4.46	3.86	3.51	3.29	3.13	3.01	2.91	2.84
21	5.83	4.42	3.82	3.48	3.25	3.09	2.97	2.87	2.80
22	5.79	4.38	3.78	3.44	3.22	3.05	2.93	2.84	2.76
23	5.75	4.35	3.75	3.41	3.18	3.02	2.90	2.81	2.73
24	5.72	4.32	3.72	3.38	3.15	2.99	2.87	2.78	2.70
25	5.69	4.29	3.69	3.35	3.13	2.97	2.85	2.75	2.68
26	5.66	4.27	3.67	3.33	3.10	2.94	2.82	2.73	2.65
27	5.63	4.24	3.65	3.31	3.08	2.92	2.80	2.71	2.63
28	5.61	4.22	3.63	3.29	3.06	2.90	2.78	2.69	2.61
29	5.59	4.20	3.61	3.27	3.04	2.88	2.76	2.67	2.59
30	5.57	4.18	3.59	3.25	3.03	2.87	2.75	2.65	2.57
40	5.42	4.05	3.46	3.13	2.90	2.74	2.62	2.53	2.45
60	5.29	3.93	3.34	3.01	2.79	2.63	2.51	2.41	2.33
120	5.15	3.80	3.23	2.89	2.67	2.52	2.39	2.30	2.22
∞	5.02	3.69	3.12	2.79	2.57	2.41	2.29	2.19	2.11

DENOMINATOR DEGREES OF FREEDOM, ν_2

Source: From M. Merrington and C. M. Thompson, "Tables of percentage points of the inverted beta (F)-distribution," *Biometrika*, 1943, Vol. 33, pp. 73–88. Reproduced by permission of the *Biometrika* Trustees.

					NUMERATOR DEGREES OF FREEDOM, ν_1					
		10	12	15	20	24	30	40	60	120	∞
DENOMINATOR DEGREES OF FREEDOM, ν_2	1	968.6	976.7	984.9	993.1	997.2	1001	1006	1010	1014	1018
	2	39.40	39.41	39.43	39.45	39.46	39.46	39.47	39.48	39.49	39.50
	3	14.42	14.34	14.25	14.17	14.12	14.08	14.04	13.99	13.95	13.90
	4	8.84	8.75	8.66	8.56	8.51	8.46	8.41	8.36	8.31	8.26
	5	6.62	6.52	6.43	6.33	6.28	6.23	6.18	6.12	6.07	6.02
	6	5.46	5.37	5.27	5.17	5.12	5.07	5.01	4.96	4.90	4.85
	7	4.76	4.67	4.57	4.47	4.42	4.36	4.31	4.25	4.20	4.14
	8	4.30	4.20	4.10	4.00	3.95	3.89	3.84	3.78	3.73	3.67
	9	3.96	3.87	3.77	3.67	3.61	3.56	3.51	3.45	3.39	3.33
	10	3.72	3.62	3.52	3.42	3.37	3.31	3.26	3.20	3.14	3.08
	11	3.53	3.43	3.33	3.23	3.17	3.12	3.06	3.00	2.94	2.88
	12	3.37	3.28	3.18	3.07	3.02	2.96	2.91	2.85	2.79	2.72
	13	3.25	3.15	3.05	2.95	2.89	2.84	2.78	2.72	2.66	2.60
	14	3.15	3.05	2.95	2.84	2.79	2.73	2.67	2.61	2.55	2.49
	15	3.06	2.96	2.86	2.76	2.70	2.64	2.59	2.52	2.46	2.40
	16	2.99	2.89	2.79	2.68	2.63	2.57	2.51	2.45	2.38	2.32
	17	2.92	2.82	2.72	2.62	2.56	2.50	2.44	2.38	2.32	2.25
	18	2.87	2.77	2.67	2.56	2.50	2.44	2.38	2.32	2.26	2.19
	19	2.82	2.72	2.62	2.51	2.45	2.39	2.33	2.27	2.20	2.13
	20	2.77	2.68	2.57	2.46	2.41	2.35	2.29	2.22	2.16	2.09
	21	2.73	2.64	2.53	2.42	2.37	2.31	2.25	2.18	2.11	2.04
	22	2.70	2.60	2.50	2.39	2.33	2.27	2.21	2.14	2.08	2.00
	23	2.67	2.57	2.47	2.36	2.30	2.24	2.18	2.11	2.04	1.97
	24	2.64	2.54	2.44	2.33	2.27	2.21	2.15	2.08	2.01	1.94
	25	2.61	2.51	2.41	2.30	2.24	2.18	2.12	2.05	1.98	1.91
	26	2.59	2.49	2.39	2.28	2.22	2.16	2.09	2.03	1.95	1.88
	27	2.57	2.47	2.36	2.25	2.19	2.13	2.07	2.00	1.93	1.85
	28	2.55	2.45	2.34	2.23	2.17	2.11	2.05	1.98	1.91	1.83
	29	2.53	2.43	2.32	2.21	2.15	2.09	2.03	1.96	1.89	1.81
	30	2.51	2.41	2.31	2.20	2.14	2.07	2.01	1.94	1.87	1.79
	40	2.39	2.29	2.18	2.07	2.01	1.94	1.88	1.80	1.72	1.64
	60	2.27	2.17	2.06	1.94	1.88	1.82	1.74	1.67	1.58	1.48
	120	2.16	2.05	1.94	1.82	1.76	1.69	1.61	1.53	1.43	1.31
	∞	2.05	1.94	1.83	1.71	1.64	1.57	1.48	1.39	1.27	1.00

TABLE 10
Critical Values for the F Statistic: $F_{.01}$

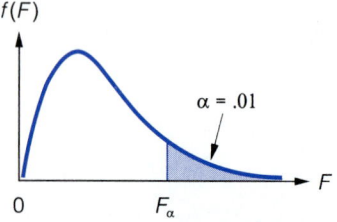

$f(F)$

$\alpha = .01$

$0 \qquad F_\alpha \qquad\qquad F$

		NUMERATOR DEGREES OF FREEDOM, ν_1							
	1	2	3	4	5	6	7	8	9
1	4,052	4,999.5	5,403	5,625	5,764	5,859	5,928	5,982	6,022
2	98.50	99.00	99.17	99.25	99.30	99.33	99.36	99.37	99.39
3	34.12	30.82	29.46	28.71	28.24	27.91	27.67	27.49	27.35
4	21.20	18.00	16.69	15.98	15.52	15.21	14.98	14.80	14.66
5	16.26	13.27	12.06	11.39	10.97	10.67	10.46	10.29	10.16
6	13.75	10.92	9.78	9.15	8.75	8.47	8.26	8.10	7.98
7	12.25	9.55	8.45	7.85	7.46	7.19	6.99	6.84	6.72
8	11.26	8.65	7.59	7.01	6.63	6.37	6.18	6.03	5.91
9	10.56	8.02	6.99	6.42	6.06	5.80	5.61	5.47	5.35
10	10.04	7.56	6.55	5.99	5.64	5.39	5.20	5.06	4.94
11	9.65	7.21	6.22	5.67	5.32	5.07	4.89	4.74	4.63
12	9.33	6.93	5.95	5.41	5.06	4.82	4.64	4.50	4.39
13	9.07	6.70	5.74	5.21	4.86	4.62	4.44	4.30	4.19
14	8.86	6.51	5.56	5.04	4.69	4.46	4.28	4.14	4.03
15	8.68	6.36	5.42	4.89	4.56	4.32	4.14	4.00	3.89
16	8.53	6.23	5.29	4.77	4.44	4.20	4.03	3.89	3.78
17	8.40	6.11	5.18	4.67	4.34	4.10	3.93	3.79	3.68
18	8.29	6.01	5.09	4.58	4.25	4.01	3.84	3.71	3.60
19	8.18	5.93	5.01	4.50	4.17	3.94	3.77	3.63	3.52
20	8.10	5.85	4.94	4.43	4.10	3.87	3.70	3.56	3.46
21	8.02	5.78	4.87	4.37	4.04	3.81	3.64	3.51	3.40
22	7.95	5.72	4.82	4.31	3.99	3.76	3.59	3.45	3.35
23	7.88	5.66	4.76	4.26	3.94	3.71	3.54	3.41	3.30
24	7.82	5.61	4.72	4.22	3.90	3.67	3.50	3.36	3.26
25	7.77	5.57	4.68	4.18	3.85	3.63	3.46	3.32	3.22
26	7.72	5.53	4.64	4.14	3.82	3.59	3.42	3.29	3.18
27	7.68	5.49	4.60	4.11	3.78	3.56	3.39	3.26	3.15
28	7.64	5.45	4.57	4.07	3.75	3.53	3.36	3.23	3.12
29	7.60	5.42	4.54	4.04	3.73	3.50	3.33	3.20	3.09
30	7.56	5.39	4.51	4.02	3.70	3.47	3.30	3.17	3.07
40	7.31	5.18	4.31	3.83	3.51	3.29	3.12	2.99	2.89
60	7.08	4.98	4.13	3.65	3.34	3.12	2.95	2.82	2.72
120	6.85	4.79	3.95	3.48	3.17	2.96	2.79	2.66	2.56
∞	6.63	4.61	3.78	3.32	3.02	2.80	2.64	2.51	2.41

DENOMINATOR DEGREES OF FREEDOM, ν_2

Source: From M. Merrington and C. M. Thompson, "Tables of percentage points of the inverted beta (F)-distribution," *Biometrika*, 1943, Vol. 33, pp. 73–88. Reproduced by permission of the *Biometrika* Trustees.

		10	12	15	20	24	30	40	60	120	∞
		NUMERATOR DEGREES OF FREEDOM, ∞_1									
DENOMINATOR DEGREES OF FREEDOM, ν_2	1	6,056	6,106	6,157	6,209	6,235	6,261	6,287	6,313	6,339	6,366
	2	99.40	99.42	99.43	99.45	99.46	99.47	99.47	99.48	99.49	99.50
	3	27.23	27.05	26.87	26.69	26.60	26.50	26.41	26.32	26.22	26.13
	4	14.55	14.37	14.20	14.02	13.93	13.84	13.75	13.65	13.56	13.46
	5	10.05	9.89	9.72	9.55	9.47	9.38	9.29	9.20	9.11	9.02
	6	7.87	7.72	7.56	7.40	7.31	7.23	7.14	7.06	6.97	6.88
	7	6.62	6.47	6.31	6.16	6.07	5.99	5.91	5.82	5.74	5.65
	8	5.81	5.67	5.52	5.36	5.28	5.20	5.12	5.03	4.95	4.86
	9	5.26	5.11	4.96	4.81	4.73	4.65	4.57	4.48	4.40	4.31
	10	4.85	4.71	4.56	4.41	4.33	4.25	4.17	4.08	4.00	3.91
	11	4.54	4.40	4.25	4.10	4.02	3.94	3.86	3.78	3.69	3.60
	12	4.30	4.16	4.01	3.86	3.78	3.70	3.62	3.54	3.45	3.36
	13	4.10	3.96	3.82	3.66	3.59	3.51	3.43	3.34	3.25	3.17
	14	3.94	3.80	3.66	3.51	3.43	3.35	3.27	3.18	3.09	3.00
	15	3.80	3.67	3.52	3.37	3.29	3.21	3.13	3.05	2.96	2.87
	16	3.69	3.55	3.41	3.26	3.18	3.10	3.02	2.93	2.84	2.75
	17	3.59	3.46	3.31	3.16	3.08	3.00	2.92	2.83	2.75	2.65
	18	3.51	3.37	3.23	3.08	3.00	2.92	2.84	2.75	2.66	2.57
	19	3.43	3.30	3.15	3.00	2.92	2.84	2.76	2.67	2.58	2.49
	20	3.37	3.23	3.09	2.94	2.86	2.78	2.69	2.61	2.52	2.42
	21	3.31	3.17	3.03	2.88	2.80	2.72	2.64	2.55	2.46	2.36
	22	3.26	3.12	2.98	2.83	2.75	2.67	2.58	2.50	2.40	2.31
	23	3.21	3.07	2.93	2.78	2.70	2.62	2.54	2.45	2.35	2.26
	24	3.17	3.03	2.89	2.74	2.66	2.58	2.49	2.40	2.31	2.21
	25	3.13	2.99	2.85	2.70	2.62	2.54	2.45	2.36	2.27	2.17
	26	3.09	2.96	2.81	2.66	2.58	2.50	2.42	2.33	2.23	2.13
	27	3.06	2.93	2.78	2.63	2.55	2.47	2.38	2.29	2.20	2.10
	28	3.03	2.90	2.75	2.60	2.52	2.44	2.35	2.26	2.17	2.06
	29	3.00	2.87	2.73	2.57	2.49	2.41	2.33	2.23	2.14	2.03
	30	2.98	2.84	2.70	2.55	2.47	2.39	2.30	2.21	2.11	2.01
	40	2.80	2.66	2.52	2.37	2.29	2.20	2.11	2.02	1.92	1.80
	60	2.63	2.50	2.35	2.20	2.12	2.03	1.94	1.84	1.73	1.60
	120	2.47	2.34	2.19	2.03	1.95	1.86	1.76	1.66	1.53	1.38
	∞	2.32	2.18	2.04	1.88	1.79	1.70	1.59	1.47	1.32	1.00

TABLE 11

Critical Values for the Wilcoxon Rank Sum Test

a. $\alpha = .025$ one-tailed; $\alpha = .05$ two-tailed

		n_1															
		3		**4**		**5**		**6**		**7**		**8**		**9**		**10**	
		T_L	T_U	T_L	T_U	T_L	T_U	T_L	T_U	T_L	T_U	T_L	T_U	T_L	T_U	T_L	T_U
	3	5	16	6	18	6	21	7	23	7	26	8	28	8	31	9	33
	4	6	18	11	25	12	28	12	32	13	35	14	38	15	41	16	44
	5	6	21	12	28	18	37	19	41	20	45	21	49	22	53	24	56
	6	7	23	12	32	19	41	26	52	28	56	29	61	31	65	32	70
n_2	7	7	26	13	35	20	45	28	56	37	68	39	73	41	78	43	83
	8	8	28	14	38	21	49	29	61	39	73	49	87	51	93	54	98
	9	8	31	15	41	22	53	31	65	41	78	51	93	63	108	66	114
	10	9	33	16	44	24	56	32	70	43	83	54	98	66	114	79	131

b. $\alpha = .05$ one-tailed; $\alpha = .10$ two-tailed

		n_1															
		3		**4**		**5**		**6**		**7**		**8**		**9**		**10**	
		T_L	T_U	T_L	T_U	T_L	T_U	T_L	T_U	T_L	T_U	T_L	T_U	T_L	T_U	T_L	T_U
	3	6	15	7	17	7	20	8	22	9	24	9	27	10	29	11	31
	4	7	17	12	24	13	27	14	30	15	33	16	36	17	39	18	42
	5	7	20	13	27	19	36	20	40	22	43	24	46	25	50	26	54
	6	8	22	14	30	20	40	28	50	30	54	32	58	33	63	35	67
n_2	7	9	24	15	33	22	43	30	54	39	66	41	71	43	76	46	80
	8	9	27	16	36	24	46	32	58	41	71	52	84	54	90	57	95
	9	10	29	17	39	25	50	33	63	43	76	54	90	66	105	69	111
	10	11	31	18	42	26	54	35	67	46	80	57	95	69	111	83	127

Source: From F. Wilcoxon and R. A. Wilcox, "Some Rapid Approximate Statistical Procedures," 1964, pp. 20–23. Reproduced with the permission of American Cyanamid Company.

TABLE 12
Critical Values for the Wilcoxon Signed Ranks Test

ONE-TAILED	TWO-TAILED	$n = 5$	$n = 6$	$n = 7$	$n = 8$	$n = 9$	$n = 10$
$\alpha = .05$	$\alpha = .10$	1	2	4	6	8	10
$\alpha = .025$	$\alpha = .05$		1	2	4	6	8
$\alpha = .01$	$\alpha = .02$			0	2	3	5
$\alpha = .005$	$\alpha = .01$				0	2	3
		$n = 11$	$n = 12$	$n = 13$	$n = 14$	$n = 15$	$n = 16$
$\alpha = .05$	$\alpha = .10$	14	17	21	26	30	36
$\alpha = .025$	$\alpha = .05$	11	14	17	21	25	30
$\alpha = .01$	$\alpha = .02$	7	10	13	16	20	24
$\alpha = .005$	$\alpha = .01$	5	7	10	13	16	19
		$n = 17$	$n = 18$	$n = 19$	$n = 20$	$n = 21$	$n = 22$
$\alpha = .05$	$\alpha = .10$	41	47	54	60	68	75
$\alpha = .025$	$\alpha = .05$	35	40	46	52	59	66
$\alpha = .01$	$\alpha = .02$	28	33	38	43	49	56
$\alpha = .005$	$\alpha = .01$	23	28	32	37	43	49
		$n = 23$	$n = 24$	$n = 25$	$n = 26$	$n = 27$	$n = 28$
$\alpha = .05$	$\alpha = .10$	83	92	101	110	120	130
$\alpha = .025$	$\alpha = .05$	73	81	90	98	107	117
$\alpha = .01$	$\alpha = .02$	62	69	77	85	93	102
$\alpha = .005$	$\alpha = .01$	55	61	68	76	84	92
		$n = 29$	$n = 30$	$n = 31$	$n = 32$	$n = 33$	$n = 34$
$\alpha = .05$	$\alpha = .10$	141	152	163	175	188	201
$\alpha = .025$	$\alpha = .05$	127	137	148	159	171	183
$\alpha = .01$	$\alpha = .02$	111	120	130	141	151	162
$\alpha = .005$	$\alpha = .01$	100	109	118	128	138	149
		$n = 35$	$n = 36$	$n = 37$	$n = 38$	$n = 39$	
$\alpha = .05$	$\alpha = .10$	214	228	242	256	271	
$\alpha = .025$	$\alpha = .05$	195	208	222	235	250	
$\alpha = .01$	$\alpha = .02$	174	186	198	211	224	
$\alpha = .005$	$\alpha = .01$	160	171	183	195	208	
		$n = 40$	$n = 41$	$n = 42$	$n = 43$	$n = 44$	$n = 45$
$\alpha = .05$	$\alpha = .10$	287	303	319	336	353	371
$\alpha = .025$	$\alpha = .05$	264	279	295	311	327	344
$\alpha = .01$	$\alpha = .02$	238	252	267	281	297	313
$\alpha = .005$	$\alpha = .01$	221	234	248	262	277	292
		$n = 46$	$n = 47$	$n = 48$	$n = 49$	$n = 50$	
$\alpha = .05$	$\alpha = .10$	389	408	427	446	466	
$\alpha = .025$	$\alpha = .05$	361	379	397	415	434	
$\alpha = .01$	$\alpha = .02$	329	345	362	380	398	
$\alpha = .005$	$\alpha = .01$	307	323	339	356	373	

Source: From F. Wilcoxon and R. A. Wilcox, "Some Rapid Approximate Statistical Procedures." 1964, p. 28. Reproduced with the permission of American Cyanamid Company.

TABLE 13
Critical Values of Spearman's Rank Correlation Coefficient

n	$\alpha = .10$ / $\alpha = .05$	$\alpha = .05$ / $\alpha = .025$	$\alpha = .02$ / $\alpha = .01$	$\alpha = .01$ ← TWO-TAILED / $\alpha = .005$ ← ONE-TAILED
5	.900	—	—	—
6	.829	.886	.943	—
7	.714	.786	.893	—
8	.643	.738	.833	.881
9	.600	.683	.783	.833
10	.564	.648	.745	.794
11	.523	.623	.736	.818
12	.497	.591	.703	.780
13	.475	.566	.673	.745
14	.457	.545	.646	.716
15	.441	.525	.623	.689
16	.425	.507	.601	.666
17	.412	.490	.582	.645
18	.399	.476	.564	.625
19	.388	.462	.549	.608
20	.377	.450	.534	.591
21	.368	.438	.521	.576
22	.359	.428	.508	.562
23	.351	.418	.496	.549
24	.343	.409	.485	.537
25	.336	.400	.475	.526
26	.329	.392	.465	.515
27	.323	.385	.456	.505
28	.317	.377	.448	.496
29	.311	.370	.440	.487
30	.305	.364	.432	.478

Source: From E. G. Olds, "Distribution of sums of squares of rank differences for small samples," *Annals of Mathematical Statistics*, 1938, Vol. 9. Reproduced with the permission of the Editor, *Annals of Mathematical Statistics*.

TABLE 14
Critical Values for the Theil Zero-Slope Test

x	4	5	8	9	12	13	16	17	20	21	24	25	28	29	32	33	36	37	40
0	.625	.592	.548	.540	.527	.524	.518	.516	.513	.512	.510	.509	.508	.507	.506	.506	.505	.505	.505
2	.375	.408	.452	.460	.473	.476	.482	.484	.487	.488	.490	.491	.492	.493	.494	.494	.495	.495	.495
4	.167	.242	.360	.381	.420	.429	.447	.452	.462	.464	.471	.472	.477	.478	.481	.482	.484	.484	.486
6	.042	.117	.274	.306	.369	.383	.412	.420	.436	.441	.451	.454	.461	.463	.468	.469	.473	.474	.477
8		.042	.199	.238	.319	.338	.378	.388	.411	.417	.432	.436	.446	.448	.455	.457	.462	.464	.468
10		.008	.138	.179	.273	.295	.345	.358	.387	.394	.413	.418	.430	.434	.442	.445	.452	.453	.459
12			.089	.130	.230	.255	.313	.328	.362	.371	.394	.400	.415	.419	.430	.433	.441	.443	.449
14			.054	.090	.190	.218	.282	.299	.339	.349	.375	.382	.400	.405	.417	.421	.430	.433	.440
16			.031	.060	.155	.184	.253	.271	.315	.327	.356	.364	.385	.390	.405	.409	.420	.423	.431
18			.016	.038	.125	.153	.225	.245	.293	.306	.338	.347	.370	.376	.392	.397	.409	.413	.422
20			.007	.022	.098	.126	.199	.220	.271	.285	.320	.330	.355	.362	.380	.385	.399	.403	.413
22			.002	.012	.076	.102	.175	.196	.250	.265	.303	.314	.341	.348	.368	.373	.388	.393	.404
24			.001	.006	.058	.082	.153	.174	.230	.246	.286	.297	.326	.334	.356	.362	.378	.383	.395
26			.000	.003	.043	.064	.133	.154	.211	.228	.270	.282	.312	.321	.344	.350	.368	.373	.386
28				.001	.031	.050	.114	.135	.193	.210	.254	.266	.298	.308	.332	.339	.358	.363	.377
30				.000	.022	.038	.097	.118	.176	.193	.238	.251	.285	.295	.320	.328	.347	.353	.369
32					.016	.029	.083	.102	.159	.177	.223	.237	.272	.282	.309	.317	.338	.344	.360
34					.010	.021	.070	.088	.144	.162	.209	.222	.259	.270	.298	.306	.328	.334	.351
36					.007	.015	.058	.076	.130	.147	.195	.209	.246	.257	.287	.295	.318	.325	.343
38					.004	.011	.048	.064	.117	.134	.181	.196	.234	.246	.276	.285	.308	.315	.334
40					.003	.007	.039	.054	.104	.121	.169	.183	.222	.234	.265	.274	.299	.306	.326
42					.002	.005	.032	.046	.093	.109	.156	.171	.211	.223	.255	.264	.290	.297	.318
44					.001	.003	.026	.038	.082	.098	.145	.159	.200	.212	.244	.254	.280	.288	.309
46					.000	.002	.021	.032	.073	.088	.134	.148	.189	.201	.234	.244	.271	.279	.301
48						.001	.016	.026	.064	.079	.123	.138	.178	.191	.224	.235	.262	.271	.293
50						.001	.013	.021	.056	.070	.113	.128	.168	.181	.215	.225	.254	.262	.285
52						.000	.010	.017	.049	.062	.104	.118	.158	.171	.206	.216	.245	.254	.277
54							.008	.014	.043	.055	.095	.109	.149	.162	.197	.207	.237	.245	.270
56							.006	.011	.037	.049	.087	.101	.140	.153	.188	.199	.228	.237	.262
58							.004	.009	.032	.043	.079	.093	.131	.144	.179	.190	.220	.229	.255
60							.003	.007	.027	.037	.072	.085	.123	.136	.171	.182	.212	.222	.247
62							.002	.005	.023	.032	.066	.078	.115	.128	.163	.174	.204	.214	.240
64							.002	.004	.020	.028	.059	.071	.108	.120	.155	.166	.197	.206	.233
66							.001	.003	.017	.024	.054	.065	.101	.112	.147	.158	.189	.199	.226
68							.001	.002	.014	.021	.048	.059	.094	.105	.140	.151	.182	.192	.219
70							.001	.002	.012	.018	.044	.054	.087	.099	.133	.144	.175	.185	.212
72							.000	.001	.010	.015	.039	.049	.081	.092	.126	.137	.168	.178	.205
74								.001	.008	.013	.035	.044	.075	.086	.119	.130	.161	.171	.199
76								.001	.007	.011	.031	.040	.070	.080	.113	.124	.155	.165	.192
78								.000	.006	.009	.028	.036	.065	.075	.107	.117	.148	.158	.186
80									.005	.008	.025	.032	.060	.070	.101	.111	.142	.152	.180
82									.004	.007	.022	.029	.055	.065	.095	.106	.136	.146	.174
84									.003	.005	.019	.026	.051	.060	.090	.100	.130	.140	.168
86									.002	.005	.017	.023	.047	.056	.085	.095	.124	.134	.162
88									.002	.004	.015	.021	.043	.052	.080	.090	.119	.129	.156
90									.002	.003	.013	.018	.039	.048	.075	.085	.114	.123	.151
92									.001	.002	.011	.016	.036	.044	.070	.080	.108	.118	.146
94									.001	.002	.010	.014	.033	.041	.066	.075	.103	.113	.140
96									.001	.002	.009	.013	.030	.037	.062	.071	.099	.108	.135
98									.001	.001	.007	.011	.027	.034	.058	.067	.094	.103	.130
100									.000	.001	.006	.010	.025	.031	.054	.063	.089	.098	.125

Continued

TABLE 14 (Continued)

x	6	7	10	11	14	15	18	19	22	23	26	27	30	31	34	35	38	39
1	.500	.500	.500	.500	.500	.500	.500	.500	.500	.500	.500	.500	.500	.500	.500	.500	.500	.500
3	.360	.386	.431	.440	.457	.461	.470	.473	.478	.479	.483	.484	.486	.487	.488	.489	.490	.490
5	.235	.281	.364	.381	.415	.423	.441	.445	.456	.458	.465	.467	.472	.473	.477	.478	.480	.481
7	.136	.191	.300	.324	.374	.385	.411	.418	.434	.438	.448	.451	.458	.460	.465	.466	.470	.472
9	.068	.119	.242	.271	.334	.349	.383	.391	.412	.417	.431	.434	.444	.446	.453	.455	.460	.462
11	.028	.068	.190	.223	.295	.313	.354	.365	.390	.397	.414	.418	.430	.433	.442	.444	.450	.452
13	.008	.035	.146	.179	.259	.279	.327	.339	.369	.377	.397	.402	.416	.420	.430	.433	.440	.443
15	.001	.015	.108	.141	.225	.248	.300	.314	.348	.357	.380	.386	.402	.407	.418	.422	.431	.433
17		.005	.078	.109	.194	.218	.275	.290	.328	.338	.363	.371	.389	.394	.407	.411	.421	.424
19		.001	.054	.082	.165	.190	.250	.267	.308	.319	.347	.355	.375	.381	.396	.400	.411	.414
21		.000	.036	.060	.140	.164	.227	.245	.289	.301	.331	.340	.362	.368	.384	.389	.401	.405
23			.023	.043	.117	.141	.205	.223	.270	.283	.316	.325	.349	.355	.373	.378	.392	.396
25			.014	.030	.096	.120	.184	.203	.252	.265	.300	.310	.336	.343	.362	.368	.382	.387
27			.008	.020	.079	.101	.165	.184	.234	.248	.285	.296	.323	.331	.351	.357	.373	.377
29			.005	.013	.063	.084	.147	.166	.217	.232	.270	.281	.310	.318	.340	.347	.363	.368
31			.002	.008	.050	.070	.130	.149	.201	.216	.256	.268	.298	.306	.329	.336	.354	.359
33			.001	.005	.040	.057	.115	.133	.186	.201	.242	.254	.286	.295	.319	.326	.345	.350
35			.000	.003	.031	.046	.100	.119	.171	.187	.229	.241	.274	.283	.308	.316	.336	.341
37				.002	.024	.037	.088	.105	.157	.173	.216	.228	.262	.272	.298	.306	.327	.333
39				.001	.018	.029	.076	.093	.144	.160	.203	.216	.251	.261	.288	.296	.318	.324
41				.000	.013	.023	.066	.082	.131	.147	.191	.204	.239	.250	.278	.286	.309	.315
43					.010	.018	.056	.072	.120	.135	.179	.192	.228	.239	.268	.277	.300	.307
45					.007	.014	.048	.062	.109	.124	.168	.181	.218	.229	.259	.267	.291	.298
47					.005	.010	.041	.054	.099	.114	.157	.170	.208	.219	.249	.258	.283	.290
49					.003	.008	.034	.047	.089	.104	.147	.160	.198	.209	.240	.249	.274	.282
51					.002	.006	.029	.040	.080	.094	.137	.150	.188	.199	.231	.240	.266	.274
53					.002	.004	.024	.034	.072	.086	.127	.141	.178	.190	.222	.232	.258	.266
55					.001	.003	.020	.029	.064	.078	.118	.132	.169	.181	.213	.223	.250	.258
57					.001	.002	.016	.025	.058	.070	.110	.123	.160	.172	.205	.215	.242	.250
59					.000	.001	.013	.021	.051	.063	.102	.115	.152	.164	.196	.206	.234	.243
61						.001	.011	.017	.045	.057	.094	.107	.144	.155	.188	.198	.227	.235
63						.001	.009	.014	.040	.051	.087	.099	.136	.147	.180	.191	.219	.228
65						.000	.007	.012	.035	.046	.080	.092	.128	.140	.173	.183	.212	.221
67							.005	.010	.031	.041	.073	.085	.121	.132	.165	.176	.205	.214
69							.004	.008	.027	.036	.067	.079	.114	.125	.158	.168	.198	.207
71							.003	.006	.024	.032	.062	.073	.107	.118	.151	.161	.191	.200
73							.003	.005	.021	.028	.057	.067	.100	.112	.144	.154	.184	.193
75							.002	.004	.018	.025	.052	.062	.094	.105	.137	.148	.177	.187
77							.001	.003	.015	.022	.047	.057	.088	.099	.131	.141	.171	.180
79							.001	.003	.013	.019	.043	.052	.083	.093	.125	.135	.165	.174
81							.001	.002	.011	.017	.039	.048	.077	.088	.119	.129	.158	.168
83							.001	.002	.010	.015	.035	.044	.072	.082	.113	.123	.152	.162
85							.000	.001	.008	.013	.032	.040	.067	.077	.107	.117	.147	.156
87								.001	.007	.011	.029	.036	.063	.072	.102	.112	.141	.150
89								.001	.006	.009	.026	.033	.059	.068	.097	.107	.135	.145
91								.001	.005	.008	.023	.030	.054	.063	.092	.101	.130	.139
93								.000	.004	.007	.021	.027	.051	.059	.087	.096	.125	.134
95									.003	.006	.019	.025	.047	.055	.082	.092	.120	.129
97									.003	.005	.017	.022	.043	.052	.078	.087	.115	.124
99									.002	.004	.015	.020	.040	.048	.074	.083	.110	.119
101									.002	.004	.013	.018	.037	.045	.070	.078	.105	.114

ASP TUTORIAL

This appendix provides an overview of the ASP program. It gives the minimal hardware requirements and start-up procedures necessary to begin an ASP session on a personal computer (PC). This tutorial is not intended to replace any of the ASP documentation manuals available from the publisher or DMC Software, Inc.

ASP must be run on an IBM-compatible PC with at least 512K of memory, two disk drives (either one hard drive and one floppy drive, or two floppy drives), and DOS 2.0 or higher. A blank formatted floppy disk is also required for data storage, unless your PC has a hard drive (i.e., fixed disk) available for storing data.

HARDWARE REQUIREMENTS

GETTING STARTED

To use the ASP program, you must first load it into the memory of the computer. To accomplish this when starting ASP from a floppy disk:

1. Insert your copy of ASP into either of your two disk drives, drive A or drive B. (Assume drive A.)
2. Type **A:** and press **ENTER** to make drive A the current drive.

    A: ⟨**ENTER**⟩
3. Type **ASP** and press **ENTER** to load the ASP program into memory.

    ASP ⟨**ENTER**⟩

The ASP disk must remain in drive A for as long as you are using the program.

To start ASP from a fixed disk or hard drive (e.g., drive C), it is first necessary to install ASP on the fixed disk. This is accomplished by placing your copy of the ASP disk into floppy drive A and entering the following commands at the DOS prompt:

```
C:          ⟨ENTER⟩
MD \ASP     ⟨ENTER⟩
CD \ASP     ⟨ENTER⟩
COPY A:*.*  ⟨ENTER⟩
```

(This sequence of DOS commands assumes the drive letter of the fixed disk is **C** and that the subdirectory in which the ASP program resides is **\ASP**.) Once ASP has been installed on the fixed disk, it need not be installed again. The ASP program can then be started at any point in the future by entering the following commands at the DOS prompt:

```
C:          ⟨ENTER⟩
CD \ASP     ⟨ENTER⟩
ASP         ⟨ENTER⟩
```

989

THE MAIN MENU

The initial screen to appear as the ASP program is loaded into memory displays copyright and licensing information. After reading this information, press any key to obtain the MAIN MENU shown in Figure H.1.

The MAIN MENU is a typical ASP "bounce bar" menu. The highlighted bar can be moved from option to option by pressing the SPACE BAR, the cursor control keys ($\rightarrow \leftarrow \uparrow \downarrow$), or the TAB key. Once your selection is made, press **ENTER** to display submenus associated with the option. (You can also make a selection by pressing the letter of the desired option.)

Table H.1 gives a brief description of each of the MAIN MENU options and the corresponding chapters in the text. Several of these options contain statistical procedures that are beyond the scope of the text. Only the statistical routines covered in the text are described in the table.

TABLE H.1
Options on the Main Menu

OPTION	DESCRIPTION	CHAPTER(S)
A. Analysis of Variance	One-way and two-way ANOVAs	13
B. Regression Analysis	Simple and multiple regression; residual analysis	11, 12
C. Correlation Matrix	Bivariate correlations	11
D. Summary Statistics	Mean, median, standard deviation, etc.	3
E. Probability Dists.	Binomial and normal distributions	5, 6
F. File Management Menu	Creating, saving, editing data	—
G. Time Series Analysis	(Beyond the scope of this text)	—
H. Hypothesis Tests	Confidence intervals and hypothesis tests for means, proportions, and variances; one-way table χ^2 test; nonparametric tests	8, 9, 10, 14, 15
I. INSTRUCTIONS	A short tutorial on the use of ASP	—
J. Factor Analysis Menu	(Beyond the scope of this text)	—
K. Miscellaneous Plots	Stem-and-leaf display, box plot, normal probability plot, scatter plot	2, 3, 6, 11
L. Crosstab/Contingency	Two-way (contingency) table χ^2 test	14
M. Auxiliary Programs	(Beyond the scope of this text)	—
N. Enter a DOS Command	Enter and execute DOS commands within an ASP session	—
O. Scr./Data Dir. Dflts.	Set the color scheme on the monitor; set the default directory and printer port	—

ALTERNATE COMMANDS MENU

All of the statistical routines in ASP are accessible through the MAIN MENU. However, additional commands can be executed through the ALT COMMANDS MENU. The ALT COMMANDS MENU is called by pressing the **F1** function key from the main menu or from any menu one level below the main menu. The ALT COMMANDS MENU appears as shown in Figure H.2.

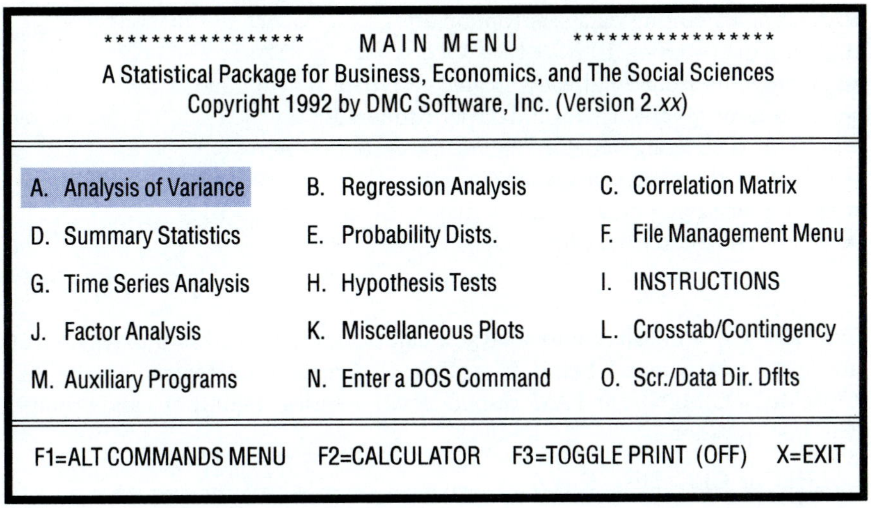

FIGURE H.1
The ASP Main Menu

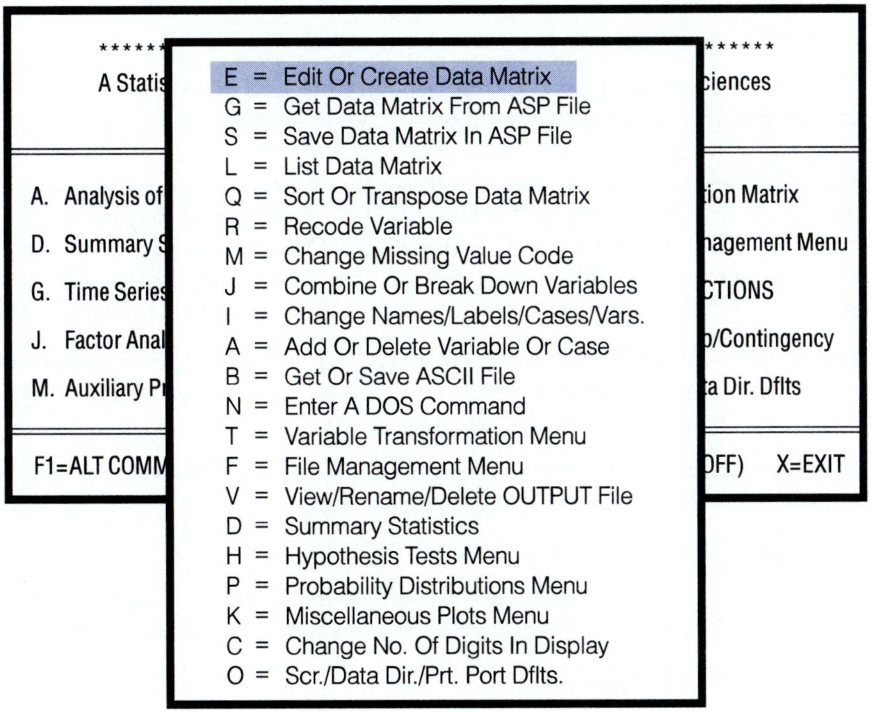

```
****************   MAIN MENU   ****************
       A Statistical Package for Business, Economics, and The Social Sciences
              Copyright 1992 by DMC Software, Inc. (Version 2.xx)


A.  Analysis of Variance      B.  Regression Analysis      C.  Correlation Matrix

D.  Summary Statistics        E.  Probability Dists.       F.  File Management Menu

G.  Time Series Analysis      H.  Hypothesis Tests         I.  INSTRUCTIONS

J.  Factor Analysis           K.  Miscellaneous Plots      L.  Crosstab/Contingency

M.  Auxiliary Programs        N.  Enter a DOS Command      O.  Scr./Data Dir. Dflts


F1=ALT COMMANDS MENU    F2=CALCULATOR    F3=TOGGLE PRINT (OFF)    X=EXIT
```

FIGURE H.2
The Alt Commands Menu

```
E  =  Edit Or Create Data Matrix
G  =  Get Data Matrix From ASP File
S  =  Save Data Matrix In ASP File
L  =  List Data Matrix
Q  =  Sort Or Transpose Data Matrix
R  =  Recode Variable
M  =  Change Missing Value Code
J  =  Combine Or Break Down Variables
I  =  Change Names/Labels/Cases/Vars.
A  =  Add Or Delete Variable Or Case
B  =  Get Or Save ASCII File
N  =  Enter A DOS Command
T  =  Variable Transformation Menu
F  =  File Management Menu
V  =  View/Rename/Delete OUTPUT File
D  =  Summary Statistics
H  =  Hypothesis Tests Menu
P  =  Probability Distributions Menu
K  =  Miscellaneous Plots Menu
C  =  Change No. Of Digits In Display
O  =  Scr./Data Dir./Prt. Port Dflts.
```

You can execute the commands on this menu by either moving the cursor to the desired option and pressing **ENTER** or by pressing the letter associated with the option. You will find this menu most useful for:

- creating or editing data sets (option **E**)
- listing data (option **L**)
- getting data from an already created ASP data set (option **G**)
- creating new variables for a data set (option **T**)
- adding or deleting variables and/or cases (option **A**)
- changing the names of variables (option **I**)
- getting or saving data in an external ASCII file (option **B**)
- saving an ASP data set (option **S**)

CREATING A DATA MATRIX

Typically, you will use ASP to analyze a data set. To do this, you must first create an ASP "data matrix." Select **E = Edit or Create Data Matrix** on the ALT COMMANDS MENU, and ASP responds with a series of questions and prompts. The first question is:

EDIT or CREATE?   E

Note that the ASP default answer is **E** for EDIT. This is used when you want to edit an existing ASP data matrix. To create a data matrix, press the letter C (for CREATE). You are now prompted with the question:

Number of Variables?   1

Change the default to the correct number and press **ENTER**. ASP creates names for the variables using the convention Var1, Var2, Var3, etc., then asks:

Are Names OK?   Y

To change the name, Press **N** (for No). ASP will then ask you to enter the new name of each variable. Once this is completed, ASP will prompt you, one case (i.e., one observation) and one variable at a time, to enter the data into the data matrix.

Important: The ASP data editor will not accept letters or special characters (e.g., dollar sign, comma) as data. Only whole numbers or numbers with decimals should be entered into the data matrix.

When data entry is complete, press **X** to exit the numerical data editor. Several questions will be asked, the most important being:

Do You Wish to Save the Data Matrix?   Y

Answer "yes" by pressing **ENTER**. ASP will then ask for the drive letter (e.g., drive A) and directory of the disk where you want to save the data:

DATA DIRECTORY: A:

If the default is correct, press **ENTER**. Otherwise, enter the correct drive/path. You will be asked to name the ASP data file, provide a file label (optional), and whether you want to save all variables and all cases.

Suppose you enter the following file name:

File Name: MYDATA

ASP will save your data matrix in the ASP file named MYDATA.ASP in the directory specified earlier. In future ASP sessions, you can access this data set by first selecting the option **G = Get Data Matrix from ASP File** from the ALT COMMANDS MENU, and then selecting MYDATA from the resulting menu-list of ASP data files.

ANALYZING A DATA MATRIX

To analyze an ASP data matrix that you have just created or accessed, return to the MAIN MENU by pressing **X** or **ESC**. From the MAIN MENU, select the desired statistical routine. Each choice will result in a series of submenus, prompts, and/or questions similar to those shown previously. After making your selections, ASP will perform the analysis and display the results immediately on the monitor screen. ASP menu selections at the bottom of the screen permit you to send the output directly to a printer or to save the output in a file for future use.

AVAILABLE DOCUMENTATION

- **ASP User's Manual** (by DMC Software, Inc.)—available free to adopters of the text from the publisher of the text, Macmillan/Dellen Publishing.
- **ASP Tutorial and Student Guide** (by George Blackford)—can be purchased directly from DMC Software, Inc., or from your campus book store.

ANSWERS TO ODD-NUMBERED EXERCISES

CHAPTER 1

1.1 a. Hawaiian properties   **b.** Quantitative; quantitative; qualitative; qualitative   **1.3 a.** Workers at a large Canadian manufacturing firm   **b.** Qualitative; quantitative; qualitative; quantitative; quantitative   **1.5 a.** Quantitative   **b.** Quantitative   **c.** Qualitative   **d.** Qualitative   **e.** Qualitative   **f.** Qualitative   **g.** Quantitative   **1.7 a.** Qualitative   **b.** Qualitative   **c.** Quantitative   **1.9 a.** Quality measures (defective or not) of all manufactured gear shifts   **b.** Quality measures of the 100 gear shifts selected for testing   **c.** No   **1.11 a.** Numbers of cavities in all school children in the United States   **b.** Numbers of cavities found in 1,000 children examined   **c.** Compute the sample average number of cavities per child; no   **1.13 a.** Opinions of all American adults concerning support of euthanasia   **b.** No   **c.** Results of survey could possibly be used to develop legislation regarding euthanasia   **1.15 a.** Diameters of all stones found in delta region   **b.** Diameters of 50 stones collected   **c.** No   **1.19** Many consumers, who would otherwise buy the product, may refuse to participate in the survey.   **1.21** Not every person will return the questionnaire.   **1.25 a.** Quantitative   **b.** Qualitative   **c.** Qualitative   **d.** Quantitative   **1.27 a.** Checkout times of all customers shopping at a Publix supermarket   **b.** No   **1.29** Method B; method A   **1.31 a.** Directions of travel of all sea turtle hatchlings released on a dark beach   **b.** Directions of travel for the 60 turtle hatchlings used in the experiment   **c.** Qualitative

CHAPTER 2

2.1 Total of class frequencies is 200; missing relative frequencies: .18, .45, .15, .14   **2.7 b.** Quest feelings much more frequent for adoptees 18 years or older   **2.9 c.** 46%   **2.13** Yes   **2.15 a.** 21, 22, 23, 24, 25, 26, 27, 28, 29, 30, 31, 32

b.

21	3, 5
22	4, 6, 7, 8
23	4, 7
24	1
25	4
26	5, 6, 7, 8
27	0, 4
28	5, 8
29	1
30	3, 3
31	6, 9
32	0

2.17 a.

1	1, 1, 2, 6, 6
2	1
3	3, 5
4	0, 3, 5
5	0, 3, 9
6	3, 4, 5, 7
7	3, 4, 6
8	2, 4, 4, 6, 9
9	4, 7

2.19 b. .923   **2.21 a.**

6	0, 2
7	0, 7, 8
8	2, 4
9	0, 1, 2, 6
10	1, 1, 3, 5, 5, 7

b. 5.88%

2.23 Brand name   **2.25 a.** 8.6   **b.** 1.8   **c.** (1.05, 2.85); (2.85, 4.65); (4.65, 6.45); (6.45, 8.25); (8.25, 10.05)   **2.27 a.** .16   **b.** .26   **2.29 a.** Frequency histogram   **b.** 250   **c.** Yes; skewed right

2.31 b.

0	0, 0, 0, 0, 0, 0
1	0, 0, 0, 0
2	0, 0, 0, 0, 0, 0, 0, 0, 0
3	0, 0, 0, 0, 0, 0, 0, 0, 0, 0, 0
4	0, 0, 0, 0, 0, 0, 0, 0, 0
5	0, 0, 0, 0, 0
6	0
7	0, 0, 0, 0, 0, 0
8	0, 0
9	0, 0
10	0
11	
12	0
13	0

2.33 a. Quantitative **b.** Frequency distribution **c.** .28 **2.35 b.** 36 smallmouth buffalo; 98 channel catfish; 10 largemouth bass **2.37 a.** 9 **b.** 5 **2.39 b.** .09 **c.** 75% **d.** Possibly; depends on whether the sample of women was randomly selected **2.41 a.** 27 **b.** 26 **c.** 84 **d.** 21 **2.43 c.** Yes **2.49 b.** Love **c.** 59% **2.51 b.** 31.1% **2.53 a.** 56% **b.** 75% **2.55 a.** 64% **b.** Skewed left **c.** 6%

CHAPTER 3

3.1 a. 12 **b.** 40 **c.** 7 **d.** 21 **e.** 144 **3.3 a.** 11.2 **b.** 12 **c.** 30 **3.5 a.** 6 **b.** 50 **c.** 42.8 **3.7** 4.6, 4 **3.9 a.** 5, 5, 5 **b.** 12, 5, 5 **c.** 0, 0, 0 **d.** 0, 4.5, 9 **3.11** Mean = 91.044; median = 92

3.13 a.

0	0, 0, 4, 4, 5, 5, 5, 7, 8, 8, 9
1	0, 6
2	
3	6

b. 8.36, 6, 5; median

3.15 a.

0	3, 4, 4, 4, 4
0	5, 5, 5, 5, 5, 6, 6, 6, 6, 7, 7, 9
1	0, 0, 0, 1, 1, 2, 3
1	5, 5, 5
2	0, 3, 4
2	

b. 9.33, 7, 5; yes

3.17 a. 12,869.15 **b.** 8,763.5 **c.** Median **3.19 a.** 10, 14, 3.74 **b.** 10, 10.4, 3.22 **c.** 0, 0, 0 **3.21** 3.66, 1.91 **3.23 a.** $\approx$70% **b.** $\approx$95% **c.** Almost all **3.25 b.** Tchebysheff's Theorem **3.27 a.** (3.95, 12.03) **b.** No; 20 falls outside $\bar{x} \pm 3s$ **3.29 b.** 159.38, 77.4, 262.5 **c.** $\frac{29}{30}$ = .967 **d.** 114.26, 76.0, 90.0 **3.31 a.** 4.04, 9.16, 3.03 **b.** Approx. 95% (Empirical Rule) **c.** .96 **d.** Decrease; decrease **3.33** 4.028, 5.506, 2.346 **3.35 b.** 15.3, 2.95 **3.37** 8.66, 373.69, 19.33 **3.39 a.** −1.6 **b.** .6 **c.** 1.6 **3.41** 228, 266.5, 291 **3.43** 3.3, 6.1, 8.2 **3.45** 75% of the starting salaries lie below $33,000; 25% lie above $33,000. **3.47 a.** 1.0 **b.** Possibly; .44 is within 2 std. dev. of the mean ($z = 1.73$) **3.49 a.** $\bar{x} = 6.03, s = 8.13$ **b.** $\bar{x} = 6.86, s = 8.47$ **c.** 1.72 **d.** 1.55 **e.** Fat tissue **3.51** Inner fences: 133.5, 385.5; Outer fences: 39, 480 **3.53 a.** 12.60; yes **b.** Measurement from a different population **3.55 a.** Suspect outlier: 8,600 **b.** Suspect outliers: 40, 43; highly suspect outlier: 61 **3.57 a.** Yes; 34,200 and 53,944 **b.** 1.77 and 3.42 **3.59 b.** .0329, .0163 **c.** .96 **3.61 b.** .169, .052 **c.** .96 **3.63** Yes, $z = 3.09$ **3.65 b.** −1.56 **c.** −2.23 **3.67 a.** 99% of the rounding errors are less than .53 **b.** 4.43 **3.69 a.** 35.06, 17.915 **b.** .64 **c.** .98 **d.** 1.00 **e.** Empirical Rule **3.71 a.** 29.45, 30, 30 **b.** 50, 87.02, 9.33 **c.** (20.12, 38.78) .70; (10.79, 48.11) .95; (1.46, 57.44) 1.00 **d.** 18 **e.** $Q_L = 23$, $Q_U = 35$, IQR = 12, inner fences: 5, 53; outer fences: −13, 71; suspect outliers: 57; highly suspect outliers: none **3.73 a.** 51.11, 66.02 **b.** 42.42, 27.04 **c.** $z = .30$; approx. 80th percentile **d.** $z = .02$; approx. 50th percentile **e.** $z = .32$; approx. 88th percentile; $z = 1.13$; approx. 92nd percentile **3.75** 1,013.85; 556.18

CHAPTER 4

4.1 a. $\frac{1}{6}$ **4.3** .55 **4.5** .06 **4.9 a.** $\frac{1}{2}, \frac{1}{2}, \frac{1}{2}$ **b.** (4, 5, 6); (2, 4, 6); (1, 3, 5) **c.** $\frac{1}{2}, \frac{1}{2}, \frac{1}{2}$ **d.** B and C **4.11 a.** $\frac{1}{36}$ **b.** $\frac{4}{36}$ **c.** $\frac{2}{36}$ **4.15 a.** $\frac{1}{10}$ **b.** $\frac{8}{10}$ **4.17 a.** (40, 300), (40, 350), (40, 400), (45, 300), (45, 350), (45, 400), (50, 300), (50, 350), (50, 400) **4.19 a.** .38 **b.** .25 **c.** .88 **4.21 a.** .18 **b.** .28 **c.** .99 **d.** .58 **4.23 a.** .97 **b.** .04 **c.** No **4.25** $\frac{8}{32} = \frac{1}{4}$ **4.27 a.** AB, AC, AD, AE, BC, BD, BE, CD, CE, DE **b.** 10 **4.31 a.** $\frac{3}{10}$ **b.** No **4.33** 1,820 **4.35 a.** AA, AB, AC; $\frac{3}{9}$ **b.** AB, BB, CB; $\frac{3}{9}$ **c.** $\frac{1}{3}$ **d.** Yes **e.** $\frac{1}{9}$ **4.37 a.** No **b.** No **c.** No **4.39 a.** $\frac{6}{60}$ **b.** $\frac{9}{40}$ **4.41 a.** .0256 **b.** .418 **4.43 a.** .0000000055 **b.** .000000005 **c.** .0000000119 **4.45 a.** $\frac{228}{234} = .974$ **b.** $\frac{3}{25} = .12$ **4.47 a.** (1/1,836)(1/1,365)(1/495) = .000000000806 **b.** Yes **c.** Yes **4.49** .7 **4.51 a.** $\frac{181}{387}$ **b.** $\frac{79}{387}$ **c.** $\frac{8}{387}$ **d.** $\frac{218}{387}$ **e.** $\frac{38}{65}$ **f.** $\frac{20}{78}$ **4.53 a.** .000025 **b.** .00002 **4.55 a.** .95 **b.** .23 **c.** .64 **d.** .95 **4.57 a.** .038 **b.** .614 **c.** .096 **d.** .103 **e.** .438 **4.59 a.** .90 **b.** .27 **c.** .198 **d.** .432 **4.61** All three pairs of events are mutually exclusive. **4.63 b.** .28 **c.** .56 **4.65 a.** .80 **b.** .20 **4.67** (B, C) and (A, B) are mutually exclusive. **4.69 a.** .0344 **b.** .1250 **c.** .1384 **4.71** .045 **4.73** 2,598,960 **4.75 a.** .57 **b.** .098 **c.** .143 **4.77 a.** $\frac{8}{36}$ **b.** $\frac{4}{36}$ **c.** $\frac{8}{36}$ **d.** $\frac{4}{36}$ **4.79 b.** No **4.81** .01782

CHAPTER 5

5.1 a. −5, 0, 2, 5 **b.** 2 **c.** .5 **d.** .2 **5.3 a.** .15 **b.** .55 **c.** .90 **5.5 a.** .8, .16, .032, .0064, .00128 **b.** .99968; no **c.** .96 **5.7 a.** Yes, probabilities sum to 1 **c.** .267 **d.** .107

5.9

x	0	1	2
p(x)	.09	.42	.49

5.11 a. Between $-\$5$ and $\$20$ **b.** $-\$15$ **5.13 a.** 24   **b.** 4   **c.** 10   **d.** .064   **e.** .2304   **5.15 a.** .0625, .25, .375, .25, .0625   **b.** .3125
c. .6875   **5.17 a.** Yes   **b.** Hotel offers shampoo in their guest rooms   **c.** .86   **d.** .383   **e.** .853   **5.19 a.** .064   **b.** .432   **c.** .648
5.21 Yes   **5.23 a.** .1515   **b.** .5125   **c.** .2027   **5.25 a.** .0648   **b.** .0144   **c.** .7735   **d.** .9154   **5.27 a.** .3487   **b.** .3874   **c.** .1937
5.29 a. .8369   **b.** .9976   **c.** .0308   **d.** .3087   **5.31 a.** .8386   **b.** .002   **5.33 a.** .0010   **b.** .6230   **c.** .9990   **5.35 a.** .8065   **b.** .0001
c. .9995   **d.** Very likely that $\pi < .70$ **5.37 a.** .0978   **b.** .1935   **5.39 a.** .8725   **b.** .5956   **c.** .0510   **5.41** Approx. .95
5.43 a. 99, .995   **b.** 80, 4   **c.** 50, 5   **d.** 20, 4   **e.** 1, .995   **5.45 a.** 891, 2.98   **b.** 720, 12   **c.** 450, 15   **d.** 180, 12   **e.** 9, 2.98
5.47 a. 25,256.57   **b.** 152.27   **c.** 36.41   **5.49 a.** .0389; .0388985   **b.** No   **5.51 a.** 20   **b.** 4.24   **c.** 20 ± 8.48 **d.** Not very likely
5.53 a. Yes   **b.** .4   **c.** 200   **d.** (178.1, 221.9)   **e.** No   **5.55 a.** .2734   **b.** .3633   **c.** .0352   **5.57 a.** .0712   **b.** Yes   **c.** Yes, claim is
most likely false   **d.** .2252; no; no   **5.59 a.** .3585   **b.** .7358   **c.** .7358   **d.** .2642   **e.** Claim is probably false   **5.61** Yes; $n = 50$, $\pi = .4$
5.63 No   **5.65 a.** .0282   **b.** .9984   **c.** .0001   **5.67 a.** .5793   **b.** .1091   **c.** .0982

CHAPTER 6

6.1 a. .3849   **b.** .4319   **c.** .1844   **d.** .4147   **e.** .0918   **6.3 a.** .25   **b.** .92   **c.** 1.28   **d.** 1.65   **e.** 1.96   **6.5 a.** .75   **b.** -1.0
c. -1.625 **d.** 2.00   **e.** -2.00 **6.7 a.** -1.28 **b.** -1.04 **c.** $-.52$ **d.** 0   **6.9 a.** .9406   **b.** .0068   **6.11 a.** .2676   **b.** No
6.13 a. .2033   **b.** No   **c.** ≈ 0 **d.** No   **6.15 a.** .0228   **b.** .9544   **c.** 38.58 seconds   **6.17 a.** Approx. 0   **b.** No   **6.19** .0080
6.21 a. .9678   **b.** .8729   **c.** .9946   **d.** Not well   **6.23 a.** Yes   **b.** .1335   **c.** .2592   **6.25** $IQR/s \approx 1.2$; data appears to be approx. normal
6.27 No; $\bar{x} \pm 2s$ includes negative numbers.   **6.29 a.** .9848   **b.** .9878   **6.31 a.** .9955   **b.** .0934   **c.** .1122   **d.** .7291   **6.33 a.** .0139
b. .4052   **c.** .0171, .4164; yes   **6.35 a.** .0017   **b.** .0823   **c.** .0823   **6.37 a.** Approx. 0   **b.** No   **6.39 a.** .8577   **b.** .0082
6.41 a. .0020   **b.** .4522   **6.43 a.** .5359   **b.** Approx. 1   **6.45 a.** .0548   **b.** 97 days   **6.47 a.** .7642   **b.** .2037   **c.** 54,175 miles
6.49 a. .6736   **b.** .3264   **c.** .5000   **d.** .3264   **6.51 a.** New   **b.** Standard

CHAPTER 7

7.1 c. Mean $= 4.73$, std. dev. $= 1.275$ **7.9 a.** 10, 2   **b.** 20, 1   **c.** 50, 30   **d.** 100, 20   **7.11 a.** .1056   **b.** .9641   **c.** .6508   **d.** .0122
7.13 a. .0043   **b.** .6065   **7.15 a.** .0166   **b.** Strong evidence that the mean is greater than 3.0   **7.17 a.** Approx. normal with $\mu_{\bar{x}} = 28.5$
and $\sigma_{\bar{x}} = 5.98$ **b.** .6203   **c.** .5987   **7.19 a.** Approx. normal with mean $= 400$ and std. dev. $= 10$ **b.** .1587   **c.** .3023
7.21 a. Approx. normal with mean $= 121.74$ and std. dev. $= 4.86$ **b.** .7348   **7.23 a.** Approx. normal with mean $= 17$ and std. dev. $= 2$
b. Approx. normal with mean $= 17$ and std. dev. $= 1$ **c.** $P(15 < \bar{x}_{100} < 19)$ **d.** .6826, .9544   **7.25 a.** Approx. normal with mean $= 9.8$
and std. dev. $= .078$ **b.** .8997   **c.** No, $z = -3.21$ **7.27 a.** Approx. normal with mean $= 90$ and std. dev. $= .52$ **b.** Approx. 0   **c.** Yes
7.31 a. .5752   **b.** .1335   **c.** 5.2 ± 1.44 **7.33 a.** .0179   **b.** Yes   **7.35** .0062   **7.37 a.** .0062   **b.** Yes   **c.** Rare event

CHAPTER 8

8.1 Proportion of similarly constructed intervals that capture the population mean   **8.3 a.** 81 ± 1.97 **b.** 81 ± 2.35 **c.** 81 ± 3.10
8.5 a. $5.7 \pm .66$ **b.** 5.7 ± 1.03 **8.9 a.** 5.4   **b.** 5.4 ± 1.09 **8.11** $29.07 \pm .86$ **8.13** 10.18 ± 2.36
8.15 c. Increase n and/or decrease the confidence coefficient   **d.** No   **8.17** $\bar{x}$ may not be normally distributed;
s may not be a good estimator of σ.   **8.19 a.** 5 ± 2.02 **b.** 5 ± 4.37 **8.23 a.** 9.9 ± 4.81 **b.** 6.7 ± 6.19 **8.25 a.** 5.99 ± 3.14
c. Distribution of TCDD levels is approx. normal   **d.** Distribution is skewed right.   **8.29 a.** .031   **b.** .016   **c.** .043   **8.31 a.** $.2 \pm .078$
b. $.2 \pm .035$ **8.33 a.** .075   **b.** $.075 \pm .028$ **8.35** $.137 \pm .079$ **8.37** $.18 \pm .035$ **8.39** $.55 \pm .02$ **8.41 a.** 10, 1.31   **b.** 13, 2
8.43 Normal with equal variances; independent and random   **8.45 a.** 9.8 ± 7.94 **b.** 9.8 ± 9.61 **c.** 9.8 ± 13.14 **8.47** $-.33 \pm .22$
8.49 a. -5 ± 1.36 **8.51 a.** 189.50 ± 156.83 **8.53 a.** -3.82 ± 4.05 **b.** Yes   **8.55 a.** 1.75, 2.63   **b.** $\mu_d = \mu_1 - \mu_2$ **c.** 1.75 ± 4.18
8.57 19.3 ± 1.47 **8.61 a.** -1.67 ± 1.36 **b.** Yes   **8.63 a.** 5.45 ± 2.49 **b.** Yes   **8.65 b.** 11.15 ± 6.80 **c.** 11.15 ± 14.21
8.67 a. .1, .057   **b.** .05, .037   **c.** $-.2$, .033   **8.69 a.** $-.08 \pm .062$ **b.** $-.08 \pm .081$ **8.71 a.** $.152 \pm .140$ **b.** Yes   **c.** Yes
8.73 a. $.13 \pm .035$ **b.** Yes; white former smokers had more difficulty quitting.   **8.75 a.** $.33 \pm .027$
b. Increase sample sizes or decrease confidence coefficient   **8.77 a.** 683   **b.** 246   **c.** 426   **8.79 a.** 57   **b.** 99   **c.** 596   **8.81** 60
8.83 86   **8.85** 4,802   **8.87** 354   **8.89 a.** (4.54, 8.81)   **b.** (.00024, .00085)   **c.** (641.86, 1,809.09)   **d.** (.95, 12.66)   **8.91** (4.27, 65.97)
8.93 (.81, .95)   **8.95 a.** (.163, 29.141)   **8.97 a.** $.333 \pm .072$ **8.99** $-.027 \pm .024$ **8.101 a.** $.9167 \pm 1.5124$ **8.103** 2.92 to 3.72; no
evidence of a difference between means   **8.105 a.** NY $.262 \pm .0084$; WI $.150 \pm .0265$; ME $.661 \pm .0833$; FL $.005 \pm .0047$; VA $.224 \pm .0937$
b. $-.511 \pm .074$ **8.107** -40 ± 95 **8.109** (.0000457, .0003216)   **8.111** $.75 \pm .1225$

CHAPTER 9

9.3 a. One-tailed **b.** One-tailed **c.** Two-tailed **d.** Two-tailed **e.** One-tailed **f.** One-tailed
9.5 a. α is probability of falsely rejecting H_0 **b.** $\beta = P(\text{Type II error})$ is unknown. **c.** $\alpha + \beta = 1$ implies we will always make an error (Type I or Type II). **9.7 c.** Type II error; Type I error **9.9 a.** -3.33 **b.** -2.36 **c.** 2.36 **d.** 1.18 **9.11 a.** $.025$ **b.** $.05$ **c.** $.01$
9.13 $z < -1.96$ or $z > 1.96$ **9.15 a.** $.0250$ **b.** $.05$ **c.** $.0038$ **d.** $.1056$ **9.17 a.** $.0022$ **b.** $.0571$ **c.** $.0139$ **d.** $.003$
9.19 Approx. 0; reject H_0. **9.23 a.** $1{,}016.31$ **b.** 74.91 **c.** 13.63 **9.25 a.** $.5080$ **b.** $.6141$ **c.** $.8186$ **9.27** Approx. 1 **9.29** Decreased
9.31 No **9.35 a.** $|z| > 2.58$ **b.** $|z| > 2.33$ **c.** $|z| > 2.05$ **9.37 a.** $H_0: (\pi_1 - \pi_2) = 0$, $H_a: (\pi_1 - \pi_2) \neq 0$
b. $H_0: \mu = 22$, $H_a: \mu < 22$ **c.** $H_0: \pi = \frac{1}{6}$, $H_a: \pi \neq \frac{1}{6}$ **d.** $H_0: (\mu_1 - \mu_2) = 0$, $H_a: (\mu_1 - \mu_2) > 0$
e. $H_0: (\mu_1 - \mu_2) = 0$, $H_a: (\mu_1 - \mu_2) > 0$ **9.39 a.** -3.13 **b.** $|z| > 1.645$ **c.** Reject H_0. **e.** Approx. 0 **9.41 a.** $\alpha > .07$ **b.** $\alpha < .07$

CHAPTER 10

10.1 a. $.33$ **b.** -10.64 **c.** -1.67 **10.3 a.** $|t| > 2.145$ **b.** $|t| > 2.977$ **c.** $t < -1.761$ **d.** $t > 1.533$ **e.** $t > 1.318$ **10.5 a.** -1.84;
do not reject H_0. **b.** -1.84; reject H_0. **c.** $.0329$ **d.** $.0329$ **10.7 a.** $z = .70$; do not reject H_0. **b.** Yes **10.9** No, $z = -1.82$
10.11 a. Yes, $t = 2.89$ **c.** p-value $= .0088/2 = .0044$; yes **10.13 a.** Yes, $z = 2.27$ **b.** No, $t = 1.07$ **c.** $.0116$ **10.15 a.** $|z| > 1.96$
b. $z < -1.645$ **c.** $z > 1.28$ **d.** $z < -2.33$ **e.** $|z| > 2.58$ **10.17 a.** $z = -1.00$; do not reject H_0. **b.** $z = 2.10$; reject H_0.
c. $z = 1.02$; do not reject H_0. **10.19 a.** $H_0: \pi = .20$, $H_a: \pi > .20$ **b.** $z = 11.00$; reject H_0. **c.** Approx. 0 **10.21** Yes, $z = -2.00$
10.23 $z = -3.28$; reject H_0. **10.25 a.** $|z| > 2.58$ **b.** $|z| > 1.645$ **c.** $z > 1.645$ **d.** $z < -1.645$ **10.27 a.** $z = 2.03$, reject H_0
b. $z = 1.02$, do not reject H_0. **10.29** $t = 1.53$, do not reject H_0. **10.31 a.** $H_0: (\mu_1 - \mu_2) = 0$, $H_a: (\mu_1 - \mu_2) < 0$ **c.** Reject H_0.
10.33 No, $t = -.58$ **10.35** $z = -2.09$, reject H_0. **10.37** $z = 7.42$, reject H_0. **10.39 a.** $|z| > 2.58$; $z = 9.20$; reject H_0.
b. $t > 2.132$; $t = 13.42$; reject H_0. **c.** $t < -1.476$; $t = -3.27$; reject H_0. **10.41 a.** $t = 3.32$, reject H_0. **b.** $t = 2.10$, do not reject H_0.
10.43 No, $t = 2.96$ **10.45 a.** -7.18 **b.** $.0001$ **10.47 a.** $H_0: (\mu_B - \mu_A) = 0$, $H_a: (\mu_B - \mu_A) \neq 0$ **b.** $t = 1.58$, do not reject H_0.
10.49 a. $|z| > 2.58$, $z = 6.17$; reject H_0. **b.** $z > 1.645$, $z = 3.65$; reject H_0. **c.** $z < -2.33$, $z = -.80$; do not reject H_0.
10.51 a. $z = 3.27$, reject H_0. **10.53** Yes, $z = -4.05$ **10.55** No, $z = 1.19$ **10.57 a.** Yes, $z = 1.96$ **b.** No; results may be invalid
10.61 a. $\chi^2 > 36.1908$ **b.** $\chi^2 < 6.84398$ or $\chi^2 > 38.5822$ **c.** $\chi^2 < 4.57481$ **d.** $\chi^2 < 5.57779$ **10.63** $\chi^2 = 35.6$, reject H_0.
10.65 No, $\chi^2 = 2.9714$ **10.67 a.** No, $\chi^2 = 6.912$ **10.69 a.** 3.18 **b.** 2.62 **c.** 2.10 **10.71** Both populations have normal distributions.
10.73 $F = 5.87$; reject H_0. **10.75 a.** No, $F = 1.09$ **10.77** Yes, $F = 1.75$ **10.79** Yes, $z = 11.14$ **10.81 a.** No, $F = 1.64$
b. No, $F = 1.42$ **10.83** Obj. 1: $z = -4.38$, reject H_0. Obj. 2: $z = 4.19$, reject H_0. Obj. 3: $z = -.40$, do not reject H_0. **10.85** Yes, $t = 3.96$
10.87 a. $z = -7.75$, reject H_0; refute claim **b.** Approx. 0 **10.89** $z = 1.83$, reject H_0. **10.91 a.** Yes, $z = 4.96$

CHAPTER 11

11.1 a. 3.5 **b.** 5.5 **d.** 4.5 **e.** Points fall exactly on the line **11.3 a.** $-.5$ **b.** -2.5 **d.** -1.5 **e.** Points fall exactly on the line
11.7 a. 15 **b.** 10 **c.** 2.2 **d.** 1 **e.** 1.5 **f.** $.7$ **11.9 b.** $\hat{y} = 4.986 - 1.934x$ **11.11 b.** $\hat{y} = 1.255 - .398x$
11.13 b. $\hat{y} = 4.4417 + .2612x$ **d.** 6.27 **11.15 a.** $\hat{y} = .2134 + 2.4264x$ **11.17 a.** Yes **b.** $\hat{\beta}_0 = 1.192$, $\hat{\beta}_1 = .987$ **11.19 a.** $.675$
b. $s^2 = .225$, $s = .474$ **11.21 a.** SSE $= 30.768$, $s^2 = 2.051$ **b.** 1.432 **11.23 a.** SSE $= 740.72$, $s^2 = 92.59$ **b.** 9.622 **11.25 a.** 4
b. 8 **c.** 23 **d.** 48 **11.27 a.** $t = 7.00$; reject H_0. **b.** $p < .01$ **c.** $1.05 \pm .353$ **11.29 a.** $.953 \pm .05$ **11.31** $2.426 \pm .473$
11.33 Yes; $t = 8.30$ **11.35 a.** $\hat{y} = -.1776 + .8991x$; $\hat{y} = 1.0179 + .9753x$ **b.** $t = 11.32$; reject H_0. **c.** $t = 11.32$; reject H_0.
11.37 $.971$ **11.39 a.** 1 **b.** -1 **11.41** Possibly; positive **11.43** Yes; $t = 3.88$ **11.45 b.** Reject H_0. **d.** Do not reject H_0.
11.47 a. No, $t = 1.36$ **b.** Yes; $t = 2.35$ **c.** Yes; $t = 2.59$ **d.** No; $t = 1.22$ **11.51** $.852$ **11.53 a.** 10 **b.** 7 **c.** 6 **d.** $.7$ **e.** 1.1
f. $.817$ **11.55** $.927$ **11.57** $.889$ **11.59** $.0049$ **11.61** $.9409$ **11.63 a.** SSE $= 4.055$; $s^2 = .225$ **b.** $10.6 \pm .22$ **c.** $8.90 \pm .32$
d. $12.3 \pm .32$ **e.** Wider **f.** 12.3 ± 1.05 **11.65 a.** $1.4 \pm .499$ **b.** 1.4 ± 1.223 **c.** Prediction interval is wider **11.67** 9.106 ± 3.198
11.69 b. $\hat{\beta}_0 = 9.15$, $\hat{\beta}_1 = .481$ **d.** Yes, $t = 8.07$ **e.** $18.77 \pm .51$ **11.71** $(13.25, 17.76)$ **11.73 d.** $.903$ **e.** $t = 20.85$; reject H_0.
f. 4.58 **11.75 a.** Negative **b.** $-.855$ **c.** Yes, $t = -4.36$ **d.** $\hat{y} = 14{,}192 - 148.978x$ **e.** $t = -4.36$; reject H_0. **f.** $8{,}977.9 \pm 1{,}388.6$
g. $x = 35$ is outside range of sample data **11.77 a.** Reject $H_0: \beta_1 = 0$, $t = 4.482$; $r^2 = .52744$; $s = 2.07383$
11.79 a. $\hat{y} = -13.622 - 0.0533x$ **c.** $r = -.923$, $r^2 = .852$ **d.** $.769 \pm .150$ **e.** $.769 \pm .474$ **11.81 a.** Possibly; negative
b. No; do not reject H_0; $t = -.79$ **c.** $.135$ **11.83 a.** $H_0: \beta_1 = 0$, $H_a: \beta_1 \neq 0$ **b.** Do not reject H_0 at $\alpha = .05$.

CHAPTER 12

12.1 $E(y) = \beta_0 + \beta_1 x_1 + \beta_2 x_2$ **12.3 b.** Parallel lines (same slope); slope $= 2$ **c.** Parallel lines **12.5 a.** $.0439$ **b.** 7.12
c. Yes; reject H_0. **12.7** Do not reject H_0; $F = 1.90$ **12.9 a.** $F = 20.914$, reject $H_0: \beta_1 = \beta_2 = \cdots = \beta_{27} = 0$; $s = 6{,}544$; $R^2 = .814$;
$R_a^2 = .7751$ **b.** $(15.13, 29.67)$ **c.** No; inflated α error; no higher-order terms in model **12.11** Reject $H_0: \beta_1 = \beta_2 = \cdots = \beta_7 = 0$
12.13 Yes; $F = 443.18$ **12.15 a.** $\hat{y} = .132 - 9.307x_1 + 1.558x_2$ **b.** Reject H_0, $F = 35.843$ (p-value $= .0005$)

c. No; do not reject H_0, $t = -1.84$ (p-value $= .1153$)   **d.** Yes; reject H_0, $t = 8.467$ (p-value $= .0001$)   **e.** .9228   **f.** .152

12.17 Do not reject H_0; $F = 1.06$ **12.19** $E(y) = \beta_0 + \beta_1 x_1 + \beta_2 x_2 + \beta_3 x_1 x_2$ **12.21 b.** Nonparallel lines   **c.** $x_2 = 0$: slope $= 2$; $x_2 = 1$: slope $= -1$; $x_2 = 2$: slope $= -4$ **12.23 a.** $F = 7.99$; reject H_0.   **c.** $t = 1.85$; reject H_0: $\beta_3 = 0$ **d.** No

12.25 a. Nuclear: $\hat{y} = -17.556 + .519x_1 + 10.889x_2 + .1322x_1 x_2$; nonnuclear: $\hat{y} = -44.682 + 2.880x_1 + 25.062x_2 - .959x_1 x_2$

b. Nuclear: yes, reject H_0 ($F = 13.22$, p-value $= .0018$); nonnuclear: yes, reject H_0 ($F = 60.85$, p-value $= .0001$)

c. Nuclear: no, do not reject H_0: $\beta_3 = 0$ ($t = .12$, p-value $= .9093$); nonnuclear: no, do not reject H_0: $\beta_3 = 0$ ($t = -1.39$, p-value $= .1904$)

d. (7.4528, 26.8903)   **f.** No; $x_2 = 1.10$ is outside the range of the sample data.

12.27 b. $\hat{y} = 54.5 + .007697x_1 + .554111x_2 + .000113x_1 x_2$ **d.** Reject H_0; $F = 44.67$, p-value $= .0005$ **e.** No, do not reject H_0: $\beta_3 = 0$; $t = .68$, p-value $= .5280$ **f.** $E(y) = \beta_0 + \beta_1 x_1 + \beta_2 x_2$ **12.29 d.** Shifts parabola along the x-axis   **e.** Controls direction of shift (right or left)   **12.31 a.** Yes; $F = 85.94$ **b.** H_0: $\beta_2 = 0$; H_a: $\beta_2 > 0$ **c.** H_0: $\beta_2 = 0$; H_a: $\beta_2 < 0$ **12.33 b.** Second-order   **c.** Different shapes   **12.35 a.** $E(y) = \beta_0 + \beta_1 x + \beta_2 x^2$ **b.** $\beta_1 < 0$ and $\beta_2 > 0$ **12.37 a.** Accountants: $F = 10.68$, reject H_0; truck drivers: $F = 22.07$, reject H_0.

b. $\hat{\beta}_2$ **c.** Yes; $t = 3.23$ **d.** Yes; $t = 4.70$ **12.39 a.** -75.71 ± 26.17 **b.** Do not reject H_0; $t = 1.38$ **12.41 b.** No; $t = -.15$

12.43 $E(y) = \beta_0 + \beta_1 x_1 + \beta_2 x_2 + \beta_3 x_3$, where $x_1 = \begin{cases} 1 & \text{if A} \\ 0 & \text{if not} \end{cases}$ $x_2 = \begin{cases} 1 & \text{if B} \\ 0 & \text{if not} \end{cases}$ $x_3 = \begin{cases} 1 & \text{if C} \\ 0 & \text{if not} \end{cases}$; $\beta_0 = \mu_D$, $\beta_1 = \mu_A - \mu_D$, $\beta_2 = \mu_B - \mu_D$,

$\beta_3 = \mu_C - \mu_D$ **12.45 a.** $\hat{\beta}_0 = \bar{x}_5 = 20$, $\hat{\beta}_1 = \bar{x}_1 - \bar{x}_5 = -5.6$, $\hat{\beta}_2 = \bar{x}_2 - \bar{x}_5 = 11.2$, $\hat{\beta}_3 = \bar{x}_3 - \bar{x}_5 = -1.7$, $\hat{\beta}_4 = \bar{x}_4 - \bar{x}_5 = -9.0$

b. H_0: $\mu_1 = \mu_2 = \mu_3 = \mu_4 = \mu_5$ **c.** $F = 2.16$; do not reject H_0.   **12.47 a.** Yes; p-value $= .001$ **b.** .187   **c.** .677   **d.** $-.046 \pm .072$

e. No, do not reject H_0: $\beta_6 = 0$ (p-value $= 1 - .10/2 = .95$)   **12.49 a.** $E(y) = \beta_0 + \beta_1 x$, where $x = \begin{cases} 1 & \text{if Hispanic} \\ 0 & \text{if Anglo} \end{cases}$

b. $\beta_0 = \mu_{\text{Anglo}}$, $\beta_1 = \mu_{\text{Hispanic}} - \mu_{\text{Anglo}}$ **c.** $E(y) = \beta_0 + \beta_1 x_1 + \beta_2 x_2 + \beta_3 x_3$, where $x_1 \begin{cases} 1 & \text{if NE} \\ 0 & \text{if not} \end{cases}$ $x_2 = \begin{cases} 1 & \text{if NC} \\ 0 & \text{if not} \end{cases}$ $x_3 = \begin{cases} 1 & \text{if S} \\ 0 & \text{if not} \end{cases}$

d. $\beta_0 = \mu_W$, $\beta_1 = \mu_{NE} - \mu_W$, $\beta_2 = \mu_{NC} - \mu_W$, $\beta_3 = \mu_S - \mu_W$ **12.51 a.** $E(y) = \beta_0 + \beta_1 x_1 + \beta_2 x_2 + \beta_3 x_3$ **b.** $F = 2.5$ **c.** $F > 3.16$

d. Do not reject H_0.   **12.53 a.** $E(y) = \beta_0 + \beta_1 x$; SSE $= 9,240,861$ **b.** 356,802   **c.** 248.99   **e.** Yes   **f.** Yes   **12.55 a.** H_0: $\beta_2 = 0$

b. Do not reject H_0; $F = .65$ **12.57 a.** H_0: $\beta_5 = \beta_6 = \beta_7 = \beta_8 = 0$; $F > 2.37$ **b.** Yes   **c.** Reject H_0.   **12.59 a.** Misspecified model; quadratic term missing   **b.** Unequal variances   **c.** Outlier   **d.** Unequal variances   **e.** Nonnormal errors   **12.61 a.** 3.197, 3.215, -2.258, .269, -1.713, -7.186, -2.659, 3.359, -2.132, 5.904   **b.** Yes; needs curvature   **c.** No outliers   **d.** Yes; needs curvature

12.63 a. .796, $-.219$, -1.004, .077, -1.071, 1.116, .868, .827, .223, -1.612 **b.** No   **c.** Possibly; needs curvature term $\beta_3 x_2^2$

12.65 a. 25.67, 18.57, -60.14, 13.41, -72.40, 64.06, 47.02, 57.93, -63.36, -75.62, 46.31, -1.43 **c.** Yes; model needs curvature term

12.67 a. $F = 83.54$; reject H_0.   **b.** Yes   **c.** No outliers   **12.69 a.** $\hat{y} = .94 - .214x$ **b.** 0, .02, $-.026$, .034, .088, $-.112$, $-.058$, .002, .036, .016   **c.** Football shape; unequal variances   **d.** Use the transformation $y* = \sin^{-1}\sqrt{y}$, and fit the model $y* = \beta_0 + \beta_1 x + \varepsilon$.

12.71 No; appears that multicollinearity exists   **12.73 a.** .0025; no   **b.** .4341; no   **c.** No   **d.** $\hat{y} = -45.154 + 3.097x_1 + 1.032x_2$; $F = 39,222$; $R^2 = .9998$ **e.** $-.8978$; x_1 and x_2 are highly correlated   **f.** No   **12.75 b.** $.331 \pm .12$ **c.** Reject H_0; $t = 3.04$

e. Reject H_0; $F = 39.70$ **12.77 a.** $\hat{y} = 66.97 + 4.03x_1 + 5.55x_2$ **b.** H_0: $\beta_1 = \beta_2 = 0$ **12.79 a.** $E(y) = \beta_0 + \beta_1 x_1 + \beta_2 x_2$

b. $E(y) = \beta_0 + \beta_1 x_1 + \beta_2 x_2 + \beta_3 x_1 x_2$ **c.** $E(y) = \beta_0 + \beta_1 x_1 + \beta_2 x_2 + \beta_3 x_1 x_2 + \beta_4 x_1^2 + \beta_5 x_2^2$ **d.** .6004   **e.** No, do not reject H_0; $F = 1.80$, p-value $= .2465$ **f.** (.1354, .5774); inadequate model   **12.81 a.** .9043   **b.** 4.5485   **c.** $F = 33.08$; reject H_0, p-value $= .0003$

d. Yes, reject H_0: $\beta_2 > 0$, p-value $= .0706/2 = .0353$ **e.** .791, -2.0062, 3.7682, 1.8362, -6.0142, -3.9582, 6.3938, 3.9718, -3.539, -1.259 **f.** No; no   **g.** No, do not reject H_0: $\beta_3 = \beta_4 = 0$; $F = 1.21$ **12.83 a.** No, at $\alpha = .01$ **b.** Means do not differ   **c.** $4,724 \pm 6,449$

12.85 a. $\hat{y} = 385.048 - 8.293x + .05982x^2$ **b.** $R^2 = .959$; $s = 5.376$ **c.** Yes, reject H_0; $F = 259.69$, p-value $= .0001$

d. Reject H_0: $\beta_2 = 0$ in favor of H_a: $\beta_2 > 0$; $t = 7.93$, p-value $= .0001/2 = .00005$ **e.** 1.924, -8.521, 3.612, 2.156, -1.449, -2.110, -2.305, 2.874, 8.78, .072, 3.651, -5.298, -2.048, 2.962, 9.727, 3.255, -1.635, .600, -1.432, 6.692, -10.348, $-.943$, -4.305, 4.960, -9.179 **f.** Mean: .068, std. dev.: 5.147   **g.** 0   **h.** No trends; assumptions appear to be satisfied

CHAPTER 13

13.1 b. 7.502   **c.** 2.381   **d.** 1   **e.** 11   **f.** 3.15   **g.**

SOURCE	df	SS	MS	F
Treatments	1	7.502	7.502	3.15
Error	11	26.190	2.381	
Total	12	33.692		

h. Reject H_0 if $F > 4.84$ **i.** Do not reject H_0.   **j.** μ_1: 9.14 ± 1.284; μ_2: 10.667 ± 1.387 **k.** 1.527 ± 1.889 **13.3 a.** df(Error) $= 30$; SSE $= 37.7$; MST $= 6.175$; MSE $= 1.257$; $F = 4.91$ **b.** 5   **c.** Yes; reject H_0.   **13.5 a.** Patients with eating disorders; depressed patients; normal eaters   **b.** H_0: $\mu_1 = \mu_2 = \mu_3$ **c.** df(Treatments) $= 2$; df(Error) $= 80$, df (Total) $= 82$ **d.** Reject H_0.   **13.7 a.** Cash compensation industries; corporate executives   **b.** No; $F = 3.556$, p-value $> .01$ **c.** $8,994.9 \pm 2,807.7$ **13.9 a.** No   **b.** Reject H_0, p-value $= .065$

13.11 No, do not reject H_0; $F = .10$ (p-value $> .10$)   **13.13 a.** df(Error) $= 15$; df(Total) $= 23$; SSB $= 74.5$; SS(Total) $= 135$; MST $= 9.033$; MSE $= 2.227$; F(Treatments) $= 4.06$; F(Blocks) $= 6.69$ **b.** No, do not reject H_0.   **c.** Yes, reject H_0 at $\alpha = .05$.   **d.** -2.4 ± 1.84

e. 2.8 ± 1.84 **13.15 a.** Treatments $= 3$ supermarkets; blocks $= 60$ items

b.

SOURCE	df	SS	MS	F
Supermarkets	2	2.64	1.32	39.23
Items	59	215.59	3.65	108.54
Error	118	3.97	.034	
Total	179	222.20		

c. Reject H_0, p-value $= .0001$ **d.** $-.260 \pm .066$ **e.** Yes

13.17 No evidence of a difference among the three plant session means; $F = .019$ **13.19** Yes; reject H_0, $F = 52.81$ **13.21 a.** $df(A) = 2$; $df(Error) = 18$; $df(Total) = 23$; $SS(B) = 559$; $SS(AB) = 5$; $SSE = 36$; $MS(A) = 50$; $MS(B) = 559$; $F(A) = 25$; $F(B) = 279.5$; $F(AB) = 1.25$ **b.** Do not reject H_0; $F = 1.25$ **c.** Reject H_0; $F = 25.0$ **d.** Reject H_0; $F = 279.5$ **13.23 a.** Type of game; position   **b.** Game: Missile Command, Pac-Man, Control; Position: Player, Observer   **c.** Missile Command/Player; Missile Command/Observer; Pac-Man/Player; Pac-Man/Observer; Control/Player; Control/Observer

d.

SOURCE	df
Game	2
Position	1
Game × Position	2
Error	78
Total	83

e. Mean difference between aggression of player and observer depends on type of game

13.25 No evidence of interaction ($F = .27$); no evidence of form main effect ($F = 0$); evidence of a difference between the two majors ($F = 6.25$)

13.27 a.

SOURCE	df	SS	MS	F
Sex	1	114.005	114.005	4.04
Weight	1	41.405	41.405	1.47
Sex × Weight	1	18.605	18.605	.66
Error	28	789.968	28.213	
Total	31	963.983		

b. No   **c.** No, do not reject H_0; $F = .66$ **d.** No, do not reject H_0; $F = 4.04$ **e.** When no interaction exists   **f.** -5.30 ± 5.44 **g.** -2.25 ± 5.44 **13.29 a.** Problem type and group independently affect proportion of ideas recalled   **b.** No evidence of a difference between the means of the two problem types   **c.** Evidence of differences among the means of the four groups   **13.31 a.** 3.930   **b.** 3.355   **c.** Approx. 3.00   **d.** 2.845   **13.33** $(\mu_1 - \mu_2)$: 2.53; $(\mu_1 - \mu_3)$: 2.70; $(\mu_2 - \mu_3)$: 2.53 (using $t \approx 2.79$)   **13.35** 5.26 (using $t = 3.930$) for all pairwise comparisons   **13.37** The treatment means for depression and eating disorder are not significantly different; the mean for normal is significantly larger than the other two means.   **13.39** The following treatment pairs are significantly different: (jail correctional, jail mental health), (jail correctional, county mental health) and (jail correctional, other mental health).   **13.41** There is a significant difference between the two groups of means (μ_5, μ_3, μ_0) and (μ_{10}); no significant differences exist within the group (μ_5, μ_3, μ_0).   **13.43 a.** Complete model:

$$E(y) = \beta_0 + \beta_1 x_1 + \beta_2 x_2 + \beta_3 x_3, \text{ where } x_1 = \begin{cases} 1 & \text{if consumer products} \\ 0 & \text{if not} \end{cases}, x_2 = \begin{cases} 1 & \text{if utilities} \\ 0 & \text{if not} \end{cases}, \text{ and } x_3 = \begin{cases} 1 & \text{if industrial} \\ 0 & \text{if not} \end{cases}; \text{ reduced model:}$$

$E(y) = \beta_0$ **13.45 a.** Complete model: $E(y) = \beta_0 + \beta_1 x_1 + \beta_2 x_2 + \beta_3 x_3 + \beta_4 x_4 + \cdots + \beta_{61} x_{61}$, where $x_1 = \begin{cases} 1 & \text{if Winn Dixie} \\ 0 & \text{if not} \end{cases}$,

$x_2 = \begin{cases} 1 & \text{if Publix} \\ 0 & \text{if not} \end{cases}$, and x_3 through x_{61} are dummy variables for grocery items; reduced model: $E(y) = \beta_0 + \beta_3 x_3 + \beta_4 x_4 + \cdots + \beta_{61} x_{61}$

13.47 a. Complete model: $E(y) = \beta_0 + \beta_1 x_1 + \beta_2 x_2 + \beta_3 x_3 + \beta_4 x_4 + \beta_5 x_5 + \beta_6 x_6 + \beta_7 x_1 x_4 + \beta_8 x_1 x_5 + \beta_9 x_1 x_6 + \beta_{10} x_2 x_4 + \beta_{11} x_2 x_5 + \beta_{12} x_2 x_6 + \beta_{13} x_3 x_4 + \beta_{14} x_3 x_5 + \beta_{15} x_3 x_6$, where x_1 through x_3 are dummy variables for antimony amount and x_4 through x_6 are dummy variables for cooling method; reduced model: $E(y) = \beta_0 + \beta_1 x_1 + \beta_2 x_2 + \beta_3 x_3 + \beta_4 x_4 + \beta_5 x_5 + \beta_6 x_6$

13.57 a. $E(y) = \beta_0 + \beta_1 x_1 + \beta_2 x_2 + \beta_3 x_1 x_2$, where $x_1 = \begin{cases} 1 & \text{if high quality} \\ 0 & \text{if low quality} \end{cases}$, $x_2 = \begin{cases} 1 & \text{if high credibility} \\ 0 & \text{if low credibility} \end{cases}$ **b.** Evidence of $A \times B$ interaction (p-value $= .002$)   **13.59** Evidence of differences among the means of the six exercises ($F = 30.21$ exceeds $F_{.05} = 2.37$)   **13.61** $\mu_{ND-H} > (\mu_{LD-H}, \mu_{ND-L}) > \mu_{LD-L}$

13.63 b.

SOURCE	df	SSE	MSE	F
Periods	2	9,726	4,863	1.04
Months	3	68,258	22,753	4.88
Error	6	27,947	4,658	
Total	11	105,932		

c. No; $F = 1.04$ **d.** 68.25 (compared to 85.30)   **e.** $F = 4.88$; reject H_0.

13.65 a. Evidence of differences among the means of the five organization types   **b.** $\mu_R < \mu_P$; $\mu_P < (\mu_D, \mu_A, \mu_B)$; $\mu_D < \mu_B$

13.67 a.

SOURCE	df	SS	MS	F
Sex, S	1	1.21	1.21	.58
Position, P	1	.01	.01	.005
$S \times P$	1	6.25	6.25	2.98
Error	96	201.60	2.10	
Total	99	209.07		

d. No; do not reject H_0.   **e.** Yes; conduct the tests for main effects.

f. S: do not reject H_0; P: do not reject H_0.   **13.69 b.** No, do not reject H_0; p-value $= .3272$ **c.** Main effects
d. Time: reject H_0 ($p = .0002$); Material: reject H_0 ($p = .0001$).   **e.** Using $B = 1.79$, the means of all charging time pairs are significantly different.   **f.** Complete model: $E(y) = \beta_0 + \beta_1 x_1 + \beta_2 x_2 + \beta_3 x_3 + \beta_4 x_1 x_3 + \beta_5 x_2 x_3$, where $x_1 = \begin{cases} 1 \text{ if 0 days} \\ 0 \text{ if not} \end{cases}$, $x_2 = \begin{cases} 1 \text{ if 25 days} \\ 0 \text{ if not} \end{cases}$, $x_3 = \begin{cases} 1 \text{ if cold rolled} \\ 0 \text{ if not} \end{cases}$; reduced model: $E(y) = \beta_0 + \beta_1 x_1 + \beta_2 x_2 + \beta_3 x_3$

CHAPTER 14

14.1 a.

CATEGORY	1	2	3	4
EXPECTED NUMBER	10	30	30	30

b. 6.25139   **c.** H_a: At least two of the cell probabilities differ from .1, .3, .3, .3, respectively.   **d.** No; $\chi^2 = 3.933$, do not reject H_0.
14.3 Yes; $\chi^2 = 40.7$ **14.5** $\chi^2 = 39.27$; reject H_0.   **14.7** Do not reject H_0; $\chi^2 = 2.6$

14.9 a.

	1	2	3
1	15.70	30.95	27.36
2	19.30	38.05	33.64

b. 3.74   **14.11 a.** 1   **b.** 3   **c.** 4   **d.** 6

14.13 a.

	1	2	3	4
1	14.25	11.73	20.95	41.07
2	10.20	8.40	15.00	29.40
3	9.55	7.87	14.05	27.53

b. 3.49   **c.** Do not reject H_0.

14.15 Reject H_0; $\chi^2 = 6.652$ **14.17 a.** Reject H_0; $\chi^2 = 5.61$ **c.** $.087 \pm .062$ **14.19 a.** H_0: "most important value" and "nation" are independent.   **b.** Reject H_0 at $\alpha = .05$ (p-value $= .033$)   **c.** $-.0001 \pm .081$ **14.21** Reject H_0; $\chi^2 = 18.91$
14.23 a. Reject H_0; $\chi^2 = 105.258$ **b.** $.0769 \pm .0512$ **14.25 a.** Reject H_0, $\chi^2 = 20.43$ **b.** Yes; $\chi^2 = 4.63$ **c.** $(.00048, .40027)$
14.27 a. .25, .25, .25, .25   **b.** 9, 9, 9, 9   **c.** 14.667   **d.** Yes, reject H_0.   **14.29** Reject H_0, $\chi^2 = 61.14$ **14.31** Do not reject H_0; $\chi^2 = .32$
14.33 a. Yes; $\chi^2 = 313.15$ **b.** $.181 \pm .069$ **14.35 a.** Reject H_0; $\chi^2 = 4,755$ **b.** $.354 \pm .012$

14.37

	HIGH-TECH	NON–HIGH-TECH
CONVERTERS	80.2	296.8
NONCONVERTERS	80.8	299.2

b. .021   **c.** No; do not reject H_0; p-value $> .10$

14.39 a.

	NATIVE	NATURALIZED	OTHER
FM	648.01	17.43	29.56
OCC	137.99	3.71	6.29
IL	69.00	1.86	3.15

b. $\chi^2 = 2.11$; do not reject H_0.

CHAPTER 15

15.1 a. .812 **b.** .012 **c.** .055 **15.3** $S = 25$, $z = 3.65$; reject H_0. **15.5 a.** Reject H_0; $z = 2.83$ (p-value $= .0023$) **15.7 a.** No **b.** No; assumption of normality is violated **c.** $S = 18$, p-value $= .0002$; reject H_0 **15.9 a.** $T_1 \leq 24$ or $T_1 \geq 56$ **b.** $T_1 \geq 84$ **c.** $T_1 \leq 22$ **15.11 a.** $z = -3.16$; reject H_0 if $z < -1.645$ **b.** $z = 2.18$; reject H_0 if $z > 1.28$ **c.** $z = 2.33$; reject H_0 if $z < -1.96$ or $z > 1.96$ **15.13** $T_1 = 71$; do not reject H_0. **15.15** No; $T_1 = 66$ **15.17** $z = 3.91$; yes **15.19** No, do not reject H_0; $T_2 = 29$ **15.21** No, do not reject H_0; $T_1 = 88.5$ **15.23 a.** Reject H_0 if $T^- \leq 2$ **b.** Reject H_0 if T (smaller of T^+ or T^-) ≤ 4 **15.25** Yes; reject H_0, $T^+ = 11$ **15.27 a.** Reject H_0; $T^+ = 2$ **15.29 a.** $z = 1.78$; reject H_0. **b.** $T^+ = 73.52$; do not reject H_0. **15.31 a.** 18.3070 **b.** 38.0757 **c.** 15.0863 **d.** 15.9871 **e.** 7.81473 **f.** 12.8381 **15.33** No, do not reject H_0; $H = .98$ **15.35 a.** Samples not normal; unequal variances **b.** $H = 16.23$; reject H_0. **15.37** $H = 11.17$; reject H_0 (test is approximate because of inadequate sample size) **15.39 a.** H_0: The three population relative frequency distributions are identical; H_a: At least two of the three population relative frequency distributions differ in location. **b.** $F_r > 5.99147$ **c.** $T_1 = 9.5$, $T_2 = 9.5$, $T_3 = 17$ **d.** 6.25 **e.** Reject H_0. **15.41** Reject H_0; $\chi^2 = 54.3$, p-value $= .0000$ **15.43** Do not reject H_0; $F_r = 1.18$ **15.45 a.** $F_r = .15$; do not reject H_0. **15.47 b.** $r_s = .893$ **c.** Yes, reject H_0; $r_{.025} = .786$ **15.49** $C = 15$, reject H_0. **15.51** Yes; reject H_0: $r_s = .745$ **15.53** Yes; $r_s = -.529$ **15.55 a.** $C = -13$ **b.** No, do not reject H_0. **15.57** $S = 14$, reject H_0; p-value $= .058$ **15.59 a.** H_0: The distributions of changes in bond prices for the four underwriters are identical; H_a: At least two of the four distributions differ in location. **b.** 1.21 **c.** No, do not reject H_0; $\chi^2_{.01} = 11.3449$ **15.61** No, $S = 5$; p-value $= 1.00$ **15.63** $r_s = 1.0$, yes **15.65 a.** No; do not reject H_0, $F_r = 6.15$ **b.** No; do not reject H_0, $F_r = 1.6$ **15.67** Yes, reject H_0; $T_1 = 78.5$

Index